Solid Oxide Fuel Cells
From Materials to System Modeling

RSC Energy and Environment Series

Editor-in-Chief:
Professor Laurence Peter, *University of Bath, UK*

Series Editors:
Professor Heinz Frei, *Lawrence Berkeley National Laboratory, USA*
Professor Ferdi Schüth, *Max Planck Institute for Coal Research, Germany*
Professor Tim S. Zhao, *The Hong Kong University of Science and Technology, Hong Kong*

Titles in the Series:
1: Thermochemical Conversion of Biomass to Liquid Fuels and Chemicals
2: Innovations in Fuel Cell Technologies
3: Energy Crops
4: Chemical and Biochemical Catalysis for Next Generation Biofuels
5: Molecular Solar Fuels
6: Catalysts for Alcohol-Fuelled Direct Oxidation Fuel Cells
7: Solid Oxide Fuel Cells: From Materials to System Modeling

How to obtain future titles on publication:
A standing order plan is available for this series. A standing order will bring delivery of each new volume immediately on publication.

For further information please contact:
Book Sales Department, Royal Society of Chemistry, Thomas Graham House, Science Park, Milton Road, Cambridge, CB4 0WF, UK
Telephone: +44 (0)1223 420066, Fax: +44 (0)1223 420247
Email: booksales@rsc.org
Visit our website at www.rsc.org/books

Solid Oxide Fuel Cells
From Materials to System Modeling

Edited by

Meng Ni
Hong Kong Polytechnic University, Hung Hom, Kowloon, P.R. China
Email: meng.ni@polyu.edu.hk

Tim S. Zhao
The Hong Kong University of Science and Technology, Hong Kong, P.R. China
Email: metzhao@ust.hk

RSCPublishing

RSC Energy and Environment Series No. 7

ISBN: 978-1-84973-654-1
ISSN: 2044-0774

A catalogue record for this book is available from the British Library

Published by The Royal Society of Chemistry,
Thomas Graham House, Science Park, Milton Road,
Cambridge CB4 0WF, UK

Registered Charity Number 207890

For further information see our web site at www.rsc.org

Preface

Fuel cells are efficient energy conversion devices that convert the chemical energy of a fuel into electricity via electrochemical reactions. Proton Exchange Membrane Fuel Cells (PEMFCs), Alkaline Fuel Cells (AFCs), and Direct Methanol Fuel Cells (DMFCs) are typically operated at room temperature and thus are featured by fast startup and are suitable for portable and mobile applications. In contrast, Solid Oxide Fuel Cells (SOFCs) are typically operated at a high temperatures ranging from 400 °C to 1000 °C. The high-temperature operation enables the direct use of alternative fuels in SOFCs as internal reforming of hydrocarbons can be achieved in the porous anode of SOFC. In addition, the waste heat from a SOFC stack is of high quality and can be recovered by integrating the stack with other thermodynamic cycles for co- or tri-generation to achieve enhanced system efficiency. Hence, SOFCs are particularly suitable for large-scale stationary applications.

This book aims to provide an overview of the SOFC technology with a focus on the recent developments in new technologies and new ideas for addressing the key issues associated with SOFC developments. The contributing authors consist of both young, promising experts and renowned senior scientists in the field. Key issues related to SOFC performance improvements, long-term stability, mathematical modelling, as well as system integration/control are addressed, including material developments, novel infiltration technique for nano-structured electrode fabrication, advanced focused ion beam- scanning electron microscopy (FIB-SEM) technique for microstructure reconstruction, Lattice Boltzmann Method (LBM) simulation for understanding the fundamental transport and reaction phenomena at pore-scale, multi-scale modeling and simulation, SOFC integration with buildings and other cycles for stationary applications.

RSC Energy and Environment Series No. 7
Solid Oxide Fuel Cells: From Materials to System Modeling
Edited by Meng Ni and Tim S. Zhao
© The Royal Society of Chemistry 2013
Published by the Royal Society of Chemistry, www.rsc.org

The editors express their appreciation to the contributing authors for their contributions to this book. Last, but not least, the editors acknowledge the efforts of the professional staff at RSC for providing invaluable editorial assistance.

T.S. Zhao
M. Ni

Contents

Chapter 1   Introduction to Stationary Fuel Cells   **1**
Ibrahim Dincer and C. Ozgur Colpan

1.1   General Introduction to Fuel Cells   1
1.2   Introduction to Low-Temperature Fuel Cells   2
1.3   Introduction to Solid Oxide Fuel Cells   4
    1.3.1   Classification of SOFC Systems   5
    1.3.2   Fuel Options for SOFC   7
1.4   Integrated SOFC Systems   9
1.5   Basic SOFC Modelling   12
1.6   Case Study   14
    1.6.1   Analysis   14
    1.6.2   Results and Discussion   20
1.7   Conclusions   22
References   24

Chapter 2   Electrolyte Materials for Solid Oxide Fuel Cells (SOFCs)   **26**
Yu Liu, Moses Tade and Zongping Shao

2.1   A General Introduction to Electrolyte of SOFCs   26
2.2   The Requirements of Electrolyte   27
2.3   Classification of Electrolytes   28
    2.3.1   Oxygen-ion Conducting Electrolyte   28
    2.3.2   Proton-conducting Electrolyte   37
    2.3.3   Dual-phase Composite Electrolyte   45
2.4   Future Vision   46
References   47

RSC Energy and Environment Series No. 7
Solid Oxide Fuel Cells: From Materials to System Modeling
Edited by Meng Ni and Tim S. Zhao
© The Royal Society of Chemistry 2013
Published by the Royal Society of Chemistry, www.rsc.org

Chapter 3   Cathode Material Development   56
Yao Wang, Yanxiang Zhang, Ling Zhao and Changrong Xia

3.1   Introduction   56
3.2   Cathodes for Oxygen Ion-Conducting Electrolyte
      Based SOFCs   57
      3.2.1   Electron Conducting Cathodes   57
      3.2.2   Mixed Oxygen Ion-Electron Conducting
              Cathodes   61
      3.2.3   Microstructure Optimized Cathodes   65
      3.2.4   Cathode Reaction Mechanisms   70
3.3   Cathodes for Proton-Conducting Electrolyte Based
      SOFCs   73
      3.3.1   Electron-Conducting Cathodes   73
      3.3.2   Mixed Oxygen Ion-Electron Conducting
              Cathodes   74
      3.3.3   Mixed Electron-Proton Conducting Cathodes   77
      3.3.4   Microstructure Optimized Cathodes   78
      3.3.5   Cathode Reaction Mechanisms   80
3.4   Summary and Conclusions   82
Acknowledgements   82
References   83

Chapter 4   Anode Material Development   88
Shamiul Islam and Josephine M. Hill

4.1   Required Properties of Anode Materials   88
4.2   Hydrogen Fuel   89
4.3   Methane Fuel   90
      4.3.1   Conventional Ni/YSZ Anodes   91
      4.3.2   Alternative Anodes   92
4.4   Higher Hydrocarbon Fuels (Propane and Butane)   94
4.5   Fuels from Biomass   95
      4.5.1   Biomass-Simulated Gas   96
      4.5.2   Biomass – Actual Gas   97
4.6   Liquid Fuels   98
4.7   Ammonia Fuel   100
4.8   Conclusions   101
References   101

Chapter 5   Interconnect Materials for SOFC Stacks   106
Xingbo Liu, Junwei Wu and Christopher Johnson

5.1   Introduction   106

5.2 Lanthanum Chromites as Interconnect 107
  5.2.1 Conductivity 108
  5.2.2 Thermal Expansion 111
  5.2.3 Gas Tightness, Processing and Chemical Stability 113
  5.2.4 Other Ceramic Interconnect 114
  5.2.5 Applications 114
5.3 Metallic Alloys as Interconnect 116
  5.3.1 Selection of Metallic Materials 116
  5.3.2 Problems for Metallic Materials as Interconnect 120
  5.3.3 Interconnect Coatings 123
  5.3.4 Applications of Metallic Interconnects 126
5.4 Concluding Remarks 130
References 130

Chapter 6 Nano-structured Electrodes of Solid Oxide Fuel Cells by Infiltration 135
San Ping Jiang

6.1 Introduction 135
6.2 Infiltration Process 136
  6.2.1 The Technique 136
  6.2.2 Factors Affecting Infiltration Process and Microstructure 140
6.3 Nano-structured Electrodes 142
  6.3.1 Performance Promotion Factor 142
  6.3.2 Nano-structured Cathodes 143
  6.3.3 Nano-structured Anodes 150
6.4 Microstructure and Microstructural Stability of Nano-structured Electrodes 155
  6.4.1 Microstructure Effect 155
  6.4.2 Microstructural Stability of Nano-structured Electrodes 158
6.5 Electrocatalytic Effects of Infiltrated Nanoparticles 162
6.6 Conclusions 168
Acknowledgement 169
References 169

Chapter 7 Three Dimensional Reconstruction of Solid Oxide Fuel Cell Electrodes 178
P. R. Shearing and N. P. Brandon

7.1 The Importance of 3D Characterisation and the Limitations of Stereology 179

7.2   Focused Ion Beam Characterisation   184
      7.2.1   The FIB-SEM Instrument   184
      7.2.2   Application of FIB-SEM Techniques to SOFC
              Materials   186
7.3   Microstructural Characterisation using X-rays   189
      7.3.1   X-ray Microscopy and Tomography   189
      7.3.2   Lab X-ray Instruments   191
      7.3.3   Synchrotron X-ray Instruments   192
      7.3.4   4-Dimensional Tomography   193
7.4   Data Analysis and Image Based Modelling   195
      7.4.1   Data Analysis   195
      7.4.2   Image Based Modelling   196
7.5   Conclusions   196
References   197

Chapter 8   Three-Dimensional Numerical Modelling of Ni-YSZ Anode   200
             Naoki Shikazono and Nobuhide Kasagi

8.1   Introduction   200
8.2   Experimental   201
      8.2.1   Button Cell Experiment   201
      8.2.2   Microstructure Reconstruction Using
              FIB-SEM   202
8.3   Numerical Method   202
      8.3.1   Quantification of Microstructural Parameters   202
      8.3.2   Governing Equations for Polarization
              Simulation   207
      8.3.3   Computational Scheme   211
8.4   Results and Discussions   212
8.5   Conclusions   215
Acknowledgements   216
References   216

Chapter 9   Multi-scale Modelling of Solid Oxide Fuel Cells   219
             Wolfgang G. Bessler

9.1   Introduction and Motivation   219
9.2   Modelling Methodologies: From the Atomistic to the
      System Scale   220
      9.2.1   Overview   220
      9.2.2   Molecular Level: Atomistic Modelling   220
      9.2.3   Electrode Level (I): Electrochemistry with
              Mean-field Elementary Kinetics   222
      9.2.4   Electrode Level (II): Porous Mass and Charge
              Transport   224

9.2.5   Cell Level: Coupling of Electrochemistry with
Mass, Charge and Heat Transport   225
9.2.6   Stack Level: Computational Fluid Dynamics
Based Design   226
9.2.7   System Level   226
9.3   Bridging the Gap Between Scales   227
9.3.1   General Aspects   227
9.3.2   Electrochemistry   228
9.3.3   Transport   232
9.3.4   Structure   234
9.4   Multi-scale Models for SOFC System Simulation and
Control   237
9.4.1   Pressurized SOFC System for a Hybrid
Power Plant   237
9.4.2   Tubular SOFC System for Mobile APU
Applications   237
9.5   Conclusions   240
Acknowledgements   241
References   241

Chapter 10 Fuel Cells Running on Alternative Fuels   **247**
Xinwen Zhou, Ning Yan and Jing-Li Luo

10.1   Introduction   247
10.2   Fuel Cell Reactor Set-up   248
10.3   SOFCs Running on Sourgas   248
10.4   SOFCs Running on C_2H_6 and C_3H_8 256
10.4.1   Development of Electrolyte of PC-SOFCs   258
10.4.2   Development of Anode Materials of
PC-SOFCs   262
10.5   SOFCs Running on Syngas Containing H_2S 269
10.6   SOFCs Running on Pure H_2S 276
10.7   Summary   281
Acknowledgements   282
References   282

Chapter 11 Long Term Operating Stability   **288**
Haruo Kishimoto, Teruhisa Horita and Harumi Yokokawa

11.1   Introduction   288
11.2   Durability of Stacks/Systems   289
11.2.1   Determination of Stack Performance   289
11.2.2   Performance Degradation and Materials
Deteriorations   289
11.2.3   Impurities and their Poisoning Effects on
Electrode Reactivity   292

11.3 Deteriorations of Electrolytes   294
      11.3.1 Destabilization of Mn Dissolved YSZ   296
      11.3.2 Conductivity Decrease in Ni-dissolved YSZ   302
11.4 Performance Degradations of Cathode and Anodes   309
      11.4.1 Cathode Poisoning   309
      11.4.2 Sintering of Ni Cermet Anodes   316
11.5 For Future Work   320
11.6 Conclusions   321
Acknowledgement   321
References   321

**Chapter 12 Application of SOFCs in Combined Heat, Cooling and
            Power Systems**   **327**
R. J. Braun and P. Kazempoor

12.1 Introduction   327
      12.1.1 Drivers for Interest in Co- and
              Tri-generation Using Fuel Cells   328
      12.1.2 Overview of CHP and CCHP   329
12.2 Application Characteristics & Building Integration   331
      12.2.1 Commercial Buildings   332
      12.2.2 Residential Applications   334
      12.2.3 Building Integration & Operating Strategies   335
12.3 Overview of SOFC-CHP/CCHP Systems   338
      12.3.1 SOFC System Description for CHP
              (Co-generation)   339
      12.3.2 SOFC System Description for CCHP
              (Tri-generation)   340
12.4 Modelling Approaches: Cell to System   342
      12.4.1 System-level Modelling and Performance
              Estimation   344
      12.4.2 Cell/Stack Modelling for SOFC System
              Simulation   349
      12.4.3 System Optimization Using Techno-economic
              Model Formulations   355
12.5 Evaluation of SOFC Systems in CCHP Applications   356
      12.5.1 Micro-CHP   356
      12.5.2 Large-scale CHP and CCHP Applications   363
12.6 Commercial Developments of SOFC-CHP Systems   365
      12.6.1 Commercialization Efforts   366
      12.6.2 Demonstrations   367
12.7 Market Barriers and Challenges   371
      12.7.1 Energy Pricing   371
      12.7.2 SOFC Costs   372
      12.7.3 Technical Barriers   373
      12.7.4 Market Barriers and Environmental Impact   373

12.8   Summary   376
References   376

Chapter 13 Integrated SOFC and Gas Turbine Systems   383
Francesco Calise and Massimo Dentice d'Accadia

13.1   Introduction   383
13.2   SOFC/GT Prototypes   385
13.3   SOFC/GT Layouts Classification   392
13.4   SOFC/GT Pressurized Cycles   394
       13.4.1   Internally Reformed SOFC/GT Cycles   395
       13.4.2   Anode Recirculation   396
       13.4.3   Heat Recovery Steam Generator (HRSG)   402
       13.4.4   Externally Reformed SOFC/GT Cycles   411
       13.4.5   Hybrid SOFC/GT-Cheng Cycles   411
       13.4.6   Hybrid SOFC/Humidified Air
                Turbine (HAT)   414
       13.4.7   Hybrid SOFC/GT-ITSOFC Cycles   415
       13.4.8   Hybrid SOFC/GT-Rankine Cycles   417
       13.4.9   Hybrid SOFC/GT with Air Recirculation or
                Exhaust Gas Recirculation (EGR)   419
13.5   SOFC/GT Atmospheric Cycles   424
13.6   SOFC/GT Power Plant: Control Strategies   427
13.7   Hybrid SOFC/GT Systems Fed by Alternative Fuels   436
13.8   IGCC SOFC/GT Power Plants   447
References   452

Chapter 14 Modelling and Control of Solid Oxide Fuel Cell   463
Xin-jian Zhu, Hai-bo Huo, Xiao-juan Wu and Bo Huang

14.1   Static Identification Model   464
       14.1.1   Nonlinear Modelling Based on LS-SVM   464
       14.1.2   Nonlinear Modelling Based on GA-RBF   470
14.2   Dynamic Identification Modelling for SOFC   478
       14.2.1   ANFIS Identification Modelling   479
       14.2.2   Hammerstein Identification Modelling   487
14.3   Control Strategies of the SOFC   496
       14.3.1   Constant Voltage Control   497
       14.3.2   Constant Fuel Utilization Control   501
       14.3.3   Simulation   502
14.4   Conclusions   505
References   506

Subject Index   511

CHAPTER 1

Introduction to Stationary Fuel Cells

IBRAHIM DINCER*[a] AND C. OZGUR COLPAN[b]

[a] Faculty of Engineering and Applied Science, University of Ontario Institute of Technology, 2000 Simcoe Street North, Oshawa, Ontario, Canada L1H 7L7; [b] Department of Mechanical Engineering, Dokuz Eylul University, Izmir, Turkey
*Email: ibrahim.dincer@uoit.ca

1.1 General Introduction to Fuel Cells

Although humankind started with wood as the main source of energy, fossil fuels have been used as the main energy source to produce power since the beginning of the industrial revolution. These fuels have mainly been converted into electricity using technologies such as the internal combustion engine, the gas turbine, and the steam turbine. Due to the increase in the global energy demand in parallel with the increase in the population of the world and in the production of high energy consuming devices, depletion of fossil fuels, and increased concern over the impact of greenhouse gases on global warming, alternative fuel and energy systems are being sought out. Among the alternative energy systems, fuel cells have received significant attention due to the fact that they convert the fuel into electricity in an efficient, effective, and environmentally friendly manner. They also help reduce the dependency on fossil fuel resources and the greenhouse gas emissions to the atmosphere.

Fuel cells are apparently known as electrochemical devices that convert the energy in the fuel into electricity. A fuel cell has mainly three components, as shown in Figure 1.1, namely, anode, cathode, and electrolyte. Fuel and air are

RSC Energy and Environment Series No. 7
Solid Oxide Fuel Cells: From Materials to System Modeling
Edited by Meng Ni and Tim S. Zhao
© The Royal Society of Chemistry 2013
Published by the Royal Society of Chemistry, www.rsc.org

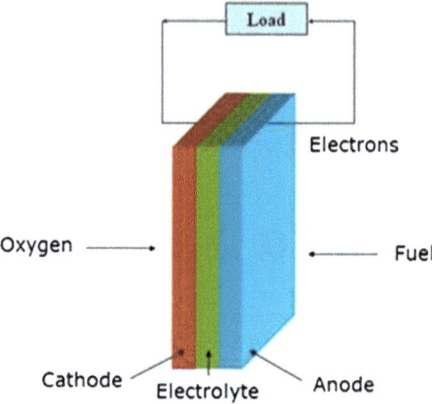

Figure 1.1   A simple schematic of a fuel cell.

continuously supplied to the anode and cathode, respectively. The ions, which are produced during the electrochemical reactions at one of the electrodes (*i.e.*, anode or cathode), are conducted to the other electrode through the electrolyte. The electrons produced during the electrochemical reactions are cycled from one electrode to the other via load. The flow of electrons forms an electric current, which effectuates work on the load. A single cell can only generate a small amount of power, which could be enough for some of the portable applications. However, for stationary applications, many single cells must be brought together to produce the required energy demand; a process referred to as 'stacking'.[1] This process is generally done by connecting single cells in series using bipolar plates or interconnects. These plates form the air and fuel channels as well as they conduct the electrons from one cell to another.

There are generally different types of fuel cells, which mainly differ from each other in terms of the electrochemical reactions that occur at the electrode and electrolyte interface and the type of ion conducting at the electrolyte. Generally, these fuel cells are categorized into two main groups: low- and high-temperature fuel cells. The most common and promising low-temperature fuel cell types and their applications are discussed in Section 1.2. There are mainly two high-temperature fuel cell types, namely, the Solid Oxide Fuel Cell (SOFC) and the Molten Carbonate Fuel Cell (MCFC). SOFC is the most employed high-temperature fuel cell type, which is the main topic of this chapter. SOFC is discussed in Section 1.3 and the subsequent sections.

1.2   Introduction to Low-Temperature Fuel Cells

Proton Exchange Membrane Fuel Cells (PEMFCs) and Direct Methanol Fuel Cells (DMFCs) are the most common low-temperature fuel cell types. PEMFC and DMFC consist of a proton conducting membrane, such as Nafion®, which is chemically highly resistant, mechanically strong, acidic, a good proton conductor, and water absorbent. The main difference between PEMFC and

Table 1.1 A comparison between PEMFCs and DMFCs.

		PEMFC	DMFC
Reactions	Anode	$H_2 \rightarrow 2H^+ + 2e^-$	$CH_3OH + H_2O \rightarrow CO_2 + 6H^+ + 6e^-$
	Cathode	$0.5O_2 + 2H^+ + 2e^- \rightarrow H_2O$	$1.5O_2 + 6H^+ + 6e^- \rightarrow 3H_2O$
	Overall	$H_2 + 0.5O_2 \rightarrow H_2O$	$CH_3OH + 1.5O_2 \rightarrow CO_2 + 2H_2O$
Benefits		• Fast startup capability • Compactness • Elimination of corrosion problems	• Using a less expensive fuel (methanol) • High energy density of methanol • Simple to use and very quick to refill
Challenges		• Need for expensive catalysts • CO poisoning problem • Water management problem	• Slow reaction kinetics at the anode • Methanol crossover • Water management problem
Application examples		• Passenger vehicles • Forklifts	• Laptops • Mobile phones

DMFC is the fuel entering the fuel cell: hydrogen in PEMFC and methanol in DMFC. The main reactions, benefits, challenges, and application examples of these fuel cells are compared in Table 1.1.

The alkaline fuel cell (AFC) has become popular particularly for powering space vehicles. However, the successful developments with PEMFC and DMFC have led to a decline in the interest in the AFC mainly due to issues related to cost, reliability, and ease of use. However, some types of AFC such as the Direct Borohydride Fuel Cell (DBFC), which uses a solution of sodium borohydride as fuel, remain promising. As the electrolyte and the fuel are mixed, it is simple to make this fuel cell. In addition, CO_2 poisoning can be prevented when highly alkaline fuel and waste borax are used.[2] However, the main challenge of this fuel cell is the side reaction known as hydrolysis reaction in which hydrogen is produced as $NaBH_4$ and reacts with water. Direct Formic Acid Fuel Cells (DFAFCs) and Direct Ethanol Fuel Cells (DEFCs), which have a proton exchange membrane, utilize formic acid and ethanol as some of the potential fuels, respectively. The main advantages of DFAFC appear to be high catalytic activity, easier water management, and minimal balance of plant. However, the performance of this fuel cell strongly depends on the feed concentration of formic acid due to mass transport limitations. DEFC may be more advantageous due to the benefits of ethanol such as high energy density, safety to use, and ease of storage. However, a lot of acetaldehyde, which is a very flammable and harmful liquid, is produced in the electrochemical reactions. In addition, DEFC reaction kinetics is very slow and ethanol crossover is a challenge.

Biofuel cells (BFCs) may be used in very low power applications. There are mainly two classes of BFC, such as microbial fuel cells and enzymatic fuel cells. The first one is more appropriate for applications such as powering underwater equipment since it has higher efficiency and complete oxidation of fuel. The latter may be used in small-scale applications such as implantable devices since it has high-power density but lower efficiency and incomplete oxidation of fuel.

1.3   Introduction to Solid Oxide Fuel Cells

A SOFC is a high-temperature fuel cell (ranging between 500 °C to 1000 °C) that contains an oxide ion-conducting electrolyte made from a ceramic material. The main application area of SOFC is stationary power and heat generation. SOFC can be used alone or integrated with other technologies such as gas turbine and gasification systems for this purpose. Another application area of the SOFC is in transportation such as being the auxiliary power unit of luxury automobiles or heavy duty commercial trucks. SOFC can also be used in military applications as this fuel cell can meet the power demands of the soldiers, which has been increasing due the new technologies (*e.g.* night-vision devices, global positioning systems, target designators, climate controlled body suits, and digital communication systems); and it could be operated with fuels such as diesel and JP-8 that are available in the battle area in any part of the world. SOFC can also be used in some specific niche applications such as miniature autonomous systems.

Compared to low-temperature fuel cells, SOFCs have important advantages including: i) simpler in concept since only solid and gas phases exist; ii) ability to utilize fuels such as carbon monoxide, methane, higher hydrocarbons, methanol, ethanol, and biomass produced gas; iii) internal reforming of the fuel; iv) efficient thermal integration with other technologies such as gas turbines and gasification systems; and v) no need for precious metal electrocatalysts. Some of the main disadvantages of SOFCs over the low-temperate fuel cells are considered to be i) challenges for construction and durability due to its high temperature; and ii) carbon deposition problem.

The operation principle of a SOFC is simple. Fuel and air are continuously fed into to the fuel and air channels, respectively. Oxygen molecules in the air stream diffuse into the cathode side and react with the electrons, which are cycling via the load, at the cathode and electrolyte interface, to form the oxide ions. These oxide ions pass through the electrolyte and react with the fuel (*e.g.* hydrogen and carbon monoxide molecules), which diffuse into the anode side, at the anode and electrolyte interface. Hence, gases such as water vapour and carbon dioxide, and electrons are formed. The flow of electrons generates the electric current. The electrochemical reactions at the anode, cathode, and the overall reaction for a hydrogen-fed SOFC are given as follows.

$$H_2 + O^{2-} \rightarrow H_2O + 2e^- \tag{1}$$

$$\frac{1}{2}O_2 + 2e^- \rightarrow O^{2-} \tag{2}$$

$$H_2 + \frac{1}{2}O_2 \rightarrow H_2O \tag{3}$$

In general, the following materials are used in a SOFC:[1] Ni-YSZ for anode, YSZ for electrolyte, LSM for cathode and magnesium-doped lanthanum chromite for interconnects. However, there is an increasingly crucial need for research to find better materials to help increase the performance of the SOFC.

For example, high-chromium containing steel, such as Crofer22APU or E-Brite, is currently considered for interconnects.

1.3.1 Classification of SOFC Systems

SOFCs may be classified according to their temperature level, cell and stack design, type of support, flow configuration, and fuel reforming type, as shown in Table 1.2.

1.3.1.1 Classification According to Temperature Levels

SOFCs may be classified as low-temperature (LT-SOFC), intermediate-temperature (IT-SOFC), or high-temperature (HT-SOFC). Increasing operating temperature decreases the resistivity of the cell components and increases the electrode kinetics. They in turn lead to an increase in the performance of the cell. In addition, a higher temperature of the exit of the SOFC will lead to a better thermal integration with other technologies, which results in better thermal efficiency. On the other hand, a higher operating temperature will cause problems such as weaker structural integrity, higher corrosion rates, higher materials costs, and longer start-up and shut-down time.

1.3.1.2 Classification According to Cell and Stack Designs

According to the cell and stack design, SOFCs may be classified as tubular, planar, and segmented-in-series (or integrated-planar). Among these design types, planar and tubular are the most common types. Planar design is more compact than the tubular design, since cells can be stacked without giving large voids. In addition, bipolar plates as used in the planar design provide simpler

Table 1.2 Classification of solid oxide fuel cells.

Classification criteria	Types
Temperature level	Low-temperature SOFC (LT-SOFC) (500 °C–650 °C)
	Intermediate temperature SOFC (IT-SOFC) (650 °C–800 °C)
	High-temperature SOFC (HT-SOFC) (800 °C–1000 °C)
Cell and stack designs	Planar SOFC (Flat-planar, radial-planar)
	Tubular SOFC (Micro-tubular, tubular)
	Segmented-in-Series SOFC (or Integrated-planar SOFC)
Type of support	Self-supporting (anode-supported, cathode-supported, electrolyte-supported)
	External-supporting (interconnect supported, porous substrate supported)
Flow configuration	Co-flow
	Cross-flow
	Counter-flow
Fuel reforming type	External reforming SOFC (ER-SOFC)
	Direct internal reforming SOFC (DIR-SOFC)
	Indirect internal reforming SOFC (IIR-SOFC)

series of electrical connection between cells and shorter current path. The manufacturing cost of planar SOFC is also lower. However, in the planar design, there is a need for gas-tight sealing; whereas in tubular design, the cells may expand and contract without any constraints. The segmented-in-series SOFC is a cross between tubular and planar geometries which have the advantages of thermal expansion freedom like the tubular and low-cost component manufacturing like the planar.

1.3.1.3 *Classification According to Support Types*

SOFCs may be manufactured as anode-supported, cathode-supported, or electrolyte-supported. The electrolyte-supported type may be suitable for HT-SOFC as the temperature of a SOFC increases, the ionic resistivity of its electrolyte decreases. For IT-SOFC and LT-SOFC, the electrolyte is generally manufactured in a very thin form and the fuel cell is either manufactured as anode or cathode-supported. These three types of manufacturing may also be called self-supporting configuration. SOFCs may also be supported externally, such as interconnect- supported and porous substrate supported SOFC.

1.3.1.4 *Classification According to Flow Configurations*

The flow configuration in a SOFC can be cross-flow, co-flow, or counter-flow. The choice of this configuration affects the temperature distribution within the cell and the stack. Recknagle *et al.*[3] showed that the co-flow configuration has the most uniform temperature distribution and the smallest thermal gradients for similar fuel utilization and average cell temperature. Schematics of planar co-, counter- and cross-flow SOFC stacks are shown in Figure 1.2.

1.3.1.5 *Classification According to Fuel Reforming Types*

Fuels that can be fed into SOFC (*e.g.* methane and syngas) are reformed into H_2 and/or CO, which are electrochemically oxidized in the SOFC. This reforming process may be outside the stack, which is called external reforming, or inside the stack, which is called internal reforming. There are two possible types of internal reforming which are called indirect internal reforming (IIR-SOFC) and direct internal reforming (DIR-SOFC). In the IIR-SOFC, the reformer section is separate from the other components inside the cell but in close thermal contact

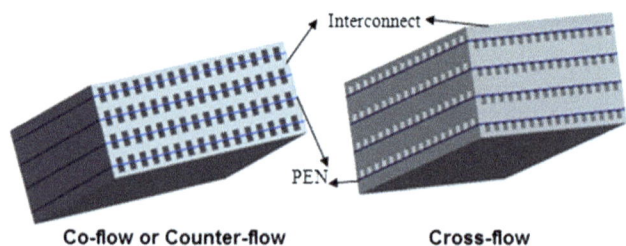

Co-flow or Counter-flow   **Cross-flow**

Figure 1.2   Planar SOFC stack with co-flow, counter-flow, and cross-flow configuration.

with the anode section; whereas in the DIR-SOFC, the reforming takes place directly on the anode catalyst. IIR-SOFC is effective in eliminating the carbon deposition problem. However, it is difficult to preserve the uniform temperature distribution in the stack in IIR-SOFC since the cells closer to the reforming section will be cooler due to the endothermic reforming reaction.

1.3.2 Fuel Options for SOFC

SOFC, when designed properly, might be operated with several fuels such as hydrogen, carbon monoxide, methane, higher hydrocarbons (*e.g.* butane), methanol, ethanol, ammonia, hydrogen sulfide, biogas, and syngas. Such fuels, excluding hydrogen and carbon monoxide, must be reformed into these gases to be electrochemically oxidized in the fuel cell. The steam reforming reactions for hydrocarbons (*e.g.* methane and butane), methanol, and ethanol are given in Eqs. (4) to (6), respectively. These reactions generally occur together with water-gas shift reaction, which is shown in Eq. (7). As can be seen in these equations, as a result of these equations, the fuel is reformed into hydrogen and carbon monoxide, as given below:

$$C_xH_y + xH_2O \rightleftharpoons xCO + \left(x + \frac{y}{2}\right)H_2 \tag{4}$$

$$CH_3OH \rightleftharpoons 2H_2 + CO \tag{5}$$

$$C_2H_5OH + 3H_2O \rightleftharpoons 2CO_2 + 6H_2 \tag{6}$$

$$CO + H_2O \rightleftharpoons CO_2 + H_2 \tag{7}$$

The other fuel option for SOFCs may be ammonia, which is an inexpensive and convenient way of storing hydrogen. A catalytic cracking of ammonia reaction, which represents the production of hydrogen from ammonia, is given in Eq. (8). One possible of usage of ammonia for SOFCs is vehicular applications. In a recent publication by Ehsan *et al.*,[4] an ammonia-fed SOFC based on proton conducting electrolyte with a heat recovery option is proposed and analyzed, as shown in Figure 1.3.

$$2NH_3 \rightleftharpoons N_2 + 3H_2 \tag{8}$$

Lu and Schaefer[5] studied the possibility of using hydrogen sulfide, which is known to be an extremely corrosive and noxious gas, as a fuel in SOFC. They suggested the usage of a H_2S decomposition reactor integrated with an SOFC as the direct use of H_2S in an SOFC causes anode deterioration over time. The decomposition reaction of H_2S is given as follows.

The decomposition of H_2S:

$$H_2S \rightleftharpoons H_2 + \frac{1}{x}S_x \tag{9}$$

Gas mixture produced from the conversion of biomass (*e.g.* wood, crops, and municipal solid waste) is another fuel option for SOFC. The conversion methods of several biomass feed stocks and their products are shown in Table 1.3. Among these products, syngas obtained from the biomass

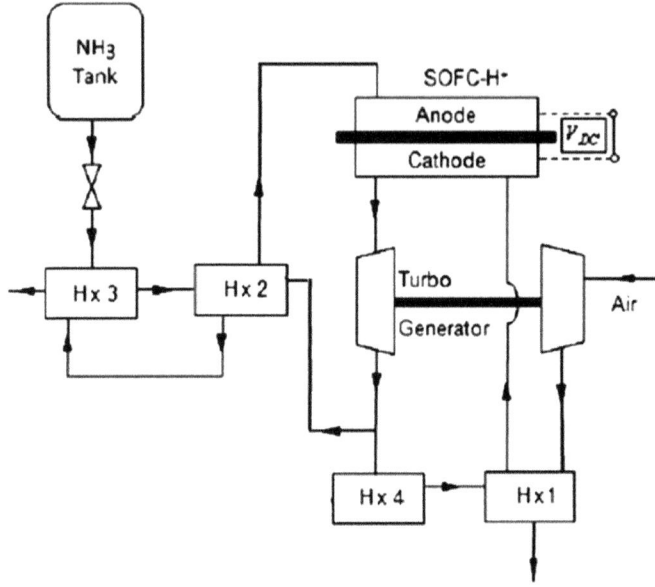

Figure 1.3 Schematic of an ammonia-fed SOFC based on proton conducting electrolyte (modified from Baniasadi and Dincer, 2011[4]).

Table 1.3 Some biomass feedstocks used as fuel in SOFC systems and their conversion methods.

Examples of Biomass Feedstock	Conversion method	Product
Wood, black liquor, municipal solid waste, dairy manure	Gasification	Syngas
Sewage sludge, animal waste	Anaerobic digestion	Biogas
Cellulosic waste, corn stover, sugarcane waste, wheat or rice straw	Fermentation	Ethanol
Wood, tyre rubber, starch, grape wastes, coconut shells	Fast pyrolysis	Bio-oil

gasification and biogas obtained from the anaerobic digestion are the most applicable fuel for biomass-fed SOFCs. Syngas produced by the system mainly consists of carbon monoxide, carbon dioxide, hydrogen, methane, water vapour, nitrogen, but also contaminants. The composition of the syngas depends mainly on the fuel, gasifier type, and gasification agent. Biogas mainly consists of methane and carbon dioxide, but also some amounts of nitrogen, oxygen, and contaminants. The product gases from gasification and anaerobic digestion need extensive cleanup before they are fed into SOFC. There are two different types of cleanup processes: a cold process involving gas cleaning at a reduced temperature and a hot process involving gas cleaning at a high temperature. The choice of the cleanup system depends on the temperature level of the SOFC and the other components in an integrated system. Bio-oil produced from pyrolysis of biomass is a liquid mixture of oxygenated compounds containing various chemical functional groups (*e.g.* carbonyl, carboxyl, and phenolic).[6] This mixture should be reformed to hydrogen with a steam reforming process before it is used in SOFC. The overall steam-reforming reaction of bio-oil is written as follows:

$$C_nH_mO_k + (2n - k)H_2O \rightleftharpoons nCO_2 + (2n + m/2 - k)H_2 \qquad (10)$$

1.4 Integrated SOFC Systems

A high exit temperature of SOFC provides an opportunity to achieve higher thermal efficiencies when SOFC is integrated with other systems. There are mainly two common types of integrated SOFC systems, namely, integrated SOFC and gas turbine systems and integrated SOFC and gasification systems. A configuration for the first system is shown in Figure 1.4. In this particular system, fuel and air compressors increase the pressure of fuel and air, respectively, according to the operating pressure level of SOFC. The unutilized

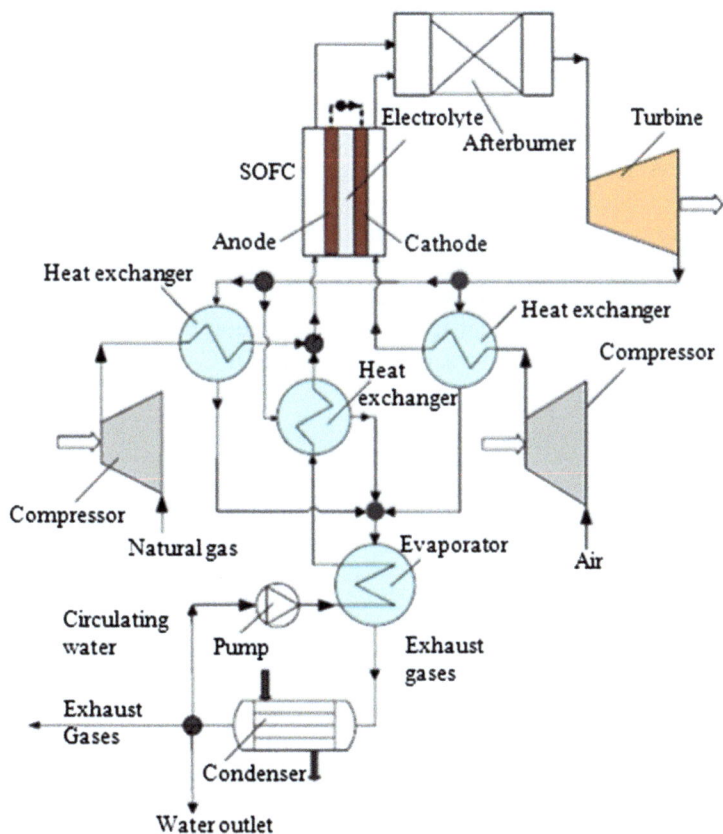

Figure 1.4 Schematic of an integrated SOFC and gas turbine system (modified from Dincer *et al.*, 2009[7]).

fuel in the SOFC exit stream is burned in an afterburner to increase the temperature of this stream. The gas mixture leaving the afterburner enters the gas turbine to generate power. The expanded gas provides the energy for increasing the temperature of the fuel (natural gas) and air compressor exits according to the SOFC inlet temperature requirement. The remaining energy of the gas mixture is used to provide the heat to generate steam in an evaporator. The steam produced in the evaporator enters the SOFC inlet to initiate the reformation process. The steam to carbon ratio should be well adjusted to prevent the carbon deposition possibility in the stack. A study by Zamfirescu et al.[7] has shown that the energy and exergy efficiencies of such a system can reach up to 70 and 80%, respectively.

Integrated SOFC and gasification systems have also been given significant attention from the SOFC community recently. Gasification systems can be fueled by either coal or biomass. As the latter one is a renewable resource, biomass is a more promising fuel for SOFCs for better environmental sustainability.

An integrated SOFC and biomass gasification are shown in Figures 1.5a and 1.5b, respectively. In both of these systems, wet biomass should be first dried as high levels of moisture in the feedstock can reduce the reaction temperature in the gasifier and lead to poorer product gas with higher levels of tar. In the conventional system, the gas mixture produced from the combustion of dried biomass process supplies heat to the HRSG where steam is produced. The steam produced in HRSG enters the steam turbine where the power is produced.

In the integrated SOFC and biomass gasification system, the dried biomass enters the gasifier where syngas is produced. A gas cleanup system cleans the syngas according to the SOFC impurity levels, in order not to cause any degradation in the SOFC. The cleaned syngas enters the SOFC, where the electricity is produced. Here, some amount of depleted fuel stream can be recirculated to the SOFC inlet to prevent the carbon deposition possibility in the SOFC. The fuel and air streams exiting the SOFC enter the afterburner to

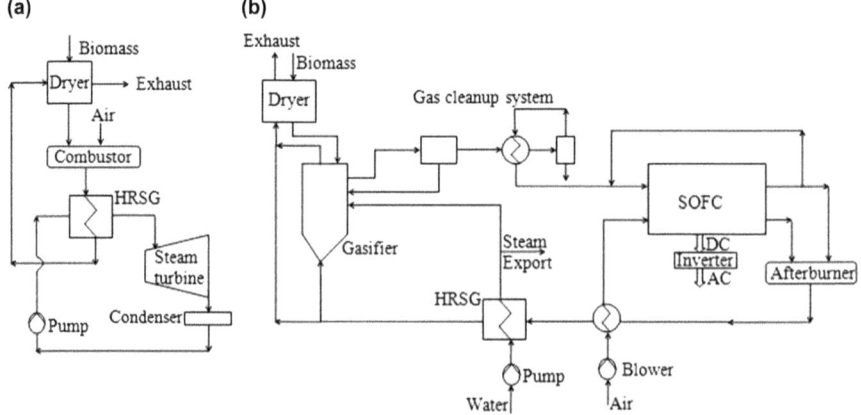

Figure 1.5 Schematic of (a) a conventional biomass fueled power production system using steam turbine and (b) an integrated SOFC and biomass gasification system (modified from Colpan et al., 2010[8] and Colpan et al., 2009[9]).

burn the unused fuel and increase the temperature of these streams. The gas mixture leaving the afterburner supplies heat to the following components, respectively: the blower, the HRSG, and the dryer. This gas mixture is emitted to the environment after exiting the dryer.

In an integrated SOFC and biomass gasification system, selection of gasifier and gasification agent plays an important role in the performance and cost of the system. Gasifiers may be classified according to the heat addition method and reactor type. Heat can be added in two ways: autothermal or allothermal. In autothermal gasification, necessary heat is provided by partial oxidation within the process; whereas in allothermal gasification, an external source is needed to supply the necessary amount of heat. There are also various reactor types that can be used in such a system, as can be seen in Table 1.4. In these gasifiers, air, oxygen, steam, or a combination of these may be used as gasification agents. A study by Colpan *et al.*[8] revealed that using steam as the

Table 1.4 Advantages and disadvantages of main biomass gasification reactor types.

Reactor type	Advantages	Disadvantages
Downdraft-fixed bed	Very simple and robust Low particulates and tar High exit gas temperature Moderate cost	Lower moisture level tolerability Scale-up limitations Feed size limitations

Updraft-fixed bed	Simple and reliable
	Higher moisture level
	  tolerability
	Low cost
	High thermal efficiency and
	  carbon conversion
		Very dirty product gas with
		  high levels of tars
		Scale-up limitations
		Intolerant to high portions
		  of fines in feed
		Low exit gas temperature

Bubbling fluid bed	Good temperature control
	Good scale-up potential
	Greater tolerance to particle
	  size range
	Large scale applications
		High particulates and
		  moderate tar
		Limited turn-down
		  capability
		Some carbon loss with ash
		Higher particle loading

Circulating fluid bed	Good temperature control
	Good scale-up potential
	Greater tolerance to
	  particle size range
	Large scale applications
		High cost at low capacity
		High particulates and
		  moderate tar
		Higher particle loading
		Difficulties with in-bed
		  catalytic processing

Entrained flow	Simple design Good scale-up potential Potential for low tar	Costly feed preparation Carbon loss with ash Limitations with particle size
Twin fluid bed	Good temperature control Greater tolerance to particle size range Large scale applications	High tar levels Difficult to scale-up High cost

Source: Adapted from Refs. 1, 10 and 11.

gasification agent yields better electrical and exergetic efficiencies for such an integrated SOFC system.

1.5   Basic SOFC Modelling

In this section, the fundamentals of SOFC modelling are discussed. The model discussed in this section may be called a zero-dimensional SOFC model. Please note that there are various other models (*e.g.* multi-dimensional, transient, and micro-level) available in the literature. A 2D and transient model is discussed in the next section.

The current generated due to the flow of electrons in a SOFC is proportional to the hydrogen utilized in the cell as shown in Eq. (11).

$$I = 2\dot{N}_{H_2,utilized} \cdot F \tag{11}$$

where F is the Faraday constant, which is approximately equal to 96485 C/mol.

The Nernst voltage (reversible cell voltage) may be shown as

$$V_N = \frac{-\Delta\bar{g}_r^\circ}{2F} - \frac{RT}{2F} \cdot \ln\left(\frac{P_{H_2O}}{P_{H_2} \cdot \sqrt{P_{O_2}/P^\circ}}\right) \tag{12}$$

The Nernst voltage may be given in terms of some non-dimensional parameters[1,12]

$$V_N = \frac{-\Delta\bar{g}_r^\circ}{2F} - \frac{RT}{2F} \cdot \ln\left(\frac{U_f}{(1 - U_f) \cdot \sqrt{\left(\frac{1 - U_a}{1/0.21 - U_a}\right) \cdot \frac{P}{P^\circ}}}\right) \tag{13}$$

where the fuel utilization ratio is the amount of hydrogen that is electrochemically reacted to the amount of hydrogen in the inlet stream; whereas air utilization ratio is the amount of oxygen that is electrochemically reacted to the amount of oxygen in the inlet stream. Fuel and air utilization ratios can be calculated using the following equations:

$$U_f = \frac{\dot{N}_{H_2,utilized}}{\dot{N}_{H_2,inlet}} \tag{14}$$

$$U_a = \frac{\dot{N}_{O_2,utilized}}{\dot{N}_{O_2,inlet}} \tag{15}$$

Here, the value of the actual cell voltage depends on the values of the ohmic, activation and concentration polarizations; and can be found using:

$$V = V_N - V_{ohm} - V_{act} - V_{con} \tag{16}$$

The ohmic polarization occurs due to the resistance to the flow of oxide ions through the electrolyte and resistance to the flow of electrons through the anode, cathode, and interconnects. The relationship between voltage drop and current density is written as follows, using Ohm's law:

$$V_{ohm} = \left(\sum_k \rho_k \cdot L_k \right) \cdot i \tag{17}$$

where the material resistivities are generally determined through conducting experimental measurements. The most significant resistance occurs at the electrolyte in a SOFC. The temperature dependence of the resistivity of YSZ, which is found using the Arrhenius equation, is given as follows:[13]

$$\rho = \left[334 \cdot \exp\left(\frac{-10300}{T} \right) \right]^{-1} (\Omega-cm)(T \text{ is in K}) \tag{18}$$

Here, it should also be noted that although there is contact resistance between the layers of the SOFC, this resistance is generally neglected in modelling.

The second type of polarization is called activation polarization which occurs due to the sluggishness of the reactions. The Butler-Volmer equation, which is shown in Eq. (19) can be used to find this polarization. Please note that charge transfer coefficient for anode and cathode is assumed as 0.5, as given in the following equation:

$$V_{act} = V_{act,a} + V_{act,c} = \frac{RT}{F} \cdot \sinh^{-1}\left(\frac{i}{2i_{o,a}} \right) + \frac{RT}{F} \cdot \sinh^{-1}\left(\frac{i}{2i_{o,c}} \right) \tag{19}$$

The third polarization is the concentration polarization, which is related to the voltage loss due to the diffusion of the gases in porous media. More specifically, when gases at the channels diffuse through the porous anode and cathode, the gas partial pressure at the electrochemically reactive sites becomes less than that in the bulk of the gas stream. Hence, a voltage drop occurs. If the microstructure is assumed not to be a function of position, this polarization may be given as follows (*e.g.*, ref. 14):

$$V_{conc,a} = -\frac{RT_s}{2F} \ln\left(1 - \frac{RT}{2F} \cdot \frac{\tau_a l_a}{D_a V_{v(a)} P_{H_2}^b} i \right) + \frac{RT_s}{2F} \ln\left(1 + \frac{RT}{2F} \cdot \frac{\tau_a l_a}{D_a V_{v(a)} P_{H_2O}^b} i \right) \tag{20}$$

$$V_{conc,c} = \frac{RT_s}{4F} \ln\left[\frac{P_{O_2}^b}{P - (P - P_{O_2}^b) \exp\left[\frac{RT}{4F} \cdot \frac{\tau_c l_c}{D_c V_{v(c)} P} i \right]} \right] \tag{21}$$

The power output of the cell may then be found as

$$\dot{W}_{FC} = I \cdot V \tag{22}$$

The electrical efficiency of the cell is calculated as

$$\eta_{el,cell} = \frac{\dot{W}_{FC}}{\dot{n}_{f,inlet} \cdot LHV} \qquad (23)$$

1.6 Case Study

As a case study, a conventional biomass fueled power production system (Figure 1.5a) is compared with an integrated biomass gasification and SOFC system (Figure 1.5b) in terms of efficiency and environmental impact. Both electrical and exergetic efficiencies, and specific greenhouse gas emissions are calculated for performance and greenhouse gas emission comparisons, respectively.

1.6.1 Analysis

In the modelling of the conventional biomass-fueled power production system (Figure 1.5a), two basic thermodynamic laws, such as the first law and the second law, are considered and the respective balance equations are written for the components of the system. In this model, it is assumed that complete combustion is achieved using 100% theoretical air. The heat recovered from the HRSG is first calculated applying an energy balance around the control volume enclosing the HRSG. Using the isentropic efficiencies of the components and the thermodynamic relations, steam produced in the HRSG is then calculated. Finally, using these findings, the power output of the steam turbine, power demand for the pump, and the net power output of the system are calculated.

For the SOFC, the transient heat transfer model developed by Colpan *et al.*[15] is used. The approach and main features of this model are as follows: A control volume around the repeat element found in the middle of a planar SOFC stack is taken. It is assumed that the other repeat elements show the same characteristics with this repeat element. The solid structure, *i.e.* electrodes, electrolyte, and interconnects, is modelled in 2D; whereas the air and fuel channels are modelled in 1D. Since the gases flow with low velocity to obtain high fuel utilization, it is assumed that fully developed laminar flow conditions are achieved at the air and fuel channels. Natural convection at the heat-up stage, forced convection at the start-up stage, conduction heat transfer between the solid parts, and all the voltage losses, *i.e.* activation, concentration, and ohmic, are taken into account in the modelling. The input parameters of this model are cell voltage, Reynolds number at the fuel channel inlet, excess air coefficient, temperature at the air and fuel channel inlets, pressure of the cell, molar gas composition at the air and fuel channel inlets, and the geometrical dimensions of the SOFC. The output parameters are the current density, temperature, molar gas composition, and carbon activity distributions, the heat-up and start-up time, the fuel utilization, the power output and the electrical efficiency of the cell. This model is validated with IEA benchmark test[16] and Braun's model.[17] The main equations of this model can be found in Tables 1.5 to 1.7. Equations to find the cell voltage and power density can be found in Section 1.5.

Table 1.5 Continuity equations considered.

Control volume	Continuity equations
Fuel channel	$\dfrac{d\dot{n}''_{CH_4}}{dx} = \dfrac{-\dot{r}''_{str}}{t_{fc}}$
	$\dfrac{d\dot{n}''_{H_2}}{dx} = \dfrac{3\dot{r}''_{str}}{t_{fc}} + \Delta\dot{n}'''_{wgs} - \dfrac{\dot{r}''_{el}}{t_{fc}}$
	$\dfrac{d\dot{n}''_{CO}}{dx} = \dfrac{\dot{r}''_{str}}{t_{fc}} - \Delta\dot{n}'''_{wgs}$
	$\dfrac{d\dot{n}''_{CO_2}}{dx} = \Delta\dot{n}'''_{wgs}$
	$\dfrac{d\dot{n}''_{H_2O}}{dx} = \dfrac{-\dot{r}''_{str}}{t_{fc}} - \Delta\dot{n}'''_{wgs} + \dfrac{\dot{r}''_{el}}{t_{fc}}$
	$\dfrac{d\dot{n}''_{N_2}}{dx} = 0$
Air channel	$\dfrac{d\dot{n}''_{O_2}}{dx} = \dfrac{-\dot{r}''_{el}/2}{t_{ac}}$
	$\dfrac{d\dot{n}''_{N_2}}{dx} = 0$

Source: Adapted from Ref. 15.

In the modelling part, the integrated SOFC and biomass gasification system, firstly, the syngas composition and the external heat needed for the gasifier are calculated by solving the set of equations derived from the thermodynamic modelling of the gasifier. These equations include three atom balances, two chemical equilibrium relations and the energy balance around the control volume enclosing the gasifier. Secondly, using the syngas composition and the heat transfer model of the SOFC, number of the SOFC stacks, molar flow rate of gases at the inlet and exit of the air and fuel channels, temperature at the exit of the air and fuel channels, and power output of the cell are found. Thirdly, combining the outputs of the gasifier and SOFC models, the molar flow rate of the dry biomass is calculated. Fourthly, applying thermodynamic principles to the components of the system, the enthalpy flow rate of all the states are calculated. Finally, using the laws of thermodynamics, work input to the auxiliary components, *i.e.* blower and pump, and net power output of the system are calculated.

Both electrical and exergetic efficiencies are selected as the performance assessment parameters. The electrical efficiency, as written in Eq. (24), is defined as the ratio of the net power output of the system to the lower heating value of the fuel. In regards to the exergetic efficiency, it is necessary to identify both a product and a fuel for the system being analyzed. The product represents

Table 1.6 Heat transfer equations.

Control volume	Heat transfer equations
Cathode interconnect	$\dfrac{1}{\alpha_{ci}}\cdot\dfrac{\partial T}{\partial t}=\dfrac{\partial^2 T}{\partial x^2}+\dfrac{\partial^2 T}{\partial y^2}$ $x=0\,\&\,x=L\Rightarrow\dfrac{\partial T}{\partial x}=0,\; y=0\Rightarrow\dfrac{\partial T}{\partial y}=0$ $y=t_{ci}\Rightarrow -k_{ci}\cdot\dfrac{\partial T}{\partial y}=\dfrac{w_{gas}}{w_{solid}}\cdot\left[h_{c,a}\cdot(T_{ci}-T_a)+h_{r,a}\cdot(T_{ci}-T_{pen})\right]+\left(1-\dfrac{w_{gas}}{w_{solid}}\right)\cdot k_{ci}\cdot\dfrac{(T_{ci}-T_{PEN})}{t_{ac}}$ $t=0\Rightarrow T=T_o$
Air channel	$\rho_{ac}\cdot c_{p,ac}\cdot\dfrac{\partial T}{\partial t}+\sum_i\dfrac{\partial}{\partial x}\left(\dot{n}_i''\cdot\bar{h}_i\right)=\dfrac{h_{c,a}(T_{PEN}-T_a)+h_{c,a}(T_{ci}-T_a)-(\dot{r}_{el}''/2)\cdot\bar{h}_{O_2}\cdot w_{solid}/w_{gas}}{t_{ac}}$ $x=0\Rightarrow\begin{array}{l}T=f(t)\,(Heat\text{-}up)\\ T=T_{w\_ac}\,(Start\text{-}up)\end{array}$ (co-flow), $\quad x=L\Rightarrow\begin{array}{l}T=f(t)\,(Heat\text{-}up)\\ T=T_{w\_ac}\,(Start\text{-}up)\end{array}$ (counter-flow) $t=0\Rightarrow T=T_o+100°C$
PEN	$\dfrac{1}{\alpha_{PEN}}\cdot\dfrac{\partial T}{\partial t}=\dfrac{\partial^2 T}{\partial x^2}+\dfrac{\partial^2 T}{\partial y^2}+\dfrac{1}{k_{PEN}}\dot{q}_{PEN}'''$ $x=0\,\&\,x=L\Rightarrow\dfrac{\partial T}{\partial x}=0$ $y=t_{ci}+t_{ac}\Rightarrow k_{PEN}\cdot\dfrac{\partial T}{\partial y}=\dfrac{w_{gas}}{w_{solid}}\cdot\left[h_{c,a}\cdot(T_{PEN}-T_a)+h_{r,a}\cdot(T_{PEN}-T_{ci})\right]+\left(1-\dfrac{w_{gas}}{w_{solid}}\right)\cdot k_{ci}\cdot\dfrac{(T_{PEN}-T_{ci})}{t_{ac}}$

$$y = t_{ci} + t_{ac} + t_{PEN} \Rightarrow -k_{PEN} \cdot \frac{\partial T}{\partial y} = \frac{w_{gas}}{w_{solid}} \cdot \left[h_{c,f} \cdot (T_{PEN} - T_f) + h_{r,f} \cdot (T_{PEN} - T_{ai})\right] + \left(1 - \frac{w_{gas}}{w_{solid}}\right) \cdot k_{ai} \cdot \frac{(T_{PEN} - T_{ai})}{t_{fc}}$$

$$t = 0 \Rightarrow T = T_o$$

Fuel channel

$$\rho_{fc} \cdot c_{p,fc} \cdot \frac{\partial T}{\partial t} + \sum_i \frac{\partial}{\partial x}(\dot{n}_i'' \cdot \bar{h}_i) =$$

$$\frac{h_{c,f}(T_{ai} - T_f) + h_{c,f}(T_{PEN} - T_f) + \left(\sum r_{prod}'' \cdot \bar{h}_{prod} - \sum r_{react}'' \cdot \bar{h}_{react}\right) \cdot w_{solid}/w_{gas}}{t_{fc}}$$

$$x = 0 \Rightarrow T = Tw\_fc\,(Start\text{-}up)(\text{co-flow}), \quad x = L \Rightarrow T = Tw\_fc\,(Start\text{-}up)(\text{counter-flow})$$

$$t = 0 \Rightarrow T = T_o$$

Anode interconnect

$$\frac{1}{\alpha_{ai}} \cdot \frac{\partial T}{\partial t} = \frac{\partial^2 T}{\partial x^2} + \frac{\partial^2 T}{\partial y^2}$$

$$x = 0 \,\&\, x = L \Rightarrow \frac{\partial T}{\partial x} = 0, y = t_{ci} + t_{ac} + t_{PEN} + t_{fc} + t_{ai} \Rightarrow \frac{\partial T}{\partial y} = 0$$

$$y = t_{ci} + t_{ac} + t_{PEN} + t_{fc} \Rightarrow -k_{ai} \cdot \frac{\partial T}{\partial y} = \frac{w_{gas}}{w_{solid}} \cdot \left[h_{c,f} \cdot (T_{ai} - T_f) + h_{r,f} \cdot (T_{ai} - T_{PEN})\right] + \left(1 - \frac{w_{gas}}{w_{solid}}\right) \cdot k_{ai} \cdot \frac{(T_{ai} - T_{PEN})}{t_{fc}}$$

$$t = 0 \Rightarrow T = T_o$$

Source: Adapted from Ref. 15.

Table 1.7 Auxiliary relations used in the modelling.

Name of the relation	Equation
Rate of electrochemical reaction	$\ddot{r}''_{el} = \dfrac{i}{2F}$
Rate of steam reforming of methane	$\ddot{r}''_{str} = 4274 \cdot P_{CH_4} \cdot \exp\left(\dfrac{-8.2 \times 10^4}{R \times T}\right)$
Chemical equilibrium constant of water-gas shift reaction	$K_{wgs} = \exp\left[-\Delta\bar{g}^\circ_{wgs}/RT\right] = \dfrac{x_{H_2} \cdot x_{CO_2}}{x_{CO} \cdot x_{H_2O}}$
Power density	$\dot{W}''_{el} = i \cdot V_{cell}$
Volumetric heat generation in PEN	$\dot{q}'''_{PEN} = \dfrac{\sum \Delta\dot{H}''_k - \dot{W}''_{el}}{t_{PEN}}$
Reynolds number at the fuel channel inlet	$Re_{D_h} = \dfrac{\dot{n}''_{k,fi} \cdot M_{mix} \cdot \left(2 \cdot t_{fc} \cdot w_{gas}\right)}{x_{k,fi} \cdot \mu_{mix} \cdot \left(t_{fc} + w_{gas}\right)}$
Excess air coefficient	$\lambda_{air} = \dfrac{\dot{n}''_{O_2,ai}}{\left(2 \cdot \dot{n}''_{CH_4,fi} + \dot{n}''_{CO,fi}/2 + \dot{n}''_{H_2,fi}/2\right)} \cdot \dfrac{t_{ac}}{t_{fc}}$
Fuel utilization	$U_f = \dfrac{\sum^m_{i=2}\ddot{r}''_{el} \cdot (\Delta x \cdot w_{solid})}{\left(4 \cdot \dot{n}''_{CH_4,fi} + \dot{n}''_{H_2,fi} + \dot{n}''_{CO,fi}\right) \cdot \left(w_{gas} \cdot t_{fc}\right)}$
Electrical efficiency	$\eta_{el} = \dfrac{\dot{W}_{SOFC}}{LHV \cdot \sum^6_{k=1} \dot{n}''_{k,fi} \cdot t_{fc} \cdot w_{gas}}$

Source: Adapted from Ref. 15.

the desired output produced by the system. The fuel represents the resources expended to generate the product. This efficiency can also be written in terms of the total exergy destructions and losses within the system, given as follows:

$$\eta_{el} = \frac{(\dot{W}_{net})_{system}}{\dot{n}_{fuel} \cdot LHV} \tag{24}$$

$$\varepsilon = \frac{\dot{Ex}_P}{\dot{Ex}_F} = 1 - \frac{\dot{Ex}_D + \dot{Ex}_L}{\dot{Ex}_F} \tag{25}$$

In addition, the environmental impact of these systems can be assessed calculating the specific greenhouse gas emissions, which is defined as the ratio of the GHG emission from the system to the net power output of the system.

From the viewpoint of energy and environment, the lower the ratio is, the more environmentally friendly the system is.

$$\sigma = \frac{\dot{m}_{GHG}}{(\dot{W}_{net})_{system}} \tag{26}$$

The input data used in this case study are tabulated in Table 1.8.

Table 1.8 Input data.

Environmental temperature	25 °C
Type of biomass	Wood
Ultimate analysis of biomass [%wt dry basis]	50% C, 6% H, 44% O
Moisture content in biomass [%wt]	30%
Exhaust gas temperature	127 °C
System-I	
Conditions of the steam entering the steam turbine	20 bar (saturated)
Pressure of the condenser	1 bar
Isentropic efficiency of the steam turbine	80%
Isentropic efficiency of the pump	80%
Electricity generator efficiency	98%
System-II	
Moisture content in biomass entering the gasifier [%wt]	20%
Temperature of syngas exiting the gasifier	900 °C
Temperature of steam entering the gasifier	300 °C
Molar ratio of steam to dry biomass	0.5
Number of cells per SOFC stack	50
Temperature of syngas entering the SOFC	850 °C
Temperature of air entering the SOFC	850 °C
Pressure of the SOFC	1 atm
Cell voltage	0.7 V
Reynolds number at the fuel channel inlet	1.2
Excess air coefficient	7
Active cell area	$10 \times 10 \, cm^2$
Number of repeat elements per single cell	18
Flow configuration	Co-flow
Manufacturing type	Electrolyte-supported
Thickness of the air channel	0.1 cm
Thickness of the fuel channel	0.1 cm
Thickness of the interconnect	0.3 cm
Thickness of the anode	0.005 cm
Thickness of the electrolyte	0.015 cm
Thickness of the cathode	0.005 cm
Pressure ratio of the blowers	1.18
Isentropic efficiency of the blowers	0.53
Pressure ratio of the pump	1.2
Isentropic efficiency of the pump	0.8
Inverter efficiency	0.95

Source: Adapted from Ref. 9.

1.6.2   Results and Discussion

The models discussed in Section 1.6 are simulated to find the performance and environmental impact of the systems studied using the data given in Table 1.1. The results and discussion of these calculations and simulations are given in this section.

For the integrated SOFC and biomass gasification system shown in Figure 1.5b, the syngas composition is first calculated as: 2.08% CH_4, 42.75% H_2, 25.80% CO, 9.44% CO_2 and 19.93% H_2O. Using this composition and the data given in Table 1.1, the SOFC model is simulated. It is found that the fuel utilization ratio of the SOFC is 82%; and the average current density is 0.253 A/cm^2 for the cell operating voltage of 0.7 V. The distribution of the current density with respect to the flow direction is also shown in Figure 1.6. The carbon activity distribution through the flow direction is found to check the carbon deposition possibility (*i.e.* the conditions where the carbon activity exceeds 1 for any location). In general, the carbon deposition possibility is more severe at the fuel channel inlet, as can be seen in Figure 1.6. It is also found that the carbon activity is less than 1 for all the locations for this case study.

Figure 1.7 shows the temperature distribution of the SOFC when the system reaches the steady-state condition. As can be seen from this figure, there is a sudden temperature drop at the x direction, *i.e.* flow direction, due to the endothermic steam reforming reaction and then the temperature increases due to exothermic electrochemical and water-gas shift reactions. This figure also shows that there is not a significant temperature change at the y direction, *i.e.* cell thickness direction. At the exit of the fuel and air channels, the temperatures of these exits are both found to be 1000 °C.

The electrical and exergetic efficiencies of the System-I (Figure 1.5a) and System-II (Figure 1.5b) are compared for the operating data given in Table 1.1.

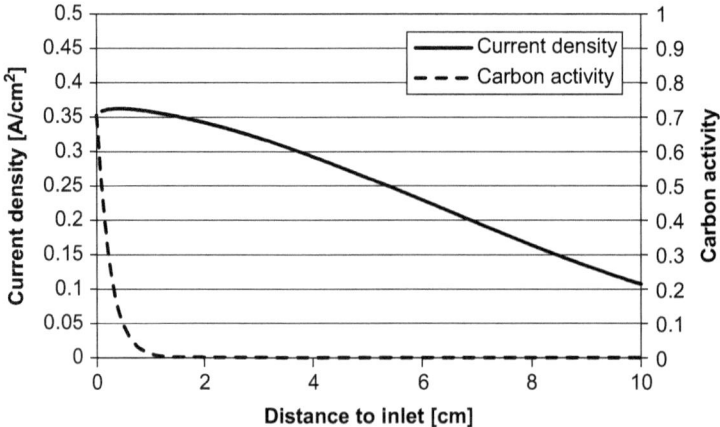

Figure 1.6   Current density and carbon activity distributions of the SOFC in the system shown in Figure 1.5b (modified from C.O. Colpan *et al.*, 2009[9]).

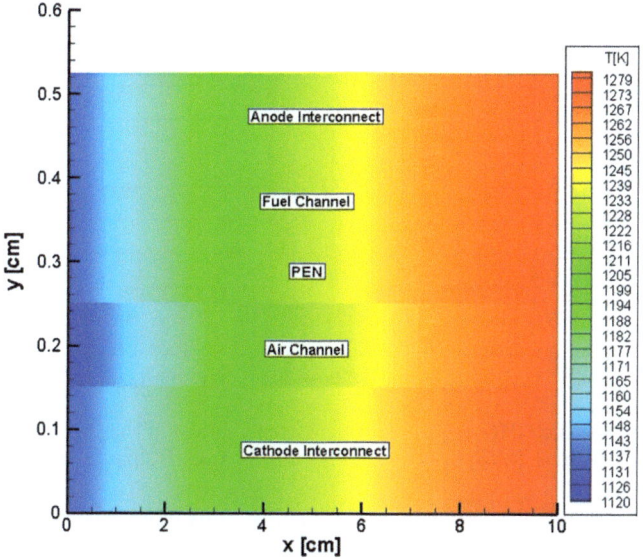

Figure 1.7 Temperature distribution of the SOFC in the system shown in Figure 1.5b (modified from C.O. Colpan *et al.*, 2009[9]).

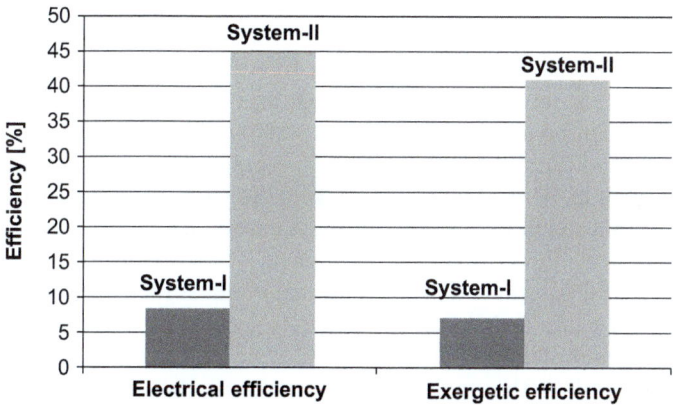

Figure 1.8 Electrical and exergetic efficiencies of the systems shown in Figure 1.5a (System-I) and Figure 1.5b (System-II) (modified from C.O. Colpan *et al.*, 2009[9]).

As shown in Figure 1.8, the electrical and exergetic efficiencies of the System-I are found as 8.3% and 7.2%, respectively; whereas the electrical and exergetic efficiencies of the System-II are found to be 44.9% and 41.1%, respectively.

The environmental impact of the systems studied is compared calculating the specific GHG emissions from these systems. It is found that System-I (Figure 1.5a) has higher GHG emissions compared to System-II (Figure 1.5b).

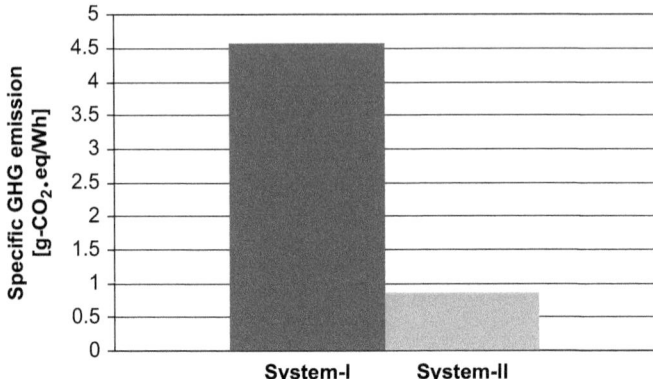

Figure 1.9   Specific GHG emissions of the systems shown in Figure 1.5a (System-I) and Figure 1.5b (System-II) (modified from C.O. Colpan *et al.*, 2009[9]).

As shown in Figure 1.9, the specific GHG emissions from System-I and System-II are found as 4.564 g-CO_2.eq/Wh and 0.847 g-CO_2.eq/Wh, respectively.

1.7   Conclusions

SOFCs are considered as one of the most feasible energy power generating devices for converting fuel into electricity, due to their high efficiency and low greenhouse gas emissions. A wide range of fuel (*e.g.* natural gas, syngas, and ammonia) can be used in these fuel cells. In addition, integrating SOFC with other systems (*e.g.* gas turbine and gasification systems), the thermal efficiency of the system can be increased significantly. However, there are challenges for construction and durability due to its high temperature. Furthermore, carbon deposition should be controlled by recirculating the depleted fuel stream or sending external steam to the SOFC inlet. A case study is presented to compare the performances and environmental impacts of an advanced biomass gasification and SOFC system with a conventional biomass fueled power production system using a steam turbine as the electricity generator. A heat transfer model for the SOFC and thermodynamic models for the rest of the components of the systems are used in the analyses. The results of the case study conducted showed that the SOFC and biomass gasification system has higher electrical and exergetic efficiencies, and lower specific GHG emissions.

Nomenclature

A	cross-sectional area, cm^2
c_p	specific heat at constant pressure, J g^{-1} K^{-1}
D	diffusivity, cm^2 s^{-1}
$\dot{E}x$	Exergy rate, kW
F	Faraday constant, C
$\bar{g}$	specific molar gibbs free energy, J $mole^{-1}$

h	heat transfer coefficient, $W\,cm^{-2}\,K^{-1}$
$\bar{h}$	specific molar enthalpy, $J\,mole^{-1}$
$\dot{H}$	enthalpy flow rate, W
i	current density, $A\,cm^{-2}$
i_o	exchange current density, $A\,cm^{-2}$
I	current, A
k	thermal conductivity, $W\,cm^{-1}\,K^{-1}$
L	length of the cell, cm
LHV	lower heating value, $J\,mole^{-1}$
$\dot{m}$	mass flow rate, $g\,s^{-1}$
M	molecular weight, $g\,mole^{-1}$
$\dot{n}$	molar flow rate, $mole\,s^{-1}$
P	pressure, bar
$\dot{q}$	heat transfer rate, W
$\dot{r}$	conversion rate, $mole\,s^{-1}$
R	universal gas constant, $J\,mole^{-1}\,K^{-1}$
Re_{D_h}	Reynolds number in an internal flow
t	time, s; thickness, cm
T	temperature, K
U_a	air utilization ratio
U_f	fuel utilization ratio
V	voltage, V; volume, cm^3
V_o	reference volume, cm^3
V_v	Porosity
w	width, cm
$\dot{W}$	power output, W
x	molar concentration

Greek letters

ε	exergetic efficiency
ρ	electrical resistivity of cell components, $ohm\,cm$; mass density, $g\,cm^{-3}$
η_{el}	electrical efficiency
λ_{air}	excess air coefficient
τ	tortuosity
μ	viscosity, $g\,s^{-1}\,cm^{-1}$
α	thermal diffusivity, $cm^2\,s^{-1}$
σ	first principal thermal stress, MPa; specific greenhouse gas emissions, $g\cdot CO_{2eq}/Wh$
σ_o	characteristic strength, MPa

Subscripts

a	anode; air
ac	air channel

act	activation
ai	anode interconnect
ave	average
c	cathode; convection
ci	cathode interconnect
conc	concentration
D	destruction
e	electrolyte
el	electrochemical; electrical
F	fuel
fc	fuel channel
fi	fuel channel inlet
ohm	ohmic
L	loss
mix	mixture
N	Nernst
o	standard
P	product
PEN	positive/electrolyte/negative
prod	product
r	reaction; radiation
react	reactant
s	solid structure
str	steam reforming reaction for methane
w	wall
wgs	water gas shift reaction

Superscripts

b	bulk
o	standard state

References

1. C. O. Colpan, Thermal modeling of solid oxide fuel cell based biomass gasification systems, PhD thesis, Department of Mechanical and Aerospace Engineering, Carleton University, 2009.
2. C. O. Colpan, I. Dincer and F. Hamdullahpur, Portable fuel cells – Fundamentals, technologies and applications, in *Mini-Micro Fuel Cells: Fundamentals and Applications*, eds. S. Kakac, A. Pramuanjaroenkij, L. Vasiliev, *NATO Science for Peace and Security Series*, Springer, Netherlands, 2008, pp. 87–101.
3. K. P Recknagle, R. E. Williford, L. A. Chick, D. R. Rector and M. A. Khaleel, Three-dimensional thermo-fluid electrochemical modeling of planar SOFC stacks, *J. Power Sources*, 2003, **113**, 109–114.

4. E. Baniasadi and I. Dincer, Energy and exergy analyses of a combined ammonia-fed solidoxide fuel cell system for vehicular applications, *Int. J. Hydrogen Energy*, 2011, **36**, 11128–11136.

5. Y. Lu and L. A. Schaefer, Solid oxide fuel cell system fed with hydrogen sulfide and natural gas, *J. Power Sources*, 2004, **135**, 184–191.

6. D. C. Dayton, Fuel cell integration-a study of the impacts of gas quality and impurities, Milestone Completion Report, *National Renewable Energy Laboratory*, U.S.A. 2001.

7. I. Dincer, M. A. Rosen and C. Zamfirescu, Exergeticperformance analysisof a gas turbine cycle integratedwith solid oxide fuel cells, *J. Energy Resour. Technol*, 2009, **131**, 032001-1–032001-11.

8. C. O. Colpan, F. Hamdullahpur, I. Dincer and Y. Yoo, Effect of gasification agent on the performance of solid oxide fuel cell and biomass gasification systems, *Int. J. Hydrogen Energy*, 2010, **35**, 5001–5009.

9. C. O. Colpan, F. Hamdullahpur, I. Dincer, Y. Yoo, Efficiency and environmental impact analyses of biomass based power production systems, *Proceedings of Global Conference on Global Warming*, Istanbul, Turkey, July 5–9, 2009.

10. A. V. Bridgwater, Renewable fuels and chemicals by thermal processing of biomass, *Chem. Eng. J.*, 2003, **91**, 87–102.

11. R. L. Bain, Biomass gasification overview, *Presentation. National Renewable Energy Laboratory*, 2004.

12. C. O. Colpan, I. Dincer and F. A. Hamdullahpur, A review on macro level modeling of planar solid oxide fuel cells, *Int. J. Energy Res.*, 2008, **32**, 336–355.

13. U. G. Bossel, *Final Report on SOFC Data Facts and Figures*, Berne, CH: Swiss Federal Office of Energy, 1992.

14. J. Kim, A.V. Virkar, K. Fung, K. Mehta and S.C. Singhal, Polarization effects in intermediate temperature, anode-supportedsolid oxide fuel cells, *J. Electrochem. Soc*, 1999, **146**(1), 69–78.

15. C. O. Colpan, F. Hamdullahpur and I. Dincer, Heat-up and start-up modeling of direct internal reforming solid oxide fuel cells, *J. Power Sources*, 2010, **195**, 3579–3589.

16. E. Achenbach, *SOFC Stack Modelling, Final Report Of Activity A2, Annex II: Modelling and Evaluation of Advanced Solid Oxide Fuel Cells*, International Energy Agency Programme on R, D&D on Advanced Fuel Cells, Juelich, Germany, 1996.

17. R. J. Braun, Optimal design and operation of solid oxide fuel cell systems for small-scale stationary applications, Phd Thesis, University of Wisconsin-Madison, 2002.

CHAPTER 2

Electrolyte Materials for Solid Oxide Fuel Cells (SOFCs)

YU LIU,[a] MOSES TADE[b] AND ZONGPING SHAO*[a,b]

[a] State Key Laboratory of Materials Oriented Chemical Engineering, Nanjing University of Technology, No.5 Xin Mofan Road, Nanjing 210009, China; [b] Department of Chemical Engineering, Curtin University, Perth, WA 6845, Australia
*Email: shaozp@njut.edu.cn

2.1   A General Introduction to Electrolyte of SOFCs

SOFCs are a kind of high-temperature fuel cell. Their name is derived from their oxide-based electrolyte, this clearly demonstrates the critical role of electrolyte in the SOFC system. The electrolytes of SOFCs, composed of a kind of solid crystalline oxide material, are pure ionic conductors with negligible electron conductivity. The electrolytes in SOFCs are typically fabricated into dense type, acting as an isolated layer to avoid the direct mixing of fuel gas and oxidizing gas. On the whole, ohmic resistance, which causes the losses of voltage according to Ohm's law, often considerably dominates the performance of SOFCs, in particular for cells with thick electrolyte. Therefore, the electrolyte exerts a vital impact on the final performance of SOFCs. Steele *et al.* set the target value for area specific resistance (ASR), a common index to indicate the resistance of materials in SOFCs, of the electrolyte as 0.15 Ω cm^2,[1] which can be reached for conventional yttria stabilized zirconia electrolyte with a membrane thickness of 15 μm at 700 °C. In practice, an operation temperature of 1000 °C was selected for Westinghouse's cathode-supported tubular cell with YSZ electrolyte membrane thickness of 30–40 μm. High operating temperature,

RSC Energy and Environment Series No. 7
Solid Oxide Fuel Cells: From Materials to System Modeling
Edited by Meng Ni and Tim S. Zhao
© The Royal Society of Chemistry 2013
Published by the Royal Society of Chemistry, www.rsc.org

however, has become a major obstacle for the widespread application of SOFC technology because of the related high costs, long start-up times, low long-term security and poor portability. Nowadays, it is generally believed that lowering the operating temperature of SOFCs to the intermediate range, *i.e.*, 500–800 °C, is of critical importance to accelerate the practical application of SOFCs. The electrolyte materials, however, essentially determine the operating temperature of SOFCs.

2.2 The Requirements of Electrolyte

A fuel cell is composed of a porous anode and a porous cathode sandwiched with a dense electrolyte. The ions, yielded at one electrode, transport to another electrode through electrolyte and the charge is balanced by the electron flow through external circuit. It means the electrolyte faces both the anode and the cathode atmospheres. To perform well as an electrolyte of SOFC, it should meet certain requirements in terms of stability, conductivity and compatbility.[2]

Stability: Since the electrolyte exposes to both atmospheres, as a basic requirement the electrolyte should possess sufficient stability to undergo no chemical changes in the reducing atmosphere on the anode side and oxidizing atmosphere on the cathode side at the wide range of operating temperatures of SOFCs; in addition, the electrolyte also needs to possess sufficient morphological and dimensional stabilities so as to prevent the phase transition and mechanical damage in the long-term operation at high temperature.

Conductivity: To minimize internal shorted current, the electrolyte should have negligible electronic conductivity, otherwise it would give birth to the voltage drop due to the electronic current flowing through the electrolyte; since the ion conductivity of electrolyte often dominates the performance of SOFC, it should be as high as possible to minimize ohmic losses; in addition, the ionic conductivity of electrolyte must also possess sufficient long-term stability.

Compatability: The chemical properties of the electrolyte should be compatible with those of other cell components, chemical interaction and elemental diffusion between the electrolyte and the other components should be limited when selecting other cell materials and the goal is to lower such unacceptable effects to a minimum level, as insulating phase formation, stability reduction, change in thermal expansion, introduction of significant electronic conductivity in the electrolyte; to maintain a sufficient mechanical strength and to prevent cracking and delamination under fabrication and operation period, electrolyte must match well with other components of SOFC in thermal expansion, in addition the thermal expansion coefficient (TEC) of the electrolyte material during the operation must be stabilized under the mutative oxygen partial pressures of the fuel and oxidant gas condition, even the exploitation of anode and cathode material must be based on the matching of their TECs with the electrolyte. In addition, the electrolyte must be dense enough to avoid the mixing of oxidant gas and fuel gas, which will lead to the losses of voltage. Other desirable properties for the SOFC electrolyte are high strength and toughness, fabrication and low cost.

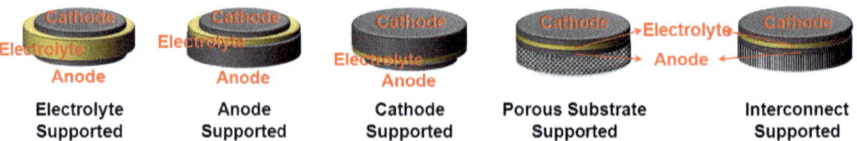

| Electrolyte
Supported | Anode
Supported | Cathode
Supported | Porous Substrate
Supported | Interconnect
Supported |

Figure 2.1   Illustration of the different types of cell support types of cell support architectures for SOFCs.

2.3   Classification of Electrolytes

Most of the research on the electrolytes of SOFCs in the 1960s was focused on the optimization of the ionic conductivity. Based on the layer that mechanically supports the fuel cells, SOFCs can be commonly divided into five categories, as shown in Figure 2.1. Anyway, the ASR of an appropriate solid-oxide electrolyte should be less than 0.15 $\Omega\,cm^2$ whatever the types of SOFCs are.[3] If the 15 μm dense impermeable electrolyte layer can be reliably produced using low-cost processing routes, the acceptable ASR can be achieved once the electrolyte conductivity reaches 1×10^{-2} $S\,cm^{-1}$. Oxide-ion and proton conducting oxide materials, classified by the conduction of their different types of ions, have been widely exploited for use as electrolytes of SOFCs.

2.3.1   Oxygen-ion Conducting Electrolyte

Most oxide-ion conducting materials possess crystal structure of the fluorite type AO_2, where A site is occupied by tetravalent cation.[4,5] Figure 2.2 shows the temperature dependence on the ionic conductivity for several typical oxygen-ion conducting oxides with the fluorite structure.[6]

The highest conductivity of fluorite-type oxide-ion conducting materials was observed in bismuth oxide compositions,[7] in particular, the ion conductivity of high-temperature δ-Bi_2O_3 phase with defective fluorite structure superlatively reaches as high as 2.3 $S\,cm^{-1}$ at 800 °C. However, it is stable only in a narrow temperature range from 730 °C to the melting point at 804 °C. Below 730 °C, the material changes to α-phase Bi_2O_3, which has ordering oxygen vacancies. By heating up again to above 730 °C, a first order vacancy transferring from order to disorder occurs, which causes the increase of conductivity by nearly three orders of magnitude.[8,9] Takahashi *et al.* demonstrated that the δ-Bi_2O_3 was stabilized to lower temperatures by partial cation substitution for Bi.[10,11] For instance, the δ-Bi_2O_3 phase was stable in the composition range of 25–43 mol% Y_2O_3, 22–27 mol% WO_3, or 35–50 mol% Gd_2O_3, 17.5–45 mol% Er_2O_3, 28.5–50 mol% Dy_2O_3, 30–40 mol% Sm_2O_3, 15–26 mol% Nb_2O_5, 20–25 mol% Ta_2O_5, 10–35 mol% $Pr_2O_{3.66}$ and 30–50 mol% $Tb_2O_{3.5}$.[12–19] Bourja *et al.* obtained the co-existence of cubic and tetragonal structures with partial substitution of Bi^{5+} by Ce^{4+}.[20] It was reported that dopants with an ionic radius smaller than that of Bi^{3+} helped to stabilize the high temperature δ-phase due to induced contraction of the open structure of δ-Bi_2O_3 by the dopants.[21] On the other hand, the introduction of dopants with a relative large

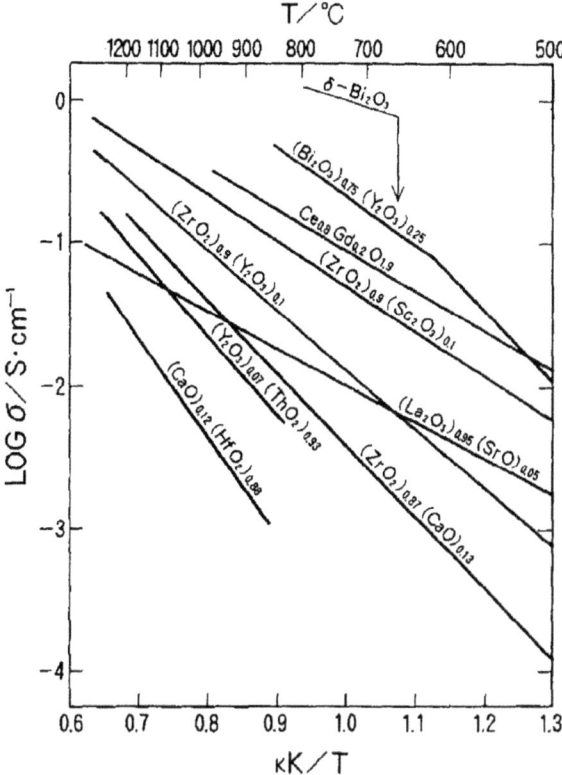

Figure 2.2 Electrical conductivity of fluorite oxides.[6]

ionic radius like La, Nd, Sm, and Gd into the Bi_2O_3 lattice was found to induce the formation of a rhombohedral structure in Bi_2O_3.[8] Compared to single-dopant systems, co-doping using two different metal oxides facilitates the stabilization of the δ-Bi_2O_3 phase down to room temperature at a lower doping concentration due to the entropy increase in the quaternary systems.[22,23] Wang *et al.* chose Gd_2O_3, Nb_2O_5, Sc_2O_3, ZrO_2, and BaO as co-dopants with Y_2O_3 to stabilize δ-phase Bi_2O_3 ceramics.[24] At 700 °C, the highest conductivity of 2.35×10^{-2} S cm^{-1} was reached for $(Ca_{0.1}W_{0.15}Bi_{0.75})_2O_{3.35}$, which may result from the higher anion vacancy concentration and more symmetrical structure.[25] Arasteh *et al.* compared lattice parameters of the cubic phases of $(Bi_2O_3)_{0.88}(Y_2O_3)_{0.06}(Er_2O_3)_{0.06}$ with that of $(Bi_2O_3)_{0.86}(Y_2O_3)_{0.07}(Er_2O_3)_{0.07}$ and revealed that the lattice parameter decreased by raising the dopant concentration.[26] The $Sm_{0.2-x}Bi_xCe_{0.8}O_{1.9}$ had much higher ionic conductivity than $Sm_{0.2}Ce_{0.8}O_{1.9}$ (SDC), and reached the highest conductivity value of 3.982 S m^{-1} at 750 °C.[27] Unfortunately, the bismuth oxide compositions are easily reduced under low oxygen partial pressure and decompose into bismuth metal at low oxygen partial pressure. Thus, practical use of stabilized Bi_2O_3 as electrolytes in SOFC is still questionable.[28]

Up to now, the most well-investigated fluorite-type oxide-ion conductors are stabilized ZrO_2. Pure ZrO_2 is in monoclinic structure at room temperature and undergoes phase transitions to tetragonal structure above 1170 °C, and finally to the cubic fluorite symmetry structure above 2300 °C.[29] However, pure zirconia is not suitable as a good electrolyte since its negligible ionic conductivity at operating temperature range for a SOFC and the appearance of phase transition during the heating/cooling process. The properties of ZrO_2-based oxide electrolytes with a series of divalent and trivalent cation dopants have been studied for many years. The HotElly project (High Operating Temperature Electrolysis) and that of Westinghouse Electric in the 1980s previously mentioned stabilized zirconia as electrolye for SOFCs. This substitution not only stabilized the cubic fluorite structure but also produced a large concentration of oxygen vacancies by charge compensation. The dissolution of yttria into the fluorite phase of ZrO_2 can be written by the following defective equation in Kroger-Vink notation:[30]

$$Y_2O_3(ZrO_2) \rightarrow 2Y'_{Zr} + 3O^{\times}_O + V_O \qquad (1)$$

A proper oxygen vacancy concentration would lead to a high ion mobility, and consequently good oxide-ion conductivity.[28] If the dopant content is not sufficient to fully stabilize the cubic structure, the material may contain mixed phases. The minimum amount of dopant required to completely stabilize the cubic structure is 12–13 mol% for CaO, 8–9 mol% for Y_2O_3 and Sc_2O_3, and 8–12 mol% for other rare-earth oxides.[31] Yttrium-stabilized zirconia (8 mol% Y_2O_3, abbreviated to YSZ) is the state-of-art electrolyte material for high temperature SOFCs. YSZ material displayed a high ion conductivity at about 1000 °C. The resistance of grain boundaries, caused by the impurities or second phases, which are easily introduced during the fabrication process, could exert a negative influence on the ion conductivity of electrolyte. In general, the detrimental impact of grain boundaries on the conductivity of YSZ may be significant at lower temperatures, while it is negligible at around 1000 °C.[28] Such a high operating temperature, however, could cause many problems, such as electrode sintering, interfacial mass diffusion between electrolyte and electrodes and mechanical stress due to different TECs.[32,33] Several ways have been exploited to reduce the thickness of YSZ electrolyte layer, thus lowering the operating temperature of SOFC, such as electrochemical vapour deposition (EVD),[34] chemical vapour deposition (CVD),[35] sol–gel processing,[36] spray pyrolysis[37] and so on. Ge *et al.* developed a cost-effective screen-printing method to fabricate YSZ films on NiO-YSZ substrates, achieving a maximum power density of 1.30 W cm^{-2} at 850 °C.[38] Another problem of YSZ electrolyte is its chemical interactions with some cathode materials, in particular cobalt-based perovskites, producing insulation phases around interface at high operating temperatures.[39–41]

Extensive research works have also been focused on an alternative material of Sc-doped ZrO_2 (ScSZ), which displayed superior ionic conductivity at intermediate temperature.[42,43] The excellent conductivity of ScSZ was ascribed to the low association enthalpy of the defective reactions and the similarities of the ionic radii of Sc^{3+} and Zr^{4+}, as compared to that of Zr^{4+} and Y^{3+}.[44–46] At 780 °C, the conductivity of ScSZ is comparable to that of YSZ at 1000 °C.[47–48] The activation energy for oxygen-ion conduction in ScSZ tends to increase with decreasing temperature, so conductivity of ScSZ is similar to or even lower than that of YSZ below 500 °C. Yamamoto *et al.* observed that samples with a low content of Sc_2O_3 (7–9 mol%) sintered at 1700 °C for 12 h had monoclinic and tetragonal phases.[47] The samples with a higher content of Sc_2O_3 (10–15 mol%) presented rhombohedral phase, and transformed to cubic phase at 600 °C, which had the highest conductivity due to the open structure and enhanced mobility of charge carriers. The presence of a rhombohedral phase in the ScSZ system has been explained by the fact that the radius of scandium cation is approximately the same as that of zirconium.[49] During the operation, however, the instability of ScSZ in phase structure significantly affects its conductivity. The conductivity aging behavior of ScSZ was attributed to the disappearance of a distorted cubic fluorite phase, which has a higher conductivity than the rhombohedral phase.[50] ZrO_2 with 8 mol% Sc_2O_3 exhibited a significant aging effect when annealing at 1000 °C. Its conductivity at 1000 °C (as sintered) decreased from an initial value of 0.3 $S\,cm^{-1}$ to 0.12 $S\,cm^{-1}$ after the aging at 1000 °C for 1000 h. On the contrary, ScSZ electrolyte with 10 or more mol% demonstrated the highest conductivity and did not show any decrease in conductivity at 1000 °C. However, a cubic-rhombohedral phase transformation appeared at a temperature of about 500 °C. Co doping can also be used to enhance the ion conductivity of electrolyte materials. For example, the addition of Bi_2O_3 to ScSZ electrolyte stabilized the cubic crystalline phase and sharply increased the conductivity of the system containing 2 mol% of Bi_2O_3.[51] Table 1 shows the conductivity, before and after annealing, as well as bending strengths, of various zirconia-based electrolytes along with their TECs.[30]

Table 2.1 Electrical conductivity, bending strength, and thermal expansion coefficient of zirconia-based electrolytes.[30]

	Conductivity at 1000 °C ($S\,cm^{-1}$)		Bending strength (MPa)	Thermal expansion coefficient. ($K^{-1} \times 10^6$)
Electrolyte	*As sintered*	*After annealing*		
ZrO_2-3 mol%Y_2O_3	0.059	0.050	1200	10.8
ZrO_2-3 mol%Yb_2O_3	0.063	0.09	/	/
ZrO_2-2.9 mol%Sc_2O_3	0.090	0.063	/	/
ZrO_2-8 mol%Y_2O_3	0.13	0.09	230	10.5
ZrO_2-9 mol%Y_2O_3	0.13	0.12	/	/
ZrO_2-8 mol%Yb_2O_3	0.20	0.15	/	/
ZrO_2-10 mol%Yb_2O_3	0.15	0.15	/	/
ZrO_2-8 mol%Sc_2O_3	0.30	0.12	273	10.7
ZrO_2-11 mol%Sc_2O_3	0.30	0.30	255	10.0
ZrO_2-11 mol%Sc_2O_3- 1 wt%Al_2O_3	0.26	0.26	250	/

ScSZ shows a good set of mechanical properties as YSZ. ZrO_2 with 11 mol% Sc_2O_3 and 1 wt% Al_2O_3 appears as one of the best electrolyte candidates for intermediate temperature SOFCs because of its high oxide ion conductivity, phase stability and excellent mechanical properties. However, the most severe problem would be the high cost of scandium.

Another fluorite-type oxide-ion conductor is doped CeO_2, which is probably the most promising electrolyte for intermediate temperature SOFCs. Doped CeO_2 shows a higher conductivity and a lower conduction activation energy as compared to stabilized ZrO_2. Ceria is often doped with Gd_2O_3 (GDC) and Sm_2O_3 (SDC) to introduce oxygen vacancies, resulting in higher ionic conductivity than that of the YSZ.[52] Ceria possesses the same fluorite structure as stabilized zirconia. The highest oxygen ion conductivity is obtained when the aliovalent doping cation matches with the host cation on the ionic radius. For example, in the case of Ce^{4+}, Sm^{3+} and Gd^{3+} are the most proper dopants. At 750 °C, the conductivities are comparable for GDC and SDC with values of 6.7×10^{-2} $S\,cm^{-1}$ and 6.1×10^{-2} $S\,cm^{-1}$, respectively.[53,54] Like zirconia, the conductivity of ceria-based electrolyte increases to a maximum value and then decreases with the further increase in dopant concentration. The conductivity of GDC is consistently higher than that of YSZ and ScSZ below 600 °C.

GDC and SDC are currently the most widely used electrolytes in SOFCs for operation from 400 to 600 °C due to their excellent conductive performance and proper TEC compared with other cell components at intermediate temperatures. At lower temperatures, the ionic transference number of GDC ($t_o > 0.9$) is sufficiently high to ensure a good efficiency of SOFCs.[55] 10 mol% Sm-doped or Gd-doped ceria have been found to possess the maximum conductivities as high as $5.0 \times 10^{-3}\,S\,cm^{-1}$ and $3.8 \times 10^{-3}\,S\,cm^{-1}$ at 500 °C, respectively.[56] Although the mechanism was not completely understood, the high conductivity was generally attributed to strain, which enhances ionic mobility. One difficulty for application of GDC and SDC in SOFCs is to obtain dense membranes with sufficient density to prevent gas cross leakage at sintering temperature below 1650 °C in air atmosphere for materials prepared by conventional solid-state reaction method.[57] Reiss *et al.* obtained relative density of 95–98 % for GDC membrane after sintering in air at 1700 °C with its precursor powders obtained by conventional solid-state reaction of ball-milled oxide raw materials.[58] Many chemical preparation methods, for instance, glycine-nitrate process, nitrate-citrate gel-combustion synthesis, carbonate coprecipitation method, oxalate coprecipitation route, homogeneous precipitation process and hydrothermal process, have been developed for the synthesis of ultrafine homogeneous doped-ceria powders.[58–65] These methods effectively reduced the sintering temperature of SDC and GDC. For example, SDC powder prepared by an oxalate co-precipitation method realized well densification when pressed uniaxially at 200–400 Mpa and then sintered at 1350–1400 °C for 4 h.[66]

Ceria doped with other dopants in a proper amount, such as lanthanum, yttrium, ytterbium and neodymium, also exhibit similar conductivities to SDC.[67] Figure 2.3 shows the conductivities of selected doped ceria oxides. Increasing the concentration of Gd_2O_3, YbO_3 or La_2O_3 dopants beyond

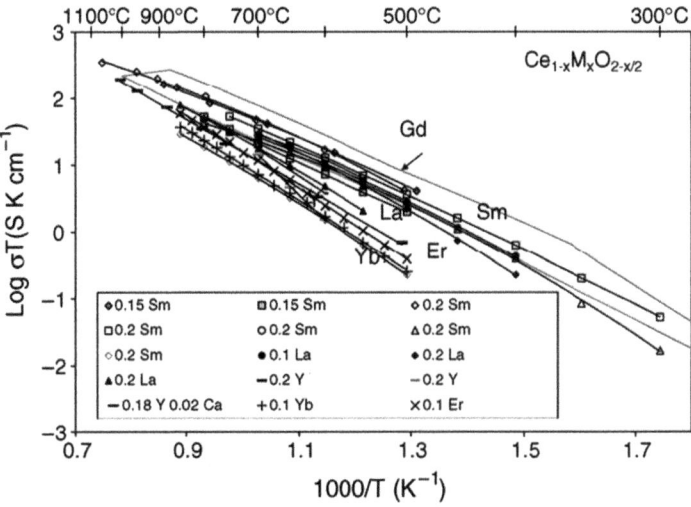

Figure 2.3 Conductivity of $Ce_{1-x}M_xO_{2-x/2}$ in air.[68]

11–14 mol% in ceria, however, lowered the ion conductivities.[68] To improve the ion conductivity of doped CeO_2 electrolyte materials, co-doping method has also been taken. Several co-doped ceria electrolytes have been investigated, for instance, the partial substitution of magnesium, praseodymium, yttrium, gadolinium and samarium for Sm or Gd increased the conductivities of SDC and GDC oxides.[69–73] Yeh and Chou have systematically studied ceria co-doped by samarium and strontium. It was found that the material with the composition of $Ce_{0.78}Sm_{0.20}Sr_{0.02}O_{1.88}$ had the highest ionic conductivity to reach 6.1×10^{-2} $S\,cm^{-1}$ at 800 °C. The increase in ion conductivity through co-doping was attributed to the increase in the number of oxygen vacancies and the decrease in association enthalpy and also the increase in radii of oxygen vacancies which widen the channels for easy movement of oxygen ions.[74] The effect of strontium doping on ionic conductivity of lanthanum-doped ceria has also been investigated. The samples co-doped with La and Sr exhibited higher ionic conductivity and lower activation energy than those of ceria doped with La only at the intermediate temperature range. The concept of average oxygen vacancy radius has been found to be useful in calculating the lattice parameter in this system.[75] Co-doped ceria with the composition of $Ce_{0.8}Gd_{0.1}Ca_{0.1}O_{2-\delta}$ achieved a maximum conductivity of $7.42 \times 10^{-2}\,S\,cm^{-1}$ at 700 °C with the minimum activation energy of 0.58 eV.[76] $CeSm_{0.13}Co_{0.07}O_{2-\delta}$ exhibited the higher conductivity of $5.38 \times 10^{-2}\,S\,cm^{-1}$ at 600 °C with activation energy of 0.567 eV, which suggests that cobalt oxide could be used not only as an effective sintering aid but as a favourable dopant to ceria as well.[77] Y_2O_3 and ZrO_2 co-doped CeO_2 possess much better mechanical and electrical properties and the mutual solubility of these two fluorite oxides leads to high ion conductivity, due to the superior chemical and mechanical properties of zirconia.[78] All these results showed that co-doped ceria with alkaline earth and rare earth ion exhibited higher conductivity than the singly doped ceria in most cases. However, it was found that ionic conductivity of a co-doped system could also be lower than that of a singly doped ceria.[74]

As mentioned previously, ceria-based ceramics were difficult to be sintered to full density below 1650 °C for the powder prepared by conventional solid state reaction technique. Adding a small amount of transition metal oxides as a sintering aid to lower the sintering temperature is of current interest. MnO_2, Bi_2O_3, CuO, MoO_3, Fe_2O_3, Li_2O and CoO_x were found to be fairly effective sintering aids for the reduction of sintering temperature of ceria-based ceramics. These sintering aids not only improved the relative density, but also exerted positive influences on the conductivity of the final ceramics. For example, Zhao et al.[79] discovered that the $Sm_{0.2-x}Bi_xCe_{0.8}O_{1.9}$ compacts could obtain its theoretical density at 1300 °C, with the usage of bismuth oxide as sintering aid, meanwhile higher ionic conductivity could be witnessed as compared with $Sm_{0.2}Ce_{0.8}O_{1.9}$. Xu et al.[80] identified that by adding a small amount of Fe_2O_3, the grain-boundary resistance of $Ce_{0.85}Sm_{0.15}O_{1.925}$ could drop considerably. The processes of densification and grain growth of $Sm_{0.2}Ce_{0.8}O_{0.9}$ with 2 mol% Li_2O as a sintering aid showed that the relative density of SDC reached 99.5% at 898 °C,[81] Moreover, $Ce_{0.79}Gd_{0.2}Cu_{0.01}O_{2-\delta}$ demonstrated improved conductivity to reach 2.6×10^{-2} S cm^{-1} at a testing temperature of 600 °C.[82] Nevertheless, further in-depth studies should be dedicated to the specific roles of transition metal oxides, for the reason that various discrepant reports exist, particularly for cobalt oxides as sintering aids. For instance, Pérez-Coll et al.[83] notified that the grain boundary conductivity of $Ce_{0.8}Sm_{0.2}O_{2-\delta}$ ceramics could be greatly enhanced with the introduction of cobalt, whereas Zhang et al.[84] suggested a delicate effect on the grain boundary conductivity by adding cobalt in $Ce_{0.8}Gd_{0.2}O_{2-\delta}$ samples. Some other researchers,[85,86] however, pointed out that the addition of cobalt vastly improved the grain boundary as well as bulk conductivity. On the other hand, literatures described considerable difference in the solubility of cobalt oxide in doped-ceria ceramics. For example, Sirman et al.[87] mentioned that the solubility limit of CoO in GDC was less than 0.5 mol% at 1400 °C, while Kharton et al.[88] held the view that the solubility of cobalt oxide in GDC was up to 10–15 mol%. In addition to the solubility, due to the co-existence of CoO, Co_3O_4 and Co_2O_3,[77,89–92] the oxidation state of Co in the ceramics is still uncertain. In the light of this inconsistency or uncertainty, more research is needed to exploit the accurate manner of Co additive in the doped-ceria ceramic. A common impurity in ceria electrolytes is silica, which has a detrimental effect on the ionic conductivity by blocking at grain boundaries.[7] The SiO_2 contamination can also be introduced from furnace refractories during high temperature sintering, or even from the silicone grease used in the apparatus to establish input gas mixtures for fuel cell test assemblies. Zhang et al. reported that the small addition of Fe_2O_3 had a scavenging effect on SiO_2 impurity, and significantly improved the grain boundary conduction of the impure GDC.[84] It may be an effective way to reduce the detrimental effect of SiO_2 on the conductivity of doped ceria.

The main reason for operating SOFCs based on GDC or SDC electrolytes in the temperature range of between 400 °C and 600 °C is the appearance of significant electronic leakage above 600 °C, which could reduce the open circuit voltage (OCV) and overall fuel cell efficiency. This is due to the fact that the partial reduction of Ce^{4+} to Ce^{3+} would occur at temperatures above 600 °C and at low oxygen partial pressures, resulting in the non-negligible proportion of n-type electronic conductivity in the overall conductivity. When the cell is under electrical load, however, the electronic current is reduced, as shown in Figure 2.4, thus the fuel cell efficiency is increased.[3] Up to now, considerable efforts have been devoted to minimize the electronic conductivity of doped ceria under reducing conditions. Introducing an additional ultra-thin interfacial electrolyte layer to prevent the electronic transport is an effective approach to suppress the reduction of ceria under reducing atmosphere.[93–95] Below 500 °C, the electronic conductivity is so small it is negligible. The adoption of bi-layered electrolyte is also an available approach to enhance the performance of SOFC. Both GDC|YSZ and SDC|YSZ bi-layered electrolyte SOFCs exhibited significantly higher performance than the cell only using YSZ, SDC or GDC electrolyte.[96–98]

In addition to fluorite structure electrolytes such as stabilized zirconia and doped ceria, there are many other structured oxides which are also attractive for SOFC electrolyte application. In particular, the ABO_3 perovskite-based systems have been considered as promising alternative options. Perovskite oxides can take on a number of different symmetries, and they can be doped with aliovalent cations on both their A and B sites. They can also accommodate very large concentrations of anion vacancies into the structures. Lanthaum

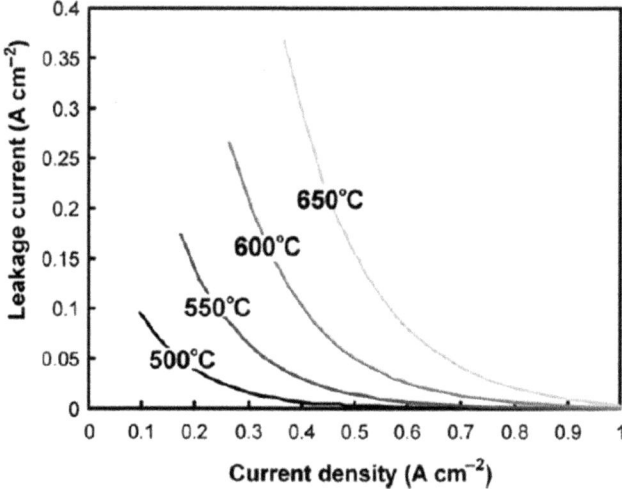

Figure 2.4 Electronic leakage current with electronic current density and temperature. R_a and R_c are the area specific resistances of the anode and cathode, respectively.[3]

gallate ($LaGaO_3$), which has received considerable attention in recent years, is a poor ionic conductor with perovskite structure. The Sr and Mg co-doped $LaGaO_3$ (LSGM), firstly reported in 1994 by Ishihara *et al.* and Goodenough *et al.*, exhibits high ionic conductivity and low electronic conductivity even at low oxygen partial pressure.[99,100] The following defective equation in Kroger-Vink notation can explain the creation of oxygen vacancies to promote rapid oxide-ion conductivity after the acceptor doping on La site and Ga site:

$$2SrO + 2La_{La}^{\times} + O_O^{\times} \rightarrow 2Sr'_{La} + V_O^{\bullet\bullet} + La_2O_3 \qquad (2)$$

Although its conductivity was slightly lower than GDC at 500 °C, its ionic domain was wider and it could be more appropriate to use this electrolyte at temperatures higher than 600 °C when the reduction of Ce^{4+} in GDC became significant.[1] Even this class of material potentially offered adequate performance at temperatures as low as 400 °C.[3] X-ray diffraction and neutron diffraction studies, respectively, demonstrated a cubic crystal structure or a monoclinic structure of LSGM.[99,101] The simultaneous Sr and Mg doping led to a reduction of tilt degree of GaO_6 octahedra and gave birth to great promotion of oxide-ion conductivity relative to the parent compound.[102] Computational techniques have extensively examined the defective chemistry and migration path of oxide ions in $LaGaO_3$.[103,104] Vacancies hopping between oxygen sites along a GaO_6 octahedron edge result in the migration of oxide ions.[105] This mechanism was also confirmed by Yashima *et al.*, who used the maximum entropy method based on neutron diffraction data.[106] In studying the defective clustering in doped $LaGaO_3$, it was suggested that the observed high oxide-ion conductivity could be promoted if the negligible binding energy for Sr dopant-vacancy clusters was considered as a major factor. On the contrary, significant binding energies among Mg-vacancy clusters referred to greater vacancy trapping. This was in accordance with the observed growth in activation energy for ion migration at higher Mg doping levels in $La_{1-x}Sr_xGa_{1-y}Mg_yO_{3-\delta}$.[105]

Systematic study showed that the optimal compositions for maximum conductivity are $La_{0.8}Sr_{0.2}Ga_{0.85}Mg_{0.15}O_3$ and $La_{0.8}Sr_{0.2}Ga_{0.8}Mg_{0.2}O_3$. The former achieved the highest ion conductivity values of 8×10^{-2} and 17×10^{-2} $S\,cm^{-1}$ at 700 °C and 800 °C, respectively.[7] The latter also exhibited a good performance. For example, Huang *et al.* reported that less ion conductivity of $11 \times 10^{-2}\,S\,cm^{-1}$ at 800 °C was derived for $La_{0.8}Sr_{0.2}Ga_{0.8}Mg_{0.2}O_3$ prepared from a sol-gel technique.[107] Another approach to enhance the conductivity of LSGM is to introduce metal oxide dopants, such as cobalt and iron.[108–110] However, the introduction of dopants may be detrimental to fuel cell performance due to increased hole conductivity, which varies with dopant levels. Thus, the optimal dopant depends on a balance between reducing impedance by increasing dopant level and reducing leakage current by decreasing dopant level.[111–113] Nickel has also been tried as a dopant for LSGM, but it also reacts with LSGM to produce impurity phase, which will be discussed later. In addition, it is difficult to fabricate pure single phase ceramic electrolytes, and second phases, most of them are detrimental to the properties of LSGM, such as $SrLaGa_3O_7$ and $La_4Ga_5O_9$, were often detected around the grain boundaries.[1]

As an excellent electrolyte of SOFCs, it should not only possess outstanding ion conductivity, but also retain chemical compatibility with anode and cathode. However, in the case of LSGM, NiO from anode reacted easily with LSGM to form ionically-insulating oxide $LaNiO_3$ at the anode-electrolyte interface during the fabrication process. Although it is possible to suppress such reaction by applying a SDC interlayer between anode and electrolyte, the diffusion of fairly mobile La^{3+} could still lead to the formation of secondary phases during long-term operation, which could block the oxide-ion transportion, thus reducing the cell performance.[114]

2.3.2 Proton-conducting Electrolyte

For many years, research into electrolyte materials has been mainly focused on oxygen-ion conducting oxides such as YSZ, SDC or GDC. As an eminent representation for electrolyte materials, proton-conducting ceramics has also caught great attention from scientists in physics, chemistry and the material science fields. The proton-conducting oxide electrolyte, which was named as high-temperature proton conductor, could be used for SOFCs and distinguished from proton-conducting polymers which had applications in proton exchange membrane fuel cells. In 1981, Iwahara *et al.* first observed that some perovskite oxides, such as $SrCeO_3$ and $BaCeO_3$, had proton conductivity at high temperatures in the presence of water vapour.[115,116] Since then, high temperature proton-conducting materials (HTPCs) have drawn special attention due to their potential applications in hydrogen sensors, hydrogen pumps, membrane reactors, solid oxide electrolysis cells and SOFCs.[117–121]

Most of the perovskite-type materials that were originally studied as anionic conductors, are also potential protonic conductors after hydration. The general formula for a perovskite structure is ABO_3, where A is a large cation 12-fold coordinated to the O-anion while B is a smaller cation occupying the centre of an octahedral unit surrounded by 6 O-anions. Figure 2.5 displays the lattice structure of a typical ABO_3 perovskite. [121] To improve protonic conductivity, it is paramount to dope the B-site with suitable trivalent element, such as Ce, Zr, Y, In, Nd, Pr, Sm, Yb, Eu, Gd and so on. The purpose of doping with a trivalent element is to form oxygen-ion vacancies, which exert a positive influence on the formation of mobile protons. The mobile protons incorporate into the perovskite structure as hydrogen defects in the presence of water vapor and in hydrogen-containing atmospheres. The formation of proton defects occurs via the following reaction:

$$H_2 + 2O_o^{\times} \rightarrow 2(OH)_o^{\bullet} + 2e \qquad (3)$$

The most important reaction leading to the formation of protonic defects at high temperatures is the dissociative absorption of water, which requires the

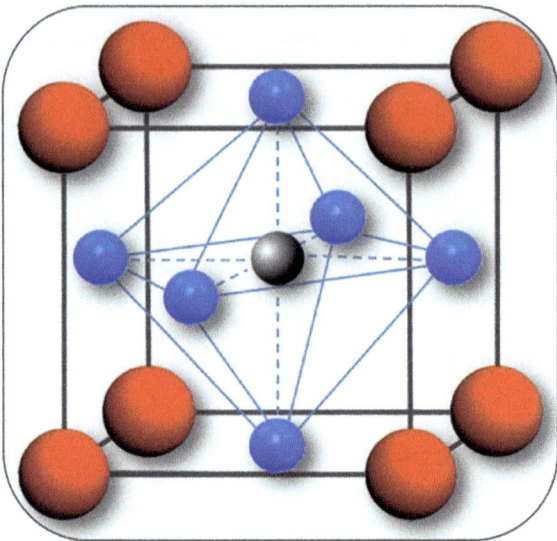

Figure 2.5 Perovskite crystal structure (ABO$_3$), where the red spheres are the A cations, the grey sphere the B cation, and the blue spheres are the oxygen ions.[121]

presence of oxide-ion vacancies. In Kroger-Vink notation this reaction is given as:

$$H_2O + V_O^{\bullet\bullet} + O_O^{\times} \rightarrow 2(OH)_O^{\bullet} \tag{4}$$

One aspect is that varying the ratio of the main constituents intrinsically leads to the formation of vacancies, while the compensation of an acceptor dopant may also extrinsically give birth to the same consequences. In order to form protonic defects, water vapour first dissociates into a hydroxide ion and proton, the hydroxide ion then incorporates into an oxygen vacancy while the proton forms a covalent bond with a lattice oxygen. The adsorption of water is an exothermic reaction, consequently protons dominate the conduction mechanism at low temperatures and oxygen vacancies at high temperatures. Protons are not part of the nominal structure, but are present as defects in equilibrium with ambient hydrogen or water vapour. Due to the negative enthalpy and entropy of hydration, higher temperatures result in a reversible loss of protons. The concentration of protonic defects can be regarded as a function of not only temperature, but also water partial pressure. The proton concentration enhanced with water partial pressure up to a certain level that corresponded to the saturation limit. For the reaction (4), it has been found that the lower electronegativity of compounds the larger equilibrium constant of hydration would be, thus protonic defects are better stabilized in the case of oxides with high basicity.[122,123]

The proton transport mechanism has been categorized by different research groups, with slight disagreements on the rules. For the vehicle mechanism, protons diffuse together with a "vehicle" such as H_3O^+ and commonly encounter in aqueous solution and other polymer membranes. Grotthuss mechanism, opposed to the vehicle mechanism, explains that protons rotationally diffuse through a combination of molecular re-orientation and they hop from one oxygen ion to an adjacent one. A recent re-analysis of quantum MD simulation revealed that proton transfer reaction and proton rotational diffusion had similar probabilities to be the rate-determining step, which resolved the contradictions of different experimental computational data.[124–126] Figure 2.6 describes Grotthuss mechanism of proton transport. Because the size of a proton is much smaller than that of an oxide ion, higher ionic conduction of proton conductor at lower temperature than conventional oxygen-ion conductors is expected by taking advantage of relatively low activation energy of 0.3–0.6 eV for proton transportation.[127–129] Furthermore, the other advantage of using proton conductors as electrolytes compared with conventional oxygen-ion conductors is that water is produced at the cathode side instead of the anode side. Water, the product of the overall reaction engendered at the cathode, can easily be removed by the oxidation ambient, while water at the anode dilutes the fuel gases so that completely obstructs the use of hydrogen.

Perovskite-type oxides (*e.g.* $SrCeO_3$, $BaCeO_3$, $KTaO_3$) doped with cations of lower valence state have exhibited protonic conductivity in an atmosphere of hydrogen or water vapor at high temperature.[130–132] In general, proton conductivity increased in the order $BaCeO_3 > SrCeO_3 > SrZrO_3 > CaZrO_3$. Lattice distortions from doping elements can influence the conducting properties of proton conductor as well. This infers that, in general, the choice of dopant cations, their ionic radius are considerably larger than B-site cation, or high dopant concentrations, can affect the electrical performance strongly. Generally, with respect to the undoped material, the strain induced by doping could be measured by the change in unit cell volume. The Goldschmidt tolerance factor (t) (as shown in Formula (5)) could illustrate the degree that a perovskite structure distort from the ideal cubic structure.

$$t = \frac{(R_A + R_O)}{\sqrt{2(R_B + R_O)}} \tag{5}$$

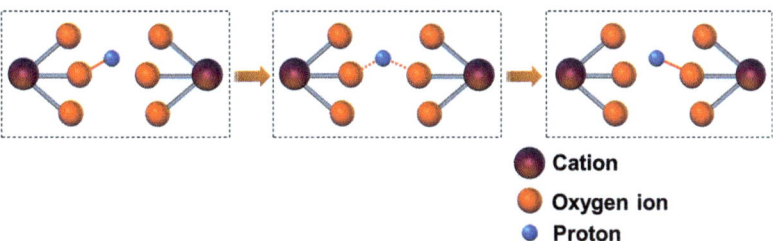

Figure 2.6 Grotthuss mechanism of proton transport.

where R_A and R_B are the ionic radii of the cations occupying A and B site, respectively, while R_O is the oxygen ionic radius. Stable perovskite structure can be obtained when the tolerance factor (t) ranges between 0.75 and 1.0; cubic symmetry has been observed for $0.95 \leq t \leq 1.04$, while compounds having a tolerance factor in the range of 0.75–0.90 typically have orthorhombic symmetry. The perovskite oxide $SrCeO_3$ doped with various lower valent cations has been studied in detail by Uchida *et al.* and has been shown to exhibit appreciable protonic conductivity at high temperature in a hydrogen or moist atmosphere.[133–136] Electrical properties of Eu, Ho, Mg, Sc, Sm, Tm, Y, or Yb doped $SrCeO_3$ have been studied.[130,137] $SrCe_{0.95}Er_{0.05}O_{3-\delta}$ ceramic was a pure protonic conductor with a maximal conductivity of 10^{-2} S cm^{-1} in wet hydrogen at 1000 °C.[138] Among these materials, $SrCe_{0.95}Yb_{0.05}O_3$ showed the highest protonic conductivity of 7×10^{-3} S cm^{-1} at 900 °C in pure hydrogen.[139] The high proton-conductivity of Tb doped $SrCeO_3$ promotes our interest in further studying the electrical properties of this group of materials in reducing atmospheres and the effects of dopant concentration on the structural and electrical properties of the materials. The total conductivity for $SrCe_{0.95}Tb_{0.05}O_{3-\delta}$ is about 3×10^{-3} S cm^{-1} in reducing atmosphere at 900 °C with an activation energy of 0.40 eV.[140] Among these HTPCs, $BaCeO_3$-based materials were regarded as benchmark materials for their high and nearly pure proton conductivity of more than 10^{-2} S cm^{-1} at 600 °C. Y-doped $BaCeO_3$ was shown to achieve a higher proton conductivity than any other rare-earth doped $BaCeO_3$ oxide systems, thus raising some interesting questions about the factors that affect proton conductivity in these compounds. Iwahara reported a high total conductivity of 5.3×10^{-2} S cm^{-1} for $BaCe_{0.8}Y_{0.2}O_{3-\delta}$ at 800 °C in hydrogen.[141] The proton conductivity in $BaCe_{1-x}Y_xO_{3-\delta}$ increases with yttrium concentration up to $x = 0.2$.[142] Moreover, in $BaCe_{0.9}Y_{0.1}O_{3-\delta}$, proton conductivity is significantly higher than that of the oxygen-ion conducting counterpart over the temperature range of 600–800 °C. Slade *et al.* studied the proton conduction of $BaCe_{1-x}M_xO_{3-\delta}$ (M = La, Nd, Gd and Y) in a moist atmosphere and found that Gd doped $BaCeO_3$ gave the highest conductivity among the tested dopants.[143] A similar result was also observed by Stevenson *et al.*[144] By electrochemical hydrogen pumping and AC impedance spectroscopy, Chen *et al.* investigated the proton conduction of $BaCe_{1-x}Gd_xO_{3-\delta}$ ($0.05 \leq x \leq 0.20$) samples, prepared by a microemulsion method, and they verified that the ceramic samples were almost pure protonic conductors at 300–600 °C in hydrogen atmosphere.[145] Many papers in the past reported that Nd was one of the effective dopants to show high conductivity in $BaCeO_3$ based ceramics, indeed, Liu and Nowick reflected that Nd-doped $BaCeO_3$ was the best proton conductor amongst Yb, Eu and Gd-doped $BaCeO_3$.[146,147] However, Stevenson *et al.* displayed that Nd-doped $BaCeO_3$ showed lower water uptake compared to that in Yb and Gd doped $BaCeO_3$. Nd-doped $BaCeO_3$ also exhibited a slower response to the changes in the atmosphere between dry and wet state and a lower weight gain in wet atmosphere than Gd and Yb doped $BaCeO_3$.[144] It was widely accepted that proton conductivity of Eu-doped $BaCeO_3$ oxides was not as high as Gd or Y doped cerate. Eu^{3+} and

Gd^{3+} have nearly the same ionic size; however, electron clouds shell size of Eu^{3+} is less than that of Gd^{3+}, which determines their different chemical behaviours. According to the ionic sizes, larger Eu^{2+} ions, with an ionic radius close to that of Sr^{2+}, prefer A-sites, while smaller Eu^{3+} ions take up both A- and B-sites. However, the EPR results demonstrated that all of Eu ions were trivalent in Eu doped $BaCeO_3$ oxides.[148] Therefore, the lower conductivities of Eu^{3+} doped $BaCeO_3$ oxides may be due to the fact that some Eu^{3+} ions were doped onto the Ba-sites, resulting in the decrease of oxygen vacancy concentration. At $300\,°C$, the conductivity of Eu-doped $BaCeO_3$ was as high as $2.36 \times 10^{-4}\,S\,cm^{-1}$, similar to Gd-doped $BaCeO_3$.[149] In the work of Iwahara *et al.*, it was found that the total conductivity of the $BaCe_{1-x}Sm_xO_{3-\delta}$ system in hydrogen atmosphere increased with samarium content and reached its maximum approximately at $x = 0.2$.[150] Furthermore, its values were higher than those measured in $BaCe_{0.9}Nd_{0.1}O_{3-\delta}$ and $SrCe_{0.95}Yb_{0.05}O_{3-\delta}$. Sharova *et al.* found that the conductivity of $BaCe_{1-x}Sm_xO_{3-\delta}$ was two orders of magnitude greater than that of pure $BaCeO_3$.[151]

Although protonic defects were stabilized in doped $BaCeO_3$ electrolyte materials, poor stability was shown in CO_2, H_2O and other trace species (SO_2, SO_3, and H_2S). Barium cerate reacts easily with acidic gases, such as CO_2, and water vapour to form carbonates and hydroxides, respectively. Such behaviour can be explained in terms of the following reaction:

$$BaCeO_3 + CO_2 \rightarrow BaCO_3 + CeO_2 \tag{6}$$

$$BaCeO_3 + H_2O \rightarrow Ba(OH)_2 + CeO_2 \tag{7}$$

$BaCeO_3$ reacts with pure CO_2 at temperature below $1150\,°C$, while the reaction with water vapour occurs below $400\,°C$.[152] Although different thermodynamic data were proposed by different authors, all of them predicted that the reaction of $BaCeO_3$ with CO_2 and H_2O was favourable at lower temperatures.[153,154] To enhance their chemical stability against CO_2 or H_2O, various doping strategies were tried. Not only high conductivity but also sufficient chemical stability under a wide range of SOFC operating conditions could be gained by these materials via proper doping. As a whole, the chemical stability can be strengthened by partially replacing Ce with elements of higher electronegativity. Take $BaCe_{0.8}Gd_{0.2}O_{3-\delta}$ as electrolyte, the cell voltage degradation rate of single cell was 24%/1000 h with 80% H_2/ and 20% CO_2 as fuel but when a discharge current density was $100\,mA\,cm^{-2}$ at $800\,°C$ it was only 7%/1000 h with pure H_2 as fuel.[155] The 20 mol% Gd-doped $BaCeO_3$ was seen to be stable in water vapour at 600 and $700\,°C$ for 1000 h but unstable when heated in water at $85\,°C$.[156] By using M (M = Ta, Ti, and Sn) and Y as the co-dopants, $BaCe_{0.8-x}M_xY_{0.2}O_{3-\delta}$ proved to maintain high conductivity as well as good stability.[157-159] Chemical stability of $BaCeO_3$ in the presence of CO_2 was improved if titanium was incorporated into its lattice and its total and protonic conductivities was achieved relatively high with $Ba(Ce_{0.95}Ti_{0.05})_{0.8}Y_{0.2}O_{3-\delta}$ composition.[160] Consequently, it is expected that maintaining high Ce content, a unique composition with high proton conductivity might be offered and stability in CO_2 and H_2O-rich environments could be promoted by co-doping the B-site with In and Y. $BaIn_{0.1}Y_{0.2}Ce_{0.7}O_{3-\delta}$ in wet CO_2 atmosphere boasts both relatively high conductivity and acceptable stability.[161] Exposing to wet 3% CO_2 at 700 °C for 24 h, the $BaCe_{0.7}In_{0.3}O_{3-\delta}$ membrane still maintained sufficient stability.[162] Nb-doped $BaCe_{0.9}Y_{0.1}O_{3-\delta}$ compounds, showed an increased chemical stability with higher Nb content under CO_2 containing atmosphere, confirming the positive impact on reducing oxide basicity.[163] It was suggested that Sm was a good dopant for $BaCeO_3$ in term of high protonic conductivity. Both Sm- and In- co-doped proton conductor $BaCe_{0.8-x}Sm_{0.2}In_xO_{3-\delta}$ ($x = 0.0–0.2$) reacted with CO_2 and boiling water to generate barium carbonate and hydroxide. However, the In doping, to some extent, decreased the CO_2 resistance but enhanced the liquid water resistance of the co-doped samples.[164] Matskevich demonstrated that $BaCe_{0.8}Nd_{0.2}O_{2.9}$ could react with water at room temperature and the reactions were more thermodynamically favoured than those of $BaCe_{0.8}Nd_{0.2}O_{2.9}$.[165]

The partial substitution of Ce with Zr can reduce the tendency of decomposition in CO_2 at high temperatures; however, it also results in the decrease of ionic conductivity. Recently, Zuo *et al.* developed a new composition of proton conductor, $BaZr_{0.1}Ce_{0.7}Y_{0.2}O_{3-\delta}$ (BZCY1), which displayed proton conductivity as high as 10^{-2} $S cm^{-1}$ at 600 °C, as well as sufficient chemical and thermal stability over a wide range of conditions relevant to SOFC operation.[166] Zhong and Guo, respectively, studied the stability and conductivity of the $BaCe_{0.9-x}Zr_xY_{0.1}O_{3-\delta}$ and $BaCe_{0.8-x}Zr_xY_{0.2}O_{3-\delta}$ series, and observed that the chemical stability improved while the electric conductivities of sintered samples reduced with the Zr content.[167,168] Due to the fact that Zr^{4+} has smaller ionic radius than Ce^{4+}, the increase of Zr content in barium cerate generally resulted in the decrease of unit cell volume. The optimal zirconium content was found to be around 40% of the B-site cations. However, $BaZr_{0.4}Ce_{0.4}Y_{0.2}O_{3-\delta}$ (BZCY4) possessed a poor sintering capability with a relative density of only 84.9% after sintering at 1500 °C for 5 h. Katahira *et al.* drew similar conclusions through investigation of the electrical conductivity and chemical stability of $BaCe_{0.9-x}Zr_xY_{0.1}O_{3-\delta}$ systems.[169] The phase stability and protonic conductivity of BZCY1, a specific composition among the $BaCe_{0.8-x}Zr_xY_{0.2}O_{3-\delta}$ series, was studied by Zuo *et al.* A fuel cell with a 65 μm thick BZCY1 electrolyte was fabricated and delivered attractive OCVs of 1.01 and 1.05 V and PPDs of 148 and 56 mW cm^{-2} at 600 °C and 500 °C, respectively.[166] However, later experiments illustrated that BZCY1 still had insufficient stability under a reforming gas atmosphere. Fabbri *et al.* researched the effects of Zr doping amount on the sintering, CO_2 resistivity and fuel cell performance of $BaCe_{0.8-x}Zr_xY_{0.2}O_{2.95}$.[170] The optimal Zr doping content was around 40–50% of the B-site cations. Sintered at 1550 °C for 8 h with Pt electrodes, a thick $BaZr_{0.5}Ce_{0.3}Y_{0.2}O_{3-\delta}$ electrolyte-supported cell delivered a PPD of only 18 mW cm^{-2} at 700 °C. A synergetic effect between the electrode and electrolyte was created. The participation of Zr and Zn effectively enhanced the chemical stability with no vital deterioration on the protonic conductivity. Take $BaCe_{0.5}Zr_{0.3}Y_{0.16}Zn_{0.04}O_{2.88}$ as an example, it maintained high stability after the TGA measurement in pure CO_2 up to 1200 °C, while its total conductivity in wet 5% H_2 was 3.14×10^{-3} S cm^{-1} at 400 °C and over 10×10^{-3} S cm^{-1} above 600 °C.[171] $BaCe_{0.2}Zr_{0.6}Gd_{0.2}O_{3-\delta}$ and $BaCe_{0.4}Zr_{0.4}In_{0.2}O_{3-\delta}$ remained stable for more than 200 h in boiling water and for more than 4000 h surrounded by steam.[172]

Among this class of proton conductors, yttrium-doped barium zirconate (BZY) has obtained growing attention in recent years due to its good chemical stability and high proton conductivity. Figure 2.7 compares ionic conductivity of the typical oxygen-ion conducting electrolytes with the BZY electrolyte at the temperature range of 300–600 °C.[173] The earliest studies of doped barium zirconate, which appeared in the early 1990s, suggested poor proton conductivity of this material, compared to doped barium cerate. However, Babilo demonstrated that the total conductivity of $BaZr_{0.2}Y_{0.8}O_{3-\delta}$ could still reach as high as 7.9×10^{-3} S cm^{-1} at 600 °C under humidified nitrogen.[174] However, the requirement of a rather high sintering temperature (about 1600 °C) for the densification of BZY could engender some detrimental chemical reactions at the interface. High sintering temperature and/or long annealing times could also lead to BaO evaporation, causing a detrimental effect on the conductivity since the lower occupancy of the A sites favours dopant substitution, reducing the concentration of oxygen vacancies. Iguchi *et al.* presented systematic research on the influence of sintering time on BZY electrical conductivity, showing that both bulk and grain conductivities decreased with the sintering time.[175] Recently, several investigations have been focused on the synthesis of ultrafine barium zirconate powders by wet chemical routes so as to facilitate the sintering at lower temperatures. In addition, a large variety of wet chemistry

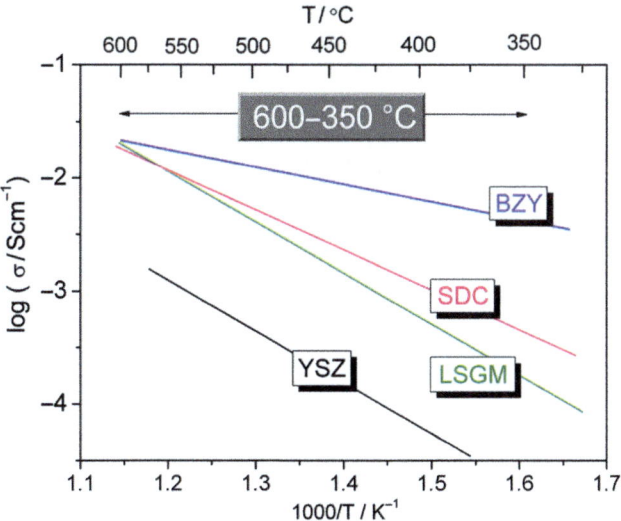

Figure 2.7 Comparison between the Arrhenius plots of the electrical conductivity of the proton-conducting Y-doped barium zirconate (BZY) and the best performing oxygen-ion conducting electrolytes.[173]

methods, such as co-precipitation and modifications of the Pechini method, glycine-nitrate combustion, sol-gel synthesis, polyacrylamide gel process, use of molten salts and hydrothermal syntheses,[176–183] have been widely employed, presenting their respective advantages and disadvantages. Drawbacks include difficulty in controlling the rate of hydrolysis of different metal alkoxides in the sol-gel process, a non-homogeneous compositional distribution and agglomeration of particles prepared by co-precipitation, the thermodynamic instability of barium cerate in the molten salt reaction medium and inhomogeneous particle size and irregular morphology in hydrothermal methods. Soft chemical processes allow one to reduce the synthesis temperature of barium zirconate, thereby giving birth to highly homogeneous and pure ultrafine powders. 20% yttrium doped **BZY** synthesized by a wet chemical route and sintered at 1500 °C reached up to 95% of the theoretical density.[184]

Anyway, from a materials point of view, $BaZrO_3$ doped with trivalent rare earth or bivalent alkaline earth metallic oxides is featured by the capabilities of high proton conductivity and excellent chemical stability.[185] If the tetravalent cation Zr^{4+} at the B-site is partially replaced with the trivalent rare-earth metallic cation, oxygen vacancies may be formed as the following equation:

$$2Zr_{Zr}^{\times} + Y_2O_3 + O_O^{\times} \rightarrow 2Y'_{Zr} + V_O^{\bullet\bullet} + 2ZrO_2 \qquad (8)$$

A strong correlation between conductivity and dopant ionic radius has not been clearly observed. A case in point: even though the ionic radii of six-coordinated Sc^{3+} (0.745 Å) and In^{3+} (0.8 Å) matched fairly well with that of Zr^{4+} (0.72 Å), (while the ionic radius of Y^{3+} (0.9 Å) is much larger), BZY reflected much larger proton conductivity than Sc or In-doped barium zirconates.[186] Compared to $BaZr_{0.9}Y_{0.1}O_{3-\delta}$ (BZY10), a drop in conductivity has been found when Ga or In was used as the dopant, even though pellet densification was improved.[187] Similar results as doping barium zirconate with 15% Sc were also obtained, while barium zirconate co-doped with Y and Sc (a Sc concentration below 5%) showed almost the same conductivity as $BaZr_{0.85}Y_{0.15}O_{3-\delta}$ (BZY15) but a slightly improved densification capability.[188] However, barium zirconate doped with Pr or Nd element displayed much higher sinterability and protonic conductivity than those of $BaZr_{0.8}Y_{0.2}O_{3-\delta}$ (BZY20).[189,190] Han et al. observed $BaZrO_3$ doped with various amounts of rare-earth dopants, like Sc, Y, Sm, Eu or Dy. All the examined dopants mainly occupied the B-site; nevertheless, a partial A-site occupation of Sm and Eu was also recognized. The bulk conductivity of $BaZr_{0.8}Dy_{0.2}O_{3-\delta}$ in wet atmosphere was higher than that in dry atmosphere, suggesting the generation of protonic conductivity in $BaZr_{0.8}Dy_{0.2}O_{3-\delta}$ after humidification of the atmosphere.[191] The measured total conductivities of $BaZr_{0.9}Dy_{0.1}O_{3-\delta}$ ceramic in wet hydrogen and in air were 7.90×10^{-3} S cm^{-1} and 7.31×10^{-3} S cm^{-1} at 800 °C, respectively.[192] Moreover, sintering aids have been also used to lower **BZY** densification temperature and to enlarge grain size. TiO_2, MgO, MoO_3, Al_2O_3, and Bi_2O_3 additives did not improve BZY10 densification, while it impaired its conductivity.[193] The conductivity of **BZY** was substantially improved by some other additives, such as ZnO and CuO, both in humidified hydrogen and humidified air atmospheres. However, the formation of a $Ba_2YCu_3O_x$ impurity phase by introducing CuO as a sintering aid was observed by Babilo *et al.*[194] In real fuel cell applications, the reduction of CuO to metallic copper under a reduced atmosphere should also be taken into consideration. The addition of 1 wt% ZnO to BZY20 led to a density of 96% of the theoretical density at 1350 °C.[195] A solid solution was formed with zinc entering into the lattice after firing and the metal possibly occupied B-site of the perovskite structure, giving rise to a decrease in bulk conductivity of Zn-doped BZY20 with respect to the values. Using NiO as a sintering aid, a total conductivity as large as 3.3×10^{-2} $S\,cm^{-1}$ was reached at 600 °C under wet argon atmospheres for the sintered BZY20 pellets.[173] Up to now, BZY20 with lithium sintering-aid was in possession of larger proton conductivity values than that of reported BZY sintered together with other metal oxides, for the reason that the protons transported rapidly in the "clean" grain boundaries and grain interior. The total conductivity of BZY20-Li in wet argon gas reached $4.45 \times 10^{-3}\,S\,cm^{-1}$ at 600 °C.[196]

2.3.3 Dual-phase Composite Electrolyte

All the above electrolytes, both oxide-ion and proton conductors, are single-phase oxide materials. However, current SOFCs based on these electrolytes still did not satisfy the demands for intermediate temperature operation because the basic requirement for ionic conductivity of $10^{-2}\,S\,cm^{-1}$ at 600 °C was hardly reached. Recently, dual-phase composite electrolyte materials were proposed for SOFCs, which were investigated firstly by Zhu *et al.*[197] CeO_2-based material, acting as a main parent compound to combine with a dispersed phase such as carbonates, sulphates, halides chlorides, fluorites and hydrates, showed mixed ionic transport properties under fuel cell operation conditions. These composite materials can also effectively suppress the electronic conduction. Extensive studies have showed that dual-phase composite electrolyte enjoyed superior ionic conductivity to reach as high as $10^{-1}\,S\,cm^{-1}$ at 600 °C and excellent fuel cell performance of $300-1100\,mW\,cm^{-2}$ at 400–600 °C was reported.[198–200] The high ionic conductivity of these materials could substantially reduce the operating temperature of SOFCs from 800–1000 °C to 300–600 °C. Development on ceria-salt composite electrolyte has thus opened up a new horizon in the low temperature SOFC (LT-SOFC) research field.

The composite electrolyte system typically contains a molten phase, which highly distributes and incorporates in the interfacial regions of the ceria oxide grains/phases. A certain amount of molten carbonate could form composites with a controlled microstructure. Under this circumstance, it can also lead to extremely high conductivity based on interfacial effects, instead of weakening the mechanical strength. So the system can remain a solid-like state. The SDC-carbonate composites have proved to be hybrid oxygen ion and proton conductors, and the oxygen-ion conduction was predominant over the proton conduction.[201] Oxygen ion and proton conductions took place respectively in the SDC and carbonate or in the two-phase interfaces based on different mechanisms.[201] The past decade has witnessed remarkable achievements in the exploration of ceria-based composite electrolyte, consisting of the ceria phase and the salt phase in low-temperature SOFCs.[202–204] Fuel cell based on SDC-Li_2CO_3-K_2CO_3 composites electrolyte reached the maximum power density of 400 and 500 mW cm^{-2} at 550 and 600 °C, respectively.[205] Fuel cells with SDC-Li_2CO_3-Na_2CO_3 composite electrolytes containing 25 and 20 wt.% carbonates achieved cell outputs of 1085 mW cm^{-2} at 600 °C and 690 mW cm^{-2} at 500 °C, respectively.[202] These performances are much higher than the best performance ever reported for the SOFC based on a doped CeO_2 thin-film electrolyte at similar temperature. Recently, novel nanocomposite electrolyte of SDC-Na_2CO_3 has been successfully developed which exhibited distinguished conductivity (10^{-1} S cm^{-1} at 300 °C) and excellent performance in fuel cell (800 mW cm^{-2} at 550 °C). Additionally, Gao *et al.* demonstrated the feasibility of direct utilization of methanol in low-temperature SOFCs with the SDC-Na_2CO_3 nanocomposite electrolyte to achieve a fairly high PPD of 512 mW cm^{-2} at 550 °C.[206,207] Meanwhile, Zhu and co-workers also studied the oxyacid salt oxide composite electrolyte.[208] Up to now, most work for oxyacid salt oxide composite focused on sulphate-alumina, among them Li_2SO_4-AL_2O_3 system is a typical example, which enjoys high proton conductivity in hydrogen-containing atmosphere. However, Li_2SO_4 remained chemically unstable in hydrogen atmosphere, resulting in poor stability under SOFCs operating conditions.[209] Tao and Meng first found that NaCl had proton and oxygen ion conduction, but it exhibited such poor performance that could not be used as the second phase in SDC electrolyte.[210] GDC-LiCl-$SrCl_2$ was developed and exploited as the electrolyte for intermediate temperature fuel cells, and it demonstrated much higher electrical conductivity than pure GDC electrolyte, so that the composition of GDC and chloride salts significantly improved the current and power density output at intermediate temperature range.[211] Further improvement in OCV and power density output could be anticipated if improved preparation techniques can be realized.

Recently, it was found that some inorganic salt phases were also potential additives to increase the proton conductivity of high temperature proton conducting materials. For example, Guo *et al.* prepared the BZCY-NaCl composite proton conductor materials by conventional solid-state method, which reached a total conductivity reached 1.26×10^{-2} S cm^{-1}, much higher than that of pure BZCY.[212] Recently, Schober *et al.* also investigated the properties of $BaCe_xY_{1-x}O_{3-\delta}$ combined with Na_2SO_4 composite materials.[213] By applying ZnO as sintering aid, BZY-NaOH and BZY-Na_2SO_4 system further increased total conductivity by 0.5 magnitude.[214] Up to now, the information about operation stability of such composite electrolyte, however, is still lacking. More research works are needed before their practical application can be realized.

2.4   Future Vision

Appropriate electrolyte materials are a pre-requisite for the improvement of fuel utilization efficiency and power output of SOFCs. Among novel and conventional electrolyte materials, there still flows a river of knowledge for us to exploit and optimize. Overall, long-term stability, reducing cost and pure ion conduction are on the foundation of existing materials development and novel compounds and structure types exploration. Development of electrolytes with low resistance at intermediate or low temperature, *i.e.*, increasing the ion conductivity or/and decreasing the membrane thickness, remain the mainstreams of research in the current SOFC electrolyte world. Doping with multiple valent elements is considered a promising approach to enhance conductivity up to several orders of magnitude, while interfacial conduction could also exert a dramatic impact on the total conductivity. The poor sinterability of zirconate-based proton conductors, however, limits their application as electrolytes in SOFCs, irrespective of their high stability against CO_2 and water vapour. Consequently, some advanced fabrication techniques, such as pulsed laser deposition, should be applied to process thin and dense electrolyte membranes with small grain boundary volumes. On the other hand, poor ionic conductivity of single phase oxygen-ion conducting electrolyte resists the rapid development of SOFCs. In this regard, oxygen ion-based dual-phase nanocomposites and proton conducting oxides emit their brilliant sparkles as electrolyte materials for reduced temperature SOFCs. Future works should be more focused on their long-term operation stability.

References

1. B. C. H. Steele and A. Heinzel, *Nature*, 2001, **414**, 345–352.
2. N. Q. Minh and Takahashi, *Science and Technology of Ceramic Fuel Cells*, Elsevier, 1995.
3. D. J. L. Brett, A. Atkinson, N. P. Brandon and S. J. Skinnerd, *Chem. Soc. Rev.*, 2008, **37**, 1568–1578.
4. A. Orera and P. R. Slater, *Chem. Mater.*, 2010, **22**, 675–690.
5. A. J. Jacobson, *Chem. Mater.*, 2010, **22**, 660–674.
6. H. Inaba and H Tagawa, *Solid State Ionics*, 1996, **83**, 1–6.
7. H. A. Harwig, *Z. Anorg. Allg. Chem.*, 1978, **444**, 151–166.
8. P. Shuk, H. D. Wiemhöfer, U. Guth, W. Göpeld and M. Greenblatt, *Solid State Ionics*, 1996, **89**, 179–196.
9. H. A. Harwig and A. G. Gerard, *Thermochim. Acta*, 1979, **28**, 121–131.
10. T. Takahashi, H. Iwahara and Y. Nagai, *J. Appl. Electrochem.*, 1972, **2**, 97–104.
11. T. Takahashi, H. Iwahara and T. Arao, *J. Appl. Electrochem.*, 1975, **5**, 187–195.
12. T. Takahashi, T. Esaka and H. Iwahara, *J. Appl. Electrothem.*, 1975, **5**, 197–207.
13. M. J. Verkerk, K. Keizer and A. J. Burggraaf, *J. Appl. Electrochem.*, 1989, **10**, 81–90.
14. M. J. Verkerk and A. J. Burggraaf, *J. Electrochem. Soc.*, 1981, **128**, 75–82.
15. H. Iwahara, T. Esaka, T. Sato and T. Takahashi, *J. Solid State Chem.*, 1981, **39**, 173–180.

16. T. Takahashi, H. Iwahara and T. Esaka, *J. Electrochem. Soc.*, 1977, **124**, 1563–1569.
17. T. Esaka, H. Iwahara and H. Kunieda, *J. Appl. Electrochem.*, 1982, **12**, 235–240.
18. P. Shuk, S. Jakobs and H. H. Mobius, *Z. Anorg. Allg. Chem.*, 1985, **524**, 144–156.
19. T. Esaka and H. Iwahara, *J. Appl. Electrochem.*, 1985, **15**, 447–451.
20. L. Bourja, B. Bakiz, A. Benlhachemi, M. Ezahri, S. Villain, C. Favotto and J. R. Gavarri, *Adv. Mater. Sci. Eng.*, 2012, 452383.
21. D. W. Jung, K. L. Duncan and E. D. Wachsman, *Acta Mater.*, 2010, **58**, 355–363.
22. G. Meng, C. Chen, X. Han, P. Yang and D. Peng, *Solid State Ionics*, 1988, **28**, 533–538.
23. D. Mercurio, M. E. Farissi and B. Frit, *Solid State Ionics*, 1990, **39**, 297–304.
24. S. F. Wang, Y. F. Hsu, W. C. Tsai and H. C. Lu, *J. Power Sources*, 2012, **218**, 106–112.
25. C. Y. Hsieh, H. S. Wang and K. Z. Fung, *J. Eur. Ceram. Soc.*, 2011, **31**, 3073–3079.
26. S. Arasteh, A. Maghsoudipour and M. Alizadeh, *A. Nemati, Ceram. Int.*, 2011, **37**, 3451–3455.
27. W. G. Zhao, S. L. An and L. Ma, *J. Am. Ceram. Soc.*, 2011, **94**, 1496–1502.
28. N. Q. Minh, *J. Am. Ceram. Soc.*, 1993, **76**, 563–588.
29. H. G. Scptt, *J. Mater. Sci.*, 1975, **10**, 1572–1585.
30. S. C. Singhal and K. Kendall, High Temperature Solid Oxide Fuel Cell: Fundamentals, Design and Applications, Elsevier, 2003.
31. S. P. S Baswal, *Solid State Ionic*, 1992, **52**, 23–32.
32. T. M. He, Z. Lu, Y. L. Huang, P. F. Guan, J. Liu and W. H. Su, *J. Alloys Compd.*, 2002, **337**, 231–236.
33. S. D. Kim, S. H. Hyun, J. Moon, J. H. Kim and R. H. Song, *J. Power Sources*, 2005, **139**, 67–72.
34. A. Mineshige, K. Fukushima, K. Tsukada, M. Kobune, T. Yazawa, K. Kikuchi, M. Inaba and Z. Ogumi, *Solid State Ionics*, 2004, **175**, 483–485.
35. K. W. Chour, J. Chen and R. Xu, *Thin Solid Films*, 1997, **304**, 106–112.
36. H. J. Hwang and M. Awano, *J. Eur. Ceram. Soc.*, 2001, **21**, 2103–2107.
37. M. Gaudon, E. Djurado and N. H. Menzler, *Ceram. Int.*, 2004, **30**, 2295–2303.
38. X. D. Ge, X. Q. Huang, Y. H. Zhang, Z. Lu, J. H. Xu, K. F. Chen, D. W. Dong, Z. G. Liu, J. P. Miao and W. H. Su, *J. Power Sources*, 2006, **159**, 1048–1050.
39. D. Kuščer, J. Hole, M. Hrovat, S. Bernik, Z. Samardžija and D Kolar, *Solid State Ionics*, 1995, **78**, 79–85.
40. D. J. Chen, F. C. Wang and Z. P. Shao, *Int. J. Hydrogen Energy*, 2012, **37**, 11946–11954.

41. A. Mitterdorfer and L. J. Gauckler, *Solid State Ionics*, 1998, **111**, 185–218.
42. S. P. S. Badwal, F. T. Ciacchi and D. Milosevic, *Solid State Ionics*, 2000, **136**, 91–99.
43. M. A. Laguna-Bercero, S. J. Skinner and J. A. Kilner, *J. Power Sources*, 2009, **192**, 126–131.
44. D. Lee, I. Lee, Y. Jeon and R. Song, *Solid State Ionics*, 2005, **176**, 1021–1025.
45. M. Okamoto, Y. Akimune, K. Furuya, M. Hatano, M. Yamanaka and M. Uchiyama, *Solid State Ionics*, 2005, **176**, 675–680.
46. C. Varanasi, C. Juneja, C. Chen and B. Kumar, *J. Power Sources*, 2005, **147**, 128–135.
47. O. Yamamoto, *Electrochim. Acta*, 2000, **45**, 2423–2435.
48. Y. Mizutani, K. Hisada, K. Ukai, H. Sumi, M. Yokoyama, Y. Nakamura and O. Yamamoto, *J. Alloys Compd.*, 2006, **408**, 518–524.
49. O. Yamamoto, Y. Arati, Y. Takeda, N. Imanishi, Y. Mizutani, M. Kawai and Y. Nakamura, *Solid State Ionics*, 1995, **79**, 137–142.
50. L. J. Gauckler and K. Sasaki, *Solid State Ionics*, 1995, **75**, 203–210.
51. S. Sarat, N. Sammes and A. Smirnova, *J. Power Sources*, 2006, **160**, 892–896.
52. D. J. L. Brett, A. Atkinson, N. P. Brandon and S. J. Skinner, *Chem. Soc. Rev.*, 2008, **37**, 1568–1578.
53. J. V. Herle, T. Horita, T. Kawasa and N. Sakai, *J. Eur. Ceram. Soc.*, 1996, **16**, 961–973.
54. T. Kudo and H. Obayashi, *J. Electrochem. Soc.*, 1975, **122**, 142–147.
55. C. Milliken and S. Guruswamy, *J. Am. Ceram. Soc.*, 2002, **85**, 2479–2486.
56. M. Mogensen, N. M. Sammes and G. A. Tompsett, *Solid State Ionics*, 2000, **129**, 63–94.
57. I. Reiss, D. Braunshtein and D. S. Tannhauser, *J. Am. Ceram. Soc.*, 1981, **64**, 479–485.
58. R. Peng, C. Xia, Q. Fu, G. Meng and D. Peng, *Mater. Lett.*, 2002, **56**, 1043–1047.
59. R. Peng, C. Xia, Q. Fu, D. Peng and G. Meng, *Mater. Lett.*, 2004, **58**, 604–608.
60. C. Peng, Y. Zhang, Z. Cheng, X. Cheng and J. Meng, *J. Mater. Sci.*, 2002, **13**, 757–760.
61. X. Song, J. Peng, Y. Zhao, W. Zhao and S. An, *J. Rare Earths*, 2005, **23**, 167–171.
62. T. Moria, J. Drennanb, A. Y. Wang, G. Auccheerlonie, J. Li and A. Yago, *Sci. Technol. Adv. Mater.*, 2003, **4**, 213–219.
63. J. V. Herle, T. Horita, T. Kawasa, N. Sakai, H. Yokokana and M. Dokiya, *Solid State Ionics*, 1996, **86**, 1255–1258.
64. B. Djuricic and S. Pickering, *J. Eur. Ceram. Soc.*, 1999, **19**, 1925–1934.
65. S. Dikmen, P. Shuk, M. Greenblatt and H. Gocmez, *Solid State Sci.*, 2002, **4**, 585–590.
66. R. F. Gao and Z. Q. Mao, *J. Rare Earths*, 2007, **25**, 364–367.
67. J. W. Fergus, *J. Power Sources*, 2006, **162**, 30–40.

68. D. J. Seo, K. O. Ryu, S. B. Park, K. Y. Kim and R. H. Song, *Mater. Res. Bull.*, 2006, **41**, 359–366.
69. F. Wang, S. Chen, Q. Wang, S. Yu and S. Cheng, *Catal. Today*, 2004, **97**, 189–194.
70. F. Y. Wang, S. Y. Cheng and S. Cheng, *Electrochem. Commun.*, 2004, **6**, 743–746.
71. S. Lubke and H. D. Wiemhofer, *Solid State Ionics*, 1999, **117**, 229–243.
72. X. Q. Sha, Z. Lu, X. Q. Huang, P. J. Miao and L. Jia, *J. Alloys Compd.*, 2006, **424**, 315–321.
73. X. Guan, H. Zhou, Y. Wang and J. Zhang, *J. Alloys Compd.*, 2008, **464**, 310–316.
74. T. H. Yeh and C. C. Chou, *Phys. Scr.*, 2007, **T129**, 303–307.
75. N. Jaiswa, S. Upadhyay, D. Kumar and O. Parkash, *J. Power Sources*, 2013, **222**, 230–236.
76. X. Zhao, J. Liu, T. Xiao, J. Wang, Y. Zhang, H. Yao, J. Wang and Z. Li, *J. Electroceram.*, 2012, **28**, 149–157.
77. H. Yao, X. Zhao, X. Zhao, J. Wang, Q. Ge, J. Wang and Z. Li, *J. Power Sources*, 2012, **205**, 180–187.
78. T. Yeh and C. Chou, *Solid State Ionics*, 2009, **180**, 1529–1533.
79. W. G. Zhao, S. L. An and L. Ma, *J. Am. Ceram. Soc.*, 2011, **94**, 1496–1502.
80. D. Xu, X. M. Liu, S. F. Xu, D. T. Yan, L. Pei, C. J. Zhu, D. J. Wang and W. H. Su, *Solid State Ionics*, 2011, **192**, 510–514.
81. S. R. Le, S. C. Zhu, X. D. Zhu and K. N. Sun, *J. Power Sources*, 2013, **222**, 367–372.
82. Y. C. Dong, S. Hampshire, J. E. Zhou and G. Y. Meng, *Int. J. Hydrogen Energy*, 2011, **36**, 5054–5066.
83. D. Pérez-Coll, D. Marrero-López, P. Núñez, S. Piñol and J. R. Frade, *Electrochim. Acta*, 2006, **51**, 6463–6469.
84. T. S. Zhang, J. Ma, Y. J. Leng, S. H. Chan, P. Hing and J. A. Kilner, *Solid State Ionics*, 2004, **168**, 187–195.
85. D. Pérez-Coll, P. Núñez, J. C. C. Abrantes, D. P. Fagg, V. V. Kharton and J. R. Frade, *Solid State Ionics*, 2005, **176**, 2799–2805.
86. J. Ayawanna, D. Wattanasiriwech, S. Wattanasiriwech and P. Aungkavattana, *Solid State Ionics*, 2009, **180**, 1388–1394.
87. J. D. Sirman, D. Waller and J. A. Kilner, Proceedings of the Fifth International Symposium on Solid Oxide Fuel Cells (SOFC-V), Aachen, Germany 1997, 1159–1167.
88. V. V. Kharton, F. M. Figueiredo, L. Navarro, E. N. Naumovich, A. V. Kovalevsky, A. A. Yaremchenko, A. P. Viskup, A. Carneiro, F. M. B. Marques and J. R. Frade, *J. Mater. Sci.*, 2001, **36**, 1105–1117.
89. C. Kleinlogel and L. J. Gauckler, *Solid State Ionics*, 2000, **135**, 567–573.
90. G. S. Lewis, A. Atkinson, B. C. H. Steele and J. Drennan, *Solid State Ionics*, 2002, **152**, 567–573.
91. D. P. Fagg, J. C. C. Abrantes, D. Pérez-Coll, P. Núñez, V. V. Kharton and J. R. Frade, *Electrochim. Acta*, 2003, **48**, 1023–1029.

92. C. J. Fu, Q. L. Liu, S. H. Chan, X. M. Ge and G. Pasciak, *Int. J. Hydrogen Energy*, 2010, **35**, 11200–11207.
93. E. D. Wachsman, P. Jayaweera, N. Jiang, D. M. Lowe and B. G. Pound, *J. Electrochem. Soc.*, 1997, **144**, 233–236.
94. H. Yahiro, Y. Baba, K. Eguchi and H. Arai, *J. Electrochem. Soc.*, 1988, **135**, 2077–2080.
95. A. V. Virkar, *J. Electrochem. Soc.*, 1991, **138**, 1481–1487.
96. Z. G. Lu, J. Hardy, J. Templeton, J. Stevenson, D. Fisher, N. J. Wu and A. Ignatiev, *J. Power Sources*, 2012, **210**, 292–296.
97. P. K. Lohsoontom, N. Laosiripojana and J. Bae, *Curr. Appl Phys.*, 2011, **11**, S223–S228.
98. Y. J. Niu, W. Zhou, J. Sunarso, L. Ge, Z. H. Zhu and Z. P. Shao, *J. Mater. Chem.*, 2010, **20**, 9619–9622.
99. T. Ishiham, T. Kudo, H. Matsuda and Y. Takita, *J. Am. Ceram. Soc.*, 1994, **77**, 1682–1684.
100. M. Feng and J. B. Goodenough, *Eur. J. Solid State Inorg. Chem.*, 1994, **31**, 663–672.
101. M. Lerch, H. Boysen and T. Hansen, *J. Phys. Chem. Solids*, 2001, **62**, 445–455.
102. M. Kajitani, M. Matsuda, A. Hoshikawa, S. Harjo, T. Kamiyama, T. Ishigaki, F. Izumi and M. Miyake, *Chem. Mater.*, 2005, **17**, 4235–4243.
103. M. S. Islam and R. A. Davies, *J. Mater. Chem.*, 2004, **14**, 86–93.
104. A. Kuwabara and I. Tanaka, *J. Phys. Chem. B*, 2004, **108**, 9168–9172.
105. L. Malavasi, C. A. J. Fisher and M. S. Islam, *Chem. Soc. Rev.*, 2010, **39**, 4370–4387.
106. M. Yashima, K. Nomura, H. Kagayama, Y. Miyazaki, N. Chitose and K. Adachi, *Chem. Phys. Lett.*, 2003, **380**, 391.
107. K. Huang, M. Feng and J. B. Goodenough, *J. Am. Ceram. Soc.*, 1996, **79**, 1100–1104.
108. J. W. Stevenson, K. Hasinska, N. L. Canfield and T. R. Armstrong, *J. Electrochem. Soc.*, 2000, **147**, 3213–3218.
109. B. A. Khorkounov, H. Näfe and F. Aldinger, *J. Solid State Electrochem.*, 2006, **10**, 479–487.
110. M. Enoki, J. W. Yan, H. Matsumoto and T. Ishihara, *Solid State Ionics*, 2006, **177**, 2053–2057.
111. T. Ishihara, Y. Tsuruta, T. Todaka, H. Nishiguchi and Y. Takita, *J. Electrochem. Soc.*, 2003, **150**, E17–E23.
112. T. Ishihara, M. Ando, M. Enoki and Y. Takita, *J. Alloys Compd.*, 2006, **408**, 507–511.
113. T. Ishihara, S. Ishikawa, C. Y. Yu, T. Akbay, K. Hosoi, H. Nishiguchi and Y. Takita, *Phys. Chem. Chem. Phys.*, 2003, **5**, 2257–2263.
114. J. B. Goodenough, *Annu. Rev. Mater. Res.*, 2003, **33**, 91–128.
115. H. Iwahara, H. Uchida, K. Ono and K. Ogaki, *J. Electrochem. Soc.*, 1988, **135**, 529–533.
116. H. Iwahara, *Solid State Ionics*, 1988, **28**, 573–578.

117. P. H. Chiang, D. Eng and M. Stoukides, *Solid State Ionics*, 1993, **61**, 99–103.
118. P. A. Stuart, T. Unno, J. A. Kilner and S. J. Skinner, *Solid State Ionics*, 2008, **179**, 1120–1124.
119. K. D. Kreuer, *Annu. Rev. Mater. Res.*, 2003, **33**, 333–359.
120. D. Hirabayashi, A. Tomita, S. Teranishi, T. Hibino and M. Sano, *Solid State Ionics*, 2005, **176**, 881–887.
121. E. Fabbri, D. Pergolesi and E. Traversa, *Chem. Soc. Rev.*, 2010, **39**, 4355–4369.
122. T. Norby, M. Widerøe, R. Glöckner and Y. Larring, *Dalton Trans.*, 2004, **19**, 3012–3018.
123. K. D. Kreuer, *Annu. Rev. Mater. Res.*, 2003, **33**, 333–359.
124. K. D. Kreuer, *Solid State Ionics*, 2000, **136**, 149–160.
125. M. Pionke, T. Mono, W. Schweika, T. Springer and H. Schober, *Solid State Ionics*, 1997, **97**, 497–504.
126. W. Münch, G. Seifert, K. D. Kreuer and J. Maier, *Solid State Ionics*, 1997, **97**, 39–44.
127. H. G. Bohn and T. Schober, *J. Am. Ceram. Soc.*, 2004, **83**, 768–772.
128. E. Fabbri, A. D'Epifanio, E. Di Bartolomeo, S. Licoccia and E. Traversa, *Solid State Ionics*, 2008, **179**, 558–564.
129. K. Katahira, Y. Kohchi, T. Shimura and H. Iwahara, *Solid State Ionics*, 2000, **138**, 91–98.
130. H. Iwahara, T. Esaka, H. Uchida and N. Maeda, *Solid State Ionics*, 1981, **3**, 359–363.
131. U. N. Shrivastava, K. L. Duncan and J. N. Chung, *Int. J. Hydrogen Energy*, 2012, **37**, 15350–15358.
132. W. Lee and A. S. Nowick, *Solid State Ionics*, 1986, **18**, 989–993.
133. H. Uchida, N. Maeda and H. Iwahara, *J. Appl. Electrochem.*, 1982, **12**, 645–651.
134. H. Uchida, H. Yoshikawa and H. Iwahara, *Solid State Ionics*, 1989, **34**, 103–107.
135. H. Uchida, H. Yoshikawa, T. Esaka, S. Ohtsu and H. Iwahara, *Solid State Ionics*, 1989, **36**, 89–95.
136. H. Uchida, H. Yoshikawa and H. Iwahara, *Solid State Ionics*, 1989, **35**, 229–234.
137. T. L. Wen, Z. Y. Lu and Z. Xu, *J. Mater. Sci. Lett.*, 1994, **13**, 1032–1034.
138. X. H. Kang, J. Yu and G. L. Ma, *Chin. J. Inorg. Chem.*, 2006, **22**, 738–742.
139. K. D. Kreuer, *Solid State Ionics*, 1997, **97**, 1–15.
140. X. T. Wei and Y. S. Lin, *Solid State Ionics*, 2008, **178**, 1804–1810.
141. H. Iwahara, *Solid State Ionics*, 1995, **77**, 289–298.
142. K. Takeuchi, C. K. Loong, J. W. Richardson Jr., J. Guan, S. E. Dorris and U. Balachandran, *Solid State Ionics*, 2000, **138**, 63–77.
143. R. C. T. Slade and N. Singh, *Solid State Ionics*, 1991, **46**, 111–115.
144. D. A. Stevenson, N. Jiang, R. M. Buchanan and F. E. G. Henn, *Solid State Ionics*, 1993, **62**, 279–285.

145. C. H. Chen, S. J. Chang, S. P. Chang, M. J. Li, I. C. Chen, T. J. Hsueh and C. L. Hsu, *Chem. Phys. Lett.*, 2009, **476**, 69–72.
146. J. F. Liu and A. S. Nowick, *Solid State Ionics*, 1992, **50**, 131–138.
147. J. F. Liu and A. S. Nowick, *Mater. Res. Soc. Symp. Proc.: Solid State Ionics II*, 1991, **210**, 657–661.
148. Y. K. Liu, D. D. Hou and G. H. Wang, *Mater. Chem. Phys.*, 2004, **86**, 69–73.
149. J. X. Wang, W. H. Su, D. P. Xu and T. M. He, *J. Alloys Compd.*, 2006, **421**, 45–48.
150. H. Iwahara, T. Yajima, T. Hibino and H. Ushida, *J. Electrochem. Soc.*, 1993, **140**, 1687–1691.
151. N. V. Sharova, V. P. Gorelov and V. B. Balakireva, *Russ. J. Electrochem.*, 2005, **41**, 665–670.
152. H. Matsumoto, Y. Kawasaki, N. Ito, M. Enoki and T. Ishihara, *Electrochem. Solid-State Lett.*, 2007, **10**, B77–B80.
153. S. L. Sorokina, Y. Y. Skolis, M. L. Kovba and V. A. Levitskii, *Russ. J. Phys. Chem.*, 1986, **60**, 186–196.
154. C. W. Tanner and A. V. Virkar, *J. Electrochem. Soc.*, 1996, **143**, 1386–1389.
155. N. Taniguchi, E. Yasumoto and T. Gamo, *J. Electrochem. Soc.*, 1996, **143**, 1886–1890.
156. Z. L. Wu and M. L. Liu, *J. Electrochem. Soc.*, 1997, **144**, 2170–2175.
157. L. Bi, S. Q. Zhang, S. M. Fang, Z. T. Tao, R. R. Peng and W. Liu, *Electrochem. Commun.*, 2008, **10**, 1598–1601.
158. K. Xie, R. Q. Yan and X. Q. Liu, *J. Alloys Compd.*, 2009, **479**, L40–L42.
159. K. Xie, R. Q. Yan and X. Q. Liu, *J. Alloys Compd.*, 2009, **479**, L36–L39.
160. P. Pasierb, M. Osiadły, S. Komornicki and M. Rekas, *J. Power Sources*, 2011, **196**, 6205–6209.
161. F. Zhao, Q. Liu, S. W. Wang, K. Brinkman and F. L. Chen, *Int. J. Hydrogen Energy*, 2010, **35**, 4258–4263.
162. L. Bi, S. Q. Zhang, L. Zhang, Z. T. Tao, H. Q. Wang and W. Liu, *Int. J. Hydrogen Energy*, 209, **34**, 2421–2425.
163. E. Di. Bartolomeo, A. D'Epifanio, C. Pugnalini, F. Giannici, A. Longo, A. Martorana and S. Licoccia, *J. Power Sources*, 2012, **199**, 201–206.
164. C. J. Zhang and H. L. Zhao, *J. Electrochem. Soc.*, 2012, **159**, F316–F321.
165. N. I. Matskevich, T. Wolf, M. Y. Matskevich and T. I. Chupakhina, *Eur. J. Inorg. Chem.*, 2009, **11**, 1477–1482.
166. C. D. Zuo, S. W. Zha, M. L. Liu, M. Hatano and M. Uchiyama, *Adv. Mater.*, 2006, **18**, 3318–3320.
167. Z. M. Zhong, *Solid State Ionics*, 2007, **178**, 213–220.
168. Y. M. Guo, Y. Lin, R. Ran and Z. P. Shao, *J. Power Sources*, 2009, **193**, 400–407.
169. K. Katahira, Y. Kohchi, T. Shimura and H. Iwahara, *Solid State Ionics*, 2000, **138**, 91–98.
170. E. Fabbri, A. D'Epifanio, E. Di. Bartolomeo, S. Licoccia and E. Traversa, *Solid State Ionics*, 2008, **179**, 558–564.

171. S. W. Tao and J. T. S. Irvine, *Adv. Mater.*, 2006, **18**, 1581–1584.
172. N. Taniguchi, C. Nishimura and J. Kato, *Solid State Ionics*, 2001, **145**, 349–355.
173. E. Fabbri, L. Bi, D. Pergolesi and E. Traversa, *Adv. Mater.*, 2011, **24**, 1–14.
174. P. Babilo, T. Uda and S. M. Haile, *J. Mater. Res.*, 2007, **22**, 1322–1330.
175. F. Iguchi, T. Tsurui, N. Sata, Y. Nagao and H. Yugami, *Solid State Ionics*, 2009, **180**, 563–568.
176. C. D. Sagel-Ransijn, A. J. A. Winnubst, A. J. Burggraaf and H. Verweij, *J. Eur. Ceram. Soc.*, 1996, **16**, 159–166.
177. Y. M. Guo, R. Ran and Z. P. Shao, *Int. J. Hydrogen Energy*, 2010, **35**, 10513–10521.
178. G. Taillades, M. Jacquin, Z. Khani, D. J. Jones, M. Marrony and J. Rozière, *ECS Trans.*, 2007, **7**, 2291–2298.
179. Y. Liu, Y. M. Guo, W. Wang, C. Su, R. Ran, H. T. Wang and Z. P. Shao, *J. Power Sources*, 2011, **196**, 9246–9253.
180. I. Luisetto, S. Licoccia, A. D'Epifanio, A. Sanson, E. Mercadelli and E. Di Bartolomeo, *J. Power Sources*, 2012, **220**, 280–285.
181. S. F. Liu and W. T. Fu, *Mater. Res. Bull.*, 2011, **36**, 1505–1512.
182. S. Komarneni, *J. Am. Ceram. Soc.*, 1998, **81**, 3041–3043.
183. R. Ran, Y. M. Guo, Y. Zheng, K. Wang and Z. P. Shao, *J. Alloys Compd.*, 2010, **491**, 271–277.
184. A. Magrez and T. Schober, *Solid State Ionics*, 2004, **175**, 585–588.
185. K. H. Ryu and S. M. Haile, *Solid State Ionics*, 1999, **125**, 355–367.
186. V. P. Gorelov, V. B. Balakireva, Y. N. Kleshchev and V. P. Brusentsov, *Inorg. Mater.*, 2001, **37**, 535–538.
187. N. Ito, H. Matsumoto, Y. Kawasaki, S. Okada and T. Ishihara, *Solid State Ionics*, 2008, **179**, 324–329.
188. S. Imashuku, T. Uda, Y. Nose, K. Kishida, S. Harada, H. Inui and Y. Awakura, *J. Electrochem. Soc.*, 2008, **155**, B581–B586.
189. E. Fabbri, L. Bi, H. Tanaka, D. Pergolesi and E. Traversa, *Adv. Funct. Mater.*, 2011, **21**, 158–166.
190. Y. Liu, Y. M. Guo, R. Ran and Z. P. Shao, *J. Membr. Sci.*, 2012, **415**, 391–398.
191. D. L. Han, Y. Nose, K. Shinoda and T. Uda, *Solid State Ionics*, 2013, **213**, 2–7.
192. Y. J. Gu, Z. G. Liu, J. H. Ouyang, Y. Zhou and F. Y. Yan, *Electrochim. Acta*, 2012, **75**, 332–338.
193. S. B. C. Duval, P. Holtappels, U. Stimming and T. Graule, *Solid State Ionics*, 2008, **179**, 1112–1115.
194. P. Babilo and S. M. Hailew, *J. Am. Ceram. Soc.*, 2005, **88**, 2362–2368.
195. S. W. Tao and J. T. S. Irvine, *J. Solid State Chem.*, 2007, **180**, 3493–3503.
196. Z. Q. Sun, E. Fabbri, L. Bi and E. Traversa, *Phys. Chem. Chem. Phys.*, 2011, **13**, 7692–7700.
197. B. Zhu, *J. Power Sources*, 2001, **93**, 82–86.

198. J. B. Huang, F. C. Xie, C. Wang and Z. Q. Mao, *Int. J. Hydrogen Energy*, 2012, **37**, 877–883.
199. B. Zhu, *J. Power Sources*, 2003, **114**, 1–9.
200. B. Zhu, X. R. Liu, M. T. Sun, S. J. Ji and J. C. Sun, *Solid State Sci.*, 2003, **5**, 1127–1134.
201. B. Zhu and M. D. Mat, *Int. J. Electrochem. Sci.*, 2006, **1**, 383–402.
202. J. B. Huang, Z. Q. Mao, Z. X. Liu and C. Wang, *Electrochem. Commun.*, 2007, **9**, 2601–2605.
203. A. Bodén, J. Di, C. Lagergren, G. Lindbergh and C. Y. Wang, *J. Power Sources*, 2007, **172**, 520–529.
204. J. B. Huang, L. Z. Yang, R. F. Gao, Z. Q. Mao, Z. X. Liu and C. Wang, *Electrochem. Commun.*, 2006, **8**, 785–789.
205. J. B. Huang, Z. Q. Mao, L. Z. Yang and R. R. Peng, *Electrochem. Solid-State Lett.*, 2005, **8**, A437–A440.
206. Z. Gao, R. Raza, B. Zhu, Z. Q. Mao, C. Wang and Z. X. Liu, *Int. J. Hydrogen Energy*, 2011, **36**, 3984–3988.
207. J. B. Huang, Z. Q. Mao, Z. X. Liu and C. Wang, *J. Power Sources*, 2008, **175**, 238–243.
208. B. Zhu, *Solid State Ionics*, 1999, **119**, 305–310.
209. S. W. Tao, Z. L. Zhan, P. Wang and G. Y. Meng, *Solid State Ionics*, 1999, **116**, 29–33.
210. S. W. Tao and G. Y. Meng, *J. Mater. Sci. Lett.*, 1999, **18**, 81–84.
211. Q. X. Fu, S. W. Zha, W. Zhang, D. K. Peng, G. Y. Meng and B. Zhu, *J. Power Sources*, 2002, **104**, 73–78.
212. T. Schober and H. Ringel, *Ionics*, 2004, **10**, 391–395.
213. Z. Z. Peng, R. S. Guo, Z. G. Yin and J. Li, *Journal of Wuhan University of Technology-Mater. Sci. Ed.*, 2009, **24**, 269–172.
214. R. S. Guo, Y. P. Deng, Y. Y. Gao and L. X. Zhang, *J. Alloys Compd.*, 2011, **509**, 8894–8900.

Cathode Material Development

YAO WANG, YANXIANG ZHANG, LING ZHAO
AND CHANGRONG XIA*

CAS Key Laboratory of Materials for Energy Conversion, Department
of Materials Science and Engineering, University of Science and Technology
of China, Hefei, Anhui 230026, China
*Email: xiacr@ustc.edu.cn

3.1   Introduction

The cathode serves as the catalyst layer for oxygen reduction. Partially owing to the high activation energy and the slow kinetics of the cathode reaction, an operation temperature of 800–1000 °C is traditional for typical SOFCs that, for example, consist of strontium-doped lanthanum manganates (LSM) cathode, yttria-stabilized zirconia (YSZ) electrolyte, and Ni-YSZ cermet anode. High operation temperature gives rise to serious problems, such as the cost of fabrication and maintenance and the stability of cell components, accordingly setting rigorous requirements for cathode materials.[1] Reducing the operation temperature to an intermediate range of 600–800 °C or even lower is an effective strategy, however, this decreases the cell's performance because of the increased ohmic resistance of the electrolyte and electrode polarization resistances arising from the thermally activated processes of ionic transport and electrochemical reactions. It is found that the overall cell loss is dominated by the cathode process,[2] which can be 65% of the total voltage loss.[3] Accordingly, the development and optimization of cathodes with high

RSC Energy and Environment Series No. 7
Solid Oxide Fuel Cells: From Materials to System Modeling
Edited by Meng Ni and Tim S. Zhao
© The Royal Society of Chemistry 2013
Published by the Royal Society of Chemistry, www.rsc.org

performance and stability operating at intermediate temperatures is of critical significance for commercial and widespread application of SOFCs.

For the oxygen ion-conducting electrolyte based SOFCs (O-SOFCs), LSM is the most commonly used and studied cathode material, due to its high electrical conductivity ($\sim 500\,\mathrm{S\,cm}^{-1}$), good thermal expansion coefficient (TEC, $11 \sim 12 \times 10^{-6}\,\mathrm{K}^{-1}$) and chemical compatibility with YSZ, and excellent stability.[4] However, its negligible ionic conductivity limits the application at decreased temperatures. The perovskite-type mixed oxygen ion-electron conductors (MIECs) such as doped $LaCoO_3$, $BaCoO_3$ and $LaFeO_3$ have also been extensively studied as possible cathodes for their high surface exchange and oxygen ion diffusion coefficients, which are required to achieve the desired area-specific resistance (ASR) at intermediate temperatures. However, some technical issues need to be resolved. For example, the Co containing MIECs readily react with the YSZ electrolyte and form resistive $La_2Zr_2O_7$ and $SrZrO_3$ phases at high temperatures during the fabrication process. And the TECs of these MIECs are usually much higher than the typical electrolyte materials.

On the other hand, tremendous progress has been made in cathode development for the proton-conducting electrolyte based SOFCs (H-SOFCs), due to their unique characters, such as great efficiency in fuel utilization, high ionic transfer numbers and low activation energies for proton conduction.[5–7] Moreover, compared with those proton exchange membrane fuel cells operated below 100 °C, no high-cost precious metal catalyst is needed and the CO poisoning of electrodes is avoided for H-SOFCs, also the easy water/thermal management system can be achieved.[8,9]

In this chapter, the progress of cathodes for both O-SOFCs and H-SOFCs will be reviewed briefly in terms of the conducting category of cathode materials, including electron-conducting cathodes, mixed ion-electron conducting cathodes (and mixed electron-proton conducting cathodes for H-SOFCs), and microstructure optimized cathodes. Some fundamental understanding of cathode reaction mechanisms is also described.

3.2 Cathodes for Oxygen Ion-Conducting Electrolyte Based SOFCs

3.2.1 Electron Conducting Cathodes

Perovskite-type manganites $La_{1-x}Sr_xMnO_{3-\delta}$ are typical electron-conducting cathode materials for the high-temperature SOFCs. The undoped $LaMnO_3$ is orthorhombic at room temperature and exhibits an orthorhombic/rhombohedral transition relating to the oxidation of Mn^{3+} to Mn^{4+}.[10] The transition is sensitive to Mn^{4+} content and could be induced at lower temperatures by the doping of divalence cations. For $La_{1-x}Sr_xMnO_{3-\delta}$, the lattice structures are complicated, and are reported to be: rhombohedral ($0 < x < 0.5$), tetragonal ($x = 0.5$), and cubic ($x = 0.7$), depending on the Sr^{2+} content.[11]

LaMnO$_3$ is an intrinsic p-type conductor, and the electronic conductivity is performed by a small polaron hopping, generally expressed by,[12]

$$\sigma T = (\sigma T)^\circ \exp\left(-\frac{E_a}{kT}\right) = A\left(\frac{h\nu^\circ}{k}\right)c(1-c)\exp\left(-\frac{E_a}{kT}\right) \qquad (1)$$

where $(\sigma T)^\circ$ is the pre-exponential constant, E_a the activation energy, and c the ratio of the carrier occupancy of the level on which the hopping conduction takes place. Figure 3.1a shows the temperature dependence of electronic conductivity of La$_{1-x}$Sr$_x$MnO$_{3-\delta}$ with Sr^{2+} doping.[13] The electronic conductivity increases with low Sr^{2+} contents, and then decreases with further Sr^{2+} doping. The peak conductivity is obtained at x = 0.5, with the value of 300 S cm^{-1} at 1000 °C. The electronic conductivity is also a function of the oxygen partial pressure, P_{O_2},[14] as shown in Figure 3.1b, the electronic conductivity exhibits nearly constant irrespective of the P_{O_2} in the high P_{O_2} region (>10^{-8} Pa), while it decreases exponentially with P_{O_2} in the low region (10$^{-15} < P_{O_2} < 10^{-8}$ Pa). These behaviours are typical for p-typed oxide semiconductors. Compared with its electronic conductivity, the ionic conductivity of LSM is negligible. The ionic conduction of LSM is performed by oxygen vacancies, which are produced by the substitution of La^{3+} to Sr^{2+} or the partial reduction of Mn^{3+} to Mn^{2+}.[15,16] The ionic conductivity for La$_{0.8}$Sr$_{0.2}$MnO$_{3-\delta}$ has been measured to be ∼5.76 × 10^{-6} S cm^{-1} at 1000 °C using the Hebb–Wagner method.[17] The low ionic conductivity also leads to a low oxygen ion diffusion coefficient, which is found to be in the range of 10^{-13} ∼ 10^{-11} cm^2 s^{-1} by the secondary ion mass spectrometry measurement.[17,18] Other factors including TEC, and surface exchange coefficient for LSM based materials are summarized in Table 3.1 for reference.

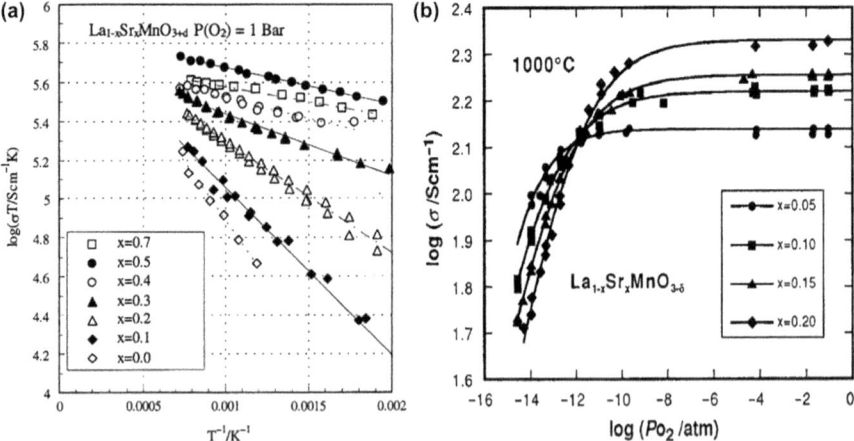

Figure 3.1 Electrical conductivity of La$_{1-x}$Sr$_x$MnO$_{3-\delta}$ as functions of (a) temperature and (b) oxygen partial pressure.[13,14]

Table 3.1 D^*, k, TEC, and electric and ionic conductiviyies of $La_{1-x}Sr_xMnO_{3-\delta}$.

Composition	Temperature (°C)	$D^*(cm^2\,s^{-1})$	$k(cm\,s^{-1})$	Conductivity (S/cm)	$\sigma_{ion}(S/cm)$	$TEC \times 10^{-6}(K^{-1})$	Reference
$LaMnO_3$	800			83		12.5	[13, 24]
	897	$2*10^{-13}$	$7.7*10^{-8}$				[25]
$La_{0.9}Sr_{0.1}MnO_3$	800		$5.9*10^{-8}$	120		11.2	[13, 24]
	1000	$4.78*10^{-12}$			$2.09*10^{-6}$		[17]
$La_{0.8}Sr_{0.2}MnO_3$	900	$1.27*10^{-12}$			$5.93*10^{-7}$	12	[17]
	1000	$1.33*10^{-11}$			$5.76*10^{-6}$		[17]
	1000	$6.6*10^{-13}$	$5.62*10^{-8}$				[18]
	900	$1.6*10^{-13}$	$1.78*10^{-8}$				[18]
	800	$4.0*10^{-15}$	$5.62*10^{-9}$	190			[18]
	700	$3.1*10^{-16}$	$1.01*10^{-9}$				[18]
$La_{0.6}Sr_{0.4}MnO_3$	800			320		12.7	[13, 24]
$La_{0.7}Sr_{0.3}MnO_3$	800			178	$6.3*10^{-4}$	11.7	[26]
$La_{0.84}Sr_{0.16}MnO_3$	800			83		11.62	[27]
$La_{0.92}MnO_3$	1000	$2.45*10^{-13}$	$7.45*10^{-8}$				[28]
$La_{0.9}MnO_3$	897	$1.2*10^{-13}$	$4.8*10^{-8}$				[25]
$La_{0.65}Sr_{0.35}MnO_3$	900	$4*10^{-14}$	$5*10^{-8}$				[29]
	800				$1.7*10^{-4}$		[30]
$La_{0.5}Sr_{0.5}MnO_3$	900	$3*10^{-12}$	$9*10^{-8}$				[29]
	800	$8*10^{-14}$	$1*10^{-7}$				[29]
	700	$2*10^{-15}$	$1*10^{-8}$				[29]
$La_{0.95}Sr_{0.05}MnO_3$	900	$2.44*10^{-13}$			$1.1*10^{-7}$		[17]

Oxygen reduction in the porous single phase LSM cathode is considered to be focused on the cathode/electrolyte interface, resulting in a high polarization loss, especially at intermediate temperatures. Thus, LSM is often used in the form of composite cathodes by introducing oxygen ionic phases, such as YSZ and SDC (samaria-doped ceria). Ji *et al.* have investigated the oxygen diffusion in LSM-YSZ composites and found that it increases from 10^{-12} cm^2 s^{-1} for pure LSM to 10^{-8} cm^2 s^{-1} for LSM-60% YSZ[19]; the oxygen ion diffusivities have greatly improved, but they are still lower than that of pure YSZ due to the decreased connectivity of the YSZ phase. Similarly, the surface exchange coefficient on LSM is increased with addition of an ionic phase, and it increases by three times after the introduction of YSZ.[20] The enhancement is sensitive to the ionic conductivity of the second phase and further improvement is observed by the coating of SDC for its larger ionic conductivity, as shown in Figure 3.2. In contrast with the oxygen ion diffusion, the surface exchange coefficients of the composites are higher than both the individual components,[19] indicating that both oxygen vacancies and electrons are important for the surface exchange process.

Electrochemical performances of LSM-YSZ composites are listed in Table 3.2. Power densities of single cells using LSM-YSZ cathodes are typical 300 mW cm^{-2} at 700 °C. This is not acceptable for low-temperature applications due to the increased cathode polarization losses. On the other hand, some insulating phases La$_2$Zr$_2$O$_7$ and SrZrO$_3$ can be formed near the interface during the long-time operation at high temperatures.[21–23] The conductivity of these zirconates is two to three orders of magnitude lower than that of YSZ,

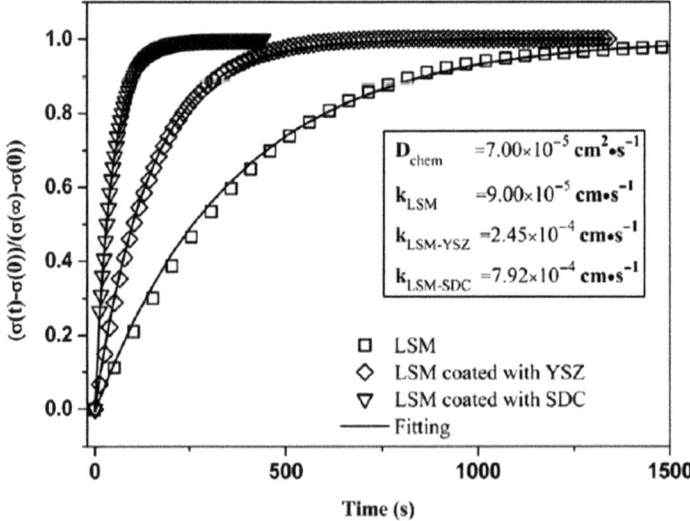

Figure 3.2 Normalized conductivity data of the LSM, LSM-YSZ and LSM-SDC specimens plotted as a function of time over the gas switching from 1% O$_2$ to CO/CO$_2$ (1:1) at 1000 °C.[20]

Table 3.2 Electrochemical performance of LSM-YSZ cathodes.

Test temperature ($°C$)	Power density ($W\,cm^{-2}$)	Polarization resistance ($\Omega\,cm^2$)	Electrolyte	Cathode composition	Ref.
800	0.85	0.071	3 µm YSZ	50 wt% LSM-YSZ	[31]
800	0.65	0.31	9 µm YSZ	60 wt% LSM-YSZ	[32]
800	1.03	—	25 µm YSZ	50 wt% LSM-YSZ	[33]
800	1.8	—	10 µm YSZ	LSM-YSZ	[34]
800	1.42	0.18	—	LSM-YSZ	[35]
800	0.41	0.46	7 µm YSZ	LSM-YSZ	[36]
750	—	1.31	—	50 wt% LSM-YSZ	[37]
750	—	0.21	—	50 wt% LSM-YSZ	[38]

which will increase the ohmic polarization resistance. Thus, it is still necessary to develop new cathode materials for SOFCs operated in intermediate and/or low temperatures.

3.2.2 Mixed Oxygen Ion-Electron Conducting Cathodes

Perovskite oxide $La_{1-x}Sr_xCoO_3$ has been widely investigated for the use of cathode materials as a MIEC. Similar to the Mn element in LSM, the transitional Co element in $LaCoO_3$ exhibits multiple valences. At a low temperature, the cobalt ions exhibit the diamagnetic low spin configuration $\left(t_{2g}^6\right)$, they transform to the paramagnetic high-spin state $\left(t_{2g}^4 e_g^2\right)$ and further produce Co^{4+}, Co^{2+} due to the delocalization of the e_g electrons when the temperature is above 930 °C.[39] The process corresponds to a metal-insulator transition which has been validated by the temperature dependence of the electrical conductivity. As shown in Figure 3.3, the electrical conductivity of pure LSC (x = 0) increases with the temperature up to about 900 °C, and this positive temperature coefficient indicates a semi-conducted behaviour. Then, the conductivity becomes less temperature dependent and decreases with temperature, indicating a metallic behaviour at temperatures above 900 °C.[40] The peak conductivity of 600 S cm^{-1} at 800 °C is observed, which is close to the value of the $LaCoO_3$ film reported by Yamamoto *et al.*[39] and Mineshige *et al.*,[41] and is much higher than that of $LaMnO_3$, 200 S cm^{-1}.[13]

Bivalent elements (such as Sr, Ca) are usually used to substitute on A-sites, permitting a fine tuning of electrical properties. In the case of Sr substitution for La, the Sr^{2+} ions act as acceptors and the strontium ions are assumed to occupy the regular La lattice sites, leading to a negative charge. According to the electrical neutrality, it should be compensated by the formation of equivalent positive charges, including $Co_{Co}^{\bullet}$ and oxygen vacancies $V_O^{\bullet\bullet}$. Accordingly, the substitution of A site La^{3+} by Sr^{2+} (or Ca^{2+}) increases the oxygen vacancy concentration and thus exhibits an increased oxygen ion conductivity. Figure 3.4a shows the oxygen ion conductivity in air for $La_{1-x}Sr_xCoO_{3-\delta}$ (680 °C) and $La_{1-x}Ca_xCoO_{3-\delta}$

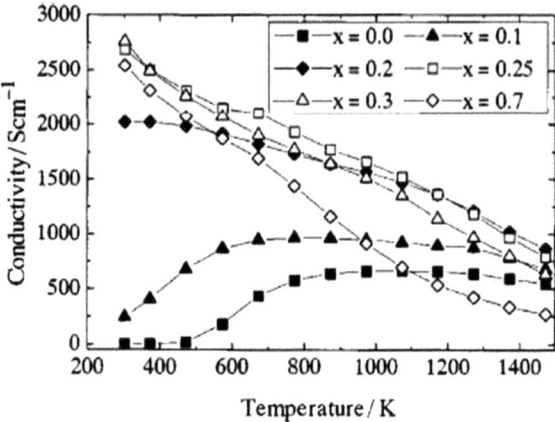

Figure 3.3   Temperature dependencies of electrical conductivity of $La_{1-x}Sr_xCoO_3$.[40]

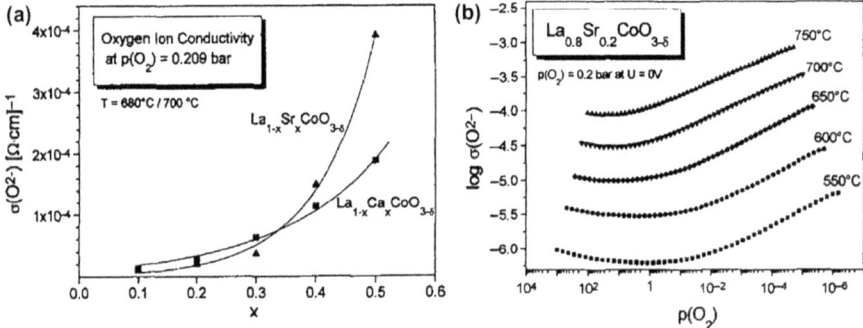

Figure 3.4   Oxygen ion conductivity of $La_{1-x}Sr_xCoO_{3-\delta}$ as functions of (a) Sr^{2+} content and (b) oxygen partial pressure at 550–750 °C.[44]

(700 °C) as a function of the earth alkaline content. The ionic conductivity of $La_{0.8}Sr_{0.2}CoO_{3-\delta}$ is 2.0×10^{-5} S cm^{-1} at 680 °C, and increases with the increment of Sr doping, since net negative charge of Sr substitution is mainly compensated by the formation of positively charged oxygen vacancies $V_O^{\bullet\bullet}$. However, much more A-sites substitution (x > 0.5) decreases ionic and electronic transport.[42,43] The effects of cation deficiency are governed by particular charge-compensation mechanisms that depend on the radius ratio of metal cations, temperature, and other factors, such as oxygen pressure.[44] Several research groups have measured the ionic-conducted properties of $La_{1-x}Sr_xCoO_{3-\delta}$. For example, Sekido *et al.*[45] measured the chemical diffusion coefficients of $La_{0.8}Sr_{0.2}CoO_{3-\delta}$ and calculated an ionic conductivity of $\sigma_{O^{2-}} = 9.2 \times 10^{-3}$ S cm^{-1} at 800 °C in air. However, Kharton *et al.*[42] reported much higher conductivities, e.g. $\sigma_{O^{2-}} = 3.3 \times 10^{-1}$ S cm^{-1} for the same composition at 830 °C in air. The discrepancy could be attributed to synthesis

conditions, pre-history, and microstructure because the cobaltite perovskites have a lower thermodynamic stability and are less tolerant to the cation non-stoichiometry in comparison with manganites. Figure 3.4b shows oxygen partial pressure dependence of the oxygen ion conductivities of $La_{0.8}Sr_{0.2}CoO_{3-\delta}$ samples at $550 \sim 750\,°C$.[44] The oxygen ion conductivity increases with decreasing oxygen partial pressure, which is attributed to the increased concentration of oxygen vacancies. The pressure dependence factor n, with respect to the relation $\sigma_{O^{2-}} = const * P_{O_2}^n$, is calculated about -0.2 at $700\,°C$, which is largely related to the doping elements, and defect concentrations.

Although these cobalt-based materials exhibit high catalytic activity and electrochemical properties, their TECs are usually much higher than the electrolyte materials. The peak TECs for $La_{0.3}Sr_{0.7}CoO_{3-\delta}$ can reach up to $26 \times 10^{-6}\ K^{-1}$, twice that of YSZ and doped ceria electrolytes ($11 \sim 12 \times 10^{-6}\ K^{-1}$). In a sequence, the TEC mismatch will result in a thermal stress, leading to the delamination of the cathode/electrolyte interface, and/or cracking of the electrolyte during the operation.[46] In addition, LSC reacts readily with YSZ to form insulated $La_2Zr_2O_7$ and $SrZrO_3$ phases at relatively low temperatures, about $1000\,°C$.[39,47]

Strontium-doped lanthanum ferrites (LSF), $La_{1-x}Sr_xFeO_{3-\delta}$, have also been studied for the use of temperature-reduced cathodes due to the sufficient p-type electrical conductivity.[48] The addition of Sr to $LaFeO_3$ creates Sr'_{La}, which may be compensated by the formation of Fe^{4+} ions or oxygen vacancies $V_O^{\bullet\bullet}$. Figure 3.5a shows the Sr addition effects on the electrical conductivity.[49] The composition of $x = 0.5$ results in the peak conductivity of $352\ S\,cm^{-1}$ at $550\,°C$ in air, which is consistent with the theoretical Sr content creating a maximum Fe^{4+}/Fe^{3+} ratio of $1:1$. Figure 3.5b presents the effects of oxygen partial pressure on conductivity. The conductivity decreases with increasing partial pressure, and then increases. Moreover, the value corresponding to the lowest conductivity increases with the temperature. Two defect equations (2) and (3)

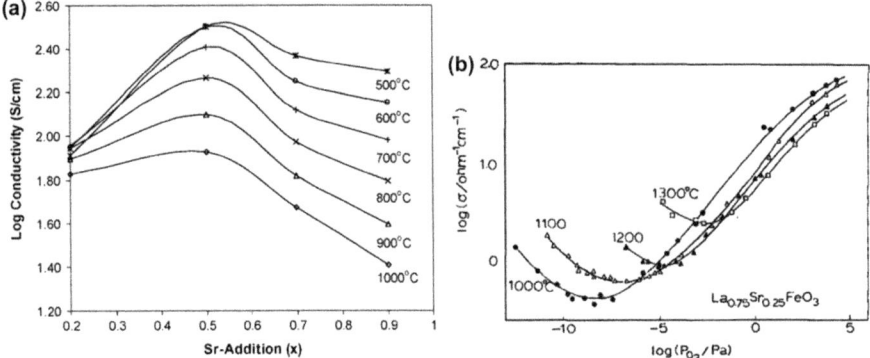

Figure 3.5 Conductivity of $La_{1-x}Sr_xFeO_{3-\delta}$ as functions of (a) Sr^{2+} content at $500 \sim 1000\,°C$ and (b) oxygen partial pressure at $1000 \sim 1300\,°C$ for $x = 0.25$.[49]

are proposed to describe these effects. At lower oxygen partial pressure, Sr'_{La} defects tend to be compensated by oxygen vacancies, which prevent the formation of electron holes, and therefore, decrease the electrical conductivity. However, at a relative higher partial pressure, the defects are dominantly compensated by holes in the form of $Fe^{\bullet}_{Fe}$, higher partial oxygen pressure resulting in larger holes concentration. Because of the much higher mobility of the electrons or holes than that of oxygen vacancies, the total conductivity in ferrites is dominated by a hole-conduction mechanism.

$$SrO + \frac{1}{2}Fe_2O_3 \overset{LaFeO_3}{\rightarrow} Sr'_{La} + Fe^{\times}_{Fe} + \frac{5}{2}O^{\times}_O + \frac{1}{2}V^{\bullet\bullet}_O \tag{2}$$

$$2SrO + Fe_2O_3 + \frac{1}{2}O_2(g) \overset{2LaFeO_3}{\rightarrow} 2Sr'_{La} + Fe^{\bullet}_{Fe} + 6O^{\times}_O \tag{3}$$

LSF is expected to be more comparable with the ionic electrolyte both in thermal and chemical stabilities in comparison with Co-based materials. TECs of LSF are about $(12 \sim 18) \times 10^{-6}$ K^{-1},[50] which is close to that of the typical electrolytes, thus the problems derived from TEC mismatch are largely avoided. Reactivity with YSZ electrolyte has also been significantly reduced since the Fe^{3+} ion has a stable electron configuration $3d^5$. Preliminary data from the reaction of iso-statically pressed $La_{0.8}Sr_{0.2}FeO_{3-\delta}$ and YSZ powder mixtures indicates that no La- or Sr-zirconate formations are observed even at $1200 \sim 1400$ °C. However, closer examination by XRD and SEM-EDS technique reveals some low angle shift in the LSF peak positions, which corresponds to a unit cell volume expansion by incorporation of Zr^{4+} cations in the perovskite, resulting in a highly A-site deficient structure.[51–53] The process has a detrimental effect on its electrical conductivity. As shown in Figure 3.6, the conductivities have decreased by 60% with the incorporation of 10% Zr

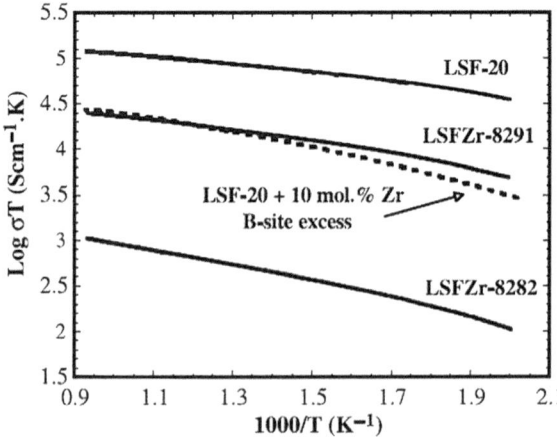

Figure 3.6 Arrhenius conductivity plots for $La_{0.8}Sr_{0.2}Fe_{1-y}Zr_yO_{3-\delta}$ ($y = 0$, 0.1, 0.2) and $La_{0.8}Sr_{0.2}Fe_{1.0}Zr_{0.1}O_{3-\delta}$ (10 mol.% Zr B-site excess).[51]

into LSF, regardless of A-site stoichiometry, and further reduction is induced with more Zr^{4+} incorporation. The decrement is possibly attributed to the fact that the fixed valence Zr^{4+} cations dispersed among the Fe cations could not participate in the conduction process but only act as blocking sites within the lattice. More serious reaction processes are also observed when the LSF cathodes are sintered at higher temperatures or doped by some other cations in A and/or B sites, or with different A/B ratio. For instance, Anderson *et al.*[52] have found that the YSZ electrolyte reacts with $Ln_{0.8}Sr_{0.2}FeO_{3-\delta}$ (Ln = La, Sr, Pr, Nd) and $La_{0.8}Ba_{0.2}FeO_{3-\delta}$ with the exception of $La_{0.8}Ca_{0.2}FeO_{3-\delta}$ below 1200 °C. Martínez-Amesti *et al.*[53] reported that higher Sr content in the LSF promoting the chemical reaction with YSZ. For x = 0.4, $SrZrO_3$ and/or $SrFe_{12}O_{19}$ are formed in the original $La_{1-x}Sr_xFeO_{3-\delta}$, while for x = 0.2, only ZrO_2 appeared at high temperatures. Thus, a layer of Sm or Gd-doped CeO_2 is still necessarily utilized between the LSF and YSZ electrolyte as a protective barrier, which also plays a very significant role in improving the electrochemical behaviour. For example, power density of 950 $mW\,cm^{-2}$ under the voltage of 0.7 V at 700 °C is reported for $La_{0.8}Sr_{0.2}FeO_{3-\delta}$ cathode on SDC interlayer, compared to 650 $mW\,cm^{-2}$ for that directly on YSZ electrolyte under the same conditions.[52] The increased performance is possibly related to the enhanced oxygen surface exchange kinetics of ceria compared to YSZ, and the ceria interlayer may prevent the formation of poor conducting phases.

$La_{1-x}Sr_xCo_{1-y}Fe_yO_{3+\delta}$ (LSCF) has also attracted much attention due to its great potential for the application in intermediate temperature SOFCs. Similar to most perovskites, the conductivity of LSCF increases with the increase of Sr content. And with the increase of Fe content, the TEC is reduced to be more compatible with that of YSZ and SDC, at the same time, the reactivity with YSZ is reduced. $La_{0.6}Sr_{0.4}Co_{0.2}Fe_{0.8}O_{3-\delta}$ is the most common composition for compromise between conductivity, catalytic activity, TEC and reactivity with the electrolyte. It offers high electrical and ionic conductivity, about 340 $S\,cm^{-1}$ at 550 °C.[54] and $10^{-2}\,S\,cm^{-1}$ at 800 °C,[55] respectively. Performances of LSCF-based cathodes in the literature are listed in Table 3.3. The polarization losses are much lower than that of LSM. The resistance of pure LSCF on gadolinia doped ceria (GDC) electrolyte is only 1.2 $\Omega\,cm^2$ at 600 °C in air. Performance can be further improved by the addition of some electrolyte phase. For example, ASR has decreased to 0.17 $\Omega\,cm^2$ at 600 °C for LSCF-60 wt% GDC composite electrode. A ceria interlayer is necessary to prevent the reactivity between the LSCF and YSZ.

3.2.3 Microstructure Optimized Cathodes

Despite of cathode materials, the electrode performance depends strongly on its microstructure. Generally, ASR can be reduced by factors of 5–10 times by composition optimization. One of the first works of composite cathodes was reported in 1991 by Kenjo *et al.*[61] It involves mixing ionic conducting phase with electrocatalytic phase (usually electronic conducting), and printing the

Table 3.3 Electrochemical performance of LSCF based cathodes.

Composition	Treatment Temperature	Performance	Reference
$La_{0.6}Sr_{0.4}Co_{0.2}Fe_{0.8}O_{3-\delta}$(LSCF6428)			
LSCF6428//GDC//LSCF6428	975 °C/2h	1.2 Ω cm^2@600 °C	[56]
LSCF6428-60wt%GDC//GDC//LSCF6428-60wt% GDC	975 °C/2h	0.17 Ω cm^2@600 °C	[56]
40%LSCF6428-60%GDC//GDC//Ni-GDC(65:35)	975 °C/2h	0.27 Ω cm^2@600 °C	[56]
		422 mW/cm^2@600 °C	
$La_{0.58}Sr_{0.4}Co_{0.2}Fe_{0.8}O_{3-\delta}$(LSCF5840)			
50wt%LSCF5840-50wt%SDC//SDC//	1000 °C/2h	0.306 Ω cm^2@700 °C	[57]
50wt%LSCF5840-50wt%SDC		0.208 Ω cm^2@750 °C	
		0.142 Ω cm^2@800 °C	
50wt%LSCF5840-50wt%SDC//SDC|YSZ//	1100 °C/2h	0.548 Ω cm^2@700 °C	[57]
50wt%LSCF5840-50wt%SDC		0.28 Ω cm^2@750 °C	
		0.21 Ω cm^2@800 °C	
Ni-8YSZ//YSZ|GDC//LSCF	1080 °C	1.2 mW/cm^2@800 °C 0.7V	[58]
Ni-8YSZ//YSZ|GDC//LSCF	GDC interlayer sintered at 1350 °C	370 mW/cm^2@750 °C	[59]
50wt%LSCF-50wt%LSGM//LSGM//	1100 °C/1h	0.69 Ω cm^2@600 °C	[60]
50wt%LSCF-50wt%LSGM		0.19 Ω cm^2@650 °C	
		0.061 Ω cm^2@700 °C	
Ni-SDC//LSGM//	1100 °C/1h	1090 mW/cm^2@750 °C	[60]
50wt%LSCF-50wt%LSGM			

(a) **(b)**

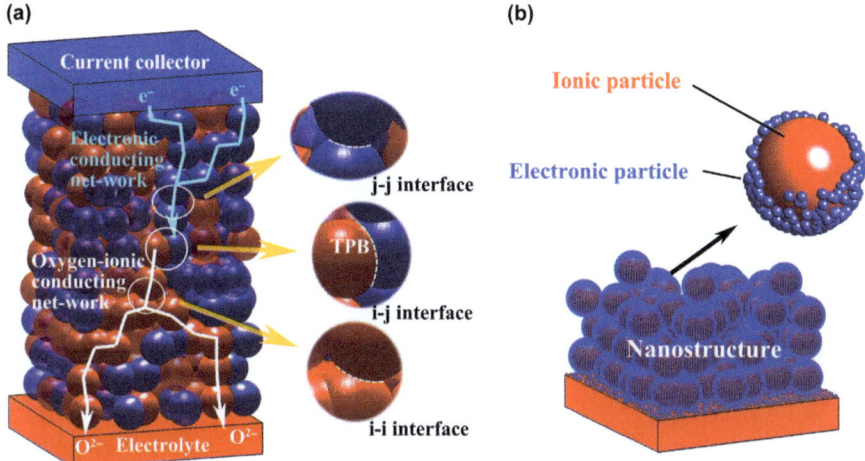

Figure 3.7 Illustrations of the microstructures of conventional composite cathodes (a) and infiltrated cathodes (b).[62,68]

composite slurry onto the electrolyte followed by high-temperature co-sintering. Until now, this strategy has become the most conventional method of cathode fabrication. The electrode microstructure is illustrated in Figure 3.7a. Compared with the porous single-phase electrodes, the composite electrodes extend the electrochemical reaction sites from the cathode/electrolyte interface to the cathode bulk. For the most commonly used composite LSM-YSZ cathodes, the reaction sites for oxygen reduction are phenomenologically considered as the three-phase boundary (TPB) lines where LSM, YSZ, and pore phases meet, because LSM/YSZ is almost a pure electronic/oxygen ionic conductor, and thus the electrode reactions are restricted to the TPB lines with a width of atomic dimensions.[4] By theoretical analysis, the cathode ASR was found to be proportional to the square root of volumetric TPB length, and the TPB length is a strong function of structure parameters, such as particle size and composition.[62] Experimental observations showed that, there exists an optimal composition corresponding to the lowest electrode ASR.[63–65] Figure 3.8 shows the ASR (800 °C) for composite LSM-YSZ cathodes at various weight fractions of YSZ, showing the optimal composition is about 50:50 wt% YSZ:LSM. The lowest ASR at 800 °C for LSM-YSZ electrodes is approximately 0.1 $\Omega\,cm^2$, marginally gratifying the need of 1 W cm^{-2} output power density for a single cell.[66] However, ASR is relatively high at operated temperatures below 800 °C. MIEC electrocatalysts, such as LSC, LSF, and LSCF perovskites are used instead of LSM for IT SOFCs. In practice, the co-sintering temperature is usually above 1000 °C to strengthen the bonding between cathodes and electrolyte and to make the cathode more conductive. Such high sintering temperature leads detrimental solid-state reactions between these perovskites (including LSM) and Zr – based electrolytes.[1] Moreover, a highly active MIEC electrocatalyst also has a large thermal expansion coefficient (TEC).

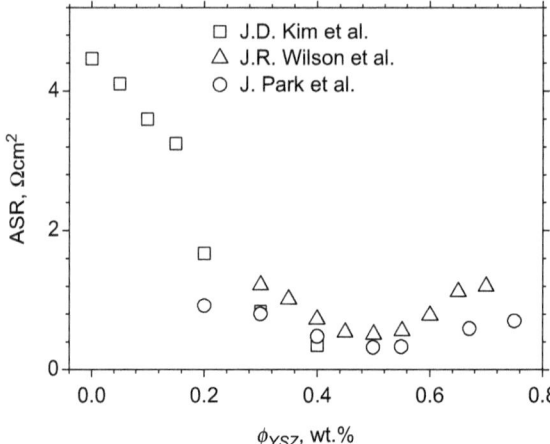

Figure 3.8 ASR at 800 °C for conventional composite LSM – YSZ cathodes at various YSZ loadings.[63–65]

For example, both the catalytic activity and TEC increase with the increase of Co content in the $La_{1-x}Sr_x(Mn\text{-}Fe\text{-}Co)O_{3-\delta}$ system, causing much higher TECs than that for electrolytes.[67] The TEC mismatch between cathode materials and electrolyte is a major consideration for choosing materials. As illustrated in Figure 3.7a, the paths for electron conducting (j–j interface), oxygen-ion conducting (i–i interface), and charge transfer reactions (TPB) can break under the internal stress by TEC mismatch. In practice, the TEC mismatch deteriorates electrode durability, and even leads to the delamination of electrodes during long-term operations, such as thermal cycles.[68]

In the last decade, the infiltration/impregnation method has drawn tremendous attention to prepare highly active electrodes. By using infiltration, the above technical barriers for conventional composite electrodes can be resolved to some extent.[66] Compared with the conventional electrodes with optimized composition, the performance of infiltrated electrodes can be enhanced by factors of several times, even dozens of times.[69] Theoretical and experimental considerations suggest that the promotion in electrode performance derives from the advanced microstructures by its distinctive fabrication processes.[62,66,68,69] In general, the infiltration approach involves making a porous backbone layer of electrolyte material onto the dense electrolyte layer by printing/casting and co-sintering processes, then infiltrating the electrocatalyst nitrate solution inside the backbone pores, followed by relative low temperature calcinations (*e.g.* 800 °C). On one hand, the multiple steps permit the separate control and optimization of the backbone structure and the electrocatalyst architect. On the other hand, the low calcination temperature can suppress the detrimental reactions between the electrolyte and electrocatalyst/electrode materials. In addition, TEC for infiltrated electrodes was found to be determined by backbone materials.[66] Thus the combinations of variable electrode materials are extended. More importantly, nanosized electrocatalytic particles are usually formed in a typical range of 50 ∼ 100 nm.[69] This unique structure results in a significant enhancement in electrochemical active sites for electrode reactions. Figure 3.7b illustrates the infiltrated microstructure, which is, in principle, considered as a sphere-packed backbone, coated with nanoparticles of another phase. The TPB length is expected to be enhanced compared with the conventional composite electrodes. By the particle-layer model,[62] Figure 3.9 shows the TPB length (a), effective electrode thickness (b), and electrode ASR (c) for conventional (CCC) and infiltrated (ICC) LSM-YSZ cathodes. It is found that the TPB length can be increased significantly by infiltration, and thus the effective electrode thickness is decreased, leading to a lower electrode ASR. For MIEC infiltrated cathodes, such as LSCF, and LSF, Barnett *et al.*[70] showed that the electrode ASR is proportional to the surface area of infiltrated phase. Most interestingly, Xia *et al.* found that LSC infiltrated SDC cathodes exhibit exceptional durability in thermal cycle processes, in spite of the unacceptable TEC mismatch between LSC and SDC.[71] Figure 3.10 shows the performance comparison between the conventional and infiltrated LSC-SDC cathodes. Two stages of thermal cycle processes – 20 500–800 °C cycles followed by 10 RT–800 °C cycles – were performed. For the conventional composite cathode, the ASR increased slightly at the first stage of the thermal cycle, and then increased from 3.5 to 12.5 Ω cm^2 at the second stage. But the ASR for the infiltrated cathode kept almost constant during the two stages of the thermal cycle. It was suggested that the promoted durability may be caused by the nanoscale size of infiltrated particles.[68] According to Weibull theory, smaller particles usually exhibit a lower failure probability under a tensile stress, which is the so-called size effect.[68] Another similar phenomenon was reported by Chen and Jiang.[72] They showed that LSM infiltrated YSZ cathode exhibited an excellent durability,

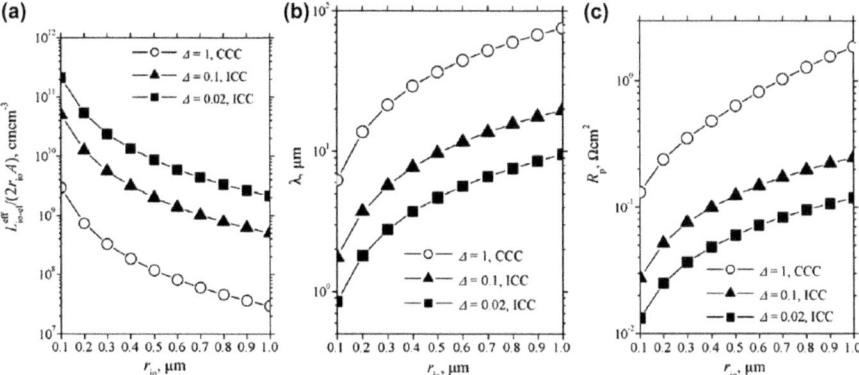

Figure 3.9 The comparison of volumetric TPB length (a), effective electrode thickness (b, 800 °C) and ASR (c, 800 °C) between conventional composite LSM – YSZ cathodes (CCC) and LSM infiltrated YSZ cathodes (ICC) by the particle – layer model. r_{io} represents the radius of YSZ particles; Δ represents the ratio of particle radius between LSM and YSZ particles.[62]

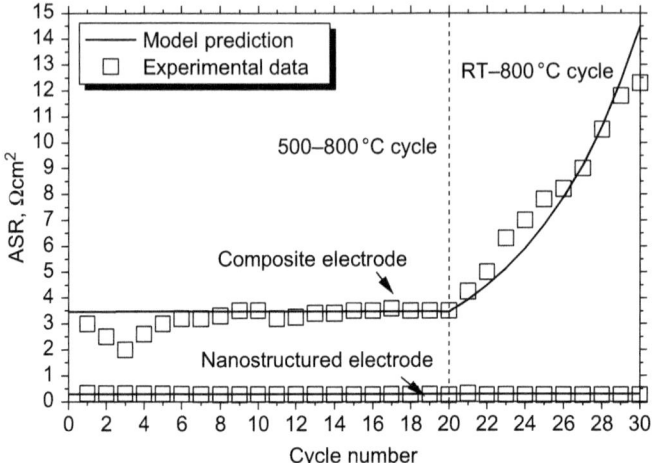

Figure 3.10   ASR at 600 °C for conventional composite LSC – SDC cathodes and
LSC infiltrated SDC cathodes under two stages of thermal cycle
treatments. Both experimental data and modelling prediction are
shown.[68,71]

while the conventional composite LSM-YSZ cathode delaminated form YSZ
electrolyte during anodic polarization, which builds up high oxygen pressure
near the LSM- YSZ interface. These works showed the significance of
geometric properties. In addition to geometric properties, catalytic effects of the
infiltrated nanoparticles may be another reason for performance promotion.[69]
For example, significant enhancement in electrode performance was observed
by fitting a small amount of isolated Pd nanoparticles into the pore structure of
composite LSM-YSZ scaffold.[73] The in-depth understanding of the geometric
properties and the catalytic effects of infiltrated nanoparticles is crucial to
identify reaction mechanisms and to optimize the microstructure for
performance enhancement.

3.2.4   Cathode Reaction Mechanisms

In the cathode, the oxygen (from air) is reduced into oxygen ions through the
following electrochemical reactions,

$$O_2 + 4e^- + 2V_O^{\bullet\bullet} = 2O_O^\times \qquad (4)$$

where, in Kröger–Vink notation, O_2 represents oxygen molecular in air,
e^- represents electrons in cathode material, $V_O^{\bullet\bullet}$ represents vacant oxygen site in
cathode material or electrolyte material, and $O_O^\times$ represents oxygen ion in
electrolyte material. As indicated, the oxygen reduction process requires the
presence of oxygen, electrons, and vacant oxygen sites. When the cathode
material possesses electronic conductivity and neglectable ionic conductivity

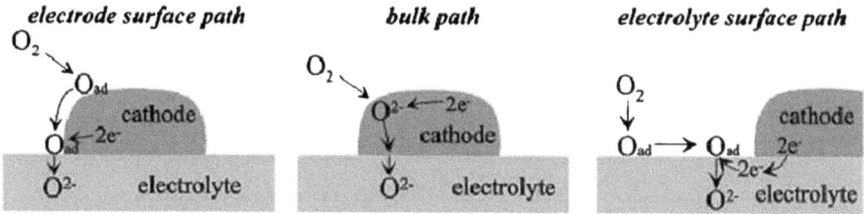

Figure 3.11 Possible reaction paths in perovskite cathodes for ion – conducting SOFCs.[74]

(*e.g.* LSM), pores, cathode material, and electrolyte material act as the carriers of oxygen, electrons, and vacant oxygen sites, respectively. And thus, the oxygen reduction process is restricted to the TPB lines. When the cathode material exhibits mixed ion-electron conductivity, such as LSCF, the sites for oxygen reduction are extended from the TPB lines to the surface of the cathode material. And the oxygen ions in the cathode material can be transferred into electrolyte material through the cathode-electrolyte interface.[66] Apparently, the cathode reaction is affected by the microstructure details for particular cathode systems. In addition, the single diatomic oxygen must be converted to some intermediate form via one or more processes.[4] For perovskite cathodes, Fleig[74] summarized three possible paths for oxygen reduction: that is, the cathode surface path, the cathode bulk path, and the electrolyte surface path as shown in Figure 3.11. One intermediate species, O_{ad} is suggested. In sequence, these three paths may be applicable to the cases of electronic conducting cathodes (*e.g.* LSM), MIEC cathodes (*e.g.* LSCF), and composite cathodes (*e.g.* composite LSM-YSZ), respectively. However, the elementary reactions were not understood clearly until now, even for the simple system- porous LSM cathode on YSZ electrolyte.

A most rational mechanism for the elementary reactions contain the following four steps,[75]

$$\frac{1}{2}O_2 \leftrightarrow O_{ad,\,cathode} \tag{5.1}$$

$$O_{ad,\,cathode} + e^- \leftrightarrow O^-_{ad,\,cathode} \tag{5.2}$$

$$O^-_{ad,\,cathode} \leftrightarrow O^-_{TPB} \tag{5.3}$$

$$O^-_{TPB} + e^- + V^{\bullet\bullet}_{O,\,electrolyte} \leftrightarrow O^{\times}_{O,\,electrolyte} \tag{5.4}$$

It is reported that step 5.2 or 5.3 maybe the rate-limiting step for the LSM cathode. Chen and Chan established a refined model, in which the step 5.4 is divided into two elementary steps,[76]

$$O^-_{TPB} + e^- \leftrightarrow O^{2-}_{TPB} \tag{5.5}$$

$$O^{2-}_{TPB} + V^{\bullet\bullet}_{O,\,electrolyte} \leftrightarrow O^{\times}_{O,\,electrolyte} \tag{5.6}$$

By the measurement and fitting of polarization curves of the porous cathode at various temperatures and environmental oxygen pressures, they proposed the rate-limiting-steps 5.3 and 5.6. As these steps take place in the vicinity of TPB lines, the area-specific current density is normalized as current density per TPB length (i, $A\,cm^{-1}$) as,

$$i = \frac{1}{1\big/ i_{0,3}\left(p_{o2}/p_{o2}^0\right)^{1/2}\exp(-f\eta) + 1\big/ i_{0,6}\left(p_{o2}/p_{o2}^0\right)^{1/2}\exp(-2f\eta)}$$
$$-\frac{1}{1\big/ i_{0,3}\exp(f\eta) + 1\big/ i_{0,6}} \tag{6}$$

where $f = F/RT$, p_{o2} the local oxygen pressure, and p_{o2}^0 the oxygen pressure in air (0.21 atm). Note this treatment eliminates the effects of microstructures, so Eq. 6 offers essential information for reaction kinetics. For LSM-based cathode on YSZ electrolyte, the exchange current densities per TPB length for steps 5.3 and 5.6 are given by Eq. 7a and 7b, respectively.

$$i_{0,3} = 78.74 \times p_{o2}^0 1/4 \exp\left(-2.05 \times 10^4 / T\right) \tag{7a}$$

$$i_{0,6} = 7.21 \times 10^{-2} \exp\left(-1.24 \times 10^4 / T\right) \tag{7b}$$

The results suggest that environmental conditions affect the cathode reactivity under polarized states through the exchange current density. However, this model cannot explain some other experimental phenomenon, such as the effects of electrolyte conductivity. It was found that the exchange current density increases with the increase in electrolyte conductivity at a given temperature. In other words, the ASR scales with electrolyte resistivity.[77] One possible reason was explained as the local polarization variations due to discrete contacts between the porous cathode and dense electrolyte.[78] Another explanation is the transfer of oxygen ions in the electrolyte near the cathode/electrolyte interface,[79] which can be considered as an elementary step after step 5.6,

$$V^{\bullet\bullet}_{O,\,electrolyte_B} \leftrightarrow V^{\bullet\bullet}_{O,\,electrolyte_S} \tag{5.7}$$

where the subscript 'S' and 'B' represent surface and bulk sites in the electrolyte. If step 5.7 is rate-limiting, the exchange current density increases linearly with the increase of electrolyte conductivity. As discussed above, the oxygen nonstoichiometry relates with temperature and environmental oxygen pressure, indicating the oxygen nonstoichiometry is also an essential factor governing the cathode reactivity.

For the MIEC cathodes, such as LSCF, the overall reaction is usually used instead of elementary reactions. Due to their relative high electronic conductivity, the electric potential in MIEC is almost uniform. And thus the diffusion of oxygen ions is induced by the concentration gradient of oxygen ions, with which the MIEC material could be in equilibrium, governed by the local oxygen ion concentration or the local oxygen nonstoichiometry.[80]

The reactivity is usually represented by two parameters related to the oxygen transfer properties of the MIEC, that is, the oxygen surface exchange coefficient (k, cm s^{-1}) and oxygen bulk diffusion coefficient (D, $\text{cm}^2\,\text{s}^{-1}$). According to the Alder or ALS model,[81] the ASR for porous MIEC cathodes is given by,

$$ASR = \frac{RT}{2F^2} \sqrt{\frac{\tau}{(1 - \varepsilon)aC_0^2 Dk}} \tag{8}$$

where τ, ε, and a represent tortuosity factor, porosity, and volumetric surface area, respectively, C_0 is the surface concentration of oxygen. It indicates that the cathode performance can be optimized by improving D and k and the cathode microstructure. When the oxygen-conducting thickness of MIEC is smaller than D/k, the cathode reaction is limited by surface exchange process (k), and thus D can be not considered. For instance, D/k for LSCF is about 100 µm at typical conditions, which is much higher than the oxygen-conducting thickness, especially for the LSCF infiltrated cathodes. This conclusion is also applicable to some other MIECs, such as LSC and LSF.[70] In literature, ASR for MIEC-based cathodes is usually simplified as,[70]

$$ASR = (A_{FP}/A_C)ASR_S \tag{9}$$

where ASR_S represents the MIEC surface resistance, A_{FP}, and A_C represent the cathode footprint area and MIEC inner surface area, respectively. All the details of electrochemical reactions are included in ASR_S, implicitly.

3.3 Cathodes for Proton-Conducting Electrolyte Based SOFCs

3.3.1 Electron-Conducting Cathodes

For proton- conducting SOFCs, water is also considered to be produced at TPB lines, where proton $(OH_O^\bullet)$, electron, and oxygen molecular are available. Thus, the TPB of electronic conductor single phase cathode is limited in the electrode/electrolyte interface. As shown in Figure 3.12a, the dissociated oxygen ion, which can transfer along the surface of the cathode to the TPB site, meets and reacts with protons to form water only at the interface. Due to the limit of the conduction species, the opportunities for contact of oxygen ion and proton is reduced, leading to the low electrochemical reaction rate.

Primal studies use the noble metal, Pt, which is an excellent electronic conductor, but its oxygen ion conductivity and proton conductivity are negligible. Therefore, though the catalytic activity for oxygen reduction reaction is excellent, Pt still exhibits large overpotential and high ASR with high-temperature proton conductors ($\sim 0.61\,\Omega\,\text{cm}^2$ at 800 °C). In addition, using precious metals is too expensive to achieve commercialization application.[82]

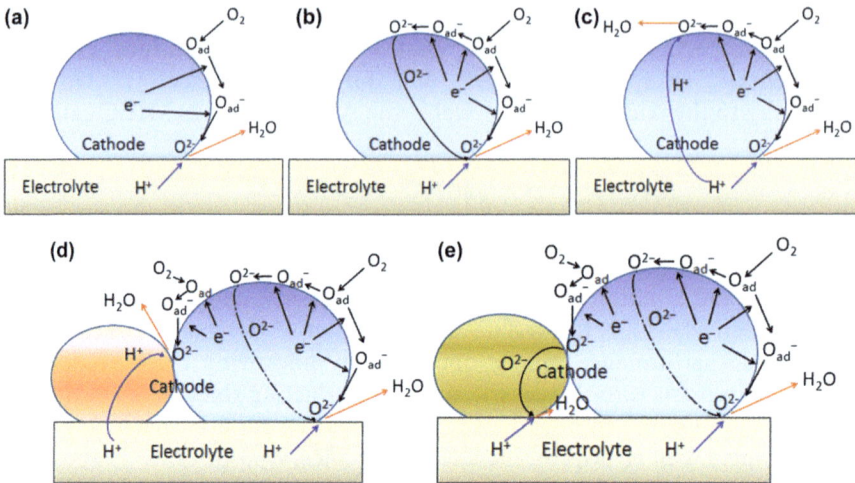

Figure 3.12 Possible reaction paths in proton – conducting SOFC cathodes, including electronic conductor single phase cathode (a), MIEC single phase cathode (b), mixed electronic/proton conductor single phase cathode (c), mixed electronic/proton conductor composite cathode (d), and MIEC composite cathode (e).

3.3.2 Mixed Oxygen Ion-Electron Conducting Cathodes

With further in-depth research, MIEC is widely used as the single phase cathode for H-SOFCs. As shown in Figure 3.12b, the possible reaction path is proposed. Oxygen ion is formed on the surface, which possibly migrates through the cathode bulk and/or surface to the cathode/electrolyte interface. Therefore, the opportunity for contact between oxygen ion and proton is promoted, leading to the enhanced reaction rate and low polarization resistance. The oxygen ion conduction process in MIECs is significantly affected by two essential characters of material. One is the oxygen surface exchange coefficient (k) and the other is the oxygen diffusion coefficient (D^*). Promotion in k and D^* gives rise to the higher oxygen ion conduction process. Figure 3.13 shows the k and D^* values of typical cathode materials, measured using O^{16}/O^{18}.[29,83–85]

Yamaura *et al.* have reported that among the $La_{0.7}Sr_{0.3}MO_3$ (M = Mn, Fe, Co) cathode materials for H-SOFCs based on $SrCe_{0.95}Yb_{0.05}O_3$ electrolytes, $La_{0.7}Sr_{0.3}FeO_3$ obtains the minimum ASR.[86] They also pointed out that $La_{0.7}Sr_{0.3}CoO_3$, $La_{0.6}Sr_{0.4}Co_{0.2}Fe_{0.8}O_3$ cathodes, containing Co element, easily react with electrolyte at 1000 °C and form the second phase. However, the $La_{0.7}Sr_{0.3}FeO_3$ basically does not react with $SrCe_{0.95}Yb_{0.05}O_3$ electrolyte.[86]

Simple perovskite oxide $Ba_{0.5}Sr_{0.5}Co_{0.8}Fe_{0.2}O_3$ (BSCF) owning high k and D^* is a fast oxygen ion conductor.[87] Lin *et al.* have shown that,[88] below 1100 °C, there seems to be no reaction occurring between $Ba_{0.5}Sr_{0.5}Co_{0.8}Fe_{0.2}O_3$ and $BaCe_{0.9}Y_{0.1}O_{2.95}$ (BCY). However, Ba element likely diffuses from BCY

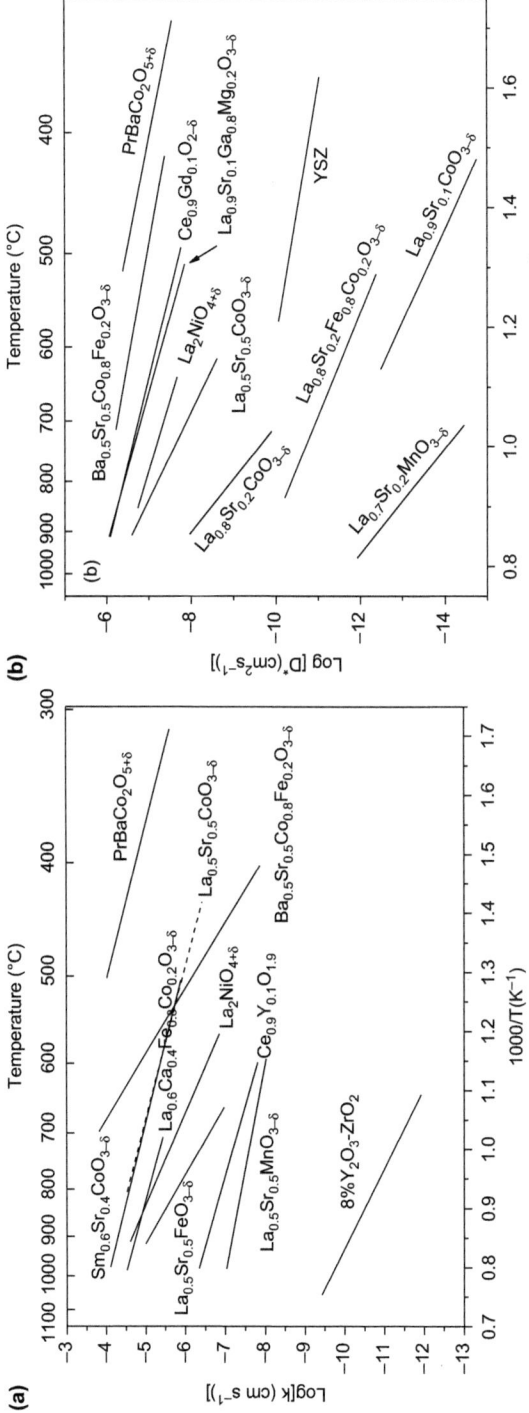

Figure 3.13 (a) oxygen surface exchange coefficient (k) and (b) oxygen diffusion coefficient (D^*) of cathode materials.[29,83–85]

electrolyte into BSCF cathode. Although the Ba diffusion does not affect the oxygen reduction reaction of the cathode, it probably blocks the transfer of the proton, reducing contact probability of proton and oxygen ion. Nevertheless, the polarization resistance of a single cell with a BSCF cathode is quite small, about $0.36\,\Omega\,cm^2$ at 500 °C. In order to further improve the stability of BSCF material, different element doping in B-site has been studied. Lin *et al.* also suggested that $Ba_{0.6}Sr_{0.4}Co_{0.9}Nb_{0.1}O_3$ perovskite oxide[89] is another good cathode with the interface impedance of a single cell at 700 °C being $0.06\,\Omega\,cm^2$. Moreover, using $BaCo_{0.7}Fe_{0.2}Nb_{0.1}O_3$ perovskite oxide[90] as the cathode material, the interface impedance of this single cell at 700 °C is $0.10\,\Omega\,cm^2$, presenting a good electrochemical performance. In addition, Co-free oxides $Ba_{0.5}Sr_{0.5}Zn_{0.2}Fe_{0.8}O_3$ ($0.08\,\Omega\,cm^2$ at 700 °C)[91] has also been used as the H-SOFC cathode materials, in order to improve the chemical stability and the thermal cycle durability.

Due to the excellent k and D^*, the cathode materials for H-SOFC are intensively focused on $LnBaCo_2O_{5+\delta}$ (Ln = La, Pr, Nd, Sm, Eu, Gd, and Y) layered perovskite oxides. The ideal structure of these family oxides can be represented by the stacking sequence, $LnO_\delta|CoO_2|BaO|CoO_2$, as shown in Figure 3.14. Transformation of a simple cubic perovskite with randomly occupied A-sites into a layered crystal with alternating lanthanide and alkali earth planes reduces the strength of oxygen binding and provides disorder-free channels for ionic motion, thereby theoretically increasing the oxygen diffusivity by many orders of magnitude.[92] Zhang *et al.*[93] have suggested that $PrBaCo_2O_{5+\delta}$ has the highest k and D^* among these $LnBaCo_2O_{5+\delta}$ oxides, which is in good agreement with the test results shown in Figure 3.13. Most of the researches about these layered perovskite cathode materials for H-SOFCs are focused on the power density and polarization resistance of single cells, instead of the ASR of the cathodes. The intensive study of the cathode interfacial impedance for H-SOFCs is limited.

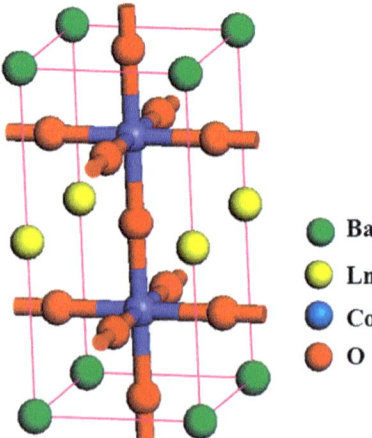

Figure 3.14   Schematic diagram of layered perovskite oxides ($LnBaCo_2O_{5+\delta}$).

For the first time, Lin *et al.* have applied GdBaCo$_2$O$_{5+\delta}$[94] and SmBaCo$_2$O$_{5+\delta}$[95] materials for H-SOFCs with BaCe$_{0.7}$Zr$_{0.1}$Y$_{0.2}$O$_3$ (BZCY) proton conductor electrolytes, At 700 °C, the interface impedance of the single cell with GdBaCo$_2$O$_{5+\delta}$ and SmBaCo$_2$O$_{5+\delta}$ is 0.16 Ω cm^2 and 0.15 Ω cm^2, respectively. Subsequently, PrBaCo$_2$O$_{5+\delta}$[96] layered perovskite oxide is used as the cathode for BZCY. At 700 °C, the interfacial impedance of the single cell with PrBaCo$_2$O$_{5+\delta}$ is 0.15 Ω cm^2. Because the impedance of a single cell includes the interface of anode-electrolyte and the interface of cathode-electrolyte, and the micro-structure of cathodes are non-uniform, so it is difficult to distinguish electrochemical activity of GdBaCo$_2$O$_{5+\delta}$, SmBaCo$_2$O$_{5+\delta}$ and PrBaCo$_2$O$_{5+\delta}$.

However, similar to the simple perovskite oxides that Co occupying the whole of B-sites, these cathodes usually cause some problems such as TEC mismatch and chemical stability during practical long-term applications. Lin *et al.*[97] have shown that, below 1000 °C, there seems to be cation diffusion occurring between PrBaCo$_2$O$_{5+\delta}$ and BZCY. Co cation spreads to the BZCY electrolyte, reducing the proton conductivity of the electrolyte BZCY. Nian *et al.*[98] reported that TEC of SmBaCo$_2$O$_{5+\delta}$ is 21.2×10^{-6} K^{-1} while it is 10.3×10^{-6} K^{-1} for BaCe$_{0.8}$Sm$_{0.2}$O$_3$ (BCS) electrolyte. The mismatch might cause low resistance to the thermal cycle. In order to reduce TEC, much attention has been paid to the B-sites ions. SmBaCuCoO$_{5+\delta}$ and SmBaCuFeO$_{5+\delta}$ oxides with the TEC of 15.5×10^{-6} K^{-1} and 14.4×10^{-6} K^{-1} are used for BCS electrolyte. At 700 °C, the interfacial impedance of the single cell with SmBaCuCoO$_{5+\delta}$ and SmBaCuFeO$_{5+\delta}$ is 0.14 Ω cm^2 and 0.20 Ω cm^2, respectively. Subsequently, Ling *et al.*[99] have reported LaBaCuCoO$_{5+\delta}$ and LaBaCuFeO$_{5+\delta}$ oxides with the single cell interfacial impedance at 700 °C being 0.15 Ω cm^2 and 0.27 Ω cm^2, respectively. As is known, the chemical stability of perovskite oxides containing Ba element might be not stable enough for high volatility and easy diffusion of BaO. Some research focus on that Sr element partly replaces the element of Ba, while not destroy the layered perovskite structure, such as, LnBa$_{0.5}$Sr$_{0.5}$Co$_2$O$_{5+\delta}$. Ding *et al.* reported that the interface impedance of the single cell with PrBa$_{0.5}$Sr$_{0.5}$Co$_2$O$_{5+\delta}$[100] is 0.12 Ω cm^2 at 700 °C, also with the similar interfacial impedance of SmBa$_{0.5}$Sr$_{0.5}$Co$_2$O$_{5+\delta}$.[101]

Except for the perovskite structure oxides, other oxides with different structures could also be used as cathode materials for H-SOFCs. For example, the hexagonal structured YBaCo$_{4-x}$M$_x$O$_7$ oxides with low thermal expansion coefficients $(<14 \times 10^{-6}$ K$^{-1})$[102] leads to an interfacial impedance of 0.16 Ω cm^2 at 700 °C.[103]

3.3.3 Mixed Electron-Proton Conducting Cathodes

Different from the MIEC single phase cathode, in which oxygen ion migrates through the cathode bulk to the cathode/electrolyte interface and reacts with proton to generate water, as shown in Figure 3.12c, the mixed electron/proton conductor single phase cathode can transfer protons from the cathode/electrolyte interface to the surface of the cathode bulk and then react with oxygen ions to produce water. Therefore, such a cathode can extend the TPB from the cathode/electrolyte interface to the cathode bulk and enhance the opportunity for contact between the oxygen ion and proton, leading to the high electrochemical reaction rate, *i.e.* low interfacial polarization resistance. B-site doped $BaCeO_3$ or $BaZrO_3$ oxides are regarded as the most potential materials for the mixed electronic/proton conductor single phase cathodes.

Fabbri *et al.*[104] have studied the electrochemical conductivity of doped $BaCeO_3$ materials. Under dry atmosphere, $BaCe_{0.9}Yb_{0.1}O_3$ (BCYb) is a mixed conductor of protons and electrons. While in the cathode environment, where H_2O exists, its electronic conductivity decreases significantly, resulting in large cell resistance. Mukundan *et al.*[105] have also pointed out that the electronic conductivities of $BaCe_{0.8-y}Pr_yGd_{0.2}O_3$ oxides are very small when H_2O exists, which agrees well with BCYb. Among the $BaCe_{0.8-y}Pr_yGd_{0.2}O_3$ oxides of various compositions, $Ba(Pr_{0.8}Gd_{0.2})O_3$ has the highest electronic conductivity, as well as a high proton conductivity, $\sim 3 \times 10^{-2}$ S cm^{-1} at 800 °C. However, the interfacial impedance of a single cell with $Ba(Pr_{0.8}Gd_{0.2})O_3$ cathode is relatively high, $\sim 0.34\,\Omega\,cm^2$ at 800 °C, implying low catalytic activity for oxygen reduction. On the contrary, the proton conductivity of $BaCe_{0.5}Bi_{0.5}O_3$ is very low, $\sim 8 \times 10^{-7}$ S cm^{-1} at 800 °C,[106] but the interfacial impedance is relatively low, $0.28\,\Omega\,cm^2$ at 700 °C for single cells on BZCY electrolytes.[107] So, it is considered that catalytic activity for oxygen reduction plays a significant role of these mixed electronic/proton conductor single phase cathodes. Tao *et al.*[108] have reported that the interface impedance based on BZCY electrolyte using $BaCe_{0.5}Fe_{0.5}O_3$ cathode is $0.17\,\Omega\,cm^2$ at 700 °C. $BaCe_{0.5}Fe_{0.5}O_3$ is an alternative cathode for H-SOFCs, but its proton conductivity is not understood clearly. Co or Fe doping on $BaCeO_3$ or $BaZrO_3$ may increase the oxygen catalytic ability and the electronic conductivity, but the doping concentration is restricted. Other cathode materials, possessing both proton and electron conductivities as well as catalytic activity for oxygen reduction, are seldom reported.

3.3.4 Microstructure Optimized Cathodes

Much attention is focused on the mixed electronic/proton conductor composite cathode, which can also extend TPB from the cathode/electrolyte interface to the surface of cathode bulk, as shown in Figure 3.12d. The composite cathode usual consists of an electronic catalytic phase and a proton conducting phase, thus exhibiting mixed electronic and proton conductivities as well as oxygen catalytic activity. In addition, composite structure could improve the resistance to thermal cycle.

Fabbri *et al.*[109] have figured out that, in the LSCF-$BaCe_{0.9}Yb_{0.1}O_3$ (BCYb) systems, the composite with a mass ratio of 1 : 1 has low interface impedance, $0.14\,\Omega\,cm^2$ at 700 °C. This result is consistent with the LSCF-BZCY system reported by Yang *et al.*[110] Detailed research has been conducted by Wu *et al.*[111] on SSC-BCS cathodes. The SSC content in the composite has a great effect on the interfacial resistance of the cell, which decreases with the ratio and then increases; while it has little effect on the bulk resistance of the cell when SSC content exceeds 20wt%. They further optimize the heating temperature and fabrication procedure. It is found that BCS backbone impregnated with 55 wt% SSC has the lowest ASR, $0.17\,\Omega\,cm^2$ at 700 °C, far less than that using ball-milling method, $1.98\,\Omega\,cm^2$ at 700 °C. In addition, they have studied the electrochemical impedance spectroscopy, and found that the low-frequency arc dominates the impedance spectra, corresponding to the oxygen dissociation-adsorption process.

Much work on composite cathodes is focused on the cell performance. BSCF-BZCY,[112] $Ba_{0.5}Sr_{0.5}Zn_{0.2}Fe_{0.8}O_3$-$BaCe_{0.5}Zr_{0.3}Y_{0.16}Zn_{0.04}O_3$[113] and $Sm_{0.5}Sr_{0.5}CoO_3$ (SSC)-BZCY composites[114] are excellent candidates for H-SOFCs cathodes. SSC-BZCY shows the highest power density among these composite cathodes. Yang et al.[115] suggested that $BaCoO_3$, possibly produced by the diffusion and/or reaction between SSC and BZCY, could be beneficial to the oxygen reduction process. They also tried to dope Co to BZCY. When the doped content exceeds, the second phase $BaCoO_3$ appears. With 20% Co doped BZCY cathode, the cell resistance at 700 °C is only $0.16\,\Omega\,cm^2$ and the peak power density is $370\,mW\,cm^{-2}$. Co and/or Ba elements are easy to spread during the cathode preparation process. The layered perovskite oxides also used in composite cathodes, such as: $PrBa_{0.5}Sr_{0.5}Co_2O_{5+\delta}$-BZCY.[116]

It is noteworthy that some mature cathode preparation technologies used in O-SOFC are gradually applied to the H-SOFC, such as impregnation technology.[117,118] The features of this technology include: (1) the cathode preparation temperature is reduced and might prevent the reaction and/or diffusion occurred; (2) nanostructured cathode is obtained leading to high catalytic activity; (3) the stability of this micro-structure can be maintained for a long time. Zhao et al.[119] successfully prepared SSC impregnated BZCY cathode for H-SOFC. At 700 °C, the interface impedance of the single cell is reduced to $0.064\,\Omega\,cm^2$.

Recently, MIEC composite cathodes, which are widely used for O-SOFCs, have also shown excellent performance for H-SOFCs. For instance, cathodes composed of SSC and SDC have shown peak power density of $665\,mW\,cm^{-2}$ on BZCY electrolytes and interface impedance of $0.066\,\Omega\,cm^2$ at 700 °C.[120] However, the stability of such cathodes needs to be further investigated. Sun et al. have used Co-free materials of $Ba_{0.5}Sr_{0.5}FeO_3$-SDC[121] and $La_{0.7}Sr_{0.3}FeO_3$-SDC[122] as the cathodes. Fundamentally, these materials could promote the cathode stability. A similar study on $Sm_{0.5}Sr_{0.5}Fe_{0.8}Cu_{0.2}O_3$-SDC[123] is also reported by Ling et al. It is noteworthy that such cathodes have excellent electrochemical performance comparable to mixed electronic/proton conductor composite cathode. As proposed in Figure 3.12e, these cathodes can transfer oxygen ion from the surface to the interface and promote the contact between the oxygen ion and proton, however, they cannot expand the length of H-TPB. But these cathodes have comparable electrochemical performance, as compared to other types of cathodes. Therefore, it is significant to reveal the cathode mechanism.

3.3.5 Cathode Reaction Mechanisms

The cathode process for H-SOFCs is much more complex than O-SOFCs due to more species being involved in the reaction. Uchida et al.[124] have proposed several elementary steps for the cathode reaction in H-SOFCs, in which the formation and evolution of water are considered in addition to the dissociative adsorption and diffusion of oxygen species. He et al.[125] and Zhao et al.[126] have proposed detailed elementary steps for the two types of composite cathodes, addressing the transfer characteristic mixed electronic/proton conductors and MIECs. Combined with the two previous models, elementary steps for different categories of cathode materials are proposed and shown in Table 3.4. The entire process of the reaction can be summarized as follows: surface dissociative adsorption and diffusion of oxygen along with charge transfer to form the oxygen ions into the lattice (step 1.5). Lattice oxygen ions migrate through bulk to the reaction sites (step 6, which happens when the oxygen ion conductor exists). At the same time, lattice protons from the electrolyte bulk transfer to the interface of electrolyte and cathode (step 7). Then lattice proton transfers to the reaction sites (step 8, which takes place when the proton conductor exists). Lattice oxygen and lattice proton meet and react to produce water (step 9.12). It is clear that elementary steps with different categories are slightly different. For electronic conductor single phase cathode, the elementary reaction steps do not include the transfer process of oxygen ion and proton (step 6 and 8). Similarly, MIEC single-phase or composite cathode does not have step 8 while the mixed electronic/proton conductor single-phase cathode is not linked with step 6.

For O-SOFCs, ASR (or R_p) is closely related to the oxygen partial pressure, P_{O2}, in the form of $R_p \propto (P_{O_2})^m$, where m is the reaction order and is linked to the rate-limiting step.[127–129] For H-SOFCs, besides gaseous oxygen, water is also involved in the cathode reactions as the product and becomes a factor in R_p, which can be in the form of $R_p \propto (P_{O_2})^m (P_{H_2O})^n$. The reaction orders with respect to P_{O_2} and P_{H_2O} can be determined for each elementary step by considering the concentration dependence of the reacting species on P_{O_2} and P_{H_2O}, and the details are shown in Table 3.4. Some steps have the same m and n factors. These steps could be identified considering the activation energy because each elementary process is, in fact, the transfer of various species, which usually corresponds to different activation energy. Uchida et al.[124] have investigated the R_p of Pt cathode at various oxygen partial pressures. It shows that R_p decreases with increasing P_{O_2} with a reaction order of $m = 1/4$. The activation energy is approximately 0.99 eV, suggesting that the probable rate-determining step for the Pt electrode is the surface diffusion of oxygen species adsorbed on the platinum to the electrochemically active site. Subsequently, Yamaura et al.[86] have studied the polarization resistances of LSF cathode on $SrCe_{0.95}Yb_{0.05}O_3$ electrolyte under dry and wet O_2–N_2 atmospheres. In both dry and wet conditions, the reaction order with respect to P_{O_2} is determined to be close to 0 (0.07–0.09). However, R_p under the wet atmosphere is 2 ~ 3 times larger than that under dry conditions, indicating that R_p depends strongly on P_{H_2O} and is almost independent on P_{O_2}. Steps 5 to 12 might be one or more

Table 3.4 Possible elementary steps for H-SOFC cathode systems, including electronic conductor single phase cathode (a), MIEC single phase cathode (b), mixed electronic/proton conductor single phase cathode (c), MIEC composite cathode (d), mixed electronic/proton conductor composite cathode (e) and the elementary reaction orders with respect to oxygen partial pressure (m) and water vapor partial pressure (n). Please note that "TPB-O" represents the three-phase interface of oxygen ion conductor, electronic conductor and gas. "TPB-OH" represents the three-phase interface of oxygen ion conductor, proton conductor and gas. "TPB-OH, O" stands for oxygen ion conductor phase of "TPB-OH", while "TPB-OH, H" proton conductor phase.

Elementary Reaction	a	b	c	d	e	$m(P_{O2})$	$n(P_{H2O})$
step1 $O_2(g) \rightarrow 2O_{ad}$	✓	✓	✓	✓	✓	1	0
step2 $O_{ad} + e^- \rightarrow O_{ad}^-$	✓	✓	✓	✓	✓	3/8	0
step3 $O_{ad}^- \rightarrow O_{TPB-O}^-$	✓	✓	✓	✓	✓	1/4	0
step4 $O_{TPB-O}^- + e^- \rightarrow O_{TPB-O}^{2-}$	✓	✓	✓	✓	✓	1/8	0
step5 $O_{TPB-O}^{2-} + V_{O,TPB-O}^{\bullet\bullet} \rightarrow O_{O,TPB-O}^{\times}$	✓	✓	✓	✓	✓	0	0
step6 $O_{O,TPB-O}^{\times} + V_{O,TPB-OH,O}^{\bullet\bullet} \rightarrow O_{O,TPB-OH,O}^{\times} + V_{O,TPB-O}^{\bullet\bullet}$	✗	✓	✗	✗/✓	✓	0	0
step7 $OH_{O,bulk,ele}^{\bullet} \rightarrow OH_{O,Inter,ele}^{\bullet}$	✓	✓	✓	✓	✓	0	1/2
step8 $OH_{O,Inter,ele}^{\bullet} \rightarrow OH_{O,TPB-OH,H}^{\bullet}$	✗	✗	✓	✓	✗	0	1/2
step9 $OH_{O,TPB-OH,H}^{\bullet} \rightarrow OH_{TPB-OH}^- + V_{O,TPB-OH,H}^{\bullet\bullet}$	✓	✓	✓	✓	✓	0	1/2
step10 $V_{O,TPB-OH,H}^{\bullet\bullet} + O_{O,TPB-OH,O}^{\times} \rightarrow O_{O,TPB-OH,H}^{\times} + V_{O,TPB-OH,O}^{\bullet\bullet}$	✓	✓	✓	✓	✓	0	0
step11 $OH_{O,TPB-OH,H}^{\bullet} + OH_{TPB-OH}^- 2O_{TPB-OH} + O_{O,TPB-OH,H}^{\times}$	✓	✓	✓	✓	✓	0	1
step12 $H_2O_{TPB-OH} \rightarrow H_2O(g)$	✓	✓	✓	✓	✓	0	1

rate-determining step(s) for LSF cathode, because the reaction orders, m, are approximately 0. The in-depth confirmation of the reaction order with respect to P_{H_2O} and the activation energy of the polarization resistances are required, in order to further determine the rate determining step. Moreover, it is clear that the rate-determining step(s) for LSF is different from Pt. He *et al.*[125] and Zhao *et al.*[126] have tried to exploit the determining steps on composite cathodes of SSC-BCS and SSC-SDC, respectively. For SSC-BCS composite cathode, the impedance spectra with two arcs are observed, implying that there might be two rate-limiting steps. The high-frequency resistances are proportionate to $(P_{H_2O})^{-1/2}$ and independent of P_{O_2} with the activation energy of 0.60 eV, suggesting that the migration of protons to TPBs (step 8) might be the corresponding step. While the low-frequency resistances are proportionate to $(P_{O_2})^{-1/4}$ and independent of P_{H_2O}, which might correspond to the surface diffusion of O_{ad}^- (step 3). For SSC-SDC composite cathode, two arcs in the impedance spectra are also obtained. The high-frequency resistances are independent of P_{O_2} and P_{H_2O} with the activation energy of 0.84 eV, suggesting that the migration of oxygen ion (step 6) might be the corresponding step. While the low-frequency resistances are proportionate to $(P_{O_2})^{-1/4}$ and independent of P_{H_2O}, which might correspond to the surface diffusion of O_{ad}^- (step 3). It is concluded that different conduction mechanism of cathode could lead to the different determining step.

3.4   Summary and Conclusions

The development of cathode materials is the key to reduce the operational temperature, and thus could accelerate the commercialization of SOFC technology. In addition to the remarkable catalytic activity for oxygen reduction, a proper cathode material should exhibit similar TEC with the electrolyte, high electric conductivity, and does not react with other components. However, despite the tremendous advances in cathode materials for both the ion-conducting and proton-conducting SOFCs, there is still not a perfect cathode material that can meet all the above requirements. MIECs are the state-of-the-art candidates for low-temperature SOFCs. However, they do provide challenges to improve the stability and durability at reduced temperatures (600–800 °C) due to their reactivity with other cell components. Cathode performance is also governed by the electrode microstructure. The electrocatalytic activity can be substantially promoted by nanostructured approached, such as infiltration. The quantification of microstructure vs. performance relationship is of practical significance to provide guidance principles for structure optimization. However, due to the limited understanding of reaction mechanisms, the effects of microstructure are still not clearly resolved.

Acknowledgements

Financial support from the Ministry of Science and Technology of China (2012CB215403) is greatly appreciated.

References

1. Z. P. Shao, W. Zhou and Z. H. Zhu, *Prog. Mater. Sci.*, 2012, **57**, 804.
2. M. Ni, M. K. H. Leung and D. Y. C. Leung, *Fuel Cells*, 2007, **07**, 269.
3. M. Koyama, C. J. Wen and K. Yamada, *J. Electrochem. Soc.*, 2000, **147**, 87.
4. C. Sun, R. Hui and J. Roller, *J. Solid State Electrochem.*, 2010, **14**, 1125.
5. T. Hibino, K. Mizutani and T. Yajima, *Solid State Ionics*, 1992, **57**, 303.
6. K. D. Kreuer, *Annu. Rev. Mater. Sci.*, 2003, **33**, 333.
7. A. S. Nowick and Y. Du, *Solid State Ionics*, 1995, **77**, 137.
8. H. F. Oetjen, V. M. Schmodt and U. Stimming, *J. Electrochem. Soc.*, 1996, **143**, 3838.
9. A. S. Patil, T. G. Dubois and N. Sifer, *J. Power Sources*, 2004, **136**, 220.
10. B. C. Tofield and W. R. Scott, *J. Solid State Chem.*, 1974, **10**, 183.
11. Z. Li, M. Behruzi, L. Fuerst, D. Stover, in *SOFC-III* ed. S. C. Singhal, H. Iwahara, The Electrochemical Society, Inc., Pennington, 1993, p 171.
12. H. Kamata, Y. Yonemura, J. Mizusaki, H. Tagawa, K. Naraya and T. Sasamoto, *J. Phys. Chem. Solids*, 1995, **56**, 943.
13. J. Mizusaki, Y. Yonemura, H. Kamata, K. Ohyama, N. Mori, H. Takai, H. Tagawa, M. Dokiya, K. Naraya, T. Sasamoto, H. Inaba and T. Hashimoto, *Solid State Ionics*, 2000, **132**, 167.
14. I. Yasuda and M. Hishinuma, *J. Solid State Chem.*, 1996, **123**, 382.
15. J. A. M. van Roosmalen and E. H. P. Cordfunke, *J. Solid State Chem.*, 1994, **110**, 109.
16. J. H. Kuo, H. U. Anderson and D. M. Sparlin, *J. Solid State Chem.*, 1990, **87**, 55.
17. I. Yasuda, K. Ogasawara, M. Hishinuma, T. Kawada and M. Dokiya, *Solid State Ionics*, 1996, **86–88**, 1197.
18. R. A De Souza, J. A Kilner and J. F Walker, *Mater. Lett.*, 2010, **43**, 43.
19. Y. Ji, J. A. Kilner and M. F. Carolan, *Solid State Ionics*, 2005, **176**, 937.
20. Y. Wang, L. Zhang and C. R. Xia, *Int. J. Hydrogen Energy*, 2012, **37**, 2182.
21. T. Kenjo and M. Nishiya, *Solid State Ionics*, 1992, **57**, 295.
22. C. Clausen, C. Bagger, J. B. Bilde-Sørensen and A. Horsewell, *Solid State Ionics*, 1994, **70–71**, 59.
23. G. Stochniol, E. Syskakis and A. Naoumidis, *J. Am. Ceram. Soc.*, 1995, **78**, 929.
24. M. Mori, Y. Hiei, N. M. Sammes, G. A. Tompsett, in *SOFC-VI*, ed. S. Singhal, M. Dokiya, The Electrochemical Society, Inc., Pennington, 1999, p 347.
25. A. V. Berenov, J. L. MacManus-Driscoll and J. A. Kilner, *Solid State Ionics*, 1999, **122**, 41.
26. Y. Sakaki, Y. Takeda, A. Kato, N. Imanishi, O. Yamamoto, M. Hattori, M. Iio and Y. Esaki, *Solid State Ionics*, 1999, **118**, 187.

27. S. Kuharuangrong, T. Dechakuptb and P. Aungkavattana, *Mater. Lett.*, 2004, **58**, 1964.
28. T. Horita, T. Tsunoda, K. Yamaji, N. Sakai, T. Kato and H. Yokokawa, *Solid State Ionics*, 2002, **152–153**, 439.
29. S. Carter, A. Selcuk, R. J. Chater, J. Kajda, J. A. Kilner and B. C. H. Steele, *Solid State Ionics*, 1992, **53–56**, 597.
30. H. Ullmann, N. Trofimenko, F. Tietz, D. Stöver and A. Ahmad-Khanlou, *Solid State Ionics*, 2000, **138**, 79.
31. C. H. Wang, W. L. Worrell, S. Park, J. M. Vohs and R. J. Gorteb, *J. Electrochem. Soc.*, 2001, **148**, A864.
32. L. Zhang, S. P. Jiang, W. Wang and Y. J. Zhang, *J. Power Sources*, 2007, **170**, 55.
33. J. Liu and S. A. Barnett, *J. Am. Ceram. Soc.*, 2002, **85**, 3096.
34. J.-W. Kim, A. V. Virkar, K.-Z. Fung, K. Mehta and S. C. Singhal, *J. Electrochem. Soc*, 1999, **146**, 69.
35. M. Zhang, M. Yanga, Z. F. Hou, Y. L. Dong and M. J. Cheng, *Electrochim. Acta*, 2008, **53**, 4998.
36. Y.-C. Chang, M.-C. Lee, W.-X. Kao, C.-H. Wang, T.-N. Lin, J.-C. Chang and R.-J. Yang, *J. Electrochem. Soc.*, 2011, **158**, B259.
37. E. P. Murray and S. A. Barnett, *Solid State Ionics*, 2001, **143**, 265.
38. A. T. Duong and D. R. Mumm, *J. Electrochem. Soc.*, 2012, **159**, B40.
39. O. Yamamoto, Y. Takeda, R. Kanno and M. Noda, *Solid State Ionics*, 1987, **22**, 241.
40. A. Mineshige, M. Kobune, S. Fujii, Z. Ogumi, M. Inaba, T. Yao and K. Kikuchi, *J. Solid State Chem.*, 1999, **142**, 374.
41. A. Mineshige, M. Inaba, T. Yao, Z. Ogumi, K. Kikuchi and M. Kawase, *J. Solid State Chem.*, 1996, **121**, 423.
42. V. V. Kharton, E. Maumovich, A. A. Vecher and A. Nikolaev, *J. Solid State Chem.*, 1995, **120**, 128.
43. V. E. Tsipis and V. V. Kharton, *J. Solid State Electrochem.*, 2008, **12**, 1367.
44. W. Zipprich, S. Waschilewski, F. Rocholl and H.-D. Wiemhöfer, *Solid State Ionics*, 1997, **101–103**, 1015.
45. S. Sekido, H. Tachibani, Y. Yamamura and T. Kambara, *Solid State Ionics*, 1990, **37**, 253.
46. X. Chen, J. Yu and S. B. Adler, *Chem. Mater.*, 2005, **17**, 4537.
47. R. Künga, S. F. Bidrawn, J. M. Vohs and R. J. Gorte, *Electrochem. Solid-State Lett.*, 2010, **13**, B87.
48. J. Mizusaki, T. Sasamoto, W. R. Cannon and H. K. Bowen, *J. Am. Ceram. Soc.*, 1983, **66**, 247.
49. E. V. Bongio, H. Black, F. C. Raszewski, D. Edwards, C. J. Mcconville and V. R. W. Amarakoon, *J. Electroceram.*, 2005, **14**, 193.
50. A. Petric, P. Huang and F. Tietz, *Solid State Ionics*, 2000, **135**, 719.
51. S. P. Simner, J. P. Shelton, M. D. Anderson and J. W. Stevenson, *Solid State Ionics*, 2003, **161**, 11.
52. M. D. Anderson, J. W. Stevenson and S. P. Simner, *J. Power Sources*, 2004, **129**, 188.

53. A. Martínez-Amesti, A. Larrañaga, L. M. Rodríguez-Martínez, A. T. Aguayo, J. L. Pizarro, M. L. Nó, A. Laresgoiti and M. I. Arriortua, *J. Power Sources*, 2008, **185**, 401.
54. L.-W. Tai, M. M. Nasrallah, H. U. Anderson, D. M. Sparlin and S. R. Sehlin, *Solid State Ionics*, 1995, **76**, 273.
55. Y. Teraoka, H. M. Zhang, K. Okamoto and N. Yamazoe, *Mater. Res. Bull.*, 1988, **23**, 51.
56. Y. J. Leng, S. H. Chan and Q. L. Liu, *Int. J. Hydrogen Energy*, 2008, **33**, 3808.
57. C. J. Fu, K. N. Sun, N. Q. Zhang, X. B. Chen and D. R. Zhou, *Electrochim. Acta*, 2007, **52**, 4589.
58. F. Tietz, V. A. C. Haanappel, A. Mai, J. Mertens and D. Stöver, *J. Power Sources*, 2006, **156**, 20.
59. W.-H. Kim, H.-S. Song, J. Moon and H.-W. Lee, *Solid State Ionics*, 2006, **177**, 3211.
60. Y. B. Lin and S. A. Barnett, *Solid State Ionics*, 2008, **179**, 420.
61. T. Kenjo, S. Osawa and K. Fujikawa, *J. Electrochem. Soc.*, 1991, **138**, 349.
62. Y. X. Zhang and C. R. Xia, *J. Power Sources*, 2010, **195**, 4206.
63. J. D. Kim, G. D. Kim, J. W. Moon, H. W. Lee, K. T. Lee and C. E. Kim, *Solid State Ionics*, 2000, **133**, 67.
64. J. R. Wilson, J. S. Cronin, A. T. Duong, S. Rukes, H.-Y. Chen, K. Thornton, D. R. Mumm and S. Barnett, *J. Power Sources*, 2010, **195**, 1829.
65. J. Park, J. Zou and J. Chung, *J. Power Sources*, 2010, **195**, 4593.
66. J. M. Vohs and R. J. Gorte, *Adv. Mater.*, 2009, **21**, 943.
67. F. Tietz, I. A. Raj, M. Zahid and D. Stover, *Solid State Ionics*, 2006, **177**, 1753.
68. Y. X. Zhang and C. R. Xia, *J. Power Sources*, 2010, **195**, 6611.
69. S. P. Jiang, *Int. J. Hydrogen Energy*, 2012, **37**, 449.
70. M. Shah, J. D. Nicholas and S. A. Barnett, *Electrochem. Commun.*, 2009, **11**, 2.
71. F. Zhao, R. Peng and C. Xia, *Mater. Res. Bull.*, 2008, **43**, 370.
72. K. F. Chen, N. Ai and S. P. Jiang, *Electrochem. Commun.*, 2012, **19**, 119.
73. F. Liang, J. Chen, S. P. Jiang, B. Chi, J. Pu and L. Jian, *Electrochem. Commun.*, 2009, **11**, 1048.
74. J. Fleig, *Annu. Rev. Mater. Res.*, 2003, **33**, 361.
75. S. H. Chan, X. J. Chen and K. A. Khor, *J. Electrochem. Soc.*, 2004, **151**, A164.
76. X. J. Chen, S. H. Chan and K. A. Khor, *Electrochim. Acta*, 2004, **49**, 1851.
77. E. V. Tsipis and V. V. Kharton, *J. Solid State Electrochem.*, 2008, **12**, 1039.
78. T. Kenjo and Y. Kanehira, *Solid State Ionics*, 2002, **148**, 1.
79. Y. L. Wang, L. Zhang, F. L. Chen and C. R. Xia, *Int. J. Hydrogen Energy*, 2012, **37**, 8582.

80. B. Ruger, A. Weber and E. Ivers-Tiffee, *ECS Trans.*, 2007, **7**, 2065e74.
81. S. B. Adler, *Solid State Ionics*, 2000, **135**, 603.
82. R. Mukundan, P. K. Davies and W. L. Worrell, *J. Electrochem. Soc.*, 2001, **148**, A82.
83. L. Wang, R. Merkle, J. Maier, T. Acartürk and U. Starke, *Appl. Phys. Lett.*, 2009, **94**, 071908–1.
84. R. Doshi, V. L. Richards, J. D. Carter, X. Wang and M. J. Krumpelt, *J. Electrochem. Soc.*, 1999, **146**, 1273.
85. G. Kim, S. Wang, A. J. Jacobson, L. Reimus, P. Brodersen and C. A. Mims, *J. Mater. Chem.*, 2007, **17**, 2500.
86. H. Yamaura, T. Ikuta, H. Yahiro and G. Okada, *Solid State Ionics*, 2005, **176**, 269.
87. Z. P. Shao and S. M. Haile, *Nature*, 2004, **431**, 170.
88. Y. Lin, R. Ran, Y. Zheng, Z. P. Shao, W. Q. Jin, N. P. Xu and J. Ahn, *J. Power Sources*, 2008, **180**, 15.
89. Y. Lin, R. Ran, D. J. Chen and Z. P. Shao, *J. Power Sources*, 2010, **195**, 4700.
90. Y. Lin, W. Zhou, J. Sunarso, R. Ran and Z. P. Shao, *Int. J. Hydrogen Energy*, 2012, **37**, 484.
91. H. P. Ding, B. Lin, X. Q. Liu and G. Y. Meng, *Electrochem. Commun.*, 2008, **10**, 1388.
92. A. A. Taskin, A. N. Lavrov and Y. Ando, *Appl. Phys. Lett.*, 2005, **86**, 091910.
93. K. Zhang, L. Ge, R. Ran, Z. P. Shao and S. M. Liu, *Acta Mater.*, 2008, **56**, 4876.
94. B. Lin, S. Q. Zhang, L. C. Zhang, X. Q. Liu and G. Y. Meng, *J. Power Sources*, 2008, **177**, 330.
95. B. Lin, Y. C. Dong, R. Q. Yan, S. Q. Zhang, M. J. Hu, Y. Zhou and G. Y. Meng, *J. Power Sources*, 2009, **186**, 446.
96. L. Zhao, B. B. He, B. Lin, H. P. Ding, S. L. Wang, Y. H. Ling, R. R. Peng, G. Y. Meng and X. Q. Liu, *J. Power Sources*, 2009, **194**, 835.
97. Y. Lin, R. Ran, C. Zhang, R. Cai and Z. P. Shao, *J. Phys. Chem. A*, 2010, **114**, 3764.
98. Q. Nian, L. Zhao, B. B. He, B. Lin, R. R. Peng, G. Y. Meng and X. Q. Liu, *J. Alloys Compd.*, 2010, **492**, 291.
99. Y. H. Ling, L. Zhao, B. Lin, Y. C. Dong, X. Z. Zhang, G. Y. Meng and X. Q. Liu, *J. Alloys Compd.*, 2010, **493**, 252.
100. H. P. Ding and X. J. Xue, *Int. J. Hydrogen Energy*, 2010, **35**, 2486.
101. H. P. Ding, X. J. Xue, X. Q. Liu and G. Y. Meng, *J. Power Sources*, 2010, **195**, 3775.
102. J. H. Kim and A. Manthiram, *Chem. Mater*, 2010, **22**, 822.
103. H. Wang, Z. T. Tao and W. Liu, *Ceram. Int.*, 2012, **38**, 1737.
104. E. Fabbri, T. Oh, S. Licoccia, E. Traversa. and E. Wachsman, *J. Electrochem. Soc.*, 2009, **156**, B38.
105. R. Mukundan, P. K. Davies and W. L. Worrel, *J. Electrochem. Soc.*, 2001, **148**, A82.

106. Z. Hui and P. Michele, *J. Mater. Chem.*, 2002, **12**, 3787.
107. Z. T. Tao, L. Bi, L. T. Yan and W. Liu, *Electrochem. Commun.*, 2009, **11**, 688.
108. Z. T. Tao, L. Bi, Z. W. Zhu and W. Liu, *J. Power Sources*, 2009, **194**, 801.
109. E. Fabbri, S. Licoccia, E. Traversa and E. D. Wachsman, *Fuel Cells*, 2009, **9**, 128.
110. L. Yang, Z. Liu, S. Z. Wang, Y. M. Choi, C. D. Zuo and M. L. Liu, *J. Power Sources*, 2010, **195**, 471.
111. T. Z. Wu, R. R. Peng and C. R. Xia, *Solid State Ionics*, 2008, **179**, 1505.
112. B. Lin, H. P. Ding, Y. C. Dong, X. Z. Zhang, D. R. Fang and G. Y. Meng, *J. Power Sources*, 2009, **186**, 58.
113. X. Lu, Y. Ding and Y. Chen, *J. Alloys Compd.*, 2009, **484**, 856.
114. L. Yang, C. D. Zuo, S. Z. Wang, Z. Cheng and M. L. Liu, *Adv. Mater.*, 2008, **20**, 3280.
115. L. Yang, S. Z. Wang, X. Y. Lou and M. L. Liu, *Int. J. Hydrogen Energy*, 2011, **36**, 2266.
116. F. Zhao, S. W. Wang and K. Brinkman, *J. Power Sources*, 2010, **195**, 5468.
117. S. P. Jiang, *Mater. Sci. Eng., A*, 2006, **418**, 199.
118. T. Z. Sholklapper, H. Kurokawa, C. P. Jacobson, S. J. Visco and L. C. De Jonghe, *Nano Lett.*, 2007, **7**, 2136.
119. F. Zhao, Q. Liu, S. W. Wang and F. L. Chen, *J. Power Sources*, 2011, **196**, 8544.
120. W. P. Sun, L. T. Yan, B. Lin, S. Q. Zhang and W. Liu, *J. Power Sources*, 2010, **195**, 3155.
121. W. P. Sun, Z. Shi, S. M. Fang, L. T. Yan, Z. W. Zhu and W. Liu, *Int. J. Hydrogen Energy*, 2010, **35**, 7925.
122. W. P. Sun, Z. W. Zhu, Y. Z. Jiang, Z. Shi, L. T. Yan and W. Liu, *Int. J. Hydrogen Energy*, 2011, **36**, 9956.
123. Y. H. Ling, J. Yu, B. Lin, X. Z. Zhang, L. Zhao and X. Q. Liu, *J. Power Sources*, 2011, **196**, 2631.
124. H. Uchida, S. Tanaka and H. Iwahara, *J. Appl. Electrochem*, 1995, **15**, 93.
125. F. He, T. Z. Wu, R. R. Peng and C. R. Xia, *J. Power Sources*, 2009, **194**, 263.
126. L. Zhao, B. B. He, J. Q. Gu, F. Liu, X. F. Chu and C. R. Xia, *Int. J. Hydrogen Energy*, 2012, **37**, 548.
127. D. J. Chen, R. Ran, K. Zhang, J. Wang and Z. P. Shao, *J. Power Sources*, 2009, **188**, 96.
128. M. J. Escudero, A. Aguadero, J. A. Alonso and L Daza, *J. Electroanal. Chem.*, 2007, **611**, 107.
129. F. H. van Heuveln, H. J. M. Bouwmeester and F. P. F. van Berkel, *J. Electrochem. Soc.*, 1997, **141**, 126.

CHAPTER 4

Anode Material Development

SHAMIUL ISLAM AND JOSEPHINE M. HILL*

Department of Chemical and Petroleum Engineering, University of Calgary,
2500 University Dr. NW, Calgary, Alberta, T2N 1N4, Canada
*Email: jhill@ucalgary.ca

4.1 Required Properties of Anode Materials

The anode of an SOFC has to meet a number of stringent criteria as outlined
below:

- Catalytic activity towards the electrochemical oxidation of the desired fuel
  for an extended period of time ($\sim 90\,000$ h).[1,2]
- Chemical stability with other cell components during fabrication and
  operation. Formation of any undesirable phases could interfere with
  electron or ion transfer from the electrolyte to the anode.[3,4]
- Good thermal stability so that the coefficient of thermal expansion of the
  anode matches with that of the other cell components to prevent structural
  damage (bending, warpage, cracking, *etc.*) during cell fabrication and
  operation.
- Sufficient porosity (> 30 vol%) to allow gas diffusion to the triple phase
  boundary (TPB).
- Good electronic and ionic conductivity. Depending on the geometry of the
  anode, the required electronic conductivity could vary from 1 to 10^2 S cm^{-1},[5]
  with the higher conductivities required for anode-supported cells in which
  the anode is thicker.[2] The ionic conductivity of the anode should be greater
  than 0.1 S cm^{-1}.[6]
- Physical support if using an anode-supported cell.

RSC Energy and Environment Series No. 7
Solid Oxide Fuel Cells: From Materials to System Modeling
Edited by Meng Ni and Tim S. Zhao
© The Royal Society of Chemistry 2013
Published by the Royal Society of Chemistry, www.rsc.org

- Carbon tolerance if using carbon-containing fuels (*e.g.* hydrocarbons, alcohols).
- Tolerance to contaminants such as tars, sulphur and chlorine (from biogas).
- Redox tolerance during start-up and shut down.

The following sections describe specific anodes that have been developed for different fuels and some of the challenges faced with these anodes and fuels.

4.2 Hydrogen Fuel

The conventional Ni/electrolyte composite anode performs well with pure H_2 as fuel. These anodes are typically prepared by mixing NiO with the electrolyte and possibly a pore former. To form the desired shape, the mixture is uni-axially pressed or tape cast, and then sintered in one or several steps with the electrolyte and cathode. Once exposed to H_2 fuel, the NiO is reduced to elemental Ni, creating or enhancing (if pore formers were used) the porous anode structure. Being a solid device, the components of the SOFC (electrodes and electrolyte) must expand during heating at a comparable rate. Thus, a composite of Ni and the electrolyte is used so that the thermal expansion coefficients of the anode and electrolyte are well matched. As described in Chapter 2, different electrolyte materials can be used. Yttria-stabilized zirconia (YSZ) is the electrolyte used for SOFC operating between 973 K and 1273 K, and so the conventional SOFC anode is a composite of Ni and YSZ (Ni/YSZ). The ionic conductivity of YSZ is on the order of 10^{-2} and $10^{-1}\,S\,cm^{-1}$ at 1123 K and 1273 K, respectively.[7–9]

Ni particles not only catalyse the electrochemical oxidation of H_2 but also provide electronic conduction from the reaction site (triple phase boundary, TPB) to the current collector. The Ni particles must, therefore, form an electronically conductive pathway through the anode. The amount of Ni required for this pathway depends on the structure of the anode (Ni to YSZ volume ratio, particle size of NiO and YSZ powders) and the processing conditions (sintering temperature, reduction time and temperature). The electrical conductivity of pure Ni is $2\times10^4\,S\,cm^{-1}$ at 1273 K. If the particles are randomly packed, $\sim30\,vol\%$ Ni is required to achieve sufficient conductivity (*e.g.* $1000\,S\,cm^{-1}$).[10] Ni foams have also been used to create the anode structure and, in this case, only 13 vol% Ni is required for sufficient conductivity.[11]

Although a number of different types of anodes, including perovskite materials have been tested in H_2, Ni/YSZ anodes are thus far the best anode material for H_2 electrooxidation. The benefits of using alternative anodes arise only when SOFC are run on fuels other than H_2 and/or with fuels containing contaminants, such as sulphur.

The performance of an SOFC depends on the electrode materials, electrolyte thickness, microstructures, fuel, and operating conditions. Power densities up to $1.16\,W\,cm^{-2}$ and $1.2\,W\,cm^{-2}$ have been achieved for Ni/YSZ anode-supported cells operated in H_2 at 1023 K[12] and 1073 K,[13] respectively. Cells with different materials have lower maximum power densities. For example, a power density of $0.9 \, W \, cm^{-2}$ was reported for a bi-layer anode, $Sr_{0.8}La_{0.2}TiO_3$ (SLT)-Ni/YSZ, at 1073 K,[14] and power densities of $0.85 \, W \, cm^{-2}$ and $0.70 \, W \, cm^{-2}$, were reported for the double perovskites, $Sr_2MgMoO_{6-\delta}$ and $Sr_2MnMoO_{6-\delta}$, respectively, at 1073 K.[7] Similarly, the introduction of impurities in the H_2 fuel decreases the maximum achievable power density. The addition of 100 ppm H_2S decreased the power density from $0.9 \, W \, cm^{-2}$ to $0.7 \, W \, cm^{-2}$ for a bi-layer anode, $Sr_{0.8}La_{0.2}TiO_3$ (SLT)-Ni/YSZ, at 1073 K.[14] The performance of this cell was restored when the fuel was switched back to H_2. Thus, perovskite anodes have lower power densities but have the advantage of being tolerant to H_2S, which may be a significant advantage depending on the source of H_2.

Annual worldwide hydrogen consumption is over 400 million m^3 with roughly one third of this amount being used for petroleum refining.[15] Processing one barrel of oil (159 L) requires between $2.8 \, m^3$ and $5.7 \, m^3$ of hydrogen.[15] Hydrogen is used to upgrade the quality of the oil but also to remove contaminants such as sulphur and nitrogen. Globally, 95% of hydrogen is produced from fossil fuel-based processes, and approximately half of the hydrogen used in refineries is produced from steam reforming of methane, which could otherwise be used as a valuable clean-burning fuel in direct applications.[16] H_2 is produced for the upgrading of heavy oil and bitumen but typically the H_2 is used on-site at the refinery and not transported and distributed in the same way as natural gas. The problems with H_2 storage and transportation have not yet been overcome and so the operation of SOFC with fuels other than H_2 has been heavily researched. The main challenge with these other fuels is deactivation of the anode from carbon formation and/or poisoning. These issues necessitate the development of materials other than Ni and/or the re-engineering of the anode.

4.3 Methane Fuel

Many locations in the world have existing natural gas distribution infra-structures, and so CH_4, which is the primary (typically >95%) component of natural gas, is a logical choice for a fuel for SOFC.[2] When CH_4 is used as fuel, the anode, particularly if it contains Ni, may catalyse the decomposition of CH_4 to form carbon. If the carbon is not removed at a similar rate at which it is formed, carbon will accumulate on/in the anode, blocking the sites for the electrochemical reactions, and blocking the pores for gas transport.[17] Ni is a particularly good catalyst for decomposing CH_4 and forming carbon (Eqn (1)). At the operating conditions of SOFC, carbon can also form by the reaction of carbon monoxide (CO) with H_2 (Eqn (2)) and the Boudouard reaction (Eqn (3)). These types of carbon are sometimes referred to as catalytic carbon.

$$CH_4 \leftrightarrow C + 2H_2 \tag{1}$$

$$CO + 2H_2 \leftrightarrow C + 2H_2O \tag{2}$$

$$2CO \leftrightarrow C + CO_2 \tag{3}$$

Carbon can also be formed through gas-phase reactions, via free-radical cracking and the polymerization of hydrocarbons. This type of carbon is called pyrolytic carbon and at typical SOFC operating conditions, only occurs for hydrocarbons that are larger than CH_4. That is, CH_4 does not decompose in the gas phase at temperatures less than 1300 K.[18,19]

Catalytic carbon forms through the reaction of hydrocarbons on a metal surface, and has been studied extensively in the steam reforming of hydro-carbons. Although the mechanism of catalytic carbon formation is complex, carbon deposition during the steam reforming of hydrocarbons over Ni catalysts includes three general steps: (a) adsorption and subsequent decomposition of hydrocarbon to carbon species on the surface of the Ni particle, (b) dissolution of carbon species into the bulk lattice of the metal and (c) precipitation of carbon from the rear metal facet. The last step separates a few particles from the bulk metal particle and support, destroying the catalyst.

For fuel cell anodes, the amount and type of carbon accumulated depends on the operating conditions such as temperature, current density, time, the type of anode and the thickness of anode.[20] For Ni/YSZ anodes, carbon fibres form at lower temperatures (<1073 K) in CH_4.[21] At higher temperatures, carbon dissolves into the Ni and expands the Ni/YSZ composite to produce sponge-like particles and eventually metal dusting. With an increase of operation time and thickness of the anode, carbon deposits become more graphitic and less hydrogenated. Carbon deposition can be eliminated by increasing the steam content of the fuel (*i.e.* promote reverse reaction in Eqn (2)) to a steam to carbon (S/C) ratio greater than two. The addition of steam dilutes the fuel, resulting in a lower maximum voltage and adds complexity to the system.

As CH_4 has been investigated more than any other hydrocarbon fuels in different types of anodes, the following sections contain several examples of direct utilization of CH_4 on different types of anodes (conventional Ni/YSZ, ceramic anodes, Cu-based anodes, and anodes promoted with BaO).

4.3.1 Conventional Ni/YSZ Anodes

As mentioned above, although Ni/YSZ anodes are excellent electrocatalysts for hydrogen oxidation, they are also excellent catalysts for non-electrochemical reactions – namely, hydrocarbon cracking, steam reforming, and water-gas shift. The cracking of CH_4 generally leads to carbon formation (Eqn (1)) if insufficient water is present in the anode. Aside from adding steam to the fuel, the water content in the anode can be increased by operating the fuel cell at a sufficiently high current density. $Ni/YSZ/(Y_2O_3)_{0.15}(CeO_2)_{0.85}$ anodes produced a stable power output for 100 h without carbon deposition in dry CH_4 at temperatures between 823 K and 923 K (the current density during operation was not reported).[22] Similar stability has also been obtained when operating Ni/YSZ anodes with humidified (3 vol% H_2O) CH_4 at 1073 K with current densities in the range of 1.4–1.8 A cm^{-2}.[23] The water generated *in situ* appears to be more effective at preventing carbon deposition than water added with the fuel.[20]

As a stack does not always operate at high current densities and the current density distribution throughout the stack is not uniform, modifications to the anode composition and structure have been studied to reduce carbon accumulation.

4.3.2 Alternative Anodes

The main strategies that have been used to prevent carbon accumulation include complete replacement of the Ni with ceramic materials, promotion of the Ni with other species, and redesign of the anode to use non-Ni containing materials in the conduction layer. Ceramic materials will be discussed first.

Although it is very difficult to obtain all the desired properties in a single ceramic material, perovskite anodes (ABO_3) have been found to have the best combination of the desired properties. In a perovskite anode, A and B are metal cations with a total charge of $+6$. One of the anode candidate materials is LST ($La_{0.3}Sr_{0.7}TiO_{3-\delta}$), where Sr is doped with La in the A site. An electronic conductivity of 1.6 $S\,cm^{-1}$ at $1073\,K$ (recall that the target range is 1–$100\,S\,cm^{-1}$) under reducing conditions (4% H_2 in wet Ar) was obtained for $La_{0.4}Sr_{0.6}T_{0.4}Mn_{0.6}O_{3-\delta}$ (LSTM).[24] An LSTM/YSZ anode with a $20\,\mu m$ thick electrolyte was tested in CH_4 at $1131\,K$ and a maximum power density of $0.06\,W\,cm^{-2}$ was obtained. The LSTM/YSZ anode did not show any deactivation as the maximum power density returned to its original value when the fuel was switched from CH_4 to H_2.[24]

Another perovskite that has been studied is LSCM-Cu-YSZ in an anode-supported cell (thickness of electrolyte: $50\,\mu m$) but a low power density of $0.05\,W\,cm^{-2}$ in dry CH_4 at $1073\,K$ was obtained, indicating that the addition of a co-catalyst to improve the performance may be required.[25] Thus, a Pd/CeO_2 catalyst was added to an LSCM/YSZ anode (thickness of electrolyte: $60\,\mu m$) and the power density was significantly higher $-0.71\,W\,cm^{-2}$ at $1073\,K$ in humidified CH_4.[26]

Double perovskites with compositions of $Sr_2Mg_{1-x}Mn_xMoO_{6-\delta}$ (SMMO), were used as anodes with $La_{0.8}Sr_{0.2}Ga_{0.83}Mg_{0.17}O_{2.815}$ (LSGM) electrolytes (thickness: $300\,\mu m$). A peak power density of $0.4\,W\,cm^{-2}$ in dry CH_4 at $1073\,K$ was reported.[7] It is not clear whether the SMMO anode will be chemically compatible with the more commonly used YSZ electrolyte.

Besides perovskite anodes, non-coking metals such as Cu, which does not catalyse carbon formation,[17] have been used in SOFC anodes operated with CH_4. In Cu-based anodes, the role of Cu is that of an electronic conductor.[17] A Cu/YSZ anode was stable in dry CH_4 for at least three days but the performance of the Cu/YSZ anode (thickness of electrolyte: $230\,\mu m$) was found to be low ($0.005\,W\,cm^{-2}$) in dry CH_4 at $1073\,K$.[27] Addition of CeO_2 to the Cu/YSZ anode imparted catalytic activity and increased the maximum power density to $0.230\,W\,cm^{-2}$.[27] $Cu/CeO_2/YSZ$ anodes were also found to be tolerant to sulphur content up to $\sim 400\,ppm$ without significant loss in performance. However, the thermal stability of Cu is low, which resulted in poorly connected Cu particles and deactivation of the cell at temperatures above $973\,K$.

Rather than completely replacing the Ni, additional metals such as Ru, Pd, Pt, Au, Ag, Sn and oxides can be added to the Ni to reduce carbon accumulation.[28–34] Alkaline earth oxides have been found to be beneficial in reducing carbon accumulation on supported catalysts.[35–37] The basicity of BaO contributes to the removal of carbon in steam and dry reforming processes.[38–40] Similarly, carbon deposition was reduced when such oxide materials (CaO, SrO) were added to Ni/YSZ anodes of SOFC. The addition of CaO, however, caused a slight degradation in the electrochemical performance for Ni/YSZ anodes when CaO and SrO were added to a Ni/SDC anode and tested in dry CH_4 at 973 K.[41] The loss in activity was attributed to the dissolution of CaO in the lattice of cubic ZrO_2, which possibly affected the reaction rate at the electrocatalytic sites. Furthermore, as alkaline oxides are known to be electronic insulators, the addition of such materials to Ni/YSZ could potentially increase the ohmic resistance and further reduce the electrochemical performance of the Ni/YSZ anodes. Hence, the addition of alkaline oxide to Ni/YSZ anode should only be implemented in the dopant level.

In the recent studies, BaO was incorporated in Ni/YSZ anode by chemical vapour deposition method.[12] BaO-impregnated anodes performed similarly to conventional Ni/YSZ anodes in dry H_2 at 1023 K. The addition of BaO did not impact the charge transfer process of Ni/YSZ anodes. Additionally, the cells were tested in C_3H_8, wet CO and gasified carbon fuels at 1023 K and were effective at preventing the accumulation of carbon with all of these fuels.[12] The improved carbon tolerance was attributed to nano BaO/Ni interfaces that were able to adsorb H_2O preferentially and subsequently gasify the adjacent carbon deposits. The presence of H_2O at the anode is critical for BaO-modified anodes to be effective in the removal of carbon. BaO addition to Ni/YSZ anodes could also help in reducing carbon deposits with CH_4. A reduction in carbon deposition was reported for BaO-doped Ni/Cu/GDC (gadolinia doped ceria) anode in dry CH_4 at 1023 K.[42] The effect of BaO addition to conventional Ni/YSZ anodes operated with CH_4 is still being studied.

Engineering of the microstructures of the anode has resulted in improved performance in dry CH_4. For example, a bi-layer anode contains one layer (*i.e.* functional layer) in proximity to the electrolyte that is optimized for the electrochemical oxidation of the fuel, and another external layer that acts as a mechanical support for the cell and current collector (*i.e.* conduction layer). Bi-layer anodes contain Ni/YSZ and Cu/YSZ, as functional and conduction layers, respectively, (Cu/YSZ + Ni/YSZ|YSZ|Pt) and these anodes exhibited improved tolerance towards carbon when operated in dry CH_4 at 100 mA cm^{-2} over 100 h at 1023 K.[43] During the testing period, the cell voltage degraded by less than 1%, whereas a typical Ni/YSZ anode-supported cell was completely destroyed by carbon within 10 h under the same conditions. The superior stability of the Cu/YSZ + Ni/YSZ|YSZ|Pt cell to dry CH_4 was attributed to a number of factors. First, Cu has been shown to be inert to the C–H bond scission resulting in little or no carbon build-up at the conduction layer. Second, the Ni of the functional layer that is part of the TPB has less propensity for carbon deposition due to a higher water partial pressure as a result of the electrochemical oxidation of H_2. Furthermore, if any carbon is formed at this Ni, it is still possible to remove the carbon electrochemically by applying a current. Although these results are promising, operating the cell at lower temperatures, and improving the thermal stability of Cu would be necessary for the further development of this bi-layer anode.

Cu in the conduction layer was replaced by $La_{0.3}Sr_{0.7}TiO_{3-\delta}$ and the anodes were found to be much more stable in dry CH_4 at 1023 K.[44] While the Ni/YSZ anode-supported cell degraded more quickly and the cell voltage dropped to zero within 10 h, the bi-layer anode was operated at a lower current ($50\,mA\,cm^{-2}$) for 24 h with a cell voltage degradation of $1.8\,mV\,h^{-1}$. The superior tolerance of the bi-layer anode cell was attributed to the type and amount of carbon that formed in the anode. The carbon was more reactive and less formed ($17.8\,\mu mol$ versus $38.3\,\mu mol$) in the functional layer of the bi-layer anode.

In summary, CH_4 has been studied quite thoroughly in different types of anode-electrolyte assemblies and at different operating conditions. A number of results show that SOFC can operate with CH_4 for hundreds of hours in laboratory scale set-ups. Obtaining sustained power output for a prolonged period of time ($>40\,000\,h$), however, is expected in a stack operation and remains a challenging task unless a sufficient amount of steam accompanies the CH_4 in the feed.

4.4   Higher Hydrocarbon Fuels (Propane and Butane)

Propane (C_3H_8) is a high energy density (46.4 MJ/kg, 22.8 MJ/L) liquid fuel that can be used for portable and remote fuel cell power generation. In order to avoid coking according to thermodynamic analysis, the oxygen to propane ratio should be greater than 1.75 at temperatures above 1006 K. The predominant gas phase species from the partial oxidation of propane are H_2 and CO.[45] When Ni-YSZ anode-supported cells were tested under these conditions in a mixture of C_3H_8 (79%) and O_2 (21%), a stable power output of $0.7\,W\,cm^{-2}$ was achieved at 1006 K. The Ni/YSZ anode effectively catalysed the partial oxidation reaction and the cell operated by the electrochemical oxidation of H_2.

The direct utilization of 5% C_3H_8 in Ar has also been investigated with Ni/YSZ anode-supported cells at 1073 K and the major gas products were H_2, CO, CH_4, C_2H_4 and CO_2 at $120\,mA\,cm^{-2}$.[46] The reforming of C_3H_8 occurred with CO_2 instead of H_2O as the anode-surface was covered by carbon and H_2O production was depleted over time. The direct utilization of 100% C_3H_8 was tested over a Ni- $La_{0.6}Sr_{0.4}Fe_{0.8}CO_{0.2}O_3$ (LSFCO) anode on a GDC electrolyte-supported (250 µm) cell at 1073 K.[47] Although no significant carbon deposition was observed on the anode, carbon was still observed on the alumina tube and tar was detected in the outlet stream, which could be due to the gas phase decomposition of C_3H_8 at 1073 K. During SOFC operation, LSFCO reacted with Ni to form La_2NiO_4, which promoted the bond scission of C–H, and GDC worked as an oxidation catalyst to remove the deposited carbon. Additionally, LSFCO was partially modified into a lanthanum-depleted $SrFe_{1-x}Co_xO_{3-y}$

(SFCO) perovskite structure. Thus, LSFCO is not chemically stable with other components of the anode.

Butane (n-butane) has been used as a fuel in porous YSZ anodes impregnated with 10 wt% CeO_2 at 973 K.[48] The ohmic resistance of the CeO_2/YSZ anode decreased over the testing period of 24 h and this decrease was attributed to the build-up of an electronically conductive carbonaceous film. The carbonaceous deposits were characterized using temperature-programmed oxidation (TPO), which showed that the reactivity of the deposits decreased with exposure time as the oxidation temperature increased from 600 K to 920 K after 24 h of exposure to n-butane. In another study, anodes of both Ni/YSZ and Ni/YSZ with 5% CeO_2 were tested in internally reformed butane with various oxygen to butane (1.4 to 6.5) ratios.[49] At a ratio of 1.4, carbon deposition could not be prevented. However, at a ratio 1.7 (well below the stoichiometric ratio for complete combustion), the cell produced 100 mA cm^{-2} and was stable for 100 h.

In situ Raman spectroscopy has been used to characterize the carbon formed on Ni/YSZ anodes in SOFC operating at 988 K.[50] The hydrocarbon fuels used in these studies contained a simulated gas composition from the pyrolysis of butane: namely, methane, ethylene, and propylene. These gas components were chosen because they represent the primary components of butane pyrolysis (accounting for nearly 90% of the carbon products found in a butane feed that has passed through a 973 K quartz reactor in 5 s). The three main conclusions from this study were as follows. First, the methane produced from pyrolysis of higher-weight hydrocarbons was unlikely to initiate carbon deposition on Ni/YSZ anodes. Second, the initial stages of carbon deposition from ethylene began with the formation of ordered graphite. Finally, propylene formed disordered carbon deposits that appeared structurally similar to the deposits formed by butane.

4.5 Fuels from Biomass

Biomass fuels are derived from renewable energy sources. Wood chips, municipality and agricultural wastes can be gasified to produce fuels such as syngas that can be fed into an SOFC. A typical composition of syngas produced in a gasifier is provided in Table 4.1.[51,61]

Table 4.1 Typical properties of biomass gasification syngas.

Compound	Gas (vol. %)	Dry gas (ave. vol %)[a]
H_2	14.5	15
CO	21.0	24
CO_2	9.7	11
H_2O	4.8	–
CH_4	1.6	2
N_2	48.4	48

[a] Average from literature data.

Table 4.2   Elemental analysis of tar sample.

Element	Wt. %
Carbon	83.33
Hydrogen	7.53
Oxygen[a]	6.83
Nitrogen	1.36
Sulfer	0.45

[a] By difference.

SOFC can utilise the major components of syngas (H_2, CO, CH_4) as fuels to produce electricity. Additionally, the high quality of heat from gases can be further integrated with gas turbines for system integration and process heating. When a gasification unit is coupled with an SOFC system, theoretical studies show that electrical efficiencies $>40\%$ and overall system efficiencies $>80\%$ can be achieved for simultaneous heat and material integration between the gasification and SOFC systems.[52,53]

Biomass contains a number of impurities such as dust, tars, H_2S, HCl, alkali metals, which can cause performance degradation if the fuel is supplied directly into the fuel cell.[54] Thus, the fuel stream requires cleaning up before feeding the fuel into the anode of SOFC to prevent interaction between the anode catalyst and the impurities. Among these impurities, tar is a major contaminant, and is an oxygenated aromatic compound formed during the thermal decomposition of biomass.[55,56] The composition of tar depends on the feedstock, operating conditions and gasifying agent.[57,58] The composition of a real tar obtained from a coal gasifier is given in Table 4.2.[59]

Tar can decompose on Ni and form carbon in the anode of SOFC. Besides tar, the product gases (CO, CH_4) can also cause carbon deposition and hence it is important to understand the cause of carbon deposition from biomass-derived syngas, so that the deactivation of SOFC anodes due to carbon deposition can be avoided.

4.5.1 Biomass-Simulated Gas

Most of the published work on SOFC run with biomass was obtained using simulated biomass fuels instead of actual product gases from a gasifier, likely because of the simplicity of running a SOFC using bottled gases instead of integrating a gasifier with a SOFC unit. In order to understand the effect of tars, benzene and toluene have been used as model fuels in Ni/YSZ anodes of SOFC at 1048 K.[60] Lorente *et al.* have shown that the use of a model fuel (toluene) can overestimate the negative impact of tars on anode deactivation.[60] Thus, the results from a model fuel need to be considered as a worst-case scenario.

Both Ni/YSZ and Ni/GDC anodes have been tested using a benzene model tar at 1038 K with various S/C ratios.[61] The authors used thermodynamic calculations to predict the threshold current density (determined to be 365 mA cm^{-2} for the conditions used), above which carbon deposition can be avoided. Indeed carbon deposition was reduced as the load was increased towards the threshold current density value and as the S/C ratio was increased up to three. Despite being operated in the thermodynamically predicted carbon free region, the Ni/YSZ anode underwent irreversible damage from carbon deposition. Compared to the Ni/YSZ anode, the Ni/GDC anode had a higher tolerance to carbon deposition, which was attributed to the capacity of GDC in reforming the deposited carbon. Ni-GDC anodes were also tested at lower operating temperatures (T = 923 K), with simulated gasification mixtures that corresponded to the gas composition of wood chips.[62] The addition of CO decreased the performance of the cell compared to the performance with H$_2$, as did the addition of CH$_4$. The addition of 10% CO$_2$ to a mixture of moist 15% H$_2$ and 25% CO, however, did not affect the cell's performance. Similarly, carbon deposition was observed on Ni-CeO$_2$-YSZ anodes in a gas mixture of H$_2$, CO, CO$_2$ and N$_2$ at 1073 K but no carbon deposition was observed at 1173 K.[63]

Carbon deposition at 1073 K was attributed to the Boudouard reaction (Eqn (3)). Besides tar, H$_2$S is the other major contaminant present in the syngas, and the effect of this contaminant (9 ppm H$_2$S in a simulated product gas) on Ni-GDC anodes has been tested at 1123 and 1139 K.[64] H$_2$S caused deactivation of the anode for methane reforming but did not affect the hydrogen or carbon monoxide electrochemical oxidation reactions. The effect of even higher H$_2$S concentrations from 5 to 240 ppm on planar anode-supported SOFC was studied in a gas mixture of 67% H$_2$ and 33% CO$_2$, which is similar to the gas composition from a biomass steam gasification unit.[65] At 1073 K, the performance degraded significantly at lower concentration of H$_2$S (5 ppm) while at higher H$_2$S concentration (10 ppm), the additional performance degradation was minimal. This phenomenon was attributed to the possibility of the SOFC anode becoming saturated with sulphur poisoning at the lower concentration of H$_2$S.

4.5.2 Biomass – Actual Gas

The actual gas derived from wood gasification in a commercial gasifier, after being cleaned to remove sulphur and tar (leaving H$_2$, CO, CO$_2$, CH$_4$, N$_2$, H$_2$O), was tested in a commercial planar type fuel cell assembly consisting of a Ni/GDC anode.[66] An S/C ratio of 0.5 was used to avoid the thermodynamically favoured carbon deposition region at 1123 K. The cell was operated for 150 h without any visible carbon deposition with a fuel utilisation (U$_f$) of $\sim$30% and a current density of 260 mA cm^{-2} resulting in an average power density of 0.21 mW cm^{-2}. Similarly, humidified biomass gas produced by gasifying rice husk has been tested with Ni/SDC and Ni$_{1-x}$-Cu$_x$ (x = 0 – 0.3)/SDC (samaria doped ceria) anodes at 873 K.[67] The addition of Cu was found to be beneficial as it reduced carbon deposition substantially. Cu, however, is not stable at higher temperatures and reduces the catalytic activity of Ni.

4.6 Liquid Fuels

Liquid fuels such as iso-octane, jet fuel, kerosene, diesel, methanol and ethanol hold good promise as fuels for SOFC. These fuels are easily transportable and can be used in transportation or in portable-type fuel cells. Several factors determine the suitability of a liquid fuel for direct utilization in SOFC. The main factor is the propensity of the fuel to form coke at the high temperatures (>873 K) and reducing conditions found in the anode compartment. Other important factors are the energy density and the physical state of the fuel at standard conditions, which determine how easily the fuel can be stored and fed to the SOFC, as well as the availability and cost of the fuel, which are related to the abundance of the feedstock from which the fuel is produced and to the costs and capacity of production. Toxicity and environmental impact are also factors to consider.

Recent studies show utilization of transportation fuels, such as iso-octane (C_8H_{18}), in a SOFC bi-layer anode assemblies.[68] The functional layer of the bi-layer consisted of a NiO/YSZ composite while the conduction layer was made of the NiO/YSZ composite with 5 wt% $BaCO_3$. The $BaCO_3$ in the functional layer transformed to a thin coating of $BaZr_{0.8}Y_{0.2}O_{3-\delta}$ on YSZ during the anode firing stage at 1723 K. This coating of $BaZr_{1-x}Y_xO_{3-\delta}$ promotes reforming of octane and oxidation of the reformed fuels. Maximum power densities of ~ 1.2 and 0.6 W cm^{-2} were obtained in dry H_2 and wet iso-octane fuels at 1073 K, respectively. The fuel stream needed to be humidified for water adsorption and subsequent gasification of carbon on the anode.

Iso-octane has been directly fed to a Cu/CeO_2 anode in a tubular YSZ-based SOFC at a range of temperatures from 1023 K to 1123 K.[69] The fuel underwent both thermal and catalytic decomposition to form CH_4, H_2, solid C and small quantities of C_2/C_3 compounds at OCV and under working conditions. The anode performed poorly, resulting in a power density of 10 mW cm^{-2} at 1073 K. Partial oxidation studies of iso-octane (air bubbled through iso-octane to produce a stream containing 6.2% iso-octane-94% air) over Ni/YSZ and Ni/SDC anodes indicated that both anodes deactivated due to carbon deposition even under sufficiently high current densities (0.6 or 0.8 A cm^{-2}) at 1063 and 863 K, respectively.[70] However, when a $Ru-CeO_2$ catalyst was placed in between the fuel electrode and the fuel path, the catalyst reformed the incoming fuel to syngas (H_2 and CO) and thereby reduced the chance of cracking reactions at the Ni sites. Similarly, a catalytic pre-reformer has been used to decompose model jet and diesel fuels from C8–C13 into lighter hydrocarbons – C1, C2, CO, and H_2 – that can be used as fuel in an SOFC.[71]

Kerosene fuel has been tested with a Ni/ScSZ cermet at 1073 K and with a S/C ratio of two. An average current density of 140 mA cm^{-2} was obtained over a period of five days under potentiostatic conditions without carbon deposition.[72] The fuel utilization in this experiment was high (55%), which ensured that the deposited carbon was removed by electrochemically produced water. Although carbon deposition was prevented on the Ni sites of the

Ni/ScSZ anode, carbon accumulation was observed in the fuel inlet tube, likely from the pyrolytic decomposition of kerosene.

Methanol (MeOH) has been considered both as a hydrogen-source and as a direct fuel since the first development of low temperature fuel cells.[73] The advantages of directly using MeOH in SOFC for small-scale portable applications include: MeOH is a liquid with high volumetric energy density that can be easily stored and transported; the content of impurities in MeOH is low; and the amount of carbon predicted at equilibrium is significantly lower compared to ethanol, LPG, gasoline or diesel.[74–76] Pyrolysis and catalytic decomposition of MeOH can be readily achieved at the typical SOFC anodic conditions favouring subsequent electrochemical oxidation of the products. The energy densities of different fuels are shown in Figure 4.1.

Ni/YSZ, Ni/CeO$_2$, Cu/CeO$_2$ and Cu–Co/CeO$_2$ catalysts have been tested to understand the importance of pyrolysis and catalytic decomposition of MeOH at high temperatures.[77] The major gas products from thermal decomposition of MeOH were H$_2$, CO and HCHO. The minor species was 'soot', which formed due to the pyrolysis of MeOH at a temperature lower than 973 K. The presence of the catalysts allowed the gas phase compositions to reach equilibrium. All of the catalysts tested in the study accumulated carbon, which resulted in the deactivation of the catalyst, and the gas composition ultimately reverted to that of pyrolysis.

The direct utilization of MeOH in Ni/YSZ anodes has been tested with high power densities of 1.3 and 0.6 W cm^{-2} for anode-supported cells at 1073 and 923 K, respectively, obtained with no visible carbon.[78] Information on the long-term stability or on the testing protocols (*e.g.* overall time of exposure to MeOH), however, were not provided. Electrolyte-supported Ni/YSZ anodes have been tested with pure MeOH at 1273 K.[79] The performance obtained in MeOH was similar to the performance obtained in a mixture of H$_2$ and CO. Again no carbon deposition was observed on the anode, in agreement with the thermodynamic predictions at 1273 K.[80] Our group has tested both Ni/YSZ

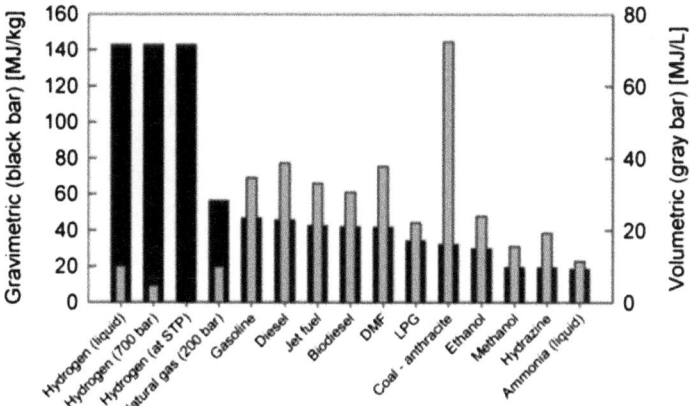

Figure 4.1 Energy density of different fuels.

and $Ni/Zr_{0.35}Ce_{0.65}O_2$(ZDC) anodes in MeOH under OCV conditions and load (0.6 V) at 1073 K.[81] The results showed that the Ni/ZDC anode accumulated less carbon. Additionally, the Ni/ZDC anode was more stable than the Ni/YSZ anode, which was likely due to the higher conductivity of ZDC. Besides conventional Ni/YSZ anodes, lanthanum cobaltite ($LaCoO_3$), copper/gadolina-doped ceria (Cu/GDC) and copper/ceria (Cu/CeO_2) have also been tested in MeOH.[82–84] Again, the issue with Cu-based anodes is sintering at high temperature and, thus, poor stability at the SOFC operating conditions.

The utilization of ethanol (EtOH) in fuel cells has been considered only recently. The advantages are that EtOH is an ultra-clean liquid fuel with reasonable energy density (Figure 4.1). EtOH, however, is harder to oxidize than H_2 and MeOH. In general, the reforming of EtOH is more problematic than natural gas and MeOH, particularly because of coking. Both steam reforming and direct utilization of EtOH have been explored and below is a summary of the findings.

Steam reforming of EtOH over Ni/YSZ catalysts at 1173–1273 K resulted in incomplete reforming and produced a gas mixture of H_2, CO, CO_2, CH_4, C_2H_6, C_2H_4 and carbon.[85] Significant amounts of C_2H_6 and C_2H_4 in the fuel could cause significant carbon formation, as these components act as promoters for carbon formation. In order to reduce carbon deposition, the S/C ratio was varied (up to five) but carbon deposition could not be prevented. In contrast, no significant carbon deposition was observed when ethanol was reformed over a $Ni/Ce–ZrO_2$ catalyst, prior to feeding the resulting product gases over a Ni/YSZ catalyst in an annular reactor at 1123–1173 K.[85] The reduction in carbon deposition was attributed to the conversion and decomposition of EtOH to lighter hydrocarbons (C_2H_6, C_2H_4) on the $Ni/Ce–ZrO_2$ catalyst producing mainly CH_4, CO and H_2, which were then further reformed over Ni/YSZ.

Direct utilization of EtOH on Ni/YSZ and Ni/GDC anodes resulted in carbon deposition and deactivation of the anodes at 1173 K.[86] The ceramic phase of GDC was not sufficiently catalytically active to prevent carbon accumulation during EtOH conversion. Similar observations of coking were also made when EtOH was directly utilized on alternative anodes such as Cu–Co(Ru)/ZDC at 1073 K.[87] Carbon deposition was found to be more severe during the direct utilization of EtOH than MeOH and the cells were less stable. Thus, limited success has been achieved in preventing carbon accumulation for both reforming and direct utilization of EtOH on the anode of SOFC.

4.7 Ammonia Fuel

Since H_2 production and storage is difficult, NH_3 has been considered as a H_2-carrier fuel. The advantage of NH_3 is that it does not contain any carbon and hence, carbon deposition from the fuel is not an issue. Additionally, utilization of NH_3 does not produce any greenhouse gases unlike other hydrocarbon fuels. Tubular Ni/YSZ cells have been tested with pure H_2 and various concentrations of NH_3 (10–100%) in H_2.[88] The Gibbs free energy of the reaction for NH_3-oxidation increases with increasing temperature (948 K to 1148 K) and hence, theoretically the OCV should also increase with temperature. The authors observed the opposite trend experimentally, however. That is, the OCV decreased with increasing temperature, which is consistent with the electrochemical oxidation of pure H_2 and not NH_3. Based on these observations, the authors concluded that NH_3 did not oxidize directly, rather it catalytically decomposed to N_2 and H_2 first, and then H_2 was electrochemically oxidized to set up the OCV. The OCV in NH_3 was 1.02 V at 1073 K, which was lower than the OCV in humidified H_2 (1.105 V). The lower OCV in NH_3 was attributed to the dilution of H_2 fuel by N_2. Compared with pure H_2 (0.12 W cm^{-2}), the peak power density in NH_3 (0.90 W cm^{-2}) was lower, which was again attributed to the lower partial pressure of H_2 in NH_3.

An anode-supported Ni-YSZ/YSZ (30 µm)/LSM-YSZ cell was tested at 923–1123 K, with the direct utilization of liquid NH_3.[89] For the range of temperatures, the performance in NH_3 was only slightly lower than that in H_2. As in the previous study, the mechanism of NH_3 oxidation was suggested to involve a two-stage process: cracking of NH_3 to H_2 and subsequent oxidation of H_2 to H_2O. NH_3 was also used as a fuel with a proton-conducting electrolyte, $BaCe_{0.9}Nd0.1O_{3-\delta}$, and the performance of the anode in NH_3 was quite similar to the performance in H_2.[90] The utilization of NH_3 does not significantly impact the electrochemical performance of the anode and averts the problem of carbon accumulation or sulphur poisoning. NH_3 does not occur naturally, however, and is produced from natural gas and air, and the environmental impact during the production of NH_3 should be considered.

4.8 Conclusions

There are many criteria for SOFC anodes and it is a challenge for one material to meet all of these criteria, especially for operation of SOFC with fuels other than hydrogen. Nonetheless, there is much activity in the development and study of anode materials. A thorough analysis (*e.g.* lifecycle analysis) of the use of different fuels would be beneficial to determine the economic and environmental viability of using these fuels in SOFC.

References

1. A. Dicks and J. Larminie, in *Fuel Cell Systems Explained*, ed. W. Vielstich, A. Lamm and H. A. Gasteiger, John Wiley & Sons Ltd, West Sussex, 2nd edn., 2003, p. 1.
2. A. Atkinson, S. Barnett, R. J. Gorte, J. T. S. Irvine, A. J. McEvoy, M. Mogensen, S. C. Singhal and J. Vohs, *Nat. Mater.*, 2004, **3**, 17.
3. J. B. Goodenough and Y. H. Huang, *J. Power Sources*, 2007, **173**, 1.
4. S. P. Jiang and S. H. Chan, *J. Mater. Sci.*, 2004, **39**, 4405.
5. S. Tao and J. T. S. Irvine, *Nat. Mat.*, 2003, **2**, 320.
6. J. T. S. Irvine and A. Sauvet, *Fuel Cells*, 2001, **1**, 205.

7. Y. H. Huang, R. I. Dass, Z.-L. Xing and J. B. Goodenough, *Science*, 2006, **312**, 254.
8. S. Primdahl and M. Mogensen, *Solid State Ionics*, 2002, **153**, 597.
9. E. P. Murray, T. Tsai and S. A. Barnett, *Solid State Ionics*, 1998, **110**, 235.
10. N. Q. Minh, *J. Am. Ceram. Soc.*, 1995, **76**, 563.
11. S. F. Corbin, R. M. C. Clemmer and Q. Yang, *J. Am. Ceram. Soc.*, 2009, **92**, 331.
12. L. Yang, Y. Choi, W. Qin, H. Chen, K. Blinn, M. Liu, P. Liu, J. Bai, T. A. Tyson and M. Liu, *Nat. Commun.*, 2011, **2**, doi:10.1038/ncomms1359, Article number: 357.
13. F. Zhao and A. Virkar, *J. Power Sources*, 2005, **141**, 79.
14. M. R. Pillai, I. Kim, D. M. Bierschenk and S. Aa. Barnett, *J. Power Sources*, 2008, **185**, 1086.
15. www. xebecinc. com/ applications- industrial- hydrogen. php (last accessed October 2012).
16. H. Balat and E. Kırtay, *Int. J. Hydrogen Energy*, 2010, **35**, 7416.
17. S. McIntosh and R. J. Gorte, *Chem. Rev.*, 2004, **104**, 4845.
18. C. Guéret, F. Billaud, B. Fixari and P. L. Perchec, *Carbon*, 1995, **33**, 159.
19. C. Guéret, M. Daroux and F. Billaud, *Chem. Eng. Science*, 1997, **52**, 815.
20. V. A. Restrepo and J. M. Hill, *Appl. Catal. A*, 2008, **342**, 49.
21. H. He and J. M. Hill, *Appl. Catal. A*, 2007, **317**, 284.
22. E. P. Murray, T. Tsai and S. A. Barnett, *Nature*, 1999, **400**, 649.
23. Y. Lin, Z. Zhan, J. Liu and S. Barnett, *Solid State Ionics*, 2005, **176**, 1827.
24. Q. X. Fu, F. Tietz and D. Stöver, *J. Electrochem. Soc.*, 2006, **153**, D74.
25. M. K. Bruce, M. van den Bossche and S. McIntosh, *J. Electrochem. Soc.*, 2008, **155**, B1202.
26. G. Kim, G. Corre, J. T. S. Irvine, J. M. Vohs and R. J. Gorte, *Electrochem. Solid-State Lett.*, 2008, **11**, B16.
27. H. He, R. J. Gorte and J. M. Vohs, *Electrochem. Solid-State Lett.*, 2005, **8**, A279.
28. T. Takeguchi, R. Kikuchi, T. Yano, K. Eguchi and K. Murata, *Catal. Today*, 2003, **84**, 217.
29. I. Gavrielatos, V. Drakopoulos and S. Neophytides, *J. Catal.*, 2008, **259**, 75.
30. N. Triantafyllopoulos and S. Neophytides, *J. Catal.*, 2006, **239**, 187.
31. I. Gavrielatos, D. Montinaro, A. Orfanidi and S. G. Neophytides, *Fuel Cells*, 2009, **9**, 883.
32. H. Kan, S. H. Hyun, Y.-G. Shul and H. Lee, *Catal. Commun.*, 2009, **11**, 180.
33. H. Kan and H. Lee, *Appl. Catal. B*, 2010, **97**, 108.
34. E. Nikolla, J. Schwank and S. Linic, *J. Electrochem. Soc.*, 2009, **156**, B1312.
35. C. E. Quincoces, S. Dicundo, A. M. Alvarez and M. G. González, *Mater. Lett.*, 2001, **50**, 21.
36. F. Frusteri, F. Arena, G. Calogero, T. Torre and A. Parmaliana, *Catal. Commun.*, 2001, **2**, 49.

37. V. R. Choudhary, B. S. Uphade and A. S. Mamman, *J. Catal.*, 1997, **293**, 281.
38. A. P. E. York, T. Xiao, M. L. H. Green and J. B. Claridge, *Catal. Rev*, 2007, **49**, 511.
39. J. R. Rostrup-Nielsen, *J. Catal.*, 1984, **85**, 31.
40. A. M. Gadalla and M. E. Sommer, *J. Am. Ceram. Soc.*, 1989, **172**, 683.
41. M. Asamoto, S. Miyake, K. Sugihara and H. Yahiro, *Electrochem. Commun.*, 2009, **11**, 1508–1511.
42. D. L. Rosa, A. Sin, M. L. Faro, G. Monforte, V. Antonucci and A. S. Aricò, *J. Power Sources*, 2009, **193**, 160.
43. M. Buccheri, PhD thesis, *Development of Nickel-based Carbon Tolerant Bi-layer Anodes for Solid Oxide Fuel Cells*, University of Calgary, 2012, p. 120–143.
44. M. A. Buccheri and J. M. Hill, *J. Electrochem. Soc.*, 2012, **159**, B361.
45. Z. Zhan, J. Liu and S. A. Barnett, *Appl. Catal. A*, 2004, **262**, 255.
46. T.-J. Huang, C.-Y. Wu and C.-H. Wang, *Fuel Process. Technol.*, 2011, **92**, 1611.
47. M. L. Faro, D. L Rosa, I. Nicotera, V. Antonucci and A. S. Aricò, *Appl. Catal. B*, 2009, **89**, 49.
48. S. McIntosh, H. He, S.-I. Lee, O. Costa-Nunes, V. V. Krishnan, J. M. Vohs and R. J. Gorte, *J. Electrochem. Soc.*, 2004, **151**, A604.
49. N. M. Sammes, R. J. Boersma and G. A. Tompsett, *Solid State Ionics*, 2000, **135**, 487.
50. M. B. Pomfret, J. Marda, G. S. Jackson, B. W. Eichhorn, A. M. Dean and R. A. Walker, *J. Phys. Chem. C*, 2008, **112**, 5232.
51. T. Reed, *Handbook of Biomass Downdraft Gasifier Engine Systems*, The Biomass Energy Foundation Press, Golden, Colorado, 1988, p. 1.
52. T. Kivisaari, P. Bjornbom, C. Sylwan, B. Jacquinot, D. Jansen and A. Degroot, *Chem. Eng. J*, 2004, **100**, 167.
53. A. Abuadala and I. Dincer, *Int. J. Hydrogen Energy*, 2010, **35**, 13146.
54. F. N. Cayan, M. Zhi, S. R. Pakalapati, I. Celik, N. Wu and R. Gemmen, *J. Power Sources*, 2008, **185**, 595.
55. C. Brage, Q Yu, G. Chen and K. Sjöström, *Biomass Bioenergy*, 2000, **18**, 87.
56. Z. A. El-Rub, E. A. Bramer and G. Brem, *Ind. Eng. Chem. Res.*, 2004, **43**, 6911.
57. A. G. Collot, Y. Zhuo, D. R. Dugwell and R. Kandiyoti, *Fuel*, 1999, **78**, 667.
58. T. Damartzis and A. Zabaniotou, *Renewable Sustainable Energy Rev*, 2011, **15**, 366.
59. E. Lorente, M. Millan and N. P. Brandon, *Int. J. Hydrogen Energy*, 2012, **37**, 7271.
60. J. Mermelstein, M. Millan-Agorio and N. Brandon, *ECS Trans*, 2009, **17**, 111.
61. Reprinted from *J. Power Sources*, 195, J. Mermelstein, M. Millan and N. Brandon, The impact of steam and current density on carbon formation

from biomass gasification tar on Ni/YSZ, and Ni/CGO solid oxide fuel cell anodes, 1657–1666, Elsevier Limited (2010), with permission from Elsevier.

62. S. Baron, N. Brandon, A. Atkinson, B. Steele and R. Rudkin, *J. Power Sources*, 2004, **126**, 58.

63. R. Suwanwarangkul, E. Croiset, E. Entchev, S. Charojrochkul, M. D. Pritzker, M. W. Fowler, P. L. Douglas, S. Chewathanakup and H. Mahaudom, *J. Power Sources*, 2006, **161**, 308.

64. J. P. Ouweltjes, P. V. Aravind, N. Woudstra and G. Rietveld, *Biosyngas Utilization in Solid Oxide Fuel Cells with Ni/GDC Anodes*, Proceedings of the First European Fuel Cell Technology & Applications Conference, Rome, Italy, 2005, Paper No. EFC2005-86089.

65. A. Norheim, Proceedings of the Second World Conference and Technology Exhibition on Biomass for Energy, Industry and Climate Protection, Rome, Italy, 2004.

66. P. Hofmann, A. Schweiger, L. Fryda, K. D. Panopoulos, U. Hohenwarter, J. D. Bentzen, J. P. Ouweltjes, J. Ahrenfeldt, U. Henriksen and E. Kakaras, *J. Power Sources*, 2007, **173**, 357.

67. Z. Xie, C. Xia, M. Zhang, W. Zhu and H. Wang, *J. Power Sources*, 2006, **161**, 1056.

68. M. Liu, Y. Choi, L. Yang, K. Blinn, W. Qin, P. Liu and M. Liu, *Nano Energy*, 2012, **1**, 448.

69. N. Kaklidis, G. Pekridis, C. Athanasiou and G. E. Marnellos, *Solid State Ionics*, 2011, **192**, 435.

70. Z. Zhan and S. A. Barnett, *J. Power Sources*, 2006, **155**, 353.

71. J. Zheng, J. J. Strohm and C. Song, *Fuel Process. Technol.*, 2008, **89**, 440.

72. H. Kishimoto, K. Yamaji, T. Horita, Y. Xiong, N. Sakai, M. E. Brito and H. Yokokawa, *J. Power Sources*, 2007, **172**, 67.

73. H. A. Liebhafsky and E. J. Cairns, *Fuel Cells and Fuel Batteries - A Guide to Their Research and Development*, John Wiley & Sons, Inc., NY, 1968.

74. G. A. Olah, A. Goeppert and G. K. S. Prakash, *Beyond Oil and Gas: The Methanol Economy*, Wiley-VCH Weinheim, Germany, 2006.

75. H. H. Kung and W. H. Cheng, *Methanol Production and Use*, Marcel Dekker, Inc., New York, 1994.

76. K. Sasaki and Y. Teraoka, *J. Electrochem. Soc.*, 2003, **150**, A878.

77. M. Cimenti and J. M. Hill, *J. Power Sources*, 2010, **195**, 54.

78. Y. Jiang and A. V. Virkar, *J. Electrochem. Soc.*, 2001, **148**, A706.

79. K. E. K. Sasaki, H. Kojo, Y. Hori and R. Kikuchi, *Electrochem. (Denki Kagaku, Japan)*, 2002, **70**, 18.

80. M. Cimenti and J. M. Hill, *J. Power Sources*, 2009, **186**, 377.

81. M. Cimenti, V. Alzate-Restrepo and J. M. Hill, *J. Power Sources*, 2010, **195**, 4002.

82. E. N. Sammes and L. Varadaraj, *Electrochem. (Denki Kagaku, Japan)*, 1995, **63**, 41.

83. D. J. L. Brett, A. Atkinson, D. Cumming, E. Ramírez-Cabrera, R. Rudkin and N. P. Brandon, *Chem. Eng. Science*, 2005, **60**, 5649.

84. T. Kim, K. Ahn, J. M. Vohs and R. J. Gorte, *J. Power Sources*, 2007, **164**, 42.
85. N. Laosiripojana and S. Assabumrungrat, *J. Power Sources*, 2007, **163**, 943.
86. R. Muccillo, E. N. S. Muccillo, F. C. Fonseca and D. Z. de Florio, *J. Electrochem. Soc.*, 2008, **155**, B232.
87. M. Cimenti and J. M. Hill, *J. Power Sources*, 2010, **195**, 3996.
88. A. Fuerte, R. X. Valenzuela, M. J. Escudero and L. Daza, *J. Power Sources*, 2009, **192**, 170.
89. Q. Ma, J. Ma, S. Zhou, R. Yan, J. Gao and G. Meng, *J. Power Sources*, 2007, **164**, 86.
90. K. Xie, Q. Ma, B. Lin, Y. Jiang, J. Gao, X. Liu and G. Meng, *J. Power Sources*, 2007, **170**, 38.

CHAPTER 5

Interconnect Materials for SOFC Stacks

XINGBO LIU,*[a] JUNWEI WU[b,c] AND
CHRISTOPHER JOHNSON[d]

[a] Mechanical and Aerospace Engineering Department, West Virginia
University, Morgantown, WV, 26506, USA; [b] School of Materials Science
and Engineering, Harbin Institute of Technology Shenzhen Graduate School,
Shenzhen, 518055, China; [c] Shenzhen Key Laboratory of Advanced
Materials, Shenzhen, 518055, China; [d] National Energy Technology
Laboratory, Morgantown, WV 26507, USA
*Email: xingbo.liu@mail.wvu.edu

5.1 Introduction

Interconnects serve a number of vital functions in the SOFC stacks. Regardless
of configurations, they provide an electrical connection between the anode of
one individual cell to the cathode of the neighbouring one to enhance voltage
output, and they act as a physical barrier to avoid any contact between the
reducing and the oxidizing atmospheres. It may offer mechanical support for
some designs as well. The criteria for the interconnect materials are the most
stringent of all cell components (anode, cathode, electrolyte, and interconnect).
In general terms, the interconnect has to meet the following demands:[1–3]

- Excellent electrical conductivity. The acceptable area-specific resistance
  (ASR) level is considered to be below $0.1 \, \Omega \, cm^2$ according to DOE
  requirements.

RSC Energy and Environment Series No. 7
Solid Oxide Fuel Cells: From Materials to System Modeling
Edited by Meng Ni and Tim S. Zhao
© The Royal Society of Chemistry 2013
Published by the Royal Society of Chemistry, www.rsc.org

- Adequate stability in terms of dimensions, microstructure, chemistry and phases at operating temperature around 800 °C in both oxidising and reducing atmospheres during 40 000 h (service lifetime).
- Excellent imperviousness for oxygen and hydrogen to prevent direct combination of oxidant and fuel during operating.
- Thermal expansion coefficient (TEC), matching those of electrodes and electrolyte, around 10.5×10^{-6} K^{-1}, so that the thermal stresses developed during start-up and shut down could be minimized.
- No reaction or interdiffusion between interconnect and its adjoining components.
- Excellent oxidation, sulfidation and carbon cementation resistance.
- Adequate strength and creep resistance at elevated temperatures.
- Low cost, as well as ease to fabrication and shaping.

Only a few such oxide systems can satisfy the rigorous requirements for the interconnect materials in SOFC. Doped lanthanum chromite (LaCrO$_3$) is currently the most common candidate material for high-temperature SOFCs (~ 1000 °C), whose properties could be adjusted by doping at both sites to meet requirements. However, high cost and difficulties in processing has made metallic alloys the primary candidate in intermediate and low-temperature SOFC.

In this chapter the technological development, commercial design and applications in interconnect materials will be reviewed, particularly the required properties, modification, and fabrication. Different parts will be emphasized for ceramic and metallic interconnects. In ceramic interconnects, defect microstructure, doping effect, electronic conductivity, TEC and fabrication will be extensively discussed. For metallic ones, we will focus on materials selection, issues for metallic materials, and how to handle them. For both materials, some designs and applications from commercial manufacturers are reviewed as well.

5.2 Lanthanum Chromites as Interconnect

Due to the stringent requirement, only a few ceramic systems can satisfy interconnect applications. The traditional material used for the SOFC interconnect is lanthanum chromite (LaCrO$_3$) at high temperature (~ 1000 °C). Firstly, it exhibits a remarkable high electric conductivity under SOFC operating conditions compared with typical ceramics,[4] and the conductivity can be improved significantly by doping with Mg, Sr, or Ca. Secondly, as indicated by the high melting point of LaCrO$_3$ (2783 ± 20 K), the material remains stable at both the cathode and anode environment.[5] Thirdly, the average TEC of LaCrO$_3$ is 9.5×10^{-6} K^{-1}, is quite close to the TEC of YSZ (10.5×10^{-6} K^{-1}),[6] the typical SOFC electrolyte.

The structure of lanthanum chromite (LaCrO$_3$) may be represented by the general formula perovskite ABO$_3$ (where A is a lanthanide and B a transition metal such as Co, Mg, Fe, Cr, Cu or V). In order to modify LaCrO$_3$ properties, alkaline earth metals such as Sr, Ca or Ba can be substituted for site A (La) and a transition metal for site B because of the ionic radius similarity for the

Table 5.1   Ionic radii (Å) for Mg, Cr and V in 6 fold coordination with different valences and for 12 fold coordinated La, Ca and Sr.[8] (With kind permission from Springer Science).

	(II)	(III)	(IV)
Cr		0.615	0.55
V		0.64	0.58
Mg	0.72		
Sr	1.44		
Ca		1.34	
La		1.36	

respective sites.[7] For instance, the replacement of part of La^{3+} ions by Sr^{2+} can highly increase the electronic conductivity of perovskite in $LaCrO_3$. Ionic radii of some elements for possible replacement are displayed in Table 5.1.[8]

5.2.1   Conductivity

$LaCrO_3$ is a p-type conductor from room temperature to high temperature (>1000 °C), and becomes nonstoichiometric through the formation of cation vacancies. Electrical conduction in the undoped $LaCrO_3$ occurs by the small polaron mechanism via transport of electron holes.[9] In Kröger-Vink notation, the p-type nonstoichiometric reaction is given by equation (1):

$$\frac{3}{2}O_2 \leftrightarrow V'''_{La} + V'''_{Cr} + 3O_O^\times + 6h^\bullet \tag{1}$$

Where V'''_{La} and V'''_{Cr} refers to La and Cr vacancy, respectively, $O_O^\times$ is oxygen site, and $h^\bullet$ is electron hole.

Doping at the A site or B site beneficially improves electrical conductivity. In the case of strontium doped $LaCrO_3$ (or LSC for short), negative charge induced by can be compensated by half an oxygen vacancy or a $Cr^{3+} \rightarrow Cr^{4+}$ transition. The conductivity change with oxygen partial pressure[1,10] (from oxidizing to reducing environments) is displayed in Figure 5.1. Under oxidizing environment ($P_{O_2} > 10^{-8}$ atm), oxygen vacancies are negligible; therefore all charge compensation occurs via valence change of $Cr^{3+} \rightarrow Cr^{4+}$. The substitution reaction in $LaCrO_3$ is

$$SrCrO_3 \leftrightarrow Sr'_{La} + Cr^\bullet_{Cr} + 3O_O^\times \tag{2}$$

$$[Sr'_{La}] = [Cr^\bullet_{Cr}] \tag{3}$$

$[Cr^\bullet_{Cr}]$ has the same function as $h^\bullet$, small-polaron hopping of charge carriers determines electrical conductivity, which occurs at the Cr-sites. It is summarized that electronic conductivity will increase with Sr'_{La}.

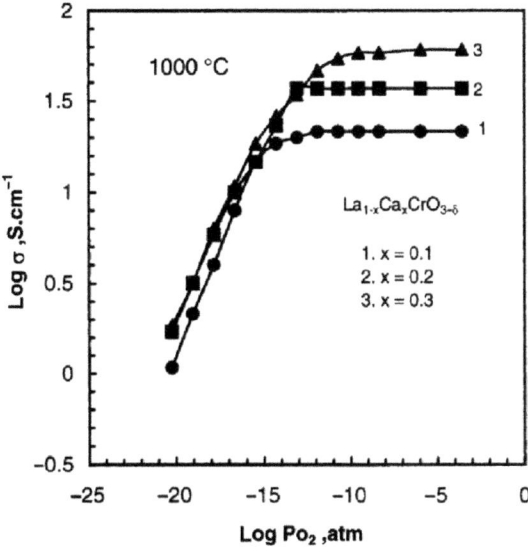

Figure 5.1 Conductivity isotherms at 1000°C for three different compositions: $x = 0.1$, 0.2, and 0.3 in $La_{1-x}Ca_xCrO_{3-\delta}$.[1,10] (Reproduced by permission of The Electrochemical Society).

$LaCrO_3$ becomes oxygen deficient at elevated temperatures under highly reducing conditions ($P_{O_2} \sim 10^{-8} - 10^{-18}$ atm). The stoichiometry should be expressed as $La_{1-x}Sr_xCrO_{3-\delta}$. In that case charge compensation is mainly achieved by oxygen vacancies and the electrical transport is mostly ionic in nature.

$$2Cr_{Cr}^{\bullet} + O_O^{\times} \rightarrow 2Cr_{Cr}^{\times} + V_O^{\bullet\bullet} + \frac{1}{2}O_2 \qquad (4)$$

The total reaction for calcium substituted lanthanum chromites is expressed by the following equation.[11]

$$La_{1-x}Ca_x(Cr^{3+})_{1-x}(Cr^{4+})_xO_3 \rightarrow La_{1-x}Ca_x(Cr^{3+})_{1-x+2\delta}(Cr^{4+})_{x-2\delta}O_{3-\delta} + \left(\frac{\delta}{2}\right)O_2 \qquad (5)$$

Figure 5.2 shows measured oxygen numbers as a function of oxygen partial pressure at different temperatures.[12] Oxygen numbers in the lattice are decreased from three with decreasing oxygen partial pressure. Therefore, the oxygen vacancies are increased correspondingly, which are the diffusion paths of oxide ions. Since the interconnect material is placed in a large oxygen potential gradient, oxygen can permeate through the $LaCrO_3$-based materials via oxygen vacancies.[13] When oxygen ions can migrate from high to low oxygen partial pressures, electrons can move in the opposite direction.

The electronic blocking electrochemical method and isotope oxygen exchange method ($^{16}O/^{18}O$ exchange) can be applied to measure oxygen

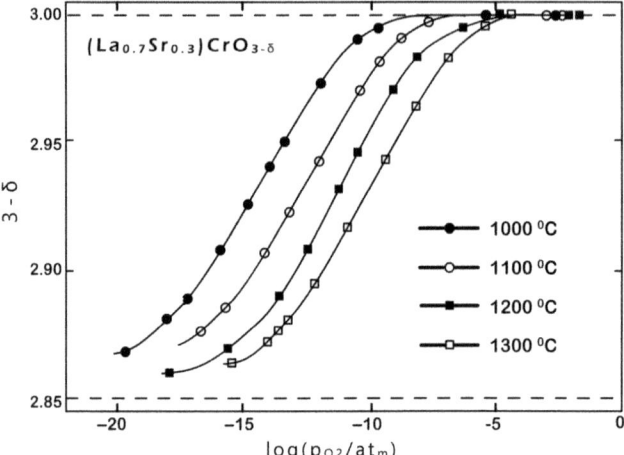

Figure 5.2   Measured oxygen numbers of $La_{0.7}Ca_{0.3}CrO_{3-\delta}$ as a function of oxygen partial pressures at 1273–1573 K.
(Reprinted from publication,[12] with permission from Elsevier).

permeation current density, which was about 3–10 mA/cm^2 at 10^{-13} Pa (about 10^{-18} atm) at the temperature of 1273 K (the thickness is assumed to be 3 mm).[13,14] The oxygen permeation needs an ionization process to oxide ion (O^{2-}) from oxygen molecules, and thus low surface reactivity can reduce the permeation flux through the LaCrO$_3$ and eventually reduce the permeation current density.

A temperature dependence of electrical conductivity in air is observed for Sr-doped LaCrO$_3$ as given in Figure 5.3, where a linear plot of Log σT vs 1/T is achieved.[10] This is often cited as convincing evidence that thermally activated hopping of small polarons is responsible for the conduction process. The relationship between the electrical conductivity σ and temperature T, is established to obey the following equation, which is typical of p-type conductor behaviour.

$$\sigma = \frac{A}{T}\exp\left(\frac{-E_a}{kT}\right) \qquad (6)$$

Where A is the pre-exponential factor, k the Boltzmann constant, and Ea the activation energy for conduction. Higher doping levels contribute to lower electrical resistance, so larger electrical conductivity, owing to the higher concentration of electron holes.[15]

Assuming that all the positive electron holes induced by the strontium are free at the higher temperatures so that the activation energy is associated with the hole mobility. It follows that the absence of a change in activation energy signifies that the carrier density is constant down to room temperature, and that the change in conductivity must be entirely due to a varying mobility, which implies localized electrons.[15]

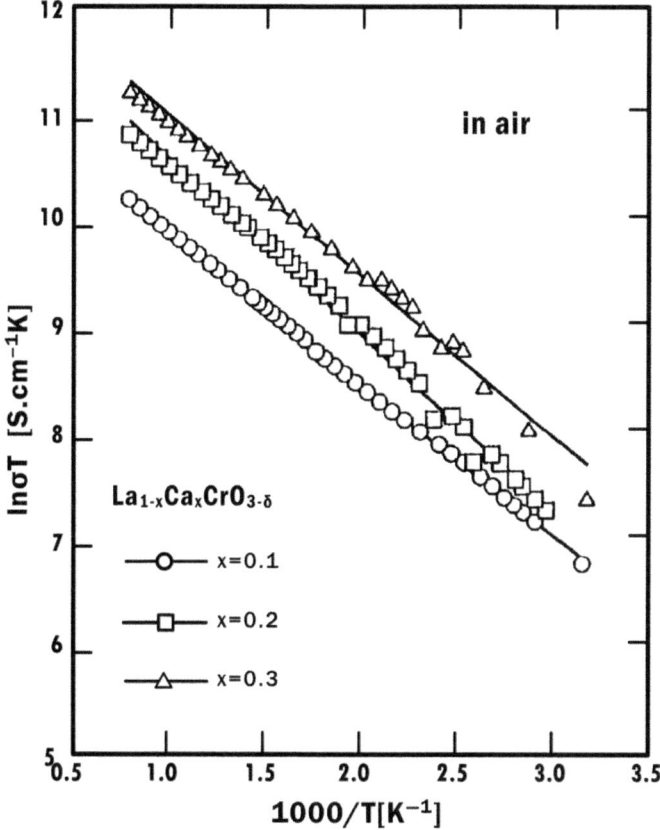

Figure 5.3 Temperature dependence of conductivity in air for three different compositions: x = 0.1, 0.2, and 0.3 in La$_{1-x}$Ca$_x$CrO$_{3-\delta}$.[10] (Reproduced by permission of The Electrochemical Society).

5.2.2 Thermal Expansion

Because interconnects contact other cell components, the thermal expansion coefficient (TEC) of the interconnect needs to be matched within an acceptable level to that of the other cell components. Large difference leads to the generation of thermal stress in the SOFC separator at the operating condition. Table 5.2 summarizes thermal expansion coefficient data for undoped and doped LaCrO$_3$ materials. The thermal expansion of LaCrO$_3$ varies with dopant content.

In air or H$_2$ atmospheres, the TEC values increase with Sr doping concentration, as shown in Figure 5.4.[16] Similar results were observed for Ca doped LaCrO$_3$. In reducing atmosphere, nonlinear expansion due to chemical reaction with the atmosphere is more common than in air. And the TEC values in reducing atmosphere are slightly larger than those in air. The lattice expansion is attributed to the increase cation size of Cr^{3+} (0.615) compared to Cr^{4+} (0.55), and to cation-cation repulsion as bridging oxygen ions are removed.[17] In the

Table 5.2 Thermal expansion coefficients of LaCrO3 and YSZ.[9] (Reprinted with permission from Elsevier).

Composition (nominal)	Thermal expansion coefficient ($\times 10^{-6}/K$)
$LaCrO_3$	9.5
$LaCr_{0.9}Mg_{0.1}O_3$	9.5
$La_{0.9}Sr_{0.1}CrO_3$	10.7
$La_{0.8}Sr_{0.2}CrO_3$	11.1
$La_{0.65}Ca_{0.35}CrO_3$	10.8
$LaCr_{0.9}Co_{0.1}O_3$	13.1
$La_{0.8}Ca_{0.2}Cr_{0.9}Co_{0.1}O_3$	11.1
8YSZ	10.5

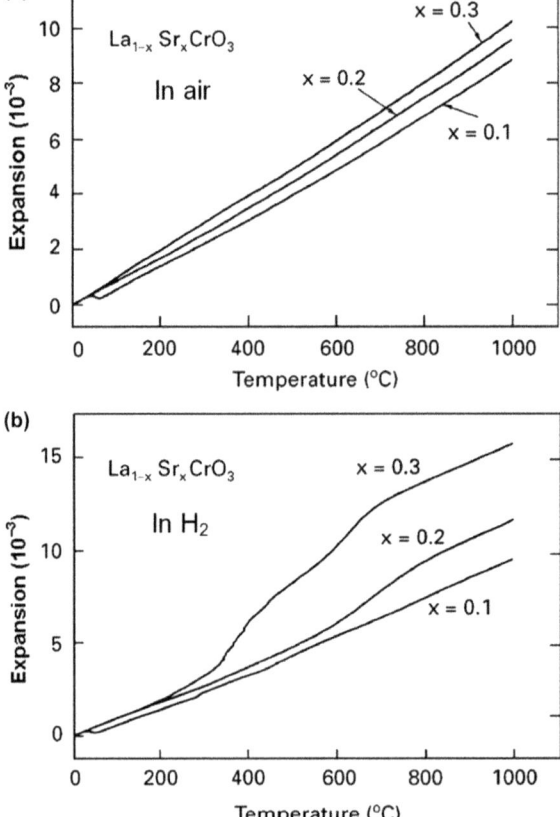

Figure 5.4 Linear thermal expansion of the Sr-doped lanthanum chromites (a) in air and (b) in H_2 atmosphere.[16]
(With kind permission from Springer Science).

case of the $La_{0.9}Sr_{0.1}CrO_3$ perovskite, a phase transformation from ortho-rhombic to rhombohedral symmetry is also observed near 80 °C.

As the temperature rises, the number of oxygen defects increase, resulting in larger thermal expansions. At the same P_{O_2}, higher doping levels lead to the generation of more oxygen vacancies, also inducing larger thermal expansions.

Substitution on B sites of $LaCrO_3$ can also be used to adjust the TEC. Doping of V decreased the TEC values to be matched with YSZ. Mori has observed Co, Al, V doping at B sites can result in TECs comparable with that of 8YSZ.[18]

5.2.3 Gas Tightness, Processing and Chemical Stability

Because interconnects separate fuel and oxidant gases from both sides, the $LaCrO_3$-based perovskite should be a dense body. Typically, it should be sintered >94% density without connected pores.[19] However, Cr starts to evaporate as gaseous CrO_3 at higher than 1000 °C in oxidizing environments, which then condenses as Cr_2O_3 at interparticle necks.[20,21] The formation of Cr_2O_3 prohibits further diffusion of other elements and therefore reduces the driving forces for further densification.[9]

To sinter $LaCrO_3$ to high densities, firing temperatures greater than 1600 °C under low oxygen partial pressures have generally been used, which is beneficial to suppress grain growth and allows maximum densification. Several other approaches have also been investigated to achieve high density, and involve the use of highly reactive powders, nonstoichiometric materials, dopants, sintering aids, and processing techniques.[9,11]

Uniform, high-surface-area powders without agglomeration lowers the sintering temperature required, due to the high free energy of the increased surface area.[22] Cr-deficient (y) 2%–3% green body would lead to a significant increase of densification as well. SaKai first found that Cr deficient $(La_{1-x}Ca_x)Cr_{1-y}O_3$ remarkably increases the sintering properties in air atmosphere.[23] The addition of dopants also improves sinterability by formation of a transient liquid phase during firing. The functions of liquid-phase sintering aids are to pull the particles closer together by surface tension forces and to enhance diffusion of the solid phase to the points of particle-to-particle contact to promote material transport via a solution/precipitation process.[24,25] The enhanced sinterability of $LaCrO_3$ doped with strontium may be due to the formation of $SrCrO_4$ melt.[26] Another more common sintering aid for $LaCrO_3$ and many other oxides is B_2O_3.[9] Furthermore, both microwave processing and spark plasma sintering of $LaCrO_3$ are beneficial to lower the sintering temperature and get dense body.[9,27]

During SOFC operation, interconnect material contacts with both electrodes at elevated temperatures, so chemical compatibility with other fuel cell components is important. Since the operation temperature of SOFC is much less than the sintering temperature of $LaCrO_3$, the vapour effect of high valence Cr is negligible, and direct reaction of $LaCrO_3$ with other components is typically not a major problem either. However, dopant addition into LaCrO$_3$ lattice may induce stability issues due to second phase reaction with other fuel cell components.[20] Reactions with LaMnO$_3$ cathode and YSZ electrolyte have been reported by Nishiyama[28] and Carter,[29] respectively. Reaction may initiate as well in Sr doped LaCrO$_3$, especially if SrCrO$_4$ formed on the interconnect.[30] Another important chemical interaction of LaCrO$_3$ involves glass sealants that are often used for gas sealing in the fiat-plate design. The interaction can alter the properties of the glass (*e.g.* changes in thermal expansion coefficient, softening temperature, *etc.*), thus degrading its sealing effectiveness.[31]

5.2.4   Other Ceramic Interconnect

LaCrO$_3$ is the most common base for SOFC interconnects, but chromites of other lanthanide elements have also been reported as an acceptable alternative. Yttrium chromite (YCrO$_3$) based perovskites exhibit several attractive features for interconnect applications over lanthanum chromites. The expansion of doped YCrO$_3$ at 1000 °C in highly reducing atmospheres is significantly less (around 40% less) than the measured expansion of similar LaCrO$_3$.[32] YCrO$_3$ is more stable in SOFC operating conditions and the formation of impurity phases can be effectively avoided.[33]

On the other hand, the electrical conductivity of yttrium chromite is known to be relatively low, but can be improved by multiple doping on A- and B-sites. Kyung Joong Yoon *et al.*, from PNNL, made a thorough study of dopant effect on electrical and thermal properties of Ca-doped yttrium chromite. Sintering and densification were facilitated by Cu substitution. Co substitution effectively enhanced the electrical conductivity in oxidizing atmosphere and Ni substitution improved stability toward reduction, additionally lowering the chemical expansion and enhancing the electrical conductivity in reducing atmospheres. With optimized amounts of Co, Ni, and Cu dopants, a good match in thermal expansion coefficients with 8 mol% YSZ was achieved. Chemical compatibility of doped yttrium chromite with YSZ, NiO, and LSM was confirmed at the standard cell processing temperatures. Based on the improved stability and thermal and electrical properties, the optimized composition of yttrium chromite, Y$_{0.8}$Ca$_{0.2}$Cr$_{0.85}$Co$_{0.1}$Ni$_{0.04}$Cu$_{0.01}$O$_3$, is recommended for use as a ceramic interconnect material for SOFCs.[34]

5.2.5   Applications

Interconnects are formed into the desired shape using ceramic processing techniques. However, material costs of perovskites are rather high and their application as ceramic interconnects is only meaningful as long as the stack design requires only small amounts of the material. By far, LaCrO$_3$ based interconnects have seldom been used for planar SOFC. All the commercial designs use very little interconnect material.

The tubular design introduced by Simens-Westinghouse uses strips of LaCrO$_3$ interconnect, which are deposited by electrochemical vapour

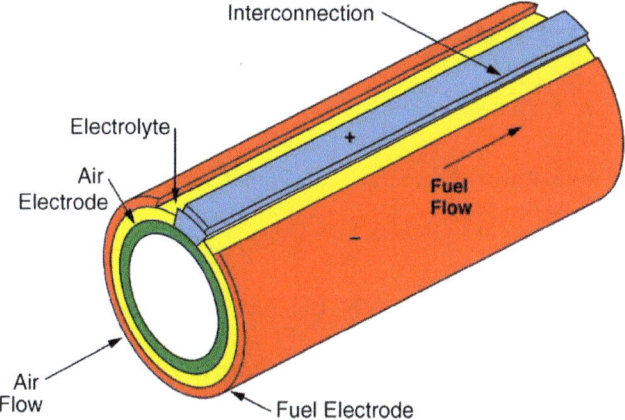

Figure 5.5 Siemens-Westinghouse tubular cells design.[35]
(With permission from Elsevier).

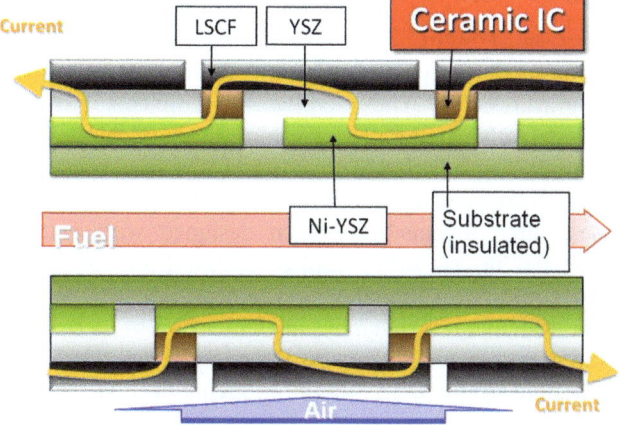

Figure 5.6 Flattened tube-type cells for a small SOFC cogeneration system by
Kyocera[36].
(courtesy of Kyocera).

deposition (EVD), as shown in Figure 5.5.[35] EVD is a modified form of CVD,
developed by Westinghouse, which utilizes an electrochemical potential
gradient to grow thin, gas-tight layers of either ionically or electronically
conducting metal oxides on porous substrates. The same technique has been
used to deposit electrolyte and anode as well.

The segmented-in-series stack concept (horizontal stripes) developed by
Tokyo Gas uses materials and manufacturing of Kyocera, pictured as shown in
Figure 5.6, also uses very little ceramic interconnect. In this case, the interconnect is applied on the sub-layer by ceramic paste mixing with an acrylic
binder and toluene.[36]

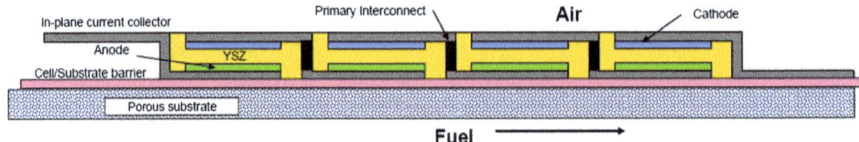

Figure 5.7   Rolls-Royce (Currently LG Chem Fuel Cells) integrated planar solid
oxide fuel cell.[37]
(Courtesy of LG Chem Fuel Cells).

Rolls-Royce fuel cells (currently LG fuel cells) developed the Integrated
Planar Solid Oxide Fuel Cell (IP-SOFC) which combines the benefits of
conventional planar and tubular approaches. Instead of being based on tubular
fuel cell, the concept is on series-connected cells fabricated on a fuel-carrying
porous support tube.[37] A cross-section view is shown in Figure 5.7. The
interconnect is based on ceramic as well.

5.3   Metallic Alloys as Interconnect

With the reduction of SOFC operation temperature to $600 \sim 800\,^{\circ}$C, metallic
materials can be used to replace $LaCrO_3$ as interconnects which are superior
at high temperatures.[1,5,38] The metallic interconnects have a number of
advantages: (1) Metallic materials have high mechanical strength. The interconnects in anode-supported SOFC, also act as a mechanical support for
ceramic parts, and constructional connection to the external inlets and outlets.
In some cases, interconnects have been designed with extra channels for
distributing the gas in co-, cross- and/or counterflow configurations. Traditional $LaCrO_3$ could not fulfill the requirement; (2) Metallic materials have
high thermal conductivity, which aides in elimination of thermal gradients both
along the interconnect plane and across the components; (3) Metallic materials
have high electronic conductivity, therefore giving decrease in resistance of the
cell increased output; (4) Ease of fabrication, low cost and readily available.

All of these alloys typically contain Cr and/or Al (Si is a third possibility, but
is much less used) to provide oxidation resistance by forming oxide scales of
Cr_2O_3 and Al_2O_3, respectively. For the chromia formers, there should be
enough chromium in the alloys to form a continuous oxide scale and to
effectively provide oxidation resistance under SOFC operating conditions. The
aluminum content in these alloys should be controlled at a minimum to avoid
formation of a continuous alumina layer, considering the insulating nature of
alumina scales.[1,38]

5.3.1   Selection of Metallic Materials

5.3.1.1   *Chromium Based Alloys*

At temperatures as high as 900–$1000\,^{\circ}$C, chromium-based oxide dispersion
strengthened (ODS) alloys have been specially used to replace $LaCrO_3$.

A representative alloy is Ducrolloy (Cr–5Fe–1Y$_2$O$_3$), which was designed by Plansee Company to match the thermal expansion coefficients of other SOFC components.[1,38] Some other chromium-based ODS alloys include: Cr–5Fe–1.3La$_2$O$_3$, Cr–5Fe–0.5CeO$_2$, Cr–5Fe–0.3Ti–0.5Y$_2$O$_3$ and others.

The reason chromium-based (chromia formation) alloys are chosen is because chromia has high conductivity compared to other oxides.[6,39] However, due to its high chromium content, chromium poisoning of cathode materials from chromia vapour and excessive chromia growth is inevitable. Excessive growth of the chromia layer will also cause spallation after thermal cycles. Additionally, ODS alloys are more difficult and costly to fabricate. Since melting can affect the dispersion of the oxides, these techniques are typically powder metallurgy based and are designed to produce near-net-shape components. For example, a powder metallurgy technique for directly producing sheets of interconnects has been reported, however, at higher cost than for typical alloys and processes.[39]

5.3.1.2 Fe-Cr-based Alloys

To get a continuous chromia layer, the substrate alloy should have enough Cr content. The critical Cr content has been summarized in the literature[38] that critical minimum Cr content is approximately 20–25% chromium in order to ensure the formation of a protective, continuous Cr$_2$O$_3$ scale. Note that low Cr content (5% and 10%) has also been used as interconnect, but the oxidation resistance was reduced significantly when lowering the Cr content.[40,41] Low-Cr steels (<5% Cr) consist of nearly pure Fe oxide accompanied by internal oxide precipitates of Cr$_2$O$_3$ and/or FeCr$_2$O$_4$ spinels. With increasing Cr content the scales become richer in spinel and chromia, which is accompanied by a decrease in the scale growth rate. Nominal composition of Fe-Cr-based alloys is shown in Table 5.3.[39]

According to their crystal and microstructures, stainless steels are usually divided into four groups: (i) ferritic steels; (ii) austenitic steels; (iii) martensitic steels; and (vi) precipitation hardening steels.[38] Among them, the ferritic stainless steels are usually the most promising because of their body-centered cubic structure, which makes the TEC quite close to that of other SOFC materials. In addition, the processing methods are quite simple.[42–44] However, the effect of substrate impurities, such as Si and Al cannot be neglected, especially silicon, which forms a continuous layer between substrate and scale. PNNL has done systemic work, for (Mn,Co)$_3$O$_4$ coated SUS430 alloy (~0.5 wt.% Si), showing ASR increases sharply at 4000 h due to the formation of continuous silica layer. Accordingly, SUS 441, with the addition of Nb and Ti based on SUS430, has also been tested. The results displayed that ASR is quite low even for bare metal, because Nb ties up Si to prevent formation SiO$_2$ layer at the scale/metal interface.[45]

During SOFC stack operation, the interconnect will face reducing and oxidizing environments on the respective sides. Test atmospheres ranging from air, to an H$_2$O + H$_2$ gas mixture simulating the anode side service gas, and to

Table 5.3 Nominal composition of Fe-based alloys.[39] (Reprinted with permission from Elsevier).

Alloys	Concentration (wt%)										
	Fe	Cr	Mn	Mo	W	Si	Al	Ti	Y	Zr	La
Fe-10Cr	Bal	10	<0.02			<0.1					
1.4724	Bal	13					1				
SUS 430	Bal	16–17	0.2–1.0			0.4–1.0	≤0.2				
Fe-17Cr-0.2Y	Bal	17							0.2		
1.4016	Bal	17									
Ferrotherm(1.4742)	Bal	17–18	0.3–0.7			0.8–0.9	0.9–1.0				
Fe-18Cr-9W	Bal	18			9						
Fe-20Cr-7W	Bal	20			7	0.3	0.6			0.3	
Fe-20Cr	Bal	20	<0.02			<0.1					
ZMG 232	Bal	21–22	0.5			0.4	0.1–0.2	0.02		0.2	0.04
AL 453	Bal	22	0.3			0.3	0.6	0.3			0.1
Fe22CrMoTiY	Bal	22	0.1	2		<0.05	<0.05	<0.05	0.4		
1.4763(446)	Bal	24–26	0.7–1.5	≤0.05		0.4–1					
FeCrMn(LaTi)	Bal	16–25	?					?			?
Fe-Cr-Mn	Bal	16–25	?								
Fe-25Cr-DIN 50049	Bal	25	0.3			0.7		0.01			
Fe-25Cr-0.1Y-2.5Ti	Bal	25						2.5	0.1		
Fe-25Cr-0.2Y-1.6-Mn	Bal	25	1.6						0.2		
Fe-25Cr-0.4La	Bal	25									0.4
Fe-23Cr-0.3Zr	Bal	25								0.3	
Fe26CrTiY	Bal	26	0.1	<0.02		<0.05	<0.05	0.3	0.4		
Fe26CrTiNbY	Bal	26	Composition not provided, but presumably same as Fe26CrTiY with Nb								
Fe26CrMoTiY	Bal	26	0.1	2		<0.05	<0.05	0.3	0.3		
E-Brite	Bal	26–27	≤0.1	1		0.03–0.2	≤0.05	≤0.05	≤0.01		
A129-4C	Bal	27	0.3	4		0.3	0.3	?			
Fe-30Cr	Bal	30	<0.02			<0.1					

Symbol '?' indicates element is present, but concentration is not specified.

air/H_2 gas dual atmosphere simulating real tests have also been carried out. In $H_2 + H_2O$ gas mixture, the chromia morphology is slightly modified and the adhesion of the scale is improved. In dual atmosphere, the scales formed in the air contained iron-rich spinel or Fe_2O_3 nodules, which were not present in the alloys exposed to air on both sides. This suggests that the mobility of iron is accelerated by hydrogen at the anode side.[39,46]

5.3.1.3 Ni-Cr-based Alloys

Compared with Fe-Cr-based alloys, Ni–Cr-base alloys always demonstrate better oxidation resistance and satisfactory scale electrical conductivity. To get a continuous Cr layer, only 15% Cr was needed to establish reasonable resistance to hot corrosion, which is lower than Fe-Cr-based alloys, and the optimum content was 18–19%. Also, Ni-based alloys are mechanically stronger. Nominal compositions of Ni-base alloy are shown in Table 5.4.[39]

Most Ni-Cr-base alloys exhibited excellent oxidation resistance in moist hydrogen, growing a thin scale that was dominated with Cr_2O_3 and $(Mn,Cr,Ni)_3O_4$ spinels or Cr_2O_3;[47] therefore, it may be used as clad metal or a plated layer in the anode side.[48,49] In air oxidation, high Cr-containing alloys, such as Haynes 230 and Hastelloy S, formed a thin scale mainly comprised of Cr_2O_3 and $(Mn,Cr,Ni)_3O_4$ spinel during high temperature exposure; on the other hand, low Cr containing alloys, such as Haynes 242, developed a thick double-layer scale consisting of a NiO outside layer above a chromia-rich substrate, raising concerns over its oxidation resistance for the interconnect applications.[47]

However, the most significant problem with Ni-Cr-base alloys is potential TEC mismatch to cell components. To take full advantage of the Ni-base alloys, novel designs of interconnects or stacks are necessary. The TECs of Ni-Cr-based alloys containing W, Mo, Al, Ti were calculated using

Table 5.4 Nominal composition of Ni-base alloys.[39] (Reprinted with permission from Elsevier).

Alloy	Concentration (wt%)									
	Ni	Cr	Fe	Co	Mn	Mo	Nb	Ti	Si	Al
Inconel 600	Bal	14–16	6–9		0.4–1			0.2–0.4	0.2–0.5	0.2
ASL 528	Bal	16	7.1		0.3				0.3	0.2
Haynes R-41(Rene 14)	Bal	19	5	11	0.1	10		3.1	0.5	1.5
Inconel 718	Bal	22	18	1	0.4	1.9				
Haynes 230	Bal	22–26	3	5	0.5–0.7	1–2				0.3
Hastelloy X	Bal	24	19	1.5	1.0	5.3				
Inconel 625	Bal	25	5.4	1.0	0.6	5.7				
Nicrofer 6025HT	Bal	25	9.5		0.1	0.5			0.5	0.15
Hastelloy G-30	Bal	30	1.5	5	5.5	5.5	1.5	1.8	1	

equation (7), a formulation derived for the TEC of Fe-free Ni alloys from room to 700 °C:

$$TEC = 13.9 + 7.3 \times 10^{-2}[Cr] - 8.0 \times 10^{-2}[W] - 8.2 \times 10^{-2}[Mo]$$
$$- 1.8 \times 10^{-2}[Al] - 1.6 \times 10^{-1}[Ti] \tag{7}$$

The bracketed terms in this equation represent the concentration (in weight percent, wt%) of the specific alloying element. According to the equation (1), most of the Ni-base alloys have higher TEC than that of fuel cell components.

NETL-Albany has done some design of Ni-Cr-based alloys containing W, Mo, Al, Ti, *etc.* After the test in moist air at room and elevated temperature for more than 1000 hours, it was found that the properties of alloy J5 were shown to be comparable to that of the commercial alloy Haynes 230. The nominal composition of J5 is Cr-12.5%, Ti-1%, Al-0.1%, Mo-22.5%, Mn-0.5%, Y-0.1%, Ni-balance.[50]

5.3.1.4   Selection Criteria for Metallic Interconnects

To further define their applicability as a SOFC interconnect, it was necessary to assess these alloys using relevant properties that correlate directly with the functional requirements of interconnect. These properties include: (i) TEC; (ii) oxidation and corrosion resistance; (iii) cost.[38]

TEC has a strong relationship with the structure.[38,51] Body-centered cubic (BCC) structure has a lower TEC than face-centered cubic (FCC) structure.[38] The structures of metallic alloys are shown in Figure 5.8.

In Wagner's theory of oxidation, it is assumed that during the oxide growth, the transport of oxygen ions and/or metallic cations through the oxide scale takes place by lattice diffusion.[38,52] Thus the growth of the surface oxide scale follows the well-known parabolic law:

$$X^2 = Kt + X_0^2 \tag{8}$$

Where X and X_0 is the thickness of the scale at time t and t = 0, respectively; K is rate constant. It has been experimentally proved that the parabolic law of the growth of scale is valid for essentially all cases in which the scale is adequately thick and homogeneous.[52]

Compared with all these results, ferritic stainless steel is the most promising candidate due to its low cost, good TEC match with YSZ, and good oxidation resistance. Comparison of key properties of different alloy groups[38] is displayed in Table 5.5.

5.3.2   Problems for Metallic Materials as Interconnect

The application of ferritic stainless steel still poses many challenges even at reduced temperatures, which includes: (1) unacceptably high oxidation rate; (2) occurrence of buckling and spallation of the oxide scale when subjected to

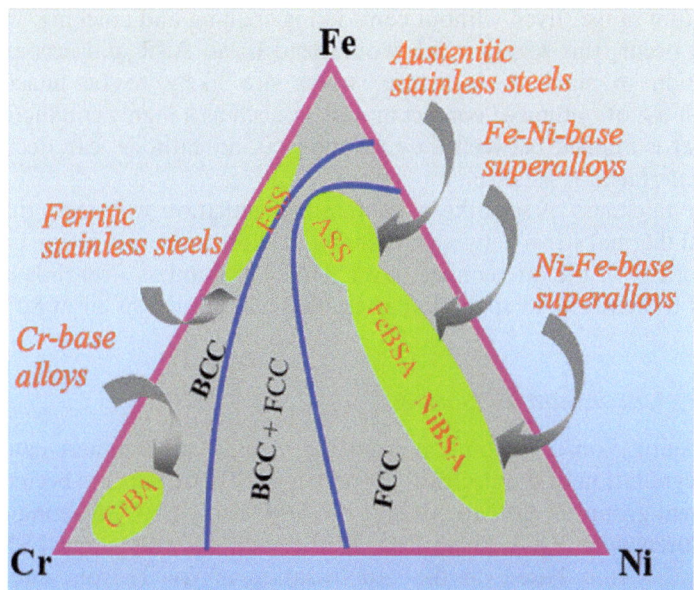

Figure 5.8 Schematic of alloy design for SOFC applications.[38] (Reproduced by permission of The Electrochemical Society).

Table 5.5 Comparison of key properties of different alloy groups for SOFC applications.[38] (Reproduced by permission of The Electrochemical Society).

Alloys	Matrix structure	$TEC \times 10^{-6} K$	Oxidation resistance	Mechanical strength	Manufactur-ability	Cost
CrBA	BCC	11.0–12.5 (RT-800 °C)	Good	High	Difficult	Very expensive
FSS	BCC	11.5–14.0 (RT-800 °C)	Good	Low	Fairly readily	Inexpensive
ASS	FCC	18.0–20.0 (RT-800 °C)	Good	Fairly high	Readily	Inexpensive
FeBSA	FCC	15.0–20.0 (RT-800 °C)	Good	High	Readily	Fairly expensive
NiBSA	FCC	14.0–19.0 (RT-800 °C)	Good	High	Readily	Expensive

thermal cycling; and (3) volatilization of high valence Cr-species in the form of CrO_3, or $Cr(OH)_2O_2$.

5.3.2.1 Excessive Growth and Spallation of Oxide Scale

Evaluation of oxidation behaviour indicates that chromia scales on ferritic stainless steels can grow to micrometers or even tens of micrometers after thousands of hours of exposure in the SOFC environment in the intermediate temperature range. Even without considering spalling and cracking, which are likely to occur, this scale growth would lead to an ASR and accompanying degradation in stack performance, which are likely to be unacceptable. Although use of optimized contact materials, such as a highly conductive oxide paste that is applied between the interconnect and cathode, can decrease the overall interfacial resistance.[53]

As the thickness of scale keeps increasing, spallation will occur due to the increased thermal stress between coatings and substrate, or, in severe cases, loss of hermeticity of the interconnect layer. UNS430 stainless steel rods had been reported to be prone to spallation after 1900 h oxidation in air at 800 °C.[54]

5.3.2.2 Chromium Poisoning

The vapourization of chromium from the metallic interconnect can lead to severe degradation of the electrical properties of SOFC. This has been observed by different groups at the cathode side of SOFC using Cr_2O_3-forming alloys as interconnects with Y_2O_3-doped ZrO_2 as the solid electrolyte and LSM as the cathode.[55–57] It is based on the vapourization of Cr_2O_3 from interconnect surface as $CrO_3(g)$ or $CrO_2(OH)_2(g)$ as major gaseous species with chromium in the 6^+ oxidation state. The equilibrium gas pressures for CrO_xH_y species as a function of oxygen partial pressure at 800 °C is displayed in Figure 5.9. The vapour pressure is becoming higher at higher oxygen partial pressure, so the cathodic side experiences a much higher Cr transport effect than anodic side. It is also clear that $CrO_2(OH)_2(g)$ shows the highest vapour pressure in the SOFC operation temperature range of 800–1000 °C.[39]

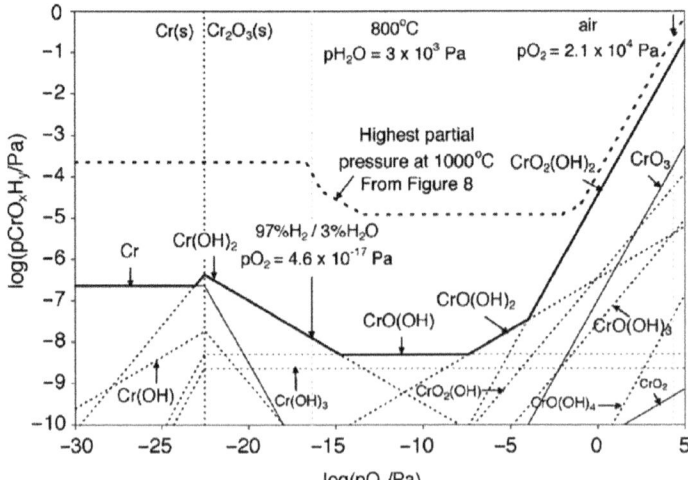

Figure 5.9   Equilibrium vapor pressures of chromium–oxygen–hydrogen gas species at 800 °C with a water vapor pressure of 3 kPa using thermodynamic data from Ebbinghaus.[39]
(Reprinted with permission from Elsevier).

Chromium-containing vapour species formed from the interconnect material will be electrochemically or chemically reduced at the three-phase boundary {electrolyte/cathode/oxidant}. The resulting deposition can block the active electrode surface and degrade cell performance. The vapour pressures are higher in air so such poisoning is most likely to occur at the cathode. This degradation can be represented by a decrease in cell voltage or an increase (*i.e.* more negative) in cell overvoltage.[39,58,59]

Both bulk alloy development and surface modifications are two important options to mitigate scale growth and Cr poisoning effect. However, the former one always introduces complicated fabrication processes and accompanying higher cost. Thus surface modifications of coatings have received more and more attention.

5.3.3 Interconnect Coatings

To qualify as viable, coating material should possess the following attributes: (1) The diffusion coefficients of Cr and O in the coating should be as small as possible so that the transport of chromium and oxygen can be effectively hindered; (2) It should be chemically compatible and stable with respect to substrate, electrodes, seal materials and contact pastes; (3) It should be thermodynamically stable in both oxidizing and reducing atmospheres over the applied temperature range; (4) It must have low ohmic resistance to maximize electrical efficiency; (5) The thermal expansion coefficient should be well matched to the substrate so that the coating is resistant to spallation during thermal cycling.[60]

5.3.3.1 Nitride Coatings

Nitride coatings have been widely used in tools coatings due to their superior wear and resistance. On the other hand, this type of coating can also supply an alternative for SOFC interconnect applications due to their low resistance and high temperature stability.[61]

Vacuum deposition, especially physical vapour deposition, has been widely used to prepare components and protective nitride coatings for SOFCs interconnect, due to the versatility of this technique, as well as the ability to control composition and morphology.[62] P.E. Gannon has applied multilayer coatings consisting of repeated sections of CrN and a CrAlN superlattice.[63] The results showed that thinner bilayer (~ 1.1 nm) CrN/AlN superlattice coatings are more favourable for the SOFC interconnect application than thicker 4.5nm bilayer. Later, he deposited Cr-Al-O-N.[64] The introduction of oxide into nitride coatings reduces the Fe and Cr migration from substrate. Both of these results suggest PVD is an effective method to fabricate high-quality coatings for metallic interconnects.

TiAlN with 30 and 50% Al[65] and SmCoN[66] coatings have been investigated in our group. Both of them revealed that nitride coatings could remain stable at 700 °C, and are helpful in inhibiting Cr and Fe migration from substrate. In addition, low ASR were obtained in the short-term test. However, these methods have limitations, such as high capital cost and low deposition rate. Most importantly, nitride is not quite stable at high temperatures.

5.3.3.2 Perovskite Coatings

Perovskite materials[67] have been widely utilized in SOFC components, such as $LaCrO_3$ for interconnect[1] and anode,[68] $LaMnO_3$ and $LaFeO_3$ for cathode.[4] Both traditional ceramic interconnect and cathode materials have been applied as coatings for metallic interconnects.

$LaCrO_3$, traditional interconnect material, had been deposited by Orlovskaya using r.f. sputtering.[69] It was found that two-step phase transformation occurred from as-deposited amorphous phase. The first step is from the X-ray amorphous state to a major intermediate phase: monoclinic $LaCrO_4$ monazite. A finite amount of La_2CrO_6 phase was also formed. The second step is the transformation of $LaCrO_4$ into the $LaCrO_3$ perovskite phase. During this transition only nanoporosity appears. These distinctive nanostructures have excellent potential for use as SOFC interconnect coating. Another option is to coat these alloys with an oxide layer that subsequently reacts with chromia to form a chromite (or chromate). Reactive formation of $LaCrO_3$ coating on SS-444 alloy via templated growth from thermally grown Cr_2O_3 and sputtered La_2O_3 layers. Reactive formation of $LaCrO_3$ coating on SS-444 alloy via templated growth from thermally grown Cr_2O_3 and sputtered La_2O_3 layers has been demonstrated.[70] The results revealed that coated samples show much lower electrical resistance compared to the uncoated samples after similar thermal exposure for 100 h at 850 °C. However, this approach suffers the obvious drawback of difficulty in controlling the composition of the reaction products. Therefore, coating with perovskite directly on the surfaces of the metallic interconnects is preferred.

LSM has been widely used as SOFC cathode materials. Due to its high electrical conductivity and thermal compatibility and stability in oxidizing environment, it has also been investigated as interconnect coatings. The presence of an LSM coating is crucial in maintaining the low level of contact resistance at elevated temperatures over extended periods of time. Slurry LSM coating was investigated on SUS430 substrate.[71] After sintering at 1200 °C for 2 h in N_2, followed by heat treatment at 1000 °C for 3 h in air, stable LSM-coated SUS 430 showed a low area-specific resistance (ASR) and maintained almost a constant value for 2600 h. Additionally, interactions between LSM coatings and RA446 were explored.[72] In the early 500 h, LSM could react with Cr diffused from substrate to form $LaCrO_3$, $(Cr,Mn)_3O_4$ and Cr_2O_3. The latter two phases have un-negligible Cr evaporation rates.

The reactions and complicated sintering process are originated from not fully dense coating and inherent porosity of perovskite, which could supply a path for Cr migration. It is necessary to deposit fully dense coatings to inhibit Cr diffusion and increase conductivity. Recently, perovskite coatings of LSM, LSCF were deposited by aerosol deposition, which is based on the impact adhesion of fine particles, and nearly 95% dense coatings could be obtained. After 100 h of oxidation at 800 °C, area-specific resistance of LSM and LSCF coated alloys was 20.6 and 11.7 $m\Omega \cdot cm^2$, respectively. And most importantly, no chromium was detected in coatings by EDX line scan.[73]

5.3.3.3 Spinels

Compared with perovskite, some spinels show better performance in preventing oxygen inward diffusion and Cr outward diffusion. However, the Cr evaporation is still a problem for Cr containing spinels. $MnCr_2O_4$ has been found as a product of reaction between chromium oxide and the Mn diffusion from Crofer 22 substrate, but the Cr evaporation is not negligible. $CoCr_2O_4$ coatings cause no cell performance degradation in 1000 hrs of testing,[74] long-term testing is necessary to prove the coating's effectiveness.

Non-chromium containing spinels, $(Mn,Co)_3O_4$ spinel are considered the most promising spinel coatings. Previous work[75] on Plansee Ducrolloy $(Cr-5\%Fe-1\%Y_2O_3)$, indicated that a $(Mn,Co)_3O_4$ spinel layer could reduce chromium migration significantly, and predicted ASR at 10 000 h is as low as $0.024 \, m\Omega \cdot cm^2$ when using LSM and LSC contact paste. Chen *et al.* reported $MnCo_2O_4$ coatings on the ferritic stainless steel AISI430 via slurry coating followed by mechanical compaction and air-heating.[76] Recently, Yang *et al.* at PNNL did a systematic study on $(Mn,Co)_3O_4$ spinel from conductivity, microstructure, TEC and long term ASR measurement of $(Mn,Co)_3O_4$ on ferritic stainless steel.[77]

(a) Microstructure. The spinel structure consists of a close-packed oxygen lattice with cations in tetrahedral and octahedral positions. Half the octahedral sites and 1/8 of the tetrahedral sites are occupied in an ordered manner. The octahedral sites alternate between empty and filled. Lattice parameters lie consistently in the range from 8.05 to 8.50 Å. Cell size depends on the ionic radius and cation distribution. Spinels containing divalent manganese cations always have the largest cell volume because of its large ionic radius.

Among the varied $Mn_{1+\delta}Co_{2-\delta}O_4$ spinel compositions, $Mn_{0.5}Co_{2.5}O_4$ ($\delta = -0.5$) and $MnCo_2O_4$ ($\delta = 0$) exhibited a cubic spinel structure, with Mn sitting on octahedral interstitial sites and Co on both tetrahedral and octahedral interstitial sites in the face-centred cubic oxygen ion lattice. While $Mn_{2.5}Co_{0.5}O_4$ ($\delta = 1.5$) and Mn_2CoO_4 ($\delta = 1.0$) demonstrated a tetragonal spinel structure. When $\delta = 0.5$, the $Mn_{1.5}Co_{1.5}O_4$ spinel was a dual phase material, containing both the cubic and tetragonal phases.[77]

During heating and cooling, the spinels $Mn_{1+\delta}Co_{2-\delta}O_4$ with δ in the range of 0.3–0.9, exhibiting a dual phase structure at room temperature, undergo a cubic $\leftrightarrow$ tetragonal spinel phase transformation. High temperature XRD analysis on $Mn_{1.5}Co_{1.5}O_4$ found that the cubic $\leftrightarrow$ tetragonal spinel phase transformation occurred around 400 °C.[77]

Table 5.6 Electrical conductivities of different spinels.[78] (Reproduced with permission from John Wiley and Sons).

	σ (S cm^{-1})	T (^{o}C)	References
$MnCr_2O_4$	0.003	800	Summarized by [78]
$Mn_{1.5}Cr_{1.5}O_4$	0.07	800	Summarized by [78]
$MgMn_2O_4$	0.6	800	Summarized by [78]
Mn_3O_4	0.1	800	Summarized by [78]
$NiMn_2O_4$	9.1	800	Summarized by [78]
$NiFe_2O_4$	1.4	800	Summarized by [78]
Co_3O_4	2.2	700	Summarized by [78]
$MnCo_2O_4$	36	800	[80]
$Mn_{1.5}Co_{1.5}O_4$	~60	800	[77]
Mn_2CoO_4	~10	850	[79]

(b) Conductivity. Spinels are generally believed to conduct by hopping off charge between octahedral sites. Thus the presence of different valence states among octahedral cations is beneficial to conduction.[78] Many $Mn_{1+\delta}Co_{2-\delta}O_4$ spinels are good electrical conductors. For example, the electrical conductivity tests on the $Mn_{1.5}Co_{1.5}O_4$ spinel indicated an electrical conductivity of ~60 S cm^{-1} at 800 °C in air, which is two to four orders of magnitude higher than Cr_2O_3, and $MnCr_2O_4$, as shown in Table 5.6.[78] By comparison, the measured electrical conductivity of Mn_2CoO_4[79] and $MnCo_2O_4$[80] was lower than that of $Mn_{1.5}Co_{1.5}O_4$ at 800 °C in air.

(c) TEC. The TEC of the phase transformation has always appeared to have negligible effects on the thermal expansion behaviour of the spinel material. Thermal expansion of $Mn_{1.5}Co_{1.5}O_4$ exhibited a good linearity with temperature (11.5×10^{-6} K^{-1}, 20–800 °C). In addition, the spinel demonstrated a good thermal expansion match to ferritic stainless steels such as Crofer22 APU and AISI 430, as well as the perovskite cathode compositions such as $La_{0.8}Sr_{0.2}MnO_3$ and $La_{0.8}Sr_{0.2}FeO_3$.[77,79] Moreover, the TEC of $MnCo_2O_4$ is very close to that of $Mn_{1.5}Co_{1.5}O_4$ below 1000 °C.

5.3.4 Applications of Metallic Interconnects

Due to its high performance, $(Mn,Co)_3O_4$ spinel has been applied to metallic interconnects by various means including slurry coating,[79] screen printing,[77] PVD,[81] aerogel coating, cathodic electroplating, anodic electrodeposition, electrophoretic deposition. Regardless of the deposition method the coatings displayed good performance in long-term operation.

Long-term $(Mn,Co)_3O_4$ spinel coatings evaluation has been done by Yang *et al.* by slurry coatings.[77,79] After the six months of 125 thermal cycling tests, the coatings still adhere well to the substrate, and no spallation has occurred. Further EDS linescan analysis of a cross-section revealed a sharp, thin Cr profile across the interface between the subscale and the spinel protection layer,

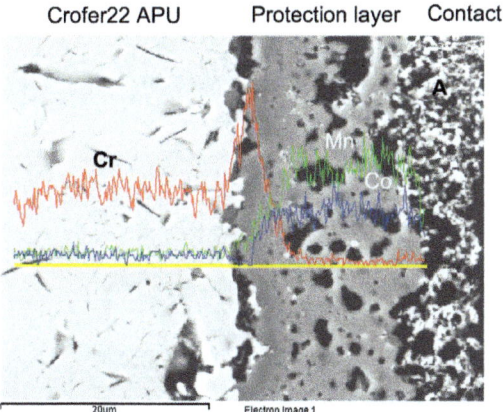

Crofer22 APU   Protection layer   Contact

Figure 5.10 Microstructural and compositional analyses on the Mn1.5Co1.5O4 protection layer subjected to a contact ASR measurement for a period of 6 months under thermal cycling.[77]
(Reproduced by permission of The Electrochemical Society).

with no chromium detectable in the spinel protection layer, as shown in Figure 5.10.

Electroplating of Mn/Co alloys followed by controlled oxidation and/or reaction to the desired phase may be a cost-effective method for the formation of dense coatings. Anodic deposition of nanocrystalline Mn-Co-O coatings has been done by Weifeng Wei *et al.*[82,83] Because Mn has higher affinity to oxygen, and the low dissociation constant (K_d) of cobalt sulfate heptahydrate in solution, diluted Mn solutions were used to get coatings with different Mn/Co ratios. Eventually, in solution of Mn : Co = 29 : 1, it is possible to obtain coatings with Mn = 18.7%, Co = 19.0%, O = 62.3%. Co and Mn bilayer deposition in two baths was done by M. Reza Bateni and co-workers.[84] $MnCo_2O_4$ spinels were achieved in the following seven-day oxidation at 750 °C.

In our group, both DC and pulse power source have been successfully deposited Mn/Co alloys on ferritic stainless steels, and various Mn/Co ratios could be controlled by adjusting depositing parameters.[85,86] Continuous 1200 h testing shows stable performance of ASR.[86] Additionally, coated interconnect had been applied to on-cell test comparing coated SUS 430[87], bare T441, preoxidized T441.[88] The cell with the MnCo coated interconnect showed stable performance even after two thermal cycles, while cells with bare and preoxidized T441 shared sharp drop in the initial 300 h,[88] suggesting coating by electroplating can achieve good results, as shown in Figure 5.11. Currently the technology is being commercialized by Faraday Technology Inc.

Electrophoretic deposition (EPD), followed by reduced-atmosphere sintering offered another option to produce the desired $(Mn,Co)_3O_4$ spinel coatings.[89] The desired spinel phase and coating thickness can be controlled by direct application of $(Mn,Co)_3O_4$ particles and simple adjustment of process

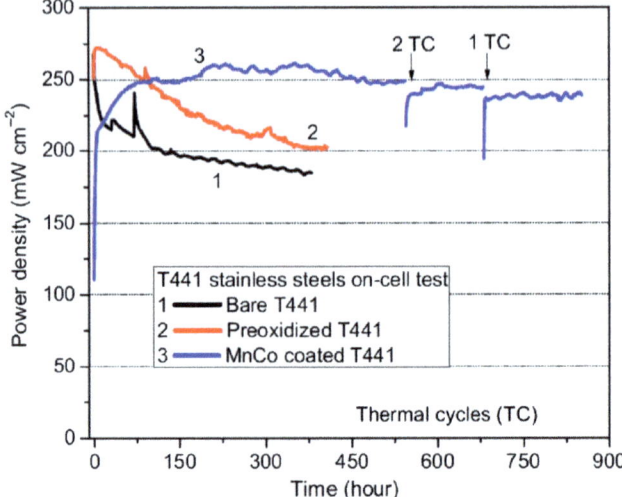

Figure 5.11   Cell performances with interconnects as cathode current collectors.[88]
(Reproduced by permission of International Association of Hydrogen
Energy).

parameters. Uniform distribution and high conductivity of the $(Mn,Co)_3O_4$
spinel coatings can be achieved by using the deposited voltage of 400 V. By
comparing reducing-atmosphere treated samples with the air-treated coatings,
it can be seen that the interfacial layer of Cr-based oxide is obviously thinner
for the spinel coating sintered in H_2/H_2O atmosphere due to the low oxygen
partial pressure. As a result, ASR tests show that the H_2/H_2O-treated coating
has a lower resistance value and much better long-term performance than the
air-treated coating.[89]

Gannon and his group have studied how two layer coatings
$La(Cr_{1-x},Mn_x)O_3$–$(Mn,Co)_3O_4$ perform at dual atmosphere.[90] La_2O_3 was
deposited by a metal organic chemical vapour deposition (MOCVD), and top
layer of Mn/Co was formed by magnetron sputtering of target alloys (Co : Mn,
1 : 1). The 'mixed' coating of $La(Cr_{1-x},Mn_x)O_3$–$(Co,Mn)_3O_4$ is formed by the
succession of a La_2O_3, annealed for 100 h at 800 °C in air atmosphere. He
found the mixed coatings effectively inhibit both chromia sub-layer formation
(compared to the single $(Co,Mn)_3O_4$ layer) and the formation of iron oxide in
dual atmosphere exposures (compared to the single La_2O_3).

NexTech has demonstrated aerosol spray deposition of SOFC interconnect
protective coatings through a flexible, high-volume process that can
cost-effectively apply protective layers to a range of interconnect geometries
and features.[91] The coatings are manganese cobalt oxides, as shown in
Figure 5.12 (a). The technology was licensed from Ceramic Fuel Cells Ltd
(CFCL). As of March 2012, NexTech continuous test on SOFC interconnects
reached the one-year milestone, the ASR profile as displayed in Figure 5.12 (b)
implies stable ASR results.

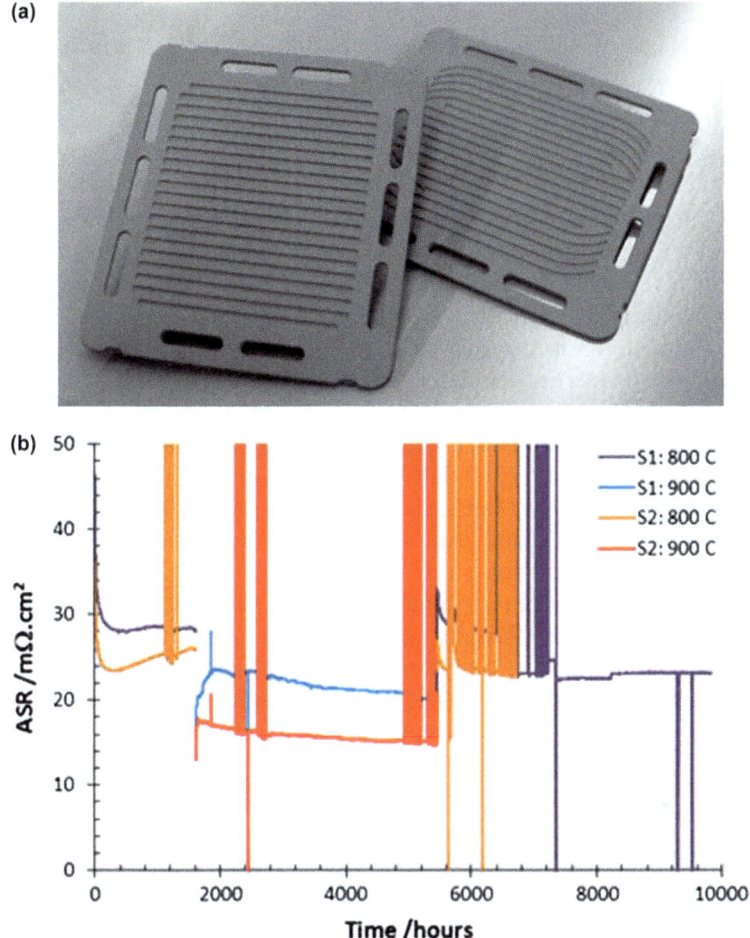

Figure 5.12   (a) Manganese cobalt oxide coated interconnect; (b) One year continuous test on coated interconnect.[91]
(Courtesy of Nextech).

Another type of interconnect material based on a CrFe PM alloy[92] is being produced by Plansee AG's High Performance Materials Division. According to Plansee, the so-called CFY- interconnects can be joined to high performance cells and hence enable the manufacture of 'robust, high performance and low cost SOFC stacks'. The CrFe alloy interconnect plates are engineered to allow the stacks to operate continuously at high temperatures (ca. 900 °C) without cracking. Plansee produces the interconnects using a powder metallurgy process in which cost effective, custom developed metal powders are pressed into the desired plate-like shape. The parts are then sintered in a sintering furnace produced by Elino specifically for these materials. Plansee reported in 2010 that its PM CrFe interconnects are being used in the Bloom ground-breaking Energy Server™ – a new class of distributed power generator, producing clean, reliable, affordable electricity at the customer site. A disadvantage of the alloy is the currently high price which could be drastically reduced by suitable production techniques.

5.4   Concluding Remarks

Ceramic-based $LaCrO_3$ and ferritic stainless steels are the most promising materials for commercial SOFC interconnect applications. Dopants at A and B site of $LaCrO_3$ are helpful to adjust the conductivity, TEC, chemical stability and so on, in order to meet the stringent requirements of SOFC interconnects. The two most common dopants to A site are strontium and calcium, which affect $LaCrO_3$ properties similarly, but to different degrees. Most $LaCrO_3$ based interconnect are applied to tubular or planar-tubular design.

Metallic-based interconnects are mainly applicable to planar SOFC. Ferritic stainless steels are the best candidate due to its good corrosion resistance, close TEC match with YSZ, easy fabrication and low cost. However, Cr volatility to poison fuel cell cathode and excessive chromia scale growth are two main challenges for ferritic stainless steels. Both bulk alloy development and surface modifications are two important alternatives. The high cost of former approach limits wide applications. Surface modifications of protective layers offer an option to achieve long term operation with adverse effects. Among all coatings reported, $(Mn,Co)_3O_4$ spinels are the most promising in good conductivity and mitigate Cr diffusion. Various deposition methods have been applied and tested. Further progress in terms of understanding elements diffusion, coating composition and thickness optimization will be necessary to achieve long term performance for SOFC commercialization.

References

1. W. Z. Zhu and S. C. Deevi, *Materials Science & Engineering*, 2003, **A348**, 227.
2. W. Z. Zhu and S. C. Deevi, *Materials Research Bulletin*, 2003, **38**, 957.
3. S. Fontana, R. Amendola, S. Chevalier, P. Piccardo, G. Caboche, M. Viviani, R. Molins and M. Sennour, *J. Power Sources*, 2007, **171**, 652.
4. N. Minh, *J. Am. Ceram. Soc.*, 1993, **76**, 563.
5. Z. Yang and S. C. Singhal, *Encyclopedia of Electrochemical Power Sources*, Volume: Solid Oxide Fuel Cells: Cell Interconnection, Elsevier, 2009, p. 63.
6. F. Tietz, *Ionics*, 1999, **5**, 129–139.
7. F. Gaillard, J. Joly, A. Boréave, P. Vernoux and J. Deloume, *Appl. Surf. Sci.*, 2007, **253**, 5876.
8. P. H. Larsen, P. V. Hendriksen and M. Mogensen, *J. Therm. Anal. Calorim.*, 1997, **49**, 1263.
9. N. Minh, T. Takahashi, *Science and Technology of Ceramic Fuel Cells*, Elsevier, 1995, Chapter 7, 165.
10. I. Yasuda and T. Hikita, *J. Electrochem. Soc.*, 1993, **140**, 1699.

11. T. Horita, *Perovskite Oxide for Solid Oxide Fuel Cells*, T. Ishihara, Springer, 2009, p. 285.
12. J. Mizusaki, S. Yamauchi, K. Fueki and A. Ishikawa, *Solid State Ionics*, 1984, **12**, 119–124.
13. N. Sakai, K. Yamaji, T. Horita, H. Yokokawa, T. Kawada, M. Dokiya, K. Hiwatashi, A Ueno and M. Aizawa, *J. Electrochem. Soc.*, 1999, **146**, 1341.
14. T. Kawada, T. Horita, N. Sakai, H. Yokokawa and M. Dokiya, *Solid State Ionics*, 1995, **79**, 201.
15. D. B. Meadowcroft, *Brit. J. Appl. Phys. (J. PHYS. D)*, 1969, **2**, 1225.
16. M. Mori, T. Yamamoto, H. Itoh and T. Watanabe, *J. Mater. Sci.*, 1997, **32**, 2423.
17. Zhenguo Yang, Jeffrey W. Fergus, in *Solid Oxide Fuel Cells: Materials Properties and Performance*, J. W. Fergus, J. Zhang, X. Li, D. P. Wilkinson and R. Hui, CRC Press, 2008, Chapter 3, p. 179.
18. M. Mori, Y. Hiei and T. Yamamoto, *J. Am. Ceram. Soc.*, 2001, **84**, 781.
19. M. Mori and N. M. Sammes, *Solid State Ionics*, 2002, **146**, 301.
20. J. W. Fergus, *Solid State Ionics*, 2004, **171**, 1.
21. H. U. Anderson, F. Tietz, in *High Temperature Solid Oxide Fuel Cells*, S. C. Singhal, K. Kendall, Elsevier, 2003, p. 173.
22. A. Chakraborty, R. N. Basu and H. S. Maiti, *Mater. Lett.*, 2000, **45**, 162.
23. N. Sakai, T. Kawada, H. Yokokawa, M. Dokiya and I. Kojima, *J. Am. Ceram. Soc.*, 1993, **76**, 609.
24. J. L. Bates, L. A. Chick and W. J. Weber, *Solid State Ionics*, 1992, **52**, 235.
25. L. A. Chick, J. Liu, J. W. Stevenson, T. R. Armstrong, D. E. McCready, G. D. Maupin, G. W. Coffery and C. A. Coyle, *J. Am. Ceram. Soc.*, 1997, **80**, 2109.
26. L. A. Chick, J. L. Bates, L. R. Pederson, H. E. Kissinger, Proceedings of the First International Symposium on Solid Oxide Fuel Cells, ed. S. C. Singhal, 1989, Electrochemical Society, USA, p. 170.
27. L. W. Tai and P. A. Lessing, *J. Am. Ceram Soc.*, 1991, **74**, 155.
28. H. Nishiyama, M. Aizawa, H. Yokokawa, T. Horita, N. Sakai, M. Dokiya and T. Kawada, *J. Electrochem. Soc.*, 1996, **143**, 2332.
29. J. D. Carter, C. C. Appel and M. Mogensen, *J. Solid State Chem.*, 1996, **122**, 407.
30. C. E. Hatchwell, N. M. Sammes, G. A. Tompsett and I. W. M. Brown, *J. Eur. Ceram. Soc.*, 1999, **19**, 1697.
31. Z. Yang, K. D. Meinhardt and J. W. Stevenson, *J. Electrochem. Soc.*, 2003, **150A**, 1095.
32. S. W. Paulik, S. Baskaran and T. R. Armstrong, *J. Mater. Sci. Lett.*, 1999, **18**, 819.
33. S. Wang, B. Lin, Y. Dong, D. Fang and H. Ding, X. Liu and G. Meng, *J. Power Sources*, 2009, **188**, 483.
34. K. J. Yoon, J. W. Stevenson and O. A. Marina, *J. Power Sources*, 2011, **196**, 8531.
35. M. C. Williams, J. P. Strakey and S. C. Singhal, *J. Power Sources*, 2004, **131**, 79.

36. T. Ito, K. Horiuchi, K. Nakamura, Y. Matsuzaki, S. Yamashita, T. Horita, H. Kishimoto, K. Yamaji and H. Yokokawa, Durability Improvement of Flatten Tubular Segmented-in-Series Type Cell-Stacks for SOFC m-CHP, European Fuel Cell 2011, Rome, Dec. 14-16. 2011, Italy.
37. T. Ohrn, Rolls Royce IP-SOSOCFC Technology Development, SECA Workshop 2009, Pittsburgh, USA.
38. Z. Yang, K. Weil, D. Paxton and J. Stevenson, *J. Electrochem. Soc.*, 2003, **150**, A1188.
39. J. Fergus, *Mater. Sci. Eng., A*, 2005, **397**, 271.
40. S. Geng, J. Zhu, M. Brady, H. Anderson, X. Zhou and Z. Yang, *J. Power Sources*, 2007, **172**, 775.
41. S. J. Geng, J. H. Zhu and Z. G. Lu, *Solid State Ionics*, 2006, **177**, 559.
42. P. Piccardo, P. Gannon, S. Chevalier, M. Viviani, A. Barbucci, G. Caboche, R. Amendola and S. Fontana, *Surf. Coat. Technol.*, 2007, **202**, 1221.
43. W. Qu, L. Jian, D. Ivey and J. Hill, *J. Power Sources*, 2006, **157**, 335.
44. I. Antepara, I. Villarreal, L. M. Rodríguez-Martínez, N. Lecanda, U. Castro and A. Laresgoiti, *J. Power Sources*, 2005, **151**, 103.
45. Z. G. Yang, G. G. Xia, G. D. Maupin, Z. M. Nie, X. S. Li, J. Templeton, J. W. Stevenson, P. Singh, Advanced Interconnect and Interconnect/Electrode Interfaces Development at PNNL, *8th Annual SECA Workshop and Peer Review*, San Antonio, TX, August 6-9, 2007, USA.
46. Z. Yang, M. S. Walker, P. Singh, J. W. Stevenson and T. Norby, *J. Electrochem. Soc.*, 2004, **151**, B669.
47. Z. Yang, G. Xia and J. Stevenson, *J. Power Sources*, 2006, **160**, 1104.
48. L. Chen, Z. Yang, B. Jha, G. Xia and J. Stevenson, *J. Power Sources*, 2005, **152**, 40.
49. K. A. Nielsen, A. R. Dinesen, L. Korcakova, L. Mikkelsen, P. V. Hendriksen and F. W. Poulsen, *Fuel Cells*, 2006, **6**, 100.
50. P. Jablonski and D. Alman, *Int. J. Hydrogen Energy*, 2007, **32**, 3705.
51. B. C. Church, T. H. Sanders, R. F. Speyer and J. K. Cochran, *Mater. Sci. Eng., A*, 2007, **452–453**, 334.
52. D. J. Young, *High Temperature Oxidation and Corrosion of Metals*, Elsevier, 2008, 17.
53. P. Jian, L. Jian, H. Bing and G. Xie, *J. Power Sources*, 2006, **158**, 354.
54. X. Deng, P. Wei, M. Bateni and A. Petric, *J. Power Sources*, 2006, **160**, 1225.
55. H. Yokokawa, T. Horita, N. Sakai, K. Yamaji, M. E. Brito, Y.-P. Xiong and H. Kishimoto, *Solid State Ionics*, 2006, **177**, 3193.
56. Y. Matsuzakiz and I. Yasuda, *J. Electrochem. Soc.*, 2001, **148**, A126.
57. Y. Matsuzaki and I. Yasuda, *Solid State Ionics*, 2000, **132**, 271.
58. J. Fergus, *Int. J. Hydrogen Energy*, 2007, **32**, 3664.
59. D. Liu, J. Almer, T. Cruses and M. Krumpelt, Chromite Doped with Strontium and Cobalt for SOFC Interconnect Applications, *ACerS 2007*, Coco beach, FL, January 22–26, 2007, USA.
60. N. Shaigan and W. Qu, *J. Power Sources*, 2010, **195**, 1529.

61. S. PalDey and S. Deevi, *Mater. Sci. Eng., A*, 2003, **324**, 58.
62. L. Pederson, P. Singh and X. Zhou, *Vacuum*, 2006, **80**, 1066.
63. P. Gannon, C. Tripp and A. Knospe, *Surf. Coat. Technol.*, 2004, **188–189**, 55.
64. A. Kayani, R. Smith and S. Teintze, *Surf. Coat. Technol*, 2006, **201**, 168.
65. X. Liu, C. Johnson, C. Li, J. Xu and C. Cross, *Int. J. Hydrogen Energy*, 2008, **33**, 189.
66. J. Wu, C. Li, C. Johnson and X. Liu, *J. Power Sources*, 2008, **175**, 833.
67. F. Gaillard, J. Joly, A. Boréave, P. Vernoux and J. Deloume, *Appl. Surf. Sci.*, 2007, **253**, 5876.
68. S. Primdahla, J. R. Hansena, L. Grahl-Madsenb and P. H. Larsen, *J. Electrochem. Soc.*, 2001, **148**, A74.
69. N. Orlovskaya, A. Coratolo, C. Johnson and R. Gemmen, *J. Am. Ceram. Soc.*, 2004, **87**, 1981.
70. J. Zhu, Y. Zhang, A. Basu, Z. Lu, M. Paranthaman, D. Lee and E. Payzant, *Surf. Coat. Technol.*, 2004, **177–178**, 65.
71. J. Kim, R. Song and S. Hyun, *Solid State Ionics*, 2004, **174**, 185.
72. Y. Zhen, S. Jiang, S. Zhang and V. Tan, *J. Eur. Ceram. Soc.*, 2006, **26**, 3253.
73. J. Choi, J. Lee, D. Park, B. Hahn, W. Yoon and H. Lin, *J. Am. Ceram. Soc.*, 2007, **90**, 1926.
74. C. Johnson, J. Wu, X. Liu, R. Gemmen, Solid oxide fuel cell performance using metallic interconnects coated by electroplating methods, *ASME Fuel Cell Conference*, New York, 2007, USA.
75. Y. Larring and T. Norby, *J. Electrochem. Soc.*, 2000, **147**, 3251.
76. X. P. Hou, C. Jacobson, S. Visco and L. De Jonghe, *Solid State Ionics*, 2005, **176**, 425.
77. Z. Yang, G. Xia, S. Simner and J. Stevenson, *J. Electrochem. Soc.*, 2005, **152**, A1896.
78. A. Petric and H Ling, *J. Am. Ceram. Soc.*, 2007, **90**, 1515.
79. J. Stevenson, Y. Chou, O. Marina, S. Simner, K. Weil, Z. Yang, P. Singh, SECA Core Technology Program: Materials Development at PNNL, *SECA Core Technology Program Review Meeting*, Lakewood, CO, October 25, 2005, USA.
80. X. Chen, P. Y. Hou, C. P. Jacobson, S. J. Visco and L. C. DeJonghe, *Solid State Ionics*, 2005, **176**, 425.
81. V. I. Gorokhovsky, P. E. Gannon, M. C. Deibert, R. J. Smith, A. Kayani, M. Kopczyk, D. VanVorous, Z. Yang, J. W. Stevenson, S. Visco, C. Jacobson, H. Kurokawa and S. W. Sofie, *J. Electrochem. Soc.*, 2006, **153**, A1886.
82. W. Wei, W. Chen and D. Ivey, *Chem. Mater.*, 2007, **19**, 2816.
83. W. Wei, W. Chen and D. Ivey, *J. Phys. Chem. C*, 2007, **111**, 10398.
84. M. Bateni, P. Wei, X. Deng and A. Petric, *Surf. Coat. Technol.*, 2007, **201**, 4677.
85. J. Wu, Y. Jiang, C. Johnson and X. Liu, *J. Power Sources*, 2008, **177**, 376.

86. J. Wu, C. Johnson, Y. Jiang, R. Gemmen and X. Liu, *Electrochim. Acta*, 2008 **54**, 793.

87. J. Wu, C. Johnson, R. Gemmen and X. Liu, *J. Power Sources*, 2009, **189**, 1106.

88. J. Wu, R. Gemmen, A. Manivannan and X. Liu, *Int. J. Hydrogen Energy*, 2011, **36**, 4525.

89. H. Zhang, Z. Zhan and X. Liu, *J. Power Sources*, 2011, **196**, 8041.

90. A. Balland, P. Gannon, M. Deibert, S. Chevalier, G. Caboche and S. Fontana, *Surf. Coat. Technol.*, 2009, **203**, 3291.

91. http://www.nextechmaterials.com/energy/.

92. http://www.ipmd.net/articles/000982.html.

CHAPTER 6

Nano-structured Electrodes of Solid Oxide Fuel Cells by Infiltration

SAN PING JIANG

Fuels and Energy Technology Institute & Department of Chemical
Engineering, Curtin University, Perth, WA 6102, Australia
Email: s.jiang@curtin.edu.au

6.1 Introduction

Fuel cells are electrochemical reactors in which the chemical energy of fuels
such as hydrogen, methane, methanol, *etc.* is converted directly into electricity
with an inherently much higher efficiency, as compared with other existing
energy conversion technologies such as internal combustion engines (ICE).
Among various types of fuel cells, the high temperature solid oxide fuel cell
(SOFC) is the most efficient for the direct conversion of chemical energy of fuels
into electricity. SOFCs operate at a temperature range of 600–1000 °C, promote
rapid kinetics with non-precious materials and offer high fuel flexibilities.

A typical SOFC consists of a dense layer of an oxygen ion conducting
electrolyte separating a cathode, on which oxygen molecules react with
electrons to produce oxygen anions, from an anode on which the fuel is
oxidized. The most common materials used in SOFCs are Y_2O_3-stabilized ZrO_2
(YSZ) electrolyte, $(La,Sr)MnO_3$ perovskite oxide cathode and Ni/YSZ cermet
anode. In practice, the actual voltage of a fuel cell is less than the theoretical
Nernst value due to the irreversible losses associated with transport of the ions

RSC Energy and Environment Series No. 7
Solid Oxide Fuel Cells: From Materials to System Modeling
Edited by Meng Ni and Tim S. Zhao
© The Royal Society of Chemistry 2013
Published by the Royal Society of Chemistry, www.rsc.org

through the electrolyte, finite rates of the electrode reactions and mass transport losses of gases in the porous electrodes. The electrolyte losses can be minimized by using thin YSZ electrolyte layers or alternative electrolyte materials such as doped ceria and $LaGaO_3$.[1-3] It is well known to the SOFC community that losses associated with the electrode processes can be reduced through careful engineering of the electrode microstructure to optimize the three phase boundaries (TPBs) where electrode, electrolyte and gas phases meet. A well-adopted strategy is to develop new cathode and anode materials with high electrochemical activity and stability for the electrode reaction and good tolerance towards impurities such as chromium in the case of metallic interconnect and carbon and sulphur in the case of hydrocarbon fuels. Despite the enormous efforts in the development of new electrode materials for SOFCs, challenges still remain.[3,4]

The nanoscale and nano-structured electrode approach by impregnation or infiltration method attracts increasing attention as the viable alternative for the development of new electrodes for SOFCs. The infiltration method is a two-step process, which effectively separates the formation temperature of the ionic and catalytic active phase from the high sintering temperature as required to establish the intimate electrode/electrolyte interfacial bonding. Due to the relatively low sintering temperature for the catalytic active phases, nano-sized particles can be formed on structurally stable and compatible scaffolds such as LSM, Ni/YSZ, YSZ, doped ceria, *etc.* The infiltration method opens a new horizon in electrode development as the technique expands the selection of variable electrode materials combinations with the minimized TEC mismatch, reduced chemical reactions between electrode and electrolyte materials, and formation of nanosized ionic and catalytic active phases. There has been significant progress in the development of nano-structured electrodes in SOFCs in last five to 10 years, as indicated by the number of review articles on this topic.[5-9] This chapter will start with a brief discussion of the infiltration process and its processing factors and the emphasis will be on the progress and achievements in the development of nano-structured cathodes and anodes for SOFCs in recent years.

6.2   Infiltration Process

6.2.1   The Technique

Due to the fact that high process temperatures are required to achieve good bonding and contact between electrode and electrolyte, *e.g.* 1000–1150 °C for LSM cathodes and 1300–1400 °C for Ni/YSZ cermet anodes,[10,11] the conventional approach to introduce nanoparticles in the green stage of the electrodes such as that commonly used in the development of Pt-based nano-structured electrocatalysts in proton exchange membrane fuel cells, PEMFCs, is not very successful for the high temperature SOFC electrodes. Thus the key consideration in the development of nano-structured SOFC electrodes is to

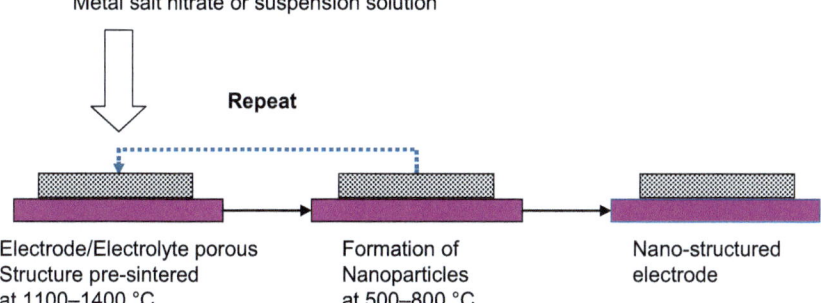

Figure 6.1 Typical process for the infiltration of metal salt nitrate solution or nanoparticle suspension into pre-sintered electrode or electrolyte porous structure/scaffold.

separate the ionic and electrocatalytic phase formation temperature from the high electrode/electrolyte interface bonding formation temperature. This can be achieved by the infiltration of catalytically and/or electrochemically active nanoparticles into a rigid and pre-fired electronic and/or ionic conducting electrode or electrolyte scaffold/framework. Figure 6.1 shows schematically a typical synthesis route for the deposition of nanosized particles into a pre-fired electrode or electrolyte scaffold or framework via infiltration of metal salt or nanoparticle suspension solutions. For example, wet infiltration of CeO_2 and Gd-doped CeO_2 (GDC) into pre-fired LSM cathode can be carried out by simply placing a drop of a $Ce(NO_3)_3$ or mixed $Gd(NO_3)_3$ and $Ce(NO_3)_3$ nitrate solution with stoichiometric composition $Gd_{0.2}Ce_{0.8}(NO_3)_x$ on top of the coating, which infiltrates the porous coating by capillary action. The infiltration process can be facilitated under vacuum.[12] Then the infiltrated LSM is heated at 500–800°C to decompose the metal salt solution, forming CeO_2 or GDC crystal phase. The phase formation temperature of the catalytic phase (*i.e.* CeO_2 or GDC) is much lower than the processing temperature of the standard LSM (*i.e.* ~1150 °C in this case). The significantly reduced phase formation temperature reduces the grain growth, resulting in the deposition of nanosized GDC particles on the porous surface of the LSM scaffold.

Infiltration technique is particularly useful to introduce low melting point phase into SOFC electrodes. For example, Ag has been introduced to the electrodes by ion impregnation/infiltration due to the low melting point (961 °C).[13,14] Infiltrated Ag nanoparticles have been shown to significantly improve the electrode performance of SOFC cathodes.[13–20] Gorte's group pioneered Cu/YSZ based anodes for direct utilization of hydrocarbon fuels in SOFCs via infiltration.[21,22] Unlike Ni, Cu does not catalyze the formation of carbon from dry hydrocarbons. However, low melting temperature of copper and copper oxides (both melt below 1150 °C) does not allow Cu to be incorporated in the anode structure using processes similar to that used for

Ni/YSZ anodes. Catalytic active phase such as ceria is generally infiltrated into Cu/YSZ anodes to promote the electrocatalytic activity of the anodes.[23]

Metallic nanoparticles such as Ni, Ag and Cu can also be infiltrated into porous electrode/electrolyte scaffold by electroless plating.[24–30] In the case of Ag plated $Ba_{0.5}Sr_{0.5}Co_{0.6}Fe_{0.4}O_3$ (BSCF) cathodes, plating bath consists of formaldehyde, silver nitrate, ammonia solution, ethanol and water.[31] Prior to the electroless plating, 5% silver nitrate solution is injected to the cathodes, followed by heat-treatment at 450 °C for ~2 min to activate the BSCF skeleton. Electroless Ag plating is then carried out in a vacuum environment (<20 kPa) for 5–180 s. The electroless plating reaction is considered as a combined result of two independent electrode reactions which are the reduction of metal ions and the oxidation of the reductant.[32] The electroless Ag plating takes place according to the following equation:

$$2Ag(NH_3)_2^+ + HCHO + 2OH^- = 2Ag + HCOO^- + NH_4^+ + 3NH_3 + H_2O$$

$$(1)$$

Reducing agent such as N_2H_4 and HCHO can also be added to reduce the Ag^+ to Ag after silver nitrate is infiltrated into porous matrix.[29,30] Figure 6.2 shows the SEM images of the BSCF cathodes with different loadings of electroless plated Ag.[31] A significant increase in the distribution and density of Ag nanoparticles is observed with the increase of Ag loading (*i.e.* the deposition time). The porous BSCF cathode is uniformly covered by a layer of Ag nanoparticles as the electroless plated Ag loading increased to 10 wt% (Figure 6.2c). In the case of copper plating on Ni/Sm-doped ceria (SDC)

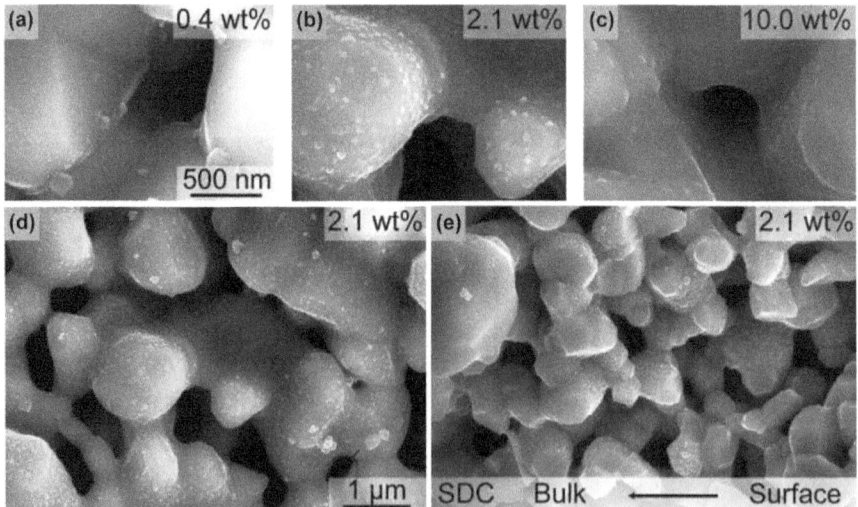

Figure 6.2 Top-view SEM images of (a) BSCF with 0.4 wt% Ag by infiltration; Ag plated BSCF with (b) 2.1 wt% and (c) 10 wt% of Ag; (d and e) Cross-section SEM images of the Ag plated BSCF cathodes with 2.1 wt% Ag.[31] Reproduced by permission from Hydrogen Energy Publications, LLC.

anodes, the anode substrates need to be reduced in hydrogen at high temperatures ($\sim 600\,^\circ$C) before the electroless plating.[28] The advantage of the electroless plating is its high efficiency of plating continuous metallic nano-particles in one plating process, in contrast to the repeated processes typically associated with the solution infiltration method. The limitation of the electroless plating technique is that porous scaffold must be electrical conducting and this has significantly limited applicability of the techniques in other non-conducting electrode systems.

The infiltrated catalytic nanoparticles can form discrete distribution or a thin and continuous network on the surface of the porous scaffold, as schematically shown in Figure 6.3. The porous scaffold can be electronic conducting electrode materials such as LSM, mixed ionic and electronic conductor such as (La,Sr)(Co,Fe)O$_3$ (LSCF) and BSCF, or ionic conducting electrolyte materials such as YSZ and doped ceria. The latter requires deposition of continuous and interconnected nanoparticles with high electronic conductivity as well as high electrocatalytic activity, and multiple infiltration steps are necessary to achieve sufficient electron conduction.[33,34] Samson *et al.* studied the electrical conductivity of La$_{0.6}$Sr$_{0.4}$Co$_{1.05}$O$_{3-\delta}$ (LSC) infiltrated GDC cathodes.[35] The measured in-plane electrical conductivity of LSC-GDC cathodes increases with an increasing number of infiltrations (see Figure 6.4). The increase in conductivity is most likely due to the increased connectivity among the infiltrated LSC nanoparticles as the LSC loading is increased. However, the need for multiple infiltrations to achieve the desired loading of electrocatalysts

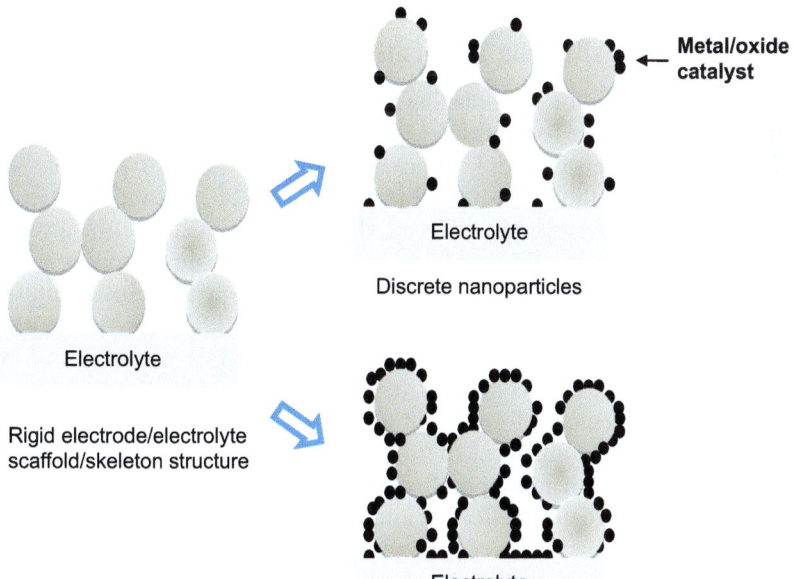

Figure 6.3 Scheme of the infiltrated nano-structured electrodes on pre-sintered porous electrode or electrolyte scaffold/skeleton.[6]
Reproduced by permission from Hydrogen Energy Publications, LLC.

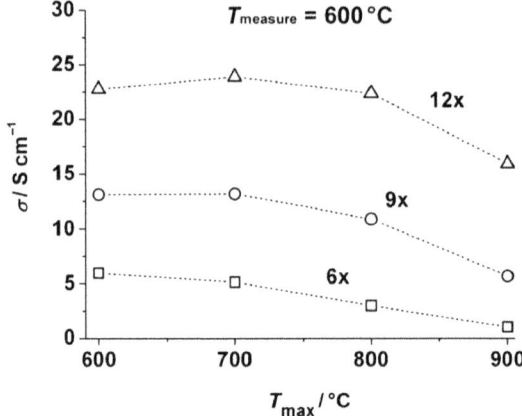

Figure 6.4   Dependence of the total conductivity of the LSC infiltrated GDC cathodes as a function of number of infiltrations and maximum firing temperature, Tmax. The in-plane conductivity of the symmetric cells was measured at 600 °C.[35]
Reproduced by permission from Elsevier.

(or connectivity of conducting phases) is time-consuming and costly, thus hindering the scale up of the infiltration process in the fabrication of SOFC electrodes.

6.2.2   Factors Affecting Infiltration Process and Microstructure

Adding surfactants or complexing agents is beneficial for the uniform distribution and phase formation of infiltrated nanoparticles. Addition of urea and polymeric dispersant can facilitate the formation of perovskite phase such as $Sm_{0.6}Sr_{0.4}CoO_3$ and LSM at low temperatures,[36,37] presumably due to the precipitation and complexing effect of the additives. Addition of glycine has also been shown to be effective to promote the uniform distribution of infiltrated yttria-stabilized bismuth (YSB) nanoparticles with reduced infiltration steps,[38] and to promote the formation of infiltrated LSM perovskite phase.[39] As shown in Figure 6.5, direct decomposition of nitrate precursors at 800 °C does not produce pure LSM perovskite phase and a pure LSM perovskite phase is formed in the presence of glycine in the nitrate solution.[39] It is suggested that glycine acts as a chelating agent to form metal ion complex in the solution, thus the oxides do not segregate upon firing of the metal nitrate precursor.[9]

Li *et al.* showed that the effect of adding urea is to enhance the distribution and electronic connectivity of Cu nanoparticles in the infiltrated Cu-YSZ composite anodes.[40] Nevertheless, the benefit of the additives appears to be related to the infiltrated nanoparticle phase and the scaffold. Adding citric acid to $AgNO_3$ solution has been shown to inhibit the grain growth of infiltrated Ag nanoparticles.[41] Nicholas and Barnett showed that adding additives such as surfactant Triton X-100 in $Sm_{0.5}Sr_{0.5}CoO_{3-x}$ (SSC) nitrite solution did not

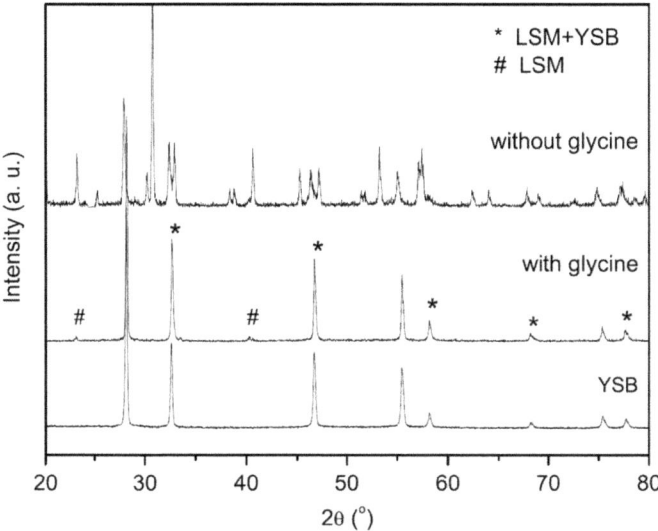

Figure 6.5 XRD patterns of pure YSB powder and YSB after infiltration with LSM nitrate precursor with and without the addition of glycine. The firing temperature of the LSM is 800 °C.[39]
Reproduced by permission from Hydrogen Energy Publications, LLC.

produce pure SSC phase while adding citric acid resulted in pure SSC phase.[42] Furthermore, addition of additives such as Triton X-100 and citric acid had little effect on the morphology or the performance of the infiltrated SSC-GDC composite cathodes.[42] On the other hand, using a concentrated LSM nitrate precursor solution with surfactant (*e.g.* Triton-X100), Sholklapper *et al.* showed that it may be possible to form a continuous and thin nano-structured LSM layer on YSZ scaffold with a single-step infiltration.[37,43,44]

As shown above, electrodeposition and electroless deposition have also been used to accelerate the infiltration process.[28,45] Distribution of infiltrated nanoparticles also depends on the wetting properties between the metal salt solution and the porous scaffold. Lou *et al.* shows that using SSC nitrate solution in a water/ethanol mixture significantly improves the wetting property between the infiltration solution and LSCF scaffold, resulting in the significantly enhanced SSC nanoparticle distribution and performance.[46]

Calcination temperature of infiltration steps affects the conductivity of infiltrated electrodes due to the grain growth and coarsening of infiltrated nanoparticles, which in turn leads to a loss of percolation.[47] Zhang *et al.*[48] studied the effect of firing temperature on the performance of Sm-doped ceria (SDC) infiltrated LSM electrode and found that there is a fine balance between the requirements of small SDC nanoparticle size and bonding between the particles. High firing temperature increases the SDC particle size but favours the strong bonding between the particles, the optimal firing temperature for SDC infiltrated LSM cathodes is 800 °C in this study. Another study on

$Y_{0.5}Bi_{1.5}O_3$ (YSB) infiltrated LSM cathodes shows that the optimal firing temperature of LSM scaffolds is related to the YSB loading.[49]

The microstructure such as porosity and surface area of the porous scaffold plays an important role in the microstructure development and optimization of infiltrated electrodes. Shah and Barnett studied the relationship between the performance of LSCF infiltrated GDC composite cathode and the sintering temperature of GDC scaffold.[50] Increasing the GDC scaffold sintering temperature from 1000 to 1100 °C effectively decreases area specific resistance (ASR, or electrode polarization resistance) due to the improved inter-connectivity between GDC particles, however, increasing the sintering temperature above 1300 °C substantially increases ASR, most likely due to the decreased porosity. The influence of size, connectivity and distribution of the pores of YSZ porous scaffolds on the three phase boundary and electrocatalytic activity of LSM infiltrated YSZ cathodes under identical total volume fraction of open pores has been studied in detail recently by Torabi et al.[51] The morphology of the porous scaffolds can be controlled to a certain degree by using various pore formers and calcination treatment and the results indicate that uniformity of the porous scaffold morphology with large specific area is essential for the infiltration of continuous and mono-layered LSM nano-particles, leading to the high electrochemical performance of infiltrated LSM-YSZ cathodes.

Organic or polymeric pore formers have been used to control the scaffold pore size and the results show the impact of the pore size on the ASR of the infiltrated LSC-YSZ cathodes[52] and infiltrated Cu-ceria-YSZ anodes.[53] Other methods to control the microstructure of scaffold are also effective. For example, Armstrong and Rich fabricated YSZ scaffold from NiO-YSZ composites, where the NiO was removed by reduction and subsequent acid leaching.[54] The advantage of this approach is to allow high sintering temperature and at the same time to retain the high surface area of the scaffold. Recent results by Küngas et al. showed that treatment of YSZ scaffold with hydrofluoric acid (HF) can significantly increase the surface of YSZ scaffold and improve the performance of the infiltrated LSF cathodes.[55]

6.3   Nano-structured Electrodes

6.3.1   Performance Promotion Factor

The enhancement in the electrochemical performance of nano-structured electrodes is truly remarkable. For instance, in the case of LSM cathodes, the electrode polarization (interface) resistance (R_E) is 11.7 Ω cm^2 at 700 °C. With the infiltration of 5.8 mg cm^{-2} GDC nanoparticles, R_E is reduced dramatically to 0.21 Ω cm^2 at 700 °C, which is 56 times smaller than that of the pure LSM cathode.[56] R_E for the O_2 reduction reaction on the nano-structured Pd+YSZ is 0.11 Ωcm^2 at 750 °C and 0.22 Ωcm^2 at 700 °C, and this is significantly lower than that of the LSM (9 to 54 Ωcm^2 at 700 °C[57,58]), LSM/YSZ (2.5 Ωcm^2 at 700 °C[59]) and LSM/GDC (1.1 Ωcm^2 at 700 °C[57]) composite cathodes.

To represent the performance enhancement of nano-structured electrode, a promotion factor, f_P, is introduced. f_p is defined as ratio of the electrode polarization resistance, R_E or the overpotential, η of the nano-structured electrodes or power density (P_w) of the cell with nano-structured electrodes to the performance of the baseline electrodes fabricated by conventional mixing and sintering processes, measured under identical conditions. For promotion effect, f_p is greater than one.

$$f_p = R_{E,\text{conventional electrode}} / R_{E,\text{infiltrated electrode}}$$

or

$$f_p = \eta_{\text{conventional electrode}} / \eta_{\text{infiltrated electrode}}$$

or

$$f_p = P_{w,\text{cell with infiltrated electrode}} / P_{w,\text{cell with conventional electrode}} \qquad (2)$$

Table 6.1 lists the performance and promotion factors of various nano-structured electrodes systems.[12,14,33,49,50,54,60–75] Infiltrated nano-structured electrodes have also been applied to metal-supported SOFCs and single-chamber SOFCs.[76–78] As shown in Table 6.1, the promotion factor varies significantly for nano-structured electrodes, 2.3 to 78 for the O_2 reduction reactions, 1.6 to 25 for the H_2 oxidation reaction and 4.6 to 26 for the CH_4 oxidation reaction. The wide variations of the promotion factors are most likely due to the significant differences in the infiltration process, microstructure, electronic and ionic properties of the scaffold, and to the distribution, loading and catalytic activities of the infiltrated nanoparticles.

6.3.2 Nano-structured Cathodes

A SOFC cathode is the material where pure oxygen or oxygen from air is reduced to oxygen ions through the combination of electrons externally from the cell, and the oxygen reduction reaction requires the presence of oxygen and electrons as well as the transportation of generated oxygen ions from the reaction sites to the bulk of the electrolyte. Thus, the materials used as the SOFC cathodes should have high electronic conductivity and oxygen ionic conductivity, and should also meet the requirement of chemical and thermal stability in oxidizing environment, chemical and thermal compatibility with the electrolyte, and high catalytic activity towards oxygen reduction reactions at intermediate temperatures.

LSM is the most common cathode materials for SOFC because of its high electrochemical activity for the O_2 reduction reaction at high temperatures, good thermal stability, good chemical stability and compatibility with the most commonly used YSZ electrolyte, and the demonstrated structural stability under fuel cell operation conditions.[79,80] LSM is an excellent electronic conductor (its electronic conductivity is $\sim 200\,\text{S cm}^{-1}$ at $800\,^\circ\text{C}$),[81] but with a negligible oxygen

Table 6.1 Performance and promotion factors of various nano-structured cathodes and anodes.

Impregnated nanoparticles	Scaffold/skeleton	Performance	Promotion factor, f_P	Reference
Cathodes				
GDC (5.8 mg cm^{-2})	LSM	$R_E = 0.21$ Ω cm^2 @ 700 °C	56 for O$_2$ reduction	60
Pd (1.8 mg cm^{-2})	LSM/YSZ	$R_E = 0.9$ Ω cm^2 @ 600 °C	78 for O$_2$ reduction	71
Pd (1.2 mg cm^{-2})	LSCF	$R_E = 2.9$ Ω cm^2 @ 600 °C	1.9 for O$_2$ reduction	69
GDC (1.5 mg cm^{-2})	LSCF	$R_E = 1.6$ Ω cm^2 @ 600 °C	3.4 for O$_2$ reduction	69
LSM ($\sim$2 mg cm^{-2})	YSZ	$R_E = 1.6$ Ω cm^2 @ 600 °C	44 for O$_2$ reduction[a]	71
LSCF (1.1 mg cm^{-2})	YSZ	$R_E = 0.54$ Ω cm^2 @ 600 °C		65
LSCF (12.5 vol%)	GDC	$R_E = 0.25$ Ω cm^2 @ 600 °C	14 for O$_2$ reduction	50
La$_{0.6}$Sr$_{0.4}$CoO$_3$ (30 vol%)	YSZ	$P = 2.1$ W cm^{-2} @ 800 °C in H$_2$/air		54
La$_{0.6}$Sr$_{0.4}$CoO$_3$ (55 wt%)	SDC	$R_E = 0.36$ Ω cm^2 @ 600 °C		72
Ag	LSCF/GDC	$P = 0.98$ W cm^{-2} @ 600 °C	3.3 for H$_2$/air	14
Pd (1.4 mg cm^{-2})	YSZ	$R_E = 0.22$ Ω cm^2 @ 700 °C		33
BSCF (1.8 mg cm^{-2})	LSM	$R_E = 1.3$ Ω cm^2 @ 700 °C	12 for O$_2$ reduction	68
Sm$_{0.6}$Sr$_{0.4}$CoO$_3$	LSM/YSZ	$R_E = 3.3$ Ω cm^2 @ 600 °C	3.0 for O$_2$ reduction	36
Y$_{0.5}$Bi$_{1.5}$O$_3$ (50 wt%)	LSM	$R_E = 0.14$ Ω cm^2 @ 700 °C		49, 74
SDC (3.3 mg cm^{-2})	GdBaCo$_2$O$_{5+\delta}$	$R_E = 0.1$ Ω cm^2 @ 600 °C	4.3 for O$_2$ reduction	75
Anodes				
Sm$_{0.2}$Ce$_{0.8}$O$_2$ ($\sim$4 mg cm^{-2})	Ni/YSZ	$R_E = 0.24$ Ω cm^2 @ 800 °C	7.3 for H$_2$ oxidation	12
GDC (4 mg cm^{-2})	(La$_{0.75}$Sr$_{0.25}$)(Cr$_{0.5}$Mn$_{0.5}$)O$_3$	$R_E = 0.44$ Ω cm^2 @ 800 °C	26 for CH$_4$ oxidation	62
GDC (4 mg cm^{-2})	(La$_{0.75}$Sr$_{0.25}$)(Cr$_{0.5}$Mn$_{0.5}$)O$_3$	$R_E = 0.12$ Ω cm^2 @ 800 °C	20 for H$_2$ oxidation	62
GDC (1.42 mg cm^{-2})	Ni	$R_E = 1.29$ Ω cm^2 @ 800 °C	25 for CH$_4$ oxidation	63
Pd (0.36-0.46 mg cm^{-2})	(La$_{0.75}$Sr$_{0.25}$)(Cr$_{0.5}$Mn$_{0.5}$)O$_3$/YSZ	$R_E = 0.88$ Ω cm^2 @ 800 °C	1.5 for H$_2$ oxidation	64
Pd (0.36-0.46 mg cm^{-2})	(La$_{0.75}$Sr$_{0.25}$)(Cr$_{0.5}$Mn$_{0.5}$)O$_3$/YSZ	$R_E = 2.0$ Ω cm^2 @ 800 °C	4.6 for CH$_4$ oxidation	64, 66
Pd (0.36-0.46 mg cm^{-2})	(La$_{0.75}$Sr$_{0.25}$)(Cr$_{0.5}$Mn$_{0.5}$)O$_3$/YSZ	$P = 0.111$ W cm^{-2} @ 800 °C	8 for C$_2$H$_5$OH/air	64
Pd (0.06 mg cm^{-2})	(La$_{0.7}$Ca$_{0.3}$)(Cr$_{0.5}$Mn$_{0.5}$)O$_3$/GDC	$R_E = 1.1$ Ω cm^2 @ 750 °C	6.5 for CH$_4$ oxidation	67
Pd (0.11 mg cm^{-2})	Ni/GDC	$R_E = 0.6$ Ω cm^2 @ 700 °C	$\sim$5 for H$_2$ oxidation	70
Pd (5 wt%)+CeO$_2$ (5 wt%)	(La$_{0.75}$Sr$_{0.25}$)(Cr$_{0.5}$Mn$_{0.5}$)O$_3$/YSZ	$P = 0.52$ W cm^{-2} @ 700 °C	$\sim$5 for H$_2$/air	73

[a] Calculated based on performance for LSM/YSZ composite cathodes.
[b] Measured after polarized at 1000 mA cm^{-2}, 800 °C for 22 h.

ionic conductivity (*e.g.* $\sim 10^{-16}$ cm^{2}s^{-1} at 700°C).[82] A number of other oxide materials have also been investigated as cathodes of SOFCs. Examples are (La,Sr)(Co,Fe)O$_3$,[83] Ba$_{0.5}$Sr$_{0.5}$Co$_{0.8}$Fe$_{0.2}$O$_3$ (BSCF),[84] lanthanum nickelate, La$_2$NiO$_{4+\delta}$ (LN),[85] La(Ni, M)O$_3$ (M = Al, Cr, Mn, Fe, Co, Ga) perovskites,[86] lanthanum strontium manganese chromite, La$_{0.75}$Sr$_{0.25}$Cr$_{0.5}$Mn$_{0.5}$O$_3$,[87] and GdBaCoO$_5$.[88] The oxygen self-diffusion coefficient of cobaltite based materials is several orders of magnitude higher than that of the manganites.[82] However, their long-term stability under SOFC operating conditions is generally unproven. Lanthanum cobalt ferrite or nickelate perovskites are in general chemically incompatible with YSZ electrolyte and they readily react with YSZ at the sintering temperatures required for the fabrication of standard SOFC electrodes.[89–91]

The infiltration process offers many advantages to the development of advanced cathodes. Since the catalytic active components are to be infiltrated in a post-firing step, the infiltrated phases can be formed at a much lower temperature. Thus, a broad range of materials can be considered including those which are highly active but prone to chemical reactions with YSZ electrolyte at conventional high electrode sintering temperatures. In this section the performance and microstructure of electronic, ionic and mixed electronic and ionic conducting scaffolds based infiltrated cathode systems will be described.

6.3.2.1 Electronic Conducting Scaffolds Based Cathode

The LSM and LSM/YSZ composite materials have been extensively studied as electronic conducting scaffolds based infiltrated cathodes.[56,60,61,92,93] In this case, the scaffolds provide the continuous electronic conductivity as well as electrocatalytic activity. We studied in detail the fabrication and performance of the GDC infiltrated LSM cathodes.[56,60,61] LSM scaffolds are generally fired at 1150 °C and GDC nitrate precursor solution is infiltrated into the LSM scaffolds by a capillary force. The electrode is then fired at 850 °C to decompose nitrates, forming GDC-type nanoparticles. At low GDC loading, the distribution of the infiltrated GDC nanoparticles is discrete and does not form a continuous network. The electrode performance of GDC infiltrated LSM cathodes increases with the GDC loading. For a 5.8 mg cm^{-2} GDC infiltrated LSM cathode, the R_E of the electrode for the O$_2$ reduction reaction is substantially lower than that of LSM and LSM/YSZ composite cathodes and is comparable with that of mixed ionic and electronic conducting oxide electrodes including LSCF and Gd$_{0.8}$Sr$_{0.2}$CoO$_3$.[60] Xia *et al.* studied the yttria-stabilized bismuth oxide (YSB) infiltrated LSM cathodes and infiltrated YSB-LSM exhibits very low electrode polarization resistance, ~ 0.14 Ωcm^2 at 700 °C.[94,95] The significant performance enhancement is most likely due to the high oxygen ionic conductivity of infiltrated YSB phase.

In addition to the infiltration of ionic conducting phases such as GDC and YSZ, infiltration of catalytic active phases such as Pd, Co$_3$O$_4$ has been found to be very effective to enhance the activity of the LSM based cathodes for the O$_2$ reduction reaction.[71,92,96] Co-infiltration of ceria and Co$_3$O$_4$ substantially enhanced the cell performance and suppressed the grain growth and

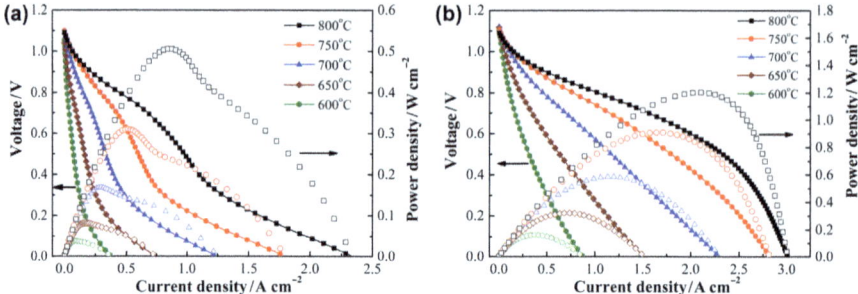

Figure 6.6   Performance of the Ni/YSZ anode-supported YSZ electrolyte film cells
with (a) pure LSM cathode and (b) 1.1 mg cm^{-2} BSCF-infiltrated LSM
composite cathode at different temperatures in H$_2$/air. H$_2$ flow rate:
200 mL min^{-1}; oxidant: stationary air.[68]
Reproduced by permission from The Electrochemical Society.

agglomeration of Co$_3$O$_4$ nanoparticles, achieving a power density of 0.58 W cm^{-2}
at 0.7 V and 700 °C on an anode-supported SOFC with ceria and Co$_3$O$_4$ co-
infiltrated LSM/YSZ composite cathode.[92] Infiltration of Pd nanoparticle
dramatically increased the power output of SOFCs, the cell with Pd infiltrated
LSM/YSZ composite cathode achieved a maximum power density of 1.42 W cm^{-2}
at 750 °C, seven times higher than 0.2 W cm^{-2} measured on the cell with
conventional LSM/YSZ composite cathodes under identical test conditions.[97]
Infiltrated Pd phase plays a significant role in the reduction of the electrode
polarization resistance of LSM based cathodes for the O$_2$ reduction reaction.
   BSCF, which is chemically reactive with YSZ at high electrode sintering
temperatures, has been incorporated into LSM cathode by infiltration.[68] In
such nano-structured BSCF-infiltrated LSM composite cathode, the
uniformly distributed BSCF nanoparticles significantly enhance the electrochemical activity of the cathode, while the LSM scaffold provides an
effective electron transfer path, TEC matching and thermal stability with the
YSZ electrolyte. The interfacial reaction between the infiltrated BSCF and
YSZ is minimized due to the low phase formation temperature of BSCF.
Figure 6.6 shows the performance of anode-supported YSZ film cells with
pure LSM and 1.1 mg cm^{-2} BSCF-infiltrated LSM cathodes at different
temperatures under H$_2$/air.[68] The cell with thin YSZ electrolyte film and the
nano-structured BSCF-LSM cathode exhibits maximum power densities of
1.21 and 0.32 W cm^{-2} at 800 °C and 650 °C, respectively, substantially higher
than 0.51 and 0.08 W cm^{-2} for the cells with the pure LSM cathode under
identical conditions.

6.3.2.2   Ionic Conducting Scaffolds Based Cathodes

Use of the ionic conducting electrolyte materials based scaffolds can increase
the scaffold's structural stability, thus the stability of the infiltrated electrodes.
One distinctive advantage of ionic conducting scaffolds is that in general, ionic

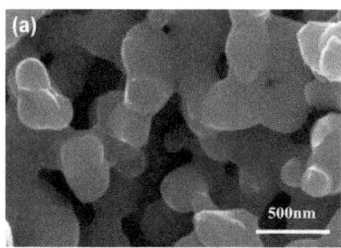

Figure 6.7 SEM micrographs of (a) conventional LSM/YSZ composite cathode with 50/50 weight ratio and (b) LSM infiltrated YSZ composite cathode. The infiltrated LSM was 2.0 mg cm^{-2}, equivalent to 25 wt%.[71]
Reproduced by permission from The Electrochemical Society.

conducting scaffold can be the same material as the electrolyte, thus, excellent bonding and perfect TEC match between the scaffold and electrolyte can be realized. The most common ionic conducting scaffold materials are YSZ and doped ceria such as GDC and SDC.

LSM infiltrated YSZ and doped ceria cathodes have been extensively studied.[37,38,43,71,98–100] Very high electrochemical performance of LSM infiltrated cathodes has been reported. For example, Jiang *et al.*[39] synthesized LSM infiltrated YSB cathodes and reported a low R_E of 0.15 Ω cm^2 at 700 °C for the O$_2$ reduction reaction. In the case of LSM infiltrated YSZ cathodes, R_E is 0.66 Ω cm^2 at 700 °C, significantly lower than 5.2 Ω cm^2 measured on conventional LSM/YSZ composite cathodes.[71] Figure 6.7 shows the SEM micrographs of conventional LSM/YSZ composite cathodes and LSM infiltrated YSZ composite cathodes.[71] The infiltrated nano-sized LSM particles are well dispersed on the surface of the porous YSZ scaffolds. The significantly high electrocatalytic performance of LSM infiltrated YSZ cathodes as compared with conventional LSM/YSZ composite cathodes is clearly due to the uniformly distributed and nano-sized infiltrated LSM nanoparticles on ionic conducting YSZ scaffolds.

Highly active cobaltite based pervoskite cathodes such as Sm$_{0.5}$Sr$_{0.5}$CoO$_{3-\delta}$ (SSC), La$_{0.6}$Sr$_{0.4}$CoO$_{3-\delta}$ (LSC), La$_{0.8}$Sr$_{0.2}$FeO$_3$ (LSF), and LSCF, and K$_2$NiF$_4$-type materials such as La$_{n+1}$Ni$_n$O$_{3n+1}$ are also developed with YSZ electrolytes via infiltration route using porous SDC or YSZ scaffolds.[34,50,52,101–104] Choi *et al.* infiltrated La$_{n+1}$Ni$_n$O$_{3n+1}$ (n = 1, 2, and 3) into porous YSZ scaffolds using citric acid as additive.[103] Single La$_{n+1}$Ni$_n$O$_{3n+1}$ (LN) phase was formed after calcination at 850 °C and SEM examination showed the formation of uniformly distributed LN nanoparticles in YSZ scaffolds (see Figure 6.8).[103] The cell with La$_{n+1}$Ni$_n$O$_{3n+1}$ (n = 3) infiltrated YSZ cathode produced a maximum power density of 0.889 W cm^{-2} at 700 °C.

6.3.2.3 Mixed Ionic and Electronic Conducting Scaffolds Based Cathodes

It has been shown that infiltration of ionic conducting and catalytic active phases to MIEC scaffolds such as LSCF, GdBaCo$_2$O$_{5+\delta}$ can further enhance

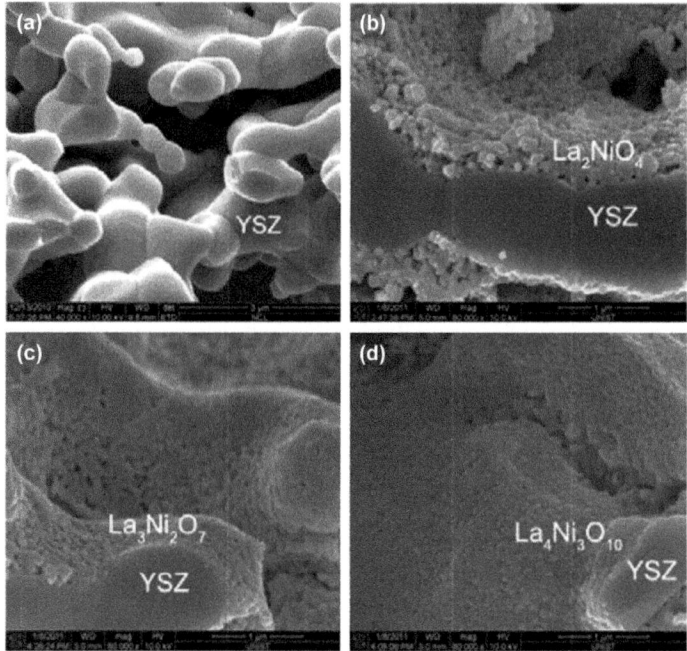

Figure 6.8 The cross-section view of microstructure with (a) YSZ scaffold and (b) La_2NiO_4-YSZ, (c) $La_3Ni_2O_7$-YSZ, (d) $La_4Ni_3O_{10}$-YSZ composites with infiltrated $La_{n+1}Ni_nO_{3n+1}$ (45 wt%) calcined at 850 °C.[103]
Reproduced by permission from The Electrochemical Society.

the electrochemical performance of the cathodes.[41,69,75,105–107] However, the enhancement of the infiltration to the performance of MIEC scaffolds based cathodes is generally smaller as compared to that observed on the performance enhancement of electronic conducting scaffold based cathodes. For example, the performance promotion factor, f_p is 1.9 and 3.4 for 1.2 mg cm^{-2} Pd infiltrated LSCF and 1.5 mg cm^{-2} GDC infiltrated LSCF measured at 600 °C, respectively.[69] In the case of 1.8 mg cm^{-2} Pd infiltrated LSM/YSZ composite cathodes, f_p is 78 for the O_2 reduction reaction at 600 °C.[71]

Infiltration technique has been used to alter and enhance the specific properties of MIEC cathodes. For example, $Ba_{0.5}Sr_{0.5}Co_{0.8}Fe_{0.2}O_3$ (BSCF) has been shown to be very active as cathodes of SOFCs, achieving the power densities of 1.01 W cm^{-2} and 402 mW cm^{-2} at 600 °C and 500 °C, respectively, when operated with hydrogen as the fuel and air as the oxidant.[84] However, BSCF is not stable and deteriorates in the presence of CO_2.[108] To protect BSCF electrodes from CO_2 poisoning, Zhou *et al.* demonstrated a novel method to fabricate a CO_2-protective shell on BSCF cathode using infiltration and microwave plasma treatment.[109] The BSCF cathode with hierarchical $La_2NiO_{4+\delta}$ (LN) shell shows a very stable performance in 10 vol% CO_2-containing air at 600 °C and high electrocatalytic activity for the oxygen reduction reaction, achieving ASR of 0.13 Ω cm^2 at 575 °C.

Another example of the MIEC scaffold based infiltrated cathodes is demonstrated by Liu's group on a well-designed study of the LSCF model electrode infiltrated with a thin layer of LSM.[110] They demonstrated that efficient electrode architecture is critical for the catalytic role of the infiltrated nanoparticles or films, see Figure 6.9.[110] The ASR of the LSM-coated LSCF cathode is larger at OCV but smaller under the cathodic bias. In contrast with the continuous degradation of the cell with LSCF model cathodes, the cell with LSM-infiltrated LSCF cathodes show a time-dependent activation phenomena. The results show that the infiltrated LSM film plays an important catalytic role in the promotion of the surface oxygen adsorption and/or dissociation and in the polarization-induced activation process. DFT calculation showed that adsorption energy is lower on LSM than that on LSC and LSF.[111] Evidently, such catalytic role of the infiltrated LSM nanoparticles would only be effective for the thin infiltrated LSM layer. Further studies show that infiltration of thin $La_{0.4875}Ca_{0.0125}Ce_{0.5}O_{2-\delta}$ (LCC) layer reduced the electrode polarization resistance to $0.076\,\Omega\,cm^2$ at $750\,°C$, significantly lower than $0.112\,\Omega\,cm^2$ measured on blank LSCF cathodes.[112] LCC surface modification also

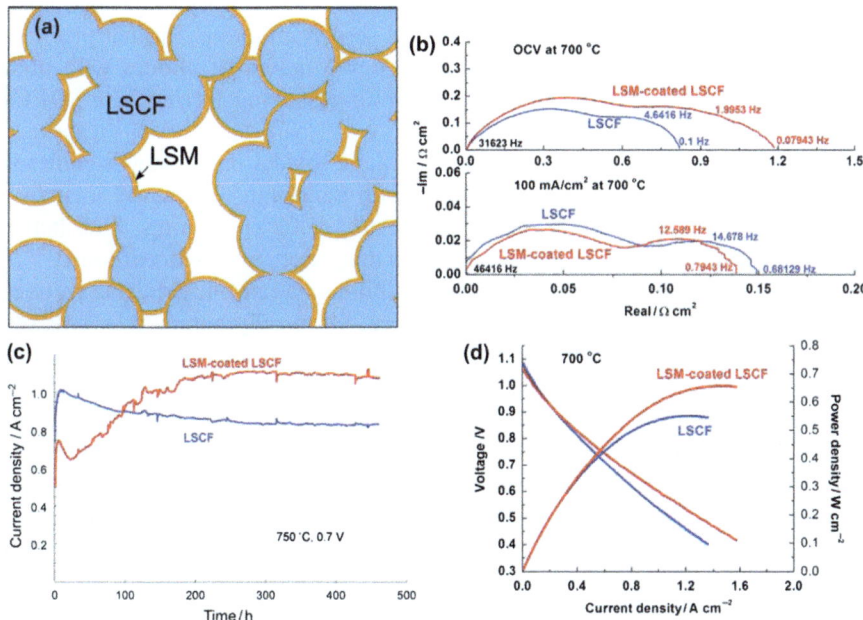

Figure 6.9 (a) Schematic diagram of the LSM-infiltrated LSCF cathode. (b) Impedance spectra of fuel cells with and without infiltration of LSM measured at OCV and at $100\,mA\,cm^{-2}$, without Ohmic portion. (c) Current density of two test cells with and without infiltration as a function of time under a constant voltage of $0.7\,V$ and approximate cathodic overpotential of $-0.12\,V$. (d) Cell voltages and power densities as a function of current density for full cells with and without infiltration of LSM after long term testing.[110]
Reproduced by permission from The Royal Society of Chemistry.

increased the stability of LSCF cathodes. The promotion effect of LCC with relatively low ionic conductivity is suggested to be on the surface kinetics of the O_2 reduction reaction.[112]

6.3.3 Nano-structured Anodes

The anode materials of SOFCs must have high electronic conductivity and oxygen ion conductivity, high activity for the oxidation reaction of fuels such as hydrogen, methane and light hydrocarbons as well as redox stability under oxidation environment.

The state-of-the-art anode materials for the H_2 oxidation reaction are the Ni/YSZ cermets due to their high electronic conductivity, high electrochemical activity for the H_2 oxidation, high structural stability and compatibility with YSZ electrolyte.[113] Ni/YSZ based composite anodes have been highly optimized and exhibit high performance for some fuels, especially H_2. They are not stable, however, when exposed to hydrocarbon fuels and have low tolerance for impurities such as sulfur.[114] For hydrocarbon fuels such as natural gas, the most abundant impurity is sulfur, existing as gaseous hydrogen sulfide, H_2S, after reforming. Sulfur is also a major impurity in gasified coal. The carbon deposition/cracking on Ni/YSZ cermet anodes is another problem in direct use of hydrocarbon fuels. Thus, development of anodes with high tolerance towards carbon deposition and sulfur poisoning is critical for SOFCs utilizing hydrocarbon fuels.

Replacing the Ni with conducting ceramic oxides can offer significant advantages if comparable electrochemical performance can be achieved, including improved redox stability, better tolerance to sulfur poisoning, and better resistance to carbon deposition and coking in the presence of hydrocarbons.[115,116] However, ceramic oxides must be stable, exhibit high electronic conductivity over the wide range of partial pressure of oxygen, chemically compatible with the electrolyte and most importantly must be catalytically active for the oxidation of fuels. The oxides must also match thermally with the electrolyte to prevent the delamination and fracture of the electrode/electrolyte interface due to the reduction and oxidation environment of fuel cells. It is a challenge to find the right ceramic oxide replacements for Ni. For example, among the ceramic oxide based materials, (LaSr)(CrMn)O_3 (LSCM) appears to be the most promising. Tao and Irvine reported a maximum power density of $0.47 \, \text{W cm}^{-2}$ on a cell with LSCM anode and LSM cathode at $900 \, ^\circ\text{C}$,[115] which is considered to be compatible to the cell based on Ni/YSZ cermet anodes. However, the electronic conductivity of LSCM is very low, $\sim 29 \, \text{S cm}^{-1}$ in air and $\sim 0.22 \, \text{S cm}^{-1}$ in $10\% H_2/N_2$.[117] The problems of the solid state reaction, differences in the thermal expansion, low catalytic activity, and conductivity of ceramic oxide anodes can be addressed in part or fully by using infiltration methods.

6.3.3.1 *Cu-infiltrated Anodes*

Cu-infiltrated anodes were initially developed for the direct oxidation of CH_4.[118] Different to Ni, Cu is relatively inert for the formation of C–C bonds,

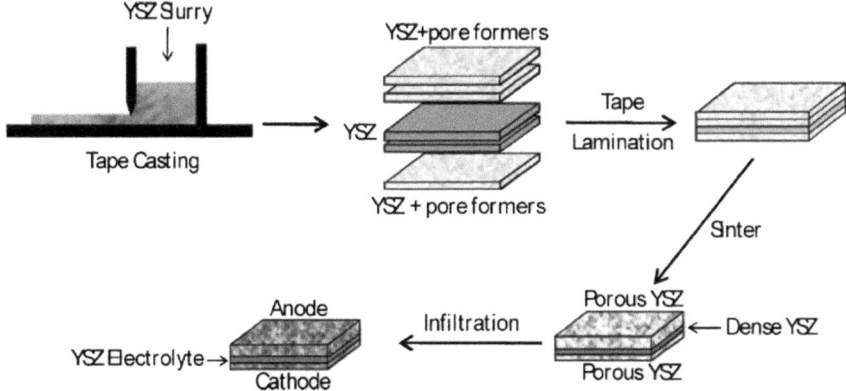

Figure 6.10 Diagram showing the steps used to fabricate an SOFC in which the electrodes are produced by infiltration of active components into a porous YSZ scaffold.[8]
Reproduced by permission from Wiley-VCH.

however, its melting temperature is 1083 °C, significantly lower than 1453 °C for Ni. Therefore, Cu-based cermet anodes cannot be produced using the same methods developed for Ni-based cermet anodes. The method developed by Prof. Gorte's group is to impregnate $Cu(NO_3)_2$ solution into porous YSZ scaffold prepared on a dense YSZ electrolyte layer, followed by calcinations to decompose the nitrate and form the oxide.[21,119] Both the dense electrolyte and porous scaffold can be fabricated by conventional tape-casting techniques, see Figure 6.10.[8] Other components such as ceria, perovskite oxides and precious metal catalysts can be added via infiltration processes to provide the ionic and electronic conductivity as well as the catalytic activity of Cu/YSZ anodes for the direct oxidation of hydrocarbons.[23,120–127] Similar to Cu-based anodes, silver with low melting temperature of 960 °C was also infiltrated into porous YSZ scaffold as potential anodes for hydrocarbon fuels, however, the activity of infiltrated Ag/YSZ composite anodes is not satisfactory, the area specific resistance, ASR is 38.4 Ωcm^2 in H_2 and 16.1 Ωcm^2 in CH_4 at 700 °C.[128]

Replacing Ni with carbon-inert Cu to form Cu/YSZ cermet anodes[129] and infiltration of porous YSZ scaffold with copper, nickel and cerium nitrate solution[121,130] were found to improve significantly the resistance towards carbon deposition in both methane and C_4H_{10}. The synthesis, characterization and performance of Cu-based anodes have been extensively reviewed elsewhere.[7,127]

6.3.3.2 Ni and Ni/YSZ Scaffolds Based Infiltrated Anodes

The promotion effect of infiltrated nanoparticles on the performance of Ni and Ni/YSZ cermet anodes is equally impressive. Infiltration of catalytic active nanoparticles can substantially enhance the electrocatalytic activity as well as the carbon and sulfur tolerance of Ni-based cermet anodes.[12,63,131–133]

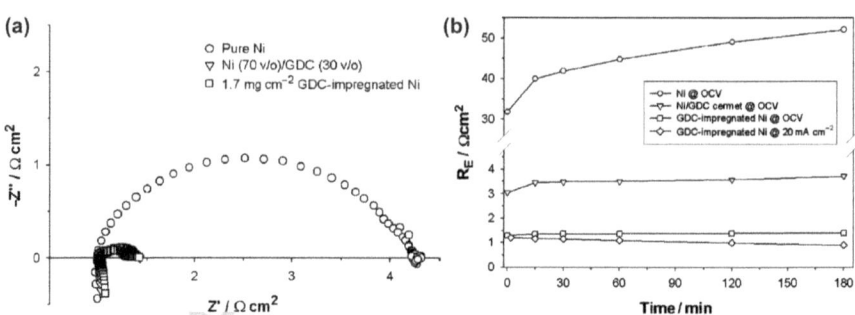

Figure 6.11   (a) Impedance curves of Ni, Ni/GDC anode and GDC infiltrated Ni
anodes in 97%H$_2$/3%H$_2$O at 800 °C and (b) RE of the anodes in
97%CH$_4$/3%H$_2$O at 800 °C.[63]
Reproduced by permission from Elsevier.

Early studies show that the infiltration of YSZ and SDC nanoparticles
significantly reduced the grain growth and agglomeration of Ni phase of the
Ni/YSZ cermet anodes during the high temperature sintering and reducing
stages of the anodes, leading to the enhancement of the electrochemical
performance of the anodes.[12] Wang *et al.* studied the electrode behaviour of Ni
anodes with and without GDC infiltration in wet methane (3% H$_2$O) at 800 °C
and found that infiltrated GDC is very effective to enhance the electrocatalytic
activity as well as the stability of the anodes, as shown in Figure 6.11.[63] R_E of
the 1.7 mg cm^{-2} GDC infiltrated Ni anode is 0.44 Ω cm^2, very close to that of
Ni/GDC cermet anodes, but significantly smaller than 3.3 Ω cm^2 measured on
Ni anode. Most important, GDC infiltrated Ni anode is very stable in wet
methane under open circuit conditions as compared to the Ni and Ni/GDC
anodes (Figure 6.11b), indicating the high carbon resistance of the GDC
infiltrated Ni anodes.

Kurokawa *et al.* added ceria nanoparticles to a conventional Ni/YSZ anodes
via a solution infiltration process and has found that the cell with ceria infiltration showed a significant increase in performance stability in H$_2$S-H$_2$ fuel
(see Figure 6.12).[132] For the cell with conventional Ni/YSZ anode, the cell
voltage drops to zero in 40 ppm H$_2$S-containing H$_2$ within several minutes,
while for cell with ceria infiltrated Ni/YSZ anode, the addition of 40 ppm H$_2$S
caused an initial cell voltage drop, but the performance was stable for over
500 h. Infiltration of Pd also increased the sulfur tolerance of Ni/GDC cermet
anodes.[131] The enhanced sulfur tolerance of infiltrated Pd may be related to the
fact that Pd nanoparticles supported on ceria is known as the sulfur tolerant
water gas shift reaction catalysts.[134]

6.3.3.3   *Ceramic Oxide Scaffolds Based Anodes*

Ceramic oxide-based materials generally have low electrical conductivity and in
some case low electrochemical activity for fuel oxidation reaction. For example,
LSCM-based electrodes show low electrocatalytic activity in methane.[135]

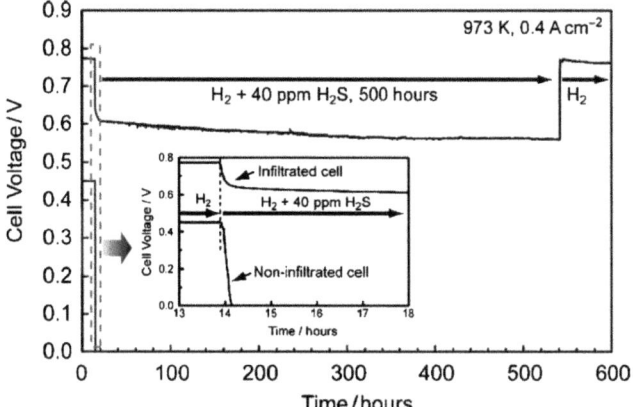

Figure 6.12 Cell voltage as a function of time for cells exposed to a fuel consisting of H$_2$ and H$_2$ plus 40 ppm H$_2$S at 700 °C. Ceria was infiltrated to the Ni/YSZ anode in one of the cells.[132]
Reproduced by permission from The Electrochemical Society.

Introduction of GDC or Pd nanoparticles by infiltration substantially enhances the activity of LSCM anodes for the electrochemical oxidation reaction in H$_2$, CH$_4$ and C$_2$H$_5$OH fuels.[62,64] Figure 6.13 shows the V-I curves and cell performance of LSCM, LSCM/YSZ, GDC- or Pd-infiltrated LSCM and LSCM/YSZ composite anodes. Pure LSCM and LSCM/YSZ composites have very poor electrocatalytic activities for the oxidation reaction of H$_2$, CH$_4$ and C$_2$H$_5$OH. With the infiltration of GDC or Pd nanoparticles, the electrocatalytic activity of the LSCM and LSCM/YSZ anodes is increased remarkably. GDC nanoparticles are equally effective for the promotion of the oxidation reaction of H$_2$ and CH$_4$ (Figure 6.13a and b). However, the effect of the infiltrated Pd nanoparticles is most pronounced for the reaction in methane and ethanol. The maximum power output is 24 mW cm^{-2} for the cell with a pure LSCM/YSZ composite anode in CH$_4$ and is almost doubled to 45 mW cm^{-2} with the infiltration of $\sim$0.4 mg cm^{-2} Pd nanoparticles (Figure 6.13c). For the ethanol oxidation reaction on a pure LSCM/YSZ anode, the cell power density is only 14 mW cm^{-2}. After the infiltration of $\sim$0.4 mg cm^{-2} Pd nanoparticles, the power density reaches 111 mW cm^{-2}, an 8-times increase in power output (Figure 6.13d).

Kim *et al.* studied the effect of infiltrated Pd on the (La$_{0.75}$Sr$_{0.25}$)(Cr$_{0.5}$Mn$_{0.5}$)O$_3$ (LSCM)/YSZ composite anodes for the H$_2$ oxidation reaction.[73] LSCM/YSZ composite anodes were prepared by infiltration of 45wt% LSCM into a 65% porous YSZ scaffold, followed by sintering at 1200 °C. For pure LSCM/YSZ composite anodes, the maximum power density was 105 mW cm^{-2} in H$_2$ at 700 °C and increased significantly to 500 mW cm^{-2} with infiltration of $\sim$0.5 wt% Pd (see Figure 6.14).[73] We also studied the effect of infiltrated Pd on the LSCM/YSZ composite anodes prepared by conventional method.[64] In the case of H$_2$, infiltration of $\sim$0.46 wt% Pd had no significant effect on the cell performance. The

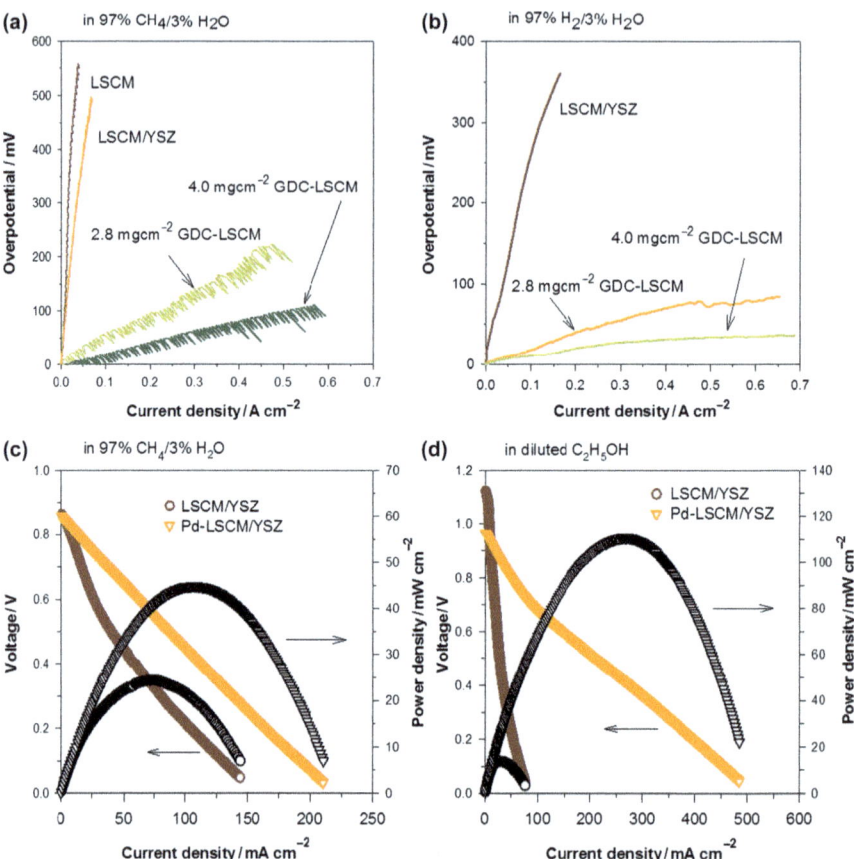

Figure 6.13 (a,b) Overpotential – current curves of LSCM, LSCM/YSZ and GDC-infiltrated LSCM anodes at 800 °C in wet CH_4 and H_2 and (c,d) performance of cells with and without $\sim 0.4\ mg\,cm^{-2}$ Pd infiltrated LSCM/YSZ anodes at 800 °C in CH_4 and C_2H_5OH. The YSZ electrolyte thickness was 1 mm.[62,64]
Reproduced by permission from The Electrochemical Society and Elsevier.

apparently significant differences in the promotion effect of infiltrated Pd on the LSCM/YSZ composite anodes for the H_2 oxidation may be related to the differences in the morphology of the scaffold of LSCM/YSZ composites prepared by infiltration and by conventional methods. As shown in Figure 6.14b, after reduction in H_2, the infiltrated LSCM film is broken into very small particles and is highly porous.[73] This may imply that Pd may be able to infiltrate to the interface between the YSZ and LSCM, promoting the electrochemical reaction at the interface regions. In the case of conventional LSCM/YSZ composites, infiltrated Pd intends to form isolated nanoparticles on the surface of the LSCM or YSZ grains. Thus, the morphology and microstructure of scaffold could also affect the distribution and thus the catalytic activity of the infiltrated nanoparticles.

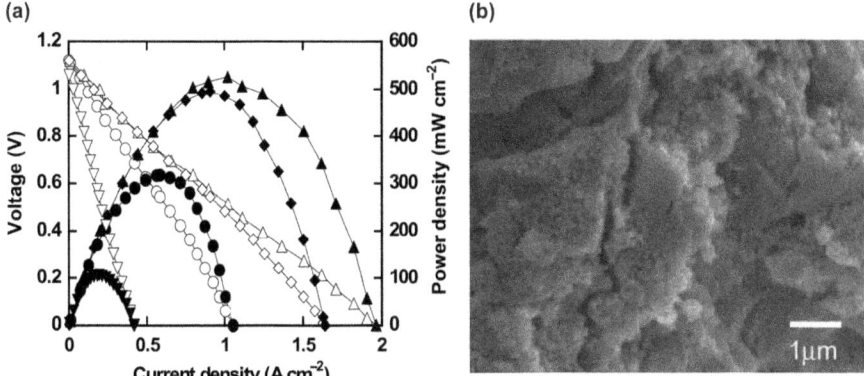

Figure 6.14 Cell performance in humidified H_2 (3% H_2O) of (a) infiltrated LSCM/YSZ composite anodes at 700 °C, where ($\triangledown$) pure LSCM/YSZ anodes, ($\bigcirc$) with 5 wt% ceria, ($\diamond$) with 0.5 wt% Pd, and ($\triangle$) with 5 wt% ceria and 0.5 wt% Pd; and (b) SEM micrograph of an infiltrated LSCM/YSZ composite anodes after reduced in humidified H_2 at 800 °C for 4 h.[73] Reproduced by permission from The Electrochemical Society.

In addition to the promotion in the electrocatalytic activity of cermic oxide anodes, infiltration of Pd has also been found to enhance the sulphur tolerance of the anodes. Lu *et al.* studied the Pd infiltrated $Sr_{0.88}Y_{0.08}TiO_{3-\delta}/La_{0.4}Ce_{0.6}O_{1.8}$ (SYT/LDC) composite anodes and the Pd infiltrated composite anodes showed no performance decay in H_2 containing up to 50 ppm H_2S at 800 °C.[136] In addition to Pd, infiltration with small amount of catalysts such as Ni, Ce, Ru into donor-substituted $SrTiO_3$ based perovskites improves their electrocatalytic activity.[126,137,138] The results presented by Karczewski *et al.* showed that the increase on the infiltrated Ni above 1 wt% does not lead to further enhancement in the power density, indicating the catalytic role of the infiltrated Ni.[138]

In addition to the LSCM infiltrated YSZ anodes,[73] other ceramic oxides infiltrated YSZ scaffolds have also been reported, including $La_{0.3}Sr_{0.7}TiO_3$,[126] $La_{0.7}Sr_{0.3}VO_{3.85}$,[139] $SrMoO_3$,[140] and $NaWO_3$.[141] Co-infiltration of catalytic promoter such as Pd has been found to be necessary to achieve a good cell performance with infiltrated ceramic oxide based anodes.[139–141] Ni/YSZ cermet anodes have also been fabricated by infiltration of Ni into porous YSZ scaffolds.[142] Busawon *et al.* infiltrated Ni into YSZ scaffold and the infiltrated Ni/YSZ composites with 12 wt% Ni showed no dimensional changes after one redox cycle, indicating the potential as the redox stable Ni-based anodes of SOFCs.[143]

6.4 Microstructure and Microstructural Stability of Nano-structured Electrodes

6.4.1 Microstructure Effect

The remarkable promotion effect of the nano-structured electrodes on the performance of anodes and cathodes of SOFCs is a direct result of the

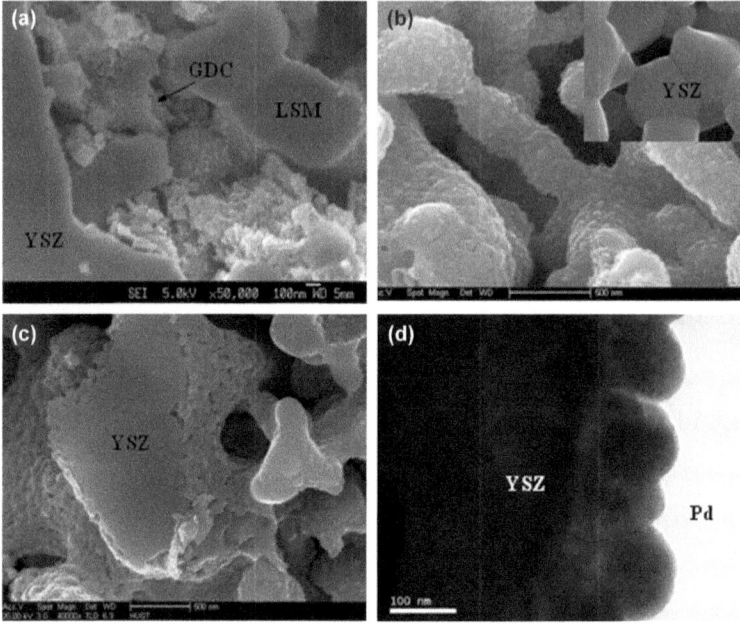

Figure 6.15 SEM micrographs of (a) GDC-infiltrated LSM, (b) LSM-infiltrated
YSZ, (c) Pd-infiltrated YSZ, and (d) high resolution TEM of (c).[97,144]
Reproduced by permission from Elsevier and Hydrogen Energy
Publications.

formation and uniform deposition of nano-sized and catalytically active phases
on the surface of porous electrode or electrolyte scaffolds. Figure 6.15 shows
typical microstructures of selected nano-structured cathodes.[97,144] The GDC
nanoparticles phase is formed after heat-treatment at 800 °C and the particle
size is in the range of ~ 50 nm, which is much smaller than 1000–1500 nm of the
LSM grains of the pre-sintered LSM electrode scaffold (Figure 6.15a). LSM
can also be infiltrated into pre-sintered YSZ electrolyte scaffold to form nano-
structured LSM+YSZ composite cathodes (Figure 6.15b). The infiltrated LSM
phase is characterized by the formation of continuous LSM nanoparticles on
the surface of YSZ grains. The formation of continuous Pd nanoparticles on
the YSZ grain surface via simple palladium nitrate solution infiltration process
is clearly demonstrated by the SEM and high resolution TEM pictures of a Pd-
infiltrated YSZ cathode (Figure 6.15c and d). The Pd nanoparticles are well
interconnected. However, infiltrated Pd intends to form discrete nanoparticles
on the surface of the anodes such as LSCM/YSZ and Ni/GDC under reducing
environment.[66,70,145]
    One of the most important conclusions from the observed formation of
continuous nanoparticles layers on highly porous YSZ scaffold (Figure 6.15b
and c) is that the resulting structure may not be entirely random. As shown
above, adding surfactant or complexing agents would help to enhance the
formation and uniform distribution of infiltrated nanoparticles. Such a

characteristic is particularly important to achieve sufficient conductivity on the dominant ionic electrolyte scaffold structure using perovskite loadings below the normal percolation threshold of 30 vol%.[8] It has been shown that the conductivity of the infiltrated LSM+YSZ composite cathodes increases at very low LSM loading and threshold for the reasonable conductivity is below 20 vol%, significantly lower than the minimum value of 30 vol% based on the percolation theory for conventional LSM/YSZ composites.[99] The benefit of the electrolyte scaffold in the development of nano-structured electrodes is the better mechanical strength of the porous scaffold as compared to that of the electrode scaffold.

The intrinsic relationship between the microstructure of the nano-structured electrodes and the performance is further demonstrated in Figure 6.16.[97] The anode-supported cell with infiltrated nano-structured LSM-YSZ cathode (electrode II) achieved a power density of 0.83 W cm^{-2} at 750 °C in H$_2$/air, more than four times higher than that on similar cells with conventional LSM-YSZ composite cathode (electrode I). With the infiltration of Pd to conventional LSM-YSZ composite cathode (electrode III), the cell performance was enhanced substantially to 1.42 W cm^{-2}. Though the microstructure of LSM-infiltrated YSZ (LSM+YSZ) and Pd-infiltrated LSM-YSZ (Pd+LSM-YSZ) composite electrodes appears similar (Figure 6.16b), the reaction paths on the electrode II and III are in fact very different. For electrode II, YSZ scaffold only provides the path for the oxygen ion conductivity and infiltrated LSM nano-particles on YSZ surface provides the TPB for the O$_2$ reduction. In the case of nano-structured Pd+LSM-YSZ (electrode III), the LSM-YSZ scaffold is both electronic and ionic conductive and the deposition of Pd nanoparticles layers on the surface of the LSM-YSZ scaffold not only provides the additional reaction sites but also significantly accelerate the reaction rate of the dissociation and diffusion of oxygen species (Figure 6.16a). This explains the very high performance of the cell with nano-structured Pd+LSM-YSZ composite cathodes.

Zhu *et al.* carried out a modelling study on the enhancement of TPB of nano-structured electrodes by infiltration.[146] The model is based on the random packing of sphere particles and electronic or ionic conducting scaffold coated with a continuous nanoparticle layer, see Figure 6.17.[146] The results predict the substantial increase in the TPB length, *e.g.*, 6.2 and 14.6 times higher as compared to conventional mixed composites when the infiltrated nanoparticle size is 50 and 20 nm, respectively (assuming the particle size of conventional composite phase is 1 μm). The finite element calculation based on a model structure consisted of micrometer-scale columns and nanoscale branches of ionically conducting materials also shows that the increase in the surface area of the infiltrated mixed conductor could result in a factor of 10 polarization resistance decrease.[147] Based on a simple model, Shah *et al.* predicted the apparent saturation in performance enhancement at higher infiltrate loading most likely due to the dominated ionic transport limitation at the scaffold.[148] The impedance responses of SOFC cathodes prepared by infiltration has also been modelled by Bidrawn *et al.*[149]

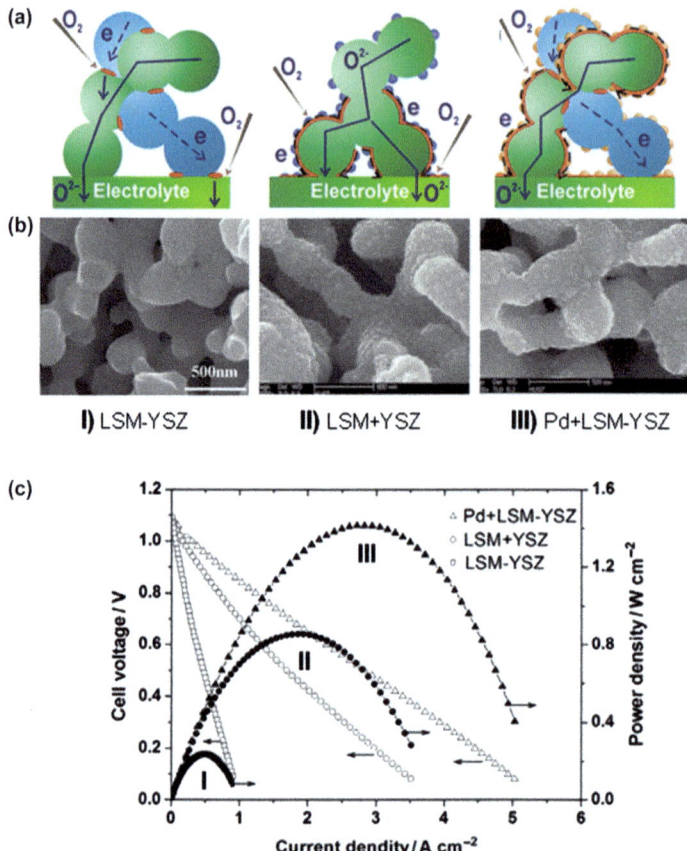

Figure 6.16 Demonstration of the performance and microstructure of the nano-
structure electrodes. (a) Schematic illustrations of electrodes I, II, and
III, showing the paths of transport for electrons and oxygen ions and
reaction sites for the O_2 reduction reaction; (b) SEM micrographs of
electrodes I, II, and III; (c) Power output of anode-supported cells with
electrode types I, II, and II, measured at 750 °C in H_2/air. Electrodes:
(I) standard LSM-YSZ composite cathode by mechanical mixing of
LSM and YSZ phases; (II) nano-structured LSM-infiltrated YSZ
(LSM+YSZ) composite cathode; and (III) nano-structured Pd-
infiltrated LSM-YSZ (Pd+LSM-YSZ) composite cathode.[97]
Reproduced by permission from Elsevier.

6.4.2   Microstructural Stability of Nano-structured Electrodes

One of the challenges associated with the application and development of nano-
structured electrodes is the long-term stability of the microstructure and
activity of the infiltrated nanoparticles. Specifically, the high surface area
nanoparticles (20–100 nm) are highly prone to sintering and grain growth at
operation temperature of SOFCs (500–800 °C). Wang *et al.* studied the stability
of $La_{0.8}Sr_{0.2}FeO_{3-\delta}$ (LSF) infiltrated YSZ composite cathodes calcined at

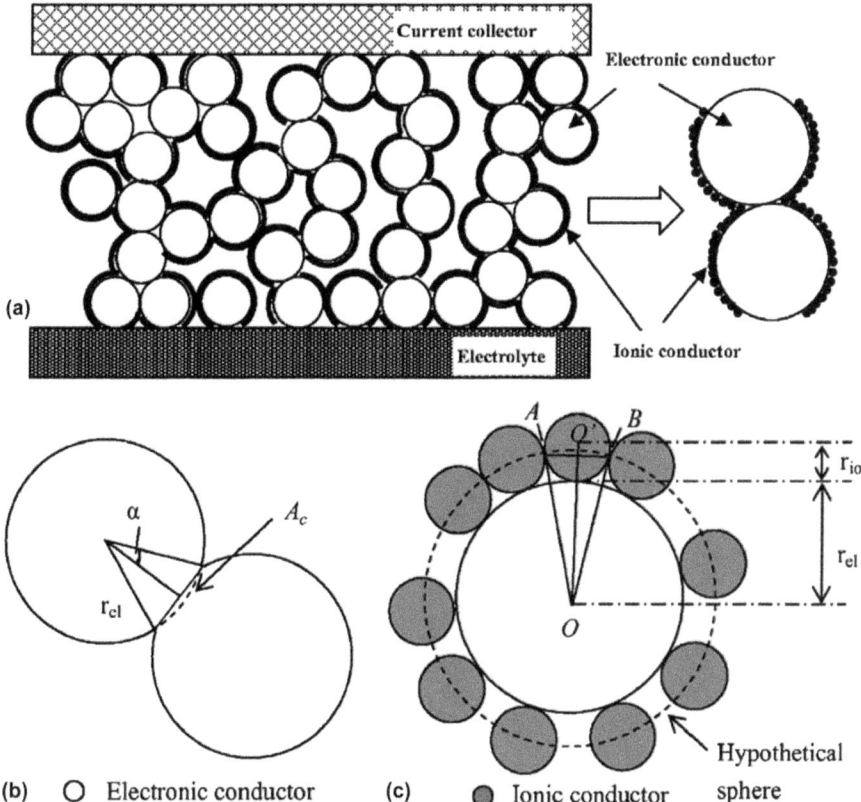

Figure 6.17 Modelling approach based on the random packing of sphere particles and electronic or ionic conducting scaffold coated with a continuous nanoparticle layer.[146]
Reproduced by permission from The Electrochemical Society.

850 °C.[150] The ASR increased linearly from 0.15 to 0.55 $\Omega\,cm^2$ after testing for 2500 h at 700 °C. In the case of infiltrated Pd nanoparticles on porous YSZ scaffold, the agglomeration and grain growth resulted in the formation of continuous and dense Pd films on the YSZ scaffold surface, leading to the increase in the polarization losses due to the blocking of the oxygen diffusion path.[151]

Alloying with cobalt, manganese and silver has been found to be effective to increase the thermal stability and to enhance the performance stability of Pd-infiltrated electrodes.[151,152] Liang *et al.* studied the effect of Mn alloying on Pd infiltrated YSZ cathode. As shown in Figure 6.18, co-infiltration of Mn and Pd with stoichiometric composition of $Pd_{0.95}Mn_{0.05}$ effectively reduced the grain growth and agglomeration of infiltrated Pd nanoparticles (Figure 6.18d).[151] $Pd_{0.95}Mn_{0.05}$ infiltrated YSZ cathode shows a significantly enhanced polarization performance stability as compared with unalloyed Pd infiltrated YSZ cathodes.

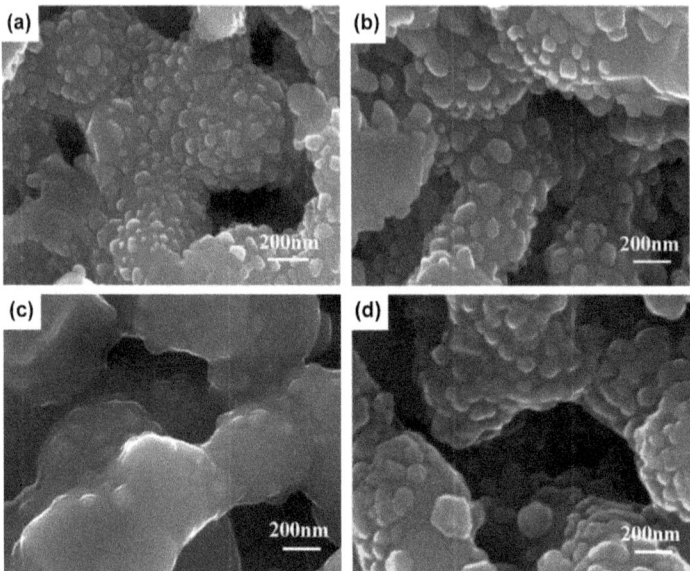

Figure 6.18   SEM micrographs of the cross sectioned (a) Pd infiltrated YSZ and
(b) $Pd_{0.95}Mn_{0.05}$ infiltrated YSZ before the polarization test and
(c) Pd infiltrated YSZ and (d) $Pd_{0.95}Mn_{0.05}$ infiltrated YSZ after the
polarization test at $200\,mA\,cm^{-2}$ and $750\,°C$ for 30 h in air.[151]
Reproduced by permission from Wley-VCH.

Kim *et al.* showed that highly active and stable Pd catalysts can be obtained
by infiltrating $Pd@CeO_2$ core-shell nanoparticles into YSZ scaffold.[145] The
core-shell nano-structure inhibits the growth of Pd nanoparticles. Co-
infiltration of mixed ceria and cobalt solutions was also found to suppress the
aggregation of Co_3O_4 nanoparticles.[92]

Operation temperature plays a critical role in the microstructural stability of
infiltrated nanoparticles. Shah *et al.* studied the grain growth of LSCF infiltrated GDC electrodes in the temperature range from 650 to $850\,°C$.[153] They
observed the significant increase of the electrode polarization resistance
particularly at high operation temperatures, which is consistent with the
significant grain growth of infiltrated LSCF nanoparticles.

There are reports of high stability of infiltrated electrodes. Zhao *et al.* showed
that excellent thermal cycle durability can be achieved by the nano-structured
$LSC-Sm_{0.2}Ce_{0.8}O_2$ cathodes and no increase in ASR was observed, see
Figure 6.19.[154,155] In this study, the electrode was fired at a maximum
temperature of $800\,°C$. The high stability of the infiltrated LSC-SDC cathodes
is most likely due to the formation of a continuous and stable LSC nanoparticles anchored on the SDC scaffold. The modelling study suggests that the
size of the nanoparticles is probably the most important factor to stabilize the
nano-structure during the thermal cycle.[156] High stability was also reported for
a cathode with LSC infiltrated GDC.[35] The cell consisted of infiltrated cathode
with Ni/YSZ anode supported scandia and yttria stabilized zirconia (ScYSZ)

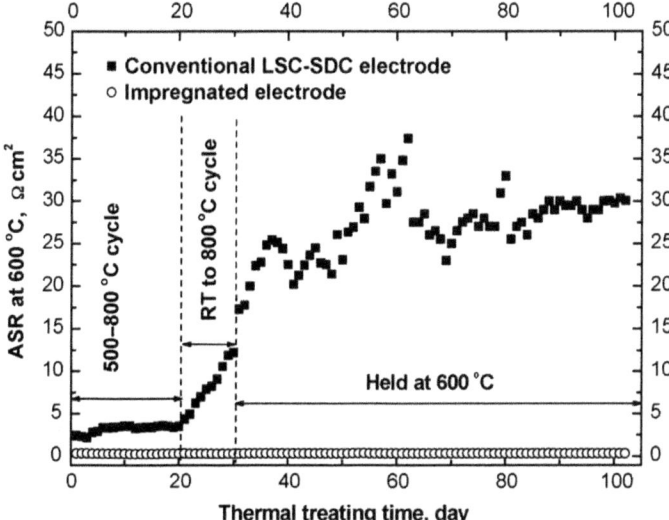

Figure 6.19 ASR at 600 °C for the infiltrated LSC-SDC cathode and conventional LSC-SDC cathodes as a function of thermal treatment.[154] Reproduce by permission from Elsevier.

and GDC barrier layer pre-tested at 850 °C showed no degradation under a current density of $0.5\,A\,cm^{-2}$ at 700 °C for 1500 h in H_2/air.

Glass and glass-ceramics are promising seals for SOFCs due to their tailorable structure and properties.[157–159] However, the volatility and reactivity of some components of sealing glass can be significant at the operating temperature of SOFCs.[160] The presence of volatile species under SOFC operation conditions such as boron from borosilicate based sealants can be another issue for the stability of nano-structured electrodes. We have reported recently that heat-treatment in the presence of volatile boron species significantly accelerates the coarsening of nano-structured GDC infiltrated LSM and LSM infiltrated YSZ cathodes and therefore leads to significant degradation of the electrochemical activity of the electrodes for the O_2 reduction reaction.[161,162] In the case of LSM electrodes, the poisoning effect of boron is related to the size of LSM, as shown in Figure 6.20.[162] SEM images were taken on conventional LSM and infiltrated LSM-YSZ composite cathodes before and after heat-treatment in the presence of borosilicate glass for 30 days. After being treated in the presence of glass powder for 30 days, there is a formation of irregularly shaped small particles (Figure 6.20b), which has been identified as $LaBO_3$. In the case of LSM infiltrated YSZ electrode, infiltrated LSM nanoparticles are no longer exist (Figure 6.20d), indicating the significant agglomeration of the nano-sized LSM particles and disintegration of the perovskite structure.

Microstructural and long term stability of infiltrated anodes and cathodes is still an open issue and requires more systematic and comprehensive studies to understand the effects of several factors such as the nanoparticle size and

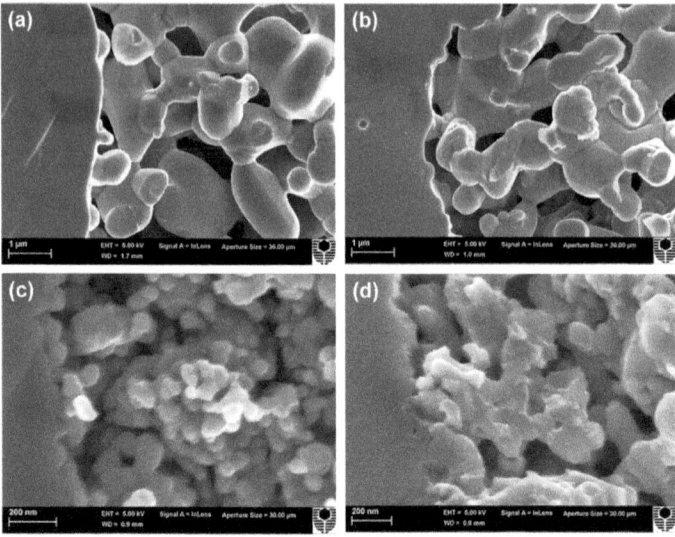

Figure 6.20   SEM micrographs of cross section of conventional LSM cathodes (a)
before and (b) after treated at 800 °C for 30 days in the presence of glass
and of infiltrated LSM-YSZ cathodes (c) before and (d) after treated at
700 °C for 30 days in the presence of glass.[162]
Reproduced by permission from The Electrochemical Society.

distribution, infiltration firing temperature, scaffold structure, operation
temperature, and current load on the structural stability of infiltrated phases.

6.5   Electrocatalytic Effects of Infiltrated Nanoparticles

The peculiar microstructure of infiltrated nano-structured electrodes as
shown in this Chapter suggests the significantly enlarged TPBs for the electrode
reactions. However, the observed significant promotion effect of the
nano-structured electrodes with discretely distributed nanoparticles (*i.e.,*
nanoparticles are not interconnected) indicates that there must be significant
electrocatalytic and catalytic effects of the infiltrated nanoparticles on the
electrochemical performance of the electrodes in addition to the obviously
enhanced TPB. This is particularly true that in the case of metallic nanoparticles (*e.g.*, Pd), the amounts of metallic nanoparticles required to provide
significant enhancement in electrode performance are so small that the effect
must be catalytic.

Early studies on the effect of infiltrated GDC nanoparticles on the
performance and electrode behavior of the LSM cathode show that the
promotion effect of infiltrated GDC is far more pronounced on the electrode
process at low frequencies as compared to that at high frequencies.[60] The
activation energy of the electrode process at high frequencies, σ_H for O_2
reduction on the GDC-infiltrated LSM is $\sim 100\,kJ\,mol^{-1}$, similar to that for
the reaction on pure LSM.[163] On the other hand, the activation energy of the

electrode process associated with low frequency arc, σ_L for the reaction on the GDC-infiltrated LSM is $140\,kJ\,mol^{-1}$, significantly lower than ~ 170–$200\,kJ\,mol^{-1}$ for the reaction on pure LSM.[163] The low activation energy of the reaction on the GDC-impregnated LSM is probably related to the low activation energy of $\sim 74\,kJ\,mol^{-1}$ for the ionic conductivity on the infiltrated GDC.[164] This indicates that the infiltration of nano-sized GDC particle phase greatly accelerates the oxygen dissociation and diffusion process in addition to the substantial reduced overall electrode polarization resistance of the nano-structured GDC-LSM cathodes for the O_2 reduction reaction. Similar promotion effect is also reported for the reaction on $Y_{0.5}Bi_{1.5}O_3$-infiltrated LSM electrodes.[74]

The effect of the nano-structured electrodes on the mechanism of the O_2 reduction reaction is also studied by AFM. Figure 6.21 is the AFM images of

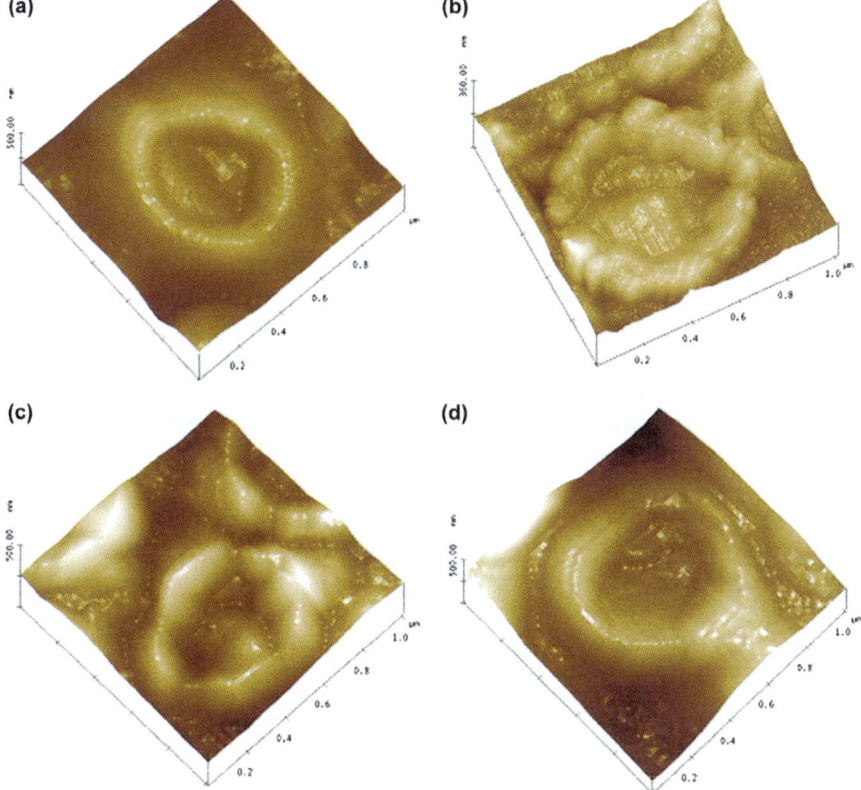

Figure 6.21 AFM micrographs of the YSZ surface in contact with the cathodes: (a) before and (b) after cathodic current passage at $200\,mA\,cm^{-2}$, $700\,°C$ for $120\,h$ for the pure LSM cathode and (c) before and (d) after cathodic current passage at $200\,mA\,cm^{-2}$, $700\,°C$ for $120\,h$ for the $1.2\,mg\,cm^{-2}$ BSCF-infiltrated LSM cathode. Electrode coatings were removed by HCl treatment.[68]
Reproduced by permission from The Electrochemical Society.

the electrode/YSZ electrolyte interface in contact with the BSCF-impregnated LSM cathodes before and after the polarization test.[68] The convex rings on the YSZ electrolyte surface are the contact interfaces between the LSM grains and YSZ electrolyte formed during the sintering of the LSM electrodes.[165] There is clearly a morphological change and broadening of the convex rings after the polarization at 200 mA cm^{-2} and 700 °C for 120 h for the reaction on the pure LSM cathode, as compared to that before the polarization (Fig 6.21a and b). The morphological change of the convex rings is the direct indications of the O_2 reduction reactions that take place at the LSM/YSZ interface.[165] In contrast to that of the pure LSM cathode, the morphological changes of the convex rings on the YSZ electrolyte surface in the case of the nano-structured BSCF-infiltrated LSM cathodes before and after polarization treatment are negligible (Figure 6.21c and d). This indicates that the O_2 reduction reaction is no longer restricted to the electrode/electrolyte interface region in the case of the nano-structured BSCF-LSM cathodes. Instead, the TPBs for the O_2 reduction reaction extend to the bulk of the electrode.

Among the metallic nanoparticles, infiltrated Pd nanoparticles are the most interesting and widely studied catalysts for the O_2 reduction and in particular the hydrocarbon oxidation reactions. The transition between nano-sized PdO and Pd occurs at temperatures of ~ 800 °C.[166] Even in the reducing environment, XPS studies indicates that $\sim 27\%$ of the surface of the infiltrated Pd nanoparticles on Ni/GDC scaffold remains as PdO.[70] The reason could be related to the presence of nickel which could retard the reducibility of palladium, forming PdO on the surface of Pd nanoparticles. Fox *et al.*[167] studied the reducibility of bimetallic PdCu/CeO$_2$ catalysts and showed that mutual interaction between Cu, Pd and CeO$_2$ components affects the reduction process. The presence of copper significantly retards the reducibility of Pd on ceria at high temperatures probably due to the greater oxygen affinity of copper over palladium. The presence of Pd/PdO redox couples appears to play a critical role in the promotion of the H_2 oxidation on the Ni/GDC anodes. The electrode polarization resistance for the H_2 oxidation reaction on Ni/GDC anodes decreases with the infiltrated Pd nanoparticles, indicating the significant promotion effect of the Pd on the H_2 oxidation reaction. The most interesting observation is that the impedance responses of the H_2 oxidation reaction on the Pd-infiltrated Ni/GDC anodes become clearly separated at low and high frequencies. The clear separation of the electrode processes indicates that the promotion effect of infiltrated Pd nanoparticle phase is also preferential towards particular electrode processes taking place on the Ni/GDC anode.

The preferential catalytic effect of infiltrated Pd on the H_2 oxidation reaction on Ni/GDC anodes is also indicated by the dependence of the activation energies of the electrode processes associated with low and high frequency arcs, R_L and R_H on the loading of the infiltrated Pd phase, as shown in Figure 6.22.[70] The activation energy of R_L decreases almost linearly with the increasing Pd loading, while for R_H it is independent of the infiltrated Pd loading. The significant reduction in the activation energy for the hydrogen dissociation and diffusion process (*i.e.*, the process associated with the low frequencies, R_L) is

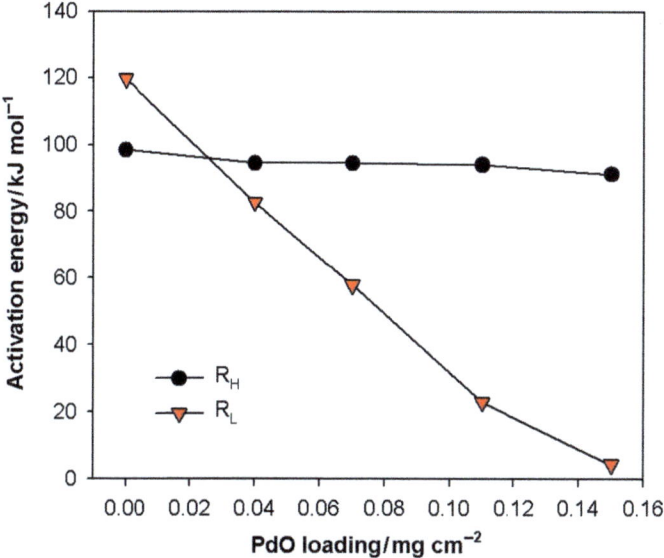

Figure 6.22   Plot of the activation energies of high frequency arc resistance (RH) and low frequency arc resistance (RL) as a function of the infiltrated PdO loading for the hydrogen oxidation reaction on Pd-infiltrated Ni/GDC cermet anodes.[70]
Reproduced by permission from The Electrochemical Society.

most likely due to the accelerated hydrogen and oxygen spillover mechanism over the Pd/PdO redox couple on the Ni/GDC anodes. The observed significant promoting effect of infiltrated Pd nanoparticles on the oxidation reaction of methane and ethanol is most likely related to the high catalytic activity of palladium on the dissociation and activation of methane and ethanol and subsequent gasification and oxidation of carbon species.[168] However, it should be pointed out here that unique catalytic activity of infiltrated nano-particles is not limited to Pd. Other metals such as infiltrated Rh, Ni and Fe also show high catalytic activity for the anode reactions in SOFCs.[73]

Infiltrated Pd nanoparticles are also very effective for the promotion of the O_2 reduction reactions. Figure 6.23 is the impedance responses of the conventional LSM/YSZ, LSM infiltrated YSZ (LSM+YSZ) and Pd infiltrated LSM/YSZ (Pd+LSM-YSZ) composite cathodes measured at 700 °C in air at open circuit.[71] The electrode polarization resistance of nano-structured LSM+YSZ cathode is substantially smaller than that on the conventional LSM/YSZ composite cathodes, clearly due to the substantially increased TPB as the results of the nanoscale engineering of the conventional LSM/YSZ composite. The R_E value of the Pd+LSM/YSZ cathode is 0.18 Ω cm^2 at 700 °C, which is comparable to those of the MIEC cathodes like LSCF (0.32 Ω cm^2 at 700 °C)[169] and $Gd_{0.8}Sr_{0.2}CoO_3$ (0.10 Ω cm^2 at 700 °C).[170] The activation energy of the O_2 reduction reaction on nano-structured Pd+LSM/YSZ composite cathodes is 96 kJ mol^{-1}, also significantly lower than 163 kJ mol^{-1} on

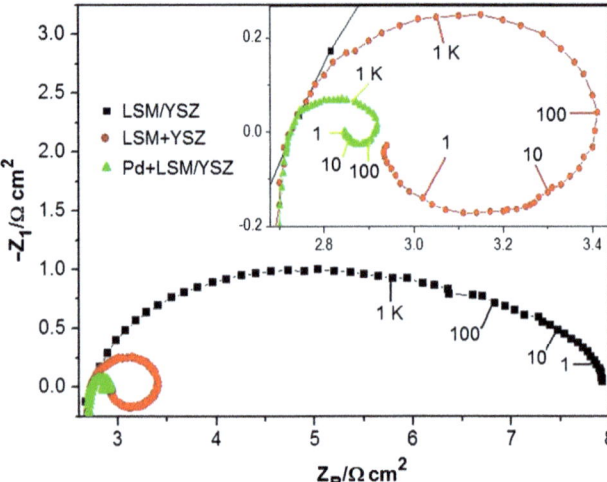

Figure 6.23   Electrochemical impedance spectra of the conventional LSM/YSZ, nano-
structured LSM+YSZ and Pd+LSM/YSZ composite cathodes measured
at 700 °C in air at open circuit. Numbers are frequencies in Hz.[71]
Reproduced by permission from The Electrochemical Society.

conventional LSM/YSZ composite cathodes.[71] This indicates that nano-
structure reduces the energy barrier for the O_2 reduction reaction as compared
to that on micro-structured conventional electrodes. The observed inductance
loop for the O_2 reduction reaction appears to be closely associated with the
nano-structured electrodes.[33,66,69,71] Pd was also shown to significantly increase
the electrochemical activity of SSC/$La_{0.8}Sr_{0.2}Mg_{0.15}Co_{0.05}O_3$ composite
cathodes.[171]

There are significant discrepancies in the electrocatalytic roles of the infiltrated nanoparticles. Yamahara et al.[96] reported a significant increase of the
power densities by infiltration of cobalt oxide to $(La,Sr)MO_3$-Y_2O_3-ZrO_2
(LSM-YSZ) cathode, while Huang et al.[172] showed that addition of 10 wt%
cobalt oxide to the LSM-YSZ composite has basically no effect on the cathode
performance. An early study by Haanappel et al. showed that LSM cathodes
are indistinguishable with or without addition of Pd.[17] On the other hand,
our work consistently showed that infiltrated Pd nanoparticles remarkably
improve the electrochemical performance of the LSM, LSM-YSZ, and
$(La,Sr)(Co,Fe)O_3$ (LSCF) electrodes for O_2 reduction reaction and the
promotion effect is primarily on the electrode processes of oxygen dissociation
and diffusion.[69,97] Søgaard et al.[173] found that addition of CeO_2 and
$Sm_{0.2}Ce_{0.8}O_{1.9}$ (SDC) nanoparticles led to similar promotion effect on the
$(La,Sr)CoO_3$ (LSC)-GDC and LSM-YSZ composite cathodes. On the other
hand, Wang et al.[174] reported that the infiltration of SDC nanoparticles
resulted in much greater enhancement in the oxygen surface exchange coefficient (k) of LSM than pure ceria. Bidrawn et al.[175] fabricated $(La,Sr)FeO_3$
(LSF) and LSM infiltrated YSZ composite cathodes and studied the influence

of various infiltrated phases such as YSZ, Pd, SDC, CaO, and K_2O on the cathode performance. The results showed that infiltrated phases had little influence on the composite cathodes sintered at 900 °C, but improved the performance of the composite cathodes sintered at 1100 °C. They claim that the enhancement associated with addition of a catalyst is simply an artifact of structure.

Though the enhancement of the addition of inert materials such as CaO and K_2O on the performance of LSF/YSZ scaffold as observed by Bidrawn *et al.*[175] is difficult to comprehend, a study by Hansen *et al.*[176] showed that infiltration of LSM in general improves the performance of LSM/YSZ cathodes, and adding inert alumina nanoparticles had a detrimental effect on the activity of the LSM/YSZ composite electrodes. Lee *et al.* studied the effect of ionic conductive LSC and electrically insulating Sr-doped $La_2Zr_2O_7$ (LSZ) infiltrated SDC/LSCF cathodes.[177] The LSC-infiltrated SDC/LSCF cathodes improved the performance, while infiltration of LSZ has a negative effect on the performance. The detrimental effect of the infiltrated LSZ increased with the increased infiltrated loading of the insulated LSZ phase. Hojberg and Sogaard showed that the co-infiltration of LSM and SDC into YSZ scaffolds reduced the activation energy for the O_2 reduction to 1.19 eV, lower than 1.35 eV on conventional LSM/YSZ composite cathodes.[178]

One important issue or factor which is generally missing in the discussion of the promotion effect of the infiltrated nanoparticle phases is the impact of the possible interaction or modification of the infiltrated phase on the surface properties and activities of the porous scaffolds. Recently, we studied the electrode behaviour of ceria and GDC infiltrated Pt electrodes.[179] Metallic platinum electrode was selected as porous scaffold structure template as Pt is widely used as a model electrode in SOFCs due to its chemical inertness, simple structure and the confinement of the electrode reaction to the TPB region.[180,181] The chemical inertness of metallic Pt is important to minimize possible interactions between the infiltrated phase/solution and the porous skeleton structure. Figure 6.24 is the *iR*-free polarization curves of pure Pt and CeO_2 and GDC infiltrated Pt electrodes (CeO_2-Pt and GDC-Pt) measured at 800 °C in air.[179] For the O_2 reduction reaction on pure Pt electrode, η increases very rapidly to 550 mV at a current density of $50 \, mA \, cm^{-2}$, and then decreases slightly with further increase in current. The rapid increase in η indicates the poor activity of Pt for the O_2 reduction reaction. Pt is a pure electrical conductor and the reaction is restricted to the TPB region. Infiltration of $0.2 \, mg \, cm^{-2}$ CeO_2 and GDC nanoparticles enhance significantly the polarization performance of the Pt electrodes. It is interesting to note that the polarization performance of CeO_2-Pt and GDC-Pt starts to differentiate at high current densities. Under a cathodic current density of $100 \, mA \, cm^{-2}$, η is 93 mV for the reaction on GDC-Pt electrode, significantly smaller than 196 mV for the reaction on the CeO_2-Pt electrode. The increase in η values for the reaction on GDC-Pt electrode is substantially smaller than that on CeO_2-Pt electrode at high currents. This shows that GDC is more effective over CeO_2 nanoparticles on the promotion of the electrocatalytic activity of Pt electrodes for the O_2

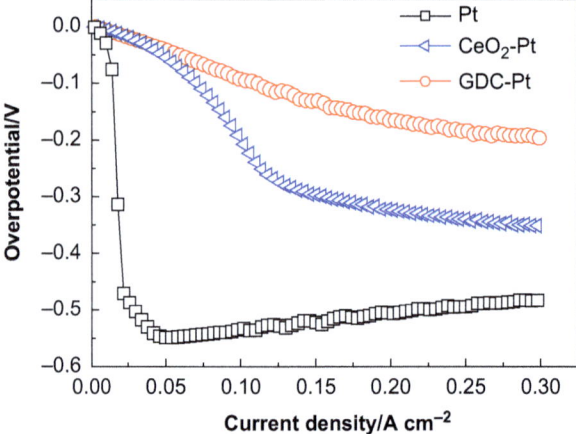

Figure 6.24   (a) IR-free overpotential curves of O_2 reduction reaction on Pt, CeO_2-Pt and GDC-Pt electrodes, measured at 800 °C in air.[179] Reproduced by permission from Elsevier.

reduction reaction. The high promotion effect of infiltrated GDC-Pt cathodes is most likely related to the higher ionic conductivity of GDC as compared with pure ceria.[182]

6.6   Conclusions

The nanoscale engineering approach via low temperature infiltration process has attracted increasing attention as the most effective and alternative techniques in the development of the nano-structured electrodes to achieve high performance and advantageous microstructure in a way that otherwise would not be possible with high temperature processing processes for standard SOFCs electrodes. The unique advantages and flexibilities of the nanoscale engineering of the conventional electrode structure via a two-step firing approach have been clearly demonstrated on wide range nano-structured electrode systems for SOFCs. Infiltration process can be scaled up for large and practical planar cells with active areas of $81\,cm^2$ (100×100 mm cells).[183] Infiltration can also be used to fabricate a doped ceria based diffusion barrier layer to minimize the problems associated with the solid state reaction of lanthanum cobalt based cathode and YSZ interface.[184] The results so far clearly demonstrate that infiltrated metal and metal oxide nanoparticles not only substantially increase the TPBs for the reaction but also play very important catalytic roles in the enhancement of the fuel cell performance.

One of the most challenging problems associated with a SOFC system over a three- to five-year lifetime is the gradual degradation and deactivation of the cathodes and anodes by contaminants which can be either in the air or fuel streams or from the volatile species of cell components, such as metallic interconnect, sealant and manifold.[185] Nano-structured electrodes already show specific tolerance towards the carbon deposition[62,63] and poisoning of sulfur[136] and chromium.[186] Nano-structured electrodes based on infiltration are not limited to one electrocatalytic active species and in principle, it is possible to develop nano-structured electrodes with high activity and tolerance towards contaminants with infiltration of multiple nanoparticles phases with specific functions and activities. However, the most significant challenge in the application and development of nano-structured electrodes is the long-term stability of the microstructure and activity of the infiltrated nanoparticles.

Fundamental understanding of the microstructure and electrocatalytic promotion effect of infiltrated nanoparticles on the performance and electrode reactions of SOFCs is important for the development of the highly efficient and durable nanoscale and nano-structured electrodes based on infiltration techniques. Part of the difficulties lie in the differences in the infiltration process such as heat treatment temperature, surfactant in the infiltration solutions, and the loading and distribution of the infiltrated phase. Another difficulty is related to the fact that the catalytic effect of the infiltrated phases depends strongly on the architect or the morphology and microstructure of the infiltrated nanoparticles or films as demonstrated by Liu *et al.*,[110] the nature of the porous scaffold, as observed for the significant different effects of infiltrated Pd on Ni/GDC and LSCM scaffolds,[64,73] and the potential interaction between the scaffold and infiltrated phase.[187] The activation and deactivation phenomena of SOFC cathodes, which depends strongly on the microstructure, polarization and presence of other phases[188–190] and the phase segregation at the electrode/electrolyte interface[191,192] also complicate the mechanism of the promotion effect of infiltrated nanoparticles.

Despite the multiple-step nature associated with the infiltration method, nanoscale engineering of electrode structures via infiltration is probably the most effective way to develop highly active and advanced electrode structures for SOFCs. However, much more systematic and extensive studies and long-term tests are needed to fundamentally understand the promotion mechanism of the infiltrated nanoparticles and to fully assess the structural and activity stability of the nano-structured electrodes under SOFCs operation conditions. The modelling simulations can be very useful and important in the fundamental and structural stability assessment of nano-structured electrodes,[146–149,156,193] particularly if the microstructural as well as electrochemical and catalytic aspects of the infiltrated nanoparticle are incorporated into the models.

Acknowledgement

This work was supported by the Australian Research Council (LP110200281).

References

1. V. V. Kharton, F. M. B. Marques and A. Atkinson, *Solid State Ionics*, 2004, **174**, 135–149.

2. T. Ishihara, *Bull. Chem. Soc. Jpn.*, 2006, **79**, 1155–1166.
3. D. J. L. Brett, A. Atkinson, N. P. Brandon and S. J. Skinner, *Chem. Soc. Rev.*, 2008, **37**, 1568–1578.
4. A. J. Jacobson, *Chem. Mat.*, 2010, **22**, 660–674.
5. S. P. Jiang, *Mater. Sci. Eng. A–Struct. Mater. Prop. Microstruct. Process.*, 2006, **418**, 199–210.
6. S. P. Jiang, *Int. J. Hydrog. Energy*, 2012, **37**, 449–470.
7. R. J. Gorte and J. M. Vohs, *Curr. Opin. Colloid Interface Sci.*, 2009, **14**, 236–244.
8. J. M. Vohs and R. J. Gorte, *Adv. Mater.*, 2009, **21**, 943–956.
9. Z. Jiang, C. Xia and F. Chen, *Electrochim. Acta*, 2010, **55**, 3595–3605.
10. A. Mitterdorfer and L. J. Gauckler, *Solid State Ionics*, 1998, **111**, 185–218.
11. S. P. Jiang, P. J. Callus and S. P. S. Badwal, *Solid State Ionics*, 2000, **132**, 1–14.
12. S. P. Jiang, Y. Y. Duan and J. G. Love, *J. Electrochem. Soc.*, 2002, **149**, A1175–A1183.
13. T. Z. Sholklapper, V. Radmilovic, C. P. Jacobson, S. J. Visco and L. C. De Jonghe, *J. Power Sources*, 2008, **175**, 206–210.
14. Y. Liu, M. Mori, Y. Funahashi, Y. Fujishiro and A. Hirano, *Electrochem. Commun.*, 2007, **9**, 1918–1923.
15. L. S. Wang and S. A. Barnett, *Solid State Ionics*, 1995, **76**, 103–113.
16. V. N. Tikhonovich, V. V. Kharton, E. N. Naumovich and A. A. Savitsky, *Solid State Ionics*, 1998, **106**, 197–206.
17. V. A. C. Haanappel, D. Rutenbeck, A. Mai, S. Uhlenbruck, D. Sebold, H. Wesemeyer, B. Rowekamp, C. Tropartz and F. Tietz, *J. Power Sources*, 2004, **130**, 119–128.
18. K. Sasaki, K. Hosoda, T. N. Lan, K. Yasumoto, S. Wang and M. Dokiya, *Solid State Ionics*, 2004, **174**, 97–102.
19. K. T. Lee and A. Manthiram, *J. Power Sources*, 2006, **160**, 903–908.
20. Y. Wang, S. Wang, Z. Wang, T. Wen and Z. Wen, *Journal of Alloys and Compounds*, 2007, **428**, 286–289.
21. S. Park, R. J. Gorte and J. M. Vohs, *J. Electrochem. Soc.*, 2001, **148**, A443–A447.
22. R. J. Gorte, J. M. Vohs and S. McIntosh, *Solid State Ionics*, 2004, **175**, 1–6.
23. S. D. Park, J. M. Vohs and R. J. Gorte, *Nature*, 2000, **404**, 265–267.
24. X. Changrong, G. Xiaoxia, L. Fanqing, P. Dingkun and M. Guangyao, *Colloids and Surfaces A: Physicochemical and Engineering Aspects*, 2001, **179**, 229–235.
25. G. Wen, Z. X. Guo and C. K. L. Davies, *Scr. Mater.*, 2000, **43**, 307–311.
26. J. Mukhopadhyay, M. Banerjee and R. N. Basu, *J. Power Sources*, 2008, **175**, 749–759.
27. N. Li, Z. Lu, B. O. Wei, X. Q. Huang, K. F. Chen, Y. H. Zhang and W. H. Su, *J. Alloys Compd.*, 2008, **454**, 274–279.
28. N. Ai, K. F. Chen, S. P. Jiang, Z. Lu and W. H. Su, *Int. J. Hydrog. Energy*, 2011, **36**, 7661–7669.

29. W. Zhou, R. Ran, R. Cai, Z. P. Shao, W. Q. Jin and N. P. Xu, *J. Power Sources*, 2009, **186**, 244–251.
30. W. Zhou, R. Ran, Z. P. Shao, R. Cai, W. Q. Jin, N. P. Xu and J. Ahn, *Electrochim. Acta*, 2008, **53**, 4370–4380.
31. R. Su, Z. Lu, S. P. Jiang, Y. B. Shen, W. H. Su and K. F. Chen, *Int. J. Hydrog. Energy*, 2013in press..
32. M. E. Ayturk and Y. H. Ma, *Journal of Membrane Science*, 2009, **330**, 233–245.
33. F. L. Liang, J. Chen, J. L. Cheng, S. P. Jiang, T. M. He, J. Pu and J. Li, *Electrochem. Commun.*, 2008, **10**, 42–46.
34. Y. Y. Huang, J. M. Vohs and R. J. Gorte, *J. Electrochem. Soc.*, 2004, **151**, A646–A651.
35. A. J. Samson, P. Hjalmarsson, M. Sogaard, J. Hjelm and N. Bonanos, *J. Power Sources*, 2012, **216**, 124–130.
36. C. Lu, T. Z. Sholklapper, C. P. Jacobson, S. J. Visco and L. C. De Jonghe, *J. Electrochem. Soc.*, 2006, **153**, A1115–A1119.
37. T. Z. Sholklapper, C. Lu, C. P. Jacobson, S. J. Visco and L. C. De Jonghe, *Electrochem. Solid State Lett.*, 2006, **9**, A376–A378.
38. Z. Y. Jiang, C. R. Xia, F. Zhao and F. L. Chen, *Electrochem. Solid State Lett.*, 2009, **12**, B91–B93.
39. Z. Jiang, Z. Lei, B. Ding, C. Xia, F. Zhao and F. Chen, *Int. J. Hydrog. Energy*, 2010, **35**, 8322–8330.
40. W. Li, Z. Lue, X. Zhu, B. Guan, B. Wei, C. Guan and W. Su, *Electrochim. Acta*, 2011, **56**, 2230–2236.
41. Y. Liu, S. Hashimoto, K. Yasumoto, K. Takei, M. Mori, Y. Funahashi, Y. Fijishiro, A. Hirano and Y. Takeda, *Curr. Appl. Phys.*, 2009, **9**, S51–S53.
42. J. D. Nicholas and S. A. Barnett, *J. Electrochem. Soc.*, 2010, **157**, B536–B541.
43. T. Z. Sholklapper, C. P. Jacobson, S. J. Visco and L. C. De Jonghe, *Fuel Cells*, 2008, **8**, 303–312.
44. T. Z. Sholklapper, H. Kurokawa, C. P. Jacobson, S. J. Visco and L. C. De Jonghe, *Nano Lett.*, 2007, **7**, 2136–2141.
45. S. W. Jung, J. M. Vohs and R. J. Gorte, *J. Electrochem. Soc.*, 2007, **154**, B1270–B1275.
46. X. Y. Lou, Z. Liu, S. Z. Wang, Y. H. Xiu, C. P. Wong and M. L. Liu, *J. Power Sources*, 2010, **195**, 419–424.
47. A. Samson, M. Sogaard, R. Knibbe and N. Bonanos, *J. Electrochem. Soc.*, 2011, **158**, B650–B659.
48. L. Zhang, F. Zhao, R. R. Peng and C. R. Xia, *Solid State Ionics*, 2008, **179**, 1553–1556.
49. Z. Y. Jiang, L. Zhang, K. Feng and C. R. Xia, *J. Power Sources*, 2008, **185**, 40–48.
50. M. Shah and S. A. Barnett, *Solid State Ionics*, 2008, **179**, 2059–2064.
51. A. Torabi, A. R. Hanifi, T. H. Etsell and P. Sarkar, *J. Electrochem. Soc.*, 2012, **159**, B201–B210.

52. Y. Y. Huang, K. Ahn, J. M. Vohs and R. J. Gorte, *J. Electrochem. Soc.*, 2004, **151**, A1592–A1597.

53. M. Boaro, J. M. Vohs and R. J. Gorte, *J. Am. Ceram. Soc.*, 2003, **86**, 395–400.

54. T. J. Armstrong and J. G. Rich, *J. Electrochem. Soc.*, 2006, **153**, A515–A520.

55. R. Kuengas, J.–S. Kim, J. M. Vohs and R. J. Gorte, *J. Am. Ceram. Soc.*, 2011, **94**, 2220–2224.

56. S. P. Jiang and W. Wang, *Solid State Ionics*, 2005, **176**, 1351–1357.

57. E. P. Murray and S. A. Barnett, *Solid State Ionics*, 2001, **143**, 265–273.

58. S. P. Jiang, J. P. Zhang and K. Foger, *J. Electrochem. Soc.*, 2000, **147**, 3195–3205.

59. E. P. Murray, T. Tsai and S. A. Barnett, *Solid State Ionics*, 1998, **110**, 235–243.

60. S. P. Jiang and W. Wang, *J. Electrochem. Soc.*, 2005, **152**, A1398–A1408.

61. S. P. Jiang, Y. J. Leng, S. H. Chan and K. A. Khor, *Electrochem. Solid State Lett.*, 2003, **6**, A67–A70.

62. S. P. Jiang, X. J. Chen, S. H. Chan and J. T. Kwok, *J. Electrochem. Soc.*, 2006, **153**, A850–A856.

63. W. Wang, S. P. Jiang, A. I. Y. Tok and L. Luo, *J. Power Sources*, 2006, **159**, 68–72.

64. S. P. Jiang, Y. M. Ye, T. M. He and S. B. Ho, *J. Power Sources*, 2008, **185**, 179–182.

65. J. Chen, F. L. Liang, L. N. Liu, S. P. Jiang, B. Chi, J. Pu and J. Li, *J. Power Sources*, 2008, **183**, 586–589.

66. Y. M. Ye, T. M. He, Y. Li, E. H. Tang, T. L. Reitz and S. P. Jiang, *J. Electrochem. Soc.*, 2008, **155**, B811–B818.

67. A. Babaei, L. Zhang, S. L. Tan and S. P. Jiang, *Solid State Ionics*, 2010, **181**, 1221–1228.

68. N. Ai, S. P. Jiang, Z. Lu, K. F. Chen and W. H. Su, *J. Electrochem. Soc.*, 2010, **157**, B1033–B1039.

69. J. Chen, F. L. Liang, B. Chi, J. Pu, S. P. Jiang and L. Jian, *J. Power Sources*, 2009, **194**, 275–280.

70. A. Babaei, S. P. Jiang and J. Li, *J. Electrochem. Soc.*, 2009, **156**, B1022–B1029.

71. F. L. Liang, J. Chen, S. P. Jiang, B. Chi, J. Pu and L. Jian, *Electrochem. Solid State Lett.*, 2008, **11**, B213–B216.

72. F. Zhao, L. Zhang, Z. Y. Jiang, C. R. Xia and F. L. Chen, *J. Alloy. Compd.*, 2009, **487**, 781–785.

73. G. Kim, S. Lee, J. Y. Shin, G. Corre, J. T. S. Irvine, J. M. Vohs and R. J. Gorte, *Electrochem. Solid State Lett.*, 2009, **12**, B48–B52.

74. Z. Y. Jiang, L. Zhang, L. L. Cai and C. R. Xia, *Electrochim. Acta*, 2009, **54**, 3059–3065.

75. B. Wei, Z. Lue, T. Wei, D. Jia, X. Huang, Y. Zhang, J. Miao and W. Su, *Int. J. Hydrog. Energy*, 2011, **36**, 6151–6159.

76. M. C. Tucker, G. Y. Lau, C. P. Jacobson, L. C. DeJonghe and S. J. Visco, *J. Power Sources*, 2007, **171**, 477–482.
77. C. M. Zhang, Y. Lin, R. Ran and Z. P. Shao, *Int. J. Hydrog. Energy*, 2010, **35**, 8171–8176.
78. X. Zhu, Z. Lue, B. Wei, Y. Zhang, X. Huang and W. Su, *Int. J. Hydrog. Energy*, 2010, **35**, 6897–6904.
79. S. P. Jiang, *Journal of Power Sources*, 2003, **124**, 390–402.
80. S. P. Jiang, *J. Mater. Sci.*, 2008, **43**, 6799–6833.
81. Y. Sakaki, Y. Takeda, A. Kato, N. Imanishi, O. Yamamoto, M. Hattori, M. Iio and Y. Esaki, *Solid State Ionics*, 1999, **118**, 187–194.
82. S. Carter, A. Selcuk, R. J. Chater, J. Kajda, J. A. Kilner and B. C. H. Steele, *Solid State Ionics*, 1992, **53–6**, 597–605.
83. A. Esquirol, N. P. Brandon, J. A. Kilner and M. Mogensen, *J. Electrochem. Soc.*, 2004, **151**, A1847–A1855.
84. Z. P. Shao and S. M. Haile, *Nature*, 2004, **431**, 170–173.
85. C. Laberty, F. Zhao, K. E. Swider–Lyons and A. V. Virkar, *Electrochem. Solid State Lett.*, 2007, **10**, B170–B174.
86. R. Chiba, F. Yoshimura and Y. Sakurai, *Solid State Ionics*, 1999, **124**, 281–288.
87. D. M. Bastidas, S. W. Tao and J. T. S. Irvine, *J. Mater. Chem.*, 2006, **16**, 1603–1605.
88. A. Tarancon, S. J. Skinner, R. J. Chater, F. Hernandez–Ramirez and J. A. Kilner, *J. Mater. Chem.*, 2007, **17**, 3175–3181.
89. A. Martinez–Amesti, A. Larranaga, L. M. Rodriguez–Martinez, A. T. Aguayo, J. L. Pizarro, M. L. No, A. Laresgoiti and M. I. Arriortua, *J. Power Sources*, 2008, **185**, 401–410.
90. M. D. Anderson, J. W. Stevenson and S. P. Simner, *J. Power Sources*, 2004, **129**, 188–192.
91. H. Y. Tu, Y. Takeda, N. Imanishi and O. Yamamoto, *Solid State Ionics*, 1999, **117**, 277–281.
92. N. Imanishi, R. Ohno, K. Murata, A. Hirano, Y. Takeda, O. Yamamoto and K. Yamahara, *Fuel Cells*, 2009, **9**, 215–221.
93. X. Y. Xu, Z. Y. Jiang, X. Fan and C. R. Xia, *Solid State Ionics*, 2006, **177**, 2113–2117.
94. Z. Jiang, L. Zhang, L. Cai and C. Xia, *Electrochim. Acta*, 2009, **54**, 3059–3065.
95. Z. Jiang, L. Zhang, K. Feng and C. Xia, *J. Power Sources*, 2008, **185**, 40–48.
96. K. Yamahara, C. P. Jacobson, S. J. Visco and L. C. De Jonghe, *Solid State Ionics*, 2005, **176**, 451–456.
97. F. L. Liang, J. Chen, S. P. Jiang, B. Chi, J. Pu and L. Jian, *Electrochem. Commun.*, 2009, **11**, 1048–1051.
98. Y. Y. Huang, J. M. Vohs and R. J. Gorte, *J. Electrochem. Soc.*, 2005, **152**, A1347–A1353.
99. H. P. He, Y. Y. Huang, J. Regal, M. Boaro, J. M. Vohs and R. J. Gorte, *J. Am. Ceram. Soc.*, 2004, **87**, 331–336.

100. Y. Huang, J. M. Vohs and R. J. Gorte, *Electrochem. Solid State Lett.*, 2006, **9**, A237–A240.
101. H. Zhang, F. Zhao, F. Chen and C. Xia, *Solid State Ionics*, 2011, **192**, 591–594.
102. J. Chen, F. Liang, L. Liu, S. Jiang, B. Chi, J. Pu and J. Li, *J. Power Sources*, 2008, **183**, 586–589.
103. S. Choi, S. Yoo, J.–Y. Shin and G. Kim, *J. Electrochem. Soc.*, 2011, **158**, B995–B999.
104. Z. Liu, Z. W. Zheng, M. F. Han and M. L. Liu, *J. Power Sources*, 2010, **195**, 7230–7233.
105. Y. Sakito, A. Hirano, N. Imanishi, Y. Takeda, O. Yamamoto and Y. Liu, *J. Power Sources*, 2008, **182**, 476–481.
106. M. Sahibzada, S. J. Benson, R. A. Rudkin and J. A. Kilner, *Solid State Ionics*, 1998, **115**, 285–290.
107. L. F. Nie, M. F. Liu, Y. J. Zhang and M. L. Liu, *J. Power Sources*, 2010, **195**, 4704–4708.
108. A. Y. Yan, M. Yang, Z. F. Hou, Y. L. Dong and M. J. Cheng, *J. Power Sources*, 2008, **185**, 76–84.
109. W. Zhou, F. L. Liang, Z. P. Shao and Z. H. Zhu, *Scientific Reports*, 2012, **2**.
110. M. E. Lynch, L. Yang, W. Qin, J.–J. Choi, M. Liu, K. Blinn and M. Liu, *Energy & Environmental Science*, 2011, **4**, 2249–2258.
111. Y. Choi, M. C. Lin and M. Liu, *J. Power Sources*, 2010, **195**, 1441–1445.
112. M. F. Liu, D. Ding, K. Blinn, X. X. Li, L. F. Nie and M. Liu, *Int. J. Hydrog. Energy*, 2012, **37**, 8613–8620.
113. S. P. Jiang and S. H. Chan, *J. Mater. Sci.*, 2004, **39**, 4405–4439.
114. S. W. Zha, Z. Cheng and M. L. Liu, *J. Electrochem. Soc.*, 2007, **154**, B201–B206.
115. S. W. Tao and J. T. S. Irvine, *Nat. Mater.*, 2003, **2**, 320–323.
116. A. Atkinson, S. Barnett, R. J. Gorte, J. T. S. Irvine, A. J. McEvoy, M. Mogensen, S. C. Singhal and J. Vohs, *Nat. Mater.*, 2004, **3**, 17–27.
117. S. P. Jiang, L. Liu, K. P. Ong, P. Wu, J. Li and J. Pu, *J. Power Sources*, 2008, **176**, 82–89.
118. S. Park, R. Craciun, J. M. Vohs and R. J. Gorte, *J. Electrochem. Soc.*, 1999, **146**, 3603–3605.
119. R. Craciun, S. Park, R. J. Gorte, J. M. Vohs, C. Wang and W. L. Worrell, *J. Electrochem. Soc.*, 1999, **146**, 4019–4022.
120. S. I. Lee, K. Ahn, J. M. Vohs and R. J. Gorte, *Electrochem. Solid State Lett.*, 2005, **8**, A48–A51.
121. R. J. Gorte, S. Park, J. M. Vohs and C. H. Wang, *Adv. Mater.*, 2000, **12**, 1465–1469.
122. S. McIntosh, J. M. Vohs and R. J. Gorte, *Electrochim. Acta*, 2002, **47**, 3815–3821.
123. S. McIntosh, J. M. Vohs and R. J. Gorte, *Electrochem. Solid State Lett.*, 2003, **6**, A240–A243.

124. G. Kim, G. Corre, J. T. S. Irvine, J. M. Vohs and R. J. Gorte, *Electrochem. Solid State Lett.*, 2008, **11**, B16–B19.
125. S. W. Jung, C. Lu, H. P. He, K. Y. Ahn, R. J. Gorte and J. M. Vohs, *J. Power Sources*, 2006, **154**, 42–50.
126. S. Lee, G. Kim, J. M. Vohs and R. J. Gorte, *J. Electrochem. Soc.*, 2008, **155**, B1179–B1183.
127. S. McIntosh and R. J. Gorte, *Chem. Rev.*, 2004, **104**, 4845–4865.
128. A. Cantos–Gomez, R. Ruiz–Bustos and J. van Duijn, *Fuel Cells*, 2011, **11**, 140–143.
129. H. Kim, C. da Rosa, M. Boaro, J. M. Vohs and R. J. Gorte, *J. Am. Ceram. Soc.*, 2002, **85**, 1473–1476.
130. H. Kim, C. Lu, W. L. Worrell, J. M. Vohs and R. J. Gorte, *J. Electrochem. Soc.*, 2002, **149**, A247–A250.
131. L. L. Zheng, X. Wang, L. Zhang, J. Y. Wang and S. P. Jiang, *Int. J. Hydrog. Energy*, 2012, **37**, 10299–10310.
132. H. Kurokawa, T. Z. Sholklapper, C. P. Jacobson, L. C. De Jonghe and S. J. Visco, *Electrochem. Solid State Lett.*, 2007, **10**, B135–B138.
133. S. P. Jiang, S. Zhang, Y. Da Zhen and W. Wang, *J. Am. Ceram. Soc.*, 2005, **88**, 1779–1785.
134. H. P. He, A. Wood, D. Steedman and M. Tilleman, *Solid State Ionics*, 2008, **179**, 1478–1482.
135. S. P. Jiang, X. J. Chen, S. H. Chan, J. T. Kwok and K. A. Khor, *Solid State Ionics*, 2006, **177**, 149–157.
136. X. C. Lu, J. H. Zhu, Z. G. Yang, G. G. Xia and J. W. Stevenson, *J. Power Sources*, 2009, **192**, 381–384.
137. Q. X. Fu, F. Tietz, D. Sebold, S. W. Tao and J. T. S. Irvine, *J. Power Sources*, 2007, **171**, 663–669.
138. J. Karczewski, B. Bochentyn, S. Molin, M. Gazda, P. Jasinski and B. Kusz, *Solid State Ionics*, 2012, **221**, 11–14.
139. J. S. Park, J. Luo, L. Adijanto, J. M. Vohs and R. J. Gorte, *J. Power Sources*, 2013, **222**, 123–128.
140. B. H. Smith and M. D. Gross, *Electrochem. Solid State Lett.*, 2011, **14**, B1–B5.
141. L. Adijanto, R. Kungas, J. Park, J. M. Vohs and R. J. Gorte, *Int. J. Hydrog. Energy*, 2011, **36**, 15722–15730.
142. J. S. Qiao, K. N. Sun, N. Q. Zhang, B. Sun, J. R. Kong and D. R. Zhou, *J. Power Sources*, 2007, **169**, 253–258.
143. A. N. Busawon, D. Sarantaridis and A. Atkinson, *Electrochem. Solid State Lett.*, 2008, **11**, B186–B189.
144. F. Liang, W. Zhou, B. Chi, J. Pu, S. P. Jiang and L. Jian, *Int. J. Hydrog. Energy*, 2011, **36**, 7670–7676.
145. J.–S. Kim, N. L. Wieder, A. J. Abraham, M. Cargnello, P. Fornasiero, R. J. Gorte and J. M. Vohs, *J. Electrochem. Soc.*, 2011, **158**, B596–B600.
146. W. Zhu, D. Ding and C. Xia, *Electrochem. Solid State Lett.*, 2008, **11**, B83–B86.

147. J. D. Nicholas and S. A. Barnett, *J. Electrochem. Soc.*, 2009, **156**, B458–B464.
148. M. Shah, J. D. Nicholas and S. A. Barnett, *Electrochem. Commun.*, 2009, **11**, 2–5.
149. F. Bidrawn, R. Kuengas, J. M. Vohs and R. J. Gorte, *J. Electrochem. Soc.*, 2011, **158**, B514–B525.
150. W. Wang, M. D. Gross, J. M. Vohs and R. J. Gorte, *J. Electrochem. Soc.*, 2007, **154**, B439–B445.
151. F. L. Liang, J. Chen, S. P. Jiang, F. Z. Wang, B. Chi, J. Pu and L. Jian, *Fuel Cells*, 2009, **9**, 636–642.
152. A. Babaei, L. Zhang, E. J. Liu and S. P. Jiang, *J. Alloy. Compd.*, 2011, **509**, 4781–4787.
153. M. Shah, P. W. Voorhees and S. A. Barnett, *Solid State Ionics*, 2011, **187**, 64–67.
154. F. Zhao, R. R. Peng and C. R. Xia, *Mater. Res. Bull.*, 2008, **43**, 370–376.
155. F. Zhao, L. Zhang, Z. Jiang, C. Xia and F. Chen, *J. Alloy. Compd.*, 2009, **487**, 781–785.
156. Y. X. Zhang and C. R. Xia, *J. Power Sources*, 2010, **195**, 6611–6618.
157. J. W. Fergus, *J. Power Sources*, 2005, **147**, 46–57.
158. R. F. Wang, Z. Lu, C. Q. Liu, R. B. Zhu, X. Q. Huang, B. Wei, N. Ai and W. H. Su, *J. Alloys Compd.*, 2007, **432**, 189–193.
159. M. K. Mahapatra and K. Lu, *J. Power Sources*, 2010, **195**, 7129–7139.
160. T. Zhang, W. G. Fahrenholtz, S. T. Reis and R. K. Brow, *J. Am. Ceram. Soc.*, 2008, **91**, 2564–2569.
161. K. F. Chen, N. Ai, C. Lievens, J. Love and S. P. Jiang, *Electrochemistry Communications*, 2012, **23**, 129–132.
162. K. F. Chen, A. Na, L. Zhao and S. P. Jiang, *J. Electrochem. Soc.*, 2013, **160**, F183–F190.
163. S. P. Jiang, J. G. Love and Y. Ramprakash, *J. Power Sources*, 2002, **110**, 201–208.
164. B. C. H. Steele, *Solid State Ionics*, 2000, **129**, 95–110.
165. S. P. Jiang and W. Wang, *Electrochem. Solid State Lett.*, 2005, **8**, A115–A118.
166. F. L. Liang, J. Chen, B. Chi, J. A. Pu, S. P. Jiang and L. Jian, *J. Power Sources*, 2011, **196**, 153–158.
167. E. B. Fox, A. F. Lee, K. Wilson and C. S. Song, *Top. Catal.*, 2008, **49**, 89–96.
168. Y. Nabae and I. Yamanaka, *Appl. Catal. A–Gen.*, 2009, **369**, 119–124.
169. S. P. Jiang, *Solid State Ionics*, 2002, **146**, 1–22.
170. J. M. Ralph, A. C. Schoeler and M. Krumpelt, *J. Mater. Sci.*, 2001, **36**, 1161–1172.
171. S. Z. Wang and H. Zhong, *J. Power Sources*, 2007, **165**, 58–64.
172. Y. Y. Huang, J. M. Vohs and R. J. Gorte, *J. Electrochem. Soc.*, 2006, **153**, A951–A955.
173. M. Søgaard, T. Z. Sholklapper, M. Wandel, L. C. De Jonghe and M. Mogensen, Proceedings (on CD–ROM) 8th European Solid Oxide

Fuel Cell Forum, Lucerne, CH European solid oxide fuel cell forum, 2008, 30 June–4 July.

174. Y. Wang, L. Zhang and C. Xia, *Int. J. Hydrogen Energy*, 2012, **37**, 2182–2186.

175. F. Bidrawn, G. Kim, N. Aramrueang, J. M. Vohs and R. J. Gorte, *J. Power Sources*, 2010, **195**, 720–728.

176. K. K. Hansen, M. Wandel, Y. L. Liu and M. Mogensen, *Electrochim. Acta*, 2010, **55**, 4606–4609.

177. S. Lee, N. Miller, H. Abernathy, K. Gerdes and A. Manivannan, *J. Electrochem. Soc.*, 2011, **158**, B735–B742.

178. J. Hojberg and M. Sogaard, *Electrochem. Solid State Lett.*, 2011, **14**, B77–B79.

179. N. Ai, K. F. Chen and S. P. Jiang, *Solid State Ionics*, 2013, **233**, 87–94.

180. J. Nielsen and T. Jacobsen, *Solid State Ionics*, 2007, **178**, 1001–1009.

181. S. B. Adler, *Chem. Rev.*, 2004, **104**, 4791–4843.

182. M. Mogensen, N. M. Sammes and G. A. Tompsett, *Solid State Ionics*, 2000, **129**, 63–94.

183. J. Chen, F. L. Liang, D. Yan, J. Pu, B. Chi, S. P. Jiang and L. Jian, *J. Power Sources*, 2010, **195**, 5201–5205.

184. R. Kungas, F. Bidrawn, J. M. Vohs and R. J. Gorte, *Electrochem. Solid State Lett.*, 2010, **13**, B87–B90.

185. T. Horita, H. Kishimoto, K. Yamaji, M. E. Brito, Y. P. Xiong, H. Yokokawa, Y. Hori and I. Miyachi, *J. Power Sources*, 2009, **193**, 194–198.

186. Y. D. Zhen and S. P. Jiang, *J. Electrochem. Soc.*, 2006, **153**, A2245–A2254.

187. J. M. Serra and H. P. Buchkremer, *J. Power Sources*, 2007, **172**, 768–774.

188. S. P. Jiang, *J. Solid State Electrochem.*, 2007, **11**, 93–102.

189. Y. J. Leng, S. H. Chan, K. A. Khor and S. P. Jiang, *J. Appl. Electrochem.*, 2004, **34**, 409–415.

190. S. P. Jiang and J. G. Love, *Solid State Ionics*, 2001, **138**, 183–190.

191. M. Backhaus–Ricoult, K. Adib, T. S. Clair, B. Luerssen, L. Gregoratti and A. Barinov, *Solid State Ionics*, 2008, **179**, 891–895.

192. J.–S. Kim, S. Lee, R. J. Gorte and J. M. Vohs, *J. Electrochem. Soc.*, 2011, **158**, B79–B83.

193. J. D. Nicholas and S. A. Barnett, *J. Electrochem. Soc.*, 2010, **157**, B536–B541.

CHAPTER 7

Three Dimensional Reconstruction of Solid Oxide Fuel Cell Electrodes

P. R. SHEARING[*a] AND N. P. BRANDON[*b]

[a] Department of Chemical Engineering, University College London, UK;
[b] Department of Earth Science and Engineering, Imperial College London, UK
*Email: p.shearing@ucl.ac.uk; n.brandon@imperial.ac.uk

SOFC electrodes are complex porous materials, the microstructure of which will have significant impact on device performance. Whilst for some time there has been widespread agreement regarding the importance of microstructure, there has been little quantitative understanding of the link between electrode microstructure and device performance.

Historically, microstructural characterisation of SOFC electrodes has been constrained to two-dimensional optical and electron techniques, which are inherently limited in their ability to describe the fundamentally three-dimensional characteristics of these functional materials. As a result, microstructural optimisation has been largely empirical and there has been little consensus as to what constitutes a "good" microstructure.[1]

In recent years, the development of more sophisticated tools for microstructural characterisation at length scales of relevance to SOFC electrodes has enabled researchers to explore these materials in three dimensions for the first time. In doing so, it has been possible to build microstructural maps of the electrodes, and to explore the complex interplay of gas and solid phases, their percolation, and their relationship to performance.

RSC Energy and Environment Series No. 7
Solid Oxide Fuel Cells: From Materials to System Modeling
Edited by Meng Ni and Tim S. Zhao
© The Royal Society of Chemistry 2013
Published by the Royal Society of Chemistry, www.rsc.org

This chapter will introduce the most widespread techniques that have been used to examine SOFC electrode microstructures in 3D and review the relevant literature to explore what these tools can reveal. Specifically the chapter will review:

1. The importance of 3D characterisation and the limitations of stereology.
2. Focused ion beam techniques.
3. X-ray techniques.
4. Data analysis and image based modelling.

7.1 The Importance of 3D Characterisation and the Limitations of Stereology

Reactions in SOFCs are supported by complex porous materials, which provide intimate contact of an ionic, electronic, and gas phase in order to effectively catalyze the hydrogen oxidation/oxygen reduction reaction.

A wide range of materials has been explored for application as SOFC electrodes, and the enhanced throughput of modern materials discovery promises to accelerate further the introduction of new electrode materials. For the purposes of this chapter, we will focus on the "technologically relevant" materials that have already found widespread adoption in research and commercial systems. These can be broadly categorised as those based on a porous composite based on a mixture of an electronic conducting material and an ionic conducting material, such as Ni-YSZ anodes, and those based on a porous mixed ionic-electronic conducting (MIEC) material, such as LSCF cathodes, see Figure 7.1.

In anodes, such as Ni-YSZ, the hydrogen oxidation reaction is thought to occur at discrete locations within the complex porous composite structure, namely, the meeting points of the electronic, ionic, and gas phases at so-called triple phased boundaries (TPBs). The successful completion of the anode half cell reaction:

$$2H_2 + 2O^{2-} \leftrightarrow 2H_2O + 4e^- \tag{1}$$

requires transport of electrons (in the Ni phase), oxide ions (in the YSZ phase), and hydrogen and steam (in the pore phase) to/from the TPB.

TPBs can be considered active or inactive: in an active TPB, each of the pore, ionic, and electronic conducting phases must provide contiguous contact to and from the active catalytic site, in turn supporting the transport of reactant species to support the reaction. A TPB can be rendered inactive because of a lack of percolation in any of the three phases, which prevents the transport of one of the essential reacting species. This is shown schematically in Figure 7.2, where the blue phase in the electronic conductor, red is an ionic conductor, and the pore space is the surrounding void.

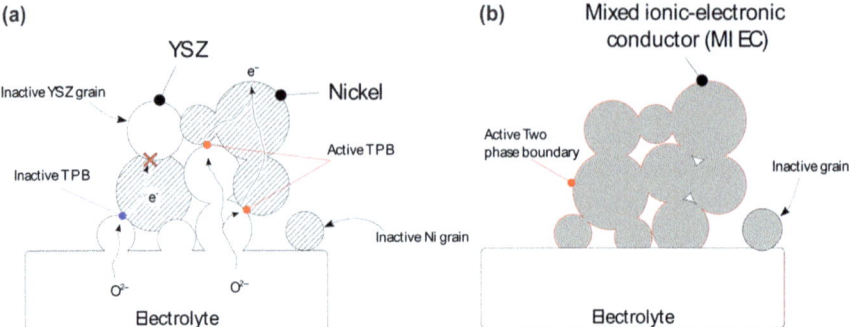

Figure 7.1 Transport of reacting species in (a) composite electrode material (Ni–YSZ) and (b) single phase MIEC electrode (reproduced with kind permission of Denis Cumming).

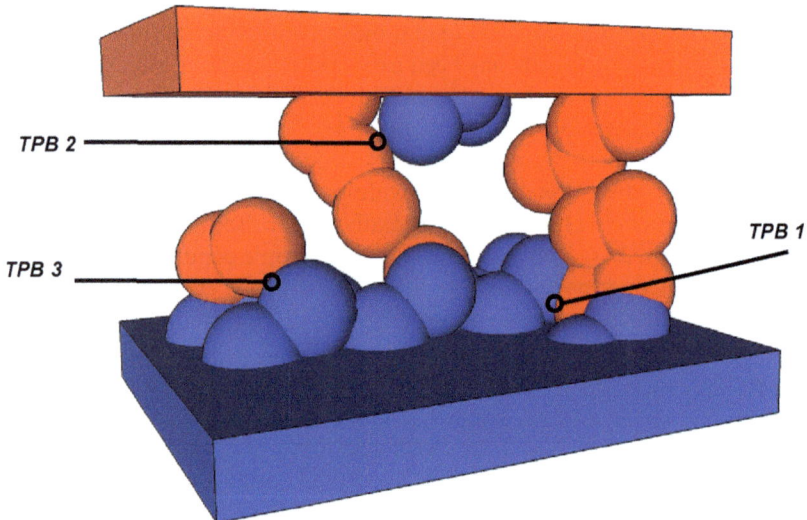

Figure 7.2 Representative composite electrode microstructure: blue relates to ion conducting phase and red to electron conducting phase.

If we assume that the above represents an anode made from a porous mixture of Ni and YSZ, then in order for the reaction to occur, electrons must be transported from the current collector to the TPB, hydrogen must be allowed to percolate through a connected pore structure from the fuel supply, steam must be allowed to percolate through a connected pore structure from the electrolyte interface into the fuel supply, and oxide ions must be transported from the electrolyte. Three TPBs have been identified in Figure 7.2. TPB1 is an example of an active TPB where there is pre-requisite contiguity in each of the three phases. TPB2 is inactive owing to a lack of percolation in the ionic phase between the electrolyte and TPB2, and TPB3 is inactive owing to a lack of percolation in the electronic phase between the current collector and TPB3.

Whilst in the above example we have considered a composite anode, the same analogy can be made for the pure electronic ionic conducting composite LSM-YSZ, where the half cell reaction:

$$O_2 + 4e^- \leftrightarrow 2O^{2-} \tag{2}$$

is supported by electronic transport in LSM, ionic in YSZ and O_2 gas in the pore structure.

The density of triple phase boundaries is a key indicator of electrochemical performance of an electrode material. In order to quantify the relationship between electrode microstructure and performance it is essential that we discriminate between the active and inactive TPBs. As their activity is dictated by percolation, characterisation of these structures in three dimensions is essential.

In mixed conducting electrodes – such as shown in Figure 7.1b, the reaction zone is extended away from the one-dimensional TPB, since the transport of ions and electrons can occur through the MIEC electrode, the active reaction zone is not isolated at TPBs. This means that the microstructural requirements can be less stringent; however, percolation between the electrolyte and current collector remains a pre-requisite for electrochemical activity. In this case an "active two-phase boundary" defines the solid-pore interface.

The microstructure of an electrode has an influence on the electrochemical performance of an SOFC over its full range of operation, see Figure 7.3. Considering the current–voltage behaviour of an SOFC electrode, operation in different regimes results in different voltage losses; at low currents, losses are predominantly due to kinetic processes, in this regime optimal microstructures should maximise triple phase contact to enhance electrochemical performance. Conversely, at high current densities, SOFC operation is limited by mass

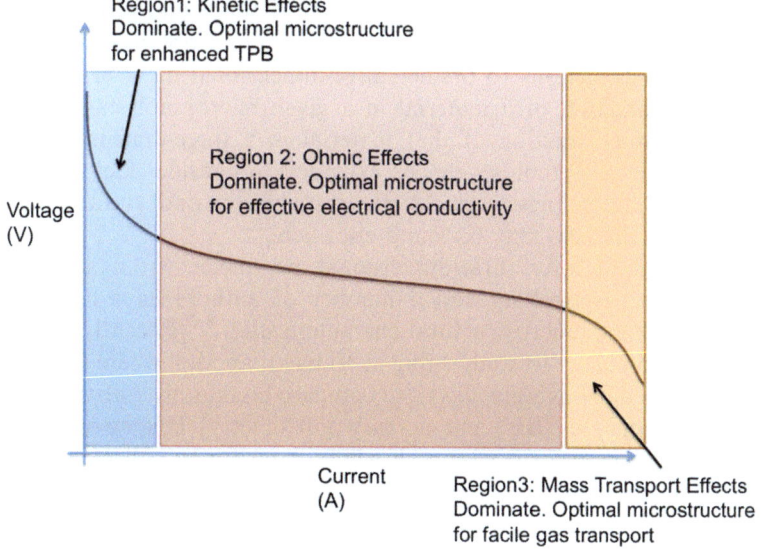

Figure 7.3 A typical SOFC polarisation curve; characteristic resistances have been associated with microstructural properties.

transport of reactant species; negative mass transport effects are mitigated by providing a pore network optimised for reactant delivery. Ohmic effects, which are a contributory loss factor at all current densities, are dictated by the nature of conductivity of the ionic and electronic phases, as well as geometric factors such as electrode thickness and current collector/cell design.

Microstructure will impact SOFC performance across all regimes of operation. However, these effects can be conflicting and as such microstructural requirements are dependent on the rate limiting step in the selected operating regime.

The implications of electrode microstructure are not constrained to electrochemical performance and attempts to design electrodes are compounded by the diverse range of processing and environmental parameters that can affect fuel cell microstructure. Processing factors include: particle size distribution, composition, deposition methods and sintering.

During operation, microstructure can be affected by redox cycling, load and thermal cycling, and long-term sintering. These microstructural evolution processes have numerous impacts, from reduced electrical or electrochemical performance to catastrophic failure by cracking or delamination. The reader is referred to a recent review paper for further discussion.[2]

The importance of electrode microstructure in determining the performance and durability of SOFCs has been demonstrated; however, the ability to utilise the sub-micron architecture of the SOFC has often been constrained by a limited understanding of existing microstructures and the effects of a wide range of processing and operating parameters.

Conventional tomography techniques (see Figure 7.4) have historically provided insufficient resolution to characterize SOFC electrodes, and hence attempts to interpret the three-dimensional structure of electrode materials have been constrained to stereology. Based on fundamental geometrical principles, stereology techniques utilise information obtained from two-dimensional cross-sections to predict three-dimensional data; in the simplest case, the area fraction of a material in a given micrograph can be used to predict the volume fraction of that material in a three-dimensional microstructure; however, stereology can be extended to consider factors including triple phase contact lines, particle size distribution and surface area. An introduction to the field may be found elsewhere.[3]

The potential to infer three-dimensional parameters from simple cross-sectional data is compelling and a number of authors have applied these techniques to SOFC microstructural characterisation.[4–7] Recently Faes et al.[7] have applied stereology to study aging in Ni based anodes; where the results of stereological predictions were shown to compare favourably with conventional porosimetry techniques (BET and mercury porosimetry). In spite of the benefits of stereology (primarily the ease with which data can be collected), the limitations of the technique are clear and have been acknowledged.[8] Furthermore, as with any 2D technique, it is not possible to make predictions regarding the nature of three-dimensional percolation, which is of significant importance when considering the electrochemical activity of electrode microstructures. For SOFC applications, Song et al. confirmed these limitations, citing the need for

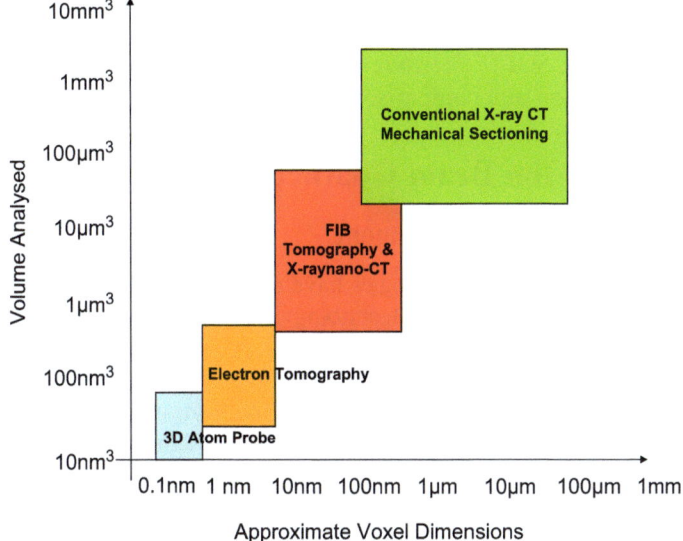

Figure 7.4 Available tomography procedures categorised as a function of the estimated analysis volume and the achievable volume pixel (voxel) resolution.

"a better technique to study microstructural features in conjunction with the electrode performance".

In recent years, improvements in tomography techniques have provided the capability to explore complex porous materials in three-dimensions at length scales varying form the atomic to the device level. The suite of available tomography techniques is shown in Figure 7.4 – where the techniques have been categorized by the resolution that can be obtained and the sample volume that can be analyzed. Clearly, both parameters must be matched to the material under investigation; sufficient resolution is necessary to capture the features of interest, whilst the sample volume analyzed must be sufficient to be considered representative of the bulk sample. This is the subject of much discussion in the literature.[9,10]

At the highest resolution, but with the correspondingly smallest sample volume, are techniques such as atom probe tomography and transmission electron tomography. These techniques provide valuable insight into the fine three-dimensional structure at sub-nm resolution and have been successfully applied to explore SOFC materials. The corresponding analysis volume is significantly smaller than can be considered representative of the bulk electrode and these techniques have been used to explore particles[11] and interfaces,[12] respectively. Owing to their limited analysis volume these techniques will not be considered further here.

For the characterisation of SOFC electrode microstructures in three dimensions, the two most widely adopted techniques are focused ion beam (FIB) tomography (also known as "slice and view") and X-ray nano computed tomography. These techniques provide the required sub-100nm resolution coupled with a field of view that can be considered representative of the bulk electrode. Over the past five years, these techniques have revolutionized the characterisation of SOFC electrode microstructures, providing access to new metric for evaluation and comparison.

7.2 Focused Ion Beam Characterisation

7.2.1 The FIB-SEM Instrument

Focused Ion Beam (FIB) technologies have developed rapidly over the past 20 years, the rapid maturation of this technology can be attributed to the wide range of disciplines to which FIB techniques may be applied. The application of FIB technologies within materials science is also very varied, including *in situ* micro-milling and ablation, etching, imaging, sample preparation, ion implantation, and material deposition. For a detailed overview of the design and application of FIB systems, two recent books by Gianuzzi and Yao provide a wealth of information.

FIB techniques employ a liquid metal ion source (LMIS), which is incident on a solid surface resulting in a number ion-solid interactions, for example, atom sputtering, electron emission, induced chemical reaction, atom displacement and ion impregnation.[13] Sputtering occurs by elastic collisions between the ion beam and target atoms resulting in momentum transfer to the solid surface. If the kinetic energy received by the target atom is sufficient to overcome the surface binding energy, the atom may be ejected as sputtered material, giving rise to the precision milling capability of FIB systems.

The LMIS, most commonly Gallium (Ga+), provides a source of ions of approximately 5 nm diameter, the extracted ions are accelerated through the ion column by potentials typically in the range 5–50 keV. The ion beam is then focused by a dual lens system before it is incident on the sample surface. The working distance from ion source to sample is relatively large, which allows for the use of samples with varying size and topography.

Gas injection systems are primarily used to introduce localised deposition of metals or insulators, for example, platinum may be introduced to the surface of a sample by ion beam assisted chemical vapour deposition – whilst there are a number of applications for GIS systems, in this context they are primarily used to provide a sacrificial metal layer to improve the quality of slice-and-view data.

Dual beam systems combine the ion beam with a conventional electron column; the ion column can be used for both milling and imaging in the same fashion as the single beam instrument, however complementary use of both beams allows the user to mill using the ion beam and simultaneously image using the electron beam. In order to facilitate simultaneous imaging and processing, the beams are separated by an angle of between 52 and 54 degrees. The capability to sequentially mill and image a sample enables the collection of sequences of 2D electron micrographs which can be effectively recombined to provide 3D microstructural maps.

Electron imaging is achieved when a surface is irradiated with incident electrons, leading to the generation of back-scattered and secondary electrons. It is of interest to consider the mechanisms by which these "imaging" electrons are generated; as this greatly affects the success of the 3D tomographic procedure. Goldstein *et al.* provide a comprehensive review of the technique for the interested reader.

Back-scattered electrons are generated by elastic collisions between the electron beam and atoms within the specimen under examination. Numerous collisions may provide sufficient deviation of the electron beam from its incident path to re-direct the electron to the surface of the specimen and subsequently escape; these are back-scattered electrons.[14] The degree of back-scattering, denoted by the back-scatter coefficient, generally increases with atomic number.

Secondary electrons are those generated as ionisation products of the primary beam. Classified as SE1 and SE2 signals, depending on whether they were generated by a beam or back-scattered electron (see Figure 7.2). Secondary electrons are much lower energy than back-scattered electrons, therefore they are very sensitive to surface conditions and the probability of escape (intensity of SE signal) decreases exponentially with distance from the surface.[14] Unlike back-scattered electrons, there is no significant relationship between secondary electron yield and atomic number. However, as the signal is based on ionization of surface atoms, the signal tends to be higher for metals where the ionization energy is correspondingly lower.

With advances in field emission electron guns, it has become possible to conduct electron imaging at very low voltages, this not only reduces the penetration of the beam into the specimen, thereby providing surface specific information,[14] but also changes the back-scatter coefficient for different materials.[15]

Figure 7.5 shows the interaction of a beam electron with a specimen surface generating backscatter, SE1 and SE2 signals. The beam penetration volume is a function of beam-accelerating voltage, beam angle of incidence and of the specimen material.

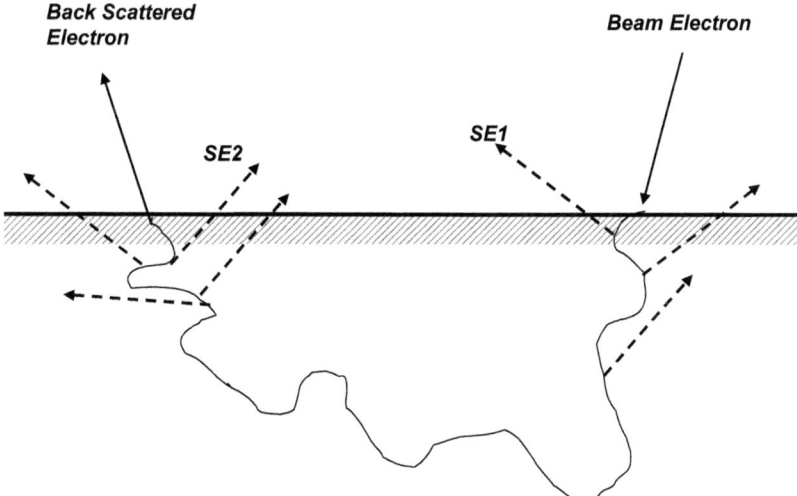

Figure 7.5   Interaction of an electron beam with a specimen surface.

FIB instruments may also combine facilities for secondary analysis, for example, energy dispersive spectroscopy (EDS) and secondary ion mass spectroscopy (SIMS) for chemical analysis and electron back scatter diffraction (EBSD), which enables crystallographic analysis.

7.2.2 Application of FIB-SEM Techniques to SOFC Materials

The pioneering work of Wilson *et al.*[16] in 2006 was the first to demonstrate the application of FIB-SEM tomography to characterise the three-dimensional structure of a Ni-YSZ SOFC electrode. Performing sequential ion beam milling with electron beam imaging, the authors were able to obtain a sequence of 2D images that could be reconstructed in three-dimensional space, providing the first comprehensive microstructural map of an SOFC electrode with direct measurement of TPB density.

The subsequent proliferation of the technique is testament to its power for characterisation of electrode microstructures, in this section we will begin by reviewing the methodologies for FIB slice and view before exploring application examples from the literature.

FIB-SEM experiments must be carefully optimised in order to maximise the quality of the raw data that is obtained. Failure to do so results in poor quality data that may prove difficult or impossible to analyze or even provide erroneous results.

Ion beam milling currents must be carefully selected to ensure even milling of the polished face, failure to do so results in the so-called "theatre curtain effect"[17] where the streaks of the face of interest prevent analysis (See Figure 7.6a). This can be mitigated by the addition of a sacrificial layer of GIS deposited platinum, which can normalise uneven topographies and also reduce the impact of non-uniform "beam tails", which result from poor ion beam alignment.[18] Figure 7.6b shows the resulting smooth, milled face of interest for a sample where a sacrificial Pt layer has been deposited.

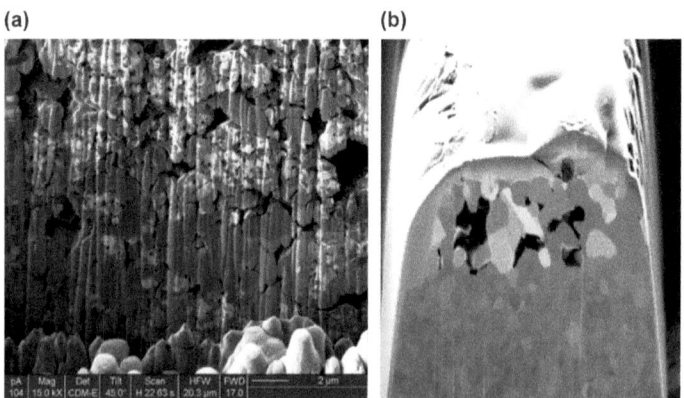

Figure 7.6 (a) The negative effects of streaking; the so-called "theatre curtain effect" (b) Elimination of streaking by the addition of a layer of sacrificial Pt.

Figure 7.7 Micrograph of electrode cross-section; shadowing caused by restricted access/egress of ions/electrons has altered the contrast differential at the far left of the image.

(a)   (b)

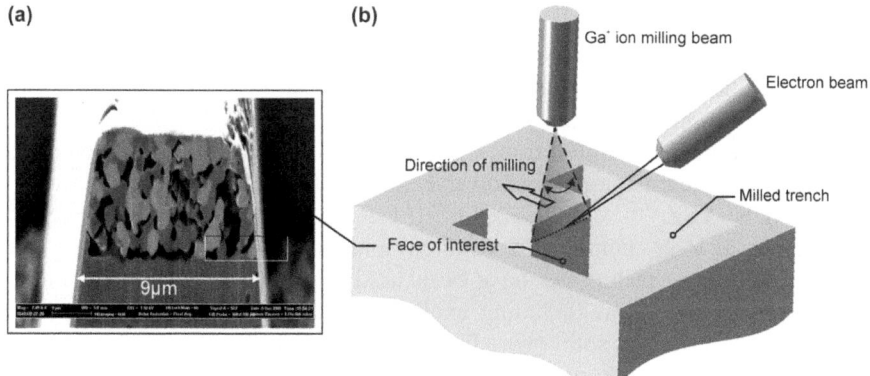

Figure 7.8 (a) an individual slice of a Ni-YSZ electrode (b) typical setup of the FIB instrument for slice and view experiments.

It is also highly desirable to maintain uniform contrast across the face of interest (although there are some software techniques which can post-process these artefacts.[19] To obtain this, incident and imaging electrons should have uniform access/egress from the face of interest. Failure to ensure this, for example, because of a milled corner will skew the contrast across the imaged face (see Figure 7.7).

In order to avoid these common pit-falls, the most widely adopted milling geometry is shown in Figure 7.8, whereby a large U-shaped trench is milled in front of the face of interest prior to the slice-and-view experiment. This trench, which can be milled at high currents, provides excellent access to the face of interest in order to prevent shadowing and also provides a large empty volume in case of any local re-deposition of sputtered material. The top-surface of the milled volume will typically be covered with a fine layer of GIS deposited Pt.

A number of groups have also adopted an impregnation technique[20,21] to infiltrate the pore structure of an electrode with an epoxy resin; this has dual advantages. Firstly, the impregnation has been generally shown to enhance the three-phase contrast. Secondly, as the FIB gun ablates a porous face of interest, the electron imaging can capture out of plane information by "seeing" through the pore structure to structures behind the current slice – whilst this can be accounted for with good image processing,[19] the impregnation approach prevents this altogether. This is particularly useful for highly porous samples.

The detector and voltage used will also affect the inter-phase contrast generated. Whilst this is less important for single phase materials (see *e.g.* Gostovic[22]), it is essential for composite materials where the reliability of phase assignation will affect the calculation of triple phase boundary contact. For Ni-based anode materials, enhanced contrast has been achieved using the SE2 signal at low electron accelerating voltages, typically below 2kV. Thyden *et al.*[15] discuss this in detail. For composite cathode materials, such as LSM-YSZ, the best contrast has been obtained using advanced back-scatter detectors[23–25] such as the Zeiss Energy Selective Back-scatter detector which provides enhanced phase contrast by filtering the energy of the back-scattered electron. The use of correlative EDS analysis has also been demonstrated to provide further confidence on phase assignation.[20]

The widespread proliferation of the FIB-SEM tomography techniques has seen characterisation of a diverse range of SOFC electrode materials, including Ni-YSZ,[16,20,26–28] Ni-CGO,[29,30] LSCF,[22] LSM-YSZ,[23–25] and LSC.[31] Alongside a large number of detailed microstructural investigations (see Figure 7.9), these techniques have been used to conduct a variety of micro-structural investigations including explorations of sintering in anodes[21] and cathodes,[24,32] effects of solid phase fractions,[30,33] novel deposition techniques,[34] alongside investigations of aging,[35] oxidation[36] and post-mortem analysis of microstructural changes in cells from operating systems.[37]

FIB-SEM techniques have also been compared with synthetic geometries[38] and combined with a variety of simulation tools.

A number of authors have also successfully correlated microstructural data with measured, macroscopic electrochemical performance – providing a direct linkage between the microscopic and device level behaviour of the SOFC.[6,20,24,39,40]

In spite of the flexibility and availability of FIB-SEM experiments, they are inherently destructive and therefore limited in their ability to characterise microstructural evolution processes. In the next section, we will look at non-destructive X-ray nano computed tomography which has demonstrated excellent correlation to FIB-SEM characterisation.[24,41]

7.3 Microstructural Characterisation using X-rays

7.3.1 X-ray Microscopy and Tomography

X-Ray computed tomography is based on the use of complex mathematics to reconstruct a sequence of X-Ray transmission micrographs into a 3D volume. In order to understand X-ray tomography, we must first understand the collection of 2-D transmission or shadow images.

In transmission X-Ray microscopy, it is useful to consider the X-ray beam as a discrete stream of photons. When a surface is irradiated with the X-ray beam, the stream of photons propagates through the material, however, when a collision occurs between the photons and atoms of the constituent matter, energy is lost and the X-ray beam is said to have been attenuated. By measuring the attenuation of incident X-rays, it is possible to probe material microstructures in two dimensions in a manner similar to medical radiography.

The extent of X-ray attenuation depends on the material with which the photon stream collides; the attenuation of the beam is defined by the Beer-Lambert Law, which can be used to calculate X-ray attenuation for different materials: the linear attenuation coefficient of a sample is a function of atomic weight and density.

In transmission X-ray microscopy (as in medical X-ray radiography), a negative image is formed on the film; therefore, areas of dense material that have attenuated the beam energy appear bright white (*e.g.* bone), whereas areas of low density which do not attenuate the beam significantly appear black (*e.g.* soft tissue).

In X-ray computed tomography (CT) the sample is rotated relative to the source and detector, with transmission images collected at discrete angular steps. The "shadow" images (see Figure 7.10b) can be converted into

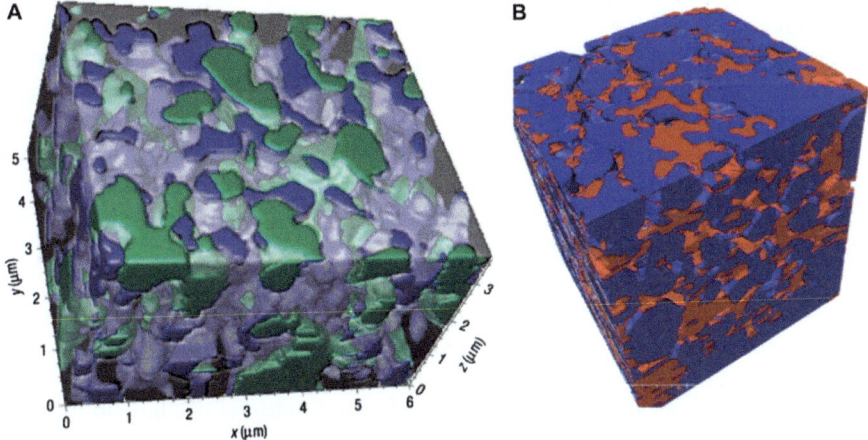

Figure 7.9   Early reconstructions of Ni-YSZ electrodes using (a) FIB Tomography reproduced from Wilson *et al.*[16] (b) X-ray nano-CT reproduced from Shearing *et al.*[9]

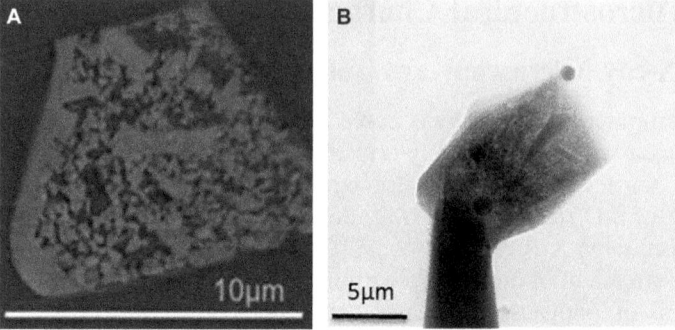

Figure 7.10   (a) Tomographic data is reconstructed into "slices" from (b) shadow or transmission images.

tomography "slice" data (see Figure 7.10a) using mathematical operations. Most commonly, this is achieved using a filtered back projection algorithm, although there are a number of alternatives – a detailed discussion is beyond the scope of this chapter, and the interested reader is referred to Banhart *et al.* for further details.

Historically, the resolution provided by conventional X-ray CT systems has limited the application to SOFC materials, which characteristically require significantly sub-micron resolution. In conventional systems, with a "cone-beam" X-ray source, the resolution that can be achieved is limited by the X-ray spot size and therefore by the working distance between the source and the sample – typically resulting in a resolution limit of >1μm.

Significant improvements in the resolution obtainable in X-ray systems have been achieved with the implementation of X-ray optics. The most widely adopted technique utilises focusing optics based on Fresnel zone plates, diffraction based lenses which can be used for both pre- and/or post-transmission focusing of the X-ray beam. Accordingly, resolution improvements of more than an order of magnitude have been demonstrated compared with conventional CT systems. A schematic of a "nano-CT" system is shown in Figure 7.11.

The improvement in resolution afforded by the introduction of X-ray optics has seen the widespread deployment of X-ray nano computed tomography in the study of SOFC materials. Two distinct techniques have emerged based on the energy, coherence and mono-chromaticity of the X-ray beam used, which can be generated by either a laboratory or synchrotron source.

Lab-based sources are typically based around the bombardment of a metal target with an electron beam. The X-ray flux that can be obtained in this manner is limited by the electron beam energy and is also a function of the target composition and is many orders of magnitude below that obtainable using synchrotron sources.

Synchrotron radiation is obtained by the radial acceleration of electrons, which causes the emission of electromagnetic radiation in the form of photons. Synchrotron facilities take advantage of this to produce high-energy X-ray

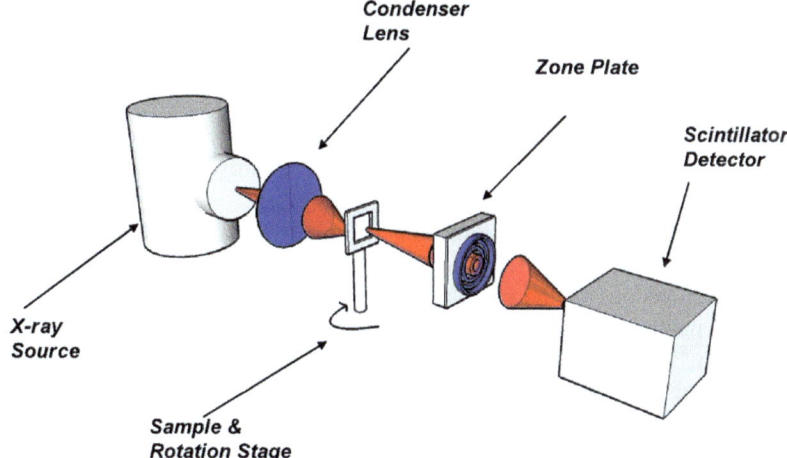

Figure 7.11   A schematic diagram of a typical X-ray nano-CT instrument utilising pre- and post-transmission optics.

beams by the continuous centripetal acceleration of electrons in a vacuum storage ring.[42] In third-generation synchrotron sources, arrays of magnets undulate the electron beam, which generates electromagnetic radiation which can be superimposed to provide brilliant radiation with minimal spot size. Banhart[43] provides a review of the historical development of X-ray sources, showing that current synchrotron sources provide X-rays over 15 orders of magnitude brighter than conventional lab sources. As well as providing enormous improvement in X-ray flux, modern synchrotrons are also equipped to provide highly monochromatic X-ray beams by use of crystal mono-chromators (see Banhart *et al.*[43]), this gives rise to complementary spectroscopy tools that will be discussed further in the next section.

7.3.2   Lab X-ray Instruments

Whilst there are a large number of commercially available laboratory CT systems, currently the only lab system utilizing focusing optics, and therefore providing sufficient resolution for SOFC electrode characterisation is manu-factured by Xradia Inc. in California.

The system uses a capillary condensing optic to increase the flux of the lab X-ray source before transmission, and a stacked Fresnel zone plate for post transmission focusing achieving sub-50 nm resolution in high resolution mode.[44] The lab source is from a Cu target providing X-rays with a peak energy of ca. 8 keV.

The first application of X-ray nano-CT to study SOFC electrodes was presented by Izzo *et al.* in 2008, who successfully characterized the pore structure of a tubular Ni-YSZ electrode and subsequently coupled the micro-structure with Lattice-Boltzmann simulations.

In order to ensure that good signal-to-noise ratios are obtained for the tomography sequences, the geometry of the sample under investigation should ideally be matched to the field of view of the X-ray microscope, which is determined by the zone plate dimensions. For the Xradia instrument, two configurations are possible, the first providing a field of view of ca. 65 µm with a ca. 150 nm spatial resolution and a second high resolution mode with ca. 15 µm field of view at sub-50 nm resolution.

During collection of the tomography data, the sample is rotated relative to the source and detector, therefore, to ensure maximum data quality, the principal dimensions of the sample must be matched to the zone plate FOV in three dimensions. This has been achieved using focused ion beam sample preparation.[41,45] For optimal geometries, excellent agreement has been demonstrated between the large FOV and high resolution mode and data has been successfully correlated to FIB[41] and mercury porosimetry.[46]

7.3.3 Synchrotron X-ray Instruments

As previously discussed, synchrotron sources provide X-ray flux many orders of magnitude greater than conventional laboratory sources, providing numerous advantages – primarily the speed at which data can be collected (owing to much reduced exposure times) and the capacity to provide a highly tunable mono-chromatic beam.

As X-rays are sensitive to electron density, the degree of X-ray attenuation is a function of atomic number. For materials with widely varying atomic number, image contrast between the two-phases can often be easily achieved, for the interested reader, the Center for X-ray Optics (CXRO) provide detailed information about the X-ray properties of the elements.[47] For composite materials demonstrating similar X-ray attenuation properties, the ability to provide mono-chromatic X-rays at varying energies can be used to provide phase discrimination.

The X-ray attenuation length of a solid (defined as the depth into the material measured along the surface normal where the intensity of X-rays falls to 1/e of its value at the surface)[47] broadly displays the trend of increasing with increasing photon energy. However, the physics of photoelectric absorption also effects the attenuation of X-rays, giving rise to discontinuities in this trend at given energies, referred to as "edges".

By taking advantage of this discontinuity, it is possible to conduct transmission X-ray imaging at photon energies above and below an absorption edge to promote contrast changes in a given material and in doing so to generate inter-phase contrast and provide material identification. This is shown in Figure 7.12, an image of a Ni-YSZ sample, where the reconstructed tomographic slices show the effects of changing X-ray energy: in Figure 7.12a, the reconstructed slice was captured below the Ni absorption edge (at ca. 8.3 keV), in this image, minimum contrast is demonstrated between the two solid phases. For the same sample location, increasing the X-ray energy to above the Ni edge (ca. 8.4 keV) significant contrast is demonstrated enabling

(a)   Below Ni edge   (b)   Above Ni edge

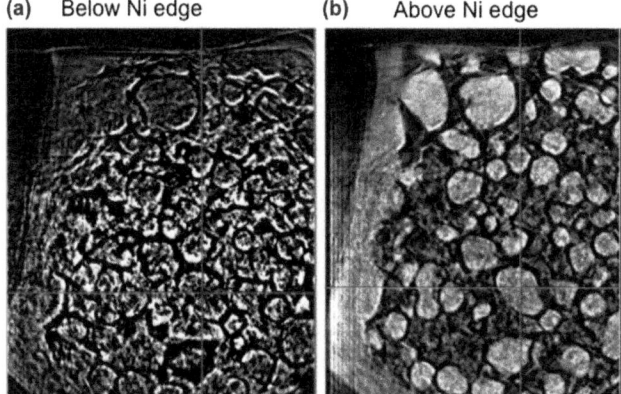

Figure 7.12   Virtual slices from a Ni-YSZ electrode structure obtained using X-ray nano-CT (a) below the Ni edge at 8.317 keV and (b) above the Ni edge at 8.357 keV. Changes in the Ni grain opacity enable phase discrimination and TPB identification (reproduced from Shearing *et al.*[9]).

discrimination of the Ni and YSZ phase and the corresponding triple phase contact.

The determination of triple phase boundary length in Ni-YSZ composite anode materials was demonstrated in 2010 both by Shearing[9] *et al.* and Grew[48] *et al.* from experiments conducted at the Advanced Photon Source, Argonne National Labs. These techniques have also been adopted by a number of groups to characterize anode[49] and cathode[25,50,51] materials. Most recently, the introduction of marker-less tomography[52] has also enabled the simultaneous characterisation of significantly larger sample volumes to include full cells[53] (*i.e.* anode/electrolyte/cathode).

Chiu and co-workers have extended this capability to explore the Ni/NiO content in model systems and working SOFC electrodes[54] – this is achieved by obtaining tomographic information at multiple X-ray energies, therefore effectively obtaining coupled imaging and XANES data. As the attenuation length of the material under investigation changes this is manifested in changes to image based contrast – by using many different X-ray energies, it is possible to isolate multiple constituent phases. This novel technique has also been applied to battery materials and is gaining considerable attention owing to its ability to simultaneously and non-destructively provide microstructural and chemical information.

7.3.4   4-Dimensional Tomography

Microstructural changes in SOFC electrodes are understood to be a key factor leading to device degradation. Whilst the effects of aging and sintering have been explored using FIB tomography,[21] the destructive nature of the technique means that all microstructural comparisons must be purely statistical.

Microstructural failure may be predicated by microscopic heterogeneities in the microstructure, therefore the inherent differences in material microstructures may cloud the interpretation of these statistical relationships. Amongst the primary benefits of X-ray nano-CT is its non-destructive nature, which facilitates the study of microstructural evolution on a grain-by-grain basis for the same sample.

X-ray nano-CT has been used to explore the effects of Ni to NiO redox cycling, which is known to cause significant microstructural degradation. A sample prepared using FIB machining was characterized at various stages during a redox cycle: the sample was scanned before and after exposure to air at high temperatures *ex situ* of the microscope stage.[55] The high temperature heat treatment promotes the Ni to NiO transition, causing microstructural change, which was subsequently characterized using the non-destructive X-ray technique. Figure 7.13 shows the progression of the microstructural change following oxidation at 300, 500, and 700 °C. Figure 7.13b and c show the resulting significant change in the Ni morphology.

Following the development of a micro furnace compatible with the X-ray microscope,[51] it is possible to carry out *in situ* heat treatments. This has recently been applied to study cathode microstructures at working temperatures for the first time. With on-going improvement to the furnace design, it is anticipated that higher temperatures will facilitate studies of sintering in the near future.

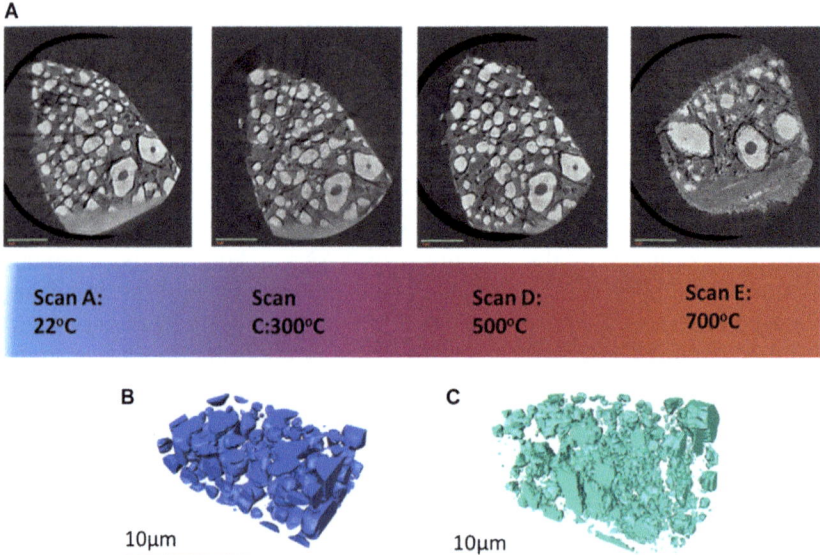

Figure 7.13 Studies of the microstructural effects of Ni to NiO transition (a) virtual slices from tomographic sequences following 22, 300, 500 and 700 °C exposure (b) Ni morphology from scan A (c) Ni morphology from scan E (reproduced from Shearing *et al.*[55]).

7.4 Data Analysis and Image Based Modelling

7.4.1 Data Analysis

In order to extract meaningful parameters from tomographic data sets, significant analysis is required. Whilst experimental techniques seek to maximise the quality of raw data obtained, analysis and segmentation of the image sequences is non-trivial. Accurate image analysis is essential to determine the fractions of each phase present and therefore the nature of triple phase contact.

FIB experiments provide raw data typically as 256 channel grey-scale "tiff" (tagged image file format) sequences, this raw data often requires alignment before it can be concatenated as a 3D volume. X-ray tomography data is reconstructed from transmission images into 3D slice-by-slice equivalent, typically also in the tiff format.

These image data sets, which can be represented as large 3D matrices, must be analyzed to assign a given label to each of the constituent phases. If care is taken during image collection, the burden of image analysis can be greatly eased as the image histogram displays distinct and separated gray-scale peaks. In practice, there is often over-lapping of the gray-scale histograms which must be separated for accurate image analysis. This can be achieved by numerous combinations of image filtering and morphological operations, the reader is referred to Gonzalez *et al.* for an overview of the field[56,57] and exemplary papers by Jorgensen[19] and Salzer[58] for exemplary application of image processing tools to FIB-SEM data.

Once image analysis has been successfully completed, there are a large number of microstructural metrics that can be computed – phase fractions, for example, can be calculated simply by summing the number of matrix elements associated with each phase as a fraction of the total number of matrix elements. Other metrics such as TPB density are more complicated; Golbert *et al.*[59] and Wilson *et al.*[60] provide methodologies for computation of TPB density based on the local contact of the three constituent phases and their respective percolation.

The availability of three-dimensional image data also gives rise to new microstructural metrics that are not obtainable from 2D data – whilst a detailed discussion of this is beyond the scope of this chapter; a few key examples are considered.

Conventional methods of two-dimensional geometrical characterisation do not necessarily translate to three-dimensional structures. For example, for a highly non-spherical convoluted pore structures, the conventional measurements of equivalent pore diameter holds little meaning. Holzer *et al.*[29,61] have developed a methodology know as "continuous particle size diameter" measurement for the interpretation of 3D particle size distribution which seeks to replicate geometrically the physical effects of mercury porosimetry.

The powerful Random Walk[62] and Fast Marching[63] methods also provide novel means for characterisation of three-dimensional pore (and solid) networks and from resulting distance maps provide access to a wide range of geometrical parameters including tortuosity, path diameters, the novel dead ends property and shape independent particle size distribution.

7.4.2 Image Based Modelling

Microstructural characterisation tools are extremely valuable for their ability to capture detailed geometrical information, enabling comparison and evaluation of SOFC electrodes. However, this microstructural data is perhaps most powerful when it is effectively combined with relevant simulation tools, which enables researchers to explore the microscopic performance of these complex materials. The literature includes numerous examples from parameter extraction for systems level models[64] to Lattice-Boltzmann models of transport and kinetics[20,27,46] The latter examples will be omitted here, as they feature as a separate chapter of this volume.

Historically, in the absence of advanced tomography techniques, electrode level models have utilised often over-simplified synthetic representations of electrode microstructures. Whilst these simulations are of clear value in providing qualitative design guidelines, they are limited in their ability to represent real-life structures, and therefore establish accurate performance/ microstructure relationships.

In the 2006 paper from Wilson[2] *et al.*, the commercial finite element package Comsol was used to compute values for tortuosity of the porous structure based on a solution to Fick's law of diffusion. The first electrochemical simulations of SOFC electrodes using image derived microstructural frameworks were presented by Shearing *et al.*[39] using a volume of fluid approach that had previously been developed for synthetic structures.[59]

The ability to accurately represent these structures *in silico*, and in particular the generation of FE meshes, is non-trivial. This is especially true for composite structures where (unlike single solid phase materials) – the phase interfaces cannot be obtained by simple Boolean subtraction. In recent years, improvements to meshing tools (for example, those introduced by Simpleware Ltd) have demonstrated the capability to generate accurate meshes without the need for extensive and labour-intensive manual repair. These tools have facilitated both mechanical[65] and electrochemical[66] simulations of SOFC electrodes and have also been routinely used for transport calculations in other porous structures such as Li-ion batteries.[67,68] The geometrical frameworks that have been extracted from tomography experiments have also been used as the basis for simulations of microstructural evolution in SOFC electrodes.[69]

7.5   Conclusions

Since 2006, the ability to characterise the three-dimensional structure of SOFC electrodes has revolutionised our understanding of the performance/ microstructure relationship. Through the application of both FIB-SEM tomography and X-ray nano-CT, the research community have collectively explored a diverse range of processing effects including phase fractions, sintering temperatures and electrode lifetime, thereby establishing a library of design criteria for optimisation of SOFC electrodes.

Increasingly, the ability to characterise microstructural evolution processes is also informing our understanding of degradation processes and the combination with relevant simulation tools is providing new insight into the microscopic performance of devices. With increasing adoption, this suite of research tools will play a key role in the future development of durable and high-performance SOFC electrodes.

References

1. D. Waldbillig, *et al.*, *J Power Sources*, 2005, **145**, 206.
2. P. R. Shearing, *et al.*, *Int Mater Rev.*, 2010, **55**.
3. R. T. DeHoff, *Quantitative Microscopy*, F. N. Rhines, McGraw Hill New York, 1968.
4. H. S. Song, *et al.*, *J. Power Sources*, 2005, **145**, 272.
5. J. R. Wilson, *et al.*, *Electrochem Solid State Lett.*, 20008, **11**, B181.
6. N. Shikazono, *et al.*, *J Power Sources*, 2009, **193**, 530.
7. A. Faes, A. Hessler-Wyser, D. Presvytes, C. G. Vayenas and J. Van herle, *Fuel Cells*, 2009, **9**, 841.
8. E. A. Holm and P. M. Duxbury, *Scripta Materialia.*, 2006, **5**, 1035.
9. P. R. Shearing, *et al.*, *Electrochem Comm.*, 2010, **12**, 1021.
10. Q. Cai, C. S. Adjiman and N. P. Brandon, *Electrochim. Acta*, 2011, **56**, 5804–5814.
11. D. G. Ivey, *et al.*, *J. Power Sources*, 2010, **195**, 6301.
12. N. J Vito, *et al.*, *217th Electrochem Soc Meeting, Abstract #728*.
13. J. Meingailis, *J Vacuum Sci. and Tech.*, 1986, **5**, 469.
14. J. I. Goldstein, *et al.*, *Scanning Electron Microscopy and X-Ray Microanalysis* 3rd ed. Kluwer Academic Press: New York, 2003.
15. K. Thyden, *Solid State Ionics*, 2008, **178**, 1984.
16. J. R. Wilson, *et al.*, *Nat Mater*, 2006, **5**, 541.
17. L. A. Gianuzzi ed, *Introduction to Focused Ion Beams*, Springer, Berlin, 2005.
18. N. Yao ed. *Focused Ion Beam System Basics and Applications*, University of Cambridge, Cambridge, UK, 2007.
19. P. S. Jørgensen, *et al.*, *Ultramicroscopy*, 2010, **110**, 216.
20. H. Iwai, *et al.*, *J. Power Sources*, 2009, **195**, 955.
21. J. S. Cronin, J. R. Wilson and S. A. Barnett, *J. Power Sources*, 2011, **196**, 2640.
22. D. Gostovic, *et al.*, *Electrochem Solid State Lett.*, 2007, **10**, B214.
23. J. R. Wilson, *et al.*, *Electrochem Comm.*, 2009, **11**, 1052.
24. J. S. Cronin, K. Muangnapoh, Z. Patterson, K. J. Yakal-Kremski, V. P. Dravid and S. A. Barnett, *J. Electrochem. Soc.*, 2012, **159**, B385.
25. G. J. Nelson, *et al.*, *Electrochem Comm.*, 2011, **13**, 586.
26. P. R. Shearing, *et al.*, *Chem Eng Sci.*, 2009, **64**, 3928.

27. N. Shikazono, *et al.*, *J. Electrochem. Soc.*, 2010, **157**, B665.
28. N. Vivet, *et al.*, *J Power Sources*, 2011, **196**, 7541.
29. L. Holzer, B. Münch, B. Iwanschitz, M. Cantoni, T. Hocker and T. Graule, *J. Power Sources*, 2011, **196**, 7076.
30. K. T. Lee, *et al.*, *J Power Sources*, 2013, **228**, 220.
31. G. Gaiselmann, M. Neumann, L. Holzer, T. Hocker, M. René Prestat and V. Schmidt, *Computational Materials Science*, 2013, **67**, 48.
32. J. R. Smith, *et al.*, *Solid State Ionics*, 2009, **180**, 90.
33. N. Vivet, *et al.*, *J Power Sources*, 2011, **196**, 9989.
34. D. Marinha, *et al.*, *Chem. Mater.*, 2011, **23**, 5340.
35. L. Holzer, B. Iwanschitz, Th. Hocker, B. Münch, M. Prestat, D. Wiedenmann, U. Vogt, P. Holtappels, J. Sfeir, A. Mai and T. Graule, *Journal of Power Sources*, 2011, **196**, 1279.
36. H. Galinski, A. Bieberle-Hütter, J. L. M. Rupp and L. J. Gauckler, *Acta Materialia*, 2011, **59**, 6239.
37. K. Eguchi, N. Kamiuchi, J.-Y. Kim, H. Muroyama, T. Matsui, M. Kishimoto, M. Saito, H. Iwai, H. Yoshida, N. Shikazono, N. Kasagi, J. Akikusa, H. Eto, D. Ueno, M. Kawano and T. Inagaki, *Fuel Cells*, 2012, **12**, 537.
38. H. Choi, D. Gawel, A. Berson, J. Pharoah and K. Karan, *ECS Trans.*, 2011, **35**, 997.
39. P. R. Shearing, *et al.*, *J Power Sources*, 2010, **195**, 4804.
40. J. R. Wilson, *et al.*, *Scripta Materialia.*, 2011, **65**, 67.
41. P. R. Shearing, *et al.*, *J Eur Cer Soc.*, 2013, **30**, 1809.
42. J. B. Eberhart, *Structural and Chemical Analysis of Materials*, John Wiley & Sons, Chichester, UK. 1991.
43. J. Banhart, ed. *Advanced Tomographic Methods in Materials Research and Engineering*. Oxford University Press, Oxford, 2008.
44. A Tkachuk., *Z. Kristallogr.*, 2007, **222**, 650.
45. J. J. Lombardo, *et al.*, *J Synchrotron Rad.*, 2012, **19**, 789.
46. J. R. Izzo, *et al.*, *J. Electrochem. Soc.*, 2008, **155**, B504.
47. CXRO: http://henke.lbl.gov/optical_constants/.
48. K. N. Grew, *et al.*, *J. Electrochem. Soc.*, 2010, **157**, B783.
49. J. Laurencin, *et al.*, *J Power Sources*, 2012, **198**, 182 15.
50. Y. Guan, *et al.*, *J Synchrotron. Rad.*, **17**, 2010, 782.
51. P. R. Shearing, *et al.*, *Electrochem Solid State Lett.*, 2011, **14**, B1117.
52. J. Wang, *et al.*, *Appl. Phys. Lett.*, 2012, **100**, 143107.
53. J. S Cronin, *et al.*, *J. Power Sources*, 2013, **233**, 174.
54. G. J. Nelson, *et al.*, *Appl Phys Lett.*, 2011, **98**, 173109.
55. P. R. Shearing, *et al.*, *Solid State Ionics*, 2012, **216**, 69.
56. M. Petrou, *Image Processing, the fundamentals*, John Wiley & Sons, Chichester, UK. 2010.
57. R. C. Gonzalez, *et al.*, *Digital Image Processing Using MATLAB*, Pearson Prentice Hall, New Jersey, 2003.
58. M. Salzer, *et al.*, *Materials Characterization*, 2012, **69**, 115.
59. J. Golbert, C. S. Adjiman and N. P. Brandon, *Ind. Eng. Chem. Res.*, 2008, **47**, 7693.

60. J. R. Wilson, *et al.*, *Microscopy and Microanalysis*, 2009, **15**, 71.
61. B. Münch, *et al.*, *J. Am. Ceram. Soc.*, 2008, **91**, 4059.
62. M. Kishimoto, *et al.*, *J Power Sources*, 2011, **196**, 4555.
63. P. S. Jørgensen, *et al.*, *J. Microscopy*, 2011, **244**, 45.
64. U. Dorawasami, P. R. Shearing, N. Droushiotis, K. Li, N. P. Brandon and G. Kelsall, *Solid State Ionics*, 2011, **192**, 494.
65. R. Clague, P. R. Shearing, P. D. Lee, Z. Zhang, D. J. L. Brett, A. J. Marquis and N. P. Brandon, *J. Power Sources*, 2011, **196**, 9018.
66. J. Joos, *et al.*, *J. Power Sources*, 2011, **196**, 7302.
67. M. Ender, J. Joos, T. Carraro and E. Ivers-Tiffée, *Electrochem. Comm.*, 2011, **13**, 166.
68. P. R. Shearing, *et al.*, *Electrochem Comm.*, 2010, **12**, 374.
69. H. Chen, H. Yu, J. S. Cronin, J. R. Wilson, S. A. Barnett and K. Thornton, *J. Power Sources*, 2011, **196**, 1333.

CHAPTER 8

Three-Dimensional Numerical Modelling of Ni-YSZ Anode

NAOKI SHIKAZONO*[a,b] AND NOBUHIDE KASAGI[c,d]

[a] Professor, Institute of Industrial Science, The University of Tokyo, Komaba 4-6-1, Meguro-ku, Tokyo 153-8505, Japan; [b] CREST, Japan Science and Technology Agency; [c] Professor Emeritus, The University of Tokyo; [d] Principal Fellow, CRDS, Japan Science and Technology Agency
*Email: shika@iis.u-tokyo.ac.jp

8.1 Introduction

Solid oxide fuel cells (SOFC) are expected to be one of the most important energy technologies in the future because of their superior efficiency and fuel flexibility.[1] However, its cost and durability must be further improved for wide market penetration. It is widely known that the electrode microstructure has significant effects upon the cell performance and durability of SOFCs. Thus, basic understanding of microscopic features of the electrode is indispensable. Quantitative investigations which relate the electrode microstructural parameters obtained from two-dimensional images and the polarization resistances have been reported. Simwonis *et al.*[2] applied the concept of contiguity (CC) theory to Ni-YSZ cermets and investigated the contiguity of the phases. Shikazono *et al.*[3] also used stereology and CC theory to investigate the relationship between the polarization characteristics and the microstructural parameters. However, a uniform and isotropic mixture is assumed in stereology and CC theory, which has to be carefully validated. For example, dead ends of the phase connections and disconnected inactive TPBs should be rationally

RSC Energy and Environment Series No. 7
Solid Oxide Fuel Cells: From Materials to System Modeling
Edited by Meng Ni and Tim S. Zhao
Published by the Royal Society of Chemistry, www.rsc.org

removed for the quantitative discussion on polarization characteristics. In addition, in order to investigate local electrochemistry in detail at the three phase boundary (TPB), which is one of the key interests of SOFC electrode research, it is indispensable to separate the effect of local electrochemistry and the transport characteristics inside a complex microstructure.[4] In order to overcome such issues, it is necessary to establish a method which can directly predict the polarization characteristics using three-dimensional microstructure.

Recently, direct measurements of three-dimensional SOFC electrode microstructure have been carried out using focused ion beam scanning electron microscopy (FIB-SEM)[5–15] and X-ray computed tomography (XCT).[16–19] As a result, useful quantitative data such as TPB length and tortuosity factor have now become availabe from the reconstructed three-dimensional micro-structures.[20] However, a difficulty still remains in removing errors which arise from discretization process and insufficient sample volume size.[11,14]

In addition to the above experimental studies, numerical simulations have the possibility to provide useful information which cannot be obtained from experiments. It is clear that numerical simulation should be made with an identical three-dimensional electrode microstructure which is directly used in the polarization experiment. The recent progress in computational methods has made it possible to conduct three-dimensional numerical simulation of SOFC electrodes.[21–33] These modelling works can provide local potential distributions inside the electrode, which affects local reaction kinetics and physical properties of materials such as ionic or electronic conductivities and Young's modulus, *etc.* Such information will be indispensable for further improving electrode performance and reliability.

In this report, the anode overpotential is calculated by the lattice Boltzmann method (LBM), which solves the species-transport coupled with the electrochemical reaction inside three-dimensional microstructures of Ni-YSZ anode measured by a dual-beam focused ion beam-scanning electron microscopy (FIB-SEM).[26,28] The dependences of TPB density and tortuosity factors on the sample volume size are evaluated by microstructure samples with different volume sizes. The accuracies of TPB length calculation methods and exchange current models are assessed. Finally, the three-dimensional ionic and electronic current distributions are presented.

8.2   Experimental

8.2.1   Button Cell Experiment

An electrolyte-supported button cell is used in this study.[28] NiO-8YSZ anode (60:40 vol%, AGC Seimi Chemical Co., Ltd.) and $La_{0.8}Sr_{0.2}MnO_3$ cathode (AGC Seimi Chemical Co., Ltd.) were screen printed on the 0.5 mm thick 8YSZ electrolyte disk (Japan Fine Ceramics Co., Ltd.), and sintered at 1400 °C and 1150 °C, respectively. Then, the anode was reduced at 800 °C for one hour. The SOFC was placed between two alumina tubes with glass seals. Platinum meshes were used as current collectors, which were mechanically pressed against the electrodes. The performance of SOFC was evaluated at different temperatures with humidified hydrogen as a fuel, and pure oxygen as an oxidant. I–V and electrochemical impedance spectroscopy measurements (frequency range $1–10^6$ Hz, AC signal strength $10\,mV$) were conducted using Solartron frequency analyzer (1255B).

8.2.2   Microstructure Reconstruction Using FIB-SEM

The samples are reconstructed by FIB-SEM, as described in our previous study.[11] The samples were infiltrated with epoxy resin (Marumoto Struers KK) under vacuum conditions so that the pores could be easily distinguished during SEM observation. Cured samples were polished using an Ar-ion beam cross-section polisher (JEOL Ltd., SM-09010) and made available for the FIB-SEM (Carl Zeiss, NVision 40) observation.

Three samples with different volume sizes were prepared.[28] The volume size of sample A is $22.3\,\mu m \times 8.56\,\mu m \times 12.7\,\mu m$ $(2.42 \times 10^3\,\mu m^3)$ in x, y and z directions, while those of samples B and C are $22.6\,\mu m \times 10.8\,\mu m \times 16.0\,\mu m$ $(3.91 \times 10^3\,\mu m^3)$ and $45.8\,\mu m \times 14.5\,\mu m \times 26.2\,\mu m$ $(1.74 \times 10^4\,\mu m^3)$, respectively. Sample C is the largest sample, which covers whole electrode thickness. The cross-sectional resolutions are 37.2 nm/pixel for samples A and B, and 55.8 nm/pixel for sample C. The FIB milling pitch distances for the three samples are 74.5, 61.7 and 74.7 nm, respectively.

Three phases (pore, Ni, YSZ) in the cross-sectional images are distinguished by their brightness values. Epoxy infiltration resulted in very clear contrast of the pore phase. The phase distinguished images are then aligned for three-dimensional reconstruction. In addition, all the samples are rearranged in 124 nm cubic voxels for the LBM calculation. Reconstructed three-dimensional structures of samples A, B, and C are shown in Figure 8.1.

8.3   Numerical Method

8.3.1   Quantification of Microstructural Parameters

The reconstructed microstructure is represented by a number of small voxels. If the neighbouring four cubic voxels comprise all three phases with different phases in two diagonal voxels, the edge surrounded by the four voxels is defined as a TPB segment.[26] One of the possible voxel arrangements and the corresponding TPB segments are shown in Figure 8.2(a). If the lengths of TPB segments are simply summed, it is apparent that the total TPB length would be overestimated because of the inevitable step-like pattern of voxel edges. Suzue *et al.*[25] assumed that TPB length was 20% smaller than the value directly calculated from the cubic voxel perimeter of reconstructed structure. Golbert *et al.*[21] counted the number of all the voxels neighboring a TPB edge (four per edge) and divided the overall number by four to obtain the total number of effective TPB edges. Wilson *et al.*[34] divided the total voxel edge length by a constant factor of 1.455, which was derived from an analytical calculation of a sphere to get total TPB. However,

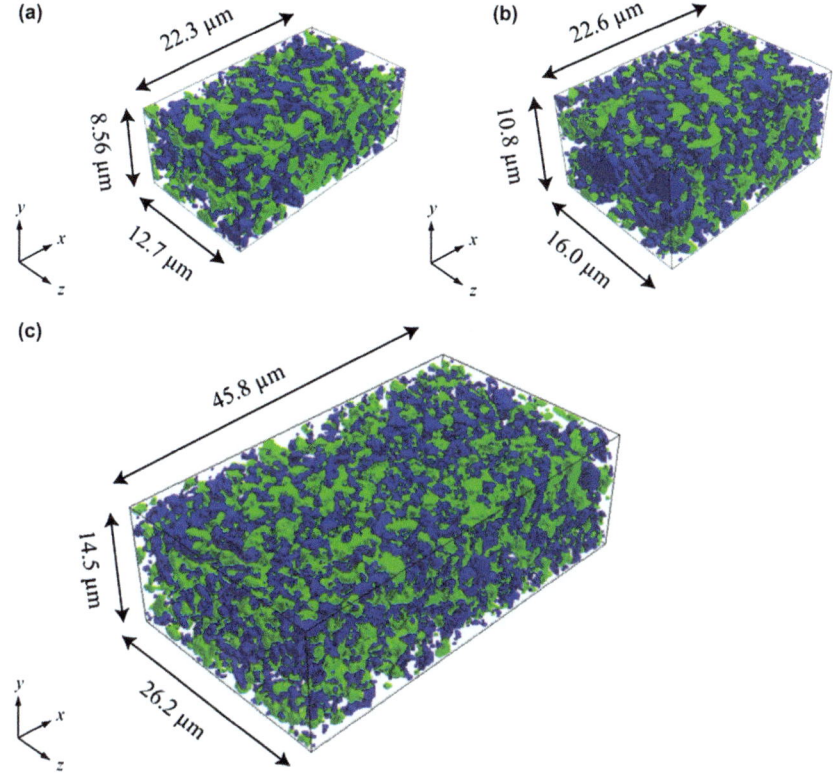

Figure 8.1 Three-dimensional reconstructed anode structures: (a) sample A, (b) sample B, (c) sample C (Blue: YSZ, Green: Ni, Transparent: pore).[28]

a method which can locally calculate TPB length accurately is indispensable in order to predict electrode polarization characteristics.

In the following, three methods to calculate local TPB length from voxel edge segments are evaluated. The first method is a simple summation of voxel edge segments as shown in Figure 8.2(b). Figure 8.2(c) shows the second method which defines the connection length of the midpoints of the TPB edge segments. The third method shown in Figure 8.2(d) counts the total distance between the centroids of the triangles defined by the neighbouring midpoints of the edge segments. These three methods are compared by applying them to a structure with well-defined TPB, *i.e.* two overlapped spheres. As shown in Figure 8.3, TPB length can be analytically obtained for this configuration. The radius of the sphere r and the distance between two centres l are varied. Overlap ratio c is defined as:

$$c = 1 - \frac{l}{2r} \tag{1}$$

Figure 8.4 shows the error values of TPB lengths calculated from three different methods. The symbols in the figure represent the mean values from 500

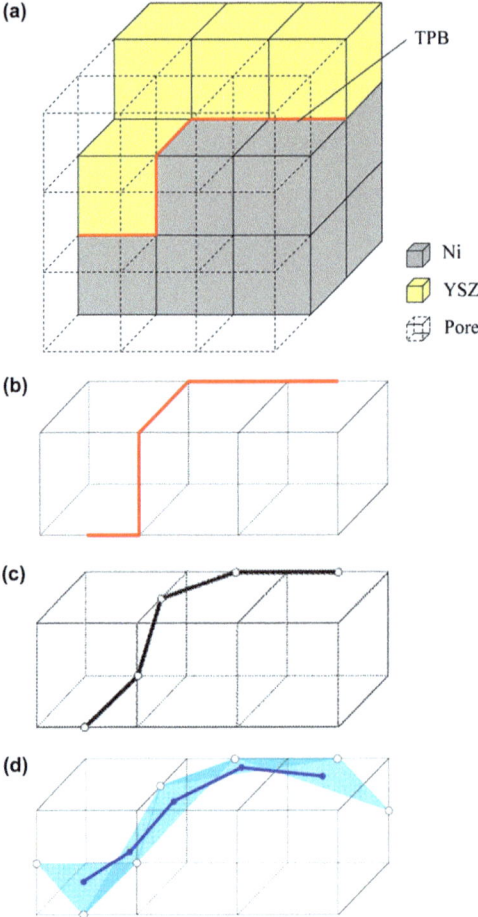

Figure 8.2 TPB length calculation methods.[26] (a) one of the possible voxel arrangements, (b) edge segment length, (c) midpoint length, and (d) centroid length.

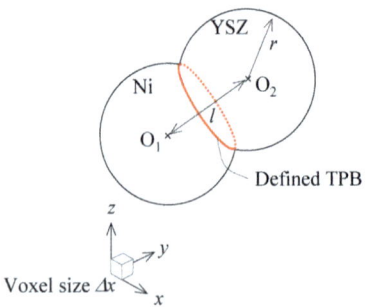

Figure 8.3 TPB defined by two spheres.[26]

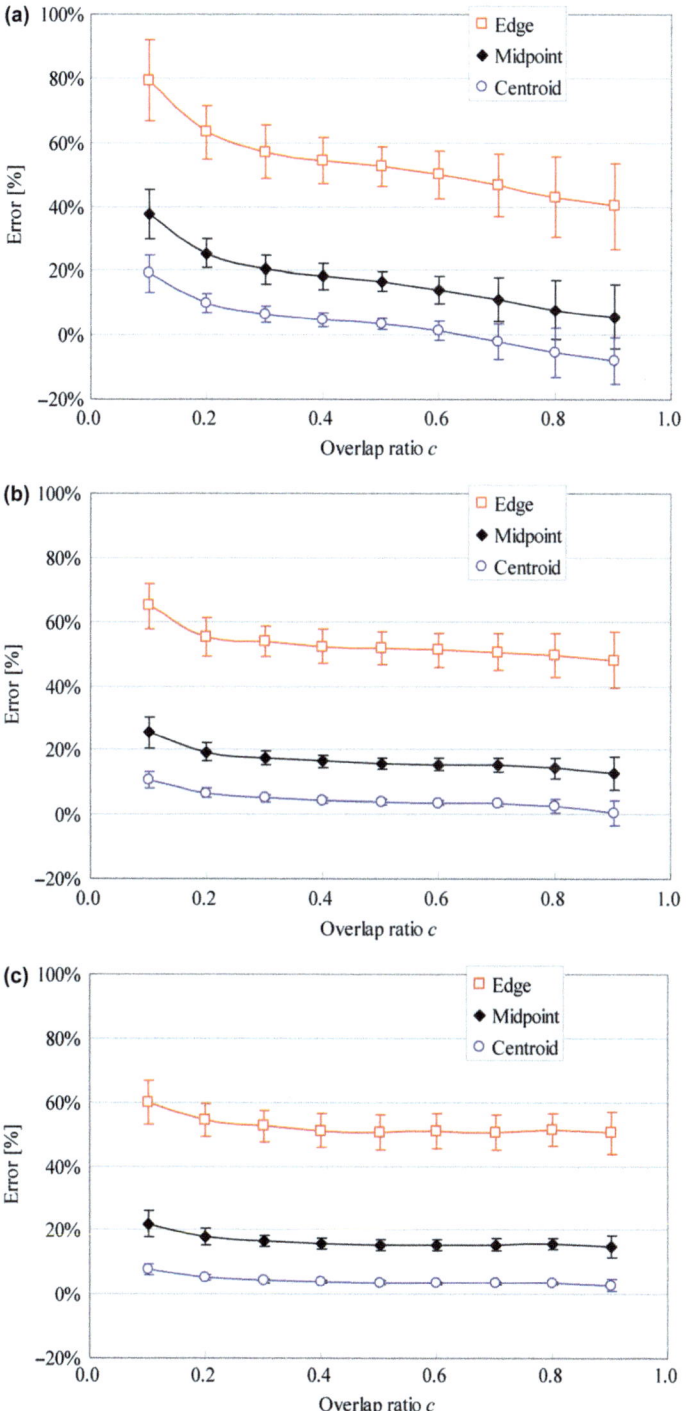

Figure 8.4 Comparison of TPB lengths by three methods.[26] (a) $r/\Delta x = 5$, (b) $r/\Delta x = 15$ and (c) $r/\Delta x = 30$.

Table 8.1 Total TPB lengths of the samples.[28]

Sample	Total TPB length [$\mu m/\mu m^3$]
A	2.11
B	1.92
C	2.05

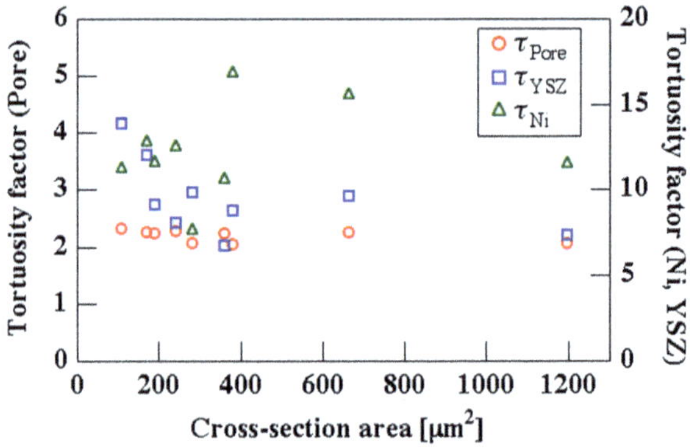

Figure 8.5 Tortuosity factors against cross-sectional area.[28]

randomly chosen sphere positions for given radius and overlap ratio, and the bars are the standard deviations. It is evident that the simple summation of edge segments overestimates TPB length more than 50%. On the other hand, the centroid method can predict TPB length within a 5% margin of error, provided that grid resolution is high ($r/\Delta x > 15$). Table 8.1 shows total TPB lengths calculated by the centroid method for the three samples in Figure 8.1. It is confirmed that the variation of TPB lengths between these three samples is relatively small.

Tortuosity factor is a measure which represents (1) the increase of actual diffusion path lengths compared to the straight-line distance in the mean diffusion direction in a porous media, and (2) the increase of diffusion flux according to the decrease of effective diffusive cross sectional area due to the inclination of the paths:

$$D_{\text{eff}} = \frac{\varepsilon_{\text{pore}}}{\tau_{\text{pore}}} D, \tag{2}$$

$$\sigma_{e^-,\text{eff}} = \frac{\varepsilon_{\text{Ni}}}{\tau_{\text{Ni}}} \sigma_{e^-}, \tag{3}$$

$$\sigma_{O^{2-},\text{eff}} = \frac{\varepsilon_{\text{YSZ}}}{\tau_{\text{YSZ}}} \sigma_{O^{2-}}, \tag{4}$$

where D_{eff} and σ_{eff} are the effective diffusivity and conductivity, and ε is the volume fraction. Diffusion inside each phase is solved by the LBM, which will be described in detail in the next section. Figure 8.5 shows tortuosity factors of all samples plotted against cross-sectional areas normal to the direction of diffusion. The pore tortuosity factor is nearly constant regardless of cross-sectional area. The YSZ tortuosity factor remains nearly unchanged when cross-sectional area exceeds $200 \, \mu m^2$. This result indicates that pore and YSZ structures are nearly isotropic and homogeneous. On the other hand, Ni tortuosity factor shows very large variation even for the samples with large cross-sectional areas. Complex nature of Ni phase network results in very large variation of tortuosity factor.

8.3.2   Governing Equations for Polarization Simulation

Gaseous, electronic and ionic diffusion equations are solved inside each of the obtained three-dimensional Ni, YSZ and pore phases. In the gaseous phase, hydrogen and steam diffusion is solved based on a dusty gas model (DGM).[35] If a constant total pressure is assumed, DGM is written as follows:

$$\frac{N_i}{D_{i,k}} + \sum_{j \neq i} \frac{y_j N_i - y_i N_j}{D_{i,j}} = -\frac{1}{RT} \nabla p_i, \tag{5}$$

where y_i is the molar fraction, N_i is the molar flux, and p_i is the partial pressure. Subscripts i and j represent gas species such as hydrogen and steam. In the present study, binary (H_2 and H_2O) equi-molar diffusion in a constant total pressure p_t environment is assumed, $i.e.$,

$$p_t = p_{H_2O} + p_{H_2}. \tag{6}$$

Therefore, only the diffusion equation of hydrogen is solved:

$$\nabla \left(\left[\frac{1 - \alpha y_{H_2}}{D_{H_2,H_2O}} + \frac{1}{D_{H_2,k}} \right]^{-1} \nabla C_{H_2} \right) = \frac{i_{reac}}{2F}, \tag{7}$$

where, C_{H_2} is the molar concentration of hydrogen gas with

$$\alpha = 1 - \left(\frac{M_{H_2}}{M_{H_2O}} \right)^{1/2}. \tag{8}$$

In Eq. (7), D_{H_2}, H_2O and $D_{H_2,k}$ represent the binary and Knudsen diffusion coefficients, respectively:

$$D_{H_2,H_2O} = 0.018833 \sqrt{\frac{1}{M_{H_2}} + \frac{1}{M_{H_2O}}} \frac{T^{3/2}}{p \Omega_D \zeta^2_{H_2,H_2O}}, \tag{9}$$

$$D_{H_2,k} = \frac{2}{3} \left(\frac{8RT}{\pi M_{H_2}} \right)^{1/2} r. \tag{10}$$

The mean pore radius r is calculated by the maximum sphere inscription method (MSI).[36] The mean radii of samples A, B and C are $r = 422$ nm, 416 nm and 462 nm, respectively.

Table 8.2   Gas properties.

Substance	M [g/mol]	ζ [Å]	ε/k [K]
H_2	2.016	2.93	37
H_2O	18.015	2.65	356

The collision integral Ω_D is given as:

$$\Omega_D = 1.1336\left(\frac{Tk}{\varepsilon}\right)^{-0.1814}. \tag{11}$$

When calculating the binary diffusion coefficient, an intermolecular force constant ζ_{H_2,H_2O} is taken as an arithmetic mean of ζ_{H_2} and ζ_{H_2O}. Geometric mean of ε_{H_2} and ε_{H_2O} is used for ε. The gas parameters are shown in Table 8.2.

Assuming that Ni and YSZ are perfect electronic and ionic conductors, respectively, the following equations are solved in the solid phases:

$$\nabla\left(\frac{\sigma_{e^-}}{F}\nabla\tilde{\mu}_{e^-}\right) = -\,i_{reac}, \tag{12}$$

$$\nabla\left(\frac{\sigma_{O^{2-}}}{2F}\nabla\tilde{\mu}_{O^{2-}}\right) = i_{reac}, \tag{13}$$

where $\tilde{\mu}_{e^-}$ and $\tilde{\mu}_{O^{2-}}$ are the electrochemical potentials of electron and oxide ion, respectively.

The reaction current i_{reac}, which is defined at the TPB, in the RHS of Eqs. (7), (12) and (13) is calculated as[37]:

$$i_{reac} = i_0 L_{TPB}\left\{\exp\left(\frac{2F}{RT}\eta_{act}\right) - \exp\left(-\frac{F}{RT}\eta_{act}\right)\right\}. \tag{14}$$

The lineal exchange current density i_0 is fitted from the patterned anode experiments of de Boer[38] and Bieberle *et al.*[39] as follows:

$$i_0 = 3.14 \times p_{H_2}^{-0.03}p_{H_2O}^{0.4}\exp\left(\frac{1.52 \times 10^5}{RT}\right) \tag{15}$$

$$i_0 = 0.0013 \times p_{H_2}^{0.11}p_{H_2O}^{0.67}\exp\left(-\frac{0.849 \times 10^5}{RT}\right) \tag{16}$$

The values for the activation energy and the exponents of P_{H_2} and P_{H_2O} in Eq. (16) are taken from the zero overpotential data of Bieberle *et al.*, but the leading coefficient 0.0013 in Eq. (16) is chosen so that the predicted results fit best to the present experimental data at 10 mol% H_2O – 90 mol% H_2, 1073 K. The numerical conditions are listed in Table 8.3.[40,41]

In the LBM calculation, the electrochemical potential of electron $\tilde{\mu}_{e^-}$ in the Ni phase and that of oxide ion $\tilde{\mu}_{O^{2-}}$ in the YSZ phase are solved. The local overpotential η_{local} is defined as the voltage difference between the virtual

Table 8.3 Numerical conditions.

Properties	Value
Total pressure p_t [Pa]	1.013×10^5
Electronic conductivity [Sm^{-1}] [40]	$3.27 \times 10^6 - 1065.3T$
Ionic conductivity [Sm^{-1}] [41]	$3.34 \times 10^4 \exp(-10300/T)$

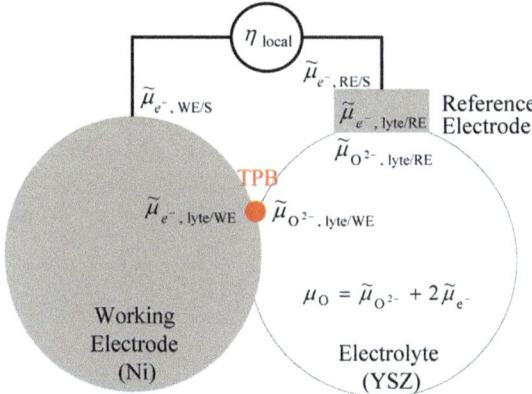

Figure 8.6 Schematic of local overpotential at the vicinity of TPB.[26]

reference electrode (RE) and the working electrode (WE) which is defined in the vicinity of TPB as shown in Figure 8.6:

$$\eta_{\text{local}} = E_{\text{WE}/\text{S}} - E_{\text{RE}/\text{S}} = \frac{1}{F}\left(\tilde{\mu}_{\text{e}^-,\,\text{RE}/\text{S}} - \tilde{\mu}_{\text{e}^-,\,\text{WE}/\text{S}} \right), \tag{17}$$

where $\tilde{\mu}_{\text{e}^-,\,\text{RE}/\text{S}}$ and $\tilde{\mu}_{\text{e}^-,\,\text{WE}/\text{S}}$ are the electron electrochemical potentials at the surfaces of RE and WE, respectively. The local activation overpotential η_{act} at TPB is obtained by subtracting the ohmic losses from the local overpotential η_{local}, which is written as follows:

$$\eta_{\text{act}} = \eta_{\text{local}} - \frac{1}{F}\left(\tilde{\mu}_{\text{e}^-,\,\text{RE}/\text{S}} - \tilde{\mu}_{\text{e}^-,\,\text{lyte}/\text{RE}} \right)$$

$$- \frac{1}{F}\left(\tilde{\mu}_{\text{e}^-,\,\text{lyte}/\text{WE}} - \tilde{\mu}_{\text{e}^-,\,\text{WE}/\text{S}} \right) - \frac{1}{2F}\left(\tilde{\mu}_{\text{O}^{2-},\,\text{lyte}/\text{RE}} - \tilde{\mu}_{\text{O}^{2-},\,\text{lyte}/\text{WE}} \right)$$

$$= -\frac{1}{2F}\left(-\mu_{\text{O},\,\text{lyte}/\text{WE}} + \mu_{\text{O,RE}} \right)$$

$$= -\frac{1}{2F}\left(2\tilde{\mu}_{\text{e}^-,\,\text{lyte}/\text{WE}} - \tilde{\mu}_{\text{O}^{2-},\,\text{lyte}/\text{WE}} + \left\{ \mu_{\text{O}}^\circ + \Delta G^\circ + RT \log\left(\frac{p_{\text{H}_2\text{O}}}{p_{\text{H}_2}} \right) \right\}_{\text{RE}} \right). \tag{18}$$

where ohmic loss in the reference electrode is zero, $\tilde{\mu}_{\text{e}^-,\,\text{RE}/\text{S}} - \tilde{\mu}_{\text{e}^-,\,\text{lyte}/\text{RE}} = 0$. In Eq. (18), the electron in the electrolyte side at the RE/electrolyte interface is assumed to be in equilibrium with that in the electrode side $\tilde{\mu}_{e^-,\,\text{lyte}/\text{RE}}$, and also a local equilibrium is assumed in the electrolyte[42]:

$$\tilde{\mu}_{O^{2-}} = 2\tilde{\mu}_{e^-} + \mu_O. \tag{19}$$

In addition, the oxygen is assumed to be in equilibrium with the gaseous phase at RE:

$$\mu_O = \mu_O^\circ + \frac{1}{2}RT\ln p_{O_2} = \mu_O^\circ + \Delta G^\circ + RT\ln\left(\frac{p_{H_2O}}{p_{H_2}}\right). \tag{20}$$

The variables in the RHS of Eq. (18) are defined at the voxels adjacent to the TPB segment, which are solved in the LBM calculation.

The total overpotential of the anode η_{anode} is obtained by subtracting the ohmic losses of current collector (CC), electrolyte and RE from the potential difference between RE and CC. Assuming that the gas compositions are the same at CC and RE, we have:

$$\eta_{\text{anode}} = \frac{1}{F}\left(\tilde{\mu}_{e^-,\,\text{RE}/\text{S}} - \tilde{\mu}_{e^-,\,\text{CC}}\right) - \frac{1}{F}\left(\tilde{\mu}_{e^-,\,\text{anode}/\text{CC}} - \tilde{\mu}_{e^-,\,\text{CC}}\right)$$

$$- \frac{1}{2F}\left(\tilde{\mu}_{O^{2-},\,\text{lyte}/\text{RE}} - \tilde{\mu}_{O^{2-},\,\text{anode}/\text{lyte}}\right) - \frac{1}{F}\left(\tilde{\mu}_{e^-,\,\text{RE}/\text{S}} - \tilde{\mu}_{e^-,\,\text{lyte}/\text{RE}}\right)$$

$$= -\frac{1}{2F}\left(-\mu_{O,\,\text{anode}/\text{CC}} + \mu_{O,\,\text{CC}}\right)$$

$$= -\frac{1}{2F}\left(2\tilde{\mu}_{e^-,\,\text{anode}/\text{CC}} - \tilde{\mu}_{O^{2-},\,\text{anode}/\text{lyte}} + \left\{\mu_O^\circ + \Delta G^\circ + RT\log\left(\frac{p_{H_2O}}{p_{H_2}}\right)\right\}_{\text{CC}}\right). \tag{21}$$

where ohmic loss in the reference electrode is zero, $\tilde{\mu}_{e^-,\,\text{RE}/\text{S}} - \tilde{\mu}_{e^-,\,\text{lyte}/\text{RE}} = 0$. The schematic of the total overpotential is shown in Figure 8.7. Again, the variables in the RHS of Eq. (21) are solved in the LBM calculation.

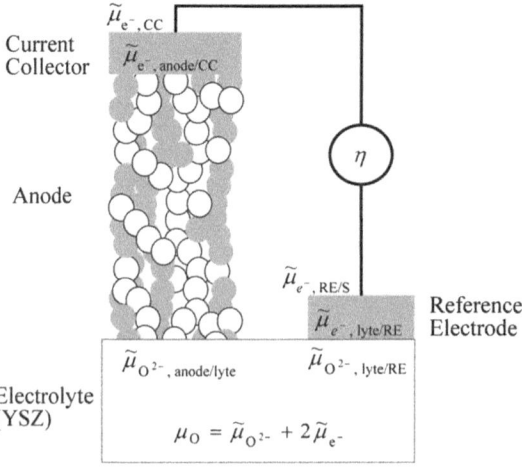

Figure 8.7 Schematic of total anode overpotential.[26]

8.3.3 Computational Scheme

The LBM is used to solve Eqs. (7), (12) and (13) in each of the three phases. Generally, for the 3D LBM simulations, D3Q15 ($i = 1$–15) or D3Q19 ($i = 1$–19) models are commonly used. However, it has been shown that, in the case of simple diffusion simulation without convection, D3Q6 ($i = 1$–6) model can be efficiently used with a slight loss of accuracy.[43] So, the D3Q6 model is used in this study. The lattice Boltzmann equation with the lattice Bhatnagar-Gross-Krook (LBGK) model in the collision term is written as follows:

$$f_i(\mathbf{x} + \mathbf{c}_i\Delta t, t + \Delta t) = f_i(\mathbf{x}, t) - \frac{1}{t^*}\left[f_i(\mathbf{x}, t) - f_i^{eq}(\mathbf{x}, t)\right] + w_i\Delta t. \quad (22)$$

In Eq. (22), f_i represents the density distribution function of gas, electron or ion with a velocity c_i in the i–th direction, and f_i^{eq} is the Maxwellian local equilibrium distribution,

$$f_i^{eq}(\mathbf{x}, t) = \frac{1}{6}\sum_{i=1}^{6} f_i(\mathbf{x}, t). \quad (23)$$

The relaxation time t^* is a function of diffusion coefficient, voxel size Δx and time step Δt, and it is given as:

$$t^* = 0.5 + \frac{3D\Delta t}{\Delta x^2}, \quad (24)$$

where

$$D = \left(\frac{1 - \alpha y_{H_2}}{D_{H_2,H_2O}} + \frac{1}{D_{H_2,k}}\right)^{-1}. \quad (25)$$

In the present study, the time step Δt is chosen so that the relaxation time becomes $t^* = 0.99$. However, the DGM diffusion coefficient is not constant in the gaseous phase. So the relaxation time is changed according to the DGM diffusion coefficient. The last term of Eq. (22) is the production term calculated from the reaction current density (Eq. (14).

For samples A and B, three mirrored FIB-SEM structures are repeated in the anode thickness direction to secure sufficient electrode thickness. On the other hand, original single structure is used in the case of sample C which covers whole anode thickness. The electrolyte and current collector layers are added at both ends of the computational volume, as shown in Figure 8.8. The thicknesses of electrolyte and CC layers are 2.60 μm and 1.24 μm, respectively. A zero gradient condition is assumed at the side boundaries. At the current collector surface, constant gas composition (Dirichlet boundary) is applied. Constant electronic and ionic current flux conditions (Neumann boundary) are imposed on the current collector and electrolyte boundaries, respectively. A no-flux boundary condition is imposed on the solid-gas interfaces in the porous media by applying the halfway bounceback scheme with a second-order accuracy.[44]

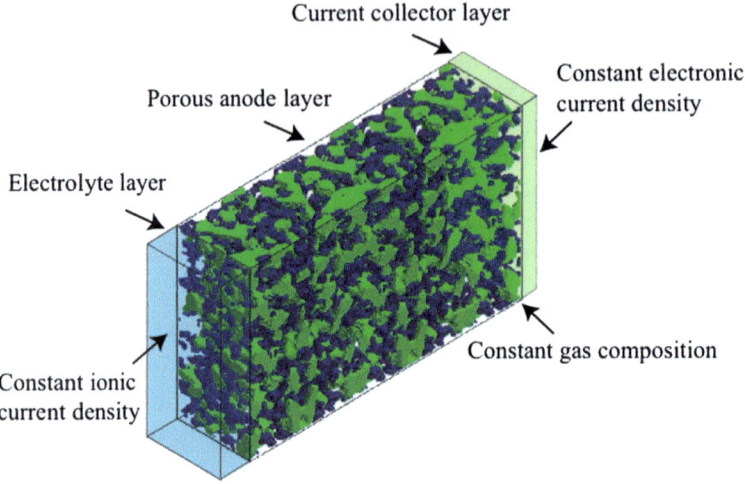

Figure 8.8   Schematic of computational domain, green: Ni, blue: YSZ.[26]

8.4   Results and Discussions

Overpotential calculations with the de Bore current density model (Eq. 15) are conducted in order to evaluate the effects of different sample volume sizes. Predicted anode overpotentials at 1073 K are shown in Figure 8.9. For sample C, calculation is conducted only for the 3 mol% H_2O – 97 mol% H_2 case because of the computational cost. The predicted overpotential of the largest sample C is slightly larger than the smaller samples A and B, but the difference is less than 0.01 V. Thus, it can be said that small volume size results in underprediction of overpotential but its influence is relatively small.

Overpotential calculations are conducted in different steam partial pressures using sample A. The results with different exchange current density models are shown in Figure 8.10. In Figure 8.10(a), prediction by de Boer's model shows weaker dependence on steam partial pressure than the experimental data. The dependence on steam partial pressure is somewhat improved by the Bieberle *et al.*'s model in Figure 8.10(b). This indicates that the exponent of P_{H_2O} in Bieberle *et al.*'s exchange current density i_0 is preferred than the de Boer's model. However, still there's a discrepancy between the prediction and experimental data. It is concluded that further validation of the exchange current density model is required.

Figure 8.11 shows overpotential predictions in sample A with different temperatures using two exchange current density models. The result indicates that the activation energy in de Boer's model somewhat overpredicts the temperature dependence. As can be seen from Figure 8.11(b), the temperature dependence for Bieberle *et al.*'s model agrees better with the experiment. This result supports the validity of the activation energy in the Bieberle *et al.*'s model.

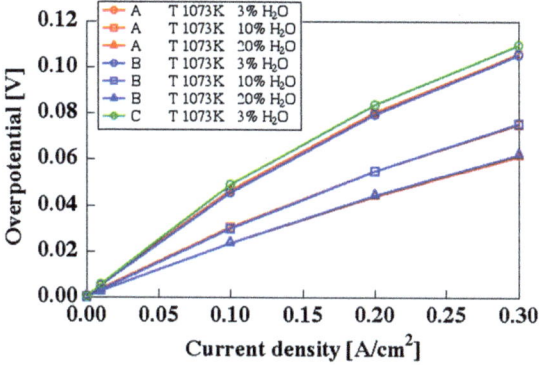

Figure 8.9   Predicted overpotential results of different samples.[28]

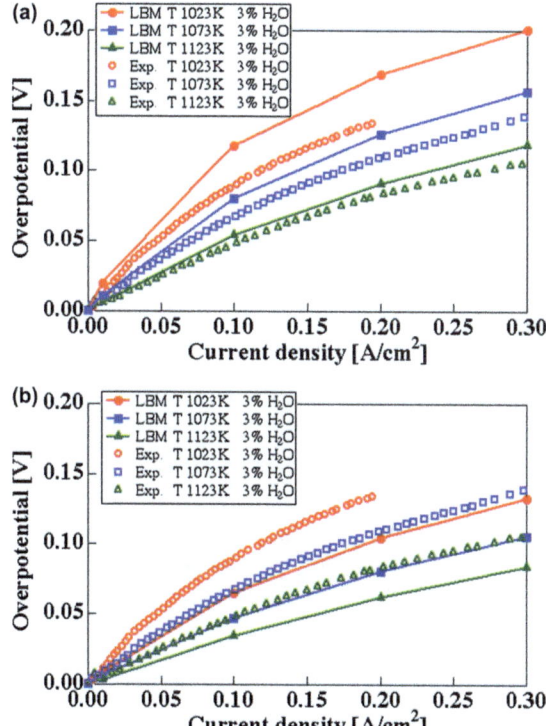

Figure 8.10   Dependence of predicted overpotentials on steam partial pressure: (a) de Boer's model and (b) Bieberle's model.[28]

Figure 8.12 shows ionic (red) and electronic (blue) current stream line distributions in samples A and C, for 3mol% $H_2O - 97$ mol% H_2, $i = 0.3$ A cm^{-2} at 1073 K case. The total thicknesses including electrolyte and current collector layers are 41.8 μm and 49.6 μm for samples A and C, respectively. As can be seen from the figure, a large portion of ionic current is changed to

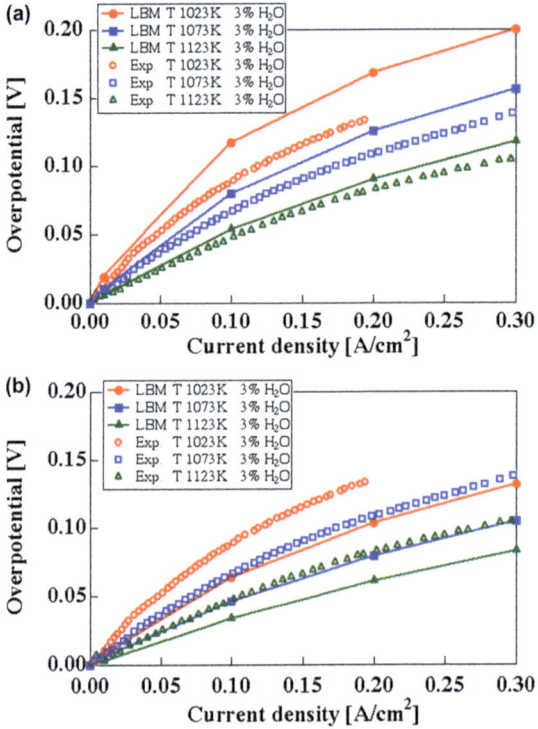

Figure 8.11 Dependence of predicted overpotentials on temperature: (a) de Boer's model and (b) Bieberle's model.[28]

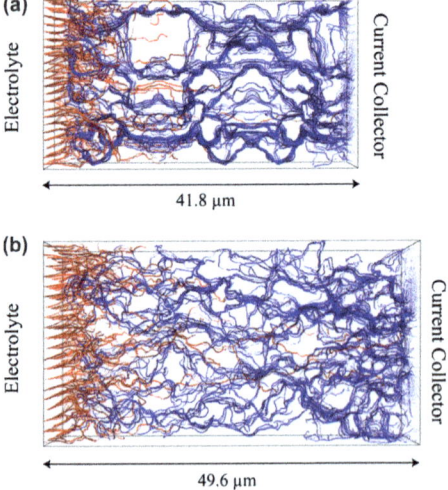

Figure 8.12 Three-dimensional current stream line distributions (red: ionic current, blue: electronic current), (a) sample A and (b) sample C.[28]

electronic current in the region close to the electrolyte. Electronic paths are more concentrated and reactive thickness is somewhat thinner for sample A than for sample C. This is considered to be due to the mirrored structures used in sample A, which resulted in factitious connections of the Ni network. In sample C, electron paths are more distributed and reactive TPBs, where ionic current is converted to electric current, distributed widely. As a result, ionic current must diffuse farther from the electrolyte to the current collector side through less conductive YSZ network. This might be the reason why sample C showed slightly larger overpotential than the smaller samples as shown in Figure 8.9. It is thus recommended to use larger volume size samples which can cover whole reactive thickness when discussing the local potential and flux distributions.

One of the merits of this numerical scheme is the possibility of reducing the difficulty which results from the microstructure complexity. For example, it will be possible to discuss local reaction rate, i.e. local current density at TPB, using the real microstructure and the macroscopic experiment. In other words, it will become possible to investigate local kinetics at TPB using electrodes which are made by the common low-cost fabrication processes instead of using artificial patterned electrodes. It is expected that this information will greatly contribute to the understanding of local reaction and transport kinetics inside complex electrodes.

Another promising future applications of the present method will be a numerical durability testing. As already shown, overpotential can be calculated from a given microstructure with the present approach. Thus, the next challenge is to predict the temporal alteration of the microstructure based on numerical simulation. Recently, several numerical works on sintering kinetics of composite electrodes have been reported.[45,46,47,48] Combining these methods with the present polarization simulation, it will become possible in the future to calculate the temporal change of the overpotential. This numerical durability testing will drastically improve the efficiency of research and development processes of SOFC.

8.5 Conclusions

A three-dimensional numerical simulation of the anode overpotential is conducted in a microstructure reconstructed by FIB-SEM. The transport of species is calculated by the LBM coupled with the electrochemical reaction at TPB. The TPB length is estimated by the centroid method, which can easily calculate the TPB length within a 5% margin error provided that grid resolution is high ($r/\Delta x > 15$). YSZ tortuosity factor remains nearly unchanged when cross-sectional area exceeds approximately $200\ \mu m^2$, while pore tortuosity factor is almost independent of the sample volume size. However, Ni tortuosity shows very large variation regardless of sample volume size. Small volume size results in underprediction of the overpotential, but its influence is relatively small. Two exchange current models from the patterned electrode experiments of de Boer and Bieberle *et al.* are evaluated. The dependence on temperature with Bieberle *et al.*'s model agrees well with the experimental data. However, both models show weaker dependence on the steam concentration than the experimental data. Further studies on the H_2O partial pressure dependence of the exchange current density should be required. From the visualized three-dimensional current stream lines, it is shown that effective reaction area becomes thinner for the mirrored computational structure because of the factitious connection of the Ni network. It is thus recommended to use larger volume size samples which can cover the whole reactive thickness when discussing the local potential and flux distributions. The present method is very promising to investigate local reaction kinetics at the three phase boundary inside actual electrodes. In addition, it will become possible in the future to predict temporal overpotential variation by combining the present method with the morphological evolution schemes. Thus, it is expected that three-dimensional electrode modelling will drastically contribute to the improvement of SOFC.

Acknowledgements

This work was supported by the New Energy and Industrial Technology Development Organization (NEDO) and Japan Science and Technology Agency (JST), CREST project.

References

1. S. C. Singhal and K. Kendall, *High Temperature Solid Oxide Fuel Cells*, Elsevier (2002).
2. D. Simwonis, F. Tiez and D. Stöver, *Solid State Ionics*, 2000, **132**, 241–251.
3. N. Shikazono, Y. Sakamoto, Y. Yamaguchi and N. Kasagi, *J. Power Sources*, 2009, **193**, 530–540.
4. C. W. Tanner, K. –Z. Fung and A. V. Virkar, *J. Electrochem. Soc.*, 1997, **144**(1), 21–30.
5. J. R. Wilson, W. Kobsiriphat, R. Mendoza, H. -Y. Chen, J.M. Hiller, D. J. Miller, K. Thornton, P.W. Voorhees, S. B. Adler and S. Barnett, *Nature Materials*, 2006, **5**, 541–544.
6. D. Gostovic, J. R. Smith, D. P. Kundinger, K. S. Jones and E. D. Wachsman, *Electrochem. Solid –State Lett.*, 2007, **10**(12), B214–B217.
7. J. R. Smith, A. Chen, D. Gostovic, D. Hickey, D. P. Kundinger, K. L. Duncan, R. T. DeHoff, K. S. Jones and E. D. Wachsman, *Solid State Ionics*, 2009, **180**, 90 –98.
8. J. R. Wilson, M. Gameiro, K. Mischaikow, W. Kalies, P. Voorhees and S. Barnett, *Microsc. Microanal.*, 2009, **15**, 71 –77.
9. J. R. Wilson, A. T. Duong, M. Gameiro, H. -Y. Chen, K. Thornton, W. Kalies, D. R. Mumm and S. Barnett, *Electrochemistry Communications*, 2009, **11**, 1052–1056.
10. P. R. Shearing, J. Golbert, R. J. Chater and N. P. Brandon, *Chemical Eng. Sci.*, 2009, **64**, 3928–933.

11. H. Iwai, N. Shikazono, T. Matsui, H. Teshima, M. Kishimoto, R. Kishida, D. Hayashi, K. Matsuzaki, D. Kanno, M. Saito, H. Muroyama, K. Eguchi, N. Kasagi and H. Yoshida, *J. Power Sources*, 2010, **195**(4), 955–961.

12. L. Holzer, B. Iwanschitz, Th. Hocker, B. Münch, M. Prestat, D. Wiedenmann, U. Vogt, P. Holtappels, J. Sfeir, A. Mai and Th. Graule, *J. Power Sources*, 2011, **196**(3), 1279–1294.

13. N. Gunda, H.-W. Choi, A. Berson, B. Kenney, K. Karan, J. G. Pharoah and S. K. Mitra, *J. Power Sources*, 2011, **196**(7), 3592–3603.

14. J. Joos, T. Carraro, A. Weber and E. Ivers –Tiffée, *J. Power Sources*, 2011, **196**(17), 7302–7307.

15. N. Vivet, S. Chupin, E. Estrade, T. Piquero, P. L. Pommier, D. Rochais and E. Bruneton, *J. Power Sources*, 2011, **196**(18), 7541–7549.

16. J. R. Izzo, Jr., A. S. Joshi, K. N. Grew, W. K. S. Chiu, A. Tkachuk, S. H. Wang and W. Yun, *J. Electrochem. Soc.*, 2008, **155**(5), B504–B508.

17. Y. Guan, W. Li, Y. Gong, G. Liu, X. Zhang, J. Jie Chen, W Gelb, Y. Yun, Y. Xiong, Tian and H. Wang, *J. Power Sources*, 2011, **196**(4), 1915–1919.

18. J. Laurencin, R. Quey, G. Delette, H. Suhonen, P. Cloetens and P. Bleuet, *J. Power Sources*, 2012, **198**, 182–189.

19. P. R. Shearing, R. S. Bradley, J. Gelb, F. Tariq, P. J. Withers and N. P. Brandon, *Solid State Ionics*, 2012, **216**, 69–72.

20. P. R. Shearing, D. J. L. Brett and N. P. Brandon, *Int. Mat. Rev.*, 2010, **55**(6), 347–363.

21. J. Golbert, C. S. Adjiman and N. P. Brandon, *Ind. Eng. Chem. Res.*, 2008, **47**, 7693–7699.

22. A. S. Joshi, K. N. Grew, A. A. Peracchio and W. K. S. Chiu, *J. Power Sources*, 2007, **164**, 631–638.

23. P. Asinari, M. C. Quaglia, M. R. von Spakovsky and B. V. Kasula, *J. Power Sources*, 2007, **170**, 359–375.

24. W. K. S. Chiu, A. S. Joshi and K. N. Grew, *Eur. Phys. J. Special Topics*, 2009, **171**, 159–165.

25. Y. Suzue, N. Shikazono and N. Kasagi, *J. Power Sources*, 2008, **184**, 52–59.

26. N. Shikazono, D. Kanno, K. Matsuzaki, H. Teshima, S. Sumino and N. Kasagi, *J. Electrocchem. Soc.*, 2010, **157**(5), B665–B672.

27. P. R. Shearing, Q. Cai, J. Golbert, V. Yufit, C. S. Adjiman and N. P. Brandon, *J. Power Sources*, 2010, **195**, 4804–4810.

28. D. Kanno, N. Shikazono, N. Takagi, K. Matsuzaki and N. Kasagi, *Electrochimica Acta*, 2011, **56**, 4015–4021.

29. K. Matsuzaki, N. Shikazono and N. Kasagi, *J. Power Sources*, 2011, **196**(6), 3073–3082.

30. M. Kishimoto, H. Iwai and H. Yoshida, *J. Power Sources*, 2011, **196**(10), 4555–4563.

31. Q. Cai, C. S. Adjiman and N. P. Brandon, *Electrochimica Acta*, 2011, **56**, 10809–10819.

32. K. N. Grew and W. K. S. Chiu, *J. Power Sources*, 2012, **199**(1), 1–13.
33. T. Carraro, J. Joos, B. Rüger, A. Weber and E. Ivers –Tiffée, *Electrochimica Acta*, 2012, **77**, 315–323.
34. J. R. Wilson, J. S. Cronin, A. T. Duong, S. Rukes, H. –Y. Chen, K. Thornton, D. R. Mumm and S. Barnett, *J. Power Sources*, 2010, **195**(7), 1829–1840.
35. R. Krishna and J. A. Wesselingh, *Chem. Eng. Sci.*, 1997, **52**, 861–911.
36. V. Novak, F. Stepanek, P. Koci, M. Marek and M. Kubicek, *Chem. Eng. Sci.*, 2010, **65**(7), 2352–2360.
37. T. Kawada, N. Sakai, H. Yokokawa and M. Dokiya, *J. Electrochem. Soc.*, 1990, **137**, 3042–3047.
38. B. De Boer, Ph. D. Thesis, University of Twente, (1998).
39. A. Bieberle, L. P. Meier and L. J. Gauckler, *J. Electrochem. Soc.*, 2001, **148**(6), A646–A656.
40. U. Anselmi-Tamburini, G. Chiodelli, M. Arimondi, F. Maglia, G. Spinolo and Z. A. Munir, *Solid State Ionics*, 1998, **110**, 35–43.
41. J. R. Ferguson, J. M. Flard and R. Herbin, *J. Power Sources*, 1996, **58**, 109–122.
42. J. Mizusaki, K. Amano, S. Yamauchi and K Fueki, *Solid State Ionics*, 1987, **22**, 313–322.
43. T. H. Zeiser, P. Lammers, E. Klemm, Y. W. Li, J. Bernsdorf and G. Brenner, *Chem. Eng. Sci.*, 2001, **56**, 1697–1704.
44. M. A. Gallivan, D. R. Noble, J. G. Georgiadis and R. O. Buckius, *Int. J. Numer. Methods Fluids*, 1997, **25**, 249–263.
45. H. –Y. Chen, H. –C. Yu, J. S. Cronin, J. R. Wilson, S. A. Barnett and K. Thornton, *J. Power Sources*, 2011, **196**, 1333–1337.
46. Y. Zhang, C. Xia and M. Ni, *Int. J. Hydrogen Energy*, 2012, **37**, 3392–3402.
47. K. Shikata, S. Hara, N. Shikazono S. Izumi, and S. Sakai, *Proc. IUMRS Int. Conf. Electronic Materials*, 2012, A-2-P26-014.
48. Z. Jiao and N. Shikazono, *J. Electrochem. Soc.*, 2013, 160(6), F709–F715.

CHAPTER 9

Multi-scale Modelling of Solid Oxide Fuel Cells

WOLFGANG G. BESSLER

Offenburg University of Applied Sciences, Badstrasse 24, 77652 Offenburg University of Applied Sciences, Institute for Energy System Technology, Germany
Email: wolfgang.bessler@hs-offenburg.de

Dedicated to David G. Goodwin who passed away on November 11, 2012

9.1 Introduction and Motivation

The present chapter introduces and reviews multi-scale modelling techniques for solid oxide fuel cells (SOFCs). Methods of mathematical modelling and numerical simulation are increasingly being used as tools to understand and optimize both fundamental electrochemical properties and technological applications of SOFCs. The overall goal of modelling activities is to identify, understand, and improve critical components and processes responsible for cell and system performance, lifetime, and cost. To this goal, a large number of different physical, chemical, and fluid mechanical processes ("*multi-physics*") that take place over a wide range of spatial and temporal scales ("*multi-scale*") have to be considered. This represents a considerable challenge for theoreticians not only during the course of model development, but also when performing numerical simulations with different software codes that are typically specialized towards particular processes and scales.

RSC Energy and Environment Series No. 7
Solid Oxide Fuel Cells: From Materials to System Modeling
Edited by Meng Ni and Tim S. Zhao

The SOFC modelling and simulation literature is vast and has been strongly growing in the past decade. Several reviews have been published before.[1–8] Here, we put a focus on multi-scale techniques, that is, techniques and examples that cross the time and length scales, and therefore typically cross different methodologies. Section 9.2 gives an overview of modelling methodologies that are used on various individual scales. Core model equations are given where appropriate. This overview forms the basis for crossing the scales. Section 9.3 consequently treats scale-combining methods. It firstly introduces general aspects of multi-scale and multi-physics modelling techniques. All SOFC models have to cover three fundamental aspects in one way or the other, that is, *chemistry* (electrochemistry and other types of chemistry such as reforming), *transport* (of mass, charge and heat), and *structure* (of electrodes, cells and stacks). Section 9.3 therefore covers scale-crossing techniques for chemistry, transport, and structure. Section 9.4 gives two specific examples for multi-scale modelling of SOFC systems, a pressurized SOFC system for stationary hybrid power plants, and a tubular SOFC system for mobile auxiliary power units. Finally, Section 9.5 concludes the chapter.

9.2 Modelling Methodologies: From the Atomistic to the System Scale

9.2.1 Overview

The fuel cell is an outstanding example of a multi-scale system. This situation is shown schematically in Figure 9.1. Electrochemical reactivity takes place on a nanometer scale and strongly depends on nano- and microstructural properties. Mass, charge, and heat transport take place from a nanometer (atomistic level) up to meter scale (system level). Time scales vary from sub-nanoseconds (electrochemical reactions) over seconds and minutes (mass and heat transport) up to days or even months (structural and chemical degradation). All processes are strongly, and often nonlinearly, coupled over the various scales. Processes on the microscale can therefore dominantly influence macroscopic behaviour. A detailed understanding of the relevant processes on all scales is required for a computer-based optimization of fuel cell design, performance and durability. Indeed, it is easy to disable an SOFC system by disabling any individual scale, but hard to optimize all individual scales to achieve the best system behaviour.

9.2.2 Molecular Level: Atomistic Modelling

The basic physicochemical processes of chemical reactions, charge transfer, and formation of space-charge layers take place on an atomistic scale on the sub-nanometer range. A number of different modelling techniques is used to address this scale, including quantum mechanics, molecular dynamics, and Monte Carlo techniques.

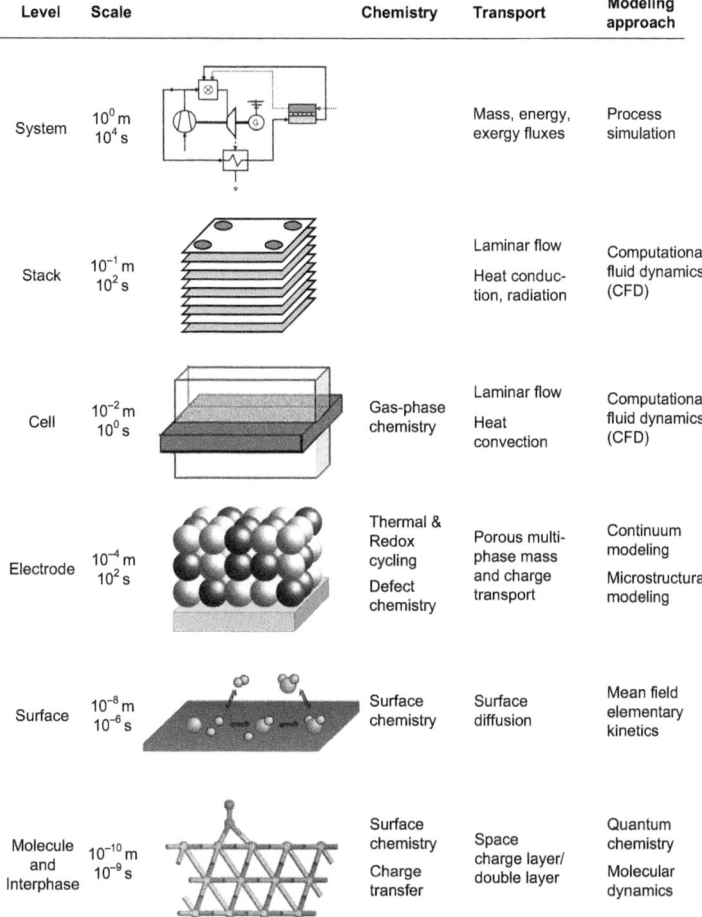

Level	Scale		Chemistry	Transport	Modeling approach
System	10^0 m 10^4 s			Mass, energy, exergy fluxes	Process simulation
Stack	10^{-1} m 10^2 s			Laminar flow Heat conduction, radiation	Computational fluid dynamics (CFD)
Cell	10^{-2} m 10^0 s		Gas-phase chemistry	Laminar flow Heat convection	Computational fluid dynamics (CFD)
Electrode	10^{-4} m 10^2 s		Thermal & Redox cycling Defect chemistry	Porous multi-phase mass and charge transport	Continuum modeling Microstructural modeling
Surface	10^{-8} m 10^{-6} s		Surface chemistry	Surface diffusion	Mean field elementary kinetics
Molecule and Interphase	10^{-10} m 10^{-9} s		Surface chemistry Charge transfer	Space charge layer/ double layer	Quantum chemistry Molecular dynamics

Figure 9.1 Relevant scales for the mathematical modelling and numerical simulation of solid oxide fuel cells.[82]

9.2.2.1 Quantum Chemistry

Quantum mechanical *ab initio* or *first-principles* calculations (referring to the fact that no adjustable model parameters are used) are based on the solution of Schrödinger's equation,

$$\hat{H} \Psi = E \Psi \qquad (1)$$

where $\hat{H}$ is the total energy operator (Hamiltonian), Ψ the wave function and E the total energy of the system. The solution for different geometries of the atoms yields the potential energy surface, from which ground-state energies, activation energies (by following reaction trajectories) and entropy contributions (*via* vibrational analysis) can be assessed. An efficient numerical method to solve Schrödinger's equation commonly used today is density functional theory (DFT).[9,10] There is a number of software tools available for DFT calculations (VASP, GAUSSIAN, CASTEP, *etc.*).

Quantum chemical calculations are computationally expensive; as computational cost scales with the cube of the number of atoms, practical systems sizes usually are in the order of tens up to hundreds of atoms for stationary models and smaller for transient models; the use of parallel computing techniques is often mandatory.

9.2.2.2   Molecular Dynamics

If larger system sizes are to be investigated, the level of detail of particle interactions has to be drastically reduced. An appropriate methodology to do so is *molecular dynamics* (MD). In this technique, the interaction between atoms or molecules is modeled based on a semi-empirical description of attractive and repulsive forces, the so-called *force field*, borrowing concepts from classical mechanics. The kinematic equations are solved using Newton's law, allowing following the motion of atoms and molecules in time. When *reactive force fields* are used, chemical reactions can be modelled.[11] Because molecular dynamics are computationally much less expensive than quantum chemical methods, they can be used to study larger systems of thousands of atoms, and dynamical simulations can be performed on the nanosecond time scale.

9.2.2.3   Kinetic Monte Carlo

A fundamental principle of quantum mechanics states that the deterministic macroscopic behaviour of a system is the result of the non-deterministic (stochastic) atomistic behavior of each of its particles. Statistical simulation methods, so-called *kinetic Monte Carlo* (KMC) methods, take advantage of this principle by studying a sufficiently large system (hundreds or thousands of particles), assigning probabilities to the processes these particles can undergo (*e.g.*, reaction, diffusion), and simulating the system's behaviour over time using random numbers.[12] The probabilities can be calculated based on quantum chemical studies of the individual processes. Thus, KMC simulations represent an elegant way to use quantum chemical results for studying the behaviour of a large system.

9.2.3   Electrode Level (I): Electrochemistry with Mean-field Elementary Kinetics

Elementary kinetics means the resolution of chemistry into single steps that represent reactivity on the molecular scale. In the so-called *mean-field approach*, chemical kinetics are modelled in an *ensemble-averaged* way. In the case of heterogeneous chemistry, the surface is described using averaged quantities like surface coverages, surface site denisities, and macroscopic thermodynamic properties of adsorbates. Chemical reactions are modelled using mass-action kinetics. The atomic-scale surface structure (terraces, steps, edges), composition (*e.g.*, impurities), adsorbate behaviour (*e.g.*, end-on or side-on geometry) and reaction trajectory are not resolved and are assumed to be included in the averaged quantities.[5] This approach allows describing chemical reactivity in a detailed way at relatively cheap computational cost.

The key governing equations in elementary kinetic modelling are the so-called rate equations, that is, the change of each species' concentration c_i (surface, gas-phase or bulk phase concentration) with time t due to chemical reactions, which can be generally formulated as

$$\frac{\partial c_i}{\partial t} = \sum_m \nu_{i,m} \left(k_{f,m} \prod_{j \in R_{f,m}} c_j^{\nu_j'} - k_{r,m} \prod_{j \in R_{r,m}} c_j^{\nu_j''} \right), \quad (2)$$

where ν' and ν'' are stoichiometric coefficients for reactants and products, and the sum m runs over all reactions that involve species i, including surface reactions, gas-phase reactions, bulk phase reactions, charge-transfer reactions, and interfacial reactions. The products j run over reactants and products, respectively, of reaction m. The reaction rate constants k_f and k_r of forward and reverse reaction, respectively, are of particular importance, as they include coefficients that can be obtained from atomistic models. A generalized Arrhenius-type expression for k can be derived from transition state theory,

$$k_f = k_f^0 \, T^{\beta_f} \, \exp\left(-\frac{E_f^{act}}{RT}\right) \, \exp\left(-(1-\alpha)\frac{zF}{RT}\Delta\phi\right) \quad (3)$$

$$k_r = k_f^0 \, T^{\beta_f/\beta_r} \, \exp\left(\frac{\Delta G}{RT}\right) \, \exp\left(\alpha\frac{zF}{RT}\Delta\phi\right), \quad (4)$$

where k_f^0, β_f and E_f^{act} are the preexponential factor, temperature coefficient, and activation energy of the forward reaction, respectively, α is the symmetry factor of a charge-transfer reaction, $\Delta\phi = \phi_{electrode} - \phi_{electrolyte}$ the electric-potential difference of the electron before and after the charge-transfer step (which can generally include double-layer effects), and z the number of electrons transferred ($z=1$ for an elementary charge transfer step). For reactions between uncharged species (non-electrochemical reactions, $z=0$) the last terms vanish. The reverse reaction rate constant (Eq. 4) should be calculated by using the free enthalpy of reaction ΔG in order to ensure thermodynamic consistency (microscopic reversibility) of the chemical reactions,[13] an issue often overlooked in elementary kinetic models.

An elementary kinetic reaction mechanism can consist of several thousand reactions (*e.g.*, gas-phase combustion chemistry).[14] A number of software tools (CANTERA, CHEMKIN, DETCHEM) have been developed to handle and evaluate such large mechanisms.

9.2.4 Electrode Level (II): Porous Mass and Charge Transport

The microstructural design of the porous electrodes is known to have a major influence on macroscopic fuel cell performance. Microstructural engineering is therefore an important part of electrode technology development. Design parameters include composition, size distribution, porosity and tortuosity, functional grading, *etc.* There are two conceptually different approaches to porous electrode modelling, *continuum approaches* and *structure-resolved approaches.* Microstructure-resolved approaches are computationally expensive and are often only used within a small, representative volume element of the porous medium. Continuum approaches are computationally cheaper and therefore can be relatively easily integrated into a larger-scale description of the complete cell or even the stack.

9.2.4.1 Continuum Approach

In the *continuum approach* (also called *homogenization* or *macrohomogeneous* approach), the porous material is described as homogenized continuum of all phases using effective structural (porosity, tortuosity), transport (diffusion coefficients, permeability), electrical (conductivity) and electrochemical (specific catalytic surface area, specific three-phase boundary length) parameters. The microstructure itself is not spatially resolved, but its properties enter the homogenized description via the effective parameters. There is a number of standard continuum models for mass transport in porous media, including Stephan-Maxwell diffusion,[15] dusty-gas model (DGM, originally developed for astrophysical applications, hence the name),[16] and mean transport pore model (MTPM);[17,18] for electrodes, they are coupled in a multi-physics approach with a governing equation for ionic and electronic charge transport (typically, Ohm's law).

9.2.4.2 Structure-resolved Approach

In the *structure-resolved approach*, the microstructure of the porous medium is directly represented in the model with high (sub-micrometer) spatial resolution. This usually requires two steps: First, the microstructure itself is generated. Second, the geometrical, electrochemical and/or transport properties of the microstructure are simulated. There are different approaches to both of these steps.

The microstructure may be generated either computationally or experimentally. Computational generation usually involves a statistical approach, such as the arbitrary placement of spheres in a given volume.[19,20] The experimental measurement of 3D microstructures are based either on processing of a set of 2D images (FIB-SEM technique: combining 2D scanning electron microscope imaging with focused-ion beam milling)[21,22] or on direct tomographic generation of 3D data (*e.g.*, via X-ray or neutron tomography).

Different methods have been applied for analyzing the geometrical, electrochemical and/or transport properties of the generated microstructure.

They can broadly be divided into discrete particle methods and continuum equation methods. In discrete particle methods, the microstructure is treated as set of objects (*e.g.*, overlapping spheres or interconnected pipes). For example, in *random resistor network models*, contact resistances and capacitances are assigned to a network of spheres to simulate electric behaviour.[23] In continuum equation methods, the microstructure is treated as continuous 3D spatial domain. A sufficiently fine computational grid is imposed over the structure. Transport and electrochemical properties are simulated by solving appropriate equations on the grid. For example, solving the Laplace equation allows to simulate the electric potential distribution.[22] The volume of fluid (VoF) method allows to simulate gas-phase transport.[24]

9.2.5 Cell Level: Coupling of Electrochemistry with Mass, Charge and Heat Transport

The cell level represents a single fuel cell membrane-electrode assembly, often including gas supply (channels, contact meshes *etc.*). This scale is governed by macroscopic mass, charge and energy transport on the millimeter to centimeter scale, which is strongly coupled to electrochemistry in the electrodes. There is a large number of modelling approaches treating the cell scale on different levels of complexity and with different goals, including studies of detailed electrochemistry,[25] internal reforming chemistry,[26] mass transport,[27] heat management,[28] mechanical stability and thermomechanical stresses,[29] degradation phenomena,[30] transient phenomena,[31] *etc.* The high level of specialization and diversification of cell-level models has been realized by either flexible commercial software such as COMSOL, MATLAB and gPROMS, or in-house software tools such as MEMEPHYS,[32] DENIS,[13] and DETCHEM.[33]

Cell-level models are typically solved on 1D or 2D (either true 2D or decoupled 1D + 1D) domains. Such a reduced-dimensional description is usually sufficient to capture major geometrical influences while still being able to include detailed models of transport and chemistry. Furthermore, the implementation of physicochemically complex models often requires home-made software, in which 1D or 2D domains can be handled with reasonable effort. Fully resolved 3D models, on the other hand, usually have to rely on commercial CFD-type software packages. 3D is required for studying geometrical effects in detail (*e.g.*, design of cell and gas supply). This will be further discussed in Section 9.2.6.

The governing equations are generally based on conservation equations for mass, momentum, species, energy, and charge (cf. Section 9.2.6). These equations are solved on the relevant computational subdomains under simplifying assumptions. For example, momentum conservation can be neglected for gas-phase transport in the pore phases (Darcy law in dusty-gas model); channel flow may be reduced to a 1D description along the channel; charge transport is not needed in the channels; *etc.*

9.2.6   Stack Level: Computational Fluid Dynamics Based Design

This level represents the fuel cell stack, that is, multiple cells, interconnectors, gas manifolds, thermal insulation, *etc.*[34,35] This scale is governed by mass and heat transport in the gas phase and solid phases. It is therefore classically modelled with methods from computational fluid dynamics (CFD), often using readily-available software such as FLUENT, STAR-CCM+, OPEN-FOAM, COMSOL, *etc.* In the gas phase, the Navier-Stokes equations are solved (conservation of mass, momentum, species and energy), which are in a general form given by:[36]

$$\frac{\partial \rho}{\partial t} = - \ (\nabla \cdot \rho \, \vec{v}) + S_{\mathrm{m}} \tag{5}$$

$$\frac{\partial (\rho \, \vec{v})}{\partial t} = - \ [\nabla \cdot \rho \, \vec{v} \vec{v}] - \nabla p - [\nabla \cdot \bar{\bar{\tau}}] + S_{\mathrm{v}} \tag{6}$$

$$\frac{\partial \rho_i}{\partial t} = - (\nabla \cdot \rho_i \vec{v}) - \left(\nabla \cdot \vec{j}_i^{\mathrm{diff}}\right) + S_i \tag{7}$$

$$\frac{\partial (\rho \, h)}{\partial t} = - (\nabla \cdot \rho \, h \, \vec{v}) - (\nabla \cdot \vec{j}_{\mathrm{q}}) - (\bar{\bar{\tau}} : \nabla \vec{v}) + \frac{\partial p}{\partial t} + (\vec{v} \cdot \nabla p) + S_{\mathrm{q}} \tag{8}$$

extended by charge conservation in solid materials,

$$0 = - \nabla \cdot (\sigma \, \nabla \cdot \phi) + S_{\phi} \tag{9}$$

The structural complexity over a relatively large, typically 3D computational domain makes stack simulations computationally expensive, and parallel computing techniques are often used. The main application of stack-level models concerns the optimum geometrical design with respect to homogenous distributions of reactants and products (maximizing electrical output) and of temperature (minimizing thermal stresses) under the constraints of materials properties and manufacturing possibilities.

In order to handle the *geometrical complexity* within reasonable computational time, stack-level models often rely on *physicochemical simplicity* by, for example, using simple linear electrochemical kinetics. This approach is opposite to typical cell-level models (Section 9.9.2.5) that combine *multiphysics complexity* with *geometrical simplicity* by, for example, using reduced-dimensional domains. Performing transient simulations as compared to stationary simulations essentially adds another dimension (the time dimension) with correspondingly increasing computational effort.

9.2.7   System Level

This level represents the complete fuel cell system, that is, the stack itself and its periphery components such as pumps, blowers, heat exchangers, fuel processing units (*e.g.*, desulphurization, reforming), exhaust gas treatment, ac/dc converter, electronic control, *etc.*[37–40] System modelling considers not only mass and energy balances between these components, but also assesses exergy fluxes (*i.e.*, the part of the energy that is available for work) in order to understand and optimize the system's overall efficiency. There are a number of commercial software tools available for system-level modelling (*e.g.*, APROS, SIMULINK, ASPEN, DYMOLA, *etc.*). There are libraries available for many standard components of the system level; furthermore, the codes usually offer the possibility to integrate custom code from lower-level languages (*e.g.*, C or Fortran).

On the system level, the component interaction is complex and nonlinear. In order to cope with computational expense and numerical stability, the individual components (subsystems) are often simulated in a simple way, that is, by analytical expressions or look-up tables. These simplifying assumptions are justified as the goal of system simulations is not the understanding of detailed component behaviour, but the optimization of component size and interaction, as well as the development of potentially real-time capable control methods. Modern model reduction techniques allow to keep both the nonlinearity and the dynamics of the reduced models at significantly increased computational performance for the use in model-predictive control scenarios.[41]

9.3 Bridging the Gap Between Scales

9.3.1 General Aspects

In the previous section, we have presented an overview of available modelling methods for the different scales. These methods are generally highly specialized and highly specific, and are usually applied to particular problems within their respective scale. This is of high importance and has led to a considerable insight into fuel cell processes at all scales. However, the coupling between the scales is often not considered. Indeed, bridging the gap between the scales is a particular challenge. This is so not only because of the complex interaction of physico-chemical processes on different scales, but also because of the specialization of different research groups on individual scales and their methods. Indeed, it is difficult to combine expert knowledge of, for example, a DFT code like VASP with a system code like DYMOLA within a single research group.

An overview of multi-scale methods for the different SOFC scales is shown in Figure 9.2. Multi-scale methods can be generally separated into two main principles: (1) *indirect coupling* (also referred to as *vertical coupling*) by first performing computations on the lower scale and subsequently using these results as input parameters to the higher scale, and (2) *direct coupling* (also referred to as *horizontal coupling*) by directly including lower-scale physics into higher-scale models. An example for case (1) is the use of DFT-based parameters in elementary kinetic cell-level models.[25] An example for case (2) is the use of detailed component models in system simulations.[42] The central level of multi-scale approaches is the single-cell level, because this is the smallest fully functional unit of an SOFC. For this reason, most scale bridges shown in Figure 9.2 either start or end at this scale. From the figure it can also be seen that most coupling methods today rely on indirect (vertical) coupling schemes. Direct coupling between stack and system scales are scarce; system models usually rely

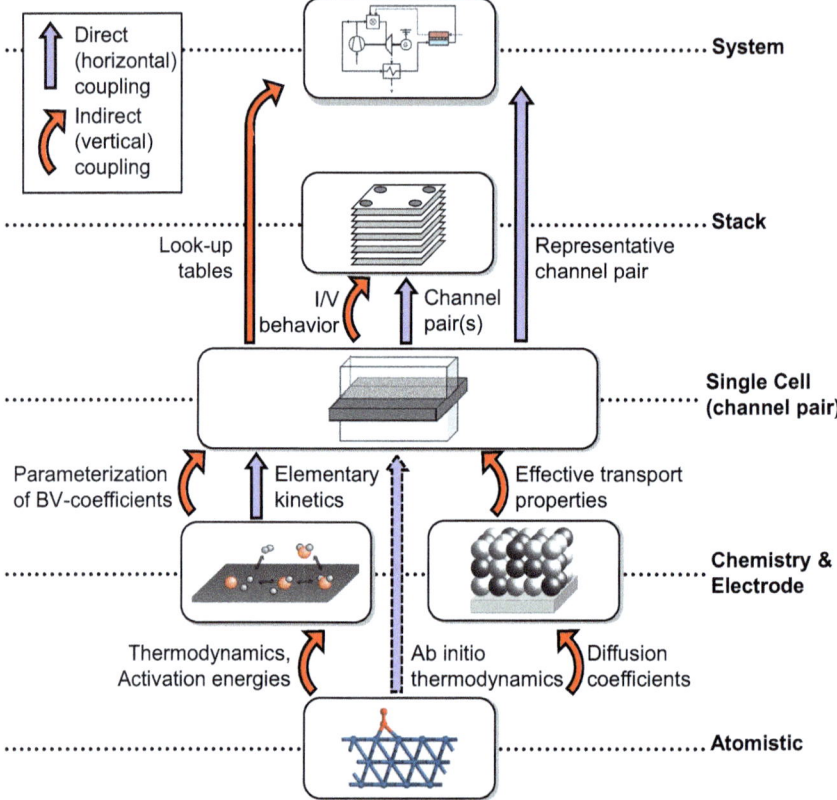

Figure 9.2   Direct and indirect coupling approaches for multi-scale modelling. The single cell as smallest fully functional unit represents a central element in multi-scale approaches.

on single-cell models as next-lower scale. The different direct and indirect multi-scale techniques will be discussed in detail in the following subsections.

Generally, each model has to describe one, two or all of the following physicochemical features: (1) Chemistry, including electrochemistry, thermodynamics, and kinetics; (2) Transport, including mass, charge, and heat transport; (3) Structure, including nano-, micro-, and macrostructure. Multi-scale modelling is possible for each of these features. This will be discussed in the following subsections. In the real system, chemistry, transport, and structure are always coupled and strongly affect each other.

9.3.2   Electrochemistry

9.3.2.1   *From the Molecule to the Electrode: Ab Inito Thermodynamics*

The use of quantum chemical calculations for determining macroscopic thermodynamic properties (chemical potentials, solubilities, equilibrium constants, phase stabilities, *etc.*) is referred to as *ab initio thermodynamics.*[43] Nørskov and co-workers have extended this method to estimate the thermo-chemistry of electrochemical reactions.[44] Enthalpies and entropies of surface-adsorbed intermediates are calculated by DFT. The influence of electrode potential is then included *a posteriori* by adding electric-potential contributions to charged intermediates. This allows us to determine free-energy landscapes of electrochemical reactions as a function of cell potential. It is important to note that the method does not include details of the charge-transfer reaction itself (*e.g.*, double layer formation, reaction barriers as a function of potential). Therefore, it is mainly applicable in trend studies, where the performance of a range of different electrode materials is compared, assuming that the neglected effects cancel out. Such studies yield the *relative* macroscopic electrode performance as a function of material.

The *ab initio thermodynamics* approach was applied to the hydrogen oxdiation at SOFC anodes.[45,46] Trends in activity were investigated for a range of transition metals (W, Mn, Mo, Fe, Co, Ir, Ru, Rh, Ni, Cu, Pt, Pd, Ag, Au). In these studies, the electrocatalytic activity was observed to correlate with oxygen adsorption strength in a volcano-type behavior, underlining the importance of oxygen intermediates in the oxygen spillover pathway. Mukherjee and Linic included an elementary kinetic analysis that allowed to consider the effect of surface coverage on electrode performance.[46] In two follow-up publications, the same approach was used to investigate methane direct oxidation.[47,48] An exemplary result of this approach is shown in Figure 9.3. Based on density-functional theoretical calculations of the thermodynamic stability of surface intermediates of the H_2, CO and CH_4 electrooxidation reactions, Ingram and Linic could demonstrate that the highest electrode activity is obtained on a family of metals including Ni, Ru,

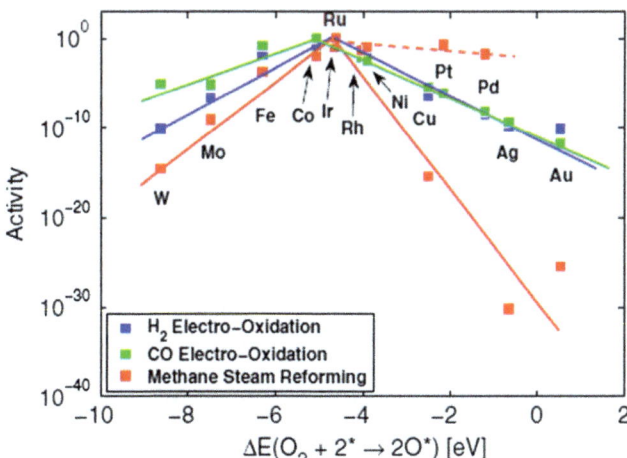

Figure 9.3 Ab initio thermodynamics calculations of the activity of different catalyst materials at the fuel electrode toward three different fuels (H_2, CO, CH_4). Reproduced from Reference.[47]

Rh, Ir, and Co.[47] These metals were predicted to activate CH_4, convert it to CO and H_2, and electro-oxidize CO and H_2 with high rates.

9.3.2.2   From the Molecule to the Cell: Elementary Kinetics

The particular strength of elementary kinetic modelling is that it allows us to bridge the large scale gap between atomistic ($\leq 10^{-10}$ m) and macroscopic ($\geq 10^{-6}$ m) modelling techniques. Thermodynamic and kinetic coefficients of elementary reaction steps (*e.g.*, thermodynamic properties of surface adsorbates; activation energies and preexponential factors of chemical reactions; diffusion coefficients of transport processes) can be obtained from atomistic studies such as DFT. Elementary kinetics is then applied in simulations with macroscopic reactive transport models on the electrode- and cell-level scale. Therefore, elementary kinetics can be used for an *indirect* scale coupling between atomistic and cell level. It can be used to predict macroscopic observables (*e.g.*, current, voltage, gas composition) and transient behavior (*e.g.*, cyclovoltammograms, electrochemical impedance spectroscopy).

The use of elementary kinetic modelling concepts in a multi-scale framework was introduced to SOFC research over the past decade by a number of groups.[13,26,49] Deutschmann and co-workers have developed and validated an elementary kinetic mechanism for methane conversion in nickel-based anodes.[49,50] Based on 21 reversible reactions between six gas-phase species (H_2, CO, H_2O, CO_2, CH_4, O_2) and 12 surface species (H, OH, H_2O, O, CH_4, CO, CO_2, CH, CH_3, CH_2, C, free Ni sites), the mechanism is able to describe steam reforming, dry reforming, water-gas shift, and partial oxidation chemistry. This mechanism has been the basis for extensive studies in the context of SOFCs operated on hydrocarbon fuels and reformate gases. Kee and co-workers have investigated the electrochemical performance of tubular SOFCs operated on methane[51] and on partially reformed logistics fuel,[52] of tubular cells for partial oxidation of methane cogenerating electricity and syngas,[53] of button cells operated on H_2 and CH_4,[54] and of planar SOFCs.[26] Goodwin and co-workers have studied the electrochemical and catalytic reactions of single-chamber SOFCs.[55,56] It was observed that the CH_4/O_2 mixture used for operating the single-chamber SOFC is rapidly converted into hydrogen-rich gases in the outermost part of the porous anode, while the electrochemically active region essentially operates on H_2. Deutschmann and co-workers studied methane conversion in button cells.[49] Bessler and co-workers included elementary charge-transfer kinetics at three-phase boundaries into multi-scale models and applied the approach to planar SOFCs[13] operated on methane, symmetric cells for studying anode impedance,[57] as well as segmented SOFCs operated on hydrogen.[58] The latter study showed that the cell may operate locally in critical operating conditions (low H_2/H_2O ratios, low local segment voltage) without notably affecting globally observed electrochemical behaviour.

An advantage of elementary kinetic models is that parameters can also be obtained and validated using a number of experimental methods. In particular, as fuel cell reactions are surface reactions, methods from heterogeneous catalysis can be used. This includes surface spectroscopy techniques for validating thermodynamic properties (LEED, AES, XPS, STM, *etc.*) as well as kinetic techniques for determining kinetic properties (TPD/O/R, *etc.*). Parameters for charge-transfer reactions are accessbile through standard electrochemical measurements (EIS, *etc.*) or through the use of particularly-designed model electrodes (patterned or point electrodes). A combined approach of surface science experiments, model electrodes, quantum chemical calculations, and elementary kinetics was applied by Yurkiv and co-workers in order to study the mechanism of CO oxidation at SOFC Ni/YSZ anodes.[25,59,60] The interaction of CO with YSZ surfaces was studied using DFT;[60] the interaction of CO and CO_2 with Ni surfaces was investigated using temperature-programmed desorption (TPD) experiments.[59] These studies served as basis for the development and parameterization of a comprehensive elementary kinetic mechanism which was subsequently refined and validated using EIS data of patterned anodes.[25] An exemplary result is shown in Figure 9.4. One of the

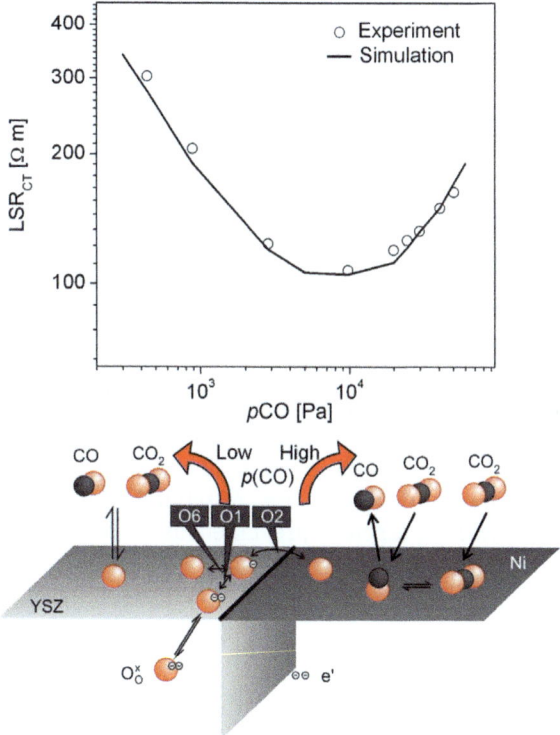

Figure 9.4 Mechanism of CO oxidation at the Ni/YSZ three-phase boundary. The experimentally observed nonlinear behavior of electrode kinetics as function of CO/CO_2 ratio can be reproduced by the elementary kinetic simulation parameterized via DFT.
Reproduced from Reference.[25]

conclusions of this study is that the CO/CO_2 ratio has strong and nonlinear influence on the reaction kinetics. Particularly, charge transfer can proceed via two different mechanisms: At high CO/CO_2 ratios, oxygen spillover from the YSZ to the Ni surface takes place, and CO is oxidized in a Langmuir–Hinshelwood type heterogeneous reaction on the Ni surface. At low CO/CO_2 ratios, the largest part of oxygen ions is fully reduced on the YSZ surface without undergoing spillover, and CO is oxidized in an Eley–Rideal type heterogeneous reaction on the YSZ surface. The study is a comprehensive example how multi-scale modelling (indirect coupling of atomistic and electrode scales) in combination with experiments can largely increase our understanding of SOFC behaviour.

9.3.2.3 From the Cell to the System: Global Kinetics and Look-up Tables

For macroscopic simulations of stacks and systems, the use of elementary kinetics is usually prohibited due to its computational expense. Instead, simple electrochemical relationships are used, for example Butler-Volmer type equations,

$$i = i^0(p, T, X) \cdot \left[\exp\left(\beta_f \frac{F}{RT} \eta_{act} \right) - \exp\left(-\beta_r \frac{F}{RT} \eta_{act} \right) \right], \qquad (10)$$

or linear kinetics,

$$i = \frac{1}{R^{CT}(p, T, X)} \eta_{act} \qquad (11)$$

where the exchange current density i^0 or the charge-transfer resistance R^{CT} may depend on pressure, temperature and gas composition. Note the use of global symmetry factors β_f and β_r (forward and reverse reaction, respectively) instead of the elementary symmetry factor α in Eqs. (3) and (4). In an indirectly-coupled multi-scale approach, the parameters i^0, β_f, β_r, and R^{CT} and their dependence on p, T, X are taken from lower-scale simulations, either as analytical approximation or in the form of tabulated numerical values. More often, however, these parameters are taken as empirical fit parameters to experimental data.

9.3.3 Transport

9.3.3.1 From the Molecule to the Component: Bulk and Surface Transport Coefficients

Microscopically, molecular transport is caused by the random movement of atoms or molecules. Macroscopically, this movement results in diffusion, migration, or convection, depending on the type of macroscopic gradient (concentration, electric potential or pressure, respectively). The microscopic phenomena can be modelled by atomistic techniques, typically DFT or MD, and macroscopic transport coefficients (*i.e.*, diffusion coefficients) can be derived. While in principle these techniques can be applied to any type of phase (bulk, surface, gas-phase), the main areas of interest in the context of fuel cells are the bulk (ionic and electronic charge transport) and the surface (surface diffusion and spillover processes), while gas-phase molecular transport is well known and binary and multi-component diffusion coefficients are readily available.[61]

To calculate self-diffusion coefficients from DFT, activation energies E_{act} and attempt frequencies ν_{at} are calculated for a movement of an atom or molecule from one stable site (surface or bulk) to the neighboring stable site at a distance l. Under the assumption that all particles are random walkers (*i.e.* they move by a sequence of spatially and temporally uncorrelated jumps from one site to the other), the macroscopic diffusion coefficient is given by the Einstein-Smoluchowsky relation,[62,63]

$$D = \frac{1}{g}\nu_{at}l^2 \exp\left(-\frac{E_{act}}{RT}\right), \tag{12}$$

where g is a geometry factor ($g=2$ for surface diffusion, $g=6$ for bulk diffusion). To calculate self-diffusion coefficients from MD, the movement of the particles is simulated over a time t (typically several picoseconds). The diffusion coefficient is calculated from the mean square displacement (MSD) of one or several particles during the simulation time,[64]

$$D = \frac{1}{gt}\left\langle [\mathbf{r}(t) - \mathbf{r}(0)]^2 \right\rangle \tag{13}$$

If MD simulations are carried out for a number of different temperatures, activation energies and preexponential factors can be obtained via an Arrhenius plot.

9.3.3.2 From the Electrode to the Cell: Porous Transport Coefficients

An important approach for indirect scale coupling is the use of microstructure-resolved models to derive effective transport coefficients, which are then used in larger-scale continuum models. Transport coefficients that can be obtained on the microstructure include the molecular diffusion coefficient (via MD simulations or by evaluating Fick's equation on the microstructure), Knudsen diffusion (random walk simulations), thermal conductivity (by evaluating the Fourier equation), and permeability (by evaluating the Darcy equation). Apart from specialized codes developed by several research groups, there are also commercial software packages available to perform these tasks (*e.g.*, GEODICT). Most of these techniques require the knowledge of the bulk transport coefficients (*e.g.*, binary molecular diffusion coefficients in the gas phase, electric or thermal conductivity in the solid phase) and provide the correction factor f between effective porous property and bulk property according to

$$D_{\text{effective}} = f \cdot D_{\text{bulk}}. \tag{14}$$

A reasonable estimate of the correction factor f can often be obtained even without requiring microstructure-resolved simulations via simple geometric consideration. The effective property depends on the porosity ε and tortuosity τ of the porous phase, for example,

$$D_{\text{effective}} = \varepsilon/\tau^2 \cdot D_{\text{bulk}}. \tag{15}$$

The factor τ^2 is referred to as *tortuosity factor*, as compared to the geometrical *tortuosity* τ defined as the ratio of real distance over linear distance. This difference has been pointed out by Epstein,[65] who showed that a rigorous derivation of the porous transport equations results in a factor of τ^2 in the transport equations or when calculating effective diffusion coefficients.[66,67] The porosity ε can be approximated from SEM measurements, and τ is often used as fit parameter. A different approach often used in the PEM fuel cell literature is to represent τ as power-law function of ε according to $\tau = \varepsilon^a$, resulting in the *Bruggemann relationship*,

$$D_{\text{effective}} = \varepsilon^\beta \cdot D_{\text{bulk}}, \tag{16}$$

where β is the *Bruggemann factor* which is often set to an empirical value of 1.5.[68] Similarly, expressions for the effective permeability can be derived as a function of ε, τ and other assumed microstructural properties like the particle diameter (*e.g.*, Kozeny–Carman relationship).

9.3.4   Structure

9.3.4.1   Homogenization

Many of the multi-scale methods discussed above apply structural homogenization approaches, that is, lower-scale structural information is embedded into the higher-scale model in an averaged (homogenized) way. This is possible on various scales. In elementary kinetic electrochemistry models, the influence of surface nanostructure like steps or kinks is implicitly included in the homogenized rate coefficients such as activation energies. For continuum porous transport models, the influence of pore structure is included in the homogenized effective transport coefficients such as effective diffusivities. For mass transport in stacks, the millimeter-scale channel/land structure can be homogenized and the transport equations solved using effective coefficients.[69,70] Although spatial homogenization is often used implicitly or based on common sense, there are also mathematically rigorous treatments of the validity of homogenization approaches.[71]

9.3.4.2 Dimension Reduction

The use of reduced-dimensional models is common in order to reduce computational effort in CFD type models. With reduced-dimensional models we refer to three different approaches. Firstly, it is common to consider only partial dimensions of the full problem according to the subject of interest. For example, instead of modelling a cell in 3D, we can consider either the 2D behaviour along a fuel cell channel, or the 2D behaviour below channel and land. Secondly, and more relevant for multi-scale approaches, spatial dimensions are often *decoupled* due their difference in scale. For example, planar fuel cells typically have large aspect ratios (cell length in the 10 cm range versus MEA thickness in the sub-millimeter range). This allows to reduce a 2D model (along length and through thickness) to a 1D + 1D model, where the MEA is described as a set of 1D models along the length of a 1D channel model. This allows to solve all transport equations in one dimension only. Thirdly, it is common to neglect dimensions due to the irrelevance of gradients for the considered problem or due to symmetry considerations. For example, the 3D laminar flow in a fuel cell channel is often modeled in 1D because radial transport is much faster than axial transport, leading to a homogeneous concentration and temperature distribution in the channel cross-sectional area.

9.3.4.3 Representative Repeat Elements

As noted in Section 9.2.6, geometrically complex 3D simulations of stacks are only possible with physicochemically simple MEA models. In order to allow the upscaling of more detailed MEA models, the concept of using *representative repeat elements* is used. This concept takes advantage of the periodic building principle of SOFC stacks, both on the cell level (where a number of structurally equivalent gas supply channels is used) and on the stack level (where a number of structurally equivalent single cells is used). The use of representative repeat elements is based on the assumption that the global stack behaviour can be described by a selection of one or more representative single channel pairs (*i.e.*, air and fuel channel coupled by a MEA), each of which can be modelled in 1D, 1D + 1D, 2D, or 3D. Indeed, system-level models often use only one single channel pair as stack submodel.[42] Although strongly simplified from the stack point of view, these models are still based on a physicochemical description of electrochemistry and transport, as compared to empirical performance maps often used in system-level models.

The concept of using representative repeat elements for SOFC stack simulations was exploited by a number of groups. Van Herle and co-workers have developed a 2D model of a stack repeat element,[72,73] including the description of degradation processes.[74] Kulikovsky studied thermal waves in SOFC stacks using a 1D analytical model.[75] Froning *et al.* compared 1D reduced models to 3D CFD models.[34]

Menon *et al.* presented an agglomerated cluster algorithm to identify representative positions in a stack.[76] An exemplary result from this study is shown in Figure 9.5. Instead of simulating every individual cell in the stack,

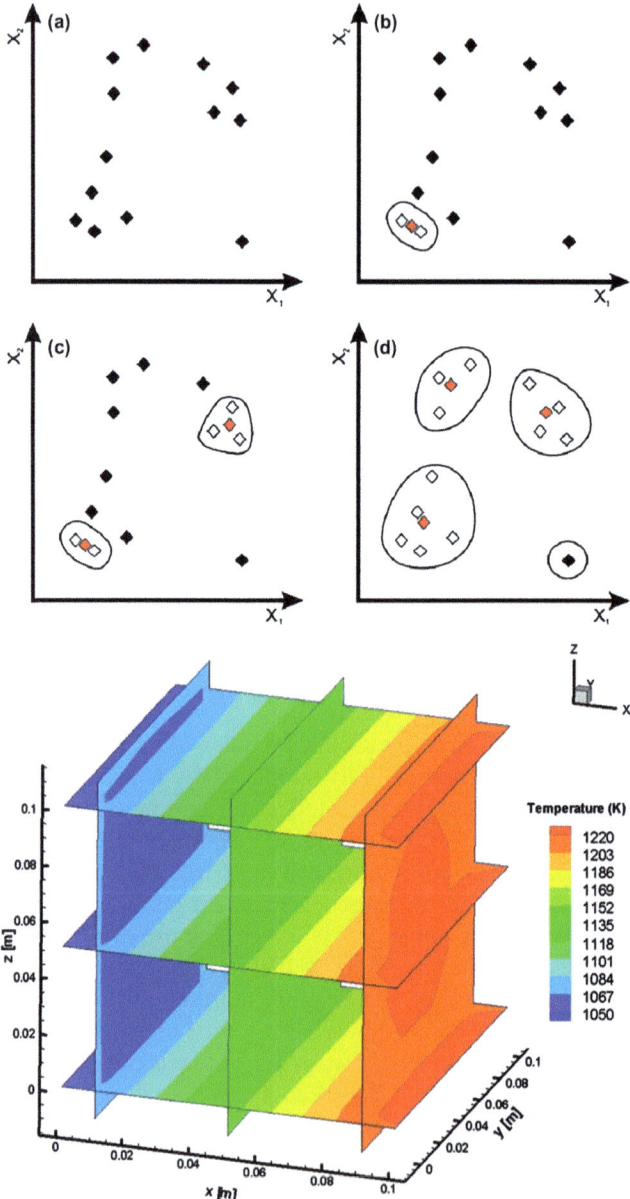

Figure 9.5   Multi-scale modelling of an SOFC stack operated on reformate gas. The upper panel illustrates the agglomerative cluster algorithm used to select representative channel pairs with detailed multiphysics for the use in the stack scale. The lower panel shows the temperature distribution in the 3D stack (solid phase) at different surface planes, along and across the flow direction, after 60 min adiabatic operation.
Reproduced from Reference.[76]

only representative ones based on the cluster algorithm are chosen for detailed simulation. The detailed model consists of a 1D + 1D model of a channel pair including elementary kinetic reforming chemistry. This approach dramatically reduces the computation time required for stack simulation.

9.4   Multi-scale Models for SOFC System Simulation and Control

The ultimate motivation for all SOFC modelling activities is to directly or indirectly contribute to the realization of a high-efficiency, environmentally-benign energy conversion technology. In this section, two examples of multi-scale models for SOFC system simulation and control will be given that demonstrate the potential of modern computational techniques.

9.4.1   Pressurized SOFC System for a Hybrid Power Plant

The combination of an SOFC and a gas turbine, the so-called hybrid power plant, allows high electric system efficiencies (up to 70%) for medium-scale stationary applications. This was firstly demonstrated by Siemens-Westinghouse in 2000.[77] In such a system, the SOFC is operated under pressurized (3–6 bar) conditions. The SOFC sub-system in a hybrid power plant was modelled by Leucht *et al.*[40,78] in a SIMULINK environment. The model is shown in the upper panel of Figure 9.6.[40] It consists of components for fuel desulphurization, humidification (partial oxidation burner and off-gas recycler), prereformer, fuel cell stack, combustor, and heat exchangers. The pressurized fuel cell stack is modelled as five-dimensional look-up table, where the current is described as function of voltage, temperature, pressure, H_2/H_2O ratio, and H_2/inert gas ratio. The look-up table was obtained from a one-dimensional single-cell model based on Butler-Volmer electrochemistry. The lower panels of Figure 9.6 show simulated system operation over 24 h for a residential application.[78] The simulation allows to quantify dynamic system states as well as time-averaged electric efficiency. It could be demonstrated that control loops are able to keep the system within a stable and safe operation range (temperature, fuel flow, air flow) avoiding conditions critical with respect to cell degradation. In total, this model is an example for an indirect coupling of the cell and the system scales via look-up tables (cf. Figure 9.2).

9.4.2   Tubular SOFC System for Mobile APU Applications

Small SOFC systems (1 kW) are being developed as high-efficiency onboard electricity source for mobile (*e.g.*, truck) applications, so-called auxiliary power units (APU). A multi-scale modelling approach for such a system was developed by Kee, Vincent and co-workers.[79,80] The model is shown schematically in the upper panel of Figure 9.7. It consists of the SOFC stack as core

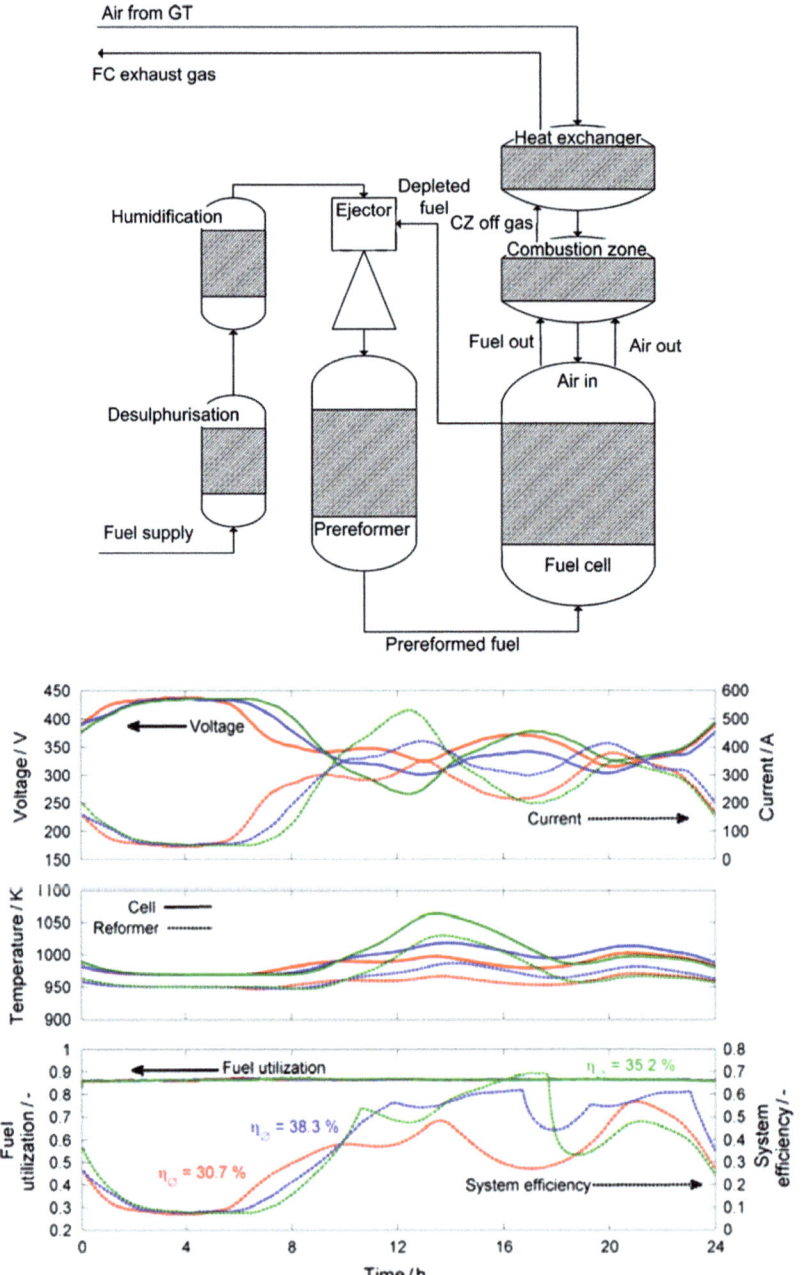

Figure 9.6   System modelling of a pressurized 150 kW SOFC system. Upper panel: System components and their coupling. The stack subsystem consists of a five-dimensional look-up table parameterized via a 1D + 1D cell-level model. Lower panels: Simulated daily load cycles for a residential area. Red shows weekdays, blue Saturdays and green Sundays. Reproduced from references.[40,78]

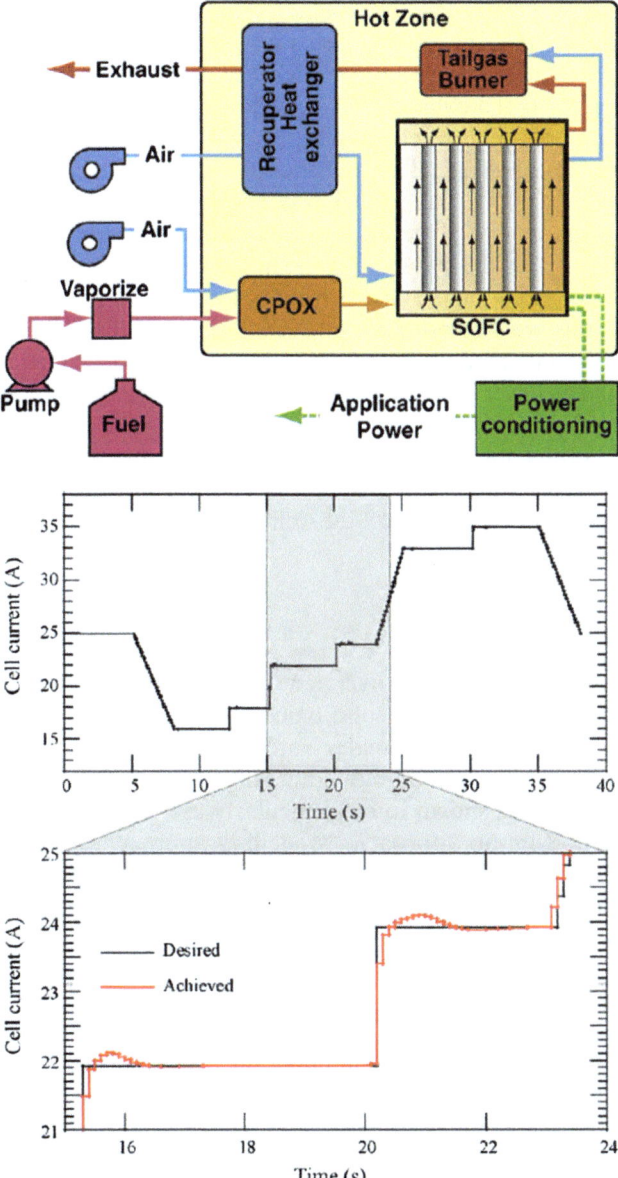

Figure 9.7 System modelling of a tubular SOFC system for mobile APU applications. Upper panel: Schematic of the SOFC system. The stack subsystem is described by a linear parameter varying (LPV) model structure derived from a detailed 1D + 1D single-cell model, maintaining full dynamics and nonlinearity. Lower panels: Overview and expanded section of cell current during dynamic system simulation. The figures show a comparison between the desired cell current and the cell current delivered by the MPC controller. Reproduced from Reference.[80]

subsystem as well as balance-of-plant components for fuel flow and air flow delivered to the stack. The stack was described as reduced version of a complex physicochemical cell-level model. The cell model used as basis for the reduction consisted of a dynamic 1D + 1D description of mass, heat and charge transport on the tubular cell domain and allowed to predict spatial and temporal profiles of chemical composition, temperature, velocity, and current density. For reduction, this model was analyzed using a linear parameter varying (LPV) method. The model was linearized around a number of operating points, and the dynamics of the linearized models were identified with small-signal perturbations. Subsequently, multiple linear models at different operating points were blended to an overall reduced nonlinear model, which was used for model-predictive control (MPC) of the stack. Exemplary results are shown in the lower panels of Figure 9.7. The controller is able to keep the system at the desired operation points even upon dynamic load changes. As in the previous subsection, this is an example for an indirect coupling of the cell and the system levels, however, using a more sophisticated reduction method that maintains the dynamics of the cell level as compared to steady-state look-up tables.

9.5 Conclusions

In this chapter we gave an overview of modelling techniques used for understanding and optimizing solid oxide fuel cells. Modelling activities are being carried out on all spatial scales – atomistic, electrode, cell, stack, system – using specialized methodology and software. Multi-scale coupling of different methods has shown to be particularly useful for describing SOFC in a consistent bottom-up approach. Modelling approaches have significantly contributed not only to our understanding of the multi-physics processes occurring in SOFCs, but also to aid technology development using system design and control. Eventually, commercial SOFC systems will be realized, to a significant extent, due to the contribution of computational techniques.

It is an open question how much multi-scale and multi-physics complexity is required in SOFC modelling. Combining all, complex multi-scale geometry, complex multi-physics processes, and time accuracy, is beyond the capabilities of even most modern computer clusters. It may, however, also well be beyond the capabilities of the researcher to interpret the resulting large and complex data set. The need to focus models on certain aspects, as imposed by researcher specialization, software handling, and computing resources, is therefore not necessarily a drawback, but helps to separate and understand the coupled effects of physicochemistry, geometry, and temporal evolution. It is therefore unlikely that full multi-scale models – with direct coupling from the atomistic scale up to the system scale – will ever be developed. Speaking with Tim Harford,[81] such a model would be no simpler than reality itself and so would add nothing to our understanding.

Acknowledgements

I had the pleasure to work with a number of excellent PhD students, postdocs, and colleagues over the past years that made multi-scale modelling of SOFCs possible. I would like to thank Marcel Vogler, Stefan Gewies, Jonathan Neidhardt, Vitaliy Yurkiv, Moritz Henke, and Florian Leucht, as well as David Goodwin, Robert J. Kee, Olaf Deutschmann, Ellen Ivers-Tiffée, Andreas Friedrich, Günter Schiller, and Josef Kallo for discussion and collaboration.

References

1. M. A. Khaleel and J. R. Selman, Cell, stack and system modelling, in *High-temperature solid oxide fuel cells: Fundamentals, design and application*, S. C. Singhal and K. Kendall, Eds. Oxford: Elsevier Science, 2003, 291–331.
2. R. J. Kee, H. Zhu and D. G. Goodwin, Solid-oxide fuel cells with hydrocarbon fuels, *Proc. Combust. Inst.*, 2005, **30**, 2379–2404.
3. V. M. Janardhanan and O. Deutschmann, Modelling of solid-oxide fuel cells, *Z. Phys. Chem.*, 2007, **221**, 443–478.
4. S. Kakac, A. Pramuanjaroenkij and X. Y. Zhou, A review of numerical modelling of solid oxide fuel cells, *Int. J. Hydrogen Energy*, 2007, **32**, 761–786.
5. S. B. Adler and W. G. Bessler, Elementary kinetic modelling of SOFC electrode reactions, in *Handbook of Fuel Cells - Fundamentals, Technology and Applications, 5*, 5, W. Vielstich, H. Yokokawa, and H. A. Gasteiger, Eds., Chichester, John Wiley & Sons, 2009, 441–462.
6. A. A. Kulikovsky, *Analytical Modelling of Fuel Cells*. Amsterdam: Elsevier, 2010.
7. M. Andersson, J. Yuan and B. Sundén, Review on modelling development for multiscale chemical reactions coupled transport phenomena in solid oxide fuel cells, *Applied Energy*, 2010, **87**, 1461–1476.
8. R. J. Kee, H. Zhu, R. J. Braun, and T. L. Vincent, Modelling the Steady-State and Dynamic Characteristics of Solid-Oxide Fuel Cells, in *Advances in Chemical Engineering, Volume 41*, Elsevier, 2012, 331–381.
9. P. Hohenberg and W. Kohn, Inhomogeneous electron gas, *Phys. Rev.*, 1964, **136**, B864–B871.
10. W. Kohn and L. J. Sham, Self-Consistent Equations Including Exchange and Correlation Effects, *Phys. Rev.*, 1965, **140**, A1133–A1138.
11. A. C. T. van Duin, S. Dasgupta, F. Lorant and W. A. Goddard III, ReaxFF: A Reactive Force Field for Hydrocarbons, *J. Phys. Chem. A*, 2001, **105**, 9396–9409.
12. R. Pornprasertsuk, T. Holme and F. B. Prinz, Kinetic Monte Carlo Simulations of Solid Oxide Fuel Cell, *Journal of The Electrochemical Society*, 2009, **156**(12), B1406–B1416.
13. W. G. Bessler, S. Gewies and M. Vogler, A new framework for physically based modelling of solid oxide fuel cells, *Electrochim. Acta*, 2007, **53**, 1782–1800.

14. J. Warnatz, U. Maas, and R. W. Dibble, *Combustion, 3.* ed. Heidelberg: Springer, 2001.
15. R. Suwanwarangkul, E. Croiset, M. W. Fowler, P. L. Douglas, E. Entchev and M. A. Douglas, Performance comparison of Fick's, dusty-gas and Stefan-Maxwell models to predict the concentration overpotential of a SOFC anode, *J. Power Sources*, 2003, **122**, 9–18.
16. E. A. Mason, A. P. Malinauskas and R. B. Evans, Flow and diffusion of gases in porous media, *J. Chem. Phys.*, 1967, **46**, 3199–3216.
17. D. Arnost and P. Schneider, Dynamic transport of multicomponent mixtures of gases in porous solids, *Chem. Eng. J.*, 1995, **57**, 91–99.
18. W. Lehnert, J. Meusinger and F. Thom, Modelling of gas transport penomena in SOFC anodes, *J. Power Sources*, 2000, **87**, 57–63.
19. S. Sunde, Monte Carlo simulations of polarization resistance of composite electrodes for solid oxide fuel cells, *J. Electrochem. Soc.*, 1996, **143**, 1930–1939.
20. G. Gaiselmann, M. Neumann, L. Holzer, T. Hocker, M. R. Prestat and V. Schmidt, *Stochastic 3D modelling of $La_{0.6}Sr_{0.4}CoO_{3-\delta}$ cathodes based on structural segmentation of FIB–SEM images*, Computational Materials Science, **67**, 48–62, Feb. 2013.
21. J. R. Wilson, W. Kobsiriphat, R. Mendoza, H.-Y. Chen, J. M. Hiller, D. J. Miller, K. Thornton, P. W. Voorhees, S. B. Adler and S. A. Barnett, Three-dimensional reconstruction of a solid-oxide fuel-cell anode, *Nature Materials*, 2006, **5**, 541–544.
22. T. Carraro, J. Joos, B. Rüger, A. Weber, and E. Ivers-Tiffée, *3D finite element model for reconstructed mixed-conducting cathodes: I. Performance quantification*, Electrochimica Acta, **77**, 315–323, Aug. 2012.
23. S. Sunde, Simulations of composite electrodes in fuel cells, *J. Electroceramics*, 2000, **5**, 153–182.
24. Q. Cai, C. S. Adjiman and N. P. Brandon, Modelling the 3D microstructure and performance of solid oxide fuel cell electrodes: Computational parameters, *Electrochimica Acta*, Jun. 2011, **56**(16), 5804–5814.
25. V. Yurkiv, A. Utz, A. Weber, E. Ivers-Tiffée, H.-R. Volpp and W. G. Bessler, Elementary kinetic modelling and experimental validation of electrochemical CO oxidation on Ni/YSZ pattern anodes, *Electrochim. Acta*, 2012, **59**, 573–580.
26. H. Zhu, R. J. Kee, V. M. Janardhanan, O. Deutschmann and D. G. Goodwin, Modelling elementary heterogeneous chemistry and electrochemistry in solid-oxide fuel cells, *J. Electrochemical Soc.*, 2005, **152**, A2427–A2440.
27. W. G. Bessler and S. Gewies, Gas concentration impedance of solid oxide fuel cell anodes. II. Channel geometry, *J. Electrochem. Soc.*, 2007, **154**, B548–B559.
28. D. Larrain, J. Van Herle, F. Maréchal and D. Favrat, Thermal modelling of a small anode supported solid oxide fuel cell, *J. Power Sources*, 2003, **118**, 367–374.

29. A. Nakajo, F. Mueller, J. Brouwer, J. van Herle and D. Favrat, Mechanical reliability and durability of SOFC stacks. Part I : Modelling of the effect of operating conditions and design alternatives on the reliability, *International Journal of Hydrogen Energy*, Jun. 2012, **37**(11), 9249–9268.

30. J. P. Neidhardt, D. Fronczek, T. Jahnke, T. Danner, B. Horstmann and W. G. Bessler, A flexible framework for modelling multiple solid, liquid and gaseous phases in batteries and fuel cells, *J. Electrochem. Soc.*, 2012, **159**, A1528–A1542.

31. H. Zhu and R. J. Kee, Modelling electrochemical impedance spectra in SOFC button cells with internal methane reforming, *J. Electrochem. Soc*, 2006, **153**, A1765–A1772.

32. A. A. Franco, P. Schott, C. Jallut and B. Maschke, A multi-scale dynamic mechanistic model for the transient analysis of PEFCs, *Fuel Cells*, 2007, **7**, 99–117.

33. O. Deutschmann, S. Tischer, C. Correa, D. Chatterjee, S. Kleditzsch, and V. M. Janardhanan, *DETCHEM Software Package, Version 2.0*. Karlsruhe: http://www.detchem.com, 2004.

34. D. Froning, L. Blum and A. Gubner, L. G. J. de Haart, M. Spiller and D. Stolten, Experiences with a CFD based two stage SOFC stack modelling concept and its application, *ECS Transactions*, 2007, **7**, 1831–1840.

35. M. Peksen, A coupled 3D thermofluid–thermomechanical analysis of a planar type production scale SOFC stack, *International Journal of Hydrogen Energy*, Sep. 2011, **36**(18), 11914–11928.

36. R. B. Bird, W. E. Stewart and E. N. Lightfoot, *Transport Phenomena*, 2nd ed. New York: John Wiley & Sons, 2001.

37. J. Van Herle, F. Maréchal, S. Leuenberger, Y. Membrez, O. Bucheli and D. Favrat, Process flow model of solid oxide fuel cell system supplied with sewage biogas, *J. Power Sources*, 2004, **131**, 127–141.

38. L. Magistri, R. Bozzo, P. Costamagna and A. F. Massardo, Simplified versus detailed solid oxide fuel cell reactor models and influence on the simulation of the design performance of hybrid systems, *Journal of Engineering for Gas Turbines and Power*, 2004, **126**, 516–523.

39. T. Ollikainen, J. Saarinen, M. Halinen, T. Hottinen, M. Noponen, E. Fontell and J. Kiviaho, Dynamic simulation tool APROS in SOFC power plant modelling at Wärtsilä and VTT, *ECS Transactions*, 2007, **7**, 1821–1829.

40. F. Leucht, W. G. Bessler, J. Kallo, K. A. Friedrich and H. Müller-Steinhagen, Fuel Cell System Modelling for SOFC/GT Hybrid Power Plants, Part I: Modelling and simulation framework, *J. Power Sources*, 2011, **196**, 1205–1215.

41. R. J. Braun, T. L. Vincent, H. Zhu, and R. J. Kee, Analysis, Optimization, and Control of Solid-Oxide Fuel Cell Systems, in *Advances in Chemical Engineering, Volume 41* , Elsevier, 2012, 383–446.

42. M. Eschenbach, R. Coulon, A. A. Franco, J. Kallo and W. G. Bessler, Multi-scale modelling of fuel cells: From the cell to the system, *Solid State Ionics*, 2011, **192**, 615–618.

43. D. Alfè, G. A. De Wijs, G. Kresse and M. J. Gillan, Recent Developments in ab initio Thermodynamics, *Int. J. Quantum Chem.*, 2000, **77**, 871–879.
44. J. K. Nørskov, J. Rossmeisl, A. Logadottir, L. Lindqvist, J. R. Kitchin, T. Bligaard and H. Jónsson, Origin of the overpotential for oxygen reduction at a fuel-cell cathode, *J. Phys. Chem. B*, 2004, **108**, 17886–17892.
45. J. Rossmeisl and W. G. Bessler, Trends in catalytic activity for SOFC anode materials, *Solid State Ionics*, 2008, **178**, 1694–1700.
46. J. Mukherjee and S. Linic, First-principles investigations of electrochemical oxidation of hydrogen at solid oxide fuel cell operating conditions, *J. Electrochem. Soc.*, 2007, **154**, B919–B924.
47. D. B. Ingram and S. Linic, First-Principles Analysis of the Activity of Transition and Noble Metals in the Direct Utilization of Hydrocarbon Fuels at Solid Oxide Fuel Cell Operating Conditions, *J. Electrochem. Soc.*, 2009, **156**, B1457–B1465.
48. N. Kleis, G. Jones, F. Abild-Pedersen, V. Tripkovic, T. Bligaard and J. Rossmeisl, Trends for Methane Oxidation at Solid Oxide Fuel Cell Conditions, *J. Electrochem. Soc.*, 2009, **156**, B1447–B1456.
49. V. M. Janardhanan and O. Deutschmann, CFD analysis of a solid oxide fuel cell with internal reforming: Coupled interactions of transport, heterogeneous catalysis and electrochemical processes, *J. Power Sources*, 2006, **162**, 1192–1202.
50. E. S. Hecht, G. K. Gupta, H. Zhu, A. M. Dean, R. J. Kee, J. Maier and O. Deutschmann, Methane reforming kinetics within a Ni-YSZ SOFC anode support, *Appl. Catal. A*, 2005, **295**, 40–51.
51. H. Zhu, A. M. Colclasure, R. J. Kee, Y. Lin and S. A. Barnett, Anode barrier layers for tubular solid-oxide fuel cells with methane fuel streams, *J. Power Sources*, 2006, **161**, 413–419.
52. G. K. Gupta, J. R. Marda, A. M. Dean, A. Colclasure, H. Zhu and R. J. Kee, Performance prediction of a tubular SOFC operating on a partially reformed JP-8 surrogate, *J. Power Sources*, 2006, **162**, 553–562.
53. H. Zhu, R. J. Kee, M. R. Pillai and S. A. Barnett, Modelling electrochemical partial oxidation of methane for cogeneration of electricity and syngas in solid-oxide fuel cells, *J. Power Sources*, 2008, **183**, 143–150.
54. G. M. Goldin, H. Zhu, R. J. Kee, D. Bierschenk and S. A. Barnett, Multidimensional flow, thermal, and chemical behavior in solid-oxide fuel cell button cells, *J. Power Sources*, 2009, **187**, 123–135.
55. Y. Hao and D. G. Goodwin, Numerical modelling of single-chamber SOFCs with hydrocarbon fuels, *J. Electrochem. Soc.*, 2007, **154**, B207–B217.
56. Y. Hao and D. G. Goodwin, Numerical study of heterogeneous reactions in an SOFC anode with oxygen addition, *J. Electrochem. Soc.*, 2008, **155**, B666–B674.
57. S. Gewies and W. G. Bessler, Physically based impedance modelling of Ni/YSZ cermet anodes, *J. Electrochem. Soc.*, 2008, **155**, B937–B952.
58. W. G. Bessler, S. Gewies, C. Willich, G. Schiller and K. A. Friedrich, Spatial distribution of electrochemical performance in a segmented

SOFC: A combined modelling and experimental study, *Fuel Cells*, 2010, **10**, 411–418.

59. V. Yurkiv, D. Starukhin, H.-R. Volpp and W. G. Bessler, Elementary reaction kinetics of the $CO/CO_2/Ni/YSZ$ electrode, *J. Electrochem. Soc.*, 2011, **158**, B5–B10.

60. V. Yurkiv, A. Gorski, W. G. Bessler and H. Volpp, Density functional theory study of heterogeneous CO oxidation over an oxygen-enriched yttria-stabilized zirconia surface, *Chem. Phys. Lett.*, 2012, **543**, 213–217.

61. R. J. Kee, M. E. Coltrin, and P. Glarborg, Chemically reacting flow. *Theory and practice*. John Wiley & Sons, 2003.

62. T. Zambelli, J. Wintterlin, J. Trost and G. Ertl, Identification of the 'Active Sites' of a Surface-Catalyzed Reaction, *Science*, 1996, **273**, 1688–1690.

63. R. Gomer, Diffusion of Adsorbates on metal surfaces, *Reports on Progress in Physics*, 1990, **53**, 917–1002.

64. M. Kilo, C. Argirusis, G. Borchardt and R. A. Jackson, Oxygen diffusion in yttria stabilized zirconia - experimental results and molecular dynamics calculations, *Phys. Chem. Chem. Phys.*, 2003, **5**, 2219–2224.

65. N. Epstein, On tortuosity and the tortuosity factor in flow and diffusion through porous media, *Chem. Eng. Sci.*, 1989, **44**(3), 777–779.

66. S. DeCaluwe, H. Zhu, R. J. Kee and G. S. Jackson, Importance of anode microstructure in modelling solid oxide fuel cells, *J. Electrochem. Soc.*, 2008, **155**, B538–B546.

67. V. H. Schmidt and C.-L. Tsai, Anode-pore tortuosity in solid oxide fuel cells found from gas and current flow rates, *J. Power Sources*, 2008, **180**, 253–264.

68. J. Newman and K. E. Thomas-Alyea, *Electrochemical Systems*, 3rd ed. Hoboken, New Jersey: John Wiley & Sons, 2004.

69. M. Roos, E. Batawi, U. Harnisch and T. Hocker, Efficient simulation of fuel cell stacks with the volume averaging method, *J. Power Sources*, 2003, **118**, 86–95.

70. M. Kvesić, U. Reimer, D. Froning, L. Lüke, W. Lehnert and D. Stolten, 3D modelling of a 200 cm^2 HT-PEFC short stack, *Int. J. Hydrogen Energy*, 37, 2430–2439, Feb. 2012, **3**.

71. M. Neuss-Radu and W. Jäger, Effective transmission conditions for reaction-diffusion processes in domains separated by an interface, *SIAM J. Math. Anal*, 2007, **39**, 687–720.

72. D. Larrain, J. Van Herle, F. Maréchal and D. Favrat, Generalized model of planar SOFC repeat element for design optimization, *J. Power Sources*, 2004, **131**, 304–312.

73. D. Larrain, F. Maréchal, N. Autissier, J. van Herle and D. Favrat, Multi-scale modelling methodology for computer aided design of a solid oxide fuel cell stack, *Computer Aided Chemical Engineering*, 2004, **18**, 1081–1086.

74. D. Larrain, J. van Herle and D. Favrat, Simulation of SOFC stack and repeat elements including interconnect degradation and anode reoxidation risk, *J. Power Sources*, 2006, **161**, 392–403.

75. A. A. Kulikovsky, Thermal waves in SOFC stacks, *J. Electrochem. Soc.*, 2008, **155**, A693–A698.
76. V. Menon, V. M. Janardhanan, S. Tischer and O. Deutschmann, A novel approach to model the transient behavior of solid-oxide fuel cell stacks, *J. Power Sources*, Sep. 2012, **214**, 227–238.
77. S. E. Veyo, L. A. Shockling, J. T. Dederer, J. E. Gillett and W. L. Lundberg, Tubular Solid Oxide Fuel Cell / Gas Turbine Hybrid Cycle Power Systems: Status, *ASME J. Eng., Gas Turbines Power*, 2002, **124**, 845–849.
78. F. Leucht, S. Seidler, M. Henke, J. Kallo, W. G. Bessler, U. Maier and K. A. Friedrich, Solid Oxide Fuel Cells: From Pressurized Cell Tests to System Dynamics Assessment, *International Colloquium on Enviromentally Preferred Advanced Generation (ICEPAG)*, 2011, ICEPAG2011–4406.
79. A. M. Colclasure, B. M. Sanandaji, T. L. Vincent and R. J. Kee, Modelling and control of tubular solid-oxide fuel cell systems. I: Physical models and linear model reduction, *J. Power Sources*, Jan. 2011, **196**(1), 196–207.
80. B. M. Sanandaji, T. L. Vincent, A. M. Colclasure and R. J. Kee, Modelling and control of tubular solid-oxide fuel cell systems: II. Nonlinear model reduction and model predictive control, *Journal of Power Sources*, Jan. 2011, **196**(1), 208–217.
81. T. Harford, *The undercover economist revised and updated*, Oxford University Press, 2012.
82. W. G. Bessler, Electrochemistry and Transport in Solid Oxide Fuel Cells, *Habilitation Thesis*, Heidelberg University, 2007.

CHAPTER 10

Fuel Cells Running on Alternative Fuels

XINWEN ZHOU, NING YAN AND JING-LI LUO*

Department of Chemical and Materials Engineering, University of Alberta, Edmonton, Alberta Canada T6G 2V4
*Email: jingli.luo@ualberta.ca

10.1 Introduction

As part of worldwide fuel cells research, solid oxide fuel cells (SOFCs) running on alternative fuels have attracted considerable attention in recent years. Besides the advantages of conventional fuel cell technology, SOFCs fuelled by industrial effluent gas or impure feed offer more significant economic benefits, optimized industrial processes and environmental friendliness. However, before successful experimental implementation of the conceptual vision on SOFCs running on alternative fuels, we need to overcome challenges of materials development, structure optimization and process improvement. For the past decades, a variety of fuels have been examined as potential fuel candidates in SOFCs. In 2008, we have briefly summarized the evolution of fuel cells powered by H_2S-containing gases.[1] In this chapter, we present the overview of our group's work on SOFCs running on alternative fuels, including methane containing H_2S ($CH_4 + H_2S$), ethane (C_2H_6), propane (C_3H_8), syngas containing H_2S ($CO + H_2 + H_2S$), and pure hydrogen sulfide (H_2S).

RSC Energy and Environment Series No. 7
Solid Oxide Fuel Cells: From Materials to System Modeling
Edited by Meng Ni and Tim S. Zhao
© The Royal Society of Chemistry 2013
Published by the Royal Society of Chemistry, www.rsc.org

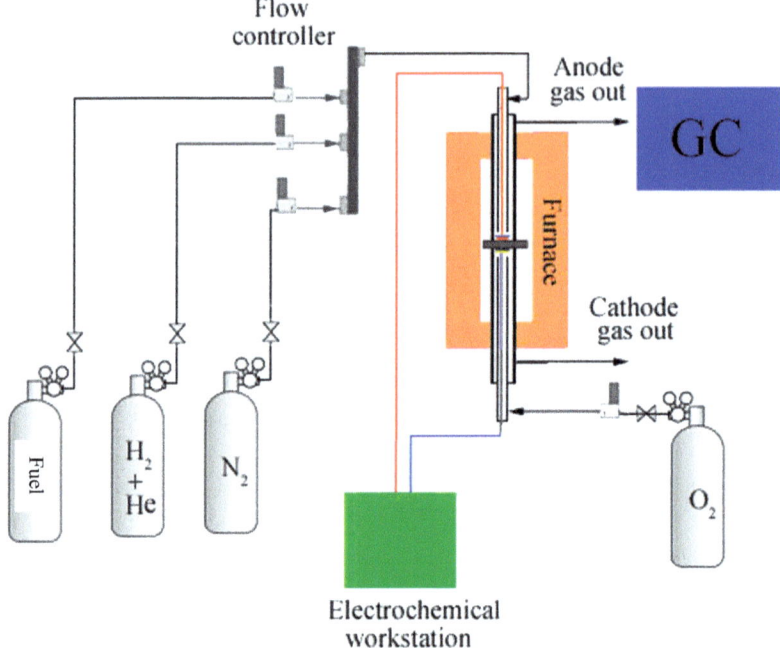

Figure 10.1   Schematic of SOFC reactor set-up and test system.

10.2   Fuel Cell Reactor Set-up

The schematic of SOFC reactor set-up and test system is shown in Figure 10.1. The fuel cell reactor was set up by securing the membrane electrode assembly (MEA) between coaxial pairs of alumina tubes and sealed using ceramic sealant, which was then cured by heating in a vertical tubular furnace. The cathode catalyst, anode catalyst, electrolyte (including proton and oxide ion conducting electrolytes), current collector and the type of fuel could be chosen according to our experimental requests. The temperature could be adjusted using temperature controller. The conductivity of the electrolyte and the electrochemical performance of fuel cell reactor were measured using an electrochemical workstation. The outlet gases from the anode chamber were analyzed using a gas chromatography (GC) equipped with a packed bed column and a thermal conductivity detector.

10.3   SOFCs Running on Sourgas

In recent years, there has been an increasing interest in improving the efficiency of power generation from sour gas and other types of biogas by use of SOFCs. Minor and trace impurities (such as H_2S) have a significant effect on the performance and durability of SOFCs. H_2S is recognized as a problem in operating conventional SOFCs as it can poison most catalysts and is corrosive. For example, H_2S can poison Ni-yttria-stabilized-zirconia (Ni/YSZ), the most efficient H_2-fueled SOFC catalyst, by either forming a sulfide or poisoning the anode surface even at low concentrations. Carbon deposition and redox cycling stability are the other major problems observed on Ni/YSZ anode cermets. In this part, we focus on the development of the sulfur and coking tolerant anode materials for SOFCs using CH_4 containing H_2S as the fuel.

The performance of a series of perovskite oxides $La_{0.75}Sr_{0.25}Cr_{0.5}X_{0.5}O_{3-\delta}$ (X = Co, Fe, Ti, Mn) as SOFC anode electrocatalysts has been studied by Danilovic *et al.*[2] Figures 10.2a and b show the cross-sectional SEM micrographs of prepared MEA of $La_{0.75}Sr_{0.25}Cr_{0.5}Fe_{0.5}O_{3-\delta}$ (LSCFe). It was found that the performance of the perovskite oxides depended on the nature of the substituted element X. Temperature programmed reaction (TPR) results of CH_4 under O_2-free conditions showed that the catalytic activity of the oxides for conversion of CH_4 was in the order of Co > Mn > Fe > Ti. The total conductivities of the materials in air decreased in the order of X = Co > Fe > Mn > Ti. Among the catalysts investigated in this work, the order of maximum fuel cell power density depended on the feed: the trend was Fe > Mn > Ti for CH_4; Fe > Mn > Ti for H_2; Ti > Fe > Mn for 0.5% $H_2S + CH_4$ (Figure 10.2c). Although all of the catalysts are mixed conductors in theory, they are primarily electronic conductors. Hence they are not ideal for use as

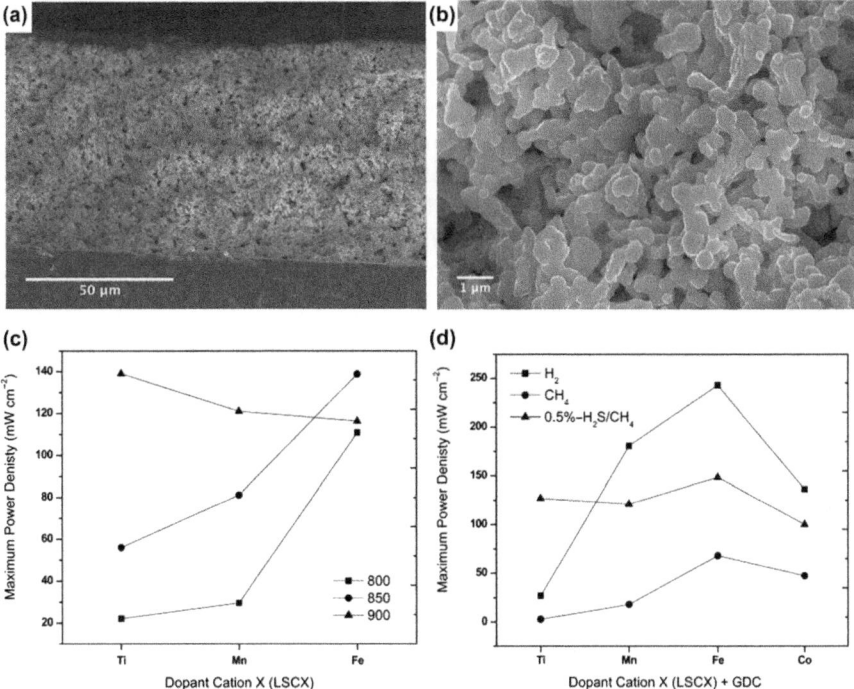

Figure 10.2 (a) and (b) cross-sectional SEM images of MEAs containing LSCFe; (c) dependence of performance on substituted cation using 0.5% $H_2S + CH_4$ feed at 800 °C, 850 °C and 900 °C; and (d) the influence of GDC on performances at 850 °C.[2]

single phase electrode materials. To improve applicability, they can be used in a composite anode with an ionic conducting component, such as gadolinia doped ceria (GDC) (Figure 10.2d).

Besides the B site doping, Vincent *et al.*[3] also found that partial substitution of Sr with Ba site in the strontium titanate structure can increase its ionic conductivity, catalytic activity to the oxidation of CH_4 and stability of the strontium titanate structure with high amount of lanthanum. $La_{0.4}Sr_{0.6}TiO_{3-\delta}$ (LST) and $La_{0.4}Sr_{0.6-x}Ba_xTiO_3$ $(0 < x \leq 0.2)$ (LSBT) were prepared using conventional solid state method. Fuel cells were fabricated using commercial YSZ disks (300 μm in thickness and 25 mm in diameter) as the electrolyte and an intimate mixture of equal weights of YSZ and strontium doped lanthanum manganite (LSM) as the cathode. After sintering the combination of anode and electrolyte, platinum paste was painted on the cathode side and gold paste on the anode side (1 cm^2 per side), then both pastes were sintered *in situ* to form current collectors. Single cell tests were performed in a vertical furnace in a coaxial two-tube (inlet and outlet) set-up. CH_4, and CH_4 with 0.5% H_2S ($CH_4 + H_2S$), H_2 and H_2 with 0.5% H_2S ($H_2 + H_2S$) were used as the fuels and fed dry at 200 mL min^{-1}. Fuels were always supplied sequentially in the following order $H_2 \rightarrow CH_4 \rightarrow CH_4 + H_2S \rightarrow H_2 + H_2S$. We found that both LST and LSBT had definite activity for conversion of hydrogen or methane, and the activity increased with the level of substitution by Ba. Most importantly, the fuel cell performance was significantly enhanced when H_2S was present in either CH_4 or H_2 fuel. SEM images of LST and LSBT after electrochemical test are shown in Figures 10.3a and b. The interfaces between LST and YSZ or LSBT and YSZ each showed good contact and good adhesion after the fuel cell tests. The grain size range was 1–5 μm for all LST and LSBT. We also found that LST had strong surface activity for conversion of methane, which was enhanced by substitution by Ba. Considering the overall performances, the maximum power densities were quite low. The performance could be improved through controlled synthesis of LST based materials of uniform shape and size.

In reference to the Ba-doped (La, Sr)TiO₃ materials, Li *et al.*[4] reported that pure $BaTiO_3$ was also a good anode catalyst for conversion of CH_4 in SOFCs, especially in H_2S-containing atmospheres. A maximum power density of 135 mW cm^{-2} was achieved at 900 °C with 0.5% $H_2S + CH_4$ in a fuel cell having a 300 μm thick YSZ electrolyte (Figure 10.4a). Compared to $SrTiO_3$ and $La_2Ti_2O_7$ anodes, the $BaTiO_3$-based fuel cell had higher tolerance to carbon deposition from the results of X-ray photoelectron spectroscopy (XPS) (Figure 10.4b), better electrochemical performance and much higher stability during long-term operation. High catalytic activity for methane conversion, mixed electronic and ionic conductivity, and the surface basicity of $BaTiO_3$ unexpectedly provided promising performance as anodes for SOFCs.

Different amounts of LSBT ($La_{0.4}Sr_{0.5}Ba_{0.1}TiO_3$) impregnated with porous YSZ were reported by Vincent *et al.*[5] The porous YSZ was obtained using graphite or polymethylmethacrylate (PMMA) to have generated about 55 vol% porosity after sintering at 1350 °C. The SEM images of the anode matrices using PMMA as pore former prior to impregnation were shown in

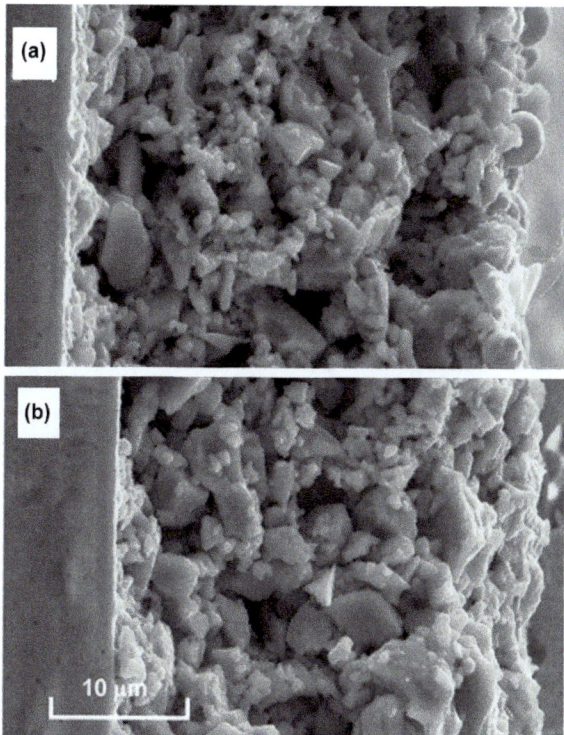

Figure 10.3 Cross-sectional SEM images of the interfaces of YSZ with (a) LST and (b) LSBT after electrochemical tests.[3]

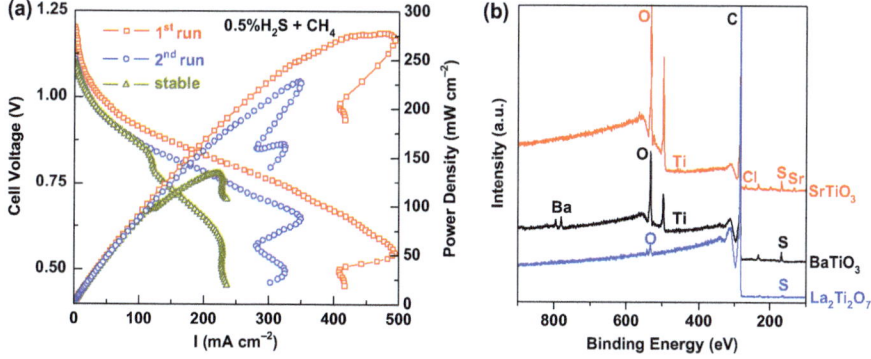

Figure 10.4 (a) Voltage and power density curves as a function of current density for a BaTiO$_3$-based cell with 0.5% H$_2$S + CH$_4$ as the anode fuel. (b) XPS spectra for various anode catalysts after calcining in 0.5% H$_2$S + CH$_4$ at 950 °C for 10 h.[4]

Figure 10.5a. Fuel cells were fabricated using commercial YSZ disk of 300 μm thickness as the electrolyte and composite YSZ/LSM 50/50 wt% as the cathode. Figure 10.5b shows the fuel cell performance variation *vs.* the number

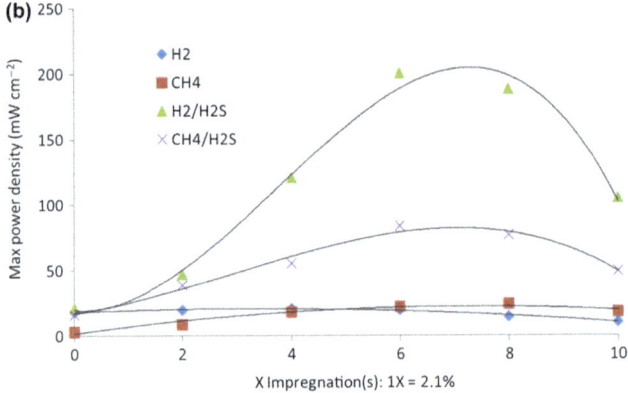

Figure 10.5   (a) SEM images of anode matrix produced using PMMA as pore formers, with inserts of their respective pore former; (b) Dependence of compensated maximum power density on the number of times LSBT impregnated into YSZ obtained at 850 °C.[5]

of times the LSBT being impregnated into porous YSZ. The temperature for each test was 850 °C and the sequence of the feeds was: H_2, $H_2 + 0.5\%$ H_2S, $CH_4 + 0.5\%$ H_2S and CH_4. It was clearly shown that when the feed was $CH_4 + 0.5\%$ H_2S, the power density was considerably higher than those in pure H_2 and pure CH_4. Six impregnations provided maximum power density of 84 mW cm^{-2}, which was about 3.5 times higher than that achieved using pure CH_4 as the fuel. When the impregnations exceeded six times, the performance decreased dramatically. This trend was much less obvious than using pure H_2. Using $H_2 + 0.5\%$ H_2S as the fuel provided the best performance, achieving 200 mW cm^{-2} for the cell using the anode with six LSBT impregnations.

A series of experiments were designed to systematically investigate the electrochemical performance and compositions of anode gases to explore the mechanism that the fuel cell performance was significantly enhanced when H_2S

was present in either CH_4 or H_2 fuel. LST was synthesized by a solid state method and was mixed with commercial YSZ power with a 50/50 weight ratio to form the anode materials.[6] Platinum paste was used as the cathode. The electrolyte was a commercial 300 μm thick YSZ dicks. The power density improved dramatically from 2 mW cm^{-2} with pure CH_4 as the fuel to more than 450 mW cm^{-2} for CH_4 containing 20% H_2S (Figure 10.6a). From thermodynamic calculations (Figure 10.6b) and mass spectroscopic data obtained at different stages during a regular potentiodynamic run (Figure 10.6c), we concluded that H_2S was not the only fuel converted, and its interaction with CH_4 was the key factor for the enhancement of performance. The anode effluent gas mixture included H_2S, SO_2, H_2 and CS_2, which was consistent with thermodynamic predictions. It was concluded that H_2S had a synergistic effect on CH_4 oxidation, *i.e.*, CH_4 and H_2S must not be considered as two separate fuels. H_2S actually ran as a powerful catalyst for fuel electro-oxidation. At low current density there was a rapid conversion of surface species, whereas at high current density there was a mass transfer blocking effect from build-up of surface intermediates. The predominant form of sulfur in the gas phase was H_2S provided there was at least 3% H_2 in the mixture. CH_4 conversion proceeded via rapid formation of syngas, and the reaction path probably included CS_2 and other sulfur species. The selectivity of the obtained products could be controlled by adjusting the applied voltage.

In order to improve the performance of the fuel cell using the LST-based anode materials, two methods were used. One was to dope LST with another element (*e.g.*, Ba for Sr), and the other was to mix LST with an ionic conductor (*e.g.* GDC in Figure 2d). Recently, LST-based composite anodes were synthesized *via* wet chemical impregnation of different amounts of $La_{0.4}Ce_{0.6}O_{1.8}$ (LDC) and LST into porous YSZ.[7] Figure 10.7 shows TEM analysis of YSZ-LDC-LST powders. Figure 10.7a is a bright field micrograph of a few relatively large particles. The simulations of the ring patterns for both LST and LDC phases are shown in Figure 10.7b. It can be seen that presence of both phases caused ring overlapping in the diffraction pattern. The corresponding selected area electron diffraction (SAED) pattern is shown in Figure 10.7c. Similar to the case of YSZ-LDC, the SAED pattern contained both bright spots and rings. The bright spots were ascribed to the YSZ matrix. Figure 10.7d shows the dark field micrograph taken from a portion of the rings corresponding to {111} and {002} rings of LDC and {011} ring of LST. As this image indicates, the LST and LDC nanocrystallites were densely dispersed across the surface of the YSZ particles.

Table 10.1 compares the peak power densities of the cells with different anode materials when different feeds fuelled the cells. Compared to LST impregnation alone, the co-impregnation of LDC with LST improved the electrochemical oxidation of the different fuels. For example, the peak power density of the cell containing LDC-LST was approximately three times of that with LST alone when using H_2S-containing fuel at 800 °C. It was found that the LST-LDC composite anode materials showed a higher performance to the oxidation of $0.5\%H_2S + CH_4$.

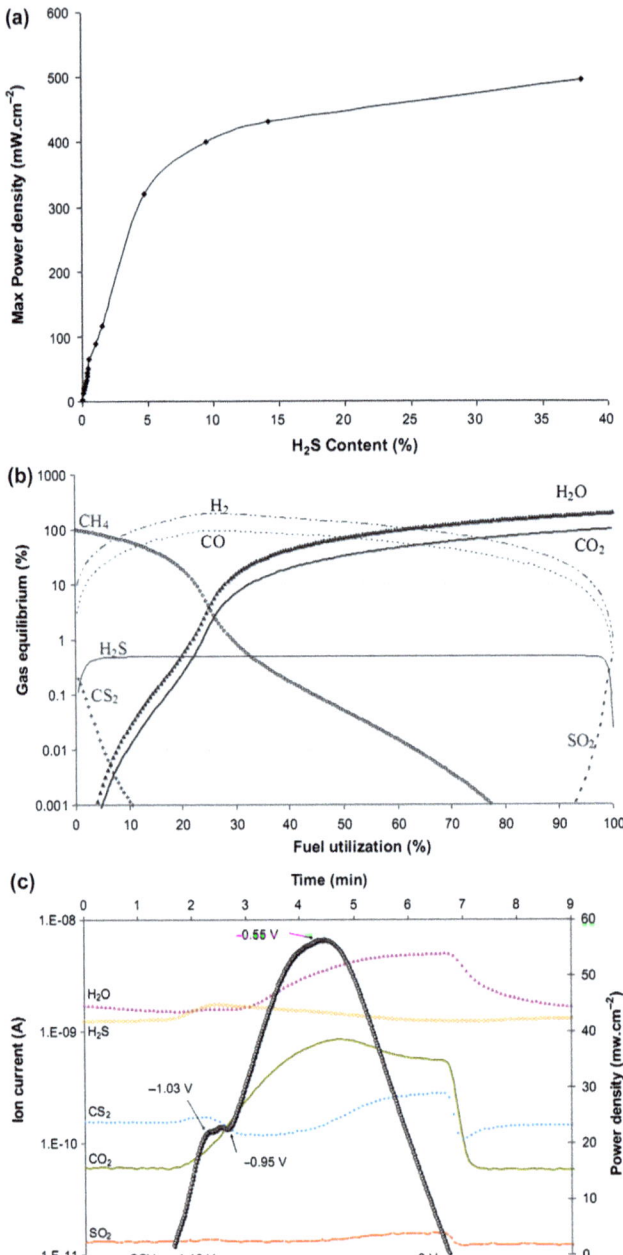

Figure 10.6 (a) Maximum power density as a function of H_2S concentration at 850 °C; (b) Calculated gas evolution from sour gas oxidation as a function of fuel utilization at 850 °C. Formation of carbon and other hydrocarbons is suppressed; (c) Correlation of the power density curve and the mass spectroscopic data for the effluent from the anode during a potentiodynamic cycle at $5\,mV\,s^{-1}$ from the open circuit voltage to $0\,V$ at 850 °C.[6]

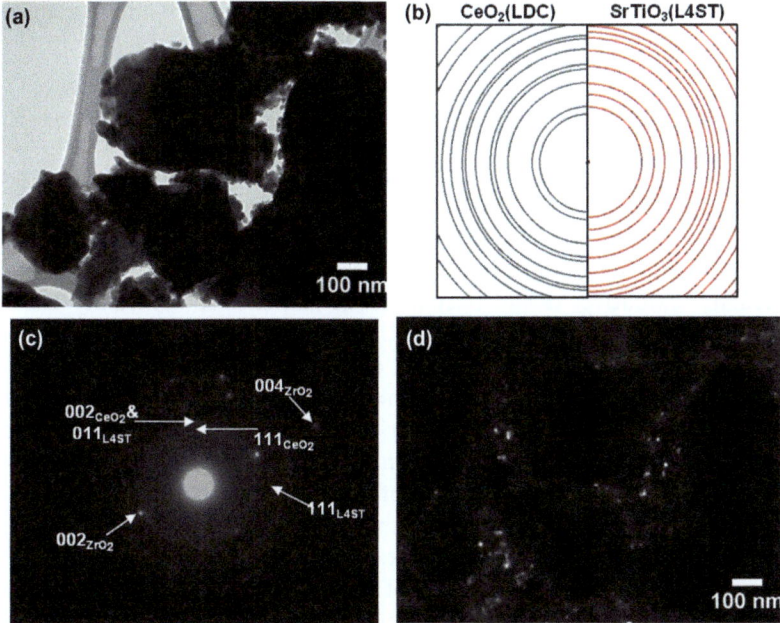

Figure 10.7 TEM micrographs of LDC-LST impregnated samples: (a) Bright field image; (b) The simulated ring patterns of CeO_2 and $SrTiO_3$ phases, highlighting the overlapping of rings; (c) Corresponding SAED pattern; and (d) Dark field micrograph.[7]

Table 10.1 Peak power densities ($mW\ cm^{-2}$) for different anode materials and fuels at $800\ °C$.[7]

	H_2	$0.5\%H_2S + H_2$	CH_4	$0.5\%H_2S + CH_4$
LST	20	36	3	30
LDC	50	25	6	30
LDC-LST	76	110	7	75

From the above analysis, we can see that LST-based materials have a great potential to be used as the anode materials of SOFCs using H_2S-containing CH_4 as the feed directly because of their excellent sulfur tolerance and oxidation enhancement effect. The main problems of the LST-based anodes are the low ionic conductivity and poor catalytic performance. Doping and the addition of active species to form composite are two common ways to increase the ionic conductivity and improve the activity of electrocatalyst. In addition to the Ba doping in Sr site and Cr doping in Ti site mentioned above, other researchers also studied Ca doping in Sr site,[8] Ce doping in La site,[9] Mn,[10] Fe, Ni,[11] Co,[12] Mg,[13] doping in Ti site of the LST-based materials and LST-based materials composite mixed with CeO_2,[14] Ni-YSZ,[15] $GDC + Cu$,[16] Bi_2O_3[17] as anode materials of SOFCs. Besides, we also applied $Ce_{0.9}Sr_{0.1}VO_3$-based

($Ce_{0.9}Sr_{0.1}VO_3$ and $Ce_{0.9}Sr_{0.1}VO_4$,[18] $Ce_{0.9}Sr_{0.1}Cr_{0.5}V_{0.5}O_3$[19]) anode materials in the SOFCs using CH_4 containing H_2S as the fuel.

10.4 SOFCs Running on C_2H_6 and C_3H_8

Ethylene is a major intermediate product in the chemical industry which is most economically produced by steam cracking. However, in this process, more than 10% of ethane feed is oxidized to CO_2, and NO_x pollutant is produced due to the high flame temperature. Some alternative methods are available. In particular, oxidative dehydrogenation of ethane to ethylene has been intensively researched in recent years. Unfortunately, ethane is readily oxidized to CO_2 and the chemical energy is not efficiently recovered from the oxidative reaction. Furthermore, oxidative methods readily produce acetylene which is detrimental to manufacture of polymers as it poisons the catalysts and must be removed to form high purity ethylene, which is an expensive process. In contrast, electrochemical conversion of ethane using proton conducting solid oxide fuel cells (PC-SOFCs) is potentially more selective, allows recovery of high-grade energy as power, and generates little or no pollutants.

The schematic working principles of SOFCs with oxide ion electrolyte and proton conducting electrolyte are shown in Figure 10.8. In the conventional oxide ion electrolyte SOFC (Figure 10.8a), O_2 is reduced to oxide ion in the cathode area. The oxide ion will migrate through the oxide ion electrolyte and move to anode area under the force of the electric field. In the anode area, the fuel is oxidized completely under the action of anode catalyst and will integrate with the oxide ion to form H_2O and carbon dioxide. The anode reactions are totally different in PC-SOFCs wherein protons are necessarily produced catalytically from fuels to enable their operation (Figure 10.8b). The hydrocarbon proton conducting SOFCs can function as membrane reactors if the anode catalyzes selective dehydrogenation of fuel to provide H^+ as shown in Figure 10.8b. Hydrocarbon fuels (for examples, C_2H_6 and C_3H_8) can be dehydrogenated under the effect of the selective anode catalysts to form H^+ and C_2H_4. The H^+ will drill through the proton conducting electrolyte and move to anode area under the force of the electric field. In the cathode area, O_2 is reduced to oxide ion and will react with the incoming H^+ to form H_2O. Thus, in the hydrocarbon PC-SOFCs, when choosing appropriate proton conducting electrolyte and highly selective anode catalysts, electricity can be generated with the production of high value-added products.[20-23] Using C_2H_6 as the feed in the PC-SOFCs was initially reported by Iwahara *et al.*[24] in 1986. They used the fuel cell with $SrCe_{0.95}Yb_{0.05}O_{3-\alpha}$ as the solid electrolyte, porous Pt as the cathode and Pt or Ni as the anode. The major limitation of the fuel cell was the large ohmic resistance of the solid electrolyte. The maximum current density (at the terminal voltage of 0 V) was less than 0.1 A cm^{-2} even at 800 °C. The deposition of solid carbon was observed at the outlet of the anode. Therefore, the development of the proton conducting electrolyte and highly selective anode catalysts in PC-SOFCs is the foremost task in this field. In this section, we will outline our work on the development of PC-SOFCs for the conversion of ethane (C_2H_6) to ethylene (C_2H_4) with co-generation of power. Our group also successfully

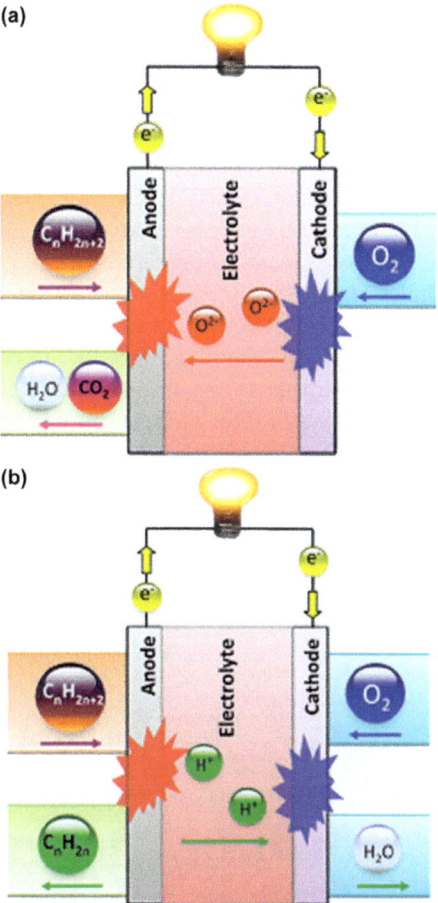

Figure 10.8 Schematic working principles of hydrocarbon SOFCs with (a) oxide ion electrolyte and anode for hydrocarbon deep oxidation, (b) proton conducting electrolyte and dehydrogenation anode.[29]

achieved the co-generation of propylene (C_3H_6) and power using propane (C_3H_8) as fuel using PC-SOFCs,[25–28] which is not elaborated on in detail in this chapter.

In order to compare the performance of the different anode or electrolyte materials, the following definitions were used to calculate the conversion, selectivity and yield (*e.g.* using C_2H_6 as fuel):

$$C_2H_6 \text{ } conversion = \left[\frac{moles \text{ } of \text{ } C_2H_6 \text{ } converted}{moles \text{ } of \text{ } C_2H_6 \text{ } introduced} \right] \times 100\% \qquad (1)$$

$$C_2H_4 \text{ } selectivity = \left[\frac{moles \text{ } of \text{ } C_2H_4 \text{ } produced}{moles \text{ } of \text{ } C_2H_6 \text{ } converted} \right] \times 100\% \qquad (2)$$

$$C_2H_4 \text{ } yield = \left[\frac{moles \text{ } of \text{ } C_2H_4 \text{ } converted}{moles \text{ } of \text{ } C_2H_6 \text{ } introduced} \right] \times 100\% \qquad (3)$$

10.4.1   Development of Electrolyte of PC-SOFCs

Commercial $BaCe_{0.85}Y_{0.15}O_{3-\alpha}$ powder (BCY15) (SCI engineered materials, Inc.) was calcined in air at 1400 °C for 10 h. The compound was ground, and then dried. The obtained powder was uniaxially pressed at 30 MPa into disks (~2.54 cm in diameter and 2 mm in thickness) which were further sintered at 1550 °C for 15 h to obtain high-density membranes. The disk surfaces were polished using abrasive papers until the thickness was reduced to about 0.8 mm. Platinum paste (Heraeus Inc., CL11 5100) was used to prepare both anode and cathode electrodes.[30] No visible holes or cracks in the membranes were found from the SEM image (Figure 10.9a). The XRD peaks (Figure 10.9b) appeared very sharp and were attributable to a single perovskite phase for BCY15 of cubic unit cell. The additional smaller peaks were due to small amounts of $BaCO_3$ added before sintering to compensate for evaporation of BaO during high-temperature calcination. The dense BCY15 membranes thus produced were stable and served as good proton conductors for use under fuel cell operating conditions.

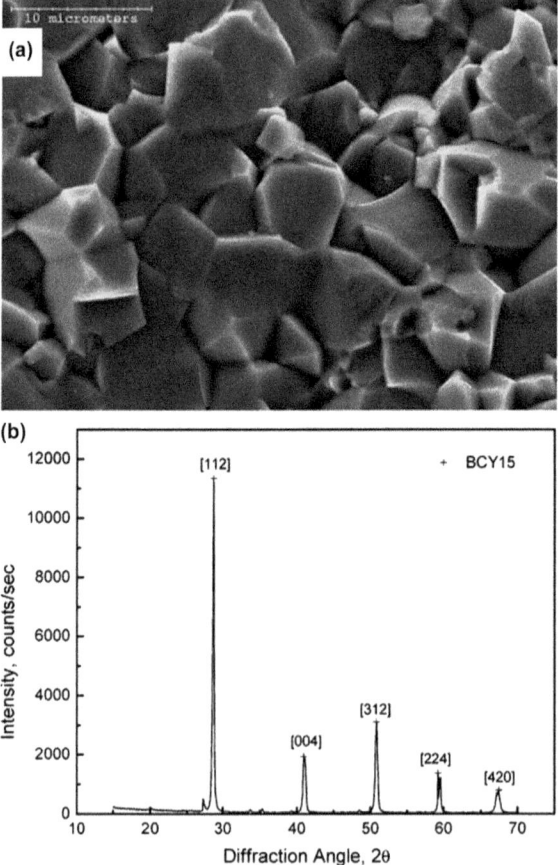

Figure 10.9   (a) Cross-sectional SEM, (b) XRD pattern of BCY15 electrolyte.[30]

Inlet and outlet anode gas streams were analyzed using online GC with helium and argon as the carrier gases. The cell operation was at 650 and 700 °C with constant ethane flow rate at 150 mL min^{-1} unless otherwise stated. The major products in the effluent were H_2, C_2H_4, and CH_4. Conversion of ethane and selectivity to ethylene at different ethane flow rates and temperatures using the vertical fuel cell are summarized in Table 10.2. It was found that the conversion was higher at 700 °C for all flow rates. The conversion of ethane was reduced as the flow rate increased, and the selectivity to ethylene increased slightly. Thus the residence time affected the composition of the effluent stream. A better seal and consequently a better performance were obtained when the sealant was under compression, achieved by operating the cell in vertical attitude. Power density of 120 mW cm^{-2} was attained at the current densities in the range 300–650 mA cm^{-2} at 700 °C and ethane flow rate of 100 mL min^{-1}. We also found that the protonic current arose in two ways: by conversion of hydrogen arising from dehydrogenation of ethane, and by electrochemical conversion of ethane. At 650 °C, the former was the major reaction path, while at 700 °C, the latter prevailed.

In order to improve the fuel cell performance and the C_2H_6 conversion, we proceeded with a successful approach to synthesize BCY15 by solid-state reactions.[31] BCY15 membranes prepared with this method showed good conductivity, 15 mS cm^{-1} at 700 °C and 20 mS cm^{-1} 750 °C. Using the fuel cell of C_2H_6, Pt/BCY15/Pt, O_2, the maximum power density was 174 mW cm^{-2} at 700 °C, with corresponding current density of 320 mA cm^{-2} and 34% ethane conversion (96% ethylene selectivity). This process had higher energy conversion efficiency when compared with current commercial technologies for direct catalytic dehydrogenation of ethane to ethylene in chemical reactors.

Besides the solid-state reactions, we have also explored other method to synthesize BCY15 power. For example, uniform spherical BCY15 nano-particles were synthesized using a citric acid-nitrate combustion method (Figure 10.10a).[32] The BCY15 powders consisted of spherical particles with an average diameter of ~50 nm. During the combustion, dense BCY15 discs were formed when being sintered at 1400 °C for 10 h (Figure 10.10b). We also found that both the electrical conductivity and resistance to CO_2 increased with the sintering temperature. The performances of the SOFC with BCY15 electrolyte and Pt electrodes are shown in Figure 10c. The maximum power density of the fuel cell was 118 mW cm^{-2} at current density of 276 mA cm^{-2} at 650 °C. The corresponding ethylene selectivity was 95.2% at 22.7% ethane conversion, and

Table 10.2 Conversion of ethane and selectivity to ethylene at different ethane flow rates and temperatures using the vertical fuel cell.[20]

| | Ethane conversion (%)/ethylene selectivity (%) | | |
| | Ethane flow rate (mL/min) | | |
Temp (°C)	100	150	200
650	8.0/98	6.8/98	5.5/98
700	27.9/95	23.7/96	20.5/96

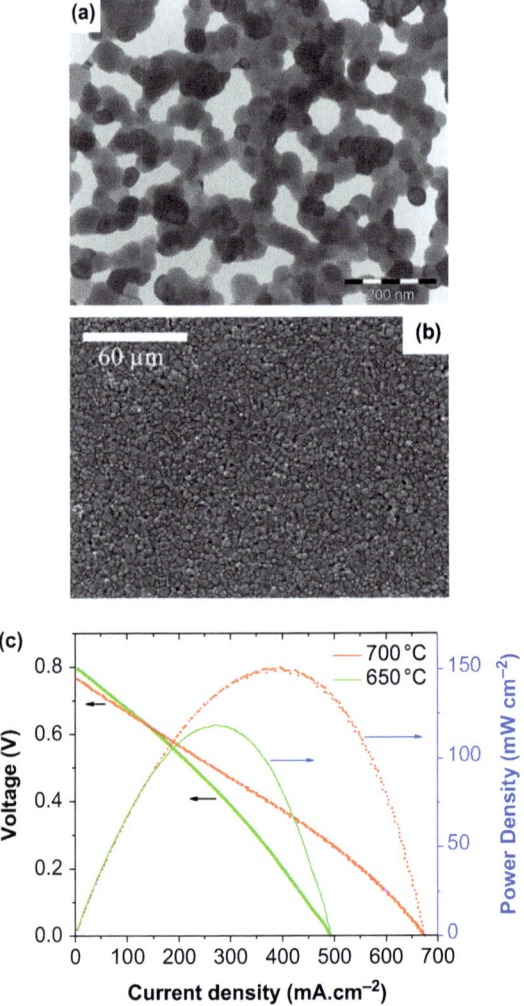

Figure 10.10   (a) TEM images of calcined BCY15 powders. (b) Surface SEM image of
BCY15 disc sintered at 1400 °C. (c) Current density–voltage and power
density curves of ethane/oxygen fuel at 650 °C and 700 °C. The
thickness of BCY15 electrolyte is about 0.8 mm. [32]

the major by-product was methane. At 700 °C, the maximum power density of
the fuel cell increased to 151 mW cm^{-2} at the current density of 393 mA cm^{-2},
with ethane conversion enhanced to 35.1%. However, ethylene selectivity
decreased to 91.6% and selectivity to methane increased. Increasing the
discharging current density increased the conversion of ethane as the rate of
proton removal from the anode increased. In addition, the ethane conversion
was higher at 700 °C than at 650 °C at the same current density, showing that
electrochemical dehydrogenation was enhanced relative to chemical dehydrogenation at the higher temperature.

Based on the combustion method, about 20 nm precursor powders for BCY15 were synthesized.[33] A bi-layered proton conducting membrane having a thick porous BCY substrate and an integrally supported dense BCY thin film were co-fabricated facilely by pressing two layers comprising the precursor powder and its mixture with starch, followed by co-sintering at 1600 °C (Figure 10.11a, b, c). Pt was impregnated into the porous BCY layer matrix as the anode catalyst for

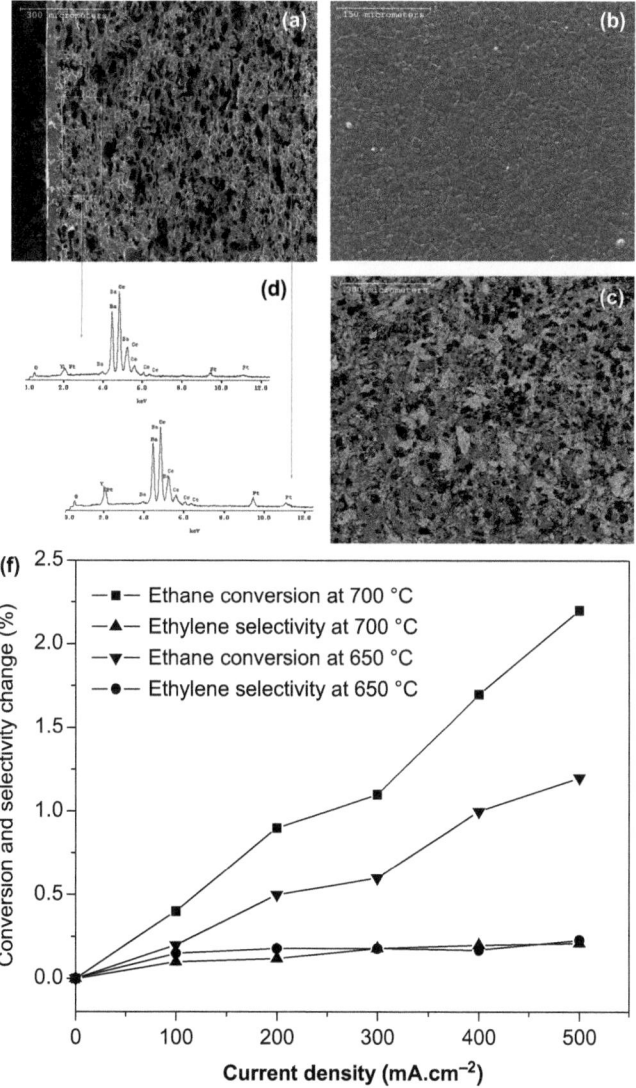

Figure 10.11 (a) Cross-sectional and (b) surface SEM images of bi-layered BCY membrane, (c) Surface SEM image of BCY porous substrate, and EDS patterns of Pt impregnated porous substrate area close to the (d) thin film and (e) porous substrate surface. (f) Change in ethane conversion, ethylene selectivity with increase in current density.[33]

dehydrogenation of ethane to ethylene (Figure 10.11d, e). The hydrocarbon SOFC with the BCY15 thin film electrolyte and Pt electrodes demonstrated high selectivity (90.5%) to ethylene at 36.7% ethane conversion (Figure 10.11f) with co-generation of $216\,mW\,cm^{-2}$ electrical energy output at $700\,^{\circ}C$. The ethane conversion and ethylene selectivity both increased with the current density.

In addition to the proton-conducting BCY15 electrolyte, we have also explored nanostructured, doped BCY15 protonic electrolyte. Very fine and loose nanostructured $BaCe_{0.85}Y_{0.1}Nd_{0.05}O_{3-\delta}$ (BCYN) powers were obtained using the citrate-nitrate combustion method (Figure 10.12a).[34] A novel tri-layered proton conducting membrane, comprising a dense thin film constructed integrally between two porous thick layers impregnated with Pt, was readily fabricated by layering, co-pressing and sintering three layers of material. The cross-sectional SEM image (Figure 10.12b) showed that the BCYN membrane comprised three layers. The centre layer was a dense thin film of ca. 50 mm thick, free of cracks and well-bonded to the other two porous layers with no delamination.

Fuel cells were assembled using Pt/BCYN(porous)-BCYN-Pt/ BCYN(porous) tri-layered membranes with Pt current collectors.[34] The anode and cathode feeds were ethane and oxygen, respectively, each at the rate of $100\,mL\,min^{-1}$. At elevated temperatures, ethane was dehydrogenated catalytically to ethylene at the anode, with co-generation of electricity. At $650\,^{\circ}C$, ethylene selectivity was 95.2% at 21.7% ethane conversion. At $700\,^{\circ}C$, ethane conversion increased to 35.5% while ethylene selectivity decreased slightly to 92.6% and selectivity to methane increased. The maximum power density of $173\,mW\,cm^{-2}$ was obtained at the current density of $355\,mA\,cm^{-2}$ when operating at $650\,^{\circ}C$. At $700\,^{\circ}C$, the maximum power density of $237\,mW\,cm^{-2}$ was achieved at the current density of $502\,mA\,cm^{-2}$ (Figure 10.12c). The PC-SOFCs thus converted ethane to value-added ethylene with high selectivity and co-generation of electricity with high power density.

Replacing BCY15 with BCYN electrolyte, the fuel cell performance was improved. At the same time, we found that the performance was also enhanced when the particle size of BCY15 power was decreased. So for future work, nanostructured, doped BCY15 electrolyte or other kind of proton conducing electrolyte obtained by other non-solid state methods with nanoscale should be explored to further improve the performance of the PC-SOFCs.

10.4.2 Development of Anode Materials of PC-SOFCs

Cr_2O_3 nanoparticles and BCYN perovskite oxides were synthesized by the combustion method and conventional solid state reaction, respectively.[35] The Cr_2O_3 consisted of very fine particles, mainly smaller than 20 nm (Figure 10.13a). For the fuel cell using Cr_2O_3 nanoparticles as the anode catalyst, BCYN perovskite oxide as the proton conducting ceramic electrolyte, and Pt as the cathode catalyst, the maximal power density increased from $51\,mW\,cm^{-2}$ to $118\,mW\,cm^{-2}$ and the ethylene yield increased from $\sim 8\%$ to 31% when the operating temperature of the solid oxide fuel cell reactor increased from $650\,^{\circ}C$ to $750\,^{\circ}C$. The temperature programmed oxidation (TPO) technique was used to

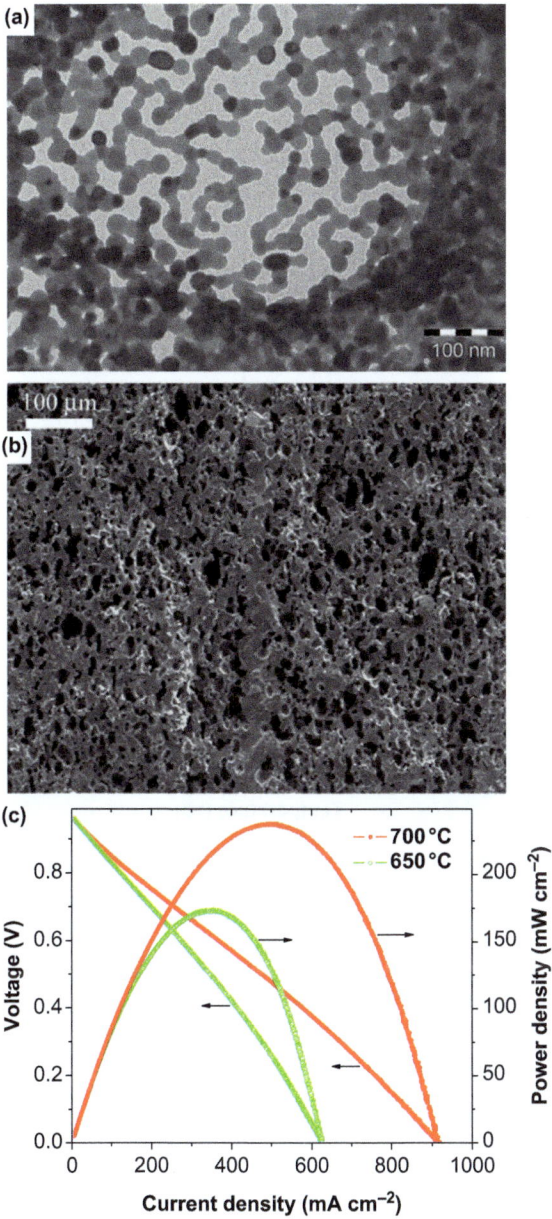

Figure 10.12 (a) TEM image of as-prepared BCYN precursor powder. (b) Cross-sectional SEM image of integral BCYN membrane SOFC. (c) Current density-voltage and power density curves of the PC-SOFCs.[34]

compare the carbon deposition on Cr_2O_3, Pt and Ni anode catalysts. Figure 10.13b illustrates that there was almost no CO_2 emission when the treated Cr_2O_3 was heated in the air, comparing to the high CO_2 peaks when using Ni and

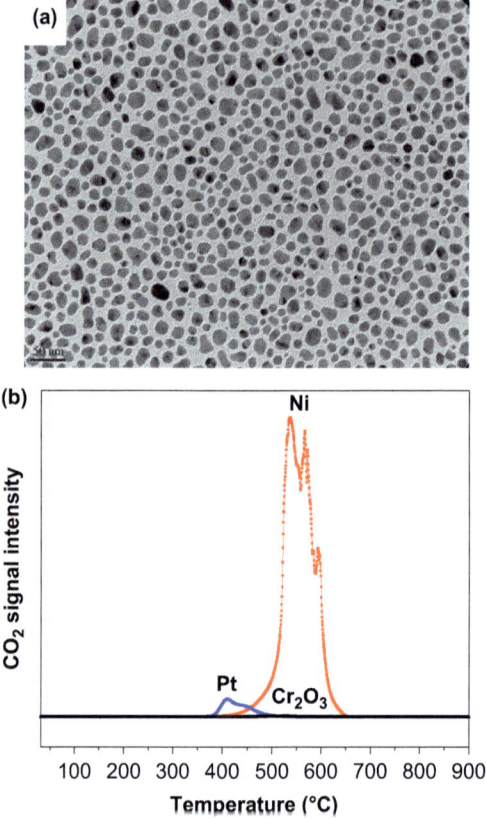

Figure 10.13   (a) TEM image of Cr_2O_3 powder. (b) O_2-TPO curves of Pt and Cr_2O_3 anode catalysts after treatment in dry ethane at 700 °C for 10 h. [35]

Pt catalysts. The results suggested that Cr_2O_3 anode material had much better coke resistance than Pt and Ni anode catalysts in ethane fuel at 700 °C.

Based on the nano-Cr_2O_3 anode power, we also explored Cu-Cr_2O_3 nano-composite used as the anode materials.[29] The anode catalyst precursor $CuCrO_2$ nanopowders with the particle sizes smaller than 10 nm were synthesized using the Pechini method (Figure 10.14a). Uniform Cu-Cr_2O_3 nanocomposites with the particle sizes smaller than 50 nm were obtained by heating $CuCrO_2$ particles at elevated temperatures in reducing gas (10% H_2 balanced with He) (Figure 10.14b). High magnification Cu (Figure 10.14c) and Cr (Figure 10.14d) auger electron spectroscopy (AES) mapping images also showed that Cu and Cr_2O_3 particles were evenly distributed within the anode, with the consequence that Cu was able to efficiently conduct electrons and was not easily sintered. The Cu-Cr_2O_3 anode catalyst had excellent resistance to carbon deposition, sintering resistance, and good electro-catalytic dehydrogenation activity. For the SOFC reactors with a Cu-Cr_2O_3 nanocomposite anode and a $BaCe_{0.7}Zr_{0.1}Y_{0.2}O_{3-\delta}$ (BCZY) proton conducting membrane, the maximum power density increased

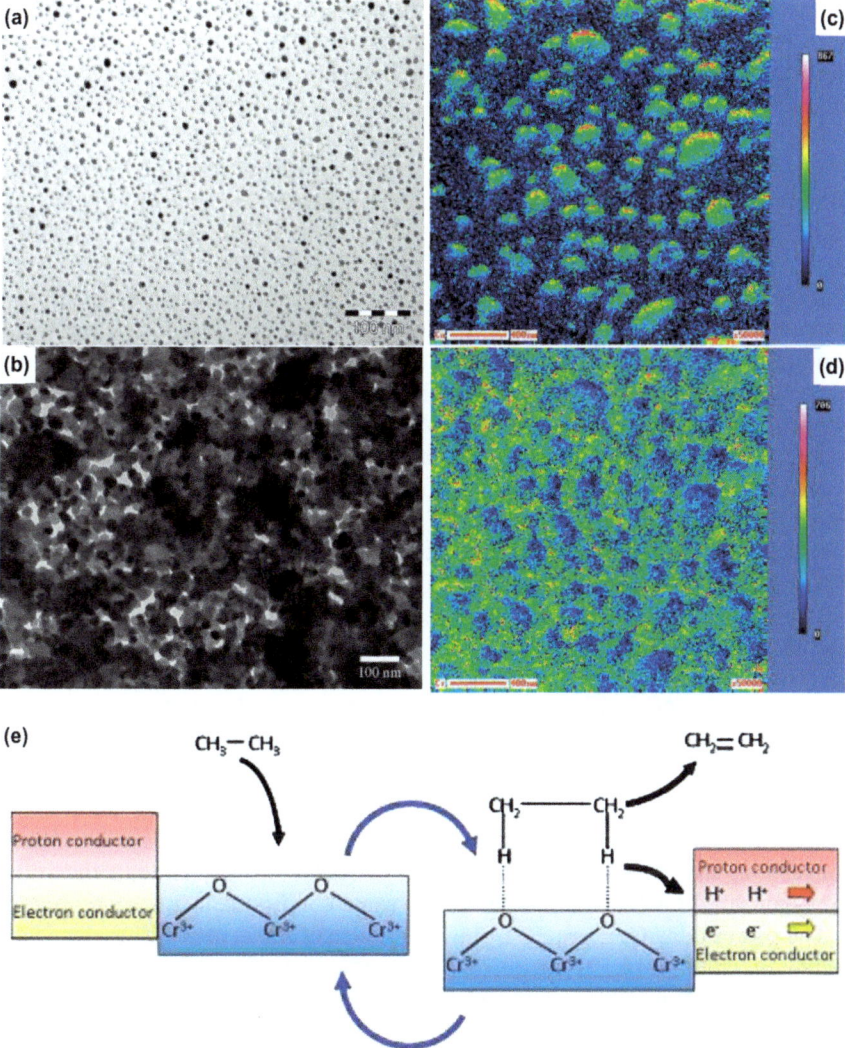

Figure 10.14   TEM images of (a) as-prepared CuCrO$_2$ powder. (b) Cu-Cr$_2$O$_3$ nano-composite powders. High magnification (c) Cu and (d) Cr AES mapping images of a Cu-Cr$_2$O$_3$ composite anode catalyst after heating at 800 °C in 10% H$_2$ as a reducing gas atmosphere for 10 h. (e) Proposed mechanism for ethane electro-catalytic dehydrogenation to ethylene over a Cu-Cr$_2$O$_3$ composite anode in PC-SOFCs.[29]

from 81 mW cm^{-2} to 170 mW cm^{-2} and the ethylene yield increased from about 9% to 39%, with high selectivity from 99% to 90% as the operating temperature arose from 650 °C to 750 °C at the same flow rate of 150 mL min^{-1}.[29] The main by-product from the SOFC reactor process of proton conducting membrane was CH$_4$, the proportion of which increased with increasing the operating temperature. There was no CO$_2$ greenhouse gas emission from the process.

However, a trace of CO was detected in the anode exhaust gas when operating at over 700 °C due to the small amount of oxide ion conductivity in the BCZY electrolyte. The PC-SOFC reactor offered highly stable performance for co-production of ethylene and power during 10 days of operation at 700 °C.

Figure 10.14e shows a possible mechanism for the electrochemical dehydrogenation of ethane over a Cu-Cr_2O_3 composite anode deposited on a proton conducting electrolyte. Sequentially, an ethane molecule chemisorbed onto the anode catalyst surface, the C-H bond was activated and broken by reaction with a coordinatively unsaturated Cr^{3+} site, then ethylene and two hydrogen ions and electrons were formed and the electron released. The released electrons migrated to the current collector through the metallic Cu and Au electronic conductor component of the anode, and the released protons (H^+) transported to the electrolyte membrane through the proton conducting Cr_2O_3. When the ethylene desorbed from the surface of the anode catalyst, the Cr^{3+}-O active site was regenerated for further dehydrogenation reaction.

Nanostructured $FeCr_2O_4$ was also explored as the anode material used in PC-SOFCs for cogenerating ethylene and electrical power. PC-SOFCs were constructed using $FeCr_2O_4$ nanoparticles as the anode catalyst, $La_{0.7}Sr_{0.3}FeO_{3-\delta}$ (LSF) as the cathode material, and BCZY perovskite oxide as the electrolyte. $FeCr_2O_4$, BCZY and LSF were synthesized by the sol-gel combustion method.[36] The BCZY electrolyte was dense, thus it separated the anode and cathode feeds very well (Figure 10.15a). The SEM image also showed the typical porous microstructure of the electrodes, required to achieve good diffusion of the feed to the electrochemically active triple phase boundary (TPB) sites. The conversion of ethane increased from 14.6% to 29% and 43.7%, while the selectivity to ethylene decreased from 96.6% to 94.2% and 90.8% as the operating temperature increased from 650 °C to 700 °C and then 750 °C (Figure 10.15b). The ethane conversions over this nanosized $FeCr_2O_4$ catalyst were higher than the values over nanosized Cr_2O_3 catalysts under the same conditions.[25]

The power density increased from 70 to 240 mW cm^{-2}, and the ethylene yield increased from ~14.1% to 39.7% when the operating temperature of the proton-conducting fuel cell reactor increased from 650 °C to 750 °C (Figure 10.15c).[36] The high values of the ohmic resistance were attributable to the overall configuration of the single cell, in particular the large thickness of the BCZY electrolyte (about 0.9 mm), suggesting that higher performance should be attainable from the cells with much thinner electrolyte layers.

Transition metal carbides (TMCs) are attracting attentions as promising alternative electrocatalysts for low temperature proton exchange membrane fuel cells (PEMFCs) and direct methanol fuel cells (DMFCs). TMCs have exceptionally high catalytic activity under hydrotreating conditions. Therefore, we also constructed PC-SOFC using Mo_2C as the anode, BCZY as the electrolyte and LSF as cathode.[37] The BCZY and LSF perovskite nanopowders were prepared using a citric acid-nitrate combustion method. Mo_2C was purchased from Aldrich with particle size of 325 mesh and used without further treatment.

The XRD results showed that Mo_2C had pure phase without any evidence of any other crystalline oxide materials.[37] The XRD pattern for BCZY showed

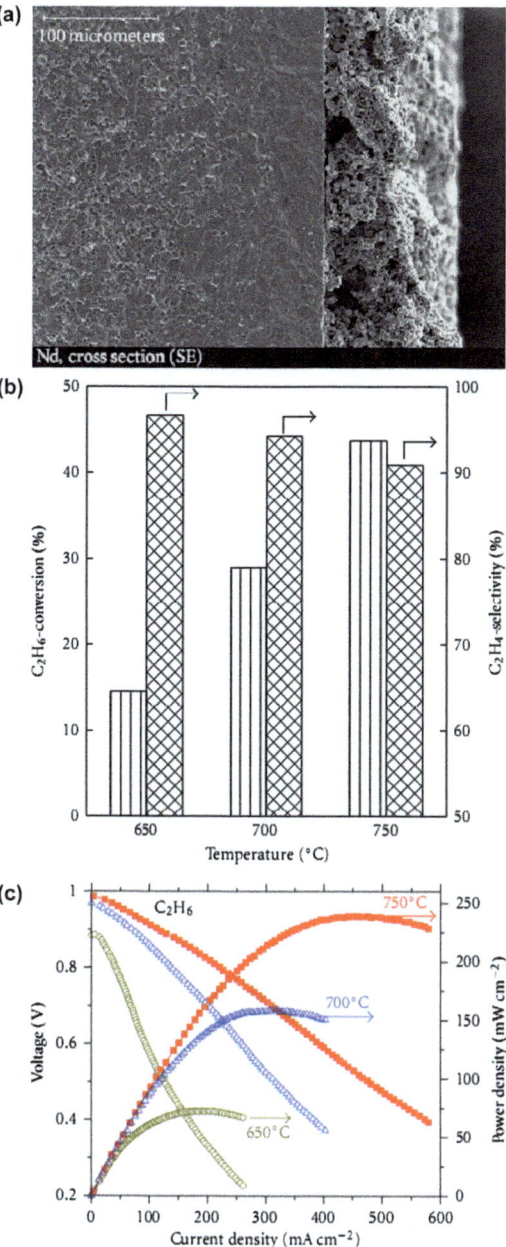

Figure 10.15   (a) SEM image of a cross section of the BCZY electrolyte-supported cell with $FeCr_2O_4$ catalyst. (b) Ethane conversion and ethylene selectivity and (c) I-V curves and power density output of a $FeCr_2O_4$-BCZY|BCZY|LSF-BCZY single cell at different temperatures.[36]

only perovskite phase. The diffraction patterns for the mixture of Mo_2C and BCZY calcined in 10% H_2/He at 950 °C for 10 h were nearly identical to those for the original pure Mo_2C and BCZY (Figure 10.16a). Ethane was

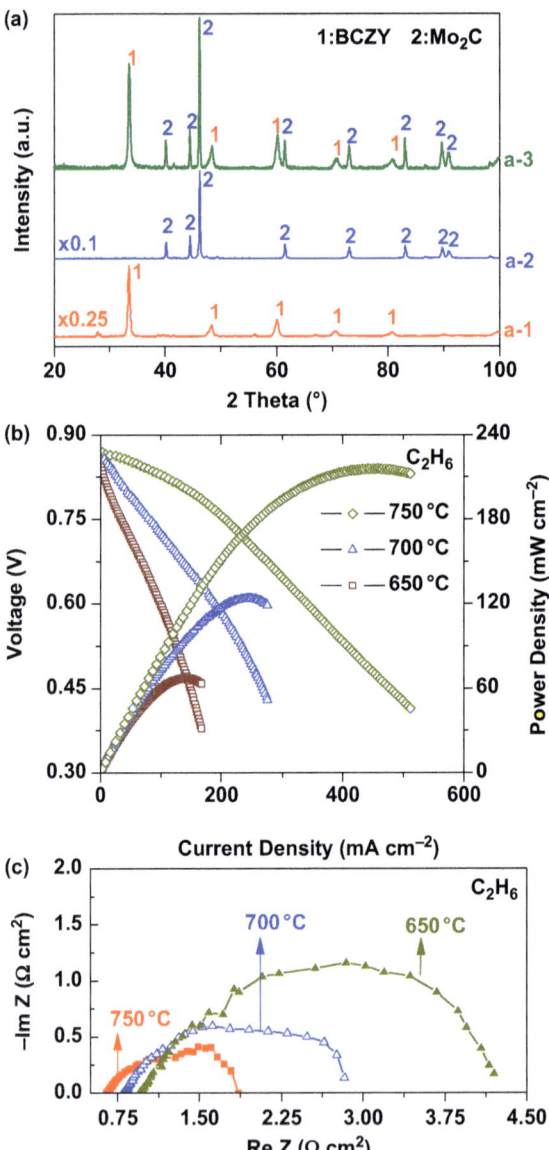

Figure 10.16 (a) XRD patterns of a-1: BCZY; a-2: Mo_2C; a-3: mixture of $Mo_2C + BCZY$ calcined in 10% H_2/He at 950 °C for 10 h. (b) Current density-voltage and current density-power density curves and (c) impedance spectra of a single cell with C_2H_6 as anode feed at different temperatures.[37]

dehydrogenated selectively to ethylene at the anode at elevated temperatures, with co-generation of electricity. Ethane conversion increased from 7.7% to 18.5% and 42.6% when the temperature increased from 650 °C to 700 °C and 750 °C, while the corresponding selectivity to ethylene decreased from 97.5% to 96.1% and 87.2%. The maximum power density was 65 mW cm^{-2} at 650 °C, and increased to 215 mW cm^{-2} at 750 °C (Figure 10.16b).[37] The polarization resistances when using ethane-fuel were considerably greater than those using H_2-fuel (Figure 10.16c).[37] At the same time, the results of XRD and weight change showed that Mo_2C anode catalyst had much better carbon deposition resistance than Ni for using ethane as the fuel. Further optimization of the fuel cell configuration, especially decreasing the thickness of electrolyte, could boost the cell performance further. Nanostructured Mo_2C obtained by other methods (*e.g.* citrate-nitrate combustion method,[22,24] Pechini method[26]) will be also explored in future work.

In summary, our group developed a new and appropriate proton conducting electrolyte (*e.g.* BCYN and BCZY[38–40]) and highly selective anode catalysts (*e.g.* nano-$FeCr_2O_4$), which enabled the more effective co-generation of electricity and the production of high value-added products (*e.g.* C_2H_4). We have also developed PC-SOFCs to convert C_3H_8 to C_3H_6 and simultaneously generate electricity.[25–28,41] Other analogous experiments are currently under way in our group.

10.5 SOFCs Running on Syngas Containing H_2S

Syngas is an important and relatively inexpensive energy source. Normally, syngas is obtained from reforming of hydrocarbons or/and carbon (*e.g.* coke). Table 10.3 shows the typical coal gas compositions from four gasification processes.[42] Syngas thus manufactured may contain H_2S derived from the sulfur compounds in the parent hydrocarbons, and is not appropriate to be used in fuel cells having conventional anodes such as Pt or Ni. It is expensive and not economical to purify the syngas by removal of H_2S just for the purpose of using conventional SOFCs anode catalysts. Therefore, the purification of the syngas is the main obstacle to overcome for developing syngas SOFCs. We investigated the feasibility of utilizing H_2S-containing syngas as the fuel in a SOFC. The obvious economic benefit is that if H_2S-containing syngas could be used directly in a fuel cell, the overall capital and operating costs could be considerably reduced by eliminating the production of sulfur free syngas.

In this part, syngas (40% H_2 and 60% CO) containing 5000 ppm H_2S was used to fuel SOFCs. The reason for choosing the 5000 ppm H_2S was because this value represented the typical sulfur concentration of syngas from coal (Table 10.3).

To develop anode material for SOFC systems fuelled with impure hydrogen, V_2O_5 was first reduced to VO_x which was then mixed with $LaCrO_3$ and YSZ to form $LaCrO_3$-VO_x-YSZ catalyst.[43,44] Commercial YSZ was used as the electrolyte and porous platinum as the cathode catalyst. It was found that the presence of CO in the anode feed enhanced the initial fuel cell performance

Table 10.3 Typical coal gas composition (Vol. %).[34]

	Air blown	*British*	*Shell*	*Texaco*
Component				
CO	15–30	54–57	62–65	45
H_2	18–25	29	28–30	30
CO_2	2–15	4.5	1.6–2	5–8
H_2O	3–11	*	0.4–2	18
N_2	45–55	3	0.7–3.1	0.8
Typical impurities (ppm)				
H_2S	750–7000	6000–7000	1260–4000	1100–3500
NH_3/HCN	50–500	*	300	800
HCl/HF	<50	*	200–830	170–690

*Not reported.

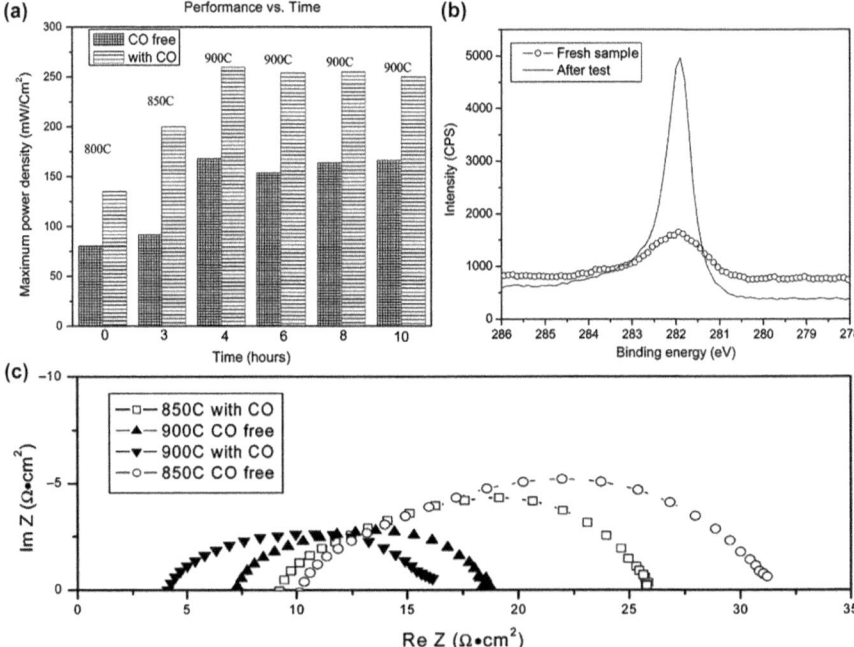

Figure 10.17 (a) Maximum power densities as functions of temperature and time. (b) Comparison of binding energy peaks of C 1s before and after 24 h fuel cell tests. (c) Impedance spectra for $LaCrO_3$-VO_x-YSZ in syngas and H_2 feeds, each containing 5000 ppm H_2S[43].

when compared to the use of CO-free H_2 containing 5000 ppm H_2S (Figure 10.17a). Impedance spectra showed that the anode polarization resistance decreased in the presence of CO (Figure 10.17c). Current density of $450\,mA\,cm^{-2}$ at $0.6\,V$ and maximum power density of $260\,mW\,cm^{-2}$ were obtained at $900\,^\circ C$. Fuel cell performance was stable when using either pure H_2 or H_2S-containing H_2 as the feed. However, the performance started to decrease more rapidly after 10 h of operation when CO existed in the feed. Figure 10.17b compares the XPS analyses of binding energy of carbon electrons on the anode surface of a fresh sample and the sample operated under the fuel cell conditions for 24 h. The amount of carbon deposited on the anode surfaces increased dramatically. Later, the carbon deposition on vanadium-based anode catalyst for syngas fuelled SOFC was studied in depth using auger electron spectroscopy and XPS technique.[45] It was found that the carbon deposition was more severe in CO than in syngas feed. The form of carbon and preferable location of deposition were dependent on the gas environment. Humidification of the anode feed could suppress the carbon deposition. However, the use of a humidified feed would decrease the initial fuel cell performance.

The use of CoS-MoS$_2$ as the anode catalyst was also investigated. MoS$_2$ and CoS, 8% YSZ nanopowder, and silver powder were mixed together in 90 : 5 : 5 weight ratio.[46] Commercial YSZ was used as the electrolyte and porous platinum as the cathode catalyst. The performance of CoS-MoS$_2$ as the anode catalyst was significantly better when compared with MoS$_2$ alone. When changing the fuel from H$_2$ containing 5000 ppm H$_2$S to H$_2$S-containing syngas (60%CO, 40%H$_2$, 5000 ppm H$_2$S), it was found that as soon as the anode catalyst was exposed to CO in the feed the cell performance dropped correspondingly and the electrochemical polarization rapidly increased, which indicated that the effect was due to strong adsorption of CO on the catalytic sites.

The above results showed that the CO in the syngas could reduce the fuel cell performance very significantly. If the CO could be oxidized directly under the operating condition of SOFC, then the degradation in performance would be much minimized. To develop the anode catalyst for electrocatalytic oxidation of all components of H$_2$S-containing syngas, nano-Au catalyst supported on MoS$_2$ (Au/MoS$_2$) was explored.[47] It was found that Au/MoS$_2$ had catalytic activity for conversion of CO, thus preventing poisoning of MoS$_2$ active sites by CO. In contrast to the use of MoS$_2$ as the anode catalyst, the performance of Au/MoS$_2$ anode catalyst improved when CO was present in the feed. Current density over 600 mA cm^{-2} and maximum power density over 70 mW cm^{-2} were obtained at 900 °C.

The perovskite oxide La$_{0.7}$Sr$_{0.3}$VO$_3$ (LSV) was synthesized using the nitrate-citrate combustion method.[48] The phase formation process of LSV was studied using thermogravimetric analysis (TGA) and XRD. LSV was active as the anode material for SOFCs operating in H$_2$S-containing syngas (40% H$_2$, 60% CO, and 5000 ppm H$_2$S) with a high chemical and electrochemical stability (Figure 10.18a). The maximum power density, 210 mW cm^{-2}, was obtained at 900 °C at the current density close to 400 mA cm^{-2} (Figure 10.18b). The potentiostatic study showed only a slight decline in output over a 24 h test period (Figure 10.18c). The degradation rate of 1.3% h^{-1} was attributable to coarsening of the electrode and not to poisoning of the catalyst; no sulfur or carbon was deposited on the anode. Impedance measurements showed that the ohmic and polarization resistance of the cell decreased with the increase in temperature (Figure 10.18d). Therefore, LSV appeared as a promising candidate for anode material in SOFCs operating with H$_2$S-containing syngas as the fuel.

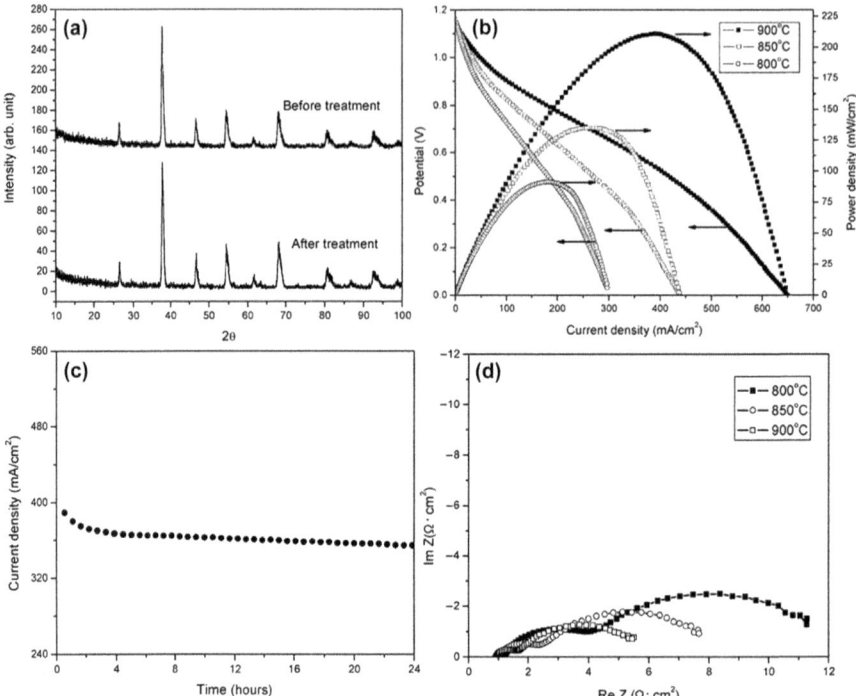

Figure 10.18 (a) XRD patterns of LSV catalyst before and after exposure to
H₂S-containing syngas at 900 °C for 72 h. (b) Performance of the fuel
cell using H₂S-containing syngas as the fuel gas operated at different
temperatures. Flow rates of the fuel gas and air are 50 standard-state
cubic centimeter per minute. (c) Electrochemical stability test
for LSV/YSZ/Pt single cell. Cell was operated at 0.5 V, 900 °C.
(d) Impedance spectra of the fuel cell using H₂S-containing syngas as
the fuel operated at different temperatures.[48]

Recently, we found that LST and $Y_{0.2}Ce_{0.8}O_{2-\delta}$ (YDC) composite proved to be
a stable and of high performance material for use as the anode in SOFC fed by
0.5% H₂S-containing syngas.[49] The power density of the cell was enhanced by
the addition of H₂S to syngas. Oxidation of CO to CO_2 was improved
significantly in the presence of H₂S. H₂S itself was not consumed because
formation of SO_2 was not detected at any potential. Thus electrochemical
oxidation of H₂S itself was not the cause of the enhancement in performance.
The role of H₂S could be the modification of the active sites of the anode catalyst.

Both LST and YDC were also synthesized using citrate-nitrate gel
combustion method.[50] MEA was prepared to have the following configuration
LST:YDC 50:50/YSZ (0.3 mm)/LSM:YSZ 50:50/LSM. It was found that the
performance of LST-YDC composite anodes in SOFCs significantly improved
when 0.5% H₂S was present in syngas (40% H_2 + 60% CO) or hydrogen. The
maximum power densities were 169, 102 and 39 mW cm⁻² when the cell was
fuelled with syngas having 0.5% H₂S, pure syngas and 0.5% H₂S balanced with

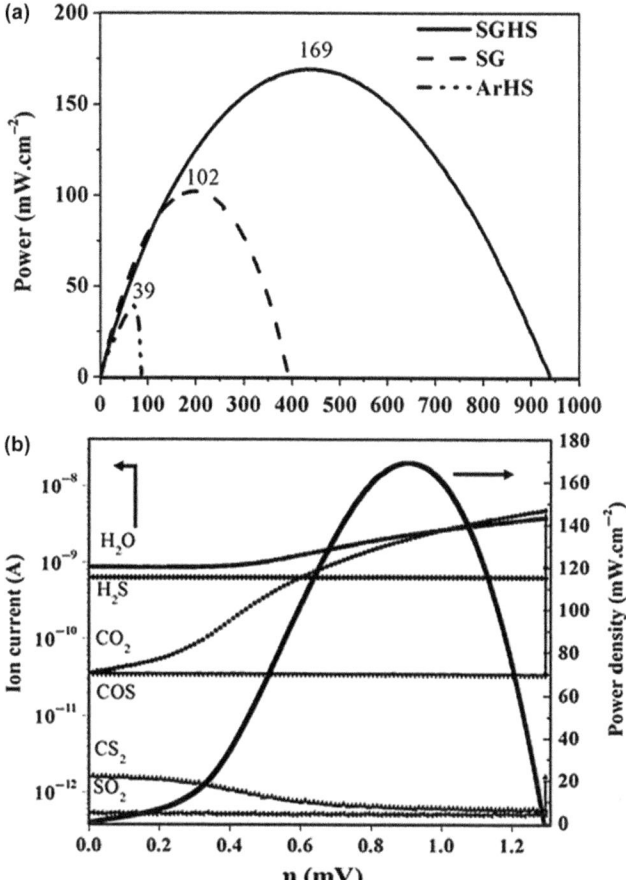

Figure 10.19 (a) Electrochemical performance of an MEA having the structure LST:YDC 50:50/YSZ/LSM:YSZ 50:50/LSM fed with different fuels (SG: syngas, SGHS: 0.5%H₂S in syngas, ArHS: 0.5%H₂S in Ar) at 850 °C. (b) Correlation of mass spectrometric peaks for the effluent and power density as a function of overpotential of the cell fed with dry syngas containing 0.5% H₂S at 850 °C.[50]

Ar at 850 °C, respectively (Figure 10.19a). However, the H₂S concentration remained unchanged during potentiodynamic and potentiostatic runs. Therefore the conversion of H₂S itself was not the reason for the enhanced performance. GC and mass spectrometric analyses revealed that the presence of H₂S could improve the rate of electrochemical oxidation of all fuel components (Figure 10.19b). At the same time, the influence of the presence of H₂S in different feeds on the power degradation could be negligible, which was proved by the electrochemical stability tests.

The results of chemical stability showed that LST-YDC composite was electrochemically stable as an anode material in syngas containing H₂S in high concentration. To explore the exact mechanism of the H₂S enhancement effect for electrochemical oxidation of syngas requires *in situ* H_2S tolerant high temperature analytic technique which is not presently available.

Other researchers also explored the cofiring CeO_2 in Ni/YSZ anode support layers,[51] $Ni-Al_2O_3$ cermet supported tubular SOFC reactor[52] and a novel compact catalytic nanoparticle bed micro-fabricated reactor[53] to enhance the performance of the SOFCs using syngas as the feed. Yu *et al.* explored the carbon formation mechanism[54] and Cayan *et al.* developed the degradation model due to impurities in coal syngas.[55]

In a fuel cell, the current collector also plays an important role in the fuel cell performance. Thus, we explored the coking resistant current collectors for syngas SOFCs.[56] Cu surface modified nickel foam was obtained by heating copper coated nickel foam in a reducing atmosphere, which led to the formation of CuNi alloy. Figures 10.20a and 10.20b show that the microstructure of $La_{0.75}Sr_{0.25}Cr_{0.5}Mn_{0.5}O_{3-\delta}$ (LSCM) anode had same porous microstructure before and after the fuel cell test. New diffraction peaks were not found in the XRD patterns after each exposure (Figure 10.20c), suggesting

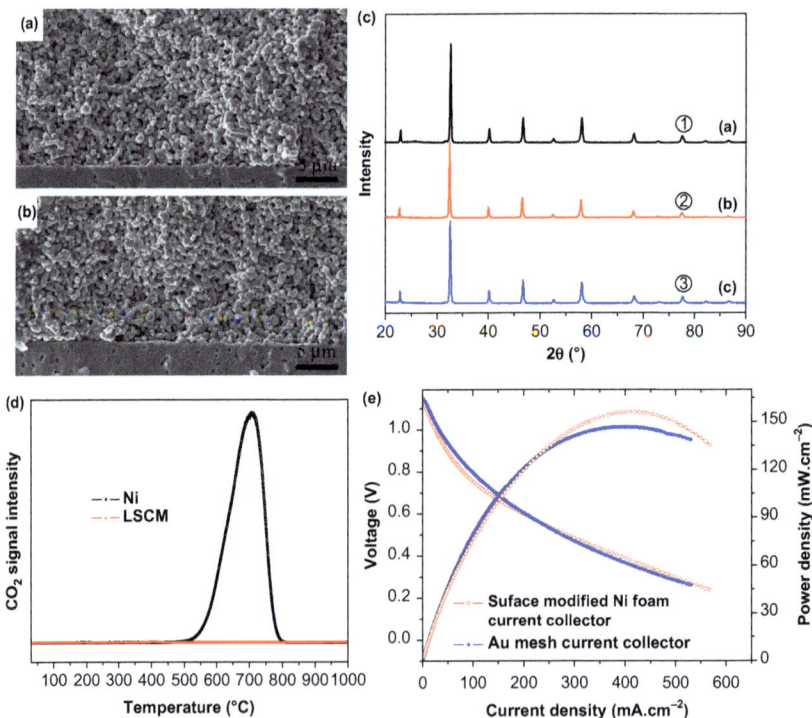

Figure 10.20   Cross-sectional SEM images of LSCM anode on YSZ electrolyte (a) before and (b) after fuel cell test. (c) XRD patterns of ① as-prepared LSCM, after heat treatment in ② syngas, and ③ CO at 900 °C for 10 h. (d) O_2-TPO curves of LSCM and Ni after the treatment in syngas at 900 °C for 10 h. (d) Curves of current density-voltage and current density-power densities for syngas SOFCs with Au and surface modified Ni foam anode current collectors after two days of operation at 900 °C.[56]

that LSCM perovskite anode catalyst had good structural and chemical stability in either syngas or carbon monoxide. The results of TPO experiments indicated that there was no obvious carbon deposition on the LSCM anode catalyst while a considerable amount of carbon was deposited on the Ni sample (Figure 10.20d). Syngas SOFCs assembled using Cu modified Ni foam as the anode current collector, LSCM as the anode catalyst, YSZ as the electrolyte, and porous Pt as the cathode provided similar power density to the one with gold anode current collector and had excellent stability during operation at 900 °C (Figure 10.20e). Low *et al.*[57] further described the properties of Cu coated Ni foam as the current collector for practical applications of SOFCs in H_2S-containing syngas at 750 °C. The Cu coated Ni foam exhibited good stability of the electronic conductivity and mechanical strength during more than 150 h exposure to the 500 ppm H_2S-containing syngas at 750 °C.

Based on the above results, we further developed Ni-P amorphous alloy coated onto porous Ni foam. The plated Ni foam exhibited excellent carbon deposition resistance and structure stability in syngas at the elevated temperature when P content was higher than 6.5 wt% in the coating.[58] SEM images (Figures 10.21 a, b) showed that the plated Ni foam was still shiny with no detectable carbon deposits on the surface and no apparent mechanical strength loss. Figure 10.21c shows the SEM image of a fuel cell cross-section after fuel cell test. The respective

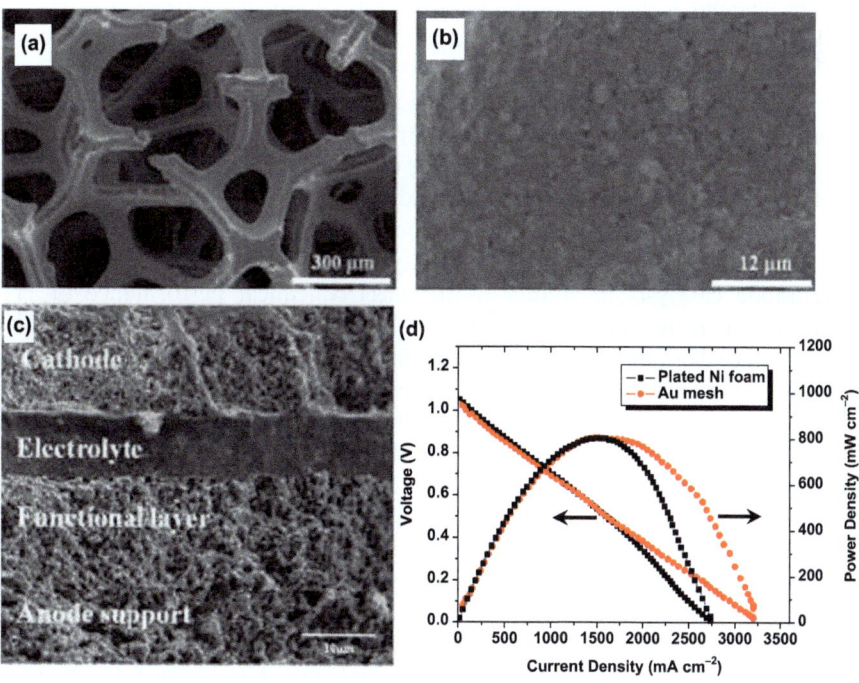

Figure 10.21   (a, b) SEM images of Ni-P coated Ni foam after treatment in syngas at 750 °C for 24 h. (c) SEM cross-sectional image of a typical tested fuel cell. (d) Comparison of current voltage and power density for SOFCs with Ni foam and Au mesh anode current collectors in syngas at 800 °C.[58]

layers included Ni-YSZ anode support substrate, Ni-YSZ anode functional layer, YSZ electrolyte and LSM-YSZ composite cathode. The syngas-fuelled cells with Au mesh and the plated Ni foam current collector reached almost identical power density and internal resistance. The maximum power density for both current collectors was $\sim 800\,mW\,cm^{-2}$ in syngas at 800 °C (Figure 10.21d). The plated Ni foam had no apparent degradation of power output during extended fuel cell stability test. These results indicated that both the Cu coated Ni foam and Ni-P alloy coated Ni foam are promising anode current collectors for SOFCs using H_2S-containing syngas due to their excellent stability and low cost.

10.6   SOFCs Running on Pure H_2S

H_2S is a by-product of many industrial operations such as processing of natural gas, coking, and hydrodesulfurization of crude oil or coal. It is anticipated that additional large quantities of H_2S will be produced when coal liquefaction attains commercial importance. As a major industrial toxic and corrosive pollutant, H_2S also serves as an increasingly significant source for sulfur production. Many processes developed to remove and/or recover H_2S include adsorption, absorption, hydrogen production and conversion to elemental sulfur *via* the two-step Claus process. High capital investments are required for such H_2S processing units. Also, as the final products are of low commercial value, there is a high demand to find more economical and efficient ways to remove H_2S or collect it for producing hydrogen and sulfur. H_2S has a high chemical potential (0.742 V if sulfur is produced and 0.758 V for SO_2 as the product, at 750 °C and 1 atm). Direct electrochemical conversion of H_2S in SOFC will offer both economical and environmental benefits. In the previous sections, we found that the presence of H_2S in the CH_4 or syngas could enhance the performance of the SOFCs. However, H_2S is not directly oxidized in those fuel cells. In this section, we focus our efforts on investigating the potential use of H_2S as a fuel in SOFCs.

Due to its stability and catalytic activity, Pt was initially used as the cathode and anode catalysts in studies of the electrochemical performance of H_2S SOFCs.[59] During the fuel cell test, the formation of PtS led to quick performance degradation and contaminated the anode surface and increased the interface resistance between Pt and YSZ, which ultimately led to detachment of the Pt anode from the YSZ membrane. The low levels of H_2 detected suggested that the primary mechanism was the electrochemical conversion of H_2S, which implied that the thermal decomposition of H_2S to H_2 only accompanied the primary electrochemical conversion as a side reaction of lesser significance.

In order to decrease the reversible formation and decomposition of PtS on the Pt-YSZ interface, the membrane structure and performance were both stabilized by interposing TiO_2 islands between the Pt anode and YSZ electrolyte.[60] The stability of Pt anode performance was thereby significantly improved by the intermediate layer of TiO_2, which helped anchor the Pt anode and increased the activity.

Four different metal sulfide catalysts were subsequently investigated for electrochemical oxidation of H_2S in SOFCs at temperatures up to 850 °C.[61]

All catalysts exhibited good electrical conductivity and catalytic activity at all temperatures investigated. MoS_2 and its composite catalysts were found to be more active than Pt anode. However, MoS_2 itself sublimes at high temperature. In contrast, composite catalysts (M-Mo-S), prepared from mixed metal sulfides $MS-MoS_2$ (M = Fe, Co, Ni), appeared to be stable and effective for electrochemical conversion of H_2S in SOFCs up to 850 °C. In order to overcome the poor electrical contact between the Pt current collecting layer and the anode layer, Ag powders were mechanically admixed into the anode material as the current collector. It was found that the anode catalyst comprising Co-Mo-S admixed with up to 10% Ag powders had much better performance and longevity, and improved electrical contact when compared with Pt/M-Mo-S anode system.

Based on the metal sulfide catalysts mentioned above, we also studied the influence of gas flow rate on the performance of H_2S/air SOFCs with MoS_2-NiS-Ag anode.[62,63] It was found that the cell OCV of H_2S, $(MoS_2 + NiS + Ag)$/YSZ/Pt, air was independent of air flow rate but increased with increasing H_2S flow rate. OCV variation followed a linear relation with the logarithm of H_2S flow rate, with the slopes being 172, 162, and 160 mV dec^{-1} at 750, 800, or 850 °C, respectively. Although an anode of $MoS_2 + NiS + Ag$ was used in the present study, the same phenomena were observed using alternative anode systems. The performances of current-voltage and power density were improved by increasing either air flow rate or H_2S flow rate. Based on three-phase boundary (TPB) theory, we developed the anode material of Mo-Ni-S with Ag and YSZ powders.[64] The Mo-Ni-S catalyst was prepared from MoS_2 and NiS (1 : 1 weight ratio). The optimum composition was ~90 wt% Mo-Ni-S, 5 wt% Ag, and 5 wt% YSZ. A fuel cell using Mo-Ni-S + Ag + YSZ anode, YSZ electrolyte, Pt cathode and pure H_2S as the feed produced the maximum sustainable current densities of over 480 mA cm^{-2} at 750 °C and over 800 mA cm^{-2} at 850 °C. The corresponding maximum power densities were 50 mW cm^{-2} at 750 °C and over 200 mW cm^{-2} at 850 °C.

We also investigated the performance and stability of MoS_2 and $Ni_{3 \pm x}S_2$-based anodes with different weight ratios of $Ni_{3 \pm x}S_2$ to MoS_2 (2 : 1, 1 : 1, 1 : 2 and 1 : 4) to find optimum anode composition for SOFC operating on pure H_2S at 700-850 °C.[65] The results showed that the disappearance of MoS_2 during the fuel cell operation was due to its oxidation to MoO_3 which sublimated at temperatures above 600 °C. The results of differential scanning calorimetry (DSC) (Figure 10.22a) showed that the addition of $Ni_{3 \pm x}S_2$ could enhance the thermal stability of the MoS_2. All anode compositions with various ratios of nickel sulfide to molybdenum sulfide had high initial electrochemical activities for H_2S oxidation in SOFC. The highest power density ca. 300 mW cm^{-2} at 850 °C was achieved with 1 : 1 weight ratio anode composition (Figure 10.22b). However, the performances of the anode materials for H_2S oxidation were not stable because of agglomeration of nickel sulfide particles and change in the anode composition.

A series of ternary transition metal sulfides with general composition of AB_2S_4 (A = Mo, Cr, Ni, Fe, Co, and Cu; B = V, Cr, and Mo) were prepared and investigated as the anode catalysts for use in H_2S-powered SOFCs.[66] The relative activities of these materials for H_2S oxidation were in the order of

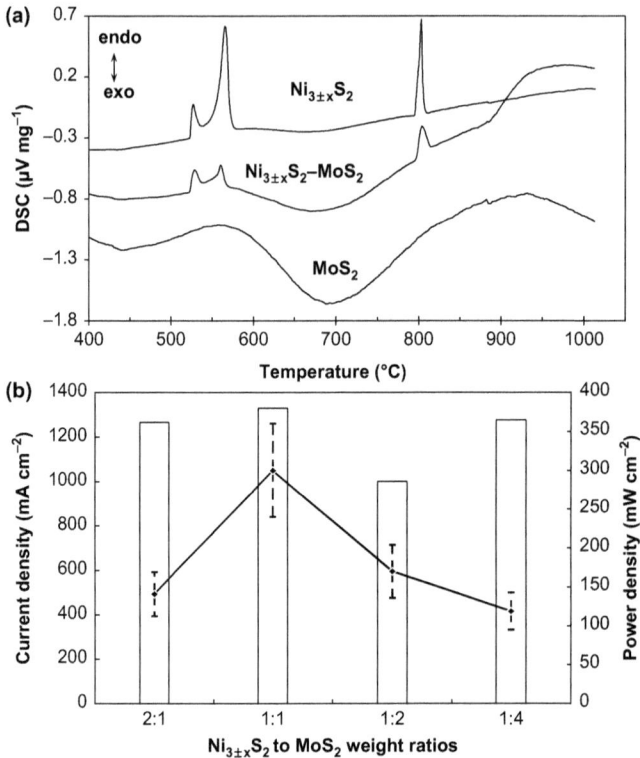

Figure 10.22 (a) DSC analysis of $Ni_{3\pm x}S_2$, MoS_2 and mixture of both in $1:1$ weight ration N_2. (b) Fuel cell performance of composite $Ni_{3\pm x}S_2$-MoS_2-Ag-YSZ anode at 850 °C (bars represent current density on the left and curve shows power density on the right).[65]

$MoV_2S_4 > NiV_2S_4 > CrV_2S_4 > CuCr_2S_4 > MoS_2 > MoS_2$-$Ni_{3\pm x}S_2 > CoMo_2S_4 >$ $FeCr_2S_4$ within the range of 700–850 °C, which was consistent with results from the anodic polarization resistances in symmetrical cell (Figure 10.23a). Polarization curves indicated that MoV_2S_4 had the lowest potential drop, with up to a $200\,mA\,cm^{-2}$ current density at 800 °C. The highest power density of ca. $275\,mW\,cm^{-2}$ was obtained at 800 °C. The power density was increased to $400\,mW\,cm^{-2}$ when the temperature was increased to 850 °C (Figure 10.23b). We also found that the MoV_2S_4 was chemically, structurally and electrochemically stable (Figure 10.23c) during prolonged exposure to a pure H_2S stream at 850 °C, which made it a potential candidate anode for use in highly concentrated H_2S SOFC at 700–850 °C.

From the above mentioned working principles of SOFCs, we know that the nature of the electrolyte membrane determines the manufacturing requirements and mode of operation, which is also true for H_2S-air fuel cells. The ions (H^+ or O^{2-}) conducted within the membrane determine the electrochemical processes that occur in the fuel cell, as shown in Table 10.4. When proton-conducting

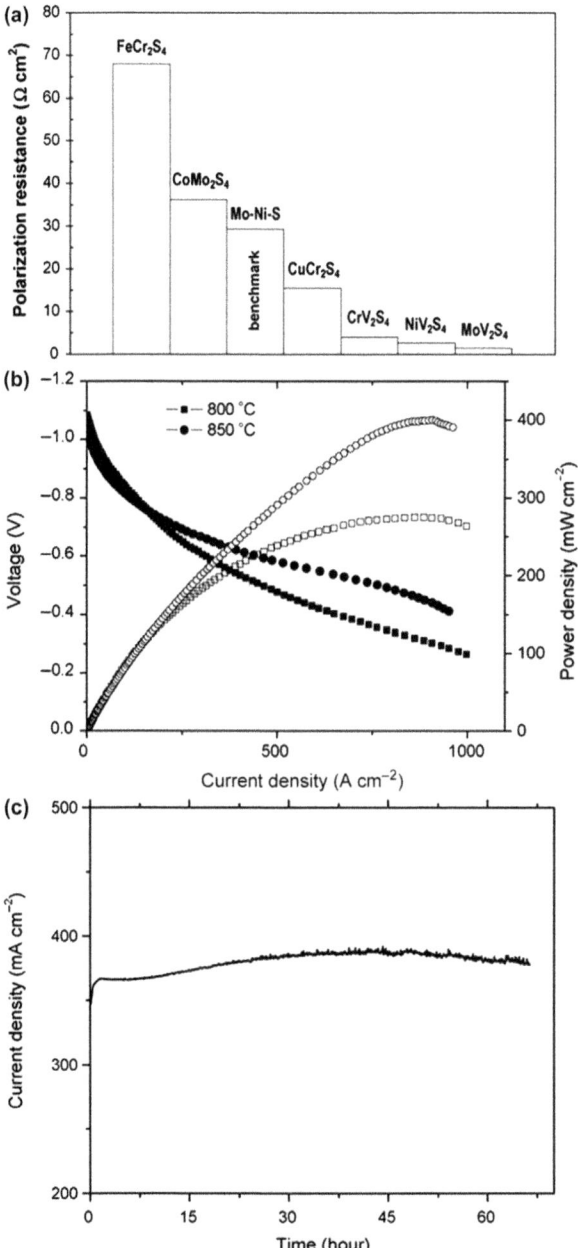

Figure 10.23 (a) Anodic polarization resistances in symmetrical cell at 800 °C. (b) Maximum attainable fuel cell performance at 800 °C and 850 °C (Electrolyte: YSZ, Cathode: Pt, Feed: H_2S). Current density-voltage (solid symbols) and current density-power density (open symbols). (c) Fuel cell current density as a function of exposure time to H_2S stream at 850 °C (at constant anodic overpotential of 0.6 V applied across anode-reference electrodes).[66]

Table 10.4 Electrochemical reactions in H_2S-air fuel cells with different membranes.[68]

Electrolyte type	Anode reaction(s)	Electrolytic conduction	Cathode reaction	Overall cell reaction(s)
Proton-conducting	$H_2S - 2e^- \rightarrow 2H^+ + 1/2S_2$ $H_2 - 2e^- \rightarrow 2H^{+\,a}$	Protons (H^+) transfer from anode side to cathode side	$O_2 + 4H^+ + 4e^- \rightarrow 2H_2O$	$2H_2S + O_2 \rightarrow 2H_2O + 1/2S_2$ $2H_2 + O_2 \rightarrow 2H_2O^a$
Oxide ion-conducting	$H_2S + 2O^{2-} -6e^- \rightarrow 2H^+ + SO_2$ $H_2S + O^{2-} -2e^- \rightarrow H_2O + 1/2S_2$ $2H_2S + SO_2 \rightarrow 2H_2O + 3/2S_2$ $H_2 + O^{2-} -2e^- \rightarrow H_2O^a$	Oxide ions (O^{2-}) transfer from cathode side to anode side	$O_2 + 4e^- \rightarrow 2O^{2-}$	$2H_2S + 3O_2 \rightarrow 2H_2O + 2SO_2$ $2H_2S + O_2 \rightarrow 2H_2O + 1/2\,S_2$ $2H_2 + O_2 \rightarrow 2H_2O^a$

^{a}H_2 results from the internal reforming of H_2S at temperatures in excess of 700 °C according to the following reaction: H_2S (g) $= H_2$ (g) $+ 1/2S_2$ (g).

membranes are used in an H_2S-air fuel cell, high-purity sulfur is the only product obtained in the anode chamber and water is the only product formed in the cathode chamber. With oxide ion-conducting membranes, elemental sulfur, H_2O and SO_2 are formed in the anode chamber, thus requiring further processing of the stream to remove SO_2. The most challenging task for the development of H_2S-air fuel cells using a proton-conducting membrane is to develop a membrane which is superior in chemical/thermal stability, mechanical strength, electrical conductivity and gas-impermeability. Our group also studied the Li_2SO_4-based proton-conducting membrane[67] incorporated with Al_2O_3 and H_3BO_3,[68] Li_2SO_4-Al_2O_3 composite electrolyte.[69,70]

Our results show that the H_2S-fuelled SOFCs can potentially generate useful electrical energy on site while disposing of H_2S, a toxic byproduct of the fossil fuel industry. We have developed several chemical-thermodynamic models for $H_2S/H_2/H_2O/N_2$ mixtures and used them in predicting equilibrated fuel composition as well as the equilibrium cell voltage.[71] Our experiments and analyses suggested that H_2S, not H_2 produced from H_2S dissociation, was preferentially electro-oxidized on the anode in our experiments. Efforts elsewhere also investigated additional anode materials, including $Sm_{0.9}Sr_{0.1}Cr_{0.5}Fe_{0.5}O_3$,[72] $Ce_{0.9}Sr_{0.1}Cr_{0.5}Fe_{0.5}O_{3\pm\delta}$,[73] $Y_{0.9}Sr_{0.1}Cr_{1-x}Fe_xO_{3-\delta}$[74] and electrolytes, such as modified barium cerate perovskite proton conductor[75] in the SOFCs using pure H_2S as the feed.

10.7 Summary

In this chapter, we summarized our work on the performance of SOFCs running on alternative fuels including $CH_4 + H_2S$, C_2H_6, C_3H_8, $CO + H_2$, $CO + H_2 + H_2S$ and H_2S. For the $CH_4 + H_2S$ fuel, we focused on the LST-based anode materials [$La_{0.75}Sr_{0.25}Cr_{0.5}X_{0.5}O_{3-\delta}$ (X = Co, Fe, Ti and Mn), $La_{0.4}Sr_{0.6-x}Ba_xTiO_{3-\delta}$, $BaTiO_3$, $La_{0.4}Sr_{0.5}Ba_{0.1}TiO_{3-\delta} + YSZ$, and $La_{0.4}Sr_{0.6}TiO_{3-\delta} + La_{0.4}Ce_{0.6}O_{1.8}$]. For the C_2H_6 and C_3H_8 fuels, we developed PC-SOFCs for co-generation of electrical energy and value-added products (*e.g.* C_2H_4 and C_3H_6). Two types of proton conducting electrolytes made of nanopowers (BCY15, BCYN) and the different anode materials (nano-Cr_2O_3, Cu-Cr_2O_3 nanocomposites, nano-$FeCr_2O_4$, Mo_2C) were employed. For the syngas fuel, anode materials ($LaCrO_3$-VO_x-YSZ, CoS-MoS_2, Au-MoS_2, $La_{0.7}Sr_{0.3}VO_3$, $La_{0.4}Sr_{0.6}TiO_{3\pm\delta} + Y_{0.2}Ce_{0.8}O_{2-\delta}$) and anode current collectors (Cu modified Ni foam, Ni-P amorphous alloy coating Ni foam) were developed. For the pure H_2S fuel, the anode materials (M-Mo-S, MoS_2-NiS-Ag, Mo-Ni-S + Ag + YSZ, MoV_2S_4) were developed.

In summary, the development of new materials for anode, cathode, electrolyte, current collector is critical for the SOFCs running on alternative fuels. The structure of the SOFCs and the reaction mechanism occurred in the SOFCs are also crucial in determining the directions of future research. Our future work shall focus on the following research areas: (1) Develop new methods to synthesize different materials with different nanostructures. (2) Establish the relationship between fuel cell performance and materials' microstructure and chemical composition. (3) Reduce the thickness of electrolyte. (4) Improve compatibility between electrolyte and electrodes by optimizing the architectures of anode, electrolyte, cathode and current collector. (5) Explore the reaction mechanism using *in situ* experimental technology, such as *in situ* XPS, *in situ* Raman spectroscopy *etc.*

Acknowledgements

We gratefully acknowledge generous financial support from Alberta Energy Research Institute (AERI) (COURSE program), AERI/ASRIP/WED, AERI/NOVA (COURSE program), Canada Foundation for Innovation, Micro Systems Technology Research Institute (MSTRI), Natural Science and Engineering Research Council of Canada (NSERC) /CVRD INCO Ltd. (CRD program), NSERC/ NGPDG (Greenhouse Gas Mitigation), NSERC/Westaim Ambeon (Strategic projects), NSERC/NOVA Chemicals CRD/Alberta Innovates Energy and Environment Solutions, NSERC (Strategic Projects), NSERC Strategic Research Network (Solid Oxide Fuel Cells Canada), Shell Global Solutions International.

Our achievements in fuel cell research would not be possible without our long-term productive collaborations with Professors K.T. Chuang and A.R. Sanger, University of Alberta. Our sincere appreciation is also devoted to our collaborators for their contributions (alphabetical order): T.H. Etsell (University of Alberta), R. Hu (NRC Institute for Fuel Cell Innovation), A. Krzywicki (NOVA), D. Mitlin (University of Alberta), P. Sarkar (Alberta Innovates e Technology Futures), K. Nandakumar (University of Alberta), Q.M. Yang (CVRD INCO Ltd.)

The advances in materials and processes reported herein were achieved through the development of innovative approaches and fine laboratory work by the following graduate students, post-doctoral fellows and research associates at the University of Alberta (alphabetical order): W. An, N. Danilovic, H. Chen, M. Chen, Y. Feng, X.Z. Fu, A. Garcia, A. R. Hanifi, P. He, J.H. Li, J. X. Li, J.Y. Li, W.S. Li, M. Liu, Q.X. Low, J. Melnik, D. S. Monder, M. Roushanafshar, A. Tsyganok, C. Peng, Z. Shi, S.V. Slavov, A. Vincent, V. Vorontsov, S. Wang, G.L. Wei, Z.R. Xu, N. Yan, G.H. Zhou and L. Zhong.

References

1. K. T. Chuang, J. L. Luo and A. R. Sanger, Evolution of fuel cells powered by H_2S-containing gases, *Chem. Ind. Chem. Eng. Q.*, 2008, **14**, 69–76.
2. N. Danilovic, A. Vincent, J. L. Luo, K. T. Chuang, R. Hui and A. R. Sanger, Correlation of fuel cell anode electrocatalytic and ex situ catalytic activity of perovskites $La_{0.75}Sr_{0.25}Cr_{0.5}X_{0.5}O_{3-\delta}$ (X = Ti, Mn, Fe, Co), *Chem. Mater.*, 2010, **22**, 957–965.
3. A. L. Vincent, J. L. Luo, K. T. Chuang and A. R. Sanger, Effect of Ba doping on performance of LST as anode in solid oxide fuel cells, *J. Power Sources*, 2010, **195**, 769–774.

4. J. H. Li, X. Z. Fu, J. L. Luo, K. T. Chuang and A. R. Sanger, Application of $BaTiO_3$ as anode materials for H_2S-containing CH_4 fueled solid oxide fuel cells, *J. Power Sources*, 2012, **213**, 69–77.
5. A. L. Vincent, A. R. Hanifi, J. L. Luo, K. T. Chuang, A. R. Sanger, T. H. Etsell and P. Sarkar, Porous YSZ impregnated with $La_{0.4}Sr_{0.5}Ba_{0.1}TiO_3$ as a possible composite anode for SOFCs fueled with sour feeds, *J. Power Sources*, 2012, **215**, 301–306.
6. A. L. Vincent, J. L. Luo, K. T. Chuang and A. R. Sanger, Promotion of activation of CH_4 by H_2S in oxidation of sourgas over sulfur tolerant SOFC anode catalysts, *Appl. Catal. B: Env*, 2011, **106**, 114–122.
7. M. Roushanafshar, Dissertation for the doctoral degree in University of Alberta, 2012.
8. M. C. Verbraeken, B. Iwanschitz, A. Mai and J. T. S. Irvine, Evaluation of Ca doped $La_{0.2}Sr_{0.7}TiO_3$ as an alternative material for use in SOFC anodes, *J. Electrochem. Soc.*, 2012, **159**, F757–F762.
9. C. Périllat-Merceroz, G. Gauthier, P. Roussel, M. Huvé, P. Gélin and R. N. Vannier, Synthesis and study of a Ce–doped La/Sr titanate for solid oxide fuel cell anode operating directly on methane, *Chem. Mater.*, 2011, **23**, 1539–1550.
10. J. H. Kim, D. Miller, H. Schlegl, D. McGrouther and J. T. S. Irvine, Investigation of microstructural and electrochemical properties of impregnated (La, Sr)(Ti, Mn)$O_{3 \pm \delta}$ as a potential anode material in high-temperature solid oxide fuel cells, *Chem. Mater.*, 2011, **23**, 3841–3847.
11. G. Tsekouras, D. Neagu and J. T. S. Irvine, Step-change in high temperature steam electrolysis performance of perovskite oxide cathodes with exsolution of B-site dopants, *Energy Environ. Sci.*, 2013, **6**, 256–266.
12. K. B. Yoo, B. H. Park and G. M. Choi, Stability and performance of SOFC with $SrTiO_3$-based anode in CH_4 fuel, *Solid State Ionics*, 2012, **225**, 104–107.
13. D. N. Miller and J. T. S. Irvine, 'B–site doping of lanthanum strontium titanate for solid oxide fuel cell anodes', *J. Power Sources*, 2011, **196**, 7323–7327.
14. X. Sun, S. Wang, Z. Wang, X. Ye, T. Wen and F. Huang, Anode performance of LST-$xCeO_2$ for solid oxide fuel cells, *J. Power Sources*, 2008, **183**, 114–117.
15. M. R. Pillai, I. Kim, D. M. Bierschenk and S. A. Barnett, Fuel-flexible operation of a solid oxide fuel cell with $Sr_{0.8}La_{0.2}TiO_3$ support, *J. Power Sources*, 2008, **185**, 1086–1093.
16. C. D. Savaniu and J. T. S. Irvine, La-doped $SrTiO_3$ as anode material for IT-SOFC, *Solid State Ionics*, 2011, **192**, 491–493.
17. X. H. Zhang, J. Zhang, C. Yuan and E. J. Liang, Synthesis and properties of LST-x%Bi_2O_3 anode materials for solid oxide fuel cells, *Adv. Mater. Res.*, 2012, 347–353.
18. N. Danilovic, J. L. Luo, K. T. Chuang and A. R. Sanger, $Ce_{0.9}Sr_{0.1}VO_x$ (x = 3, 4) as anode materials for H_2S–containing CH_4 fueled solid oxide fuel cells, *J. Power Sources*, 2009, **192**, 247–257.

19. N. Danilovic, J. L. Luo, K. T. Chuang and A. R. Sanger, Effect of substitution with Cr^{3+} and addition of Ni on the physical and electrochemical properties of $Ce_{0.9}Sr_{0.1}VO_3$ as H_2S-active anode for solid oxide fuel cells, *J. Power Sources*, 2009, **194**, 252–262.
20. J. L. Luo, K. T. Chuang and A. R. Sanger, Paraffin Fuel Cell, US Patent No. 8,377,606 B2 (Feb. 19, 2013).
21. J. L. Luo, K. T. Chuang and A. R. Sanger, Paraffin Fuel Cell, US Patent No. 8,039,167 (Oct. 18, 2011).
22. J. L. Luo, K. T. Chuang and A. R. Sanger, Paraffin Fuel Cell, US Patent No. 7,977,006 B2 (July 12, 2011).
23. K. T. Chuang, A. R. Sanger, J. Luo and S. V. Slavov, Electrochemical Process for Oxidation of Alkanes to Alkenes, US Patent No. 7,338,587 B2 (March 4, 2008).
24. H. Iwahara, H. Uchida and S. Tanaka, High temperature-type proton conductive solid oxide fuel cells using various fuels, *J. Appl. Electrochem.*, 1986, **16**, 663–668.
25. C. K. Cheng, J. L. Luo, K. T. Chuang and A. R. Sanger, Propane fuel cells using phosphoric–doped polybenzimidazole membranes, *J. Phys. Chem. B*, 2005, **109**, 13036–13042.
26. Y. Feng, J. L. Luo and K. T. Chuang, Conversion of propane to propylene in a proton-conducting solid oxide fuel cell, *Fuel*, 2007, **86**, 123–128.
27. Y. Feng, J. L. Luo and K. T. Chuang, Carbon deposition during propane dehydrogenation in a fuel cell, *J. Power Sources*, 2007, **167**, 486–490.
28. Y. Feng, J. L. Luo and K. T. Chuang, Propane dehydrogenation in a proton-conducting fuel cell, *J. Phys. Chem. C*, 2008, **112**, 9943–9949.
29. X. Z. Fu, J. Y. Lin, S. Xu, J. L. Luo, K. T. Chuang, A. R. Sanger and A. Krzywicki, CO_2 emission free co–generation of energy and ethylene in hydrocarbon SOFC reactors with a dehydrogenation anode, *Phys. Chem. Chem. Phys.*, 2011, **13**, 19615–19623.
30. S. Wang, J. L. Luo, A. R. Sanger and K. T. Chuang, Performance of ethane/oxygen fuel cells using yttrium-doped barium cerate as electrolyte at intermediate temperatures, *J. Phys. Chem. C*, 2007, **111**, 5069–5074.
31. Z. Shi, J. L. Luo, S. Wang, A. R. Sanger and K. T. Chuang, Protonic membrane for fuel cell for co-generation of power and ethylene, *J. Power Sources*, 2008, **176**, 122–127.
32. X. Z. Fu, J. L. Luo, A. R. Sanger, N. Luo and K. T. Chuang, Y-doped $BaCeO_{3-\delta}$ nanopowers as proton–conducting electrolyte materials for ethane fuel cells to co–generate ethylene and electricity, *J. Power Sources*, 2010, **195**, 2659–2663.
33. X. Z. Fu, J. L. Luo, A. R. Sanger, N. Luo, Z. R. Xu and K. T. Chuang, Fabrication of bi-layered proton conducting membrane for hydrocarbon solid oxide fuel cell reactors, *Electrochem. Acta*, 2010, **55**, 1145–1149.
34. X. Z. Fu, J. L. Luo, A. R. Sanger, N. Danilovic and K. T. Chuang, An integral proton conducting SOFC for simultaneous production of ethylene and power from ethane, *Chem. Commun.*, 2010, **46**, 2052–2054.

35. X. Z. Fu, X. X. Luo, J. L. Luo, K. T. Chuang, A. R. Sanger and A. Krzywicki, Ethane dehydrogention over nano-Cr_2O_3 anode catalyst in proton ceramic fuel cell reactors to co-produce ethylene and electricity, *J. Power Sources*, 2011, **196**, 1036–1041.

36. J. H. Li, X. Z. Fu, G. H. Zhou, J. L. Luo, K. T. Chuang and A. R. Sanger, $FeCr_2O_4$ Nanoparticles as anode catalyst for ethane proton conducting fuel cell reactors to coproduce ethylene and electricity, *Adv. Phys. Chem.*, 2011, 407480–407485.

37. J. H. Li, X. Z. Fu, J. L. Luo, K. T. Chuang and A. R. Sanger, Evaluation of molybdenum carbide as anode catalyst for proton-conducting hydrogen and ethane solid oxide fuel cells, *Electrochem. Commun.*, 2012, **15**, 81–84.

38. C. Peng, J. Melnik, J. L. Luo, A. R. Sanger and K. T. Chuang, $BaZr_{0.8}Y_{0.2}O_{3-\delta}$ electrolyte with and without ZnO sintering aid: preparation and characterization, *Solid State Ionics*, 2010, **181**, 1372–1377.

39. C. Peng, J. Melnik, J. X. Li, J. L. Luo, A. R. Sanger and K. T. Chuang, ZnO-doped $BaZr_{0.85}Y_{0.15}O_{3-\delta}$ proton-conducting electrolytes: Characterization and fabrication of thin films, *J. Power Sources*, 2009, **190**, 447–452.

40. J. X. Li, J. L. Luo, A. R. Sanger and K. T. Chuang, Chemical stability of Y-doped Ba(Ce, Zr)O_3 perovskites in H_2S–containing H_2, *Electrochim. Acta*, 2008, **53**, 3701–3707.

41. W. S. Li, D. S. Lu, J. L. Luo and K. T. Chuang, Chemicals and Energy Co-Generation from Direct Hydrocarbons/Oxygen Proton Exchange Membrane Fuel Cell, *J. Power Sources*, 2005, **145**, 376–382.

42. Z. Xu, Dissertation for the doctoral degree, University of Alberta, 2009.

43. Z. R. Xu, J. L. Luo, K. T. Chuang and A. R. Sanger, $LaCrO_3$-VO_x-YSZ anode catalyst for solid oxide fuel cell using impure hydrogen, *J. Phys. Chem. C*, 2007, **111**, 16679–16685.

44. J. L. Luo, K. T. Chuang, Z. Xu and A. R. Sanger, Anode Catalyst and Methods of Making and Using the Same, US Patent 8,318,384 B2 (Nov. 27, 2012).

45. Z. R. Xu, X. Z. Fu, J. L. Luo and K. T. Chuang, Carbon deposition on vanadium-based anode catalyst for SOFC using syngas as fuel, *J. Electrochem. Soc.*, 2010, **157**(11), B1556–B1560.

46. Z. R. Xu, J. L. Luo and K. T. Chuang, CoS-Promoted MoS_2 catalysts for SOFC using H_2S-containing hydrogen or syngas as fuel, *J. Electrochem. Soc.*, 2007, **154**(6), B523–B527.

47. Z. R. Xu, J. L. Luo and K. T. Chuang, The study of Au/MoS2 anode catalyst for solid oxide fuel cell (SOFC) using H_2S-containing syngas fuel, *J. Power Sources*, 2009, **188**, 458–462.

48. C. Peng, J. L. Luo, A. R. Sanger and K. T. Chuang, Sulfur-tolerant anode catalyst for solid oxide fuel cells operating on H_2S-containing syngas, *Chem. Mater.*, 2010, **22**, 1032–1037.

49. M. Roushanafshar, J. L. Luo, K. T. Chuang and A. R. Sanger, Effect of hydrogen sulfide on electrochemical oxidation of syngas for SOFC applications, *ECS Transactions*, 2011, **35**(1), 2799–2804.

50. M. Roushanafshar, J. L. Luo, A. L. Vincent, K. T. Chuang and A. R. Sanger, Effect of hydrogen sulfide inclusion in syngas feed on the electrocatalytic activity of LST-YDC composite anodes for high temperature SOFC applications, *Int. J. Hydrogen Energy*, 2012, **37**, 7762–7770.

51. S. Patel, P. F. Jawlik, L. Wang, G. S. Jackson and A. Almansoori, Impact of cofiring ceria in Ni/YSZ SOFC anodes for operation with syngas and n-butane, *J. Fuel Cell Sci. Tech.*, 2012, **9**, 041002–041009.

52. C. X. Li, L. L. Yun, Y. Zhang, C. J. Li and L. J. Guo, Microstructure, performance and stability of Ni/Al_2O_3 cermet-supported SOFC operating with coal-based syngas produced using supercritical water, *Int. J. Hydrogen Energy*, 2012, **37**, 13001–13006.

53. A. J. Santis-Alvarez, M. Nabavi, B. Jiang, T. Maeder, P. Muralt and D. Poulikakos, A nanoparticles bed micro-reactor with high syngas yield for moderate temperature micro-scale SOFC power plants, *Chem. Eng. Sci.*, 2012, **84**, 469–478.

54. J. Yu, Y. Wang and S. Weng, Numerical Analysis of the possibility of carbon formation in planar SOFC fueled with syngas, *J. Fuel Cell Sci. Tech.*, 2012, **9**, 021011–021017.

55. F. N. Cayan, S. R. Pakalapati, I. Celik, C. Xu and J. Zondlo, A degradation model for solid oxide fuel cell anodes due to impurities in coal syngas: part I theory and validation, *Fuel cell*, 2012, **12**, 464–473.

56. X. Z. Fu, J. Melnik, Q. X. Low, J. L. Luo, K. T. Chuang, A. R. Sanger and Q. M. Yang, Surface modified Ni foam as current collector for syngas solid oxide fuel cells with perovskite anode catalyst, *Int. J. Hydrogen Energy*, 2010, **35**, 11180–11187.

57. Q. X. Low, W. Huang, X. Z. Fu, J. Melnik, J. L. Luo, K. T. Chuang and A. R. Sanger, Copper coated nickel foam as current collector for H_2S-containing syngas solid oxide fuel cells, *Appl. Surf. Sci.*, 2011, **258**, 1014–1020.

58. N. Yan, X. Z. Fu, J. L. Luo, K. T. Chuang and A. R. Sanger, Ni-P coated Ni foam as coking resistant current collector for solid oxide fuel cells fed by syngas, *J. Power Sources*, 2012, **198**, 164–169.

59. M. Liu, P. He, J. L. Luo, A. R. Sanger and K. T. Chuang, Performance of a solid oxide fuel cell utilizing hydrogen sulfide as fuel, *J. Power Sources*, 2001, **94**, 20–25.

60. P. He, M. Liu, J. L. Luo, A. R. Sanger and K. T. Chuang, Stabilization of Platinum anode catalyst in a H_2S-O_2 solid oxide fuel cell with an intermediate TiO_2 layer, *J. Electrochem. Soc.*, 2002, **149**(7), A808–A814.

61. M. Liu, G. L. Wei, J. L. Luo, A. R. Sanger and K. T. Chuang, Use of metal sulfides as anode catalysts in H_2S-Air SOFCs, *J. Electrochem. Soc.*, 2003, **150**(8), A1025–A1029.

62. G. L. Wei, M. Liu, J. L. Luo, A. R. Sanger and K. T. Chuang, Influence of gas flow rate on performance of H_2S/air solid oxide fuel cells with MoS_2-NiS-Ag anode, *J. Electrochem. Soc.*, 2003, **150**(4), A463–A469.

63. K. T. Chuang, J. L. Luo, G. Wei and A. R. Sanger, Electrode Catalyst for H_2S Fuel Cell, P. C.T. Patent Application WO 03/096452 (2002), *US Patent No.*, 7014, 941, B2 (March 21, 2006).

64. G. L. Wei, J. L. Luo, A. R. Sanger and K. T. Chuang, High-performance anode for H_2S-air SOFCs, *J. Electrochem. Soc.*, 2004, **151**(2), A232–A237.

65. V. Vorontsov, W. An, J. L. Luo, A. R. Sanger and K. T. Chuang, Performance and stability of composite nickel and molybdenum sulfide–based anodes for SOFC utilizing H_2S, *J. Power Sources*, 2008, **179**, 9–16.

66. V. Vorontsov, J. L. Luo, A. R. Sanger and K. T. Chuang, Synthesis and characterization of new ternary transition metal sulfide anodes for H_2S–powered solid oxide fuel cell, *J. Power Sources*, 2008, **183**, 76–83.

67. J. X. Li, J. L. Luo, K. T. Chuang and A. R. Sanger, Proton conductivity and chemical stability of Li_2SO_4 based electrolyte in a H_2S-air fuel cell, *J. Power Sources*, 2006, **160**, 909–914.

68. G. L. Wei, J. L. Luo, A. R. Sanger, K. T. Chuang and L. Zhong, Li_2SO_4–based proton–conducting membrane for H_2S–air fuel cell, *J. Power Sources*, 2005, **145**, 1–9.

69. G. L. Wei, J. Melnik, J. L. Luo, A. R. Sanger and K. T. Chuang, Cathodes for fuel cells using proton–conducting Li_2SO_4-Al_2O_3 electrolyte, *J. Electroanal. Chem.*, 2005, **575**, 183–193.

70. J. Melnik, W. An, G. L. Wei, J. L. Luo, A. R. Sanger and K. T. Chuang, Use of nano–composite component to improve the conductivity and mechanical properties of Li_2SO_4–Al_2O_3 solid electrolyte, *Mater. Res. Bull.*, 2006, **41**, 1806–1816.

71. D. S. Monder, V. Vorontsov, J. L. Luo, K. T. Chuang and K. Nandakumar, An investigation of fuel composition and flow-rate effects in a H_2S fuelled SOFC: experiments and thermodynamic analysis, *Can. J. Chem. Eng.*, 2011, **90**, 1033–1042.

72. D. Xu, X. Zhu, Y. Bu, H Yan, W. Tan and Q. Zhong, Synthesis and performance of $Sm_{0.9}Sr_{0.1}Cr_{0.5}Fe_{0.5}O_3$ as anode material for SOFCs running on H_2S-containing fuel, *Ionics*, 2013, **19**, 491–497.

73. X. Zhu, Q. Zhong, D. Xu, H. Yan and W. Tan, Further investigation of $Ce_{0.9}Sr_{0.1}Cr_{0.5}Fe_{0.5}O_{3\pm\delta}$ as anode for solid oxide fuel cell fuelled with H_2S, *J. Alloy Compd.*, 2013, **555**, 169–175.

74. D. Xu, Y. Bu, W. Tan and Q. Zhong, Structure and redox properties of perovskite $Y_{0.9}Sr_{0.1}Cr_{1-x}Fe_xO_{3-\delta}$, *Appl. Surf. Sci.*, 2013, **268**, 246–251.

75. W. Tan, M. Miao and H. Qu, H_2S Solid oxide fuel cell based on a modified barium cerate perovskite proton conductor, *Ionics*, 2009, **15**, 385–388.

CHAPTER 11

Long Term Operating Stability

HARUO KISHIMOTO,* TERUHISA HORITA AND
HARUMI YOKOKAWA

National Institute of Advanced Industrial Science and Technology (AIST),
Energy Technology Research Institute, Higashi 1-1-1, AIST Central No.5,
Tsukuba, Ibaraki 305-8565, Japan
*Email: haruo-kishimoto@aist.go.jp

11.1 Introduction

The most important characteristic features of solid oxide fuel cells (SOFCs) are
conversion efficiency, life and cost. Those features are commonly used for other
energy converters such as gas engine, gas turbines or polymer electrolyte fuel
cells (PEFCs) to be utilized in similar situations. When compared with other
energy convertors, it is highly expected for SOFCs to achieve higher energy
conversion rate from fuels to electricity, while technologies associated with
durability and cost of SOFCs have been less matured because of a lack in
accumulating knowledge on operation of SOFC systems. This indicates a key
issue in SOFC technologies that it is highly needed to achieve simultaneously
cost reduction as well as long life.[1,2] In recent years, the construction and
operations of SOFC systems have been significantly progressed in the field of
stationary applications such as the residential micro CHP systems[3,4] and the
Distributed Generators.[5] This implies that the requirement of achieving low
cost and long life becomes urgent to promote further marketing of such SOFC
systems and to facilitate the wider penetration of SOFCs in various energy
conversion fields.

Degradation phenomena are closely related with the fabrication technology
of SOFC stacks, while fabrication of SOFC stacks in turn strongly depends on

RSC Energy and Environment Series No. 7
Solid Oxide Fuel Cells: From Materials to System Modeling
Edited by Meng Ni and Tim S. Zhao
© The Royal Society of Chemistry 2013
Published by the Royal Society of Chemistry, www.rsc.org

various strategies on how to construct stacks. This is because materials compatibilities are essential in construction of SOFC stacks; that is, chemical compatibility among various cell components becomes crucial in selecting materials, materials processing and stack designs.[6–8] In addition, mechanical instability becomes critical when the volume changes take place in assembly of cell components. To achieve excellent mechanical compatibility, matching in thermal expansion is crucial, while the volume change caused by the valence change of the transition metal ions or cerium ions in cell components should be also carefully taken into account.

This situation of fabrication technology suggests that the investigation of degradation of SOFC systems should be carefully examined by using SOFC stacks instead of small button cells and analyses should be also made by taking into account the differences in materials selected, adopted materials processing, stack design and fabrication strategies.

Recent progress in the residential micro CHP systems in Japan made it necessary to establish the high durability/reliability of such SOFC stacks used in the micro CHP systems. Since tight cooperation between stack developers and research groups in research institutes and universities is highly requested, the NEDO project on the durability/reliability started and enlarged to establish high durability in the framework of such cooperation.[9–12]

In this chapter, the general features of degradations of SOFC stacks/systems are first given on the basis of the recent results of the NEDO project on durability/reliability. More detailed descriptions will be given for the deterioration of electrolytes and also for the performance degradation of cathodes and anodes. And finally, the remaining issues will be given for achieving long durability within the shortest period of examination time.

11.2 Durability of Stacks/Systems

11.2.1 Determination of Stack Performance

The stack performance was measured by changing gas compositions in cathode or anode together with measurement of the electrical potential at an OCV condition. This makes it possible to separate the contributions from cathode and anode. Current interruption was also adopted to separate the ohmic losses. When this set of measurements were made in every given period of time, increases in cathode or anode overpotential and ohmic losses can be given. After a long-term operation, any parts of stacks were examined in details to detect any changes in composition or microstructure of electrode or any chemical changes particularly at interfaces of cell components. Typical results are given in Table 11.1.[12]

11.2.2 Performance Degradation and Materials Deteriorations

The most important finding in the NEDO durability/reliability project is that degraded parts of one particular stack are closely related with the fabrication sequence of the stack.

Table 11.1 Results of determination of increases in cathode/anode overpotential and ohmic losses in several different stacks after some improvements.

(a) Materials of respective stacks

Stack makers	Gen	Materials			IC	Connect	Fabrication sequence
		Cathode	Electrolyte	Anode			
Flatten tubes (Kyocera)	2nd	LSF-base	YSZ	Ni/YSZ	LC-base	Ferritic ss	Anode support
Segment-in-series (Tokyo gas)	2nd	LSCF	YSZ	Ni/YSZ		LC-base	Substrate/anode
Sealless tubular (TOTO)	1st	LSM	SSZ	Ni/YSZ		LCC	Cathode support
Micro tubular (TOTO)	2nd	LSCF	LSGM	Ni/CeO$_2$	ferritic		Anode support
Disk-type (MMC & KEPCO)	2nd	SSC	LSGMC	Ni/CeO$_2$	ferritic		Electrolyte support
Segment-in-series (MHI)	1st	LSCM	YSZ	Ni/YSZ	LST		Substrate

(b) Performance degradation data for respective stacks

Stack makers	Operation Condition	Electrical potential lowering rate (%/1000 h)				Lowering after Thermal cycle	Remarks
		Cathode	Anode	Ohmic	Total		
Flatten tube (Kyocera)	750 °C	0–11kh	0.03	0.05	0.22	0.30	0.42%/120C	Longer life
Flatten tube 2 (Tokyo gas)	775 °C	0–4kh	0.17	0.13	0.35	0.64	0.8%/100C (after 2000 h operation)	Initial degradation
		1–4kh	0.09	0.07	0.19	0.35	High durability after 120 SD	
Micro tube (TOTO)	650 °C	2010 type				0.6		
		2011 type				0.3		
Segment-in-series (MHI)	900 °C	Type 1 (Cr)	0.6	−0.07	0.36	0.89	No degradation after 50 TC + 200 Load C	SDC Cr
		Type 2	0.84	−0.15	0.07	0.76		
		Type 3	−0.02	−0.03	−0.04	−0.09		
		Type 5 (Cr)	0.10	−0.02	0.05	0.13		

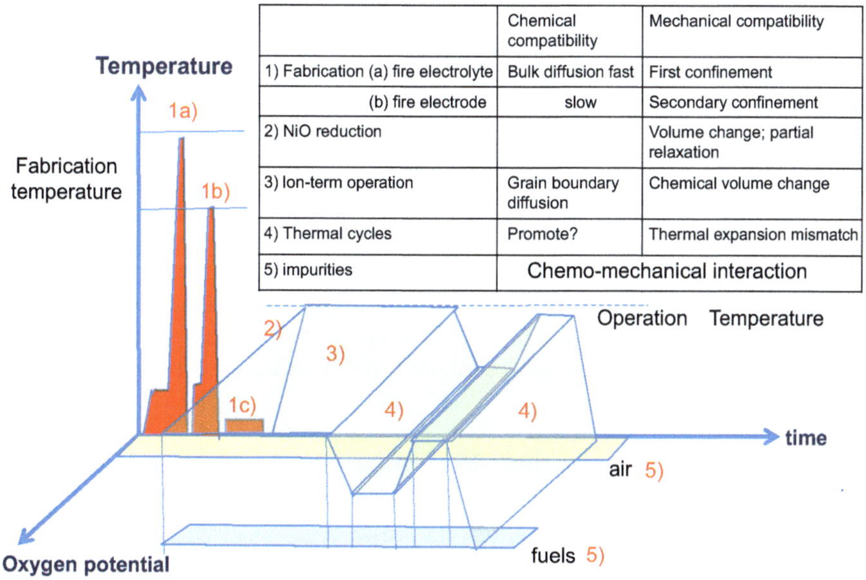

	Chemical compatibility	Mechanical compatibility
1) Fabrication (a) fire electrolyte	Bulk diffusion fast	First confinement
(b) fire electrode	slow	Secondary confinement
2) NiO reduction		Volume change; partial relaxation
3) Ion-term operation	Grain boundary diffusion	Chemical volume change
4) Thermal cycles	Promote?	Thermal expansion mismatch
5) impurities	Chemo-mechanical interaction	

Figure 11.1 Fabrication/operation procedures and corresponding changes in temperature and oxygen potential as a function of time.

Figure 11.1 shows a typical fabrication procedure of SOFC stacks consisting of at least two stages; the first one is the densification of electrolyte plate to be fired at a highest temperature, the second being the subsequent step of preparing a remaining electrode at a lower temperature. After assembling cells/stacks, nickel oxide is reduced to nickel metal and at the same time, the oxygen potential distribution is developed inside electrolytes and interconnects. During an operation period of time, thermal cycles are made depending on requirements in respective applications.

These conditions during fabrication and operations are important in considering the chemical compatibility among cell components and the mechanical compatibility in the assembled stacks. Actually, what was found in the NEDO durability project is that degraded parts are those which were fabricated in the second fabrication step in Figure 11.1. For example, the segment-in-series cell made by Mitsubishi Heavy Industries is fabricated in a sequence of substrate, anode, electrolyte and cathode so that cathode is lastly fired at a lower temperature; for this cell, significant changes in cathode microstructure were observed during operation for 10 000 h.[13] The flatten tubular cell made by Kyocera is anode-support so that cathode is the secondary fabricated electrode; this $LaFeO_3$-based cathode showed some degradation due to the $SrZrO_3$ formation at the interfaces of doped-ceria interlayer and YSZ electrolyte.[11,12] The sealless tubular cell made by TOTO is cathode-support and therefore the anode is the secondary fabricated electrode; in special cases (not always), this cell exhibited serious changes in microstructure of anodes. In the electrolyte-supporting disk-type cell made by Mitsubishi Materials Corp, nickel sintering was frequently observed even though the operational temperature was low (650–700 °C) compared with other cells.

Another interesting trend found during the NEDO project is that after knowing where the responsible part is for performance degradation in respective stacks, it was fast to improve those parts of cell components and to achieve higher durability. Particularly, the secondary fabricated electrode must be improved by changing the fabrication conditions in this second stage without modifying the main part of fabrication procedure. On the other hand, when something wrong was found in electrolyte, it was hard to overcome this issue within the already established fabrication procedure. One such electrolyte is the Mn dissolved YSZ which is utilized in the sealless cathode-support tubular cells.[14–16] This will be described in detail in the next section.

11.2.3 Impurities and their Poisoning Effects on Electrode Reactivity

Impurities are important in understanding the performance degradation of real stacks. Some comes from other cell components than electrodes themselves, others being from the system components or fuels/air. Typical values of impurities observed by the secondary ion mass spectrometry (SIMS) on selected parts of cell components are shown in Figure 11.2 on anodes of four different

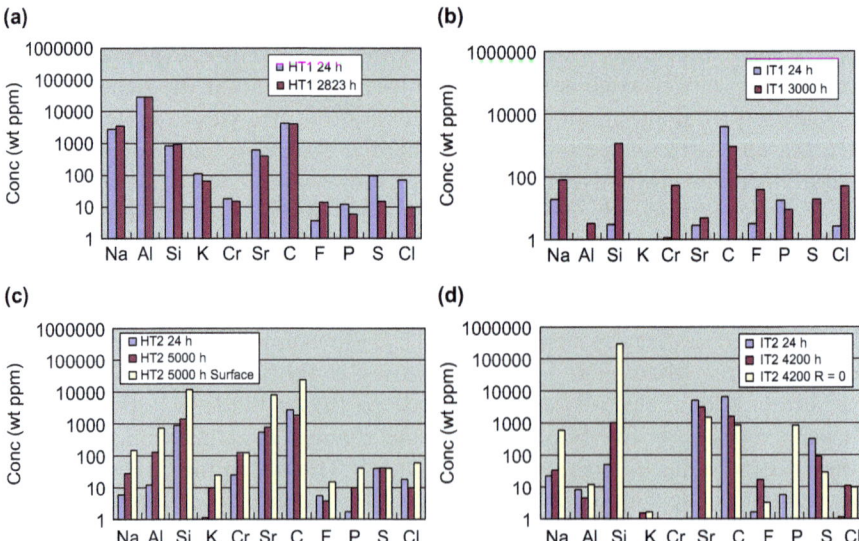

Figure 11.2 Results of SIMS Analyses on anode impurities at 24 h operation (left bar) and after 3000–5000 h operation (right bar): (a) HT segment-in-series cells by MHI, (b) IT flatten tubular cells by Kyocera, (c) HT sealless tubular cells by TOTO, (d) IT Disk-type cells by Mitsubishi Materials.

stacks. Measurements were made first on anode materials just after a 24 h operation and then after a longer operation time, typically 3000–5000 h; results are compared in Figure 11.2. For example, impurities in MHI cells are relatively large but do not change in amount even after the 5000 h operation, indicating that impurities are not entered or accumulated in the anode layer during operation. For Kyocera cells, impurity level is quite small in general but sulfur contamination was detected after the long-term operation even though this test was made using pure hydrogen from cylinder, suggesting that there must be sulfur sources between gas cylinder and the anode compartment. Thus, sulfur poisoning on nickel anode can become the major degradation factor. Since those anodes having higher sulfur contamination exhibit more significant sintering after a long-term operation or frequent thermal cycles. When phosphor contamination takes place, significant changes in microstructure are developed and as a result, performance will be degraded seriously. Although the Ni-P intermetallic compound can be formed at the entrance of phosphor containing gases and the major part of phosphor will be trapped in this region, a small amount of phosphor can be transferred from such deposited area to the electrochemically active sites. Note that the Ni-S and the Ni-P systems exhibit the eutectic lowering of melting temperature, implying that the sulfur or phosphor contamination on nickel surfaces leads to increases in surface energies and surface diffusion of nickel. This will be discussed later.

Silicon is one of the major contaminants in anode environments. Even so, there is no clear correlation between the amount of silicon contamination and the performance degradation or changes in microstructure. This is because silicon tends to be located in two distinct regions; that is, the surface region of anode layer and interface region with electrolyte. The latter should be responsible for performance degradation.

One of the most significant poisoning effects on cathode performance is the chromium poisoning on the lanthanum strontium manganite (LSM) cathode.[17] In the NEDO project, the MHI utilizes the lanthanum strontium calcium manganite (LSCM) cathode and the cathode degradation was found to be the major source of performance degradation so that the chromium concentration in the vicinity of interface between the electrolyte and the cathode was carefully observed by the secondary ion mass spectrometry (SIMS). This SIMS analysis reveals that under the experimental condition for the durability test, some amount of chromium-containing gases were emitted from oxide scales of metal tubes for introducing air to the test cell and such transported chromium was partly accumulated in those electrochemically active area, as shown in Figure 11.3. The accumulated amount of Cr at the interface is about 1000 ppm (0.1 %) after 10 000 h and this value is quite small compared with normal detection limit of EDX or other analyses on composition than SIMS. Thus, SIMS is quite powerful to identify whether accumulation of impurities has been made at the interfaces which should be compared with the performance degradation in the range of 0.1–0.25%/1000 h. Actually, the segment-in-series cells by MHI have been well improved in its microstructure and chemical composition so that only Cr in an order of 10 ppm has been found at interfaces,

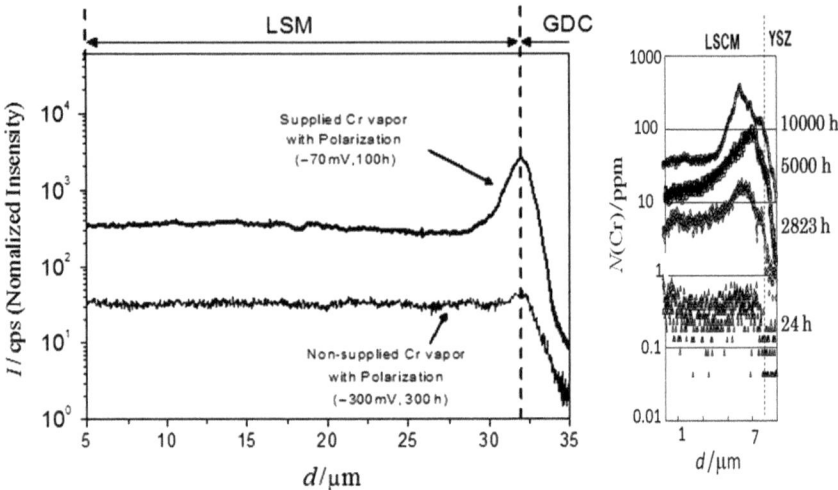

Figure 11.3   Comparison of Cr deposition profile for button cells tested in an accelerated manner (left) and for segment-in-series cells by MHI after 24, 2823, 5000, 10 000 h (right).

indicating that the improved cathode is good enough to prevent the Cr poisoning even when stainless steel will be used for introducing tubes of air.[17]

In addition, SO_2 contamination on cathodes was observed during the NEDO project. Thus, some experimental investigations have been made to clarify those effects of electrode performance. These will be described in the next sections.

11.3   Deteriorations of Electrolytes

The soundness of the electrolyte is one of the most important issues to establish not only the durability/reliability of SOFC systems but also to optimize the SOFC fabrication process. The most conventional and reliable electrolyte material is yttria stabilized zirconia (YSZ) with cubic symmetry. This is because this material is an essentially purely oxide-ion conductor with high conductivity at high temperatures and over a wide oxygen potential range and also because it has the high mechanical strength and the high chemical stability. On the other hand, the cubic YSZ is a "metastable" phase at the SOFC operating temperature, typically 800–900 °C, according to reported phase diagrams (Figure 11.4).[18] In fact, the conductivity of 8 mol% Y_2O_3 doped ZrO_2 (8YSZ), which shows the highest conductivity among YSZ electrolytes, gradually decreases with increasing the annealing time around 1000 °C, and this conductivity lowering is accompanied with the phase transformation from the cubic to the tetragonal structure (Figure 11.5).[19–22]

This phase transformation of YSZ electrolyte may lead to decrease in efficiency of SOFC systems due to conductivity lowering, and also may cause some degradation due to changes in the mechanical properties and/or the thermal expansion behaviour.

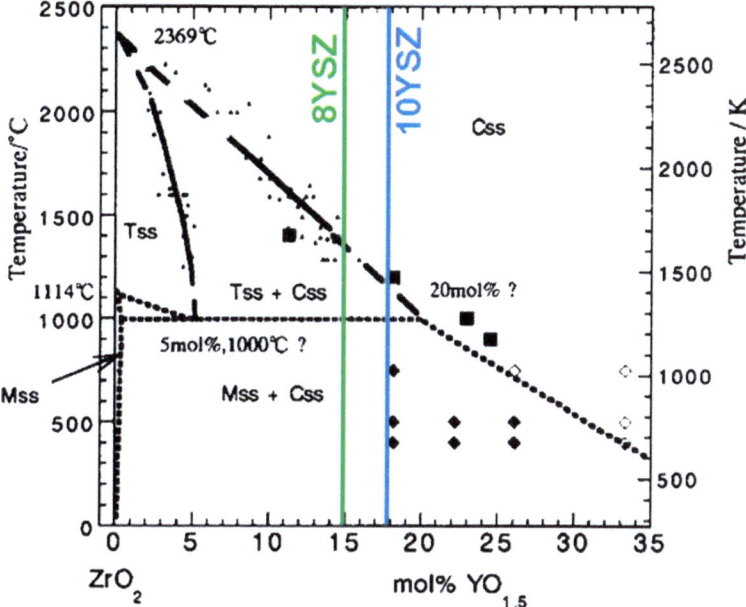

Figure 11.4 Equilibrium phase diagram for ZrO_2-$YO_{1.5}$ system.

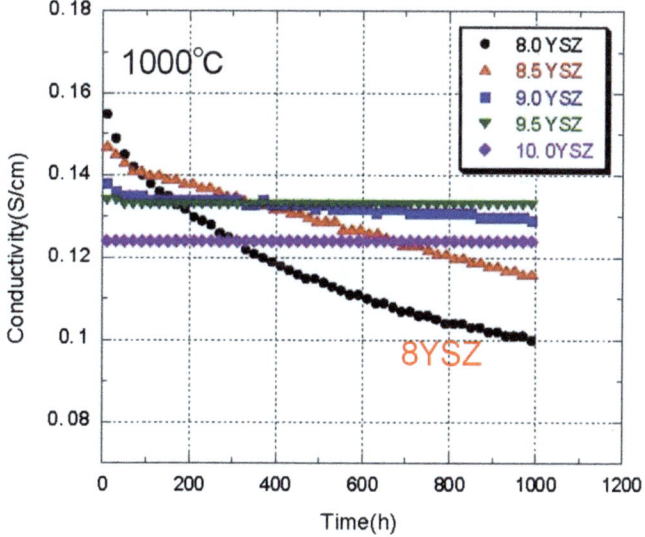

Figure 11.5 Conductivity degradation behavior of YSZ electrolytes.

A large oxygen potential distribution is extensively developed inside the electrolyte under SOFC operating conditions; that is about 15 to 20 orders of magnitude difference in oxygen partial pressure between two sides of the electrolyte. Figure 11.6 shows such an oxygen potential distribution in the

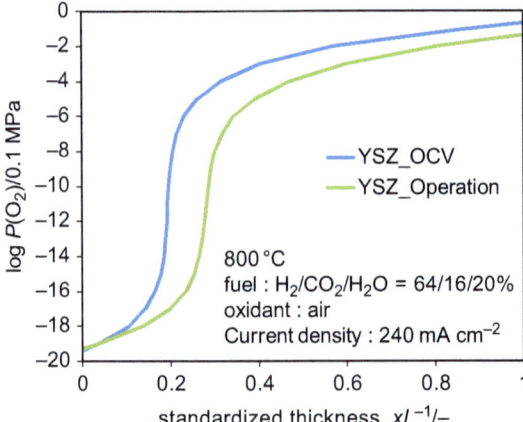

Figure 11.6 Calculated oxygen potential distribution in the 8YSZ electrolyte at 800 °C. The overpotential values of cathode and anode are assumed as 34 mV and 5 mV, respectively, under operation.

8YSZ electrolyte at 900 °C which was calculated from the oxide-ion and the electronic conductivities.[23] The oxygen potential changes significantly and steeply in a very narrow region inside the 8YSZ electrolyte. This potential distribution depends also largely on the SOFC operating conditions including fuel utilization and the overpotential for both electrodes; those conditions determine the limiting values of oxygen potential at the sides of electrolyte.

As mentioned above, electrolyte materials are co-sintered with the "base" material (cathode material for cathode-support type cells and NiO-oxide composite material for anode-supported cells) at the highest temperature to obtain the dense electrolyte films. Therefore, elemental interdiffusion between the electrolyte and the "base" electrode is unavoidable in a sintering process. For example, the Mn component in lanthanum strontium manganite (LSM) and the Ni component in precursor of Ni-YSZ anode (namely, NiO-YSZ mixture) are dissolved into the YSZ electrolyte of the cathode-support cells with the LSM cathode and the anode-support cells, respectively.[24,25] These transition metals can have different valence states according to the oxygen potential and temperature conditions. In this section, effects of dissolved transition metals into the YSZ electrolyte on durability and reliability of the YSZ electrolyte are discussed.

11.3.1 Destabilization of Mn Dissolved YSZ

For those cells with the LSM cathode, the Mn component, which occupies the B sites of the perovskite-type lattice, dissolves into the YSZ electrolyte in a high-temperature fabrication process. Figure 11.7 shows the lattice parameter of the cubic phase of Mn-8YSZ system as a function of Mn content, leading to the solubility limit at selected temperatures.[24] Thus determined solubility limit

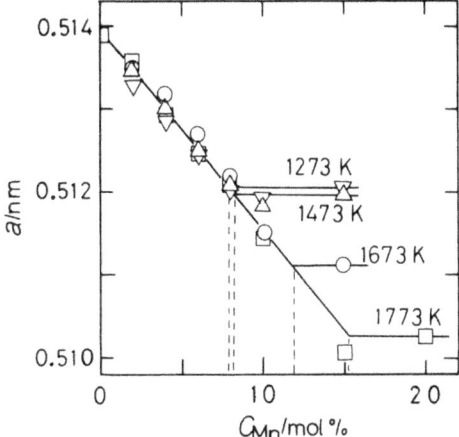

Figure 11.7 Lattice parameter of the cubic phase of Mn-YSZ (8 mol % Y_2O_3). C_{Mn} denotes the atomic concentration of Mn, $C_{Mn} = 100$ [Mn]/([Mn] + [Zr] + [Y]).

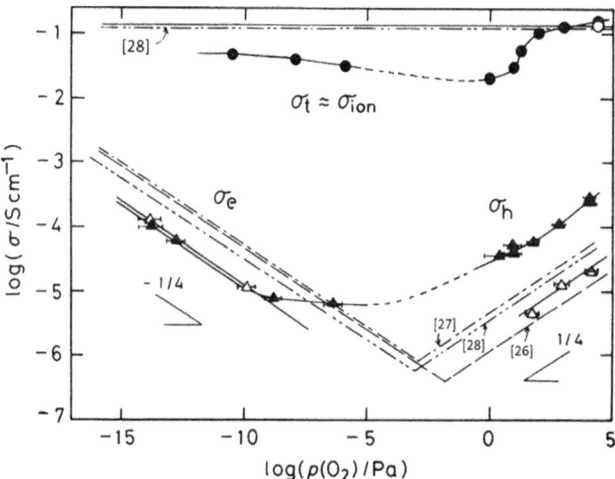

Figure 11.8 Total and electronic conductivity at 1273 K for 8YSZ and Mn (4 mol %) - 8YSZ.[24] The reference data are the electronic conductivity of CaO-ZrO_2,[26] 10 mol % Y_2O_3-ZrO_2,[27] and 8 mol % Y_2O_3-ZrO_2[28] and the ionic conductivity of 8YSZ.[28]

of Mn into 8YSZ is as high as about 12 mol % and 8 mol % at 1400 °C and 1200 °C, respectively. This 8YSZ can be regarded as essentially "*purely*" oxide-ion conductor even though Mn is dissolved into the YSZ lattice. Even so, it should be recognized that the ionic conductivity and electronic conductivity change depending on the oxygen potential (Figure 11.8).[24] These oxygen potential dependencies are related to the valence state of the Mn ions in the YSZ lattice. The concentration of oxygen vacancy and electron or electronic hole changes with changing the valence state of the Mn ions in YSZ as expressed in the following equation:

$$2Mn'_{zr} + O_O^x \leftrightarrow 2Mn''_{zr} + V_O^{\bullet\bullet} + 1/2O_2 \tag{1}$$

At 800 °C, the boundaries of MnO $(Mn^{2+})/Mn_3O_4$ (Mn^{2+}/Mn^{3+}) and Mn_3O_4 $(Mn^{2+}/Mn^{3+})/Mn_2O_3$ (Mn^{3+}) are at about $P(O_2) = 10^{-5}$ Pa and 10^4 Pa, respectively. These boundaries correspond well to the vending behaviour of the total conductivity of Mn-8YSZ electrolyte as shown in Figure 11.9.[29] This indicates that the manganese valence in YSZ gradually changes with decreasing oxygen potential as the manganese ions are reduced from trivalent to the divalent ions, in other words, as the concentration of oxide ion vacancies increases. This trend is the same as that observed from 8YSZ to 10YSZ. Solubility of Mn into YSZ also depends on the valence state of the dissolved Mn. Figure 11.10 shows the calculated solubility of some transition metals into YSZ by thermodynamic consideration.[30] It is suggested that the Mn solubility decreases with increasing the Mn valence state. This should be related to not only the valence state but also the ionic radius. This means that the Mn solubility increases with temperature due to the two effects, one is the mixing entropy term, the other being the reduction effect of manganese ions. With lowering oxygen potential, the Mn solubility increases only due to the reduction effect. Those phase diagram behaviours suggest that the Mn component dissolved during the high-temperature fabrication process can be expectedly precipitated in the air side but not in the fuel side during the lower-temperature operation.

There are some examples of the influence of dissolved Mn on electrolyte durability and reliability. One is the disintegration phenomenon of ScSZ electrolyte observed for the cathode-support tubular cell (Figure 11.11(a)).[14-16] This phenomenon was observed on the anode side surface near the anode

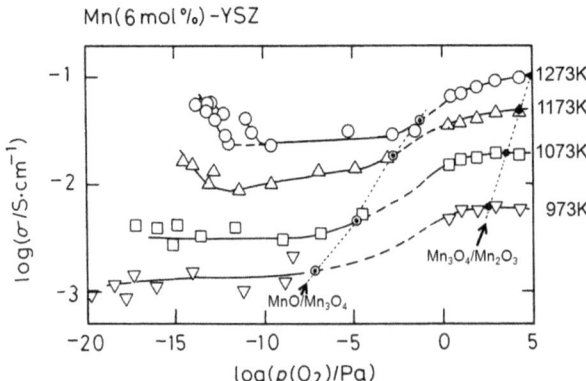

Figure 11.9 Oxygen partial pressure dependence of electrical conductivity of 6 mol % Mn-8YSZ at 1273 K, 1173 K, 1073 K and 973 K.[29] Oxygen potentials of phase transition from Mn_2O_3 to Mn_3O_4 and from Mn_3O_4 to MnO are marked on each isothermal line of the conductivity.

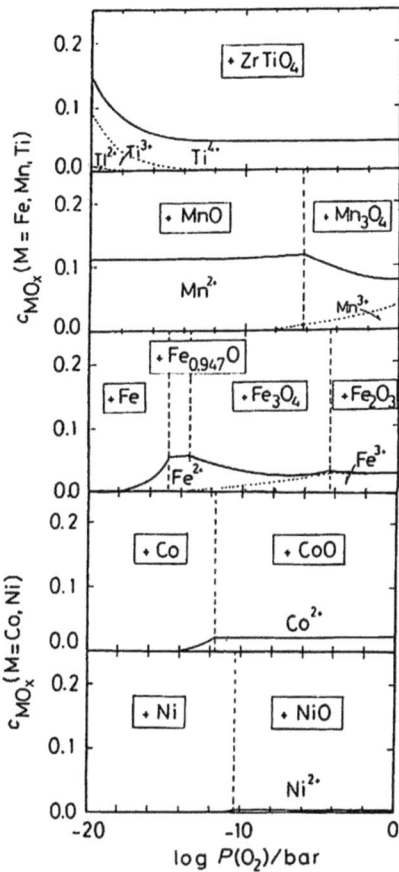

Figure 11.10 Thermodynamic calculation of solubility of transition metals into YSZ.

(zone-b in Figure 11.11(b)). In this cell, LSM is used as cathode and scandia stabilized zirconia (ScSZ) electrolyte is co-sintered with the cathode; during this process, Mn dissolves into the electrolyte.[14] Figure 11.12 shows cross-section images of the electrolyte surface especially near the edge of the anode.[15] Disintegration of the electrolyte starts from the anode side surface and the thickness of the dense (robust) electrolyte decreases with increasing operating time. When this phenomenon proceeds seriously, gas leakage and/or breaking the cell will occur. At the electrolyte surface with decomposed (pulverized) electrolyte, precipitation of Mn_3O_4 and/or related compounds are also detected (Figure 11.13).[15]

The other example is crack formation in planner cells, reported by Forschungszentrum Jülich GmbH.[31] When LSM is used as cathode in this anode support cell, the Mn concentration is initially high especially along grain boundaries in the YSZ electrolyte.[31] After a long-term operation, precipitation of the Mn component at the grain boundaries and relating micro crack formation is observed in the middle of the electrolyte (Figure 11.14).[31]

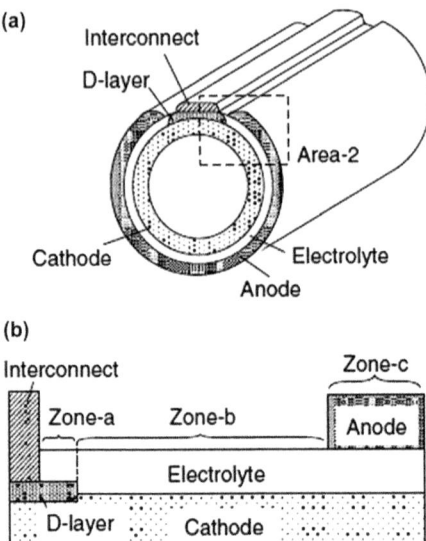

Figure 11.11   Schematic drawings of the cathode support tubular type SOFC cell. (a) Cross section image and (b) enlarged cross section image of Area 2 in (a).

There are some common points between above two examples; the initial Mn dissolution into electrolyte during fabrication and the subsequent Mn precipitation at the respectively specified areas (the electrolyte decomposition area or the crack formation parts) during long operation. In addition, the oxygen potential distribution around Mn_3O_4 precipitation area is similar between these two phenomena. Figure 11.15 shows the estimated oxygen potential distribution for the cathode support tubular cell;[32] oxygen potential at the anode side surface of the electrolyte is estimated to steeply increase near the edge of the anode even though there is exposed to the fuel atmosphere. Note that the high oxygen potential region is in accord with the decomposed part. On the other hand, for the planner anode-support cell, the oxygen potential steeply changes in the middle of the electrolyte underneath the electrode area as shown in Figure 11.15; the area corresponds to the Mn precipitation and the crack formation area in the Julich cells as well. As described above, the phase diagram behaviour predicts that the Mn precipitation should take place in the more oxidative side. But the place where the Mn precipitation takes place is in the middle of the electrolyte for the anode-support cell and in the fuel side in the cathode-support cell. Thus, the steep oxygen potential gradient should be related to these degradation behaviours induced by the Mn dissolution. Furthermore, in the cathode-support cell, catalytic effects of Mn oxide, iron oxide and silicon oxides are identified.[15] At such a large oxygen potential gradient part, oxide ions, electrons or holes diffuse steadily along their own slope of their electrochemical potentials, while other charge carriers such as cations feel different electrochemical potential depending on their valence

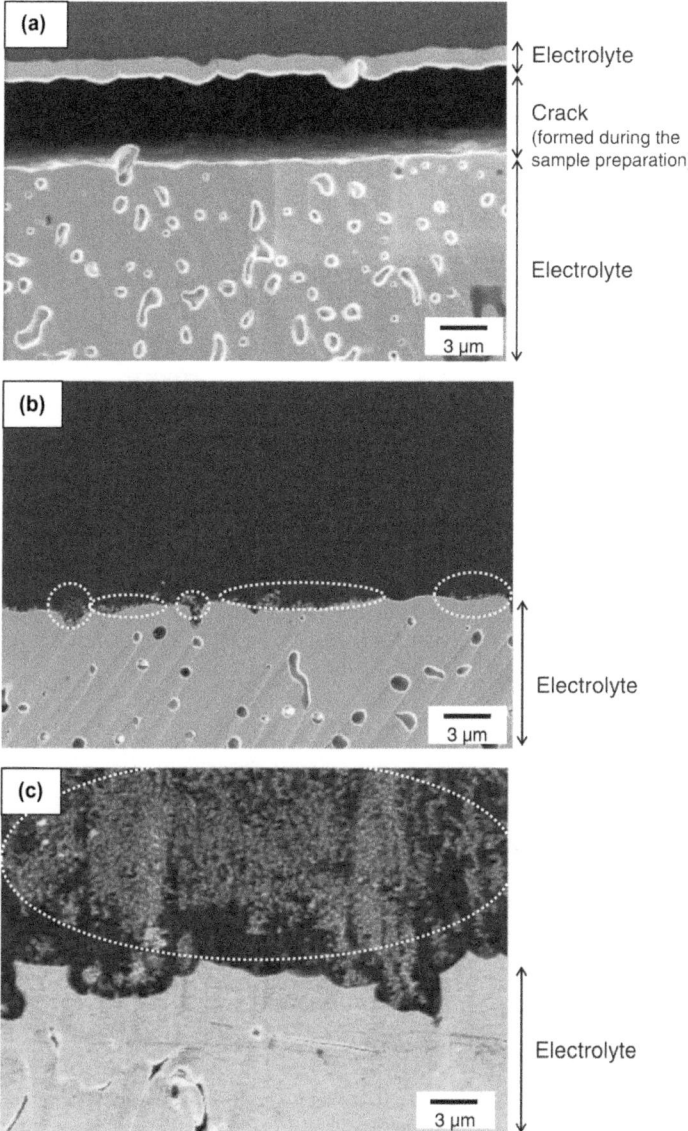

Figure 11.12 Cross section images of the electrolyte surface near the anode[15]: (a) before operation, (b) after 100 h operation and (c) after 11 000 h operation.

number. Thus, the flux of manganese (divalent or trivalent ions) may be strongly influenced by the steep oxygen potential or the electrochemical potential of electrons, leading to strong kinetic driving forces for precipitation of manganese related phase or pore formation in a similar manner to detailed discussions by Maruyama *et al.* on the pore formation inside the Fe_3O_4 based

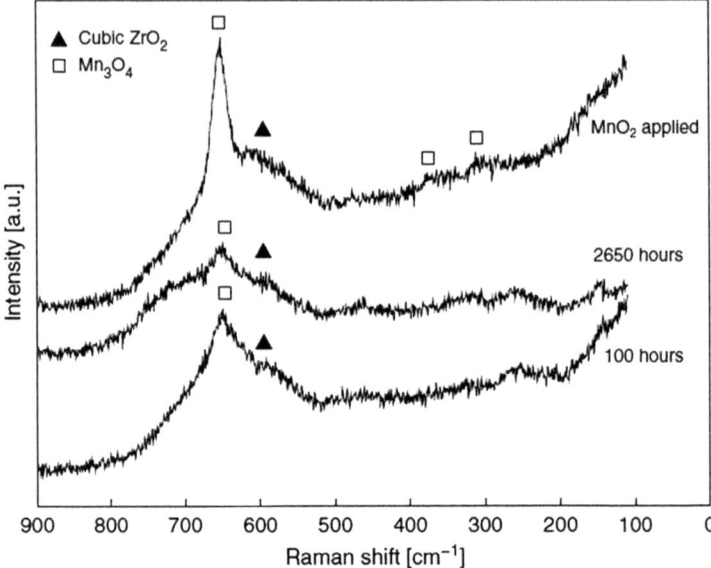

Figure 11.13   Raman spectra of the electrolyte surface after fine powders were wiped off (a) after 100 h operation, (b) after 2650 h operation and (c) the MnO₂ applied cell after 270 h operation.

oxide scale.[33] It is hoped to make a detailed analysis on this phenomena from the kinetic point of view.

11.3.2   Conductivity Decrease in Ni-dissolved YSZ

As mentioned in the introduction, the conductivity of 8YSZ gradually decreases with increasing the annealing time at a temperature of 1000 °C and simultaneously the phase transformation into the tetragonal symmetry occurs. Interestingly, however, such phase transformation phenomena are only limitedly observed in those practical SOFC cells which were used for field tests. Figure 11.16 shows the Raman spectra for the cross section of the several practical stacks with zirconia electrolyte after a long-term durability test.[34] For the anode-support cell and segment-in-series cell whose electrolyte is sintered with the anode, phase transformation is observed only for the zirconia component in the cermet anode and at the anode/electrolyte interface even after the tests for 4000 ~ 10000 h. (The cathode supported tubular cell shows the different phase transformation behaviour. It should be related to the dissolution of Mn into zirconia electrolyte.)

Nickel oxide in the anode precursor can dissolve into electrolyte in co-sintering process for anode-support cell.[25] The solubility limit of NiO into YSZ electrolyte is less than 1 mol%.[25] The conductivity of NiO dissolved YSZ is nearly the same or slightly less than that of YSZ.[35] However, the conductivity of NiO dissolved 8YSZ rapidly decreases when the electrolyte is in a reducing atmosphere as shown in Figure 11.17.[35–37] and becomes about 1/2 to1/3 of the

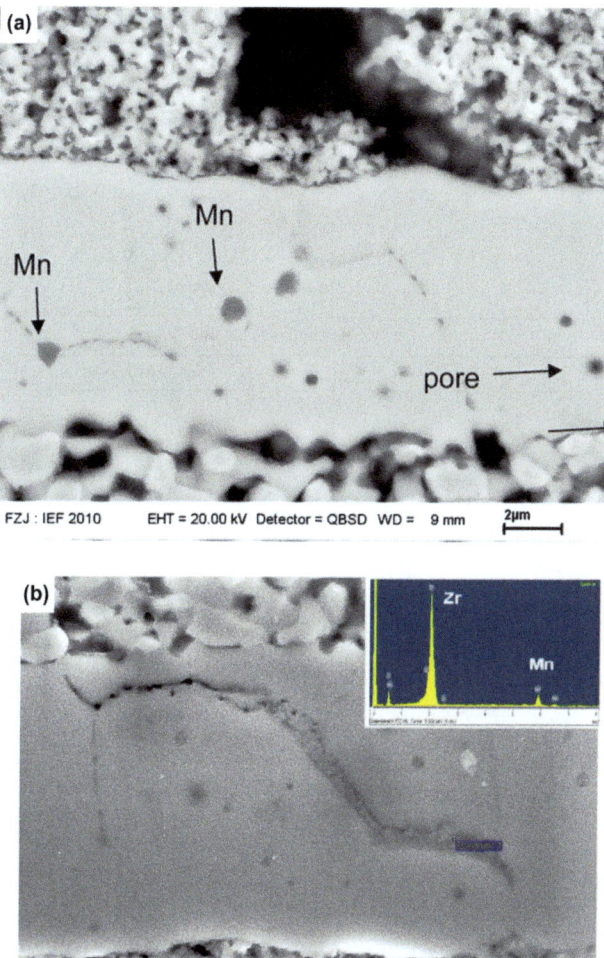

Figure 11.14 SEM images of the cross section of the electrolyte. (a) and (b) show Mn accumulation in the 8YSZ electrolyte. In (a), only an enhancement of Mn at the grain boundaries can be seen, and in (b) separation already becomes visible.

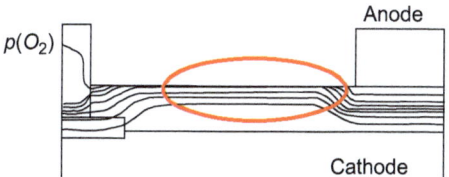

Figure 11.15 Estimated potential distribution near the anode edge for the cathode support tubular type cell.

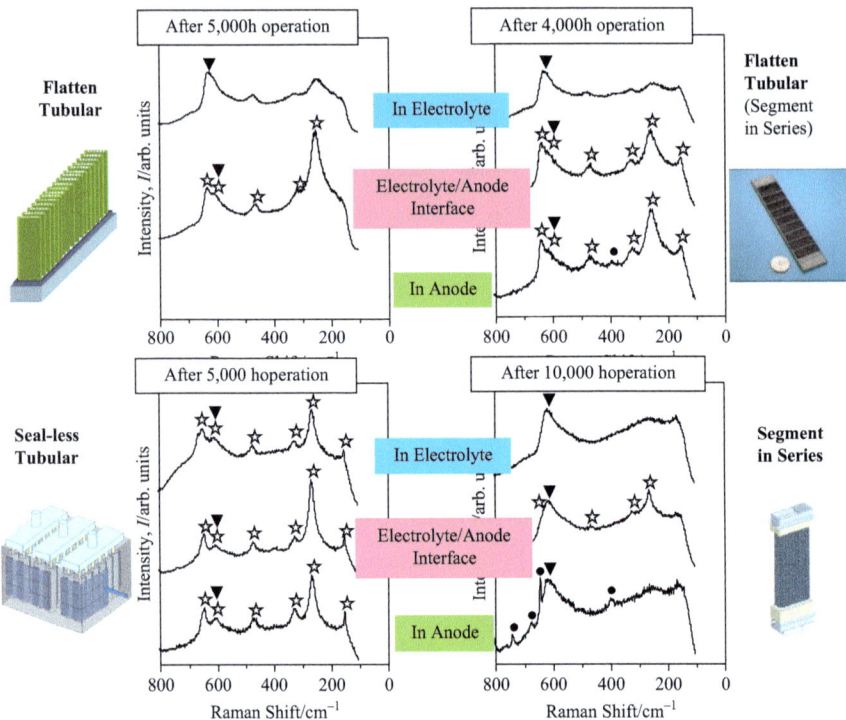

Figure 11.16   Raman spectra of cell components for several "real" stacks of flatten tubular type, flatten tubular (segment-in-series) type, sealless tubular type and segment-in-series type after being long time durability test.[34] ▼: cubic and ☆: tetragonal zirconia, and ● : resin

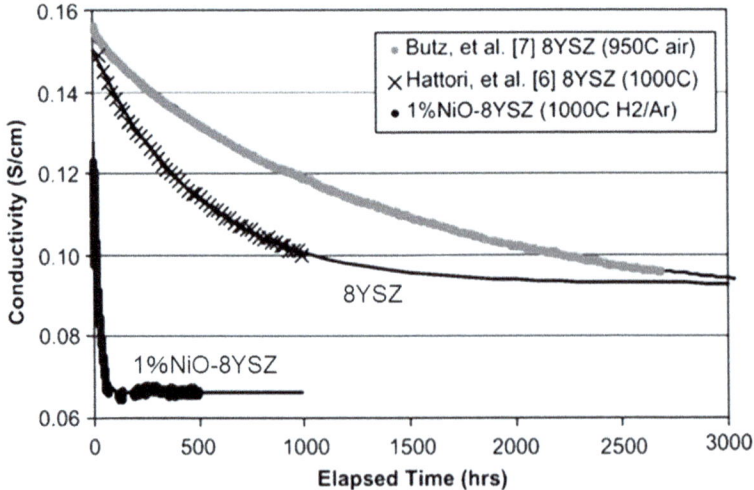

Figure 11.17   Conductivity relaxation for pure 8YSZ in air at 950 °C (Butz *et al.* [22]) and 1000 °C (Hattori *et al.*[20]) and 1 mass % NiO / 8YSZ in 4 % H₂ with Ar balance.

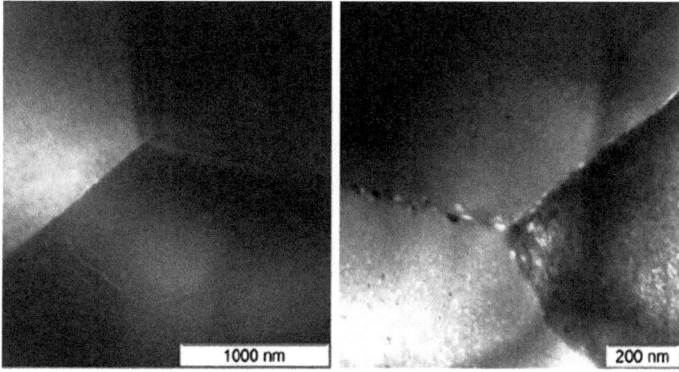

Figure 11.18 Bright field TEM of as-fired (left) and reduced (right) 1 mass %
NiO-8YSZ.

(a)   (b)   (c)

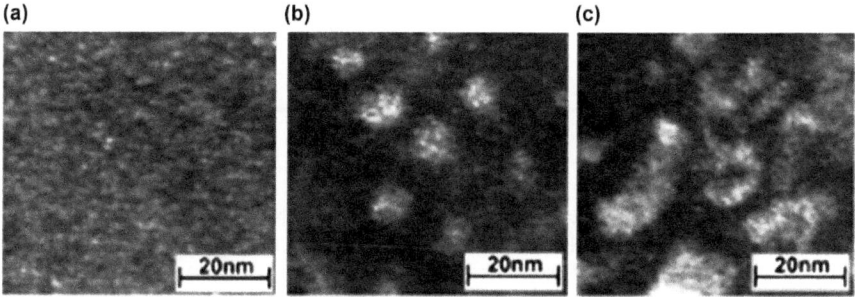

Figure 11.19 Dark field TEM images visualizing the evolution of tetragonal precipitations (detecting the {112} reflections). (a) sample directly reduction,
(b) reduced at 950 °C for 75 h and (c) for 100 h.

initial conductivity. Under the reducing condition, Ni^{2+} in the YSZ lattice
reduced to Ni^0 and its solubility is quite low. Therefore, reduced nickel
precipitates as nickel metal at grain boundaries (Figure 11.18).[35] Figure 11.19
shows the dark field TEM images of the NiO dissolved 8YSZ which visualize
the evolution of tetragonal precipitations by reducing treatment.[37] Several ten
nm size of tetragonal domains are precipitated in the cubic phase matrix and it
grows with increasing the reducing time. This domain structure is quite similar
to that for the YSZ obtained by a long-time high-temperature annealing.[22]

What phenomenon does occur under the SOFC operating condition? The
present authors' group has examined phase transformation and its related
conductivity degradation behaviours under SOFC operating conditions by
using the 1 mol % NiO doped 8YSZ as electrolyte. Figure 11.20 shows the
conductivity degradation curve of the NiO dissolved 8YSZ under the OCV
condition at 900 °C.[38] After the conductivity becomes stable during feeding dry
air to the anode, wet hydrogen is fed to the anode at $t = 0$ in Figure 11.20.
Conductivity steeply decreased at $t = 0$ h and reached nearly constant at a half

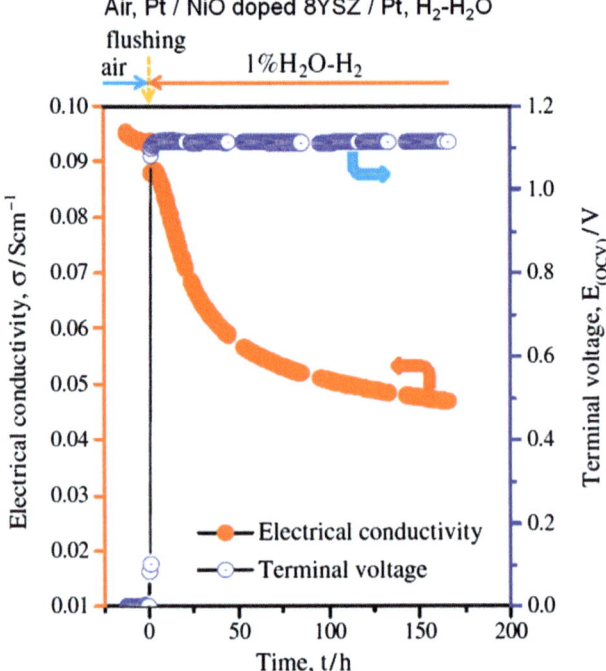

Figure 11.20 Conductivity degradation behaviour for Ni-YSZ electrolyte under the OCV condition between air and fuel gas. At $t = 0$ h, anode condition was changed from air to 1.2% H_2O–H_2 mixture gas.

of the initial value in 200 h. Figure 11.21 shows the optical image and the mapping analysis results of micro-Raman spectroscopy for the cross-section of the NiO doped YSZ electrolyte after the conductivity measurement.[38] In the mapping results, the red colour means apparently phase transformed area. The tetragonal phase was identified in a zone from the anode surface to a half of the electrolyte thickness and this transformed area completely corresponds to the dark colour area of the electrolyte in the optical image, where nickel metal is thought to be precipitated at grain boundaries or other colour centres are formed from the localized electrons surrounded by dopants and other defects. The calculated oxygen potential distribution is plotted on the Raman mapping results in Figure 11.22. Oxygen potential steeply changes in the middle of the electrolyte and exactly corresponds to the boundaries of the phase transformation observed. The Ni–NiO equilibrium oxygen potential is about 1×10^{-12} atm. It is clear that phase transformation occurs in the area where reduction of NiO occurs.

The relationship between conductivity degradation and phase transformation under the SOFC operating condition has been directly established using *in situ* Raman spectroscopy technique.[39,40] Figures 11.23 and 11.24 show results of *in situ* Raman analysis under conductivity measurement and the relationship between conductivity and phase transformation ratio,

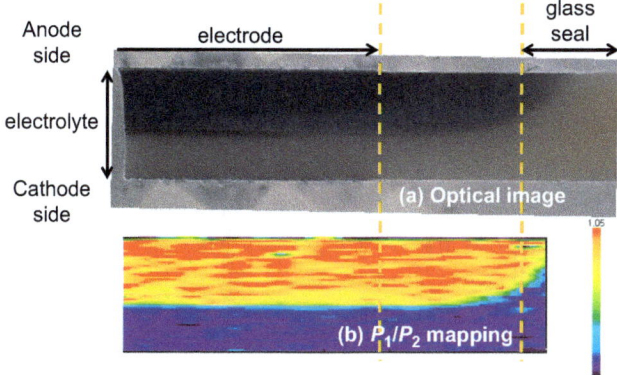

Figure 11.21 (a) Optical image and (b) mapping analysis of micro-Raman spectroscopy for the cross section of NiO doped YSZ electrolyte after conductivity measurement. Resolution of mapping analysis is carried out at room temperature in ambient air. Analyzed area is 200 μm and 25 μm for x and y axes, respectively. Spatial resolution of micro-Raman analysis is about ϕ 2 μm. In mapping analysis, peak ratio, P_1/P_2 is plotted. High P_1/P_2 value corresponds to the high phase transformation degree from cubic phase to tetragonal phase. P_1: integrated intensity of peaks from tetragonal phase at $185\,cm^{-1}$–$380\,cm^{-1}$ with background elimination, P_2: integrated intensity of peaks from both cubic and tetragonal phases at $440\,cm^{-1}$–$690\,cm^{-1}$ with background elimination.

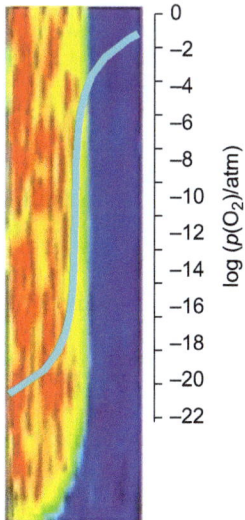

Figure 11.22 Calculated oxygen potential distribution in the NiO doped 8YSZ plotted with the result of Raman mapping analysis.

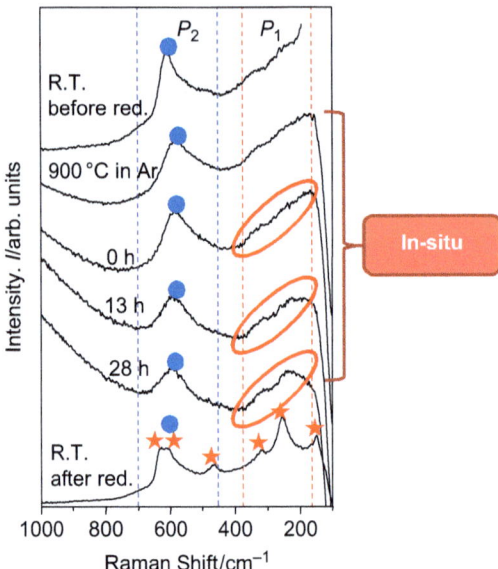

Figure 11.23   *In situ* Raman spectra for the surface of NiO dissolved 8YSZ electrolyte at the anode side surface under the SOFC operation atmosphere. Anode side and cathode side were exposed to wet Ar balanced 50 vol % H_2 and dry air, respectively.

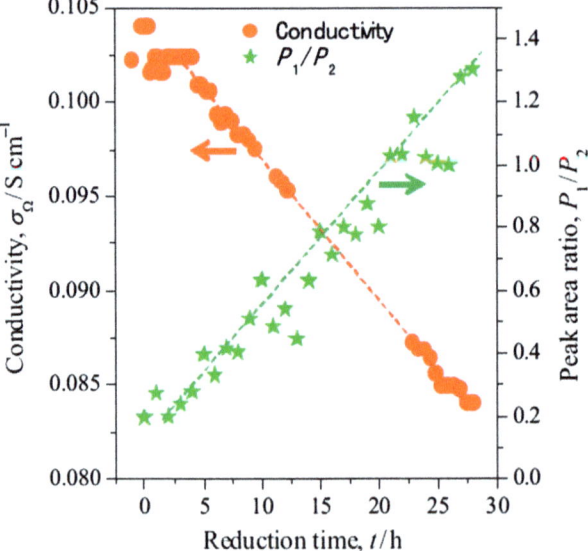

Figure 11.24   Peak ratio, P_1/P_2, and conductivity as a function of annealing time under the SOFC operation atmosphere. P_1: integrated intensity of peaks from tetragonal phase at 160–370 cm^{-1} with background elimination, P_2: integrated intensity of peaks from both cubic and tetragonal phases at 450–700 cm^{-1} with background elimination.

respectively.[39] It is clearly evident that phase transformation proceeds in the same manner as the conductivity degradation under the SOFC condition.

The effect of NiO dissolution and subsequent Ni precipitation seems to accelerate phase transformation behaviour. Since Ni is precipitated only along grain boundaries, this should be accompanied with diffusion of nickel ions to grain boundaries or back diffusion of cation vacancies from grain boundaries to inside grains. Such mass (or vacancy) transport enhances the rearrangement of cations and, as a result, more stable phases will be formed in the YSZ particles. In other words, the tetragonal phase tends to be developed with an aid of cation diffusion. For non-NiO doped YSZ, the cation vacancies are considered to be in a very small limited number to be determined for a frozen state caused by slow cation diffusivity.

11.4 Performance Degradations of Cathode and Anodes

In operating SOFC systems, oxidants (air) and fuels are continuously supplied to the cathode and the anode compartments, respectively. Therefore, impurities and/or contaminants in air and fuels are also continuously supplied and may be accumulated on electrode materials. In this section, electrode degradation behaviour induced by such impurities and contaminants are discussed.

11.4.1 Cathode Poisoning

For the cathode material of SOFCs, lanthanum strontium manganite (LSM) and lanthanum strontium cobalt ferrite (LSCF) with perovskite structure are generally used for high- and intermediate-temperature operation systems, respectively. Figure 11.25 shows schematic drawings of reaction models for cathodes.[41] For the electronic-conductive but non-oxide-ion-conductive cathode such as LSM, oxygen adsorption and dissociation occur on the electrode surface

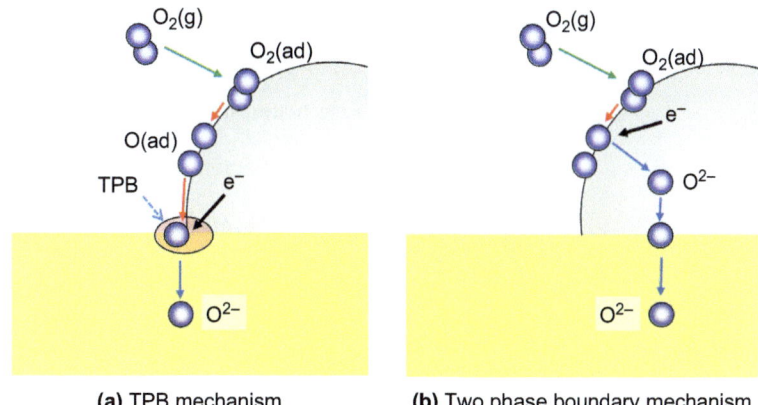

(a) TPB mechanism   **(b)** Two phase boundary mechanism

Figure 11.25 Schematic drawings for the cathode reaction: (a) electronic conductor such as LSM (TPB mechanism) and (b) mixed ionic electronic conductor such as LSCF (two phase boundary mechanism).

and then adsorbed/dissociated oxygen atom diffuses to the TPB. The oxygen atom is ionized at the TPB and it is incorporated as an oxide ion into electrolyte (Figure 11.25(a)). On the other hand, for the mixed ionic and electronic conductor (MIEC) such as LSCF, oxygen adsorption, dissociation, ionization and incorporation occurs mainly at the same place on the electrode surface (Figure 11.25(b)). In this section, the impurity-related degradation behaviour is compared mainly between LSM and LSCF cathodes and discussed mainly based on differences in electrode reaction mechanism and chemical reactivity with impurities.

11.4.1.1   Chromium Poisoning

In practical SOFC stacks and systems, the metallic components, such as interconnect materials and gas-supplying tubes, are used and can become sources of Cr containing vapours. The main volatile component of Cr has six valence (ex, CrO_3 or $CrO_2(OH)_2$).[42–44] The pressure of these Cr containing vapours is lower than 10^{-8} atm at an SOFC operating temperature of 1073 K. This Cr poisoning phenomena of cathode materials have been recognized and examined intensively.[45–61] It is well known that Cr poisoning phenomena appear in a different manner among cathode materials. The present authors' group have examined these phenomena using a specialized experimental setup

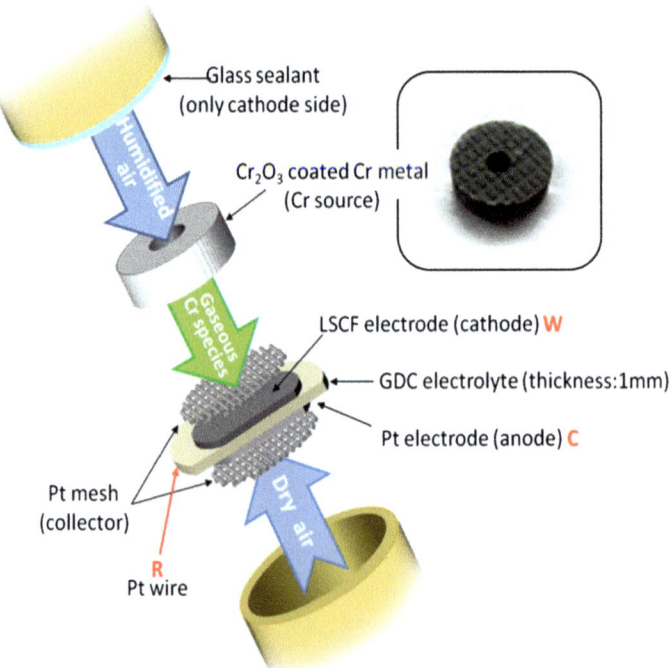

Figure 11.26   Schematic drawing of experimental configuration for Cr-poisoning test.

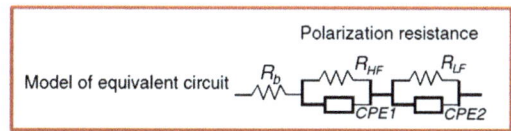

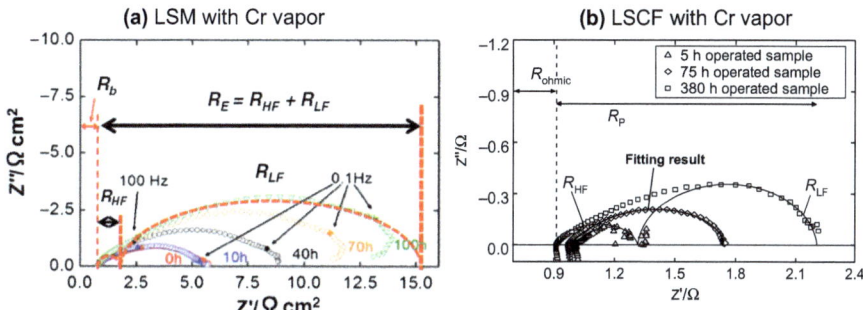

Figure 11.27 Equivalent circuit model (top) and AC impedance spectra of cathodes of (a) LSM and (b) LSCF, respectively, with Cr vapor condition. (R_b: ohmic resistance from the AC impedance intercept of Z'' axis at high frequency, R_E: polarization resistance, $R_E = R_{HF} + R_{LF}$, R_{HF}: high frequency polarization resistance and R_{LF}: low frequency polarization resistance).

for Cr poisoning test as shown in Figure 11.26,[58,60,61] where the Cr introduction from sources can be controllably switched to avoid the Cr contamination before the measurement starts. Figure 11.27(a) and (b) shows the impedance spectra change during the operating tests with Cr vapours for the LSM (($La_{0.8}Sr_{0.2})_{0.95}MnO_3$) and the LSCF (($La_{0.6}Sr_{0.4})(Co_{0.2}Fe_{0.8})O_3$) cathodes, respectively.[60,61] Polarization resistances, R_E, gradually increase with increasing operation time for both the LSM and the LSCF cathodes. It is clear that especially the low frequency component, R_{LF}, mainly increases. The electrode reaction process with long relaxation time, such as adsorption and dissociation process, is relatively inhibited by Cr vapours. The increasing rates of polarization resistance, R_E, with respect to operating time is plotted in Figure 11.28.[60] The slope of increasing rates of polarization resistance does not depend on polarization voltage for both the LSM and the LSCF cathodes. On the other hand, the slope itself is apparently different between the LSM and the LSCF cathodes; 1/2 for the LSM and 2 for the LSCF. This means that Cr poisoning mechanism must be quite different. Figure 11.29 shows the SEM / EDX images for the cross-section of the tested cathodes;[60,61] the dominant Cr deposit area is completely different between the LSM and the LSCF cathodes. Chromium is deposited at the TPB for the LSM, whereas at the top of the cathode surface for the LSCF. For the LSCF cathode, Sr is also detected with Cr, suggesting that $SrCrO_4$ is formed on the surface of the cathode.

Different degradation mechanism between the LSM and the LSCF cathodes should be related with electrode reaction mechanism based on the electrical properties and also with the chemical properties. For the cathode with the TPB

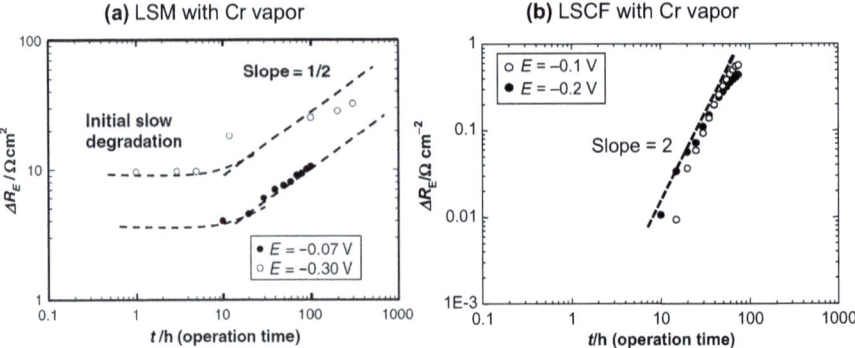

Figure 11.28 Polarization resistance increase (ΔR_E) with operation time at different cathodic voltages for (a) LSM and (b) LSCF cathodes, respectively, at 1073 K in air. ($\Delta R_E = (R_{HF} + R_{LF}) - (R_{HF} + R_{LF})_{initial}$).

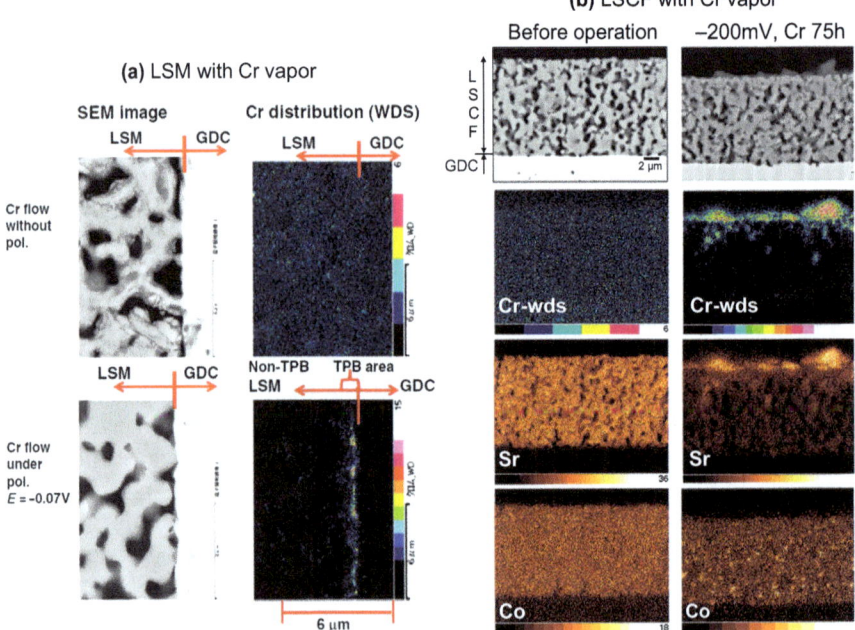

Figure 11.29 SEM images and elemental distribution maps (EDX and WDS) of cross section of (a) LSM and (b) LSCF cathodes.

electrode reaction mechanism such as LSM, Cr inhibits the cathode reaction as following reaction:

$$2CrO_3(g) + 6e^- = Cr_2O_3(s) + 3O^{2-} \tag{2}$$

Cr_2O_3 deposits at the TPB as a result of a side electrochemical reaction which should be taken place at the most active sites, leading to coverage on active sites with Cr_2O_3. As expected, this reaction does not proceed under the OCV condition as shown in Figure 11.29(a). It is also clearly shown that LSM is relatively stable against the chemical attack from Cr containing vapours as suggested from the thermodynamic consideration.[62] On the other hand, for the LSCF cathode, the $SrCrO_4$ formation is observed on the surface of the cathode under polarization condition. This can be typically written as the following chemical reaction:

$$(La_{0.6}Sr_{0.4})(Co_{0.2}Fe_{0.8})O_3 + 0.1CrO_{3(g)} = (La_{0.6}Sr_{0.3})(Co_{0.167}Fe_{0.733})O_{2.7}$$
$$+ 0.033O_2(g) + 0.1SrCrO_4 + 0.1/3\,CoFe_2O_4$$

(3)

Here, the mole number of Cr containing gas is artificially assumed to be 0.1 mole without any special reason; this is just to provide a typical example of reaction. Although this $SrCrO_4$ formation is accompanied with the $CoFe_2O_4$ precipitation and the oxygen evolution (reduction) processes, it is not necessarily regarded as electrochemical reactions; this result is also consistent with the thermodynamic calculation for perovskite cathodes with a high Sr activity.[62] The above chemical equation alone, however, cannot explain why the $SrCrO_4$ formation takes place mostly at the cathode surface not at the electrochemical active sites around the cathode/electrolyte interfaces. One possible explanation is that the SrO component migrates out of whole area of the LSCF cathode layer to the surface and reacts with the Cr-containing vapours. In other words, the above chemical reaction proceeds at the two different reaction zones; first one is the SrO evolution out of the perovskite lattice and the other is the $SrCrO_4$ formation at the surface from the evolved SrO and $CrO_3(g)$. To ensure that the $SrCrO_4$ formation takes place mainly on the surface, the supplying speed of SrO should be much faster than that of $CrO_3(g)$ in air. This implies the fast Sr diffusion to the cathode surface layer via the LSCF grain surface or through the gaseous phase. At present, it is hard to distinguish two possibilities from the experimentally observed facts. Anyway, the driving force of SrO diffusion is given by the difference in the SrO thermodynamic activity between the $SrCrO_4$ phase and the LSCF phase under polarization.

11.4.1.2 SO_2 Poisoning of Cathode

The concentration of sulfur component in air is as low as several ten ppb typically in Japan, but sulfur is one of the major contaminants in ceramic raw powders and metallic materials. Therefore, it is expected that sulfur impurities are continuously supplied to the cathode under an SOFC operating condition. Recently, the influences of sulfur on the cathode performance have been well recognized and examined intensively.[63–69] The poisoning behaviour of SO_2 on the cathode performance is also different among cathode materials. Figure 11.30 shows the results of cathode performance tests for the LSM $((La_{0.8}Sr_{0.2})_{0.95}MnO_3)$ cathode with 100 ppm of SO_2 and for the LSCF $((La_{0.6}Sr_{0.4})(Co_{0.2}Fe_{0.8})O_3)$ cathode with 1 ppm SO_2.[63,69] The LSM cathode

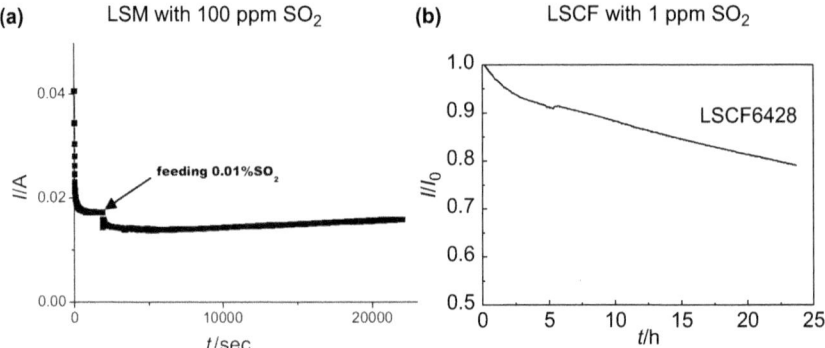

Figure 11.30   Results of cathode performance tests (a) for the LSM cathode with 100 ppm SO$_2$ and (b) for the LSCF cathode with 1 ppm SO$_2$ at a fixed applied voltage of −0.2 V.

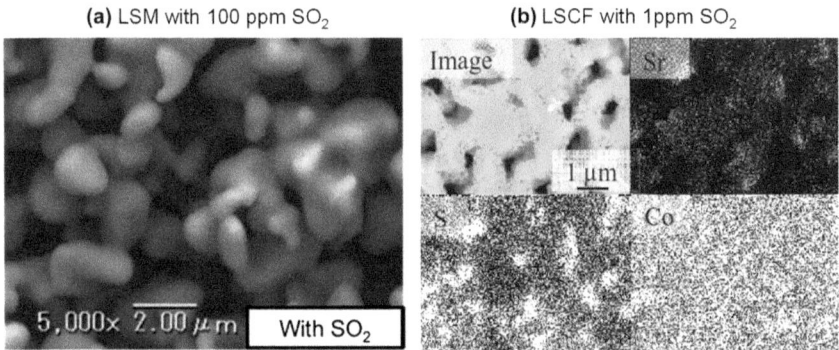

Figure 11.31   (a) SEM image of the LSM surface after the performance test with 100 ppm of SO$_2$ and (b) SEM and elemental distribution images of the cross section of the LSCF cathode after the performance test with 1 ppm SO$_2$.

shows slight stepwise performance drop at the time of SO$_2$ introduction, but after then, the performance is quite stable. On the other hand, the performance of the LSCF cathode gradually decreases even when 1 ppm of SO$_2$ is mixed into air. After the performance test, microstructure of the LSM cathode does not change (Figure 11.31(a)).[63] The LSM cathode is stable under 100 ppm of SO$_2$ containing air at an SOFC operating temperature. The morphology of the LSCF cathode changes on particle surfaces, especially around grain boundaries (Figure 11.31(b)).[68] Impurities of S and the Sr component segregate in the same area, indicating that SrSO$_4$ is formed for the LSCF cathode. This result corresponds to the thermodynamic considerations which show that SrSO$_4$ formation always takes place even under a concentration of SO$_2$ as low as 0.1ppm for the LSCF cathode.[68] This instability of LSCF cathode against SO$_2$ leads to gradual degradation of cathode performance.

The LSM cathode shows slight stepwise degradation without morphology change. It should be related to the adsorption of impurities on the electrode surface. Adsorbed SO_2 blocks adsorption of oxygen on the LSM cathode. On the other hand, the $SrSO_4$ formation and related phenomena can be regarded as the main reason for gradual performance degradation of the LSCF cathode. The $SrSO_4$ formation reaction can be written as following;

$$(La_{0.6}Sr_{0.4})(Co_{0.2}Fe_{0.8})O_3 + 0.1SO_2 + 0.017O_2(g)$$
$$= (La_{0.6}Sr_{0.3})(Co_{0.167}Fe_{0.733})O_{2.7} + 0.1SrSO_4 + 0.1/3\,CoFe_2O_4 \tag{4}$$

This is given in a similar manner to Equation (2) for the chromium poisoning. When focus is made on the SrO component, it can be written as

$$SrO\ (in\ LSCF) + SO_2 + 1/2O_2 \rightarrow SrSO_4 \tag{5}$$

This $SrSO_4$ formation reaction is basically oxidation reaction contrary to the $SrCrO_4$ formation given in Equation (3). Even so, the observed deposited sulfur distribution in the LSCF cathode is different from the expectation based on the equilibrium properties which suggest the oxidative side is preferable; actually, sulfur deposition prefers to occur near the electrode/electrolyte interface (Figure 11.32),[69] where the electrochemically cathodic reaction actively occurs and therefore the oxygen potential becomes low. This suggests that there must be some kinetic factors of controlling where this $SrSO_4$ formation takes place. This should also provide reasons for differences between the $SrCrO_4$ and $SrSO_4$ formations. One possibility is the availability of oxygen vacancies where the oxygen molecules as reactants of the above reaction can be adsorbed, dissociated and reacted with $SO_2(g)$.

The present examples of poisoning of cathode performance are determined by many factors such as the thermodynamic reactivity of the cathode components, availability of diffusion paths, or availability of reaction sites. These factors make performance degradation behaviour complicated. Even so, recent investigations have revealed interesting features to resolve such complexity.

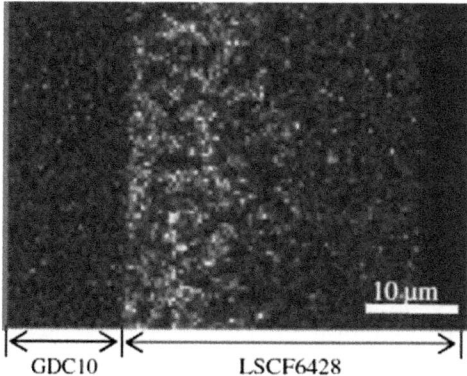

Figure 11.32 Sulfur distribution in the LSCF cathode after the performance test with 1 ppm of SO_2 for 24 h.

11.4.2   Sintering of Ni Cermet Anodes

Nickel-base cermet anodes, mainly Ni-YSZ anode, have been conventionally used for many SOFC systems; Ni metal provides catalytic activity for fuel oxidation as high as precious metals with lower cost than that of precious metals.[70] The oxide component in cermet anodes takes some roles in the electrochemical processes such as providing oxide ion paths, high mechanical strength and better fitting in the thermal expansion coefficient with the oxide electrolyte. In the practical SOFC systems, overpotential of anode reactions is relatively small compared with cathode overpotential as revealed in the NEDO durability project.

Although reliability of nickel anodes can be damaged by carbon deposition or morphological changes due to the Redox cycles, the major source against durability is come from sintering of nickel. Sintering of Ni leads to first reduction of TPB length and then breaking electronic paths. Notably, breaking electronic paths can trigger a "sudden death" of the cell. Nickel sintering is caused or enhanced by following effects:

1. Cell operation at a high temperature for a long time;[71,72] this is mainly due to bulk diffusion of nickel.
2. Redox cycle and thermal cycle;[73–76] during Redox cycles, nickel must be once pulverized and then coagulated again changing network connectivity, while sintering during thermal cycles can be caused by anomalously enhanced diffusion at lower temperatures.
3. Impurities in fuels;[77–88] this is due to impurity-enhanced diffusion.

Sulfur is usually contained as impurities in hydrocarbon fuels. There are many reports about degradation behaviour of Ni cermet anodes induced by sulfur impurities.[89–105] When concentration of H_2S in fuels is low as several ppm, active sites on the nickel surface are covered with adsorbed sulfur. Therefore, such degradation phenomena are reversible (Figure 11.33(a)[97]).

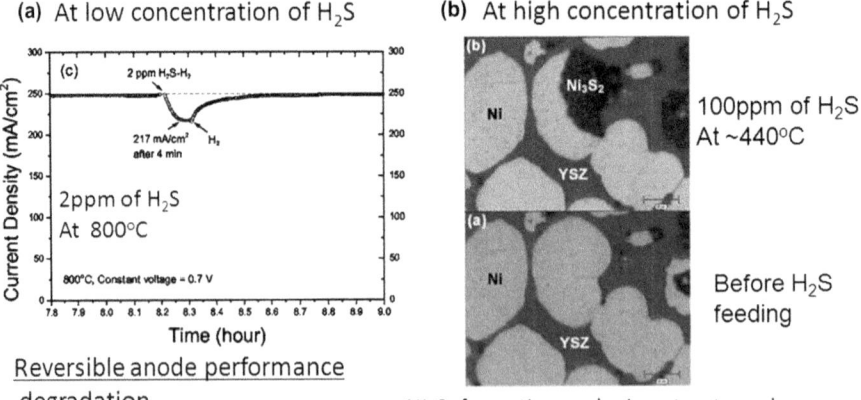

(a) At low concentration of H_2S **(b)** At high concentration of H_2S

Reversible anode performance degradation

Ni_3S_2 formation and microstructure change

Figure 11.33   Degradation behavior of the Ni cermet anode under sulfur containing fuel. (a) at low concentration of H_2S at 800 °C and (b) at high concentration of H_2S at 440 °C.

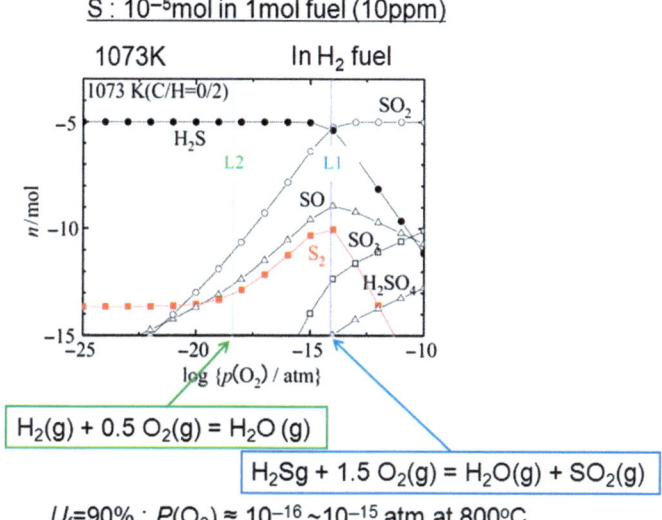

Figure 11.34 Stable form of sulfur in fuel as a function of $P(O_2)$.

In the case of high H_2S concentration at a relatively low temperature, nickel sulfide formation occurs (Figure 11.33(b)[104]). Sulfur poisoning phenomena for Ni metal can be explained in terms of results of thermodynamic calculations.[106] Figure 11.34 shows the stable form of sulfur in fuel as a function of $P(O_2)$.[106] In an anode atmosphere, major sulfur content is in the form of H_2S. Under the constant partial pressure of H_2S, however, the sulfur potential interestingly becomes oxygen-potential dependent and increases with increasing $P(O_2)$ in a typical anode environment. Figure 11.35(a) shows the calculated phase diagram for the Ni-S system.[106] The temperature that provides the Ni-S eutectic liquids decreases with increasing S activity and it reaches as low as the SOFC operating temperature of about 900 K. Figure 11.35(b) shows the calculated phase diagram for the Ni-S-O-C-H system.[106] In H_2S dominant region, the S activity increases with increasing $P(O_2)$, however, it decreases with decreasing $P(O_2)$ in SO_2 dominant region. When the concentration of S containing gases is 1 ppm, the sulfur activity increases with increasing $P(O_2)$ but Ni metal is still stable without sulfide formation under fuel condition. On the other hand, when the concentration increases to 100 ppm, liquid phase becomes stable at high $P(O_2)$ condition. Figure 11.36 also shows the calculated chemical potential diagram for the Ni-S-O-H-C system for cases where P_{SX} is (a) 100 ppm and (b) 1 ppm, respectively.[106] At low S concentrations at a high T, nickel metal is stable and reversible degradation can be expected and actually is observed in many cases. At high S concentrations at a low T, the nickel sulfide formation may occur. These considerations lead to some expectation that anomalous sintering of nickel anode is expected under cell operation with a high sulfur concentration at an intermediate temperature because approaching the Ni-S

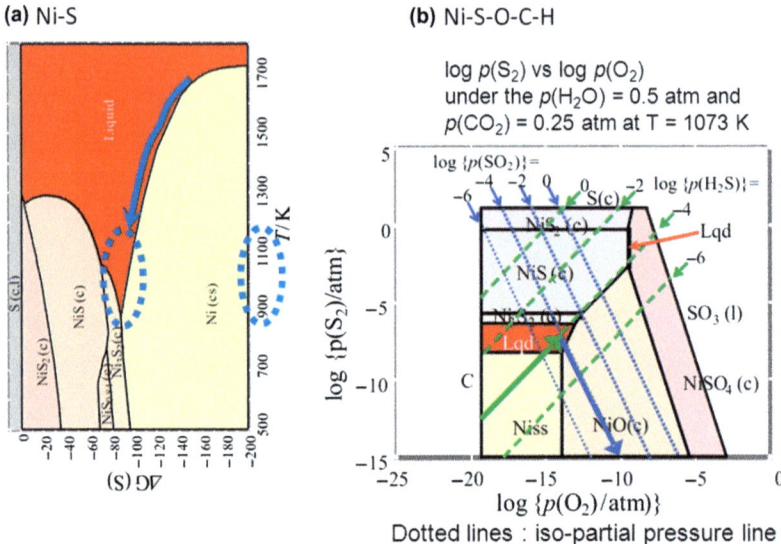

Figure 11.35   The calculated phase diagram (a) for the Ni-S system and (b) Ni-S-O-C-H system.

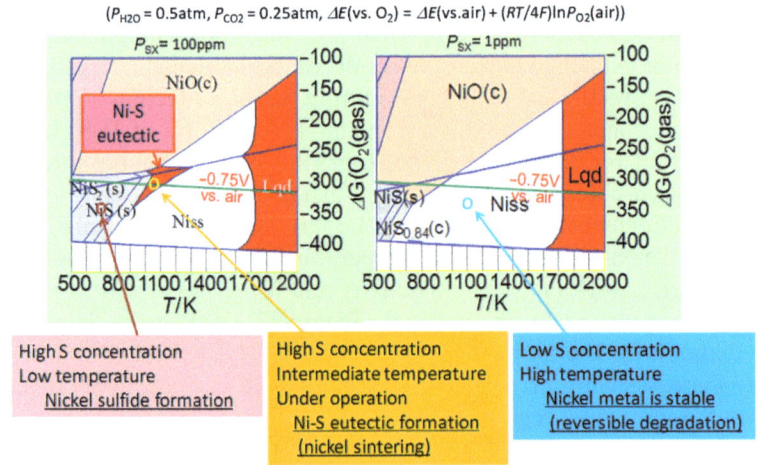

Figure 11.36   The calculated chemical potential diagram for Ni-S-O-H-C system.

eutectic formation may provide anomalously enhanced nickel diffusion even at a low temperature.

The thermodynamic considerations suggest that sulfur poisoning should be serious under high $P(O_2)$ conditions because the sulfur activity and the solubility increase with increasing $P(O_2)$ under anode conditions. It is estimated that the nickel sintering related to the sulfur impurities becomes serious at the

TPB of the anode with high over potential. In addition, morphology change should be significant by thermal cycles because sulfur adsorption becomes serious at a low temperature and the cell pass the Ni-S eutectic stable point in the equilibrium phase diagram.

Integrated Coal Gasification Fuel Cell Combined Cycle (IGFC) is promising a large-scale power plant with higher electric efficiency using the coal syngas to be fuelled to SOFC systems. One of the major impurities in coal syngas is phosphor (P) as well as sulfur. Phosphor has a high reactivity with nickel metal to form nickel phosphide or nickel phosphate and, as a result, significant nickel sintering occurs.[82–88] as shown in Figure 11.37. Figure 11.38 shows the Ellingham diagrams for the P-O-H system with (a) 1 ppm and (b) 1 ppb of phosphor.[107] Roughly speaking, under hydrogen dominant region, PH_3 is dominant, and under H_2O dominant region, oxidized phosphor becomes stable. Figure 11.39 shows Ellingham diagrams for the Ni-P-O-H system.[107] Under $P(H_2) > P(H_2O)$ region where PH_3 is dominant impurity, Ni phosphide is stable. On the other hand, under $P(H_2) < P(H_2O)$ region where $(P_2O_5)_2$ is dominant, Ni phosphate is dominant. The nickel poisoning behaviour should change according to operating condition and $P(O_2)$ distribution in the anode. For example, Ni phosphide mainly formed on the anode surface because of low $P(O_2)$. Nickel phosphate should be formed near the TPB especially for high over potential condition because $P(O_2)$ becomes high.

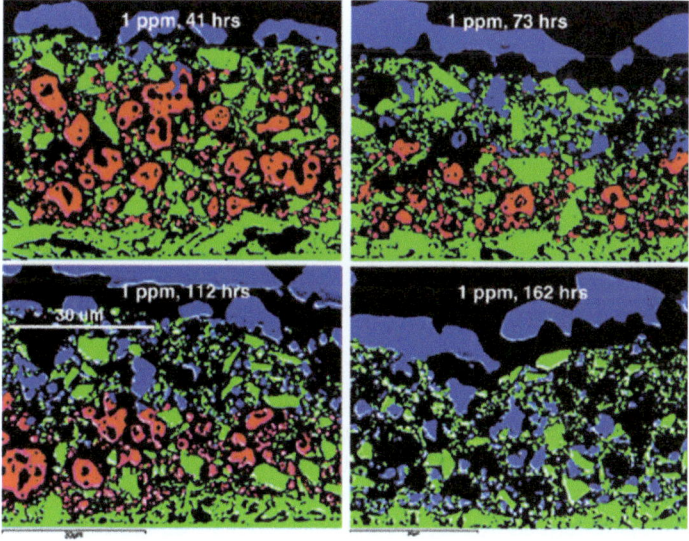

Figure 11.37 Sequence of nickel conversion to nickel phosphide in anodes of electrolyte-supported cells by 1 ppm PH_3 at 800 °C for various exposure times as determined from the elemental maps collected during cross-sectional SEM analysis. Nickel is red, YSZ is green and nickel phosphide phases are blue.

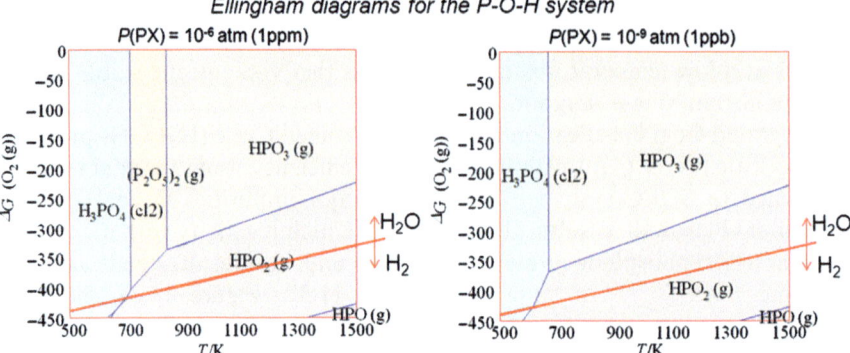

Figure 11.38   Ellingham diagrams for P-O-H system with (a) 1ppm and (b) 1ppb of phosphor.

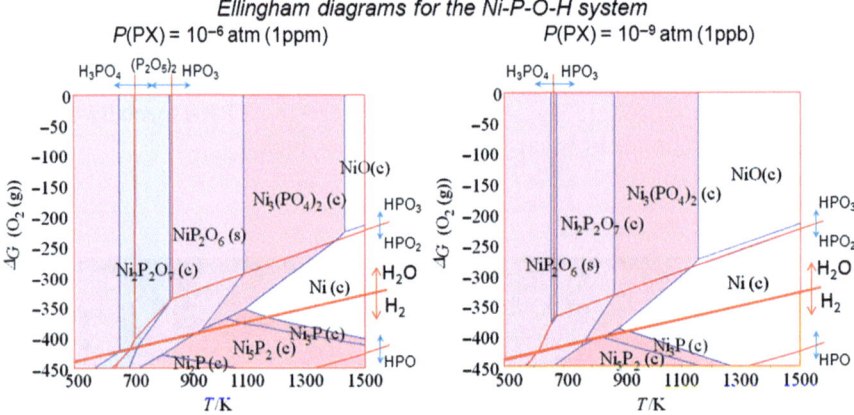

Figure 11.39   Ellingham diagrams for the Ni-P-O-H system with (a) 1ppm and (b)1ppb of phosphor.

11.5   For Future Work

Two examples for electrolyte deterioration described above suggest that the oxygen potential distribution developed inside the electrolyte is very important to examine the chemical behaviour of transition elements dissolved into the electrolyte during fabrication and operations. Particularly, redox behaviour or precipitation as second phases should be strongly influenced by the local oxygen potentials. As described above, the oxygen potential distribution itself depends largely on operation conditions such as temperature, fuel composition or overpotential of electrodes. This implies that behaviour of transition metal elements inside electrolyte may be also influenced by changes in electrode performance. This is quite an important point when material behavior on long term operation should be considered from the point of view of establishment of long life.

The present examples indicate that degradation mechanism appears very complicated. Furthermore, behaviour of degradation of one material may be related with other degradations at difference materials or places. In this sense, mutual interactions should be carefully taken into account when a long life of SOFC stacks should be considered to establish high durability and reliability. This should need a kind of cooperation of different approaches, experimental or simulation, atomistic, compound level, cell level or stack level *etc.*

11.6 Conclusions

Degradations of SOFC stacks depend first on the fabrication processes. In view of the strong correlation among materials selection, materials processing, and stack design, it is reasonable to understand that degradations of SOFC should be investigated using SOFC stacks to be considered. After extractions of what will be the major sources of degradations for particular stacks, it can be investigated using various experimental facilities for cell component test, single test etc. Among others, deterioration of electrolytes should be most severe. In this article, Mn-dissolved YSZ and Ni-dissolved YSZ have been analyzed concerning their stability against disintegrations or crack formation during operation.

Degradations also depend on the SOFC operation conditions. In this article, impurity poisoning effects on cathodes and anodes have been described with an emphasis on the strength of chemical interaction and the poisoning reasons. Since there are several different processes are involved in a poisoning effect, care must be taken to investigate such effects. Here, effects of equilibrium properties as well as kinetic properties associated with electrochemical processes are compared. From these considerations, interesting differences can be seen between LSM and LSCF cathodes.

Acknowledgement

Parts of this investigation have been made in the NEDO Project on development of the fundamentals of the SOFC systems.

References

1. H. Yokokawa and T. Horita, *Solid Oxide Fuel Cell Materials: Durability, Reliability and Cost, in Encyclopedia of Sustainability Science and Technology*, ed. R. A. Meyers, Springer Verlag, 2012, pp. 9934–9968.
2. H. Yokokawa, Performance and degradations, overview of solid oxide fuel cell degradation, in *Handbook of Fuel Cells Fundamentals Technology and Application Vol. 6, Advances in Electrocatalyst, Materials, Diagnostics, and Durability*, ed. W. Vielstich, H. Yokokawa and H. A. Gasteiger, John Wiley & Sons, 2009, pp. 923–932.
3. http://www.noe.jx-group.co.jp/newsrelease/2012/20121130_03_0960492.html (last access, February 11, 2013).

4. http://www.osakagas.co.jp/company/press/pr_2012/1196121_5712.html (last access, February 11, 2013).

5. http://www.bloomenergy.com/fuel-cell/solid-oxide/ (last access, February 11, 2013).

6. H. Yokokawa and N. Sakai, Part 4. Fuel Cell Principle, Systems and Applications, Chapter 13 History of high temperature fuel cell development, in *Handbook of Fuel Cells Fundamentals Technology and Application, Vol. 1. Fundamentals and Survey of Systems*, ed. W. Vielstich, A. Lamm and H. A. Gasteiger, John Wiley & Sons, 2003, pp. 219–266.

7. H. Yokokawa, *Annu. Rev. Mater. Res.*, 2003, **33**, 581–610.

8. H. Yokokawa, H. Tu, B. Iwanschitz and A. Mai, *J. Power Sources*, 2008, **182**, 400–412.

9. H. Yokokawa, T. Horita, K. Yamaji, H. Kishimoto and M. E. Brito, *J. Kor. Ceram. Soc.*, 2010, **47**(1), 26–38.

10. H. Yokokawa, K. Yamaji, M. E. Brito, H. Kishimoto and T. Horita, *J. Power Sources*, 2011, **196**, 7070–7075.

11. H. Yokokawa, T. Horita, K. Yamaji, H. Kishimoto and M. E. Brito, *J. Kor. Ceram. Soc.*, 2012, **49**(1), 11–18.

12. H. Yokokawa, *ECS Trans.*, 2011, **35**(1), 207–216.

13. H. Yokokawa, H. Kishimoto, K. Yamaji, T. Horita, T. Watanabe, T. Yamamoto, K. Eguchi, T. Matsui, K. Sasaki, Y. Shiratori, T. Kawada, K. Sato, T. Hashida, A. Unemoto, T. Kabata and K. Tomida, *ECS Trans.*, 2011, **35**(1), 2191–2200.

14. M. Shimazu, T. Isobe, S. Ando, K. Hiwatashi, A. Ueno, K. Yamaji, H. Kishimoto, H. Yokokawa, A. Nakajima and K. Okada, *Solid State Ionics*, 2011, **182**, 120–126.

15. M. Shimazu, K. Yamaji, T. Isobe, A. Ueno, H. Kishimoto, K. Katsumata, H. Yokokawa and K. Okada, *Solid State Ionics*, 2011, **204–205**, 120–128.

16. M. Shimazu, K. Yamaji, H. Kishimoto, A. Ueno, T. Isobe, K. Katsumata, H. Yokokawa and K. Okada, *Solid State Ionics*, 2012, **224**, 6–14.

17. H. Yokokawa, T. Horita, K. Yamaji, H. Kishimoto, T. Yamamoto, M. Yoshikawa, Y. Mugikura and K. Tomida, *Fuel cells*, 2013, DOI: 10.1002/fuce.201200164.

18. M. Yashima, M. Kakihara and M. Yoshimura, *Solid State Ionics*, 1996, **86–88**, 1131–1149.

19. K. Nomura, Y. Mizutani, M. Nakamura and O. Yamamoto, *Solid State Ionics*, 2000, **132**, 235.

20. M. Hattori, Y. Takeda, J.-H. Lee, S. Ohara, K. Mukai, T. Fukui, S. Takahashi, Y. Sakaki and A. Nakanishi, *J. Power Sources*, 2004, **131**, 247.

21. C. Haering, A. Roosen and H. Schichi, *Solid State Ionics*, 2005, **176**, 253.

22. B. Butz, P. Kruse, H. Stömer, D. Gerthsen, A. Müller, A. Weber and E. Ivers-Tiffée, *Solid State Ionics*, 2006, **177**, 3275.

23. T. Kawada and H. Yokokawa, *Key Eng. Mater.*, 1997, **125–126**, 187–248.

24. T. Kawada, N. Sakai, H. Yokokawa and M. Dokiya, *Solid State Ionics*, 1992, **53–56**, 418–425.
25. S. Linderoth, N. Bonanos, K. V. Jensen and J. B. B. Sorensen, *J. Am. Ceram. Soc.*, 2001, **84**, 2652.
26. L. Heyne and N. M. Beekmans, *Proc. Br. Ceram. Soc.*, 1971, **19**, 229.
27. W. Weppner, *J. Solid State Chem.*, 1977, **20**, 305.
28. J. H. Park and R. N. Blumenthal, *J. Electrochem. Soc.*, 1989, **136**, 2867.
29. T. Kawada, N. Sakai, H. Yokokawa, M. Dokiya and I. Anzai, *Solid State Ionics*, 1992, **50**, 189–196.
30. H. Yokokawa, N. Sakai, T. Kawada and M. Dokiya, *ISSI Letters*, 1991, **2**, 7.
31. J. Malzbender, P. Batfalsky, R. Vaßen, V. Shemet and F. Tietz, *J. Power Sources*, 2012, **201**, 196–203.
32. M. Shimazu, Doctoral dissertation, Tokyo Institute of Technology, 2011.
33. M. Ueda and T. Maruyama, *J. Kor. Ceram. Soc.*, 2012, **49**, 37–42.
34. H. Kishimoto, T. Shimonosono, K. Yamaji, M. E. Brito, T. Horita and H. Yokokawa, *ECS Trans.*, 2011, **35**, 1171–1176.
35. W. G. Coors, J. R. O'Brien and J. T. White, *Solid State Ionics*, 2009, **180**, 246–251.
36. V. Sonn and E. Ivers-Tiffee, *Proc. 8th Euro SOFC Forum*, Luzern, Switzerland, July 2008, B1005.
37. A. Lefarth, B. Butz, H. Störmer, A. Utz and D. Gerthsen, *ECS Trans.*, 2011, **35**, 1581–1586.
38. T. Shimonosono, H. Kishimoto, M. E. Brito, K. Yamaji, T. Horita and H. Yokokawa, *Solid State Ionics*, 2012, **225**, 69–72.
39. H. Kishimoto, K. Yashiro, T. Shimonosono, M. E. Brito, K. Yamaji, T. Horita, H. Yokokawa and J. Mizusaki, *Electrochimica Acta*, 2012, **82**, 263–267.
40. H. Kishimoto, K. Yashiro, T. Shimonosono, M. E. Brito, K. Yamaji, T. Horita, H. Yokokawa, and J. Mizusaki, Phase transformation related conductivity degradation of NiO doped YSZ: an in situ micro-Raman analysis, in: MRS Online Proceedings Library 2012, 1385.
41. T. Kawada and J. Mizusaki, in *Handbook of Fuel Cells – Fundamentals, Technology and Applications Volume 4*, Ed. W. Vielstich, H. A. Gasteiger and A. Lamm, 2003, John Wiley & Sons, Ltd., 987.
42. C. Gindirf, L. Singheiser and K. Hilpert, *J. Phys. Chem. Solids*, 2005, **66**, 384–387.
43. M. Stanislowski, E. Wessel, K. Hilpert, T. Markus and L. Singheiser, *J. Electrochem. Soc.*, 2007, **154**, A295–A306.
44. H. Kurokawa, C. P. Jacobson, L. C. DeJonghe and S. J. Visco, *Solid State Ionics*, 2007, **178**, 287–296.
45. S. Taniguchi, M. Kadowaki, H. Kawamura, T. Yasuo, Y. Akiyama, Y. Miyake and T. Saitoh, *J. Power. Sources*, 1995, **55**, 73–79.
46. K. Hilpert, D. Das, M. Miller, D. H. Peck and R. Weiß, *J. Electrochem. Soc.*, 1996, **143**, 3642–3647.

47. S. P. Jiang, J. P. Zhang, L. Apateanu and K. Foger, *J. Electrochem. Soc.*, 2001, **147**, 4013–4022.
48. S. P. S. Badwal, R. Deller, K. Foger, Y. Ramprakash and J. P. Zhang, *Solid State Ionics*, 1997, **99**, 297–310.
49. S. P. Jiang, J. P. Zhang and K. Foger, *J. Electrochem. Soc.*, 2001, **148**, C447–C455.
50. Y. Matsuzaki and I. Yasuda, *J. Electrochem. Soc.*, 2001, **148**, A126–A131.
51. S. C. Paulson and V. I. Birss, *J. Electrochem. Soc.*, 2004, **151**, A1961–A1968.
52. K. Fujita, T. Hashimoto, K. Ogasawara, H. Kameda, Y. Matsuzaki and T. Sakurai, *J. Power. Sources*, 2004, **131**, 270–277.
53. E. Konysheva, H. Penkalla, E. Wessel, J. Mertens, U. Seeling, L. Singheiser and K. Hilpert, *J. Electrochem. Soc.*, 2006, **153**, A765–A773.
54. E. Konysheva, J. Mertens, H. Penkalla, L. Singheiser and K. Hilpert, *J. Electrochem. Soc.*, 2007, **154**, B1252–B1264.
55. S. P. Jiang, S. Zhang and Y. D. Zhen, *J. Electrochem. Soc.*, 2006, **153**, A127–A134.
56. S. P. Jiang and Y. Zhen, *Solid State Ionics*, 2008, **179**, 1459–1464.
57. T. Komatsu, R. Chiba, H. Arai and K. Sato, *J. Power. Sources*, 2008, **176**, 132–137.
58. T. Horita, Y. Xiong, H. Kishimoto, K. Yamaji, M. E. Brito and H. Yokokawa, *J. Electrochem. Soc.*, 2010, **157**, B614–B620.
59. R. R. Liu, S. H. Kim, S. Taniguchi, T. Oshima, Y. Shiratori, K. Ito and K. Sasaki, *J. Power Sources*, 2011, **196**, 7090.
60. T. Horita, D.-H. Cho, F. Wang, T. Shimonosono, H. Kishimoto, K. Yamaji, M. E. Brito and H. Yokokawa, *Solid State Ionics*, 2012, **225**, 151–156.
61. D.-H. Cho, H. Kishimoto, M. E. Brito, K. Yamaji, M. Nishi, T. Shimonosono, F. Wang, H. Yokokawa, and T. Horita, *ECS Trans.*, submitted.
62. H. Yokokawa, T. Horita, N. Sakai, K. Yamaji, M. E. Brito, Y.-P. Xiong and H. Kishimoto, *Solid State Ionics*, 2006, **177**, 3193–3198.
63. Y. Xiong, K. Yamaji, T. Horita, H. Yokokawa, J. Akikusa, H. Eto and T. Inagaki, *J. Electrochem. Soc.*, 2009, **156**(5), B588–B592.
64. K. Yamaji, Y. Xiong, M. Yoshinaga, H. Kishimoto, M. E. Brito, T. Horita, H. Yokokawa, J. Akikusa and M. Kawano, *ECS Trans.*, 2009, **25**(2), 2853–2858.
65. A. J. Schuler, Z. Wuillemin, A. Hessler-Wyser and J. Van herle, *ECS Trans.*, 2009, **25**(2), 2845–2852.
66. R. R. Liu, S. K. Kim, Y. Shiratori, T. Oshima, K. Ito and K. Sasaki, *ECS Trans.*, 2009, **25**(2), 2859–2866.
67. J. A. Schuler, C. Gehrig, Z. Wuillemin, A. J. Schuler, J. Wochele, C. Ludwig, A. Hessler-Wyser and J. Van herle, *J. Power Sources*, 2011, **196**, 7225–7231.

68. F. Wang, K. Yamaji, D. H. Cho, T. Shimonosono, H. Kishimoto, M. E. Brito, T. Horita and H. Yokokawa, *J. Electrochem. Soc.*, 2011, **158**(11), B1391–B1397.
69. F. Wang, K. Yamaji, D. H. Cho, T. Shimonosono, H. Kishimoto, M. E. Brito, T. Horita and H. Yokokawa, *Solid State Ionics*, 2012, **225**, 157–160.
70. K. Eguchi, in *Handbook of Fuel Cells – Fundamentals, Technology and Applications*, ed. W. Vielstich, H. A. Gasteiger and A. Lamm, John Wiley & Sons, Ltd., 2003, Vol. 4, 1057.
71. T. Iwata, *J. Electrochem. Soc.*, 1996, **143**, 1521–1525.
72. A. Ioselevich, A. A. Kornyshev and W. Lehnert, *J. Electrochem. Soc.*, 1997, **144**, 3010–3019.
73. D. Fouquet, A. C. Miiller, A. Weber and E. lvers-Tiffée, *lonics*, 2003, **8**, 103–108.
74. T. Klemensø, C. Chung, P. H. Larsen and M. Mogensen, *J. Electrochem. Soc.*, 2005, **152**, A2186–A2192.
75. M. Pihlatie, A. Kaiser, P. H. Larsen and M. Mogensen, *J. Electrochem. Soc.*, 2009, **156**, B322–B329.
76. Y. Zhang, B. Liu, B. Tu, Y. Dong and M. Cheng, *Solid State Ionics*, 2009, **180**, 1580–1586.
77. H. Kishimoto, Y.-P. Xiong, K. Yamaji, T. Horita, N. Sakai, M. E. Brito and H. Yokokawa, *J. Chem. Eng. Jpn.*, 2007, **40**, 1178–1182.
78. M. Zhi, X. Chen, H. Finklea, I. Celik and N. Q. Wu, *J. Power Sources*, 2008, **183**, 485–490.
79. H. Kishimoto, K. Yamaji, Y.-P. Xiong, T. Horita, M. E. Brito and H. Yokokawa, *ECS Trans.*, 2009, **17**(1), 31–35.
80. J. E. Bao, G. N. Krishnan, P. Jayaweera, J. Perez-Mariano and A. Sanjurjo, *J. Power Sources*, 2009, **193**, 607–616.
81. C. A. Coyle, O. A. Marina, E. C. Thomsen, D. J. Edwards, C. D. Cramer, G. W. Coffey and L. R. Pederson, *J. Power Sources*, 2009, **193**, 730–738.
82. W. Liu, X. Sun, L. R. Pederson, O. A. Marina and M. A. Khaleel, *J. Power Sources*, 2010, **195**, 7140–7145.
83. O. A. Marina, C. A. Coyle, E. C. Thomsen, D. J. Edwards, G. W. Coffey and L. R. Pederson, *Solid State Ionics*, 2010, **181**, 430–440.
84. O. A. Marina, L. R. Pederson, C. A. Coyle, E. C. Thomsen and D. J. Edwards, *J. Electrochem. Soc.*, 2011, **158**, B36–B43.
85. C. Xu, J. W. Zondlo, M. Gong and X. Liu, *J. Power Sources*, 2011, **196**, 116–125.
86. O. A. Marina, L. R. Pederson, C. A. Coyle, E. C. Thomsen, P. Nachimuthu and D. J. Edwards, *J. Power Sources*, 2011, **196**, 4911–4922.
87. K. Channa, R. De Silva, B. J. Kaseman and D. J. Bayless, *Int. J. Hydrogen Energy*, 2011, **36**, 9945–9955.
88. C. Xu, J. W. Zondlo and E. M. Sabolsky, *J. Power Sources*, 2011, **196**, 7665–7672.

89. D. W. Dees, U. Balachandran, S. E. Dorris, J. J. Heiberger, C. C. McPheeter, and J. J. Picciolo, in *Solid Oxide Fuel Cells (SOFC-I)*, S. C. Singhal, 1989, Pennington, 89–11, pp. 317.
90. H. Frey, A. Kessler, W. Münch, M. Edel, and B. V. Nerlich, in *Handbook of Fuel Cells: Advances in Electrocatalysis, Materials, Diagnostics and Durability, Vol. 6*, ed. W. Vielstich, H. A. Gasteiger, and H. Yokokawa, 2009, John Wiley & Sons, New York, pp. 992–1001.
91. N. Christiansen, J. B. Hansen, H. Holm-Larsen, S. Linderoth, P. H. Larsen, P. V. Hendriksen and H. Hagen, *ECS Trans.*, 2007, **7**, 31.
92. S. Mukerjee, K. Haltiner, R. Kerr, L. Chick, V. Sprenckle, K. Meinhardt, C. Lu, J. Y. Kim and K. S. Weil, *ECS Trans.*, 2007, **7**, 59.
93. Y. Matsuzaki and I. Yasuda, *Solid State Ionics*, 2000, **132**, 261.
94. K. Sasaki, K. Susuki, A. Iyoshi, M. Uchimura, N. Imamura, H. Kusaba, Y. Teraoka, H. Fuchino, K. Tsujimoto, Y. Uchida and N. Jingo, *J. Electrochem. Soc.*, 2006, **153**, A2023.
95. K. Sasaki, K. Susuki, A. Iyoshi, M. Uchimura, H. Kusaba, Y. Teraoka, H. Fuchino, K. Tsujimoto, Y. Uchida, N. Jingo, in *Solid Oxide Fuel Cells IX (SOFC-IX)*, ed. S. C. Singhal and J. Mizusaki, The Electrochemical Society Proceedings Series, Pennington, NJ, 2005, PV 2005–07, Vol. 2, 1267–1274.
96. S. J. Xia and V. I. Birss, in *Solid Oxide Fuel Cells IX (SOFC-IX)*, ed. S. C. Singhal and J. Mizusaki, Pennington, 2005, PV 2005–07, Vol. 2, 1275–1283.
97. S. Zha, Z. Cheng and M. Liu, *J. Electrochem. Soc.*, 2007, **154**, B201.
98. J. Dong and S. Zha, and M. Liu, in *Solid Oxide Fuel Cells IX (SOFC-IX)*, ed. S. C. Singhal and J. Mizusaki, Pennington, 2005, PV 2005–07, Vol. 2, pp. 1284–1293.
99. J. Dong, Z. Cheng, S. Zha and M. Liu, *J. Power Sources*, 2006, **156**, 461.
100. J. H. Wang and M. Liu, *Electrochem. Commun.*, 2007, **9**, 2212.
101. A. I. Marquez, Y. De Abreu and G. G. Botte, *Electrochem. Solid-State Lett.*, 2006, **9**, A163.
102. J. B. Hansen, *Electrochem. Solid-State Lett.*, 2008, **11**, B178.
103. Z. Cheng and M. Liu, *Solid State Ionics*, 2007, **178**, 925–935.
104. Z. Cheng, H. Abernathy and M. Liu, *J. Phys. Chem. C*, 2007, **111**, 17997–18000.
105. S. K. Schubert, M. Kusnezoff, A. Michaelis and S. I. Bredikhin, *J. Power Sources*, 2012, **217**, 364–372.
106. H. Kishimoto, T. Horita, K. Yamaji, M. E. Brito, Y.-P. Xiong and H. Yokokawa, *J. Electrochem. Soc.*, 2012, **157**, B802–B813.
107. H. Yokokawa, K. Yamaji, M. E. Brito, H. Kishimoto and T. Horita, *J. Power Sources*, 2011, **196**, 7070–7075.

CHAPTER 12

Application of SOFCs in Combined Heat, Cooling and Power Systems

R. J. BRAUN* AND P. KAZEMPOOR

Colorado School of Mines, Department of Mechanical Engineering, College of Engineering and Computational Sciences, 1610 Illinois Street, 80401, Golden, Colorado, USA
*Email: rbraun@mines.edu

12.1 Introduction

The unique characteristics of solid oxide fuel cells (SOFCs) have encouraged their development for a wide variety of applications that range from portable, mobile and micro-combined heat and power (500 W to 20 kW) to larger-scale stationary power at both distributed generation ($\sim$100 kW –5 MW) and central utility scales (>100 MW). Attractive SOFC technology attributes include high electric efficiency, high-grade waste heat, fuel flexibility, low emissions, power scalability, and low unit capital cost potential when high production volumes are achieved. The high operating temperature of SOFCs enable production of varying grades of waste heat that can then be recovered for process heating, power augmentation via gas turbine integration, or for polygeneration of exportable products (*e.g.*, heat, cooling, power or fuels). The effective use of waste heat significantly impacts overall system efficiency, economics and environmental emissions. These attributes have accelerated SOFC technology development with the aim of replacing traditional combustion-based power

RSC Energy and Environment Series No. 7
Solid Oxide Fuel Cells: From Materials to System Modeling
Edited by Meng Ni and Tim S. Zhao
Published by the Royal Society of Chemistry, www.rsc.org

generation equipment, as well as offering solutions to emerging 21st Century energy problems.

While many research studies have been performed on larger-scale (>1 MW) SOFC power systems,[1–4] most of the industrial SOFC hardware technology development activity remains at system capacities of about 250 kW or less. In particular, the current state-of-the-art in SOFC stack power output is in the neighbourhood of 25 kW.[5–7] This technology status has contributed to the near-term focus of SOFC system development for both mobile unmanned aerial and undersea vehicle applications, auxiliary power units (APUs), and combined heat and power (CHP) systems in residential (1–5 kW) and commercial (10–250 kW) building markets. Thus, much of the commercialization activity among SOFC developers is aimed at relatively small-scale applications for mobile systems and residential and light-commercial building CHP (*i.e.*, micro-CHP); the latter of which is predominately occurring in Japan and Europe.[8,9]

12.1.1 Drivers for Interest in Co- and Tri-generation Using Fuel Cells

The development of SOFCs as a high-efficiency, low-emission stationary power generator is motivated by the recognition of the technology as a potential cross-cutting solution for a number of emerging paradigms related to energy supply and management. First, dispatchable distributed generation (DG) technologies are one component of a broader set of distributed energy resources (DERs) that are expected to be instrumental in enabling large-scale penetration of inter-mittent renewable resources, such as wind and solar.[10–13] Key DER tech-nologies envisioned to be significant players in the advancement of distributed generation include fuel cells, micro-turbines, energy storage devices and thermally activated heat recovery technologies (TAT). Large-scale energy storage development is envisioned as a key requirement in being able to both increase the flexibility of and modernize the electric grid, especially in the U.S. In particular, SOFC-derived technologies such as reversible SOFCs and solid oxide electrolysis, are receiving increased interest as candidates to offering viable grid energy storage solutions.[14–16]

Secondly, interest in microgrids as an energy supply and distribution solution to growing reliability and power quality problems associated with the centralized electric grid is also increasing. One of the potential advantages of microgrids is that as a semiautonomous power supply system, it can be controlled and operated as a single aggregated load, while also offering end-users the benefits of meeting their onsite needs for heat, uninterruptible power, and enhanced local reliability and power quality (*e.g.*, via maintaining voltage stability and minimizing harmonic distortion).[11,17] Importantly, the parallel interest and development of microgrid technology is relevant to smaller-scale DG as it can more effectively enable the utilization of waste heat by moving the production of thermal energy closer to the point of end-use.[11] The smaller scale of thermal energy production units also offers flexibility in matching the application requirements for heat and power. Figure 12.1 illustrates the range of DER technologies and their associated

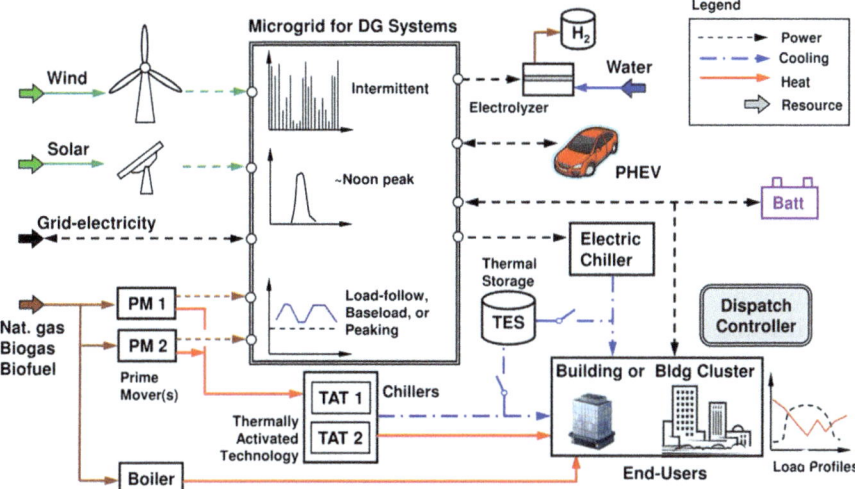

Figure 12.1 Microgrid with heterogeneous distributed energy resource technologies.

diurnal energy production characteristics that may be employed to serve building or building cluster energy demands. Renewable energy technologies, such as wind and solar, have intermittent production profiles and achieve rather low annual average energy capacity factors (~ 0.2 to 0.35). Advanced prime movers, such as fuel cells, can be integrated with TAT to provide either cooling (and/or space heating directly) to the end-user or employ thermal energy storage (TES) for later use depending on energy pricing, demand, and overall emissions considerations. Advanced DER technologies, which include fuel (hydrogen) generation via electrolyzers and plug-in hybrid vehicles (PHEV) are also shown. In such energy paradigms of the future, the ability to co- or tri-generate (or 'polygenerate') energy products becomes increasingly attractive given the relatively high costs of early production fuel cell DG technologies as it allows the allocation of capital costs to be distributed among all of the various co-products, thereby reducing the unit production costs of each commodity stream. Polygeneration encompasses CHP, combined cooling, heat, and power (CCHP), combined heat, hydrogen, and power (CHHP), and combined fuel and power (CFP). It continues to experience global interest as a means for energy supply security, efficiency enhancement, environmental impact reduction,[†] and as a response to the growing uncertainty in electricity markets due to competition and an overburdened grid infrastructure.[18]

12.1.2 Overview of CHP and CCHP

Combined cooling, heat and power is a method whereby a prime mover (such as a fuel cell or stationary engine) consumes fuel to produce power and the

---

[†]For example, through emission control and avoidance of construction of both large generating plants and the associated transmission and distribution infrastructure.

waste heat rejected from the system is recovered to provide space heating and/or hot water, and cooling which is typically (but not always) derived from a thermally activated technology such as an absorption chiller. In some cases, the term CCHP is employed for situations in which combined cooling, heat and power are not produced simultaneously; but instead heat and power are produced during the heating season, and cooling and power are produced during the cooling season. The primary difference between CHP and CCHP is that excess thermal or mechanical/electrical energy derived from the prime mover is employed to produce cooling for a CCHP application.[19] CCHP is synonymous with tri-generation or building cooling, heating and power (BCHP) and one of its main benefits is the reduction of primary energy necessary to deliver a required amount of electric power and thermal energy over separate production methods.[19,20]

Figure 12.2 depicts a generic SOFC-CHP system where a hydrocarbon fuel, such as natural gas is supplied to the system and electric power and thermal energy are exported to meet end-use requirements of a given application. Figure 12.3 depicts a schematic diagram of a typical CCHP configuration for a fuel cell. Fuel, such as natural gas, is supplied to a fuel cell sub-system and AC electric power is generated. Waste heat from the system is recovered to drive a thermally activated cooling technology (*e.g.*, absorption chiller, adsorption chiller, or dessicant dehumidifier); any remaining un-utilized thermal energy is made available for hot water production. The fuel cell-based CCHP system is comprised of the fuel cell sub-system, power conditioning, heat recovery system and thermally activated cooling hardware. In such a configuration, overall system efficiencies can range from 70–95% to produce three useful energy products compared with the 30–40% from typical coal-fired central utility power plants making electricity alone. In general, a distributed generation CCHP system can readily achieve an overall efficiency of 88–90%, while separate production methods for cooling, heat and power supplies achieve a combined efficiency of less than 60%.[19,21,22] The three commercially available thermally activated cooling technologies are absorption chillers, adsorption chillers and dessicant dehumidifiers. Direct-drive vapour-compression chilling is also available whereby steam-generated from a waste heat recovery boiler is expanded to provide the mechanical shaft power requirements of the

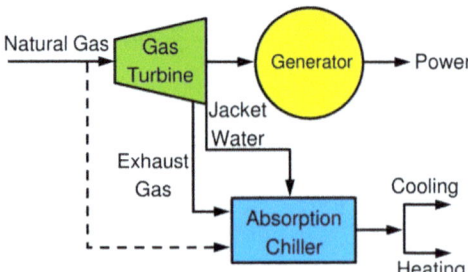

Figure 12.2   Conventional CCHP system.

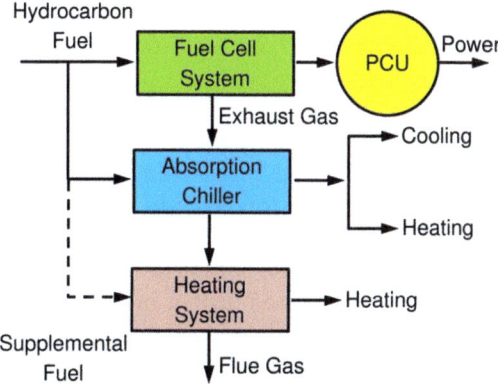

Figure 12.3 SOFC-CCHP system.

compressor in the refrigeration system. From Figure 12.3, it is apparent that an SOFC-CHP system is nearly identical to a CCHP system, except it does not contain the thermally activated components needed to supply cooling for building air-conditioning loads.

This chapter focuses on the application of SOFC technology in both CHP and CCHP systems for residential and commercial buildings. In particular, modelling approaches, integration strategies, and benefits and challenges for SOFC-based CHP and CCHP systems are presented. The chapter is organized such that a brief overview of building application requirements and economic considerations are first presented. SOFC-based CCHP system configurations and operation are then discussed. Considering the lack of SOFC-CCHP systems either commercially available or in demonstration, the presentation herein has a predominate focus on SOFC-CHP systems at relatively small scales (< 10 kW). Basic modelling approaches and techniques for system design and simulation are given next, followed by a synthesis of results and observations concerning the expected effectiveness of SOFC-CHP systems in the building energy markets. The chapter concludes with an overview of SOFC commercialization efforts, technology and economic barriers, and market outlook.

12.2 Application Characteristics & Building Integration

Building types for potential application of SOFC-CHP/CCHP systems are wide ranging and include hotels, hospitals, office buildings, educational institutions, mercantile (*e.g.*, retail), apartments, supermarkets, and single-family residential dwellings to name a few. There are many technical, economic and regulatory factors to consider when determining the suitability of a fuel cell-CHP system for application in residential and commercial buildings including:

- Building electric and thermal load characteristics
- Grid-electricity and natural gas prices and rate structures

- Utility net-metering plans
- Grid-connection requirements and regulations
- Plant siting and permitting

This section only focuses on the technical aspects of SOFC-CHP/CCHP building installations. Building load profiles and building-integrated fuel cell systems are briefly discussed in the following to introduce various application characteristics and to provide some context for subsequent discussions on system configurations, modelling, and market considerations. Economic considerations and market requirements are discussed further in Sections 12.4 and 12.7.

A useful performance characteristic for matching energy supply with energy demand is the thermal-to-electric ratio (TER). The TER of a building structure is the ratio of the thermal energy demand to the base electrical demand. A TER may be based on space heating, space cooling, or domestic hot water demands within a building and its magnitude is highly dependent on location, building type and design, usage patterns, time of day, and time of year.[23] A TER may also be expressed for an SOFC-CHP/CCHP system, in which case the ratio represents the amount of thermal energy available for export divided by the net power generated by the system. The TER of an SOFC-CHP/CCHP system depends on many factors, but in general can range from 0.5 for high electric efficiency systems to almost 2.0 for lower efficiency ones.

12.2.1 Commercial Buildings

The electrical and thermal energy demands in commercial buildings vary widely over the course of a day, season, and geographic location. Figure 12.4(a,b) illustrates an example of the diurnal electrical and thermal energy usage of a prototypical large hotel located in southern California on an hourly time-average basis. The energy demand profiles for the buildings and locations depicted in Figure 12.4 were generated using EnergyPlus software[24] and input files for standard building types defined by the U.S. Department of Energy.[25] Figure 12.4(a) shows that electric demand over the course of a winter or summer day can vary by over 220 kW where the minimum load is near 90–110 kW in the early morning hours and as high as 340 kW in the evening hours. The effect of the vapour-compression cooling load is also observable when comparing January and July days, adding as much as 100 kW to the electrical demand in the afternoon hours (which often represent a high price of electricity time period). Over the course of the day, the hourly-average space cooling TER for the building ranges from 0.35 to 1.1 in the winter and from about 1.0 to 2.1 in the summer. Building size and demand characteristics are further summarized in Table 12.1.

The thermal load profiles for the large hotel on a January day are depicted in Figure 12.4(b). In the warm climate of southern California, the building experiences significant cooling and heating loads throughout the day, where the thermal energy demand is primarily in the form of domestic hot water (DHW)

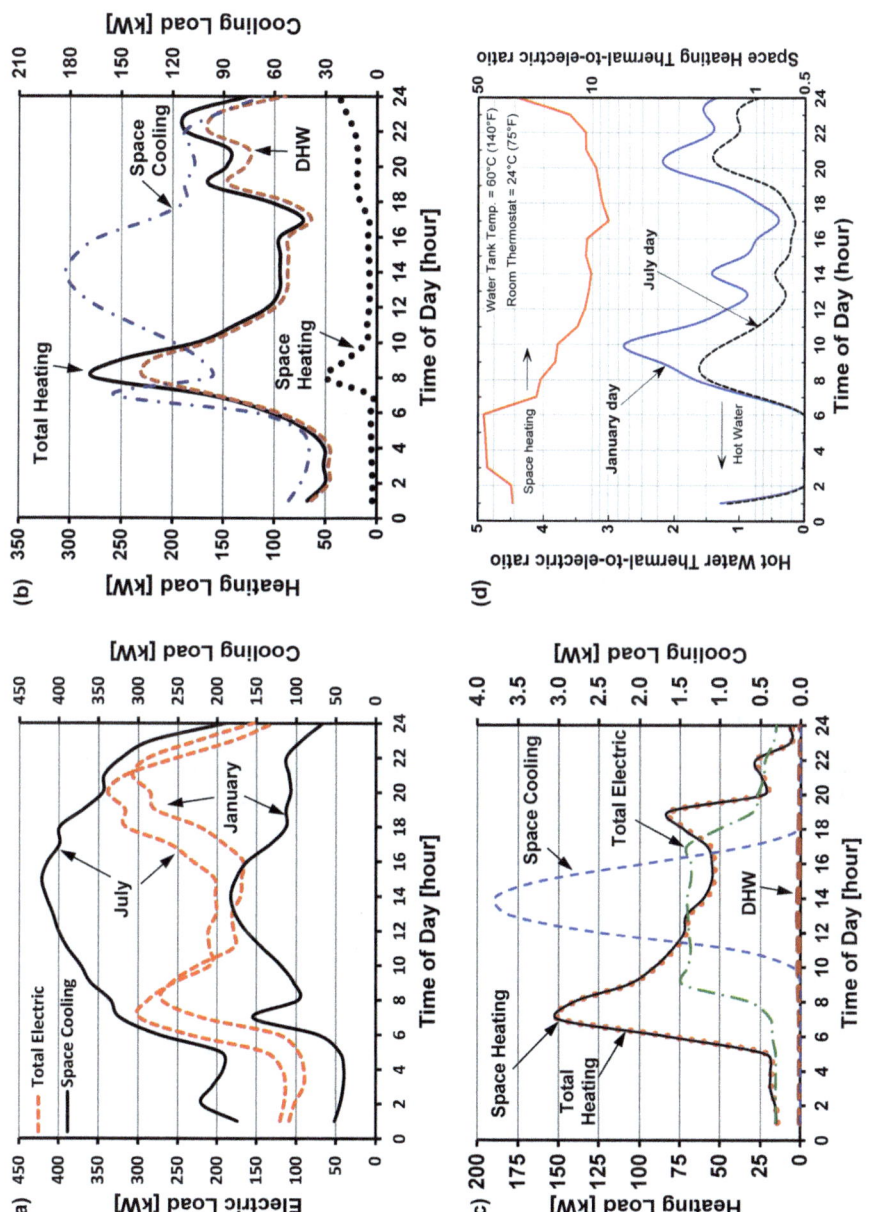

Figure 12.4 (a) Electric and cooling load profiles of prototypical large hotel in Los Angeles, CA during July and January days; (b)Thermal load profiles for prototypical large hotel in Los Angeles, CA during a January day; (c) Electric and thermal load profiles for prototypical medium office building in Boston, MA; (d) Domestic hot water and space heating TERs for a 230 m^2 residential dwelling located in Madison, WI.

Table 12.1 Summary of commercial building energy characteristics for a large hotel and medium-office in the U.S.

Statistic	Los Angeles Hotel	Boston Hotel	Los Angeles Office	Boston Office
Height (floors)	6	6	3	3
Area (thousand ft^2)	122	122	54	54
Average power demand (kW)	204	142	54	53
Maximum power demand (kW)	357	263	151	167
Minimum power demand (kW)	87	52	15	15
Average heating demand (kW)	101	294	12	51
Maximum heating demand (kW)	353	625	81	317
Minimum heating demand (kW)	29	78	0.3	0.3
Average thermal-to-electric ratio (kW)	0.49	2.16	0.29	0.96
Maximum thermal-to-electric ratio (kW)	1.24	4.23	3.1	13.6
Minimum thermal-to-electric ratio (kW)	0.16	0.95	0.01	0.02
Average cooling demand (kW)	204	137	38	29
Maximum cooling demand (kW)	599	796	228	249
Minimum cooling demand (kW)	21	0	0	0

usage. DHW-based TER values range from as low as 0.16 to over 1.2 over the course of the year (see Table 12.1).

The effect of building type and geographic location can be seen from examination of the electric and thermal loads for a medium-sized office building located in Boston, Massachusetts as presented in Figure 12.4c. The winter day loads depicted illustrate a nearly steady daytime electric load of about 65–70 kW and space heating loads exceeding 150 kW during the early morning hours. DHW heating loads are less than 5 kW and a small amount of space cooling (1–3 kW) is still required during a winter day often for cooling of the building interior. Office building TERs can range substantially over the course of a year from effectively zero to over 13 for space heating during the winter.

12.2.2 Residential Applications

Similar to commercial buildings, single-family residential applications experience significant variation in both the timing and magnitude of their energy demands. The annual hourly average domestic hot water TER for a ~230 m^2 home in the U.S. can range from 0.7–1.0.[26] Figure 12.4(d) shows the building loads for a prototypical residence located in Madison, Wisconsin during a winter and summer day in terms of TER. Load data was generated using TRNSYS[27] with typical meteorological year weather data. A peak hourly domestic hot water heating TER of less than 2.75 and a base value near 0.4 is apparent in the figure for a typical January day. The peak domestic hot water TER for a July day is about 1.6 with a base value near 0.2. Also, note both the magnitude and rate of change in domestic hot water TER during the early hours of the day. The annual hourly average domestic hot water TER is about 1.0 and this value is typical of most households in the U.S. In contrast to domestic hot water heating, the TER data for space heating is substantially higher with a peak hourly TER demand near 50 and a base load of seven. The cooling TER registers a maximum of about five during late afternoon hours. Over the course of an entire year, the annual average hourly electric load for the house is approximately 1.0 kW_e, and the average domestic hot water load is also about 1.0 kW_{th}. Residential-scale fuel cell systems typically generate TERs in the range of 0.5–2 and with the use of thermal storage, can be matched to serve domestic hot water heating loads. For micro SOFC-CHP systems, TER production in the range 0.7 to 1.0 is thus preferred for integration as it matches well with the hourly average demand of residential domestic hot water systems in the U.S.

Given the relatively slow transient response capability of SOFCs[28] and end-use load diversity, both thermal and electrical energy storage concepts may be required. In the case of grid-connected, single-family residential dwellings, electrical energy storage can be avoided by using the grid for peak power and fast dynamic power response. Operating strategy and grid-connection issues are interrelated and are discussed later in this chapter.

12.2.3 Building Integration & Operating Strategies

12.2.3.1 Building Integration

Integration of SOFC-CHP/CCHP systems within the building envelope requires interconnection with the building HVAC systems. Design and optimization of integrated systems with variable loads and environmental conditions is a complex endeavour and multi-objective in nature. The simple system diagram shown in Figure 12.5 is provided to motivate subsequent

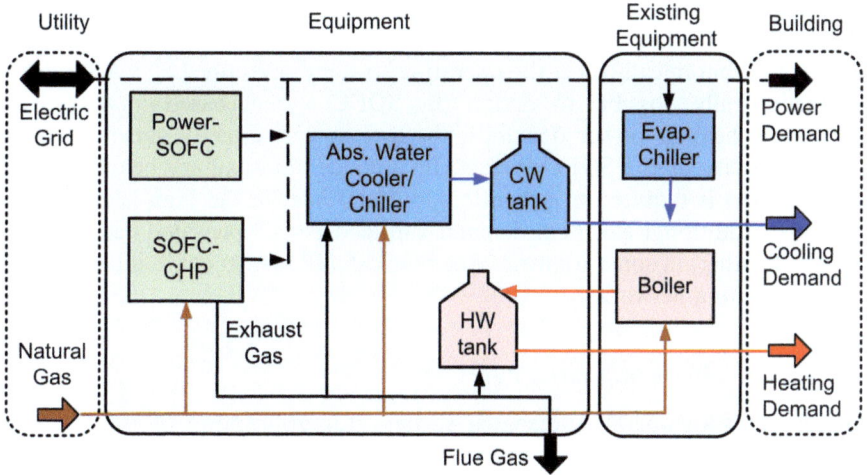

Figure 12.5   Schematic of building-integrated SOFC-CHP/CCHP System.

modelling, design, and optimization discussions. Existing equipment in commercial buildings typically involves electric-driven vapour-compression chilling and natural gas-fired heating systems. Given that most DG systems will not completely serve all thermal and electrical loads of a building, installation of SOFC-based hardware must interface with these systems, irrespective of whether the building is a new construction or a retrofit. As the previous discussion on building load profiles and SOFC TER characteristics demonstrated, some amount of thermal storage and heating capacity is preferred to be able to better match energy demands that are widely disparate. The waste heat from an SOFC power system can be recovered to provide building cooling via absorption chillers and/or to supply hot water for DHW or space heating loads as shown in Figure 12.5. The disparate timing and magnitude of building load profiles mean that thermal storage can be quite important for achieving high overall system efficiency and attractive economics. Thus, both hot and chilled water production can be stored in tanks for later use. Storage can also assist in lowering the capacity requirements of the vapour-compression systems and enable them to be 'right' sized so that they do not operate at the more inefficient part-load conditions for long durations. Hot water production from SOFC-CHP systems can also be used as boiler feedwater preheat, thereby lowering energy production costs. In addition to SOFC-CHP/CCHP systems, Figure 12.5 illustrates that combinations of power-only and CHP SOFC systems could certainly be envisioned if the application requirements are better served by lower TERs, for example.

Application of SOFC-CHP systems in smaller scale buildings, such as residential dwellings, benefit from simpler integration issues. Figure 12.6 illustrates a high-level diagram of how a micro-CHP system might interface with the equipment and loads of a single-family residence and is consistent with the few systems developed practically and/or suggested theoretically.[29–33] The residential energy system shown involves an SOFC system integrated with an auxiliary boiler and storage tank, as well as the heating and cooling systems. The SOFC is the heart of this CCHP system and must produce enough heat and power to economically meet the operating strategy envisioned.[34,35] Although it is theoretically possible to design the SOFC system based on either the maximum heat or power demands, the system cost directly depends on the rated capacity of the SOFC system. In practice, an auxiliary heat generator (*e.g.*, boiler) is required to generate additional heat in the high heat demand periods. Additional electrical demand can also be compensated easily by the grid.[36] Further, in some countries, the SOFC-CHP system could interface with district heating networks.

12.2.3.2   Operating Strategies

Selection of an operating/dispatch strategy for application of the fuel cell-CCHP system in a commercial building is important as it strongly affects the economic benefits, efficiency and overall system reliability. Furthermore, it is

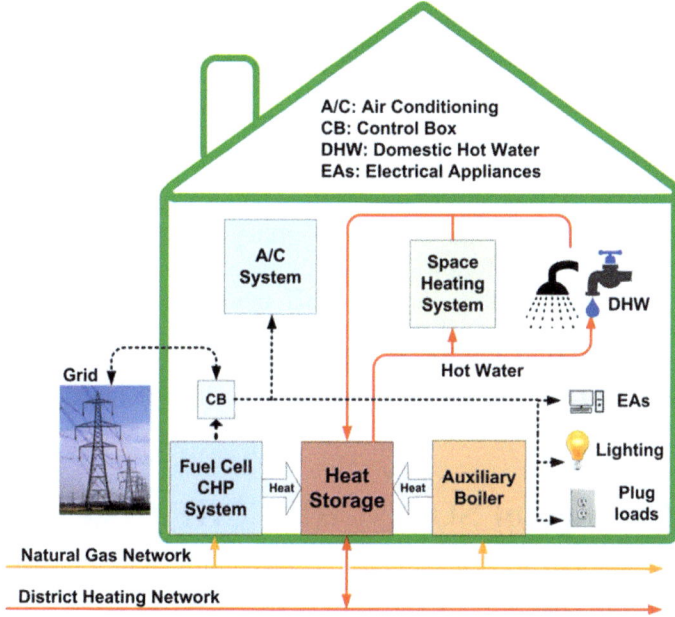

Figure 12.6   Integration of a fuel cell-CHP system into a residential building.

pre-requisite for proper sizing (*i.e.*, rated capacity) of the CCHP system and for quantifying the value proposition in any given application. The possible operating strategies include (1) electric base-loading of the SOFC system, (2) electric load-following, (3) thermal load-following, (4) seasonal load-following, or (5) peak-shaving.

In an electric base-load strategy, the SOFC-CHP system is operated in a steady-state manner at some nominal power output condition for most of the year. SOFC sizing and base-load operation can be made such that either the system power output does not exceed the expected minimum electric load of building all year or the power output exceeds the minimum building demand for a portion of the day (or year). If surplus power is generated by the SOFC system, it must be exported to the electric grid. If the SOFC-CHP system is base-loaded, then limitations on dynamic performance may be of little concern. When base-loading at rated capacity, the SOFCs are rarely turned down to part load or standby, and do not change power or exhaust gas output between time periods.

In load-following operating strategies, the SOFC system is designed to preferentially meet either the thermal or electrical demands (not both). Thus, the thermal or electrical capacity of the SOFC system exceeds either the minimum electrical or thermal requirement for the facility. With electrical load-following, the fuel cell output power changes in response to the power demand of the building. The thermal demand can be partly supported by the

SOFC system. However, an additional heat source is needed to generate more thermal energy during the high heating demand periods. With a thermal load tracking strategy, the fuel cell system is designed based on the overall heat demand and minimum amount of power required by the facility. During high electrical demand, additional power can be supplied by purchase from the grid.

Seasonal load-following involves a combination of electric and thermal load-following. Under this operating strategy, the CCHP system will operate in either load-following mode for a given month (or season) depending on the monthly (or seasonal) thermal-to-electric ratio of the building.[37] Lastly, a peak shaving CCHP system would call for fuel cell operation only during limited time periods where the time-of-day price of electricity and/or utility demand charges are so high it justifies the limited operation of the DG system.

12.3 Overview of SOFC-CHP/CCHP Systems

Much attention is often devoted to the fuel cell only, however, the SOFC is only one component of a relatively complex system. The balance-of-plant (BOP) in an SOFC system typically includes fuel pumps, air blowers, hydrocarbon fuel reformers, tail-gas combustors, and process gas heat exchangers. In fact, the chemistry and transport within components such as the reformer can be as complex as those within the SOFC.[38] Key system parameters such as performance at full and part loads (electrical, thermal and CCHP efficiencies), durability, reliability and capital cost depend strongly on the BOP and its integration with the SOFC stack. For example, it has been shown that the overall system electrical efficiency for a CCHP system is almost 20–35% lower than the stack efficiency.[23,39] The majority of these losses belong to auxiliary power consumption and inefficiency in fuel processing. Losses in the thermal components as well as the power conditioning units are also very important.[26,40] Consequently, the implementation of more efficient components and their optimal integration within the system have an appreciable effect on overall cost and benefits. This optimization is often accomplished by techno-economic modelling and design, which evaluates the most appropriate system configuration and establishes the corresponding optimal operating conditions through minimization of life cycle costs.[41]

In the following sub-section, brief overviews of system components, configurations, and operation are given. The central focus is on the SOFC system as the primary thermal energy and power generator, and thus, details regarding the building cooling and heating systems are not presented here. Importantly, the system presentation provides context for the subsequent discussion of modelling and application techniques for SOFC-CHP/CCHP systems in building applications. Additionally, the following overview proves to be useful in apprehending and evaluating the various operational and commercial SOFC-CHP systems in development as summarized in Section 12.6 of this chapter.

12.3.1 SOFC System Description for CHP (Co-generation)

A process flowsheet for a natural gas fuelled SOFC is shown in Figure 12.7.[23,39] In this system, pressurized fuel is shown to enter a desulfurizer in order to remove the sulfur compounds normally contained in utility provided natural gas at state point (2). The cleaned natural gas (3) is then mixed with super-heated steam provided by anode exhaust gas recycling (6) via an ejector to achieve the necessary steam-to-carbon ratio for fuel reforming without carbon deposition. A portion of the fuel pre-heating is also accomplished by the direct mixing of anode tail-gas with the entering fresh fuel. The fuel gas mixture (4) is then passed through an external pre-reformer, which further preheats the mixture and converts a fraction of the natural gas to hydrogen and carbon monoxide before entering the anode compartment of the cell-stack at (5). The thermal energy required to support the endothermic reforming reactions is supplied by the hot exhaust gases leaving the catalytic tail-gas combustor (15). SOFCs are air-cooled and thus, air at near-ambient conditions enters the system in excess of stoichiometric requirements via the blower (8). A portion of the pressurized air flow may be bypassed to control the burner exhaust or inlet cathode temperatures. However, the majority of the airflow is preheated by the burner exhaust gas (11). As an option, an ejector or high-temperature recycle blower may be used to recirculate a fraction of the cathode gas (13) for air pre-heating, while also reducing the size of the air preheat heat exchanger. After electrochemical oxidation of the fuel within the anode, the residual combustibles in the anode tail-gas are mixed with excess air from the cathode exhaust and catalytically oxidized in the tail-gas burner. The burner exhaust gas (15) is the highest temperature (850–1000 °C) in the system and serves as the thermal energy source stream for the downstream process gas reactors and heat exchangers. The dashed lines in Figure 12.7 indicate process flow diagrams for system concepts that may employ anode and/or cathode gas recycle.

The DC power produced by the SOFC stack must be converted into AC power (single- or three-phase, 50 or 60 Hz) for use by onsite building power demands or for export to the electric grid. This power conditioning is achieved with DC/AC inverters and is a critical component of any stationary SOFC power system. The fuel cell-CHP system necessarily requires other components too such as an air filter, flow manifolds, valves and orifices, controllers, sensors, and piping to deliver fuel and air to the stack, and remove waste gases, excess heat, and electricity. Due to the high operating temperature, selecting suitable materials and insulation for components such as pipes and valves are key to ensuring the thermal integrity of the system.

The above conceptual design is an example of system configurations intended for residential and commercial building applications in the 1–400 kW range. In larger systems (*e.g.*, > 200kW) micro-turbines can also be integrated into the

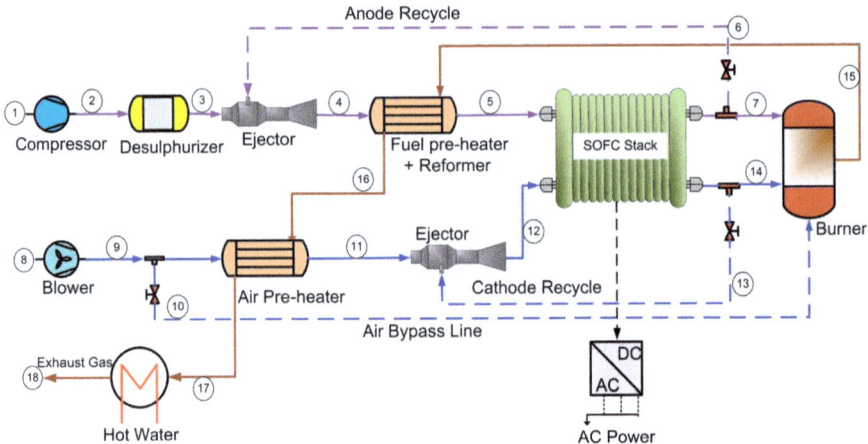

Figure 12.7   Flowsheet for SOFC-CHP system employing exhaust gas recycle concepts.

system for producing more power and consequently increasing the overall system efficiency. The coupling between gas turbine and SOFC can be done either indirectly or directly.[42] The simplest fuel cell/gas turbine integration consists of a coupling of the two components by a heat exchanger (indirect connection). In this condition, the SOFC exhaust heats compressed air in the micro gas turbine recuperator. The anode and cathode gas preheating can also be done with heat from the gas turbine exhaust gas as well as the burner exhaust flow streams. When the SOFC and the gas turbine are coupled indirectly (by a heat exchanger), TERs of in the range of 0.4 to 0.5 are realizable at SOFC operating temperatures between 950 and 1000 °C.[42]

Direct coupling of SOFC and gas turbine is typically accomplished by operating the SOFC system at pressures above 4 bar (and as high as 20 bar) and directing the exhaust from the tail-gas burner directly into a gas turbine. Direct integration configurations have been explored for both natural gas and coal gasification systems and offer electrical efficiencies of up to 70%[42] Further discussion on integrated SOFC-gas turbine cycles is provided in Chapter 13.

12.3.2   SOFC System Description for CCHP (Tri-generation)

The waste heat available in an SOFC system can be utilized to produce cooling through a thermally activated cooling technology. Figure 12.8 illustrates an SOFC system integrated with an absorption chiller and hot water heat exchanger to produce ac power, chilled water, and hot water. The high temperature burner exhaust gas can be utilized to drive a single-, double-, or triple-effect absorption chiller. If a single-effect chiller is integrated with the SOFC system, then the high-grade heat exiting the burner is first used to provide the thermal energy for fuel preheat and pre-reforming and to preheat

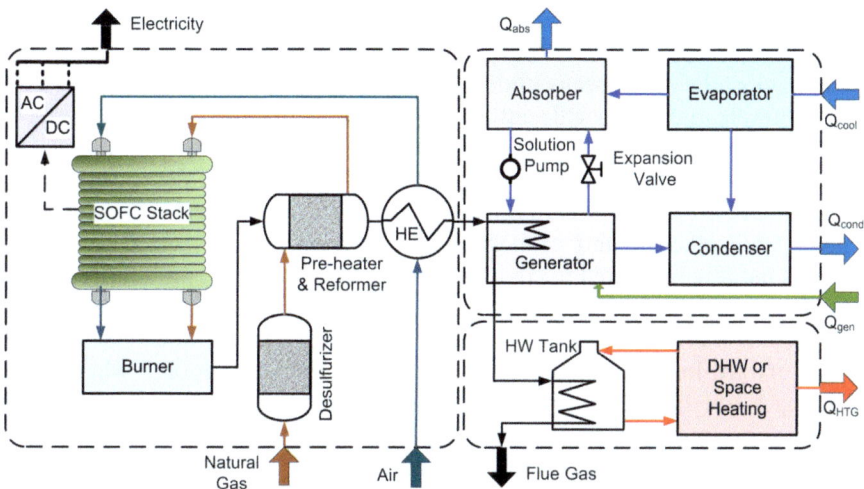

Figure 12.8 Flowsheet for SOFC-CCHP system integrated with an absorption chiller.

the cathode inlet air before waste heat is recovered in the generator section of the chiller as shown in Figure 12.8. Within the generator, the SOFC exhaust gas heats the refrigerant-absorbent mixture,[‡] resulting in a mixture separation of refrigerant vapour and a strong absorbent liquid solution. The strong liquid solution is led to the absorber where it mixes again with the refrigerant vapour generated in the evaporator. The refrigerant vapour is absorbed into the liquid solution in an exothermic process, where the resulting weak solution is pumped backed into the generator. The evaporator and condenser portions of the absorption chiller operate in the same manner as any vapour-compression system, and thus, heat rejected from both the condenser (Q_{cond}) and absorber (Q_{abs}) could be recovered for DHW or space heating purposes. Most often the heat is rejected via a cooling tower.

Double- and triple-effect absorption chillers require a higher source temperature and thus, such systems could be integrated with an SOFC system by utilizing the burner tail-gas directly in the generator. Alternatively, steam could be generated in a waste heat recovery boiler that is supplied with high-quality heat from the SOFC exhaust gas and sent to the generator of the chiller.[43]

The appropriate selection and integration of components in a fuel cell-CCHP system depends on numerous parameters such as the system size, load demands, operating strategy, utility energy pricing, climate, and the fuel infrastructure in the building zone. They also change depending on the proposed application. For example, the system can be implemented as either a co-generation system (to produce heat and power demand),[36] or as a

---

[‡]Typically mixtures of either Li/Br or ammonia-water are employed depending on the required refrigeration temperature.

polygeneration system producing cooling, heat, and power, or heat, hydrogen (or other fuel) and power.[44–46] Although currently many experimental and numerical studies have been conducted on fuel cell stacks in order to develop a more durable and highly efficient power module, only a few practical fuel cell-CHP systems have been developed. As SOFC technology is still in its relative infancy, many aspects of these systems are currently under investigation worldwide. Part of the research focuses on integrating the fuel cell in a system that is both efficient and economically attractive. Because of the emerging systems paradigms related to co- or poly-generation concepts and the interest in alternative fuel feedstocks (*e.g.*, biomass, biogas),[47] integrated system design has become a critical focus for enabling energy conversion systems congruent with forward-looking sustainable development. Furthermore, when these challenges are considered with the great variety of potential CHP/CCHP applications (many of which are unique or require custom DG solutions), identification of optimal system configurations and dispatch strategies are neither trivial nor obvious.

12.4   Modelling Approaches: Cell to System

Modelling approaches for application of SOFC-CHP/CCHP systems in commercial buildings depend largely on the purpose of the model. To enable application studies of the effectiveness of fuel cell CHP systems requires annual building energy demand profiles, CHP system models for simulation, utility energy rate structures, dispatch and control models with embedded operating strategies, and economic models for capital and operating cost estimation. Figure 12.9 depicts a model framework for modelling and simulation of DERs that are integrated with building electrical and HVAC energy systems. This structure consists of input parameter data, real-time sensor data, component and system models, decision and logic controllers, and performance analysis tasks. Input data includes weather, building construction characteristics, and costs that are utilized for design and simulation activities. Real-time data is employed to adjust demand profiles, operating strategies, and dispatch control of the various DERs. Thermodynamic component and system models are required to simulate a heterogeneous array of DG technologies. Cost models together with the overall thermodynamic system model are used for the purposes of establishing an optimal system design of the power system performance over a range of load conditions. The generated building loads are employed in a simulation of the energy system technologies dispatched to serve building energy demands. The system simulator computes instantaneous and time-averaged efficiency, economic, and environmental performance. Feedback between the distributed energy system design, operating strategy, and simulation results is necessary to assess 'optimal' application design and supervisory control schemes.

Integrated system modelling is typically carried out by dividing the system into subsystems and subcomponents, as implied in Figure 12.10. Figure 12.10

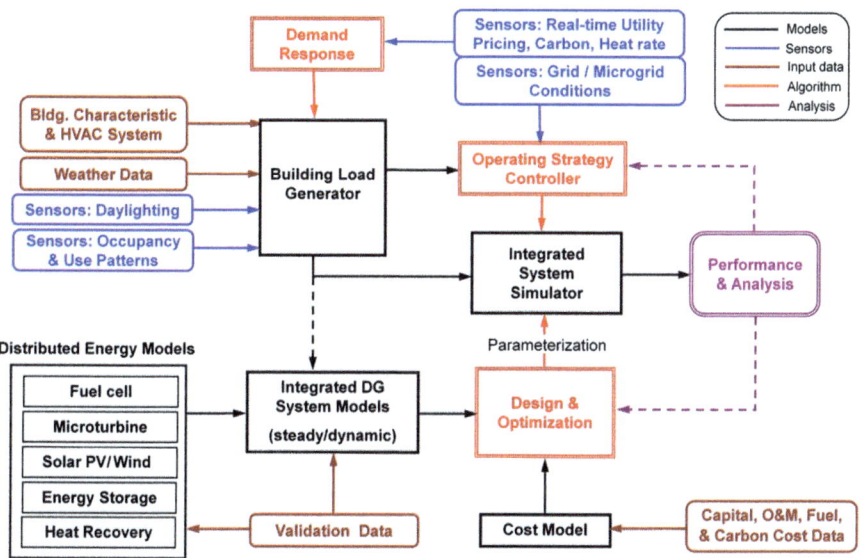

Figure 12.9   Integrated CHP-building model information flow schematic.

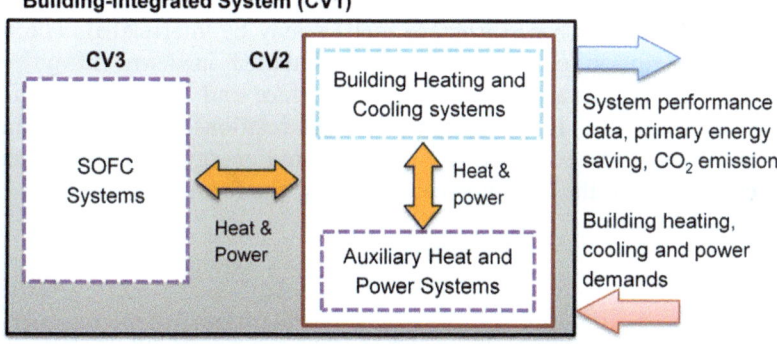

Figure 12.10   Modelling control volume schematic of building-integrated SOFC-CHP system.

shows a simplified view of a building-integrated fuel cell-CCHP system, designated as control volume 1 (CV1), that interfaces with the building energy demands. The overall system may be sub-divided into two primary subsystems: the building CCHP energy systems (CV2), and the SOFC power generator (CV3). CV2 comprises all of the CCHP system components except the SOFC system (*i.e.*, stack, BoP components including heat recovery, and power conditioning which are included in CV3). When the main objective is the conceptual design of the fuel cell system (*i.e.*, CV3), a stand-alone model of this sub-system suffices. Numerous modelling approaches that vary in fidelity from simple black box models to detailed, multi-dimensional models that have a primarily technical focus can be employed for system design and simulation.

Alternatively, model development can be based on either high-level technical or economic assessments of the whole system (*i.e.*, the level of CV1) (Figure 12.10).

Chapters 9, 13 and 14 provide substantial detail on modelling approaches for SOFC cells and systems. Thus, only a high-level overview of system and stack modelling approaches is given here with an emphasis on steady-state performance prediction of SOFC-CCHP systems. Additionally, given the emphasis on application of CCHP systems, approaches for modelling and performance estimation of the heat recovery equipment and building energy systems are summarized in this section. Approaches for system-level design and performance modelling are also provided, where the methods often used for this purpose rely on black or grey box techniques. SOFC cell/stack model formulations whose performance characteristics are semi-empirical are highlighted next. The section concludes with modelling techniques for system optimization.

12.4.1 System-level Modelling and Performance Estimation

12.4.1.1 General Modelling Overview

System-level models are typically a collection of component models that are integrated such that input and output variables are exchanged between components and whose performance metrics may be interrelated. The mathematical description of the system is formulated in terms of governing equations that are established from: (1) interface and boundary conditions, (2) conservation laws, (3) property and kinetic relations, and (4) performance characteristics of the components. The mass and energy balances written for each component in the system generally follow the form:

$$\text{Mass}: \frac{dm_{cv}}{dt} = \sum_i \dot{m}_i - \sum_e \dot{m}_e \tag{1}$$

$$\text{Energy}: \frac{dE_{cv}}{dt} = \dot{Q}_{cv} - \dot{W}_{cv} + \sum_i \dot{m}_i \left(h_i + \frac{\vec{V}^2}{2} + gz_i \right) - \sum_e \dot{m}_e \left(h_e + \frac{\vec{V}^2}{2} + gz_e \right) \tag{2}$$

where m, $\dot{m}$, t, E, $\dot{Q}$, $\dot{W}$, V, h, g, z are mass (m), mass flow rate ($\dot{m}$), time (t), energy (E), heat transfer rate ($\dot{Q}$), power ($\dot{W}$), velocity (V), enthalpy (h), gravitational acceleration (g) and elevation (z), respectively. The subscripts refer to device inlet i, outlet e, and the overall component control volume CV. Examples of device performance characteristics include fan/blower, compressor/expander, and power-conditioning efficiencies, fuel-cell polarization curves, and the effectiveness of process heat-exchangers within the system. The equations for mass and energy balances, property relationships and performance characteristics form a set of nonlinear-coupled equations incorporating design and operating variables and are common to all energy conversion devices.

12.4.1.2 Modelling Building Energy Demands, and Heating and Cooling Systems

Modelling of a building-integrated SOFC-CCHP system (*i.e.*, SOFC, CCHP, building HVAC systems, and building envelope) is mainly used to investigate the system benefits in terms of annual CO_2 emission reductions, primary energy savings, economics and overall efficiency performance compared with either utility-supplied energy or competing DG technologies.[45,48,49] The models used for these purposes usually consider the interaction between the building, CCHP system and the environment. The building (*i.e.*, the application) model input data are the building electrical, cooling and heat demands, which are strongly time dependent, as well as the building characteristics (type, location, construction, occupancy, *etc.*). Models may be developed for short- (hours), seasonal (weeks), or annual simulation of the system. Seasonal simulations are typically employed for a particular time of year, for example, to examine when the thermal or electrical demands are above the nominal ranges (*i.e.*, during winter and summer weeks).[50]

Due to the complicated nature of modelling and simulation of integrated energy systems over durations that may amount to 8,760 hours in a year, it is usually impractical to employ high-fidelity models for all sub-systems. Instead, varying fidelity component and sub-system models are integrated and employed in a fashion that largely depends on factors which relate to both the suitability (*i.e.*, flexibility, capability, availability, cost, *etc.*) of the various software platforms and the individual preferences of the modellers themselves. Currently, there are several commercial software options such as EnergyPlus[24], ESP-r[51], BeOpt[52] and TRNSYS[27] whose purpose is to generate the energy and water demand profiles in various building types (*e.g.*, see Figure 12.4(a-d)), as well as provide the means for simulating the various forms of energy supply from onsite renewable and distributed resources. Specifically, these simulation tools can model building heating, cooling, electrical, water usage, onsite DG systems and other energy flows. The models used in these tools are commonly based on quasi-steady state performance characteristics with simple approximations for transient behaviour. To simulate a building-integrated SOFC-CCHP system, it is necessary to develop simplified steady-state or dynamic models for the SOFC sub-system (CV3) since it is not typically available in standard software library components. This approach is commonly used in the literature and is described further in a subsequent section.

12.4.1.3 SOFC System-level Modelling

SOFC system-level models (as opposed to building-integrated models) may be used to establish a conceptual process design, for cost and performance analysis (in both steady-state and transient conditions), and/or for optimization purposes. These models can also be employed to evaluate the effect of some detailed parameters of the SOFC system such as fuel utilization, operating voltage, and cell operating temperature, or they can also be used to calculate the spatial distribution of one or more specific parameters inside a plant component (*e.g.*, the temperature distribution along SOFC channels).

Zero-dimensional thermodynamic models are typically employed for high-level system design and analysis purposes. In this thermodynamic modelling approach, the system can be viewed in terms of the input/output and transfer characteristics without following the details of the internal processes. This modelling approach can be sub-divided into two categories: the so-called 'grey box' and 'black box' approaches which differ from one another primarily on the basis of information resolution with respect to the system outputs. For example, in a grey box approach, given inputs of fuel type, flow, and ambient conditions, only bulk system performance, such as net power, efficiency, waste heat available, etc., is estimated. More detailed information such as stack temperature, cell temperature gradients, gas composition, *etc.* are not typically available. In black box modelling, bulk system performance estimation is accomplished by individual modelling of the entire set of integrated components which make up the system (*i.e.*, CV3-3 through CV3-7) as shown in Figure 12.11. The grey box modelling approach has been followed in various international projects focused on developing simulation tools for the conceptual design, analysis, and environmental and economic evaluation of CHP systems employing fuel cells and internal combustion (IC) engines in stationary building applications. For example, the modelling strategy employed in the Annex 42 project implemented grey box CHP (fuel cell and IC engines) models to existing whole-building simulation programs such as EnergyPlus and TRNSYS.[53] However, to solve the set of equations governed by the conservation laws (mass and energy), the performance characteristics (*e.g.*, electric efficiency and airflow versus net system power output) of the fuel cell module must be inputted. Performance curves may be approximated as simple polynomial expressions where the constant coefficients are established from either experimental data or predicted data derived from higher fidelity cell, stack, or fuel cell system models.

Modelling SOFC-CHP systems using methods consistent with black box techniques are generally more common in the technical literature. This approach involves building a model by establishing a control volume around each plant component (or set of components) and applying conservation equations, property and kinetics relations, and performance characteristics in

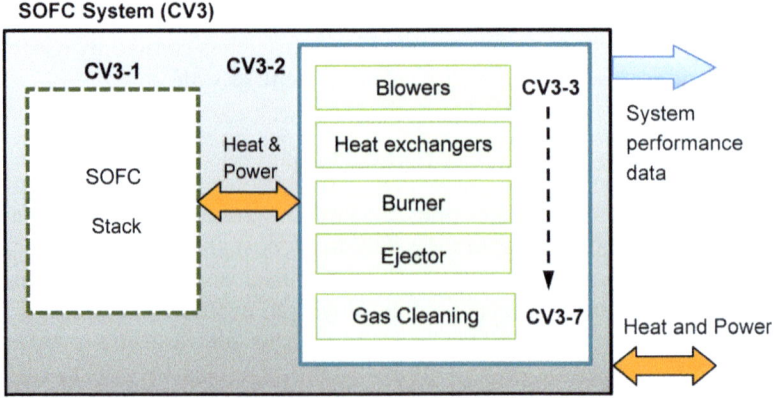

Figure 12.11   Modelling control volume schematic of SOFC and balance-of-plant.

order to generate a system of equations which characterize the physicochemical processes occurring within. One particular advantage of this modelling methodology is that it enables varying levels of model fidelity to be applied to different devices. For example, SOFC stack performance estimation can be accomplished with much higher fidelity using one- or two-dimensional models, thereby enabling cell temperature profiles and gradients, reactant utilizations, *etc.* to be resolved. SOFC system modelling and simulation can be accomplished using commercial tools such as Aspen Plus[54], gPROMS[55], and TRNSYS that have libraries for the standard (*i.e.*, BOP) components and custom models for that represent the SOFC stack and other unconventional devices. It is also not uncommon to develop system level models without using the commercial chemical engineering software platforms. Such approaches, for example, have been used by Braun *et al.*[23] and Kazempoor *et al.*[39] to develop system models in EES software environment.[56] Black box steady-state and dynamic modelling approaches for various BOP components have been provided in many studies, the details of which can be found in references.[23,29,47,49]

12.4.1.4 CCHP System Performance Metrics

Numerous system efficiencies are utilized when evaluating performance of SOFC-CCHP systems. For CCHP systems employing thermally activated cooling technologies, the absorption chiller is the most common device that is integrated with prime movers. The coefficient of performance (*COP*) is a performance metric for conventional refrigeration and absorption chiller systems and is quantified by the useful thermal energy produced divided by the energy supplied to the system as given below,

$$COP_{abs} = \frac{\dot{Q}_{cool}}{\dot{E}_{in}} \qquad (3)$$

where $\dot{Q}_{cool}$ is the cooling developed by the chiller in the form of chilled water and $\dot{E}_{in}$ is the sum of the thermal energy and auxiliary power supplied to the chiller. The nominal COPs for single-, double-, and triple-effect absorption chillers are about 0.7, 1.2, and 1.5, respectively.[19] At the system level, several useful efficiency metrics are defined as follows,

$$\text{Net system electric efficiency}: \eta_e^{sys} = \frac{P_{AC,net}}{(\dot{m}_{fuel} \cdot LHV_{fuel})_{system\ inlet}} \qquad (4)$$

$$\text{System CHP efficiency}: \eta_{CHP} = \frac{P_{AC,net} + \dot{Q}_{HR}}{(\dot{m}_{fuel} \cdot LHV_{fuel})_{system\ inlet}} \qquad (5)$$

$$\text{System CCHP efficiency}: \eta_{CCHP} = \frac{P_{AC,net} + \dot{Q}_{HR} + \dot{Q}_{gen}}{(\dot{m}_{fuel} \cdot LHV_{fuel})_{system\ inlet}} \qquad (6)$$

where $P_{AC,net}$ is the net system AC power, $\dot{Q}_{HR}$ is the amount of thermal energy from the SOFC system exhaust gas that is recovered and exported for heating purposes, $\dot{Q}_{gen}$ is the thermal energy extracted from the SOFC exhaust gas and supplied to the generator section of an absorption chiller (for example), $\dot{m}_{fuel}$ is the mass flow rate of fuel supplied to the CCHP system, and LHV_{fuel} is the fuel lower heating value. The heating and cooling efficiencies are measures of the thermal energy recovered from the system relative to the fuel energy input,

$$\text{System heating efficiency}: \eta_{HTG} = \frac{\dot{Q}_{HR}}{(\dot{m}_{fuel} \cdot LHV_{fuel})_{system\ inlet}} \qquad (7)$$

$$\text{System cooling efficiency}: \eta_{cool} = \frac{\dot{Q}_{gen}}{(\dot{m}_{fuel} \cdot LHV_{fuel})_{system\ inlet}} \qquad (8)$$

Calculating the sensible energy recovered (i.e., $\dot{Q}_{HR}$ or $\dot{Q}_{gen}$) is a straightforward thermodynamic evaluation of the change in enthalpy of the SOFC exhaust gas. For example, given the SOFC-CHP system presented in Figure 12.7, calculating the rate of heat recovered is given by,

$$\dot{Q}_{HR} = \dot{m}_{17}(h_{18} - h_{17}) = \dot{m}_{HW}c_{p,w}(T_{out} - T_{in}) \qquad (9)$$

where the h's are the enthalpy of the SOFC exhaust gas, $\dot{m}_{HW}$ is the flow rate of hot water into the heat recovery heat exchanger and T_{out} and T_{in} are the water outlet and inlet temperatures, respectively. In calculating the amount of heat recovered, the exhaust gas temperature after the heat exchanger needs to be determined and is typically above the dew point temperature of the flue gas. Prediction of the outlet gas temperature can be made once the heat exchanger performance characteristics (e.g., surface area and overall heat transfer coefficient) are known via effectiveness-NTU approaches. SOFC pre-commercial systems under development, such as the Hexis micro-CHP device, allow for condensation in the exhaust gas, thereby increasing the heating efficiency. For CCHP systems employing an absorption chiller, the generator heat exchanger effectiveness must be known.[21,43] Further, when coupling an SOFC device to a thermally driven chiller, heat normally has to be extracted and supplied to the chiller at temperatures of at least 80 °C.[19,57]

Total efficiency or CCHP efficiency of the SOFC-CCHP system is the summation of electrical (AC), heating, and cooling efficiencies and therefore can also be expressed as,

$$\eta_{CCHP} = \eta_{el}^{sys} + \eta_{HTG} + \eta_{cool} \qquad (10)$$

and similarly,

$$\eta_{CHP} = \eta_{net,el} + \eta_{HTG} \qquad (11)$$

It should be noted that although one might be inclined to employ a CCHP system efficiency that uses $\dot{Q}_{cool}$ (see Eq. 3) as opposed to $\dot{Q}_{gen}$ this could result in an overall system efficiency greater than 100% since double- and triple-effect absorption chiller COPs are greater than one. The proper efficiency expression only allows the total thermal energy extracted from the prime mover sub-system to be employed. Furthermore, industry efficiency expressions are typically based on the fuel lower heating value (LHV); but when considering CHP and CCHP systems, particularly those that utilize condensing gas heat recovery heat exchangers, a higher heating value (HHV) basis is more appropriate and reflective of the true efficiency potential of the system.

Annual capacity (or load) factors for SOFC-CCHP systems are useful performance indices considering stationary building applications. These factors are represented on electric and thermal energy-supplied bases. As given by Eq. (12), the system electric capacity factor CF_e is defined to mean the ratio of the electricity produced by the CHP system for a given time interval over the electricity that would have been produced if the plant operated 100% of the time at its rated capacity,

$$CF_e = \frac{(\text{kWh electricity supplied by CCHP system})}{(\text{Max. kWh electricity supplied at 100\% rated power})} = \frac{E_{el,actual}}{E_{el,max}} \quad (12)$$

Similarly, expressions for heating and cooling capacity factors can be written as,

$$CF_h = \frac{(\text{kWh heating supplied by CCHP system})}{(\text{Max. kWh heating supplied at 100\% rated power})} = \frac{E_{h,actual}}{E_{h,max}} \quad (13)$$

$$CF_c = \frac{(\text{kWh cooling supplied by CCHP system})}{(\text{Max. kWh cooling supplied at 100\% rated power})} = \frac{E_{c,actual}}{E_{c,max}} \quad (14)$$

12.4.2 Cell/Stack Modelling for SOFC System Simulation

Fuel cell models can be developed to meet a wide range of objectives. Achieving diverse objectives usually requires that the models incorporate significantly different levels of sophistication. For systems design and performance, the level of modelling detail required for most system components is limited to overall mass and energy balances and incorporation of component performance characteristics. However, the relative infancy of fuel cell technology requires that simulation of the fuel cell stack component be driven by a more detailed cell-level model. One-, two-, or three-dimensional cell-level models may be written depending on the requirements of the user. Two- and three-dimensional modelling are generally concerned with cell and stack design efforts. The design of SOFC stacks benefits from models that can predict gas flows through inlet and exhaust manifolds, flow distribution into multiple channel networks, and in-channel reaction chemistry and transport phenomena. Understanding and controlling thermal variations within the stack is another important stack-level design consideration. Models that focus on transport and chemistry at the microscale provide great value in assisting the optimization of MEA structures, but are not needed for stack simulation.[38] Detailed cell and stack models are outside the scope of this chapter, but some approaches are given in references.[29,38,40,58–60]

12.4.2.1   Simplified Cell-level Modelling

Relatively low-fidelity models of the fuel cell stack (see CV3-1 in Figure 12.11) can be constructed readily from modelling single-cells and then extrapolating single-cell performance to be representative of an SOFC stack of N cells. Reactant gas supply is assumed to be uniformly distributed among the cells within the cell-stack and among the channels within each repeat unit. This representation can be readily constructed as quantities such as stack voltage and stack power are scaled versions of single-cell voltage and power. Thus, in this manner, a single-cell model forms the heart of an SOFC stack model. The method is implemented by the coupling of mass and energy balances (written over anode/cathode inlets and outlets) with a polarization curve whose operating point is specified through a set of fixed operating parameters (*e.g.*, reactant utilization, cell temperature, *etc.*). Both zero- and multi-dimensional modelling can be applied with this general methodology. The simplest approach is a zero-dimensional (0-D) model of the cell which serves as a lumped, single-node thermodynamic representation that accounts for internal reforming and water-gas shift equilibrium, electrochemical polarizations and the associated heat generation, and mass transfer from cathode to anode (via cell reactions) within a single-cell repeat unit.[4,61]

A model with higher fidelity can be achieved by moving to a one-dimensional (1-D) representation of the cell, which results in a so-called channel model. One-dimensional cells models can be utilized to great effect for system studies. Depending upon the intended application and available computation resources, full stacks of cells can be represented as arrays of channels where, as noted before, each channel in the stack may be assumed to have identical behaviour.[34,35] These so-called channel models enable 1-D steady-state and dynamic cell behaviours to be simulated and are powerful tools for integration with system-level component models. These models still rely on performance extrapolation and thus, the models must be experimentally validated or calibrated for simulation and optimization purposes. In practice, of course, each channel performs differently depending upon the stack design (flow manifolds, thermal insulation, *etc.*).[38] However, the differences may not be meaningful in terms of errors in stack voltage and power prediction.[62] The 1-D modelling approach has been successfully used for the SOFC system studies using different cell geometries (*i.e.*, planar, tubular, delta, segmented-in-series, *etc.*).[23,29,47] Further discussion of one-dimensional channel-type cell models is given in references.[34,38,39,58]

In both zero- and one-dimensional approaches, the cell model is comprised of three compartments – the anode, the cathode and the electrolyte. Figure 12.16 depicts the model architecture of a 0-D, single-cell where the temperature across the cathode is typically specified, and T_{cell} and P_{cell} are the temperature and pressure at which the electrochemistry functions are evaluated. Mass balances are written individually for the anode and the cathode compartments taking into account that the consumption of H_2 in the anode and O_2 in the cathode is governed by Faraday's law and is proportional

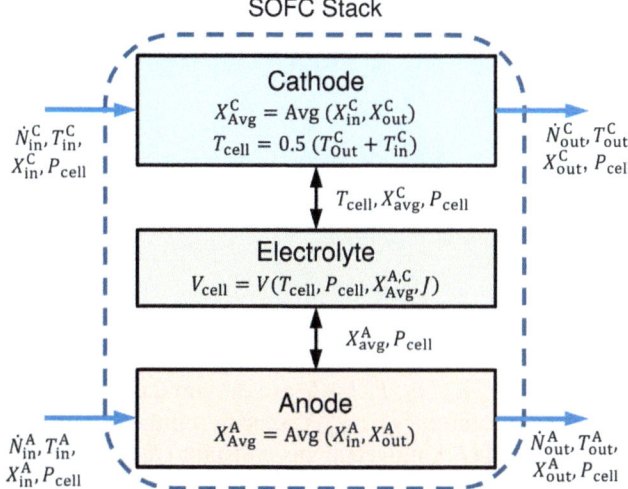

Figure 12.12 SOFC stack model overview.

to the current density. As given in the Figure 12.12, $\dot{N}^C$ and $\dot{N}^A$ are the molar flow of species into or out of the cathode and anode, respectively. The terms X^C and X^A are the molar fractions of species at the cathode and anode inlets and outlets, respectively, T is either the cathode gas, anode gas, or cell temperature, and P_{cell} is the pressure. This approach presumes that hydrocarbons and carbon monoxide are not electrochemically active but are consumed rather through reforming and water-gas shift (WGS) reactions. The produced H_2 then is the only participant in electrochemical oxidation at the triple-phase boundary and the WGS and reforming reactions are taken to be in equilibrium at the anode outlet. Mass balance equations must account for compounds consumed/produced due to the WGS, reforming and electrochemical reactions. Quantities such as fuel utilization and O_2-stoichiometry can either be calculated from the mass-balance equation framework or specified as input parameters.

An overall system energy balance (see Eq. (2)) accounts for enthalpy-flows and external heat losses from the stack. The total enthalpy-flow into the system has two components: the anode inlet flow and the cathode inlet flow. Similarly, the enthalpy-flow out of the system has the anode outlet and cathode outlet flow components. When a load is applied, the lumped system produces power and rejects thermal energy to both the surroundings and the cathode cooling air stream.

The electrochemical model that translates the charge-transfer equations into a cell voltage (and ultimately into a cell voltage-current performance map) is summarized below:

$$V_{cell} = V_{Nernst} - \eta_{Ohmic} - \eta_{Act} - \eta_{Conc} \tag{15}$$

$$V_{\text{Nernst}} = E_O + \frac{R_u T}{n_e F} \ln \left(\frac{P_{\text{H2}} P_{\text{O}_2}^{0.5}}{P_{\text{H}_2\text{O}}} \right) \qquad (16)$$

$$\eta_{\text{Ohmic}} = J \cdot R \qquad (17)$$

$$\eta_{\text{Act}} = \frac{2R_u T}{n_e F} \sinh^{-1} \left(\frac{J}{2J_{0,\,\text{Anode}}} \right) + \frac{2R_u T}{n_e F} \sinh^{-1} \left(\frac{J}{2J_{0,\,\text{Cathode}}} \right) \qquad (18)$$

$$\eta_{\text{Conc}} = \eta_{\text{act}}^{\text{anode}} + \eta_{\text{act}}^{\text{cathode}}$$

$$= \frac{R_u T}{n_e F} \sinh^{-1} \left(\frac{1}{1 - J/J_{\text{L,O}_2}} \right) + \frac{R_u T}{n_e F} \sinh^{-1} \left(\frac{1 - J/J_{\text{L,H}_2}}{1 - J/J_{\text{L,H}_2\text{O}}} \right) \qquad (19)$$

where E_0, V_{Nernst}, R, P_i, R_u, n_e, F, J_0, J_L are the standard equilibrium potential (E_0), the Nernst cell potential (V_{Nernst}), specific ohmic resistance (R), partial pressure of gas species i (P_i), universal gas constant (R_u in J k^{-1}mol^{-1}), number of electrons transferred per electrochemical reaction ($n_e = 2$), the Faraday constant ($F = 96485$ C mol^{-1}), exchange current density (J_0 in Am^{-2}) and limiting current density (J_L in Am^{-2}), respectively. In a lumped model, the fuel cell operating conditions (T, P_i) are then based on the average cathode gas temperature and the average gas compositions of the inlet and outlet fuel and air streams. These averaged quantities are employed in the electrochemical model equation set given by Eqs. (15)–(19).

Eqs. (15)–(19) are coupled, non-linear functions of the temperature and gas species inside the stack. Therefore, the set of governing equations can only be solved iteratively.[60] The main limitation with the above approach is that because the channel gas composition and temperature variations are neglected, different results are obtained depending on if the inlet, outlet or average values of these parameters are employed for the calculations. Bove et al.[60] have examined the effect of considering the three different reference values (i.e., inlet, outlet, or average) on the V-J cell polarization curve using a black box SOFC model. Their results show that the effect of fuel consumption along the gas channel cannot be estimated if the inlet gas composition is used. The cell voltage may also be underestimated if the output gas composition is considered into the calculation. A good agreement between the experimental and numerical data can be obtained by considering average values between inlet and outlet streams, however, the choice of utilizing average or outlet compositional values depends to some extent on how well the polarization characteristic can be fitted to the data. The coupled nature of the governing equations typically requires an iterative numerical solution algorithm. When using the average values, the iteration can be started by predicting the outlet gas compositions and outlet temperature of the flow streams. The set of equations from the electrochemical model (Eqs. (1), (2), (15)–(19)) can then be solved for the operational voltage. With knowledge of the cell voltage, the mass and energy balances equations can be solved simultaneously for predicting the outlet parameters, until convergence is obtained.

The 0-D model input/output and parameters are summarized in Table 12.2. The inputs to the cell model are average cell current density (or cell voltage),

Table 12.2 SOFC model inputs, outputs, and parameters.

Inputs	Outputs	Model Parameters
Avg. current density, J_{cell} (or V_{cell})	Cell voltage, V_{cell} (or J_{cell})	V-I characteristic (R, J_0, J_L)
Inlet gas temperatures, T_{in}^C, T_{in}^A	Cell power, P_{DC}	Cell geometry: Area
Reactant gas temperature rise (DT)	Fuel flow, $\dot{N}_{in}^A$, (or U_f)	
Cell temp., T_{cell} (or $\dot{N}_{in}^A$ or DT_{air})	Air flow, $\dot{N}_{in}^C$	
Fuel utilization, U_f (or $\dot{N}_{in}^A$)	Outlet fuel and air temps, T_{out}^C, T_{out}^A	
Inlet gas compositions (X_i)		

air-to-fuel ratio, inlet fuel and air temperatures, anode and cathode gas temperature rises, fuel utilization (or fuel flow), and inlet compositions. Fixed parameters (*i.e.*, geometry and performance characteristics) are the polarization curve constants, and cell area. The outputs of the model are cell voltage (or current density), power, efficiency, air flow, fuel flow (or fuel utilization, U_f) and outlet temperature of the fuel and air streams.

Although this modelling approach is usually sufficient for the preliminary design and concept studies of SOFC-CHP systems, the method is only concerned with the input and output values and cannot generate any information about the distributed parameters inside the stack.

The successful design and analysis of an SOFC system generally requires a more detailed model of the components especially the cell-stack. For example, the local temperature gradient (and local solid temperatures, fuel depletion zones, *etc.*), as well the local steam to carbon ratio (SC) are important parameters which must be maintained within specified limits in order to ensure no harmful damage occurs. Even for a well-designed SOFC system, these parameters may exceed their allowable range. The maximum allowable temperature gradient and increase are about $15\,\mathrm{K\,cm}^{-1}$ and $150\,\mathrm{K}$, respectively, SC > 2 is also necessary for protecting the cell from any carbon deposition.[63] Thus, for model objectives intent on establishing viable SOFC system designs and performance prediction, a higher-fidelity model of the SOFC stack is necessary in order to resolve the distributed functions of temperature, current density, composition, *etc.*

12.4.2.2 Simplified Stack Modelling

The use of multi-dimensional SOFC stack modelling tools is generally too computationally intensive and inefficient for system design, simulation and optimization purposes. One exception can be in designs where the SOFC power module is tightly thermally integrated and, as a result, boundary and input conditions to the various unit operations within the system are coupled due to the proximity and packaging geometry of system components. In such cases, large-scale computational fluid dynamics software augmented with special purpose software for cell electrochemistry may be employed for system simulation.[40,64] However, the more common approach is to extend the 1-D cell model to represent stack performance by considering appropriate assumptions and losses. A 1-D model usually relies on several assumptions such as uniform distribution of feed gases to each individual cell and channel, adiabatic boundaries at the cells or channels surrounding area, isopotential surfaces at each cell, *etc.* Even for a well-designed stack, these assumptions might not be valid and the associated losses must be considered in the calculation. In addition, experimental[65–67] and numerical studies[65,68,69] show that the voltage versus current density for SOFC stacks is significantly below the results presented for button- and single-cell configurations. Therefore, it is likely that besides the major cell overpotentials (*i.e.*, ohmic, activation and concentration), there are other significant losses that also must be considered to extend a 1-D cell model be more representative of stack performance estimation. Other cell polarizations may be related to an increased overall ohmic resistance of the entire stack due to: (1) the contact resistances at the interfaces between the electrodes and the electrolyte, (2) contact resistance between the electrodes and the current collectors, and (3) the resistance of stack current collection and associated wiring. The small contact area between ceramic components and resistive phases or potential barriers at the cell interfaces are two main contributions of the overall cell contact resistance. Besides the contact resistance, long in-plane conduction paths inside the electrodes can also be counted as a source of additional ohmic losses (if true, such a resistance could negate the appropriateness of isopotential electrode surface assumptions in the cell model). There are several parameters such as mechanical load, and both the temperature and size of cell components which can affect the overall cell contact and in-plane resistances.[66] In some recently reported experimental studies, 60% (or even more) of the overall stack voltage losses are estimated to be attributable to the above contact resistances.[70]

A modification to the ohmic polarization term given by Eq. (17) can then be made as follows,

$$\eta_{Ohmic} = J\ (R_{PEN} + R_{IC} + R_{contact}) \tag{20}$$

where R_{PEN}, R_{IC}, and $R_{contact}$ are the ohmic resistance of the PEN structure, interconnectors, and the resistance due to the contact between the cell components, respectively.

Evaluation of the pressure losses in the stack components are also a challenge in extrapolating cell performance prediction to full stack results. As mentioned before, the blower, pump, and compressor must be used to overcome the system pressure losses and are major contributors to part of the SOFC system auxiliary power consumption. Thus, the correct evaluation of the pressure losses in each individual component is an essential part of the system modelling. With 0- and 1-D reduced order models, the pressure losses in the fuel and air manifolds as well as the feed headers cannot generally be calculated. A generic model which accounts for the pressure losses in the stack certainly has its limitations and shortcomings, as SOFC stack analysis is very much dependent on the actual design. A number of studies concerning experimental and mathematical modelling have been presented for calculation of the pressure losses in SOFC stacks.[71,72] A general model which solves mass and momentum equations to predict pressure drop and flow uniformity within individual channels based on dimensionless groups has been developed in reference.[72]

12.4.3 System Optimization Using Techno-economic Model Formulations

Techno-economic modelling is a method whereby the technical performance characteristics of an energy conversion system translate into economic outcomes, such as the levelized cost of electricity, net present value, or other life cycle cost metrics. One objective of SOFC-CCHP system design optimization is to judiciously account for the competing objectives of capital and operating cost minimization subject to both system design and application constraints. The importance of techno-economic models is that they enable quantification of the economic benefits of CCHP system operation in a given application. Fuel-cell system performance characteristics are largely driven by cell-stack design parameters such as cell voltage, fuel utilization, operating temperature and cathode gas temperature rise. Further, the design operating point strongly influences the capital costs of the major system hardware components, such as the SOFC stack, fuel reformer, airblower and preheater, and heat recovery equipment. The operating costs are primarily associated with fuel consumption (or efficiency). Quantitatively understanding and predicting the cost–benefit trade-offs is the objective of techno-economic modelling and optimization.

Minimizing the SOFC-CCHP system life cycle costs (LCCs) is often employed as one basis for system optimization. For distributed CCHP applications, the LCC may be expressed in terms of an effective levelized cost of electricity (LCOE) where waste heat recovery for heating and cooling purposes provides value and is incorporated into the *LCOE* calculations. Cost models incorporate the forecasts for manufacturing costs of the SOFC and BOP components. The models consider capital and maintenance costs, utility energy prices (grid electricity and natural gas), interest and energy inflation rates, and system efficiency.

In a distributed CCHP application, the LCOE can be expressed as,

$$LCOE_{CCHP} = \underbrace{\frac{(CRF \cdot C_{CCHP}^{sys})}{CF_e \cdot A_{plant}}}_{\text{system capital cost}} + \underbrace{\sum MC_j}_{\text{maintenance}}$$

$$+ \underbrace{\left(\frac{F_c}{\eta_e^{sys}}\right) \cdot \left[1 - CF_h\left(\frac{\eta_{HTG}}{\eta_{B,HS}}\right) - CF_c\left(\frac{\eta_{cool}}{\eta_{B,CS}}\right)\right]}_{\text{Fuel cost} - \text{Thermal energy credits}}$$

(21)

where the first term in this expression is associated with the capital costs, the second term with maintenance costs, and the third term with fuel costs. The capital recovery factor (CRF) is defined to mean the ratio of a constant annuity and the present value of receiving that annuity for a specified period of time. The installed capital cost for an SOFC-CCHP system is expressed as C_{CCHP}^{sys} (in \$/kW). The system capacity factors CF_e, CF_h, and CF_c, have been previously defined (See Eqs. (12) to (14)). In the event that there is system downtime, the expected annual plant availability, A_{plant}, will be less than 8760 hours. The levelized annual maintenance cost, MC_j is the sum of each of the component j contributions (in \$/kWh). The unit fuel cost is F_c (e.g., \$/kJ) and $\eta_{B,HS}$ and $\eta_{B,CS}$ are the building heating and cooling system efficiencies, respectively, which would get displaced by the installation of the CCHP system. Transmission and distribution costs do not factor into the LCOE for on-site power generation.

Detailed capital cost data for SOFC systems is given elsewhere[41,73,74,75] and can be used to generate cost functions that are employed to estimate the first costs, such as in Eq. (21). The LCOE can serve as the basis for an LCC objective function that is minimized to optimize the hardware configuration in a system or to optimally select design parameters within a given system configuration.[41] Optimization of the system configuration for SOFCs has been explored parametrically[39,41] and more recently using mixed-integer linear and nonlinear programming[76,77] including system sizing and optimal dispatch of an SOFC-CHP system for a building application.[57,78] The objective function that is formulated from minimization of the system *LCC* is subject to constraints such as mass and energy conservation, property and kinetics relations, and performance characteristics of all hardware within the system. The resulting optimization problem is highly nonlinear and usually involves several independent variables to optimize on.[78]

12.5   Evaluation of SOFC Systems in CCHP Applications

12.5.1   Micro-CHP

Micro-CHP is the simultaneous generation of heat and power for small-scale building applications such as residential homes and small commercial buildings whose electric power demands are generally lower than 20 kW. The residential energy sector is one potential application for SOFC-CHP systems and is responsible for 22% of the total annual energy consumption in the U.S.[79] with over 69% of that energy consumption being used for low-efficiency space heating (43%), domestic hot water (18%), and air-conditioning (8%).[80] While the residential sector has substantial room for improvements in energy efficiency, it is also one of the most challenging markets to compete in due to the low cost and maintenance, and high durability and efficiency requirements. Despite these requirements, there are substantial potential benefits to deployment of the technology and numerous SOFC companies are developing systems for residential CHP applications as discussed further in Section 12.6.

Several studies have examined the potential benefits and challenges of grid-connected SOFC-CHP systems in this market sector. Average residential building electrical energy demands vary substantially and depend on the size and building construction of the dwelling, occupancy patterns and geographic location. Average hourly electric demand typically ranges between 0.50 kW and 2.0 kW and thus, studies involving SOFC-based micro-CHP systems are often based on 1 kW size systems.[41,49,81] The operating strategy for micro-CHP systems is dependent on several factors including the annual or seasonal TER of the building and utility energy pricing and net metering plans. In northern European countries, where annual thermal demands are relatively higher than warmer climates, operating strategies for micro-CHP are typically envisioned to be heat-led, although cost optimal operating strategies often involve a combination of heat- and electricity-led load-following modes over the course of a year.[49,82,83]

SOFC-CHP systems that employ some amount of steam methane internal reforming have been predicted to achieve net system electric efficiencies of about 45%-LHV (40%-HHV) at nominal single-cell voltages of around 0.75 V/cell, 85% fuel utilization, and a nominal operating temperature of about 750 °C.[41,23] Total CHP efficiency is estimated to range between 75 and 85%-LHV depending on the SOFC system configuration, heat recovery strategy, and the magnitude of the system heat losses. Higher electric efficiency (above 55%) is possible when operating the SOFC stack at higher cell voltages (*i.e.*, approaching 0.85 V/cell), but this also requires a larger (by as much as 3X) and more costly cell-stack to achieve the same power output. Table 12.3

Table 12.3 Basic technical characteristics of four key micro-CHP systems synthesized from published literature.

	Internal combustion engine	PEFC	SOFC	Stirling engine
Electrical efficiency (part load, full load)	10%–20%	30%, 26%	45%, 40%	5%, 10%
Overall efficiency (part load, full load)	80%, 85%	80%, 85%	75%, 80%	80%, 90%
Supplementary thermal system efficiency	86%	86%	86%	86%
Minimum operating set point (% of rated power)	20%	20%	20%	20%
Minimum up-time (min)	10	60	60	10
Maximum ramp rate (kW_e min^{-1})	0.2	0.2	0.05	0.2
Start-up energy consumption (kW_e, kW_{th})	0.008, 0.5	0.017, 1.6	0.017, 2.0	0.008, 0.5

provides a comparison of the micro-CHP system technological characteristics that have been synthesized in the literature.[84]

Simulations of 1 kW SOFC micro-CHP systems with about 35%-LHV electric efficiency (95% CHP efficiency) for residential buildings in Europe have been performed operating in a heat-following control strategy to supply hot water for DHW and space heating loads.[49] The authors found that non-renewable primary energy reductions of over 45% could be achieved if the thermal energy output of the fuel cell matched the thermal demands of the building. The capacity factor of the fuel cell was also observed to improve when applied in multi-family housing. The influence of an application in colder climates has also been noted to increase the annual CHP efficiency when in electric load-following modes due to the higher annual water heating demand.[50]

Selection of system size, operating strategy and utility electricity and gas prices are critical parameters in the successful deployment of SOFC micro-CHP systems. Simulation studies of an SOFC-CHP system configured with anode gas recycle can provide broad insights into application considerations as illustrated in Figure 12.6 where the electric power and heat produced are supplied to a prototypical U.S. residence. For example, one such study integrated the aforementioned system with a two-tank (preheat heat exchanger and a standby tank) thermal storage system to supply domestic hot water to serve the energy demands of a household located in Madison, Wisconsin (see Figure 12.4(d)).[34] Capital and maintenance costs for SOFC-CHP systems with electric power capacities of between 1 to 5 kW were estimated and simulations on an annual hourly basis were performed. Installed unit capital costs were based on high volume, mature technology cost projections and ranged from 1500 $/kW for the 5 kW system to 2450 $/kW for the 1 kW system in the analysis. The simulation studies are for grid-interconnected systems in which the electric utility acts a peaking plant, providing power to the house when the instantaneous electrical demand cannot be met with the SOFC system. Importantly, maximum system turndown for all systems simulated was 5:1 or 20% of rated electric capacity; thus a 2 kW system, for example, would either have to shut-off or move to a hot standby mode if a load of 400 W was not available. A sample electric-led, load-following operating strategy for a 2 kW SOFC-CHP system is depicted in Figure 12.13(a,b), showing both building load and fuel cell electric power and thermal energy production profiles.

Figure 12.14(a) shows that the SOFC electric capacity factor (CF_e) decreases with increasing system size. The electric capacity factor performance of the fuel cell indicates that a 2 kW size solid oxide fuel cell is under-utilized as only 43% of its annual electrical energy production capacity was used. Decreasing the fuel cell system to 1 kW increases the electric capacity factor to 77%. Other methods to increase the fuel cell system electric capacity factor include using an even smaller SOFC system, base-load operation using 'net metering,' employing lead acid batteries, and heat pumping. The heating thermal capacity factor (CF_h) of the fuel cell system is also a useful measure and Figure 12.14(a) indicates that a 1 kW SOFC system could achieve an 83% fuel cell thermal capacity factor,

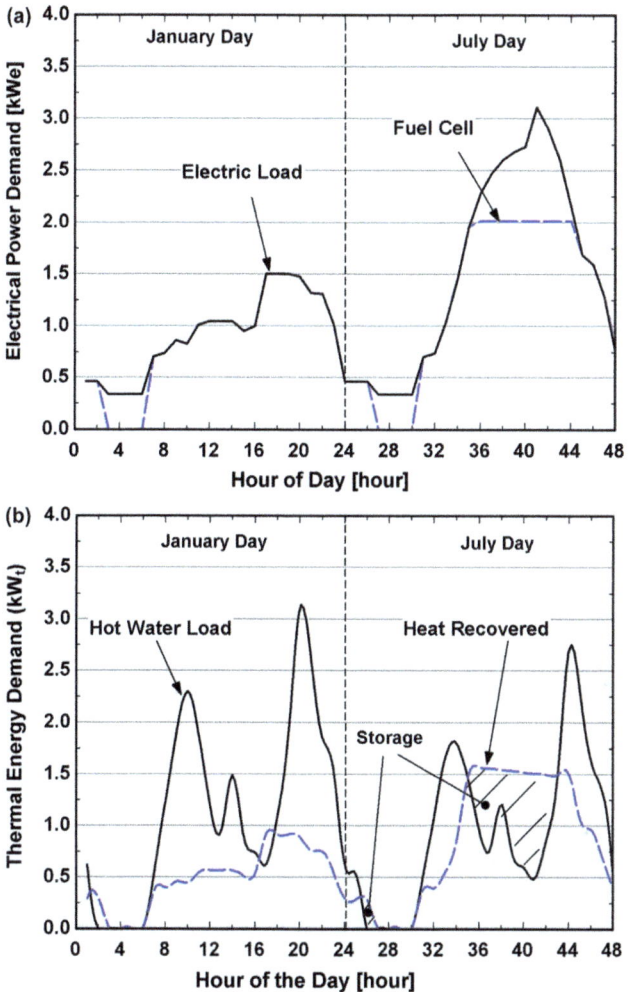

Figure 12.13 (a) System load-following power output and building demand plot on two days of the year (b) system thermal energy output and building thermal demand plot on two days of the year.

displacing 6000 kWh of thermal energy that otherwise would have been served by a conventional hot water heater.

The 'home' capacity factors are indices that indicate the effectiveness of the fuel cell system in meeting the household electric or thermal loads. That is, it is the total kWh of electricity or thermal energy supplied by the fuel cell system divided by the total kWh household electricity or heat demanded. As system capacity is increased, the house electric capacity factor passes through a maximum at 2 kW and decreases with increasing system rating due to the system turndown limitations. From the viewpoint of meeting household electrical energy demand, a 2 kW system is energetically optimal. Thus, unless the

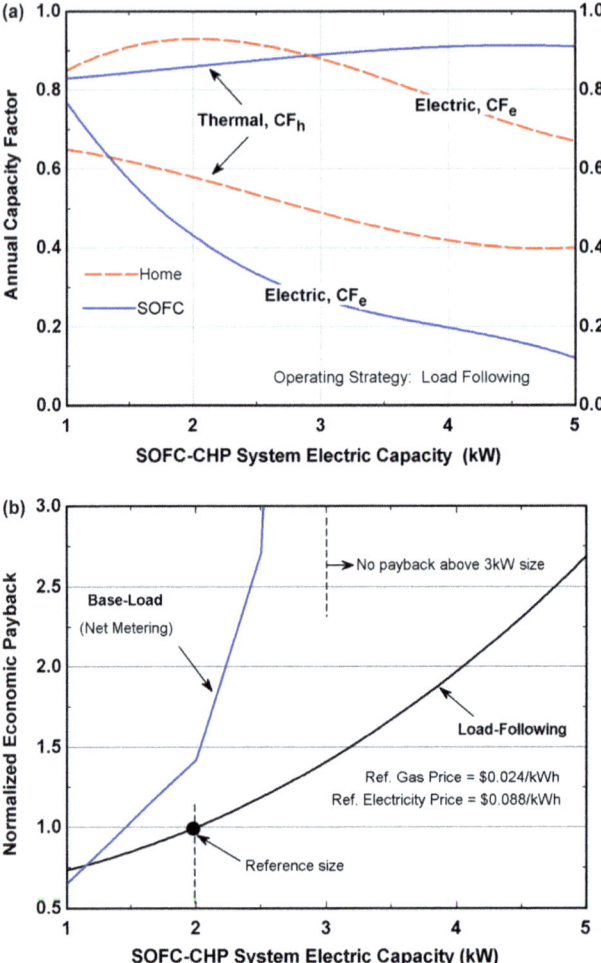

Figure 12.14 (a) Influence of power rating on SOFC-CHP system and home capacity factors (b) influence of power rating and electric-led operating strategy on normalized payback economics.

system turndown capability can be substantially increased, a 5 kW SOFC system should not be employed in a load-following scenario for both economic and household energy effectiveness reasons.

Grid-connected SOFC systems benefit from buyback of electricity from the utility when excess power is produced. Utility electric purchase price agreements for renewable energy systems, such as solar photovoltaic systems, are at retail grid-electricity prices, however, that is not always the case for CHP systems. Surveys of net metering programmes in the U.S. indicate that the average utility will repurchase excess electricity at retail rates as long as there is no net electrical energy production onsite over the billing cycle (*i.e.*, running the meter backwards).[85] Once an SOFC-CHP system becomes a net power producer, the utility will only buy back the power at avoided cost which typically ranges from 1.5 to 2.0 ¢/kWh.[34] Assuming a utility buy back rate of 2.0 ¢/kWh, Figure 12.14(b) provides a comparison of the economic effectiveness between electric-led load-following and base-load operating strategies as a function of SOFC system capacity where the payback is normalized to the 2-kW system payback period. In the base-load scenario, the SOFC system is operating in a net metering manner. As system size is increased, fewer savings are realized for the base-load scenario resulting in an infinite payback above 3 kW-sized fuel cell systems. A 1 kW-size SOFC system operating in a base-load configuration has the lowest economic payback period (*i.e.*, it is the most attractive) since it has high capacity factor and the lowest quantity of electrical energy purchased by the utility at the avoided cost. Figure 12.14(b) also shows that for single family residences, the economics grow less attractive with increasing system size as capital costs increase at a rate greater than the annual savings.

Annual simulations (8,760-hr) of a base-loaded 1-kW SOFC-CHP system produce an overall electric efficiency of 44.5%-LHV and a CHP efficiency of 84.6%-LHV. When operated in a load-following control strategy, the electric efficiency increases to 46.0%-LHV and the CHP efficiency to 85.5% due to more frequent part-load operation of the SOFC which results in higher efficiency. Despite the efficiency advantage of electric-led load-following operation, the base-loaded system produces a 10% lower payback period.

Since the SOFC system typically cannot respond to load changes very fast (slow dynamic response) without careful control and small energy storage, a thermal load tracking strategy can be attractive.[49] However, a thermally led, load-following strategy has some disadvantages. During the heating season the system produces too much electricity, which is only economic if it can be sold to the grid. Additionally, given the characteristically low TER of high-efficiency SOFC-CHP systems, sizing the system based on thermal demand requirements will result in a power capacity much higher than the application requires, thereby also resulting in much higher investment costs.

The amount of the heat recovered can strongly affect the overall economics of micro-CHP applications, but its value is also proportional to the price of natural gas. The lower the price of natural gas, the less influence the amount of waste heat recovered has on the value proposition. Moreover, a better indicator for the viability of CHP applications is the so-called 'spark spread,' which is defined as the price of grid electricity minus the price of pipeline natural gas. Plots of the normalized economic payback versus spark spread are given in Figure 12.15(a) for a 1 kW electrically-led, base-loaded SOFC system operating either as a CHP system or as a power-only system. The economic payback period of all plots is normalized against the SOFC-CHP ('cogen') system at the reference spark spread of 6.4 ¢/kWh. At this condition, the payback period is slightly over five years. There are several interesting features to note about the trends displayed in Figure 12.15(a). First, the analysis indicates that thermal energy recuperation yields a 60% lower economic payback at the reference spark spread and moreover, is more favourable than power-only systems

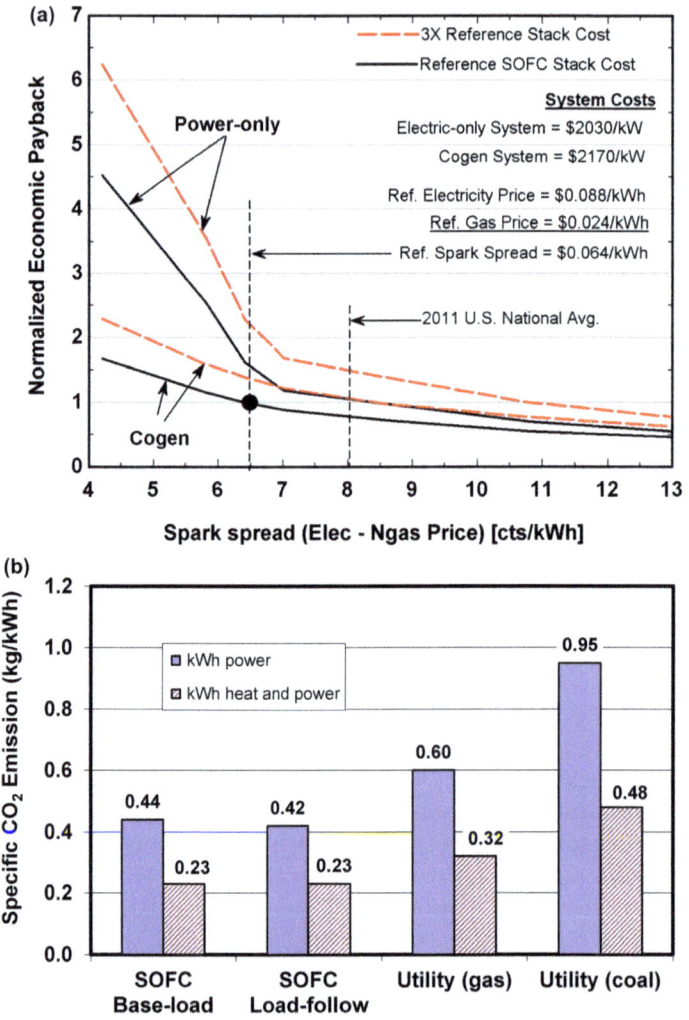

Figure 12.15   (a) Economic sensitivity to utility pricing and SOFC capital cost (b) Comparative specific CO_2 emissions for base-loaded 1–kW SOFC system versus conventionally supplied energy.

irrespective of the spark spread for the range shown. As the price of either electricity or natural gas fluctuates, cogeneration to produce domestic hot water generally decreases the sensitivity of the economic outcome to utility pricing. Second, the value of thermal energy recuperated clearly increases with increasing natural gas price and in contrast, as electricity price increases, the economic importance of cogeneration is diminished. This result is supported by other studies which show that given the option to preferentially select between heat-led or electric-led load-following operation, electric-led is always chosen as the utility power buy-back rates approach the retail (*i.e.*, grid) residential rates.[82]

The preceding results are based on a mature SOFC stack cost of \$450/kW. Results are also presented for a stack cost three times this value to evaluate the economic sensitivity of early production units. The dashed lines in Figure 12.15(a) show that the payback period for both cogeneration and electric-only systems increases by about 40%, while yielding the same trends as the mature unit stack cost.

Figure 12.15(b) presents a comparison of the specific CO_2 emissions between a 1 kW SOFC system and the average U.S. utility[86] when supplying electrical and thermal energy (domestic hot water) to the residence. The base-load SOFC system achieves a CO_2 emission output at 0.44 kg/kWh-electric, and the load-following case is slightly lower at 0.42 kg/kWh-electric due to the slight efficiency advantage. Utility CO_2 emission rates are between 35 and 115% larger than the base-loaded fuel cell system output. The improved CO_2 emission characteristics in a CHP system result because the fuel utilized to provide power also supplies thermal energy to the residence.

The annual hourly average electric load of the single-family residential application in this example was about 1.0 kW. Sizing the SOFC system at 1 kW produced the lowest payback period, the highest fuel cell electric and thermal capacity factors, and the lowest annual CO_2 emissions. Further, operating in an electric-led, base-load control mode is more attractive both economically and technologically than load-following provided net metering plans are available. Lastly, cogeneration in the form of DHW supply to the household is preferable to power-only SOFC systems. The micro SOFC-CHP simulation results presented here suggest that for single-family detached dwellings, SOFC system size should be based on the annual hourly average electric demand of the application and that cogeneration is preferred over power-only systems.

It is important to note that these results are by no means conclusive as the study has not considered many other aspects and variables, such as heat-led operating strategies, different building loads and associated TERs for other home types, sizes, and geographic locations, and the influence of different policy measures (*e.g.*, carbon taxation, grid- and utility-related externalities, *etc.*) and regulations that could affect net metering plans and interconnection standards and fees. Nevertheless, the simulation proves to be illustrative of the considerations involved in residential micro-CHP applications.

12.5.2 Large-scale CHP and CCHP Applications

The application of SOFC-CHP systems in larger commercial and industrial applications is of interest to many developers and potential end-users. Renewable fuels such as landfill gas and biogas produced in anaerobic digesters from animal waste or in wastewater treatment plants present a unique opportunity for fuel cell CHP systems. Recent studies on the use of renewable biogas as a fuel feedstock for MW-class SOFC-CHP systems show that even though the fuel gas is typically diluted such that methane represents only 60% of the gas content, high electric efficiencies in the range of 45–52%-LHV and CHP efficiencies from 85–88%-LHV could be achieved.[73,78] The potential life cycle costs of such systems have also been shown to be superior to large reciprocating IC engines, microturbines, molten carbonate fuel cell systems, and competitive with larger gas turbine technologies if SOFC technology can reach the mature costs that comes with high volume manufacturing.[73]

In large commercial building applications, the penetration of SOFC-CHP technology is challenged by many market barriers that are technology neutral (see Section 12.7). Yet recent work shows some promising options, particularly in SOFC-CHP operating strategies that favour *both* the end-user and the electric utility.[2] Business models that view the electric grid as a peaking plant are generally not attractive to electric utilities in the U.S. However, the SOFC as a dispatchable resource could be attractive to the utility if it was large enough so that it could serve the majority of the commercial building loads and export electric power to the grid during high peak demand time periods. Simulations of a 1.3 MW SOFC-CHP against experimentally measured energy demands for a large hotel indicate that CHP efficiencies over 85% can be achieved with less than 15% impact to the end-user COE.[2] Other work indicates additional promise for 1 MW SOFC-CHP systems serving residential neighbourhoods with electric power and district heating.[88]

The high-quality heat available in the exhaust gas of SOFC systems make them an attractive option for CCHP applications, especially in larger buildings (*e.g.* multifamily,[45] hospital, university,[89] governmental.)[21] The lower efficiency (and power) of conventional boiler and IC engine systems and the intermittent nature of solar power (especially the scarcity of solar radiation in winter time) make them less attractive in CCHP applications compared with SOFC systems. For example, a recent study comparing different renewable technologies for integration into a 500 kW CCHP building application found that an SOFC-based system has the highest electrical efficiency among these systems and can produce enough energy for supporting both the heating and cooling systems. In terms of maximum CCHP (*i.e.* trigeneration) efficiency, biomass- and solar-trigeneration systems are expected.[44] Integration of SOFCs with solar thermal collectors to form an integrated CCHP system for a university building located in Naples (Italy) with an electrical capacity of about 250 kW has also been explored. Simulation results indicate that the system can achieve a net electrical efficiency of nearly 47% and can overcome the severe issues regarding the thermal balance of PEM operating systems as well as the problems associated with the standalone solar systems.[89]

The potential advantages of SOFC-CCHP systems in term of both techno-economic and environmental issues for multifamily housing applications has been evaluated and compared with other competing technologies, such as the IC engine and natural gas boiler.[45] Using cell, stack and systems models,[3,28,29] a performance assessment of building-integrated SOFC-CCHP systems in different multifamily housing in the hot climate of Madrid (Spain) were assessed.[45] The system performances were determined in terms of non-renewable primary energy demand and CO_2 equivalent (CO_2-eq) emissions, and compared to a system with that employed an IC engine and with respect to the reference system. The reference system for all cases comprised a gas boiler and a mechanical chiller and grid-electricity supplied according to the selected generation mix. Compared to IC engine-based cogeneration and to traditional gas burner/mechanical chiller/grid electricity supply technology, significant savings in both energy (62%) and CO_2 emissions (35%) resulted from employing an SOFC trigeneration system, especially when an UCTE (union for the coordination of the transmission of electricity) electricity grid mix was assumed.[45]

Noteworthy studies are focusing on the technical assessment and optimization of such systems using different climates, building sizes and system configurations. For example, a CCHP system consisting of an SOFC system integrated with a double-effect water-lithium bromide (LiBr) absorption heating and cooling system was investigated[21] to assess its potential for supplying the space heating, cooling, hot water and lighting energy demands for an American office building of around 9500 m².[21] Besides proofing the technical feasibility of such a system (total efficiency more than 87%), the study illustrated that the system can become economically competitive with CHP technologies when SOFC capital costs are reduced to about \$1000/kW and the life span of at least five years are reached.[21] The SOFC trigeneration system was predicted to achieve CCHP and CHP efficiencies about 89% and 84%, respectively which is higher than the typical values (70–80%) reported for similar systems using either IC engine or micro-gas turbines.

Presently, the authors are unaware of any hardware technology demonstrations of SOFC-CCHP systems in building applications. Nevertheless, the results of these case studies add to the growing body of technical literature that demonstrates technical feasibility and potentially superior performance of SOFC-based systems compared to both renewable and conventional technologies.

12.6 Commercial Developments of SOFC-CHP Systems

During the last decade, demonstration projects on either small-scale or full-size systems have been started in various countries for verifying these characteristics under actual application conditions. The demonstration systems are largely focused on residential micro-CHP applications and although there are still various techno-economic issues, such as cost and complexity, the results show promise for the successful application of such systems in residential households. There is a growing number of successful pilot plants around the globe and several industrial companies have begun commercialization efforts of their SOFC products around the world.

The following sub-section synthesizes some of the commercialization efforts and demonstration projects related to SOFC-CHP systems. Based on the open literature, the authors are unaware of any SOFC-CCHP system currently available on the market. The only practical fuel cell-CCHP system is developed by UTC Power[90] which integrates an absorption chiller with a phosphoric acid fuel cell to produce about 50 tons of cooling. As a result, SOFC-CHP systems development is the central focus in the following summary.

12.6.1 Commercialization Efforts

There are approximately six companies offering commercial (or 'pre-commercial') SOFC power system products. Ceramic Fuel Cells Limited Co. (CFCL) has about 20 years of experience in the development of SOFC stacks and offers SOFC-CHP units in the range between 1 and 5 kW.[91,92] CFCL offers two power modules (Gennex[93] and BlueGen[94]) that are intended to support different markets and customers. Gennex is a fuel cell power module fuelled by natural gas that has been designed for integration with existing appliances such as condensing boilers (which provide hot water and space heating) by sharing common inputs and outputs from the fuel cell.[93] BlueGen is a self-contained packaged system providing heat and power. The Gennex module comprises fuel cell stack, hot BoP (integrated steam generator, burner, and fuel & air heat exchanger), and high temperature insulation. The stack operates in the temperature range between 800 and 870 °C and is modular, starting with a base manufactured cell-stack containing 28 layers and producing about 150 W of DC power at 850 °C. Seven units form a 1 kW stack and 14 are needed for a 2 kW stack. Gennex documentation indicates that it can be power modulated from 0 to 2 kW with electrical efficiencies of around 57% at 2 kW (approximately 1000 W thermal output), 60% at 1.5 kW (approximately 540 W thermal output), and 36% at 0.5 kW (approximately 400 W thermal output).[38,39] The total thermal efficiency of the system is between 60 and 85 %, depending on the amount of heat recovered. The Gennex electrical and thermal efficiencies versus the exported power are shown in Figure 12.16 (based on exhaust gas cooled to 30 °C) and the unit has a reported start-up time of about 25 hours.[93]

The Bluegen unit[94] is a small-scale heat and power generator fuelled by natural gas that is approximately the size of a washing machine (600 mm × 660 mm × 1010 mm). The waste heat generated by the power module within BlueGen is used for production of hot water and is sufficient to provide 150–200 L of DHW per day. The system is optimized as an electrical generator with a peak electrical output of 2 kW at a maximum net AC electrical efficiency of 60%. The TER is less than 0.5 and the total thermal efficiency of the system can reach up to 85%, depending on the operating condition and amount of waste heat recovered. Specific CO_2 emissions are reported to be about 340 g/kWh.

Table 12.4 provides a summary of the commercial and demonstration SOFC systems including costs and demonstrated hours of operation.

Vaillant Group, has commercialized wall-hung heating appliances based on SOFC power systems. The system is designed for single-family houses and generates 2 kW of heat and 1 kW of electricity.[95] System development has been

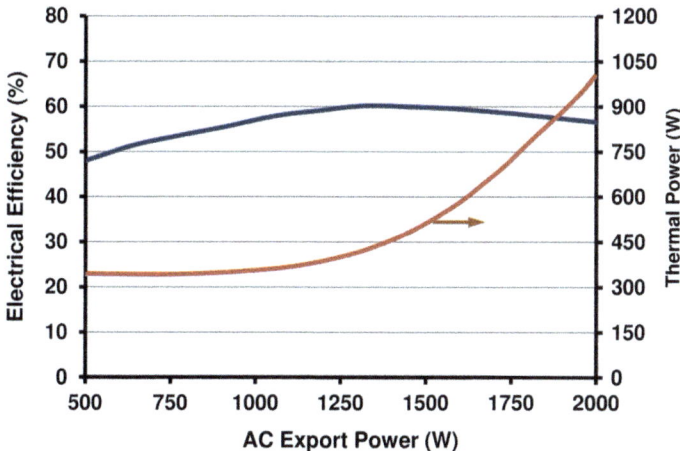

Figure 12.16 Net electrical and thermal efficiencies of Gennex power module versus AC export power. (Adapted from Ref. 93.)

in cooperation with Staxera GmbH and the Fraunhofer Institute for Ceramic Technologies and Systems (IKTS) in Dresden.[95]

In the USA, Acumentrics has developed three SOFC systems (RP20, RP 1000/1500) suitable for micro-CHP applications. The company has reported installations at various sites in states such as Colorado, West Virginia, Texas, Arkansas, Pennsylvania, and in Calgary, Canada. The RP20 SOFC stack is built from tubular cells which operate near 800 °C. The net DC output of the stack is 1000 W. RP 1000/1500 are 2.5 kW (RP1500) and 2.0 kW (RP 1000) versions of RP series.[96]

Larger SOFC systems for commercial building applications have been commercialized by Bloom Energy. Two products are offered at either 100 or 200 kW capacities – the ES5400 and ES 5700, respectively. These units are power-only systems achieving net electric efficiencies greater than 50%-LHV.[97]

In Canada, DDI energy Inc. produces an SOFC-based power generator fuelled by natural gas, propane, and methane. System products range between 3 to 40 kW (ARC1-P3 to ARC2-P40) and are suitable for both indoor and outdoor installation.[98]

12.6.2 Demonstrations

SOFC technology demonstrations are numerous and varied with support and financial incentives provided by different governments and agencies. The following synthesizes some recent SOFC technology demonstration projects in Europe and Japan, as well as some companies, such as Topsoe Fuel Cells and Versa Power, who are leaders in SOFC technology but have not yet officially commercialized systems.

Hexis GmbH is focusing on stationary applications with an electric power output below 10 kW through their Galileo 1000N system. Systems have been

Table 12.4 Summary of commercially available and demonstration SOFC systems intended for building applications.[31,33,94,95,96,97,98,103,104,106]

	Commercial products							Demonstration units	
	ARC Series (DDI Energy)	Bloom Energy	BlueGen (CFCL)	RP (Acumentrics Corp.)	ENE-FARM (Osaka Gas)	Vaillant	EBZ	Galileo 1000N Hexis	EnGen
Weight (kg)	273 to1180	19400	<200	136	94	-	-	170	17.5 (stack)
SOFC type	Planar	Planar	Planar	Tubular	FT-SIS	Planar	Planar	Circular planar	Planar
Fuel type	NG, Propane, Methane	NG	NG	NG, Propane	NG	NG	NG, LPG, biogas, Syngas	NG	NG
Max. power (kW)	3 to 40	210	2	1, 2,2.5	0.7	1	1.5	1	0.5
AC electrical efficiency (LHV) (%)	-	>50	60 (at 1500 W)	>30	46.5	30	30	25-30	30 (Stack eff.)
Thermal output (W)	-	-	300 to 1000	-	470 W	2	2.5	2.5	-
Total efficiency (%)	60	-	85	-	90	85	>90	90-105	-
Flue gas Temperature (°C)	250-482	-	< 200	-	75	-	Condensate to ambient temp.	Condensate to ambient temp.	200-500
CO₂ (g/kWh)	578.70	350.6	340	-	690	-	.05	-	-
NOₓ (g/kWh)	0.037	<4.535	No	-	-	-	0.06	<0.003	-
Cost (per kW) (US $)	~47500 to 100,000	~8000	-	-	~33700 (0.7 kW system)	-	-	-	-

*system developed by Tokyo Gas Co. has almost same characteristics.
**Topsoe Fuel Cells and Versa Power, who are leaders in SOFC technology have not yet officially commercialized systems.

tested in houses and small apartment buildings for a reported duration of more than 27 000 hours (status as of October 2010) with a power degradation rate on power of approximately 2% per 1000 h.[31] These systems convert natural gas (or bio-methane) into electricity and heat using radial planar electrolyte-supported cells and metallic interconnects (60 repeat units).[99] The Galileo unit generates an electrical output of 1 kW and a thermal output of approximately 2 kW. Additional heat up to 20 kW can be generated by an integrated burner. The system electrical and overall efficiencies are 30–35% and >90%-LHV, respectively, with low emissions of NO_x (< 30 mg/kWh) and CO (<30 mg/kWh).[99,100]

EBZ is developing planar SOFC stacks and systems (with a focus on stationary heating appliances and small CHP). Development efforts have been in conjunction with the FP7 EU-collaborative project FC-DISTRICT. Participants of this project come from 23 European organizations and companies. One of the main objectives of the FC-DISTRICT project is to develop SOFC-CHP systems which are able to produce up to 2 kW electric power and 6–8 kW thermal power with an overall efficiency up to 90%-LHV.[33] In the framework of this project, as well as the Flame SOFC project, EBZ's SOFC micro-CHP units have been designed to generate a nominal power output of 1.5 kW electric and 2.75 kW thermal at 30% net electrical efficiency. The targeted overall efficiency of this system is above 90%. The EBZ system operates with several fuels such as natural gas and biogas and employs catalytic partial oxidation to convert the fuel to a hydrogen-rich fuel gas. System emissions are reported to be NO_x < 60 mg/kWh, CO < 50 mg/kWh at 0% O_2.[101,102] The electrolyte-supported SOFC stack used in this system is developed by Staxera GmbH.[33]

The EnGen unit is a complete SOFC micro-CHP demonstration system with net power production up to 1000 W. The system is developed through collaboration between Htceramix (Switzerland) and SOFCpower (Italy). The system operates on natural gas that is reformed through a catalytic partial oxidation (CPOx) reactor and achieves electric efficiencies in the range of 30–32%.[103] The complete system is fully autonomous with a lead-acid battery for grid-independent start-up, UPS functionality, and for surge coverage. The packaged system includes heat exchanger, temperature insulation, electric heater for start-up, smart temperature control, voltage conditioning for a lead acid battery compatible output and integrated software and controls.[103]

Since 2004, the Japanese companies, Tokyo Gas, Rinnai, and Kyocera (stack manufacturer) jointly with Gastar Exploration Ltd. (USA) are focusing on CHP systems development based on flatten tubular segmented-in-series (FT-SIS) SOFCs.[104,105] This specific cell design can reduce the gas sealing problem (same as tubular design) and has also more stable electrical connections (same as segmented-in-series cells). The cell has also high durability at high temperature since it is completely made of ceramic materials.[32] The first residential CHP system using the above technology was developed in 2008 and has been implemented into several homes for data collection under realistic conditions. Each stack is composed of two bundles of 36 FT-SIS cells

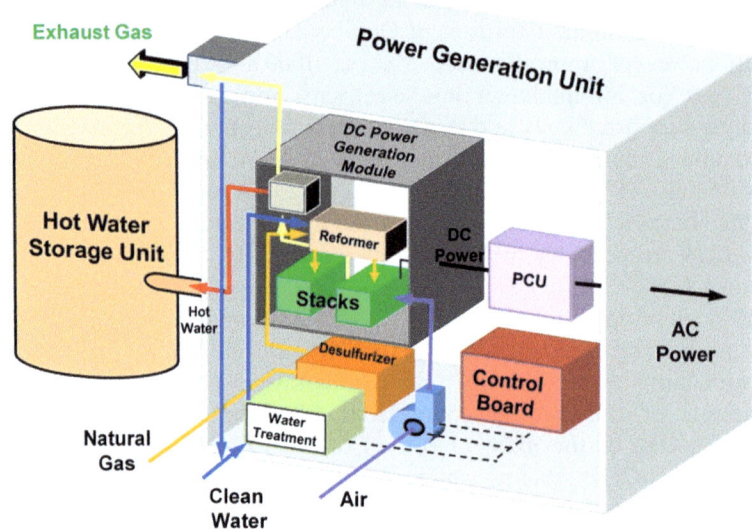

Figure 12.17 Schematic of Packaged FT-SIS SOFC system developed by several Japanese companies. (Adapted from Ref. 104.)

each and can generate almost 800 W (11 W each single cell). The electrical and heating efficiencies of this system are more than 40% and 32% at 700 W respectively. A conceptual overview of the packaged system is shown in Figure 12.17.

The ENE-FARM Type S is a SOF-CHP system developed with collaboration work between Kyocera, Osaka Gas, Aisin Seiki, Chofu Seisakusho, and Toyota Motor Corporation. This system involves the same stack as mentioned above (from Kyocera). This 700 W system costs about (¥2.75 million (US $33,700) (including taxes but excluding) costs.[106]

In Finland, the technical research centre of Finland (VTT) and Wärtsilähave been jointly working on developing of SOFC-CHP systems for both residential and industrial applications, but hardware demonstrations have been for systems in the range of 5–10 kW$_e$. VTT has also a strong participation in several European programs such as Large SOFC and SOFC600. The VTT SOFC system is designed to work with natural gas using a planar anode-supported stack (counter-flow design with internal manifold) obtained from Forschungszentrum – Jülich. The stack comprises 50 rectangular unit cells with an active area of 361 cm^2. Ambient air is supplied to the system with two blowers (3–10 air excess ratio). The maximum DC power ($P_{el,stack}$) and DC efficiency($\eta_{el,stack}$),are about 5 kW and 50% (LHV) respectively.[107] In 2009, VTT developed a 10 kW demonstration unit jointly with Versa Power Systems.[108]

Topsoe Fuel Cell (TOFC) focuses on the technical development and commercialization of SOFCs. The POWERCORE module has been developed for CHP applications with system integration in collaboration with Dantherm Power.[109,110] The TOFC POWERCORE is developed for single-family households with an average power demand of 1 kW nominal electrical power. The system operates with natural gas but additional fuels, such as methanol, LPG and biofuels are also expected in the future.[7] The system is comprised of several components such as insulation, start-up burner, heat exchanger, reformer, stack, off-gas burner and instrumentation. The DC efficiency and fuel utilization of the PowerCore unit are in the power range 750–1350W at about 60–65% and 80–85%, respectively.[7]

The potential market for small SOFC systems has also encouraged other U.S. SOFC developers and laboratories to produce demonstration units. As noted previously, Versa Power Systems jointly with VTT (Finland) is developing a 10 kW system for commercial building applications. Recently, Pacific Northwest National Laboratory (PNNL) reported the installation of a high efficiency 2 kWSOFC power-only system.[111] This system consists of four parallel planar anode-supported stacks provided by Delphi Corporation, one steam reformer, five recuperators (four in the cathode side and one in the anode loop), one condenser, and two blowers (in both anode and cathode loops). Each stack involves 30 cells, each with an active area of 105 cm^2. The system was designed so that about 83–90% of anode exhaust stream can be recycled. The reported electric efficiency record for this system is about 57%-LHVwhich is significantly higher than the performance reported elsewhere for similarly sized systems (*i.e.*, 30–50%-LHV).[21]

12.7 Market Barriers and Challenges

The widespread deployment of SOFC-based CHP and CCHP systems in DG applications provides an attractive alternative as a global energy solution to the 21st-century problems of energy shortages, security, and access to low-cost energy supplies, grid congestion, reliability, and power quality, and reduction of harmful energy-related emissions. In general, many experts observe that the increase in CHP/CCHP system deployment (inclusive of gas and steam turbines and IC engines) is a growing trend in energy supply around the world and that the potential benefits are manifold.[19,22,112] For example, over the last 15 years, Japan has increased its CCHP installations from less than 1 MW to over 10 GW; the U.S. has now over 85 GW of installed CHP/CCHP; and China has announced the goal of over 40 GW of new CCHP installations by 2020.[113] In contrast to conventional CCHP system technologies, such as gas turbines and IC engines, an SOFC-based energy technology platform is unique in its ability to serve as a scalable, fuel flexible, highly efficient, low-emission, and potentially cost-effective polygenerator of cooling, heat, fuel and power. In the following sub-sections, we discuss the outlook for application of SOFC-CHP/CCHP systems in CHP markets from the viewpoints of energy prices, capital and life cycle costs, and regulatory policy and environmental impact.

12.7.1 Energy Pricing

The long-term viability of SOFC technology is closely related to being able to provide a market solution with low life cycle costs (*i.e.*, low capital and operating costs). As previously discussed in Section 12.5, the spark spread strongly influences the life cycle costs of an SOFC-CHP/CCHP system. The larger the spark spread, the more economically attractive onsite CHP installations become. The shale gas revolution in the United States has lowered natural gas prices to their lowest levels in over a decade such that average city-gate natural gas prices ($\sim$4.55 \$/GJ in 2011) are now cheaper than coal per unit of energy.[114] Furthermore, the substantial increase in unconventional natural gas resources (over 35% of U.S. gas production in 2011) has dramatically improved gas supply stability such that natural gas prices have become independent and decoupled from petroleum prices while retail U.S. electricity prices have been relatively flat for the last five years.[114,115] The supply of cheap natural gas has created a substantial shift away from coal-based power generation from over 50% of the U.S. generation mix to less than 37% in just ten years time. Renewable biogas fuels derived from landfill and anaerobic digester gas resources also offer attractively low fuel prices. In the short-term, historically high spark spreads are promoting increased interest in CHP in general.[112] Because of the relationship between natural gas and electricity prices, it is now believed that over the long term (*i.e.*, $\sim$20-years), the forecasted increase in U.S. natural gas supply combined with reduced electricity demand (*e.g.*, due to increased 'green' building infrastructure and energy conservation efforts in the commercial building sector) will lead to substantially reduced retail electricity prices which may gradually reduce current spark-spreads in some countries.[116,117]

12.7.2   SOFC Costs

Capital cost reduction of SOFC stacks have made remarkable progress over the last ten years, dropping by nearly an order of magnitude to about \$175/kW in high volume production scenarios.[118] At the system-level, capital costs of mass-produced 1–2 kW SOFC-CHP systems have been estimated at about 2300 US\$/kW which is shown to compete with conventionally supplied grid-electricity and natural gas fired boilers when the systems are optimally designed.[41] These systems are marginally competitive without incentives when U.S. averaged utility energy pricing for electricity and natural gas are employed. Capital costs for larger-scale SOFC-CHP systems are expected to range from 1950 US\$/kW to 1120 US\$/kW for systems sized at 330 kW and 6.0 MW, respectively.[73] These capital cost estimates are consistent with other recent larger-scale SOFC-CHP manufacturing cost studies.[119] The resulting cost of electricity for such larger systems is estimated to range from 0.079 to 0.050 US\$/kWh, which is lower than commercial grid-electricity prices, and on par with industrial electricity prices in the U.S. These business-as-usual value propositions show that even mature SOFC technology in mass production scenarios, while competitive in some market sectors, are not as compelling and representative as they could be given the environmental, societal and energy reliability and security benefits associated with widespread implementation of distributed generation.

12.7.3 Technical Barriers

Interest in SOFCs for mobile APU, UAV, and UUV applications[120–122] and residential micro-CHP[33,39,47,48] has increased dramatically within the last 10 years and is partly driven by the congruency of current SOFC system capacities (1–25 kW) with the application power requirements. While the residential sector has substantial room for improvements in energy efficiency, it is also one of the most challenging markets to compete in. Despite the efficiency advantages of high temperature fuel cell systems for onsite CHP generation, the application requirements of low maintenance, high durability or lifetime, low cost, and high efficiency are severe. Section 12.6 of this chapter has already discussed current manufacturers and operational systems. However, it should be noted that SOFC-CHP systems are not strictly limited to stationary applications. There is also interest in mobile SOFC-CHP systems. In particular exploratory studies of have illustrated the potential economic benefits and system architectures if metal-supported SOFC technology at 300–400 kW scales is employed as a gas turbine APU replacement in commercial aircraft applications.[121]

Technical issues for SOFC commercialization include cost reduction, durability improvement, and dynamic operation. In general, while technical issues related to SOFC capital cost and durability are often cited as barriers to widespread adoption of the technology, system integration and viable business models in all applications remains as a key scientific challenge, particularly for CCHP systems. System integration challenges encompass establishing: (1) the optimal system architectures for effective utilization of the different grades of thermal energy for export to the building application, (2) viable operating and control strategies (*e.g.*, heat-led *vs.* power-led; load-following *vs.* base-load, *etc.*) that are mutually attractive to end-user, electric utilities, and the SOFC technology itself, (3) power conditioning topologies and the associated technology to enable power plant islanding modes when grid-connected, and (4) the various forms of energy and resources that can be integrated to maximize benefits of fuel cell systems for both electrical energy generation, thermal energy utilization, and low environmental impact.[17,23,123]

12.7.4 Market Barriers and Environmental Impact

The economic picture is increasingly favourable towards SOFC-CHP system deployment in stationary applications; yet excitement surrounding the technology is tempered by the reality of persistent market barriers in many countries to all DG technologies. Indeed, while numerous organizations and countries throughout the globe have initiated meaningful product development efforts of (mostly small-scale) SOFC-CHP systems, complex commercialization barriers remain before widespread market penetration is realized. These barriers are technology neutral and are comprised of both technological and regulatory components that include, but are not limited to:[17,22]

- high first cost for turn-key installation,
- a 'wait-and-see' approach regarding adopting an onsite CHP energy supply system,

- regulated fees and tariffs associated with grid-electricity rate structures,
- unfavourable and non-standardized governmental policies and regulations,
- utility interconnection barriers and predatory pricing,
- low electricity buy-back rates from electric utilities,
- historic natural gas price volatility and customer resistance to power purchase agreements and
- limited liberalization of electricity markets worldwide.

While many of the barriers to DG installations are technology neutral and are common throughout the world, there are regional differences. In North America, excessive utility standby and backup power charges during CHP system downtimes detrimentally affect the attractiveness of CHP, as well as difficulties in securing long-term power purchase agreements from potential customers.[112] In contrast, many utility companies in Europe are not competing for customers and therefore, are often partners for DG implementation as they believe the CHP/CCHP investor is alleviating the utility of some capital risk associated with transmission and distribution (T&D) and generating capacity expansion.[112] However, in Germany, liberalization of the electricity market initially resulted in price wars which drove the grid electricity price below production costs before government policies favouring CHP were adopted.[19] The relatively high cost of natural gas in China means that nearly 95% of CCHP is fuelled by coal and is generally 10 to 100 MW in capacity, which limits energy conversion systems to conventional boilers and steam turbines as coal does not suit small- and intermediate-scale SOFC technology. However, fuel diversification is receiving increased attention in China with growing interest in biomass and biogas resources.[19]

Even though cost reduction and durability performance have dramatically improved, a primary challenge for SOFC technology in stationary CHP/CCHP applications is the lack of policy incentives that will enable all (or even some) of the associated externalities and societal benefits from technology adoption to be internalized such that the resulting value proposition is a compelling one. Recent studies indicate that if even modest sustainability policy measures are adopted, a viable market for mass production of SOFC-CHP systems would be established in the medium- to long-term (*i.e.*, within 20 to 30 years), which would enable expected cost targets to be achieved.[75,117,124] Aggressive policy incentives for fuel cell CHP systems adoption are predicted to create a large and sustainable market in European countries such as Germany.[117] In the U.S., a government-led objective of adding 40 GW of new, cost-effective CHP by 2020 has been announced with initiation of corresponding efforts at reducing market barriers.[125] If successful, this initiative could help catalyze widespread adoption of CHP/CCHP technologies for distributed generation with positive effects for the SOFC industry.

While environmental impact awareness and health concerns are increasing internationally, challenges to internalizing the environmental and societal benefits of highly efficient and environmentally preferred advanced energy conversion systems, such as SOFC-CCHP technologies, remain. Traditional utility, industrial and commercial sector benefit-cost analyses include only transparent, market-traded monetary values; and thus, externalities, which may be significant, are largely ignored due to difficulties in value quantification.[126] The substantial benefits in terms of environmental impact of SOFC-based power generation systems for CHP applications have been evaluated through life cycle assessment studies which show that small- and large-scale SOFC technology is superior to both conventional competing technologies and the expected future utility electricity generation mix in almost all impact categories.[127–129] For example, one study found that the SOFC produces 70% less acidification than a low-NO$_X$ gas turbine and 30% less than a modern natural gas combined cycle plant on a life-cycle basis.[128] Cradle-to-grave (*i.e.*, product life cycle) energy and carbon payback times for micro SOFC-CHP systems have been estimated to range from 0.5 to 1.5 years and are even lower than renewable solar PV and micro-wind installations despite the conservative SOFC efficiency performance employed in the study.[129]

Efforts have also recently been made to quantify the benefits of fuel cell-based CHP/CCHP installations in term of the cost of electricity by assessing the value of generation-related, grid-related, and emissions- and health-related benefits.[130,131] When the benefits of fuel cell-based CHP/CCHP systems in the three afore-mentioned categories are rationally monetized, the resulting value, expressed in terms of cost of electricity, is estimated to range from 0.051–0.199 US$/kWh.[126] A breakdown of category contributions is shown in Figure 12.18, illustrating that health benefits, avoidance of grid-related costs, and avoidance of generation-related costs can amount to 8.49, 2.34, and 9.12 US$/kWh, respectively. Further, by employing waste heat recovery for CHP/CCHP purposes, the value proposition is increased by over 50% compared with power-only systems.[131]

Alternatively, another perspective when evaluating the merit of fuel cell-based CHP systems is to analyze the externalities of existing conventional

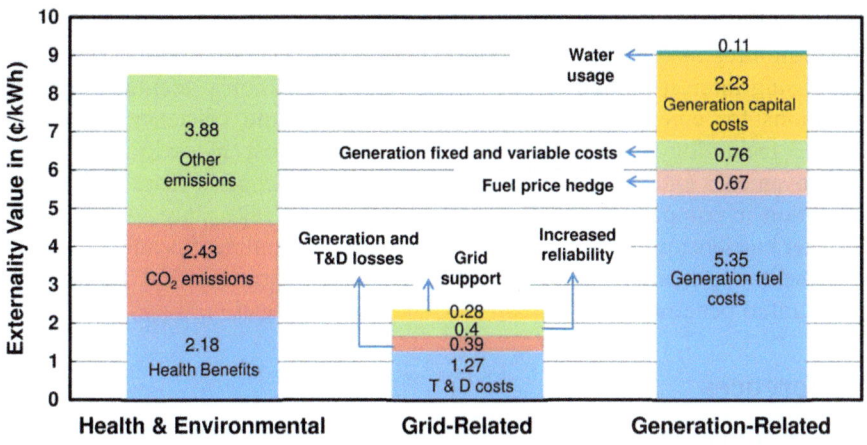

Figure 12.18 Valuation of externality benefits associated with fuel cell-CHP/CCHP systems in terms cost of electricity.

power generation and consider pricing them into the price of electricity. Coal-based power generation in the U.S. has historically accounted for over 50% of the electricity generation mix (it is less than 38% as of 2012). Moreover, at least one recent study estimates that when the costs external to the coal industry (*i.e.*, environmental and health damages associated with coal extraction, transport, processing, and combustion) are accounted for, the effective price of electricity per kWh of generation is conservatively doubled to tripled in the U.S., thereby making renewable and alternative forms of power generation 'economically competitive'.[132] Thus, either of these analysis viewpoints makes a compelling case for the deployment of SOFC-CHP/CCHP systems when the externalities associated with environmental and health effects are rationally valued and reflected in the price of energy supplies.

12.8 Summary

The unique and beneficial characteristics of SOFC technology, coupled with emerging energy production and supply paradigms related to distributed generation, hold much promise for their eventual widespread adoption in numerous residential and commercial building applications. A study of building energy demand characteristics and results from numerous SOFC system simulations indicate that the high efficiency and low thermal-to-electric ratio characteristics of SOFC-CHP systems can provide greater energetic, economic and environmental benefits than competing technologies, including renewables and other fuel cell types. Black and grey box system modelling approaches (including techno-economic modelling) are typically employed in SOFC-CHP/CCHP system performance estimations and application simulations. Performance expectations gathered from numerous studies as well as an examination of the current commercial offerings in SOFC-CHP and power-only systems indicate system electric efficiencies greater than 50%-LHV and CHP efficiencies greater than 85%-LHV are readily achievable. Hardware-based SOFC systems incorporating thermally activated cooling technologies to form an integrated SOFC-CCHP system have not yet been realized, but are of increasing interest globally as distributed polygeneration presents compelling solutions for emerging energy supply, reliability, and efficiency challenges. SOFC technology is not yet mature, and faces cost reduction, durability improvement, and robust dynamic operation challenges before widespread adoption in competitive building energy sectors can take place. Additionally, market barriers applicable to all DG technologies are numerous and varied and represent a potentially much greater challenge than technical barriers for substantial penetration of new CHP/CCHP technologies.

References

1. A. Verma and A. D. Rao, *J. Power Sources*, 2006, **158**, 417.
2. K. Nanaeda, F. Mueller and J. Brouwer, *J. Power Sources*, 2010, **195**, 3176.

3. E. Liese and J. Eng, *Gas Turbines Power*, 2010, **132**, 061703.
4. R. J. Braun, S. Kameswaran, J. Yamanis and E. Sun, *J. Eng. Gas Turbines Power*, 2012, **134**, 021801–1.
5. T. Tang, M. Pastula and B. Borglu, *ECS Trans.*, 2011, **35**, 63.
6. F. Mitlitsky, *Low-cost Co-production of Hydrogen and Electricity*, DOE Annual Merit Review Proceedings, 2008, http://www.hydrogen.energy.gov/annual_review08_fuelcells.html.
7. H. Holm-Larsen, M. J. Jørgensen, N. Christiansen, J. H. Jacobsen, Paper presented at Fuel Cell Seminar, Uncasville, CT, 2012.
8. A. Nanjou, Proceedings of the 10th European SOFC Forum, Lucerne, Switzerland, 2012.
9. K. Föger Proceedings of the 10th European SOFC Forum, Lucerne, Switzerland, 2012.
10. Quadrennial Technology Review Framing Document, U.S. DOE, Washington D.C., 2011, http://energy.gov/sites/prod/files/edg/qtr/documents/DOE-QTR_Framing.pdf.
11. R. Lasseter, A. Akhil, C. Marnay, J. Stephens, J. Dagle, R. Guttromson, A. S. Meliopoulous, R. Yinger, J. Eto, Integration of Distributed Energy Resources: The CERTS MicroGrid Concept, U.S. DOE, Washington D.C., LBNL-50829, 2002.
12. J. M. Guerrero, F. Blaabjerg and T. Zhelev, *IEEE Ind. Electron. Mag.*, 2010.
13. K. Hemmes, J. M. Guerrero and T. Zhelev, *Chem. Eng. Process*, 2012, **51**, 18.
14. C. Wendel, R. J. Braun, Proceedings of the 10th European SOFC Forum, Lucerne, Switzerland, 2012.
15. D. M. Bierschenk, J. R. Wilson and S. A. Barnett, *Energy Environ. Sci.*, 2011, **4**, 944.
16. W. L. Becker, R. J. Braun, M. Penev and M. Melaina, *Energy*, 2012, **47**, 99.
17. J. A. P. Lopes, N. Hatziargyriou, J. Mutale, P. Djapic and N. Jenkins, *Electr. Power Syst. Res.*, 2007, **77**, 1189.
18. G. Chicco and P. Mancarella, *Renewable Sustainable Energy Rev.*, 2009, **13**, 535.
19. D. W. Wu and R. Z. Wang, *Prog. Energy Combust. Sci.*, 2006, **32**, 459.
20. J. Wang, Z. Zhai, Y. Jing, X. Zhang and C. Zhang, *Appl. Energ.*, 2011, **88**, 5143.
21. F. Zink, Y. Lu and L. Schaefer, *Energy Convers. Manage.*, 2007, **48**, 809.
22. A. Shipley, A. Hampson, B. Hedman, P. Garland, P. Bautista, *Combined Heat and Power: Effective Energy Solutions for a Sustainable Future*, Oak Ridge National Laboratory, TN, 2008
23. R. J. Braun, S. A. Klein and D.T. Reindl, *J. Power Sources.*, 2006, **158**, 1290.
24. EnergyPlus Energy Simulation Software, http://apps1.eere.energy.gov/buildings/energyplus/.
25. P. Torcellini, M. Deru, B. Griffith, K. Benne, *DOE Commercial Building Benchmark Models*, National Renewable Energy Laboratory, Golden, CO, 2008.

26. R. J. Braun, S. A. Klein and D. T. Reindl, *ASHRAE Trans.*, 2004, **110**, part I.
27. *TRNSYS- A TRaNsient SYstems Simulation Program*, http://sel.me.wisc. edu/trnsys/faq/faq.htm.
28. P. Kazempoor, F. Ommi and F, V. Dorer, *J. Power Sources*, 2011, **196**, 8948.
29. P. Kazempoor, F. Dorer and F. Ommi, *Fuel Cells*, 2010, **10**, 1074.
30. CFCL BlueGen units for Virtual Power Plant Project in Netherlands. *Fuel Cells Bulletin*, 2012, **7**, 3.
31. A. Mai, B. Iwanschitz, U. Weissen, R. Denzler, D. Haberstock, V. Nerlich and A. Schuler, *ECS Trans.*, 2011, **35**, 87.
32. H. Yoshida, T. Seyama, T. Sobue and S. Yamashita, *ECS Trans.*, 2011, 97.
33. I. Frenzel, A. Loukou, D. Trimis, F. Schroeter, L. Mir, R. Marin, B. Egilegor, J. Manzanedo, G. Raju, M. D. Bruijne, R. Wesseling, S. Fernandes, J. M. C. Pereira, G. Vourliotakis, M. Founti and O. Posdziech, *Energy Procedia*, 2012, **28**, 170.
34. R. J. Braun, *Optimal Design and Operation of Solid Oxide Fuel Cell Systems for Small-Scale Stationary Applications*, PhD Thesis, University of Wisconsin-Madison, 2002.
35. S. H. Pyke, A. J. Burnett, R. T. Leah. *System Development for Planar SOFC Based Power Plant*, Harwell Laboratory, Energy Technology Support Unit, 2002.
36. I. Beausoleil-Morrison, *Annex 42 Final Report*, International Energy Agency, Canada, 2008.
37. P. J. Mago and A. K. Hueffe, *Energy and Buildings*, 2010, **42**, 1628.
38. R. J. Kee, H. Zhu, R. J. Braun, T. L. Vincent, *Advances in Chemical Engineering, Fuel Cell Engineering* K. Sundmacher, Elsevier, New York, NY, 2012, 41, 6, 331.
39. P. Kazempoor, V. Dorer and F. Ommi, *Int. J. Hydrogen Energy*, 2009, **34**, 8630.
40. K. J. Kattke and R. J. Braun, J. Fuel Cell Sci. Technol., 2011, **8**, 021009.
41. R. J. Braun, *J. Fuel Cell Sci. Technol.*, 2010, **7**, 031018.
42. D. Bohn, Micro Gas Turbines. Rhode-St-Genèse, Belgium, RTO/NATO, 2005.
43. Z. Yu, J. Han and X. Cao, *Int. J. Hydrogen Energy*, 2011, **36**, 1256.
44. F. A. Al-Sulaiman and I. Dincer, *Int. J. Hydrogen Energy*, 2010, **35**, 5104.
45. P. Kazempoor, V. Dorer and A. Weber., *Int. J. Hydrogen Energy*, 2011, **36**, 13241.
46. W. L. Becker, R. J. Braun, M. Penev and M. Melaina, *J. Power Sources*, 2012, **34**, 200.
47. S. Farhad, F. Hamdullahpur, Y. Yoo, *Int. J. Hydrogen Energy*, 2010, **35**, 3758.
48. V. Dorer and A. Weber, *Energy Convers. Manage.*, 2009, **50**, 648.
49. V. Dorer, R. Weber and A. Weber, *Energy and Buildings*, 2005, **37**, 1132.
50. K. H. Lee and R. K. Strand, *Renewable Energy*, 2009, **34**, 2839.

51. ESP-r, an integrated building/plant simulation tool, http://www.esru. strath.ac.uk/Programs/ESP-r.htm.
52. BEopt-Building Energy Optimization software, http://beopt.nrel.gov/ home.
53. N. Kelly, I. Beausoleil-Morrison, *Annex 42 Final Report: A Report of Subtask B*, International Energy Agency, Canada, 2007.
54. Aspen Plus - AspenTech, www.aspentech.com/products/aspen-plus.aspx.
55. gPROMS -*Process Systems Enterprise*, http://www.psenterprise.com/ gproms.
56. Engineering Equation Solver (EES). F-Chart Software. 2000. www. Fchart.com.
57. M. Burer, K. Tanaka, D. Favrat and K. Yamada, *Energy*, 2003, **28**, 497.
58. S. H. Chan, K. A. Khor and Z. T. Xia, *J. Power Sources*, 2001, **93**, 130.
59. H. Zhu and R. J. Kee., *J. Power Sources*, 2003, **117**, 61.
60. R. Bove, P. Lunghi and N. M. Sammes, *Int. J. Hydrogen Energy*, 2005, **30**, 181.
61. R. J Braun, T. L. Vincent, H. Zhu, R. J. Kee, *Advances in Chemical Engineering: Fuel Cell Engineering*, ed. K. Sundmacher, Elsevier, New York, NY, 2012, 41, 7, 383.
62. E. Achenbach, U. Reus, 6th International Symposium on Solid Oxide Fuel Cells (SOFC-VI), Honolulu, Hawaii, 1999.
63. S. C. Singhal, K. Kendal, *High Temperature and Solid Oxide Fuel Cells, Fundamentals, Design and Applications*, Oxford, UK, Elsevier, 2003.
64. K. J. Kattke, R. J. Braun, A. M. Colclasure and G. Goldin, *J. Power Sources*, 2011, **196**, 3790.
65. F. Zhao and A. V. Virkar, *J. Power Sources*, 2005, **141**, 79.
66. S. Koch, Contact Resistance of Ceramic Interfaces Between Materials Used for Solid Oxide Fuel Cell, PhD Thesis, Risø National Laboratory, Denmark, 2002.
67. J. E. O'Brien, C. M. Stoots and J. S. Herring, *J. Fuel Cell Sci. Technol.*, 2006, **3**, 213.
68. D. H. Jeon, J. H. Nam and C. J. Kim, *J. Electrochem. Soc.*, 2006, **153**, A406.
69. S. Liu, C. Song and Z. Lin, *J. Power Sources*, 2008, **183**, 214.
70. C. M. Stoots, J. E. O'Brien, J. J. Hartvigsen. *Test Results of High Temperature Steam/CO₂ Co-Electrolysis in a 10-Cell Stack*, Idaho National Laboratory, 2007, www.inl.gov/technicalpublications/ documents/3751099.pdf.
71. R. J. Boersma and N. M. Sammes, *J. Power Sources*, 1996, **63**, 215.
72. R. J. Kee, P. Korada, K. Walters and M. Pavol, *J. Power Sources*, 2002, **109**, 148.
73. A. Trendewicz, R. J. Braun, *J. Power Sources*, 2013, **233**, 380.
74. K. Gerdes, E. Grol, D. Keairns, R. Newby, *Integrated Gasification Fuel Cell Performance and Cost Assessment*, U.S. DOE, DOE/NETL-2009/1361, 2009.
75. J. H. J .S. Thijssen, *The Impact of Scale-up and Production Volume on SOFC Manufacturing Cost*, Final report prepared for the U.S. DOE, 2007.

76. N. Autissier, F. Palazi, F. Maréchal, J. N. Herle and D. Favrat, *J. Fuel Cell Sci. Technol.*, 2007, **4**, 123.

77. F. Palazzi, N. Autissier, F. Marechal and D. Favrat, *Appl. Therm. Eng.*, 2007, **27**, 2703.

78. K. A. Pruitt, R. J. Braun and A. M. Newman, *Appl. Energy*, 2013, **102**, 386.

79. Table A2. Energy Consumption by Sector and Source, Annual Energy Outlook, Washington D.C., Energy Information Administration, U.S. DOE, Washington D.C., 2009, http://www.eia.gov/oiaf/aeo/pdf/0383(2009).pdf.

80. Table CE3.1 Household site End-use Consumption in the U.S, Totals and Averages, Residential Energy Consumption Survey, Energy Information Administration, U.S. DOE, Washington D.C., 2009, http://www.eia.gov/oiaf/aeo/pdf/0383(2009).pdf.

81. A. D. Hawkes, P. Aguiar, B. Croxford, M. A. Leach, C. S. Adjiman and N. P. Brandon, *J. Power Sources*, 2007, **164**, 260.

82. A. D. Hawkes and M. A. Leach, *Energy*, 2007, **32**, 711.

83. T. DeValve, B. Olsommer, *Micro-CHP Systems for Residential Applications Final Report*, Prepared by United Technologies Research Center for U.S. DOE, http://www.osti.gov/bridge/servlets/purl/921640-8RSmr0/921640.pdf.

84. A. D. Hawkes, P. Aguiar, C. A. Hernandez-Aramburo, M. A. Leach, N. P. Brandon, T. C. Green and C. S. Adjiman, *Energy Environ. Sci.*, 2009, **2**, 729.

85. Network for New Energy Choices inFreeing the Grid, New York, 2010. http://www.gracelinks.org/media/pdf/freeing_the_grid_2010.pdf.

86. U. S. Environmental Protection Agency Carbon Dioxide Emissions from the Generation of Electric Power in the United States, U.S. DOE, Washington D. C., 2000.

87. S. Farhad, Y. Yoo and F. Hamdullahpur, *J. Power Sources*, 2010, **195**, 1446.

88. C. M. Colson and M. H. Nehrir, *IEEE T. Energy Conver*, 2011, **26**.

89. F. Calise, *Int. J. Hydrogen Energy*, 2011, **36**, 6128.

90. UTC Power: A United Technologies Company, http://www.utcpower.com/products/purecell400.

91. B. Godfrey, K. Föger, R. Gillespie, R. Bolden and S. P. S. Badwal, *J. Power Sources*, 2000, **86**, 68.

92. J. Love, R. Ratnaraj, Proceedings of the 6th European Solid Oxide Fuel Cell Forum, Lucerne, Switzerland, 2004.

93. *Gennex brochure*, http://www.cfcl.com.au/Assets/Files/Gennex_Brochure_(EN)_Apr-2010.pdf, Accessed Jan., 2013.

94. *BlueGen Brochure*, http://www.bluegen.info/Assets/Files/BlueGen%20Brochure_Architects-web.pdf, Accessed Jan., 2013. .

95. Vaillant Group Presents First Wall-Hung Fuel Cells Heating Appliance, http://www.vaillant.de/Presse/Press-Releases/article/Vaillant_Group_presents_first_wall-hung_fuel_cells_heating_appliance, Accessed Jan., 2013.

96. *RP1000 & RP1500 Datasheet (1,000 W & 1,500 W Outputs)*, http://www. acumentrics.com/Collateral/Documents, Accessed Jan., 2013.

97. *Bloom Energy, ES-5400/ ES-5700 Product Data Sheet*, http://www. bloomenergy.com/fuel-cell, Accessed Jan., 2013.

98. DDI Energy, ARCseriesSOFCInformation3to40kw20120316.pdf, Accessed Jan., 2013.

99. V. Nerlich, A. Schuler, A. Mai, T. Doerk, Paper presented at: 18th World Hydrogen Energy Conference, Essen, Germany, 2010.

100. Galileo – Decentralised Energy and Heat Supply with Fuel Cells, www. hexis. com/ downloads/ hexis_prospekt_englisch_web0703. pdf, Accessed Jan., 2013.

101. Information about Fuel Cell Unit Presented at Hannover Messe, http:// fc-district.eu/documents.html Accessed Jan., 2013.

102. *FC-District*, Presentation on Development of an SOFC Based μ-CHP System on Fuel Cell Day, http://fc-district.eu/documents.html, Accessed Jan., 2013.

103. O. Bucheli, M. Bertoldi, S. Modena and A. Ravagni, Paper presented at Fuel Cell Seminar, Uncasville, CT, 2012.

104. *Tokyo Gas Webpage-Challenge for the Future Society*, http://www.tokyo-gas.co.jp/techno/challenge/001_e.html, Accessed Jan., 2013.

105. T. Ishikawa, Paper presented at 23rd World Gas Conference, Amsterdam, Nethersland, 2006, http://www.igu.org/html/wgc2006/pdf/paper/add10759. pdf.

106. Japanese Group Unveils SOFC Ene-Farm Residential Cogen Unit, *Fuel Cells Bulletin*, 2012, **4**, 4.

107. M. Halinen, J. Saarinen, M. Noponen, I. C. Vinke and J. Kiviaho, *Fuel Cells*, 2010, **10**, 440.

108. M. Halinen, Paper presented at Fuel Cell Seminar, San Antonio, TX, USA, 2010.

109. H. Weineisen, J. Noe, Paper presented at Fuel Cell Seminar, Uncasville, CT, 2012.

110. N. Christiansen, H. Holm-Larsen, S. Primdahl, M. Wandel, S. Ramousse, A. Hagen, *ECS Trans.*, 2011, **35**, 71.

111. M. Powell, K. Meinhardt, V. Sprenkle, L. Chick and G. McVay, *J. Power Sources*, 2012, **205**, 377.

112. D. Engle, The Journal of Energy Efficiency & Reliability, 2012, Sep/Oct, **44**.

113. Q. Gu, H. Ren, W. Gao and J. Ren, *Energy and Buildings*, 2012, **51**, 143.

114. U.S. Energy Information Administration, *Quarterly Coal Report January – March 2012*, Washington, D.C., U.S.A., DOE/EIA-0121(2012/01Q), 2012.

115. G. Hale and B. Hobijn, R. Raina, *FRBSF Economic Letter*, 2012, **14**.

116. K. Palmer, D. Burtraw, M. Woerman, B. Beasley. The Effect of Natural Gas Supply on Retail Electricity Prices, Resources for the Future (RFF), Washington D.C., 2012.

117. W. Krewitt, J. Nitsch, M. Fischedick, M. Pehnt and H. Temming, *Energy Policy*, 2006, **34**, 793.

118. E. D. Wachsman, C. A. Marlowe and K. T. Lee, *Energy Environ. Sci.*, 2012, **5**, 5498.
119. J. Warren, S. Das, W. Zhang, Proceedings of the ASME 2012, 10th Fuel Cell Science, Engineering and Technology Conference, San Diego, CA, 2012.
120. S. B. Lee, T. H. Lim, R. H. Song, D. R. Shin and S. K. Dong., *Int. J. Hydrogen Energy*, 2008, **33**, 2330.
121. R. J. Braun, M. Gummalla and J. Yamanis, *J. Fuel Cell Sci. Technol.*, 2009, **6**, 031015–1.
122. P. Nehter, J. B. Hansen and P. K. Larsen, *J. Power Sources*, 2011, **196**, 7347.
123. J. Xu, J. Sui, B. Li and M. Yang, *Energy*, 2010, **35**, 4361.
124. TIAX LLC, Scale-up Study of 5-kW SECA Modules to a 250 kW System, Final report prepared for the U.S. DOE, National Energy Technology Laboratory, 2002.
125. Environmental Protection Agency, Combined Heat and Power: A Clean Energy Solution, U.S.DOE Washington D.C., 2012.
126. National Fuel Cell Research Center, *Build-up of Distributed Fuel Cell Value in California: 2011 Update Background and Methodology*, University of California, Irvine, California, 2011.
127. P. Pehnt, *Int. J. Life. Cycle. Assess.*, 2003, **8**, 283.
128. P. Pehnt, *Int. J. Life. Cycle. Assess.*, 2003, **8**, 365.
129. I. Staffell, A. Ingram and K. Kendall, *Int. J. Hydrogen Energy*, 2012, **37**, 2509.
130. L. S. Schell, *Economic Analysis of Large Stationary Fuel Cell Value in California*, Paper presented at International Colloquium Environmentally Preferred Advanced Power Generation (ICEPAG), Newport Beach, California, CA, 2009.
131. L.S. Schell, *The Cost Effectiveness of DG with and without CHP/CCHP*, Paper presented at International Colloquium Environmentally Preferred Advanced Power Generation (ICEPAG), Costa Mesa, California, CA, 2010.
132. P. R. Epstein, J. J. Buonocore, K. Eckerle, M. Hendryx, B. M. Stout III, R. Heinberg, R. W. Clapp, B. May, N. L. Reinhart, M. M. Ahern, S. K. Doshi and L. Glustrom, *Ann. N. Y. Acad. Sci.*, 2011, **1219**, 73.

Integrated SOFC and Gas Turbine Systems

FRANCESCO CALISE* AND
MASSIMO DENTICE D'ACCADIA

University of Naples Federico II, P.le Tecchio 80, 80125
*Email: frcalise@unina.it

13.1 Introduction

During the past decades, the world energy consumption has rapidly increased, especially due to dramatic development of the emerging countries, and further increases are expected in the near future.[1] So, it is mandatory to find a way to make such development "sustainable". First, the economic and social development model must take into account the limited availability of natural resources (*e.g.*, the fossil fuels).[2] In addition, energy conversion processes must not be harmful to the environment, and undesirable effects must be prevented, such as: global warming, emissions of pollutants, *etc.*[3]

In this framework, a special effort has been spent worldwide in order to promote a more efficient utilization of energy sources and to reduce the environmental impact of energy conversion processes.[4,5] This goal may be achieved by a double strategy. On one hand, several governments are promoting and forcing the use of renewable energy sources (solar, wind, hydro, *etc.*), that allow one to produce thermal and electrical energy without depleting natural resources and with scarce or null environmental impact.[4–8] On the other hand, innovative technologies must be used for the conventional conversion of fossil

RSC Energy and Environment Series No. 7
Solid Oxide Fuel Cells: From Materials to System Modeling
Edited by Meng Ni and Tim S. Zhao
© The Royal Society of Chemistry 2013
Published by the Royal Society of Chemistry, www.rsc.org

fuels, providing simultaneously ultra-high conversion efficiencies and ultra-low environmental impact.[9–15]

In the field of renewable energy sources, significant improvements have been done in developing cost-effective conversion systems, such as: innovative photovoltaic materials, solar power plants, solar heating and cooling systems, photovoltaic/thermal solar systems, innovative wind turbines, *etc.*[16–30] simultaneously, academia and industry joined their efforts in producing innovative highly efficient energy conversion systems using fossil fuels, such as: combined cycles,[31–33] carbon capture technology,[34–41] integrated gasification combined cycles[42–45] and fuel cells based power plants.[46–49] These latter are considered as one of the most promising energy conversion technologies.[46] Fuel cells systems can directly oxidize the fuel via an electrochemical reaction; so, their performance is not limited by the Carnot efficiency, as occurs in heat engines, and ultra-high efficiency can be reached, even in micro-scale systems.[46,50–52] Fuel cells can also benefit from low emissions, due to the intrinsic characteristics of their electrochemical reaction, whose by-products, in case hydrogen is used as a fuel, are just CO_2 and H_2O.[53,54] Their intrinsic modularity also allows one to assembly power plants from micro-scale ($<1\,kW$) to large-scale ($>10\,MW$) systems.[46,53,54] Finally, fuel cells are especially suitable for CHP, since the electrochemical reaction is exothermic and the heat released by such a reaction can be used for a number of different applications (space heating, domestic hot water, district heating, steam production, *etc.*), depending on the operating temperature of the fuel cell.[3,52,55–62] Therefore, the higher the operating temperature of the fuel cells, the wider the number of possible CHP applications. When the operating temperature of the fuel cells is sufficiently high, the heat produced by the fuel cell can be also used as a heat source for a further conventional heat engine (*e.g.*, Rankine cycle, Bryton cycle).[3,52,55–56,63–70] In this case, the overall efficiency of the overall hybrid cycle can theoretically be higher than 70%.[3,46,52]

A dramatic effort has been performed over the past 20 years by academic and industrial partners in order to develop cost-effective power systems based on fuel cell technology.[46,55,68] In particular, a number of scientific papers have been published, aiming at proposing novel power systems integrating both fuel cells and conventional technologies (gas turbines, steam turbines, combined cycles, *etc.*).[3,46,52,55,68,71–76] In fact, in case of hybrid systems, the overall electrical efficiency can be surprisingly high even for low- or medium-scale systems.[46,50,55,68,77,78]

It is well known that the electrical efficiency of a conventional heat engine increases for higher values of the temperature of the heat sources.[79] Therefore, the operating temperature of the fuel cell integrated in the hybrid system dramatically affects the overall efficiency of the cycle.[53,54] Among the different types of fuel cells, Solid Oxide Fuel Cell (SOFC) achieve the maximum operating temperature (up to 1000 °C). This makes SOFCs especially suitable for integration in hybrid systems.[3,46,52–55,68,80]

The operating temperature of the fuel cell also affects the type of bottoming cycle. In fact, when the operating temperature is around 650 °C (for example, in case of Molten carbonate Fuel Cells, MCFC, or Intermediate Temperature Solid Oxide Fuel Cells, ITSOFC), the Rankine cycle is typically selected as the bottoming cycle[52,81–84] since the inlet temperature required for a steam turbine is close to the temperature of the exhaust gases of the fuel cell. Conversely, for higher operating temperatures (around 1000 °C), the Bryton cycle is the reference choice, which can be directly or indirectly coupled with the fuel cells.[3,52–54] This last configuration, based on the integration of a high-temperature SOFC and a Gas Turbine (GT), is especially attractive, due to its potentially ultra-high efficiency: a number of research studies predicted electrical efficiencies higher than 70%.[3,52–54] Additional characteristics of SOFC/GT hybrid systems are: low emissions, long operating life, availability of high-temperature heat for CHP applications, large range of capacities. In fact, such systems were indicated as one of the best candidate to meet the goals of The Vision 21 programme of the US Department of Energy. Such a programme proposed a new approach to energy conversion that addresses pollution control as an integral part of high-efficiency energy production. So, integral pollution control, ultra-high efficiency and potential for carbon dioxide capture and sequestration are the salient features of the "Vision 21 Energy Plant" concept, aiming at achieving an electrical efficiency greater than 60% (if based on the higher heating value, HHV) and 75% (based on the lower heating value, LHV); the first limit refers to the use of coal as a fuel, the second one to natural gas. A Vision 21 plant is also required to have near-zero emissions of pollutants, including smog and acid rain-forming substances, and to reduce greenhouse gas emissions by $40 \div 50\%$.[85] In this framework, several researchers stated that SOFC/GT hybrid cycle technology is a key to reaching the Vision 21 goals, both in terms of efficiency and environmental safety.[78,85,86]

13.2 SOFC/GT Prototypes

A significant research effort was spent on the design and optimization of SOFC/GT hybrid cycles, producing a large number of papers.[3,46,50,52,68,87–92] Conversely, experimental activities on SOFC/GT hybrid power systems were scarce, mainly due to the high capital cost required to build prototypes, even at a laboratory scale.[93] Probably, the most popular prototype is the 220 kW SOFC/GT power system w manufactured by Siemens Westinghouse and installed at the University of California (Figure 13.1):[94] a similar system, with a output of 300 kW, was also tested in Pittsburgh.[95]

The main design goals of the prototype tested in California were:[94]

- increasing the power of a 140 kW atmospheric SOFC up to 200 kW by pressurizing the stack up to 3 bar, with an operation temperature of 1000 °C;
- integrating the pressurized SOFC with a commercially available 50 kW Micro Gas Turbine (MGT), operating with a pressure ratio equal to 3 and at Turbine Inlet Temperatures (TIT) ranging from 800 °C to 900 °C;

Figure 13.1   220 kW SOFC/GT unit by Siemens-Westinghouse.[94,96]

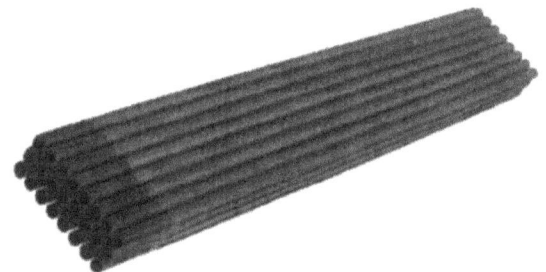

Figure 13.2   Siemens-Westinghouse tubular SOFC bundle.[94,96]

- designing and optimizing the Balance of Plant, also including recuperative heat exchanger;
- designing and optimizing the control strategies of the system.

The SOFC stack consisted in 1152 tubular fuel cells, vertically mounted, operating in a series/parallel configuration. The stack was arranged in 12 rows, consisting of bundles of 24 cells (Figure 13.2). The nominal length of the tubular SOFC is 1500 mm, whereas the diameter is 22 mm (Figure 13.3). The nominal current of the stack was 250 A.[94]

In the first design of the system, the use of a 50 kW MGT was planned. Unfortunately, no commercially available MCGT of such capacity was found. Therefore, a 75 kW MGT was used. Nevertheless, this oversized MGT was predicted to determine a reduction in the overall efficiency. In fact, the air flow of the MCGT was excessive for the SOFC stack. In addition, the thermal energy produced by the SOFC stack was not sufficient to drive the MGT. In particular, a two-shaft IRES MGT was employed. The common shaft housed a radial centrifugal compressor and a radial inflow turbine, whereas a radial

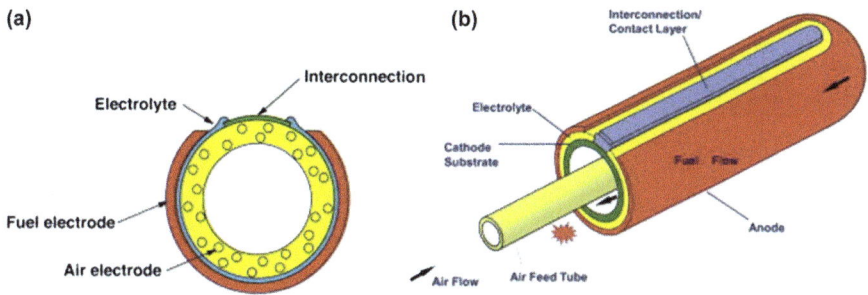

Figure 13.3 Siemens-Westinghouse tubular SOFC cross section.[94,97]

inflow free power turbine is installed on the second shaft, driving a synchronous AC generator trough reduction gear.

The simplified system layout of the 220 kW SOFC/GT prototype is shown in Figure 13.4. Air is filtered and compressed by the "Gasifier", approximately up to 3 bar. Then, it is preheated in the recuperator by the exhaust gases exiting from the low pressure turbine. Then, air enters the cathode compartment of the fuel cell. On the other side, natural gas is first compressed and then desulfurized. The gas, once compressed and desulfurized, is used for the anode compartment of the fuel cell and to feed two auxiliary combustors, placed upstream and downstream from the SOFC stack. Natural gas is reformed in the SOFC stack, where it is also converted in electricity via the electrochemical reaction. SOFC exhaust gases expand first in the high pressure turbine, driving the air compressor, and then in the low pressure turbine, driving the generator and consequently producing additional electrical power. The steady state SOFC operating temperature was around 1000 °C, whereas the operating pressure was around 3 bar. The air exits the recuperator around 550 °C. The net power produced by the MGT was about 20–30 kW AC. This is due to the low TIT which was around 700–800 °C. Such value was significantly lower than the nominal one (871 °C) as a consequence of the overcapacity of the selected gas turbine. An air bypass duct is also used in order to limit the amount of air processed by the fuel cell.[94]

First tests showed that the SOFC and MGT DC gross powers were 166 kW and 22 kW, respectively, showing an AC efficiency of 49.1%. Such values were significantly lower than the ones expected. This was also due to some gas leakage into the SOFC stack, causing about 5% of the natural gas to bypass the SOFC stack. During the first hours of operation the hybrid system encountered some failures. One of this was very severe, requiring stack disassembly. Results of the inspections showed that fuel bypass caused a combustion inside the stack and a consequent damage of some SOFC tubes. Therefore, the stack was redesigned, repaired and reinstalled. After that, SOFC and MGT DC gross powers were 172 kW and 22 kW, respectively, showing an AC efficiency of 52.1%.[94]

The researchers involved in this experimental activity concluded that the results were dramatically affected by the over-capacity of the MGT,

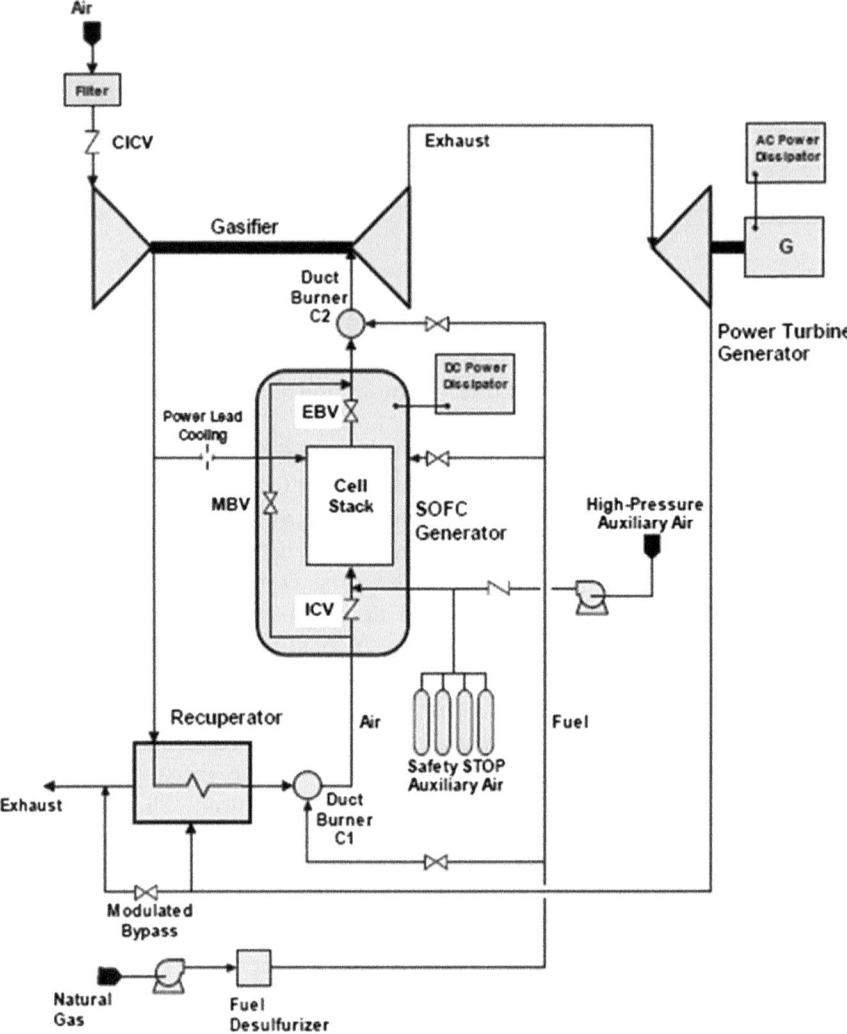

Figure 13.4 Simplified scheme of the Siemens-Westinghouse 220 kW SOFC/GT unit.[94,96]

determining a non-optimal system design. The only MGT available was too large, determining an excess air flow, accommodated by the bypass duct. However, the bypassed air, when mixed with SOFC exhaust, resulted in a low TIT value, and therefore in a poor MGT performance. The overall system efficiency was also affected by a low voltage of the second row of the stack. As a consequence, the electrical efficiency measured during the tests (52%) was significantly lower than the expected goal (>57%).[94]

Another important trial of a SOFC/GT hybrid system was performed by Mistubishi Heavy Industries, testing a 200 kW pressurized SOFC/GT power system.[52,98,99] This project was commissioned by the New Energy and

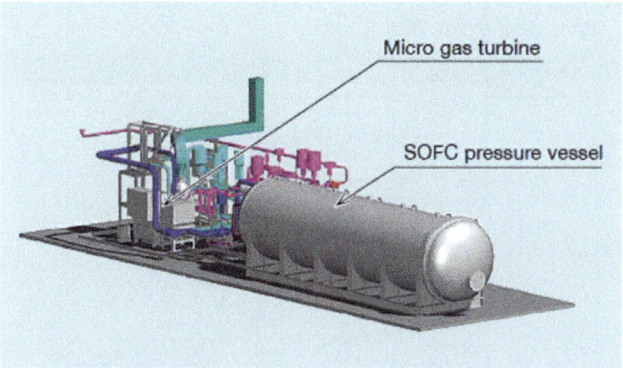

Figure 13.5 Mitsubishi 200 kW SOFC/GT prototype.[98]

Figure 13.6 Mitsubishi SOFC tube.[98]

Industrial technology Development (NEDO) in 2004 and was completed in 2006. The system consists in a pressurized vessel containing the SOFC stack, coupled with a MGT (Figure 13.5).[98]

The system operating principle was basically similar to the one discussed for the Siemens project. In fact, also in this project a tubular SOFC is used, too, with an operating temperature around 1000 °C. The length of the SOFC tube (Figure 13.6) was 1500 mm and the diameter 28 mm, providing about 143 W (maximum 151 W) at 0.65 V.

The main differences in the Mistubishi layout with respect to the previous one, are: (i) single shaft MGT, *vs.* double-shaft MGT; (ii) anode recirculation by means of a high temperature blower, *vs.* the ejector used in the previous case; (iii) anode and cathode exhaust gases burnt in an external combustor, *vs.* the internal combustion used the Siemens stack. A special effort was performed in order to design such combustor which operates with SOFC exhaust gases, whose heating value is 1/10 of that of natural gas (Figure 13.7).[98]

The cells were arranged in cartridges, each including 104 cell tubes. Cartridges were then installed in sub-modules located inside the pressure vessel, forming the SOFC stack (Figure 13.8).[98]

A total electrical efficiency of 50% was reached. This prototype was intended as a small-scale proof of concept of larger systems that the company aims at commercializing.[98]

A further SOFC/GT prototype, with smaller capacity, was recently experimented in Korea.[57] Researchers of KIER designed and constructed a pressurized 5 kW anode-supported planar solid oxide fuel cell (SOFC) power generation system with a pre-reformer for a fuel cell/gas turbine hybrid system. The SOFC stack was manufactured by Forschungszentrum Jülich (FZJ) in

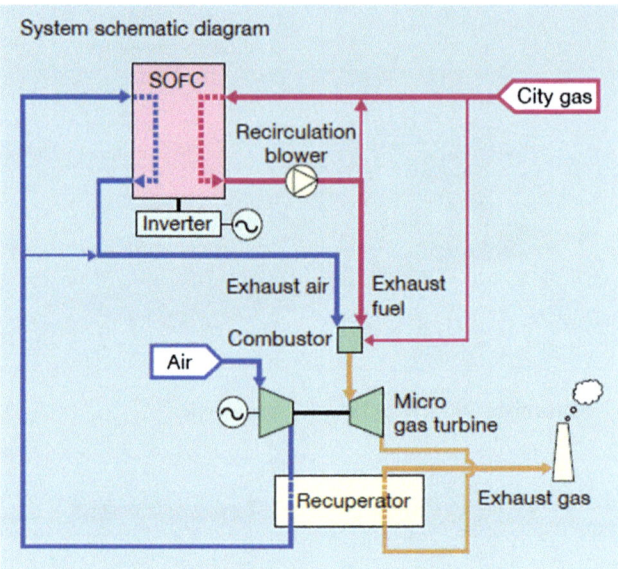

Figure 13.7   Mitsubishi 200 kW SOFC/GT system layout.[98]

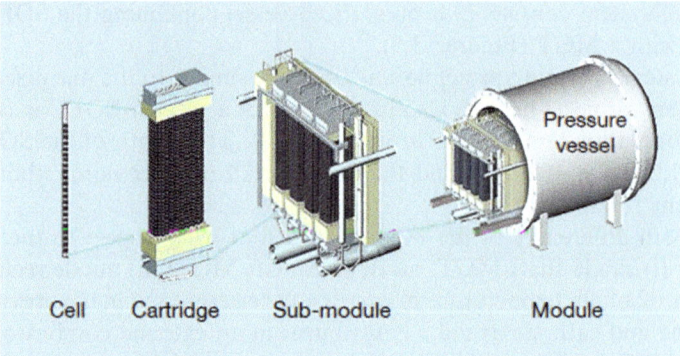

Figure 13.8   Structure of Mitsubishi module.[98]

Germany, whereas the hybrid system was assembled and installed at KIER. The authors of this study compared the performance of the SOFC stack in two operating modes: (i) atmospheric (no integration with a MGT); (ii) pressurized, integrated with a MGT. In atmospheric pressure, the output of the SOFC stack was 8.1 kW for the H_2 and 4.7 kW for the pre-reformed gas. In the hybrid configuration, the system layout was based on the integration of an externally reformed planar SOFC stack with a recuperative gas turbine, as shown in Figure 13.9.[57]

When the stack was fuelled by liquefied natural gas (LNG), the pre-reformer was operated in combination with a micro-gas turbine. In this case, the stack showed an output of 5.1 kW at about 3.5 bar. The nominal power of the micro

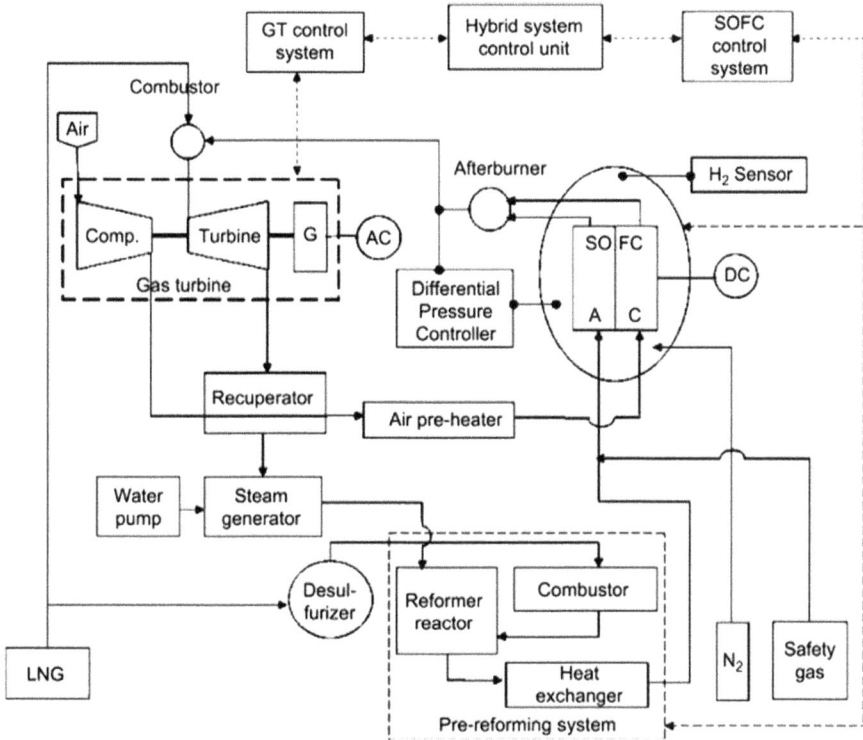

Figure 13.9 KIER hybrid SOFC/GT system layout.[57]

gas turbine was 25 kW. From the experimental results, the authors concluded that the anode-supported planar SOFC stack operated successfully in a pressurized hybrid system with a micro-gas turbine. The authors of this study reported a fuel utilization factor of 33.2%. However, no data is provided regarding the electrical efficiency of the overall system and the amount of electricity produced by the MGT.[57]

A further attempt to design a hybrid SOFC/GT system was performed by Rolls Royce Fuel Cells. This configuration uses custom-designed ejectors to recycle both anode and cathode exiting gases, avoiding the use of high temperature blowers or fans. In order to design an optimal hybrid system, Rolls Royce planned to model and design a novel two-stage/two-spool turbocharger, dedicated to this kind of application. However, due to the large pressure drops caused by the ejectors, the net power produced by the turbocharger will probably be much lower than the SOFC power output. At the present time, no data was published regarding the on-site operation of such prototype.[77] A similar project was also conducted by Allison Engine Company, evaluating a pressurized SOFC stack coupled with conventional GT plants. At design point, the pressure ratio is 7.0, showing an electrical efficiency of 67.0%. Further improvements to the system layout might determine an increase of this efficiency by 3.0%. NOx emissions were less than 1 ppm. McDermott Technology also developed a conceptual design of a hybrid system based on an atmospheric planar SOFC stack. The company predicted a power production of 700 kW with an electrical efficiency of 70%.[100] Unfortunately, no experimental data is available regarding these three projects (Rolls Royce, Allison Engine Company, McDermott Technology).

13.3 SOFC/GT Layouts Classification

The previous section showed that the few existing SOFC/GT prototypes are based on very similar system layouts. In fact, such layouts consist of a pressurized SOFC/GT stack operating in a Bryton cycle where it replaces the conventional combustor included in the basic Bryton cycle. However, this is only one of the possible configurations of SOFC/GT hybrid systems.[3,46,52,57,77,94,98] In fact, the open literature shows a number of different possible configurations. The layout of a Hybrid SOFC/GT power plant depends on several design parameters, such as:

- operating pressure of the SOFC stack;
- operating temperature of the SOFC stack;
- type of fuel;
- type of steam reforming process: internal/external, direct/indirect;
- production of steam required for the reforming process: anode recirculation or heat recovery steam generator;
- type of Bryton cycle: basic, intercooled and/or reheated.

Therefore, during the past few years, researchers developed a number of theoretical SOFC/GT configurations, aiming to improve the electrical efficiency and/or to reduce capital costs.[3,46,52,64,68,74–75,89,92,101–110]

In Figure 13.10 a possible rough classification of SOFC/GT system is proposed. Such figure summarizes the main selections which can be performed in the design of a SOFC/GT hybrid plant, neglecting a number of minor ones (*e.g.*, type of fuel, use of pre-reformers, bypass ducts, auxiliary burners, type of GT, *etc.*).

The basic selection which must be performed in the design of a hybrid SOFC/GT power plant is the operating pressure of the fuel cell. In fact, if system simplicity and reliability are assumed as the main final goals, the operation of the fuel cell under atmospheric pressure should be preferred.[110,111] In fact, in this case, the operation of the SOFC is completely independent from the Gas Turbine, and a heat exchanger represents the only connection between the two sub-systems. This allows one to achieve a safe and robust operation for both the fuel cell and the GT. In particular, the latter can operate at nominal conditions since an auxiliary burner provides additional heat to the air exiting from the SOFC heat exchanger, in order to maintain a stable value of the TIT. On the other hand, it must be pointed out that such configuration is expected to achieve lower values of the electrical efficiency with respect to the pressurized one. In fact, both SOFC and GT efficiencies are expected to be lower: the

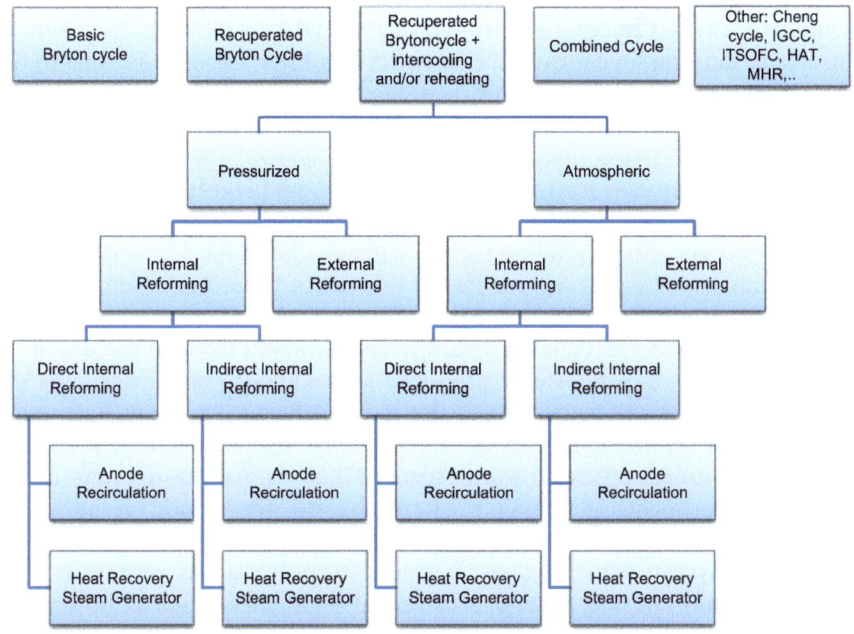

Figure 13.10 SOFC/GT hybrid cycles classification.

voltage – and consequently the efficiency – of a SOFC significantly increase with the operating pressure, as clear from the Nernst equation; in addition, the heat exchange between the SOFC exhaust and the air entering the turbine induces a lower TIT with respect to the pressurized configuration, where SOFC exhaust directly expands in the gas turbine.[3,52,110] Therefore, when the goal of the design is to achieve high values of the electrical efficiency, the pressurized configuration is the best selection. In such configuration, the SOFC stack acts as a combustor of a Bryton cycle, producing also an additional amount of electricity.[3,46,52,102] This configuration is also expected to be cheaper than the atmospheric one, since no heat exchanger is needed to transfer the heat flow of SOFC exhaust to the air entering the turbine, and such a heat exchanger is very expensive, due to the large heat exchange area required by gas-to-gas operation, characterized by very low values of the heat exchange coefficients. On the other hand, it must be considered that the direct (pressurized) coupling between the GT and the SOFC is very complex, as showed by the test trials.[94,98] In fact, the operating domain of a conventional GT power plant, in terms of pressures and mass flow rates, is very restricted, due to the intrinsic peculiarities of the turbomachinery. Therefore, the use of a SOFC stack, rather than a combustor, further restricts the operating domain of SOFC/GT power plants.[68,112–114] In addition, this coupling may also determine dangerous pressure fluctuations, that may seriously damage the SOFC stack, especially sensitive to the pressure differences between the anode and cathode compartments. Further design variables are the type of cycle (Bryton, IGCC, Cheng, *etc.*) and the solution selected for the reforming process.[3,52] From both energetic and economic points of views, an Internal Reforming (Direct, DIR or Indirect, IIR) is more attractive, showing higher overall efficiencies at a lower capital cost, due to the fact that an external reformer is not needed. On the other hand, it must be taken into account that when an internal reforming arrangement is adopted, the methane conversion rate may not always be controllable. In addition, the DIR configuration may lead to some temperature gradients inside the cells, due to the fact that such a process is endothermic and cools the first part of the cell where it occurs. Such gradients may damage the stack.[3,52] Furthermore, another important aspect regards the production of the steam required for the (internal or external) reforming process. The typical Siemens configuration is based on an anode recirculation arrangement, achieved through the use of an ejector,[94,102] that uses the steam produced at the anode compartment by the electrochemical reaction to support the internal Steam Methane Reforming Process. This arrangement can be accomplished by recirculating a part of anode outlet stream upstream with respect to the eventual pre-reformer or the anode compartment. A possible alternative to the anode recirculation arrangement is based on the external production of this steam, by using the heat of the system exhaust gases in a heat recovery steam generator. This second option is typically more expensive and slightly less efficient; however, a better control of the steam-to-carbon ratio inside the stack is possible, preventing carbon deposition.[3,28,50,52,68,102,115]

13.4 SOFC/GT Pressurized Cycles

As discussed above, the most common SOFC/GT hybrid configuration is based on the integration of a pressurized SOFC stack in a conventional Bryton cycle. Often, a recuperator is also employed in order to preheat the air entering the stack by means of the exhaust gases.[46]

The simplified layout of a SOFC/GT Bryton cycle is shown in Figure 13.11. External air (state point 1) is compressed approximately up to the SOFC operating pressure (state point 2). Then, air is preheated by the recuperative heat exchanger, up to state point 3, and enters the cathode compartment of the SOFC stack, where it participates in the electrochemical reaction. On the other side, fuel is compressed up to the SOFC operating pressure by the fuel compressor. Then, after possible preheating, the fuel enters the anode compartment of the fuel cell. Here, it is first reformed into a H_2-rich mixture and subsequently participates in the electrochemical reaction, producing electricity. Unreacted fuel and cathode exhaust air are burnt into the combustor, increasing the outlet temperature of the resulting stream (state point 5). Such flow expands in the gas turbine, driving simultaneously the air compressor and the electrical generator, and producing additional power. Turbine exhaust gases (state point 6) are used to preheat air in the recuperative heat exchanger; finally (state point 7), such exhaust are available for cogenerative purposes.[46]

The design of this type of SOFC/GT hybrid cycle is affected by a number of thermodynamic parameters. Probably, the most important is the pressure ratio

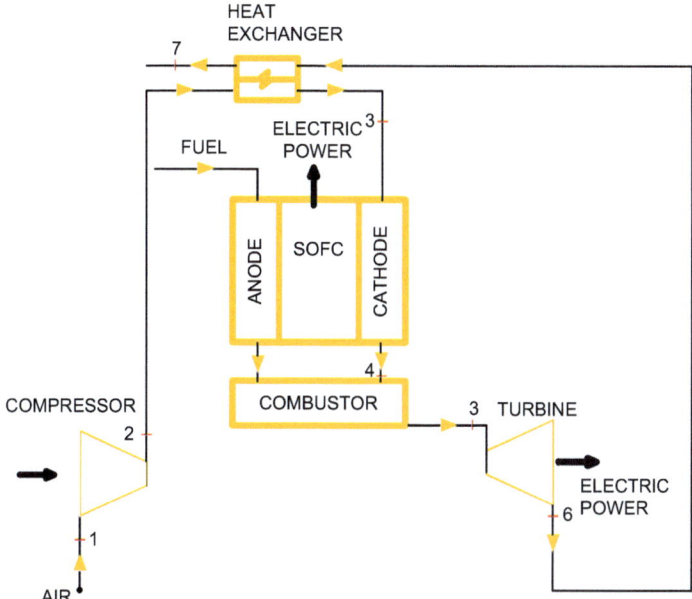

Figure 13.11   Hybrid SOFC/GT Bryton cycle.[46]

of the turbomachinery, directly affecting the SOFC operating temperature. The design operating temperature of the SOFC also dramatically affects the efficiency of both SOFC and GT subsystems. Furthermore, the Turbine Inlet Temperature (TIT) value is crucial for an efficient operation of the turbomachinery. Finally, when the final design of the system must be performed, a number of additional parameters must be also selected: geometry of turbomachinery; geometry and capacity of the reformer; geometry and size of the heat exchangers; geometry, materials and assembly of the SOFC stack, *etc.*

13.4.1   Internally Reformed SOFC/GT Cycles

In the majority of the studies published in literature regarding SOFC/GT hybrid cycles, natural gas was assumed as the fuel, internally reformed inside the SOFC stack. The Internal Reforming (IR) arrangement allows one to reduce dramatically system capital cost since no external fuel processing subsystem is required to convert hydrocarbons into hydrogen.[3,28,46,50,52,59,68,102,115–117] In some cases, a pre-reformer is also used, in order to crack the higher hydrocarbons included in the natural gas and to produce a small amount of hydrogen before the fuel enters the fuel cells; so, electrochemical reactions begin immediately, resulting in an enhanced power production. Conversely, when no external reformer is employed, the inlet sections of the SOFC typically just convert methane into hydrogen, whereas the electrochemical rate of reaction is very scarce, due to the low availability of hydrogen.[50,68,118]

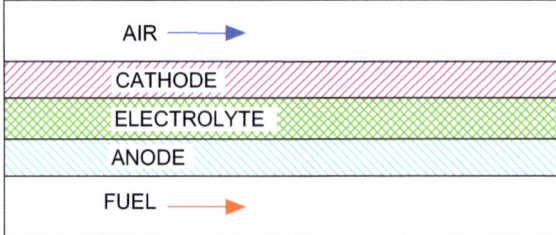

Figure 13.12   DIR arrangement.

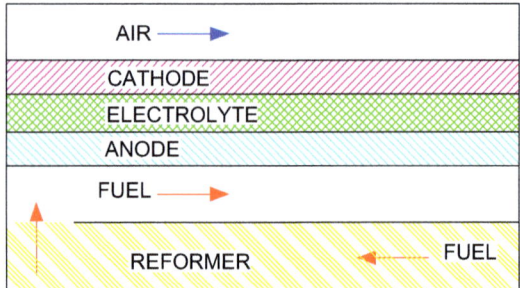

Figure 13.13   IIR arrangement.

In the Internal Reforming arrangement, two possible configurations are employed. The Direct Internal Reforming (DIR) configuration, where the fuel (natural gas) is converted into a hydrogen-rich mixture directly inside the anode compartment of the fuel cell (Figure 13.12). This configuration allows one to achieve greater system simplicity and lower capital costs. On the other hand, it must be considered that in the DIR configuration: (i) the anode compartment must be equipped with a proper catalyst for the Steam Methane Reforming (SMR) reaction; (ii) the risk of carbon deposition is higher due to larger amount of methane in the anode side of the fuel cell; (iii) the thermal balance of the fuel cell is more complex, since the SMR reaction (endothermic) is very fast and significantly cools the initial sections of the fuel cell, determining large temperature gradients inside the stack.[3,46,50,52,68,118]

All these issues may be mitigated by using the Indirect Internal Reforming (IIR) arrangement (Figure 13.13), where the SMR reactions occur in a compartment separated by the anode side of the fuel cell, which receives from the fuel cell the heat required to support the endothermic SMR reaction. Therefore, in this case only a thermal coupling between the reformer and the SOFC stack exists. Obviously, the IIR configuration results in a higher system complexity and in higher capital costs.[53,54]

13.4.2   Anode Recirculation

As mentioned above, the majority of SOFC/GT hybrid systems investigated in literature were fuelled by natural gas, which is converted in hydrogen using the

SMR process. It is well known that such a process uses steam to convert methane into hydrogen and carbon dioxide. Therefore, this steam must be supplied to the fuel processing subsystem in order to accomplish to its scope.[53,54] Therefore, a first possibility lies in the external production of steam to be supplied to the fuel processing subsystem. The steam is produced by using pressurized demineralized water, heated up to its evaporation temperature. Obviously, such a process requires a significant amount of heat, which can be supplied by the system exhaust gases. Therefore, steam production can be accomplished in a heat exchanger, a Heat Recovery Steam Generator (HRSG), designed to use a part of the heat of system exhaust gases to vapourize the demineralized water.[102–104] However, taking into account the anode electrochemical semi-reaction, it can be argued that a certain amount of steam is produced internally by the SOFC at its anode compartment. Therefore, a possible alternative to the HRSG is the recirculation (using an ejector or high-temperature blowers) of the anode exhaust gases to the fuel processing subsystem, since such gases also include the amount of steam required to support the SMR process. This configuration, pioneered by Siemens Westinghouse (Figure 13.14 and Figure 13.15), is the most common one, due to its intrinsic simplicity. In fact, the anode recirculation arrangement typically allows one to reduce system capital costs (no HRSG is needed) and it also promises high conversion efficiencies. On the other hand, the HRSG configuration allows an easier control of the system, since the amount of steam produced can be monitored and controlled also allowing a possible further use of the steam in the system (Cheng cycles) and/or for thermal purposes.[3,50,52,68,77,116,119]

A simple scheme of an internally reformed SOFC/GT power plant with an anode recirculation arrangement was presented by Song *et al.*[117] As shown in Figure 13.16, the system consists of a conventional recuperative Bryton cycle, where the conventional combustor is replaced by the SOFC stack. This stack is based on the tubular SOFC stack in which both anode recirculation and indirect internal reforming arrangements are implemented. Furthermore, a pre-reformer is also used in order to promote the SMR process. The authors assumed the operation of the cycle at constant Turbine Inlet Temperature (TIT) as 840 °C. Pressure ratio was set at 2.9. The operating values of the steam-to-carbon ratio and the fuel utilization factors were set at 2.5 and 0.85, respectively. The authors of this study calculated a net SOFC DC power of 840 kW, whereas the total power produced by the hybrid system was 220 kW. The overall electrical efficiency was 60.2%. The authors of this study also investigated temperature distributions along SOFC stack. They concluded that the stack is not isothermal and that the temperature profiles are dramatically affected by the gas flow distribution, by the cell operating parameters, and by the reforming process. In fact, the use of a pre-reformer decreases the cell operating temperature, resulting in a lower value of the electrical efficiency. However, a pre-reformer is considered very helpful since it relaxes the reforming load of the indirect internal reformer and reduces the amount of methane in the anode of the SOFC, reducing also the carbon deposition.[117]

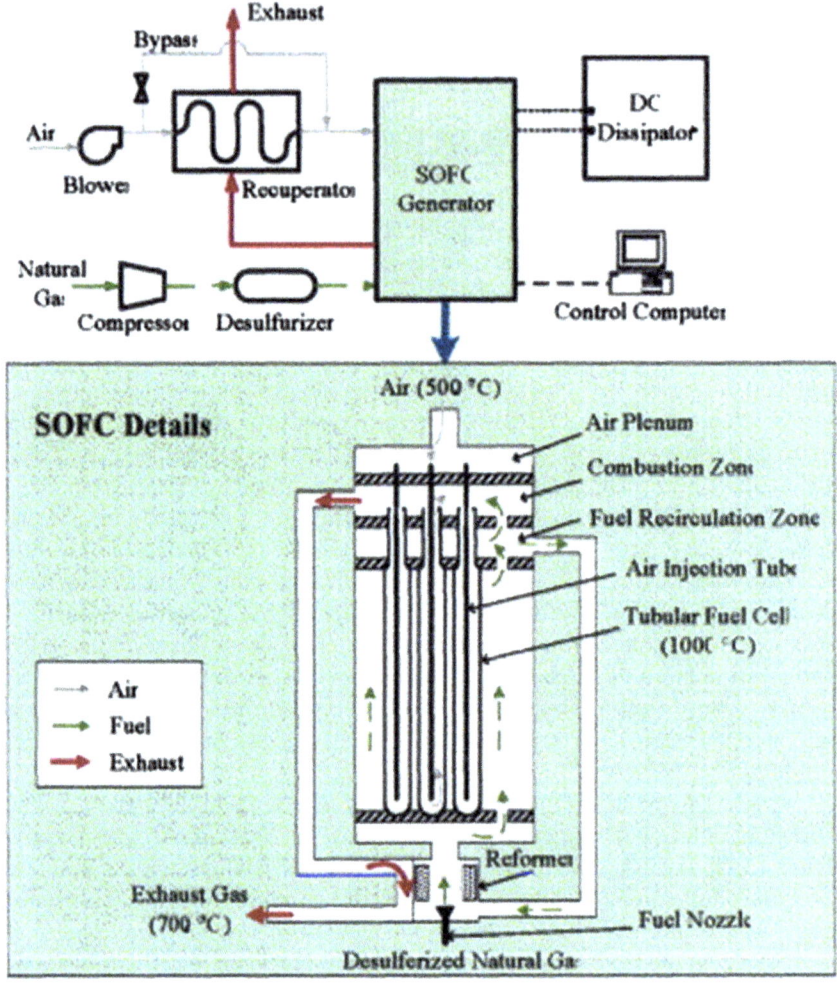

Figure 13.14   25 kW Siemens SOFC module with anode recirculation.[120]

A similar layout of the hybrid SOFC/GT power was presented by Calise *et al.*[50,68] The system layout investigated in these works is also based in a recuperative Bryton cycle, coupled with an internally reformed SOFC stack (Figure 13.17) using the Siemens anode recirculation arrangement (ejector). The main difference with respect to the previous layout lies in the use of a DIR arrangement. In fact, in this case, methane is first pre-reformed and subsequently reformed directly in the anode compartment of the fuel cell. The system is also equipped with heat exchangers for cogenerative purposes (HE3 and HE4).[50,68]

In this study, the pressure ratio was calculated by using turbomachinery performance maps on the basis of the corrected mass flow rate and rotational speed. The value calculated was around 7.8. The operating values of the steam

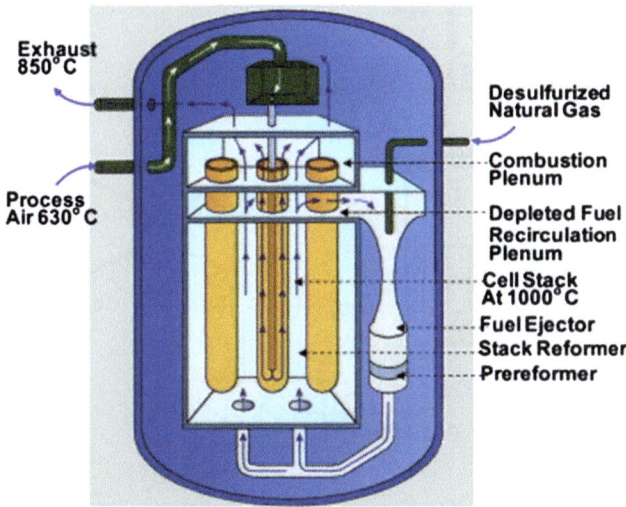

Figure 13.15 Scheme of Siemens SOFC module.[97]

to carbon ratio and the fuel utilization factor were set at 2.0 and 0.85, respectively. The authors of this study calculated a net total power of 1.5 MW and an electrical efficiency of 67.9%.

In the two previously cited studies, the gas turbine and the compressor were installed on the same shaft. However, this configuration may be also slightly different when an alternative gas turbine design is considered. In fact, it is a common selection to split the gas turbine into two. The first GT is installed on the same shaft of the compressor which is driven by the gas expansion in that turbine. The second turbine is a free power turbine which is installed on the same shaft of the electrical generator, producing an additional amount of energy compared to that produced by the SOFC.[66,77–78,116,121–123] An example of such alternative layout was presented by Chan *et al.*[116] (Figure 13.18), based on an internally reformed SOFC with anode recirculation and Direct Internal Reforming.

The study presented by Chan *et al.*[116] considered a pressure ratio around 9. The net power of the system was 381 kW. Stack temperature and TIT were respectively 1166 K and 1466 K. The calculations showed an electrical efficiency of 62.2%. The authors also performed a sensitivity analysis aiming at investigating the effect of the pressure ratio on the system. They concluded that increasing the pressure determines an increase of SOFC temperature and TIT, resulting in an improved electrical efficiency of the system.[116]

The previous configuration was also investigated in the IIR arrangement by McLarty *et al.*[78] In this study, the authors investigated the same layout adopted by Siemens in their 220 kW hybrid prototype, shown in Figure 13.4.[94] As mentioned above, this prototype was based on internally reformed tubular SOFC stack (IIR configuration), equipped with an ejector for the anode recirculation, a generator and a free power turbine. In this study, authors

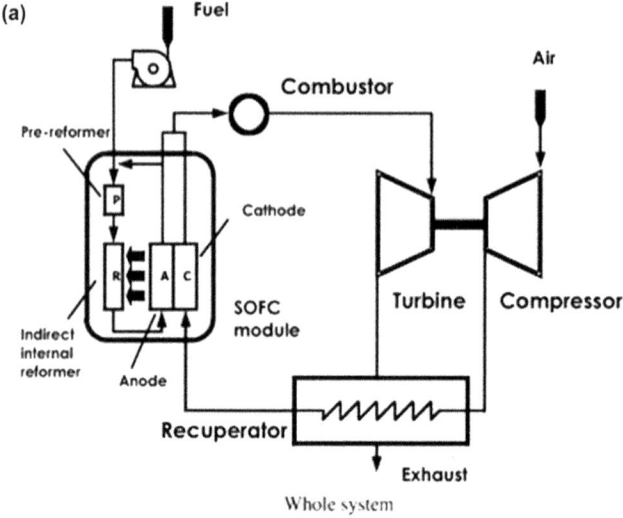

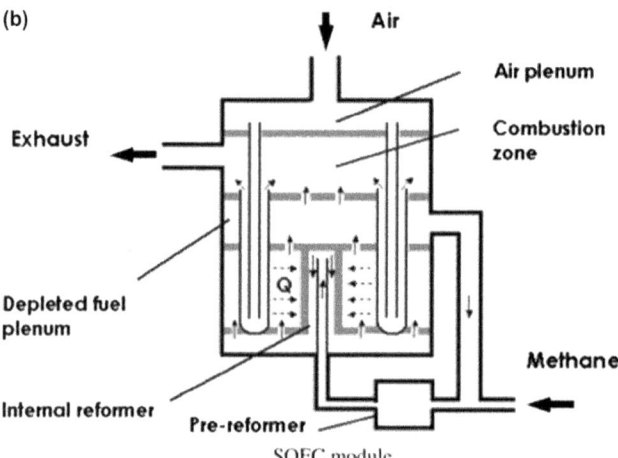

Figure 13.16 SOFC/GT hybrid system: internal reforming and anode recirculation, IIR arrangement.[117]

developed a simulation tool for designing a control strategy of the Siemens prototype. They developed control loops in order to mitigate internal temperature transients. They also observed that the monitored temperature fluctuations derived from environmental perturbation may significantly affect system efficiency and may diminish system longevity.[78] An additional system layout, reproducing the 220 kW Siemens prototype, was also presented by Bao *et al.*[66,124] The authors of this study concluded that their model fits the experimental data provided by Siemens with a very good agreement.[66]

A similar configuration was also analyzed by Burbanck *et al.*[77] (Figure 13.19). Their system layout is basically the same as that of McLarty *et al.*,[78] differing in a few peculiarities: (i) in this work,[77] the compression is divided into two stages,

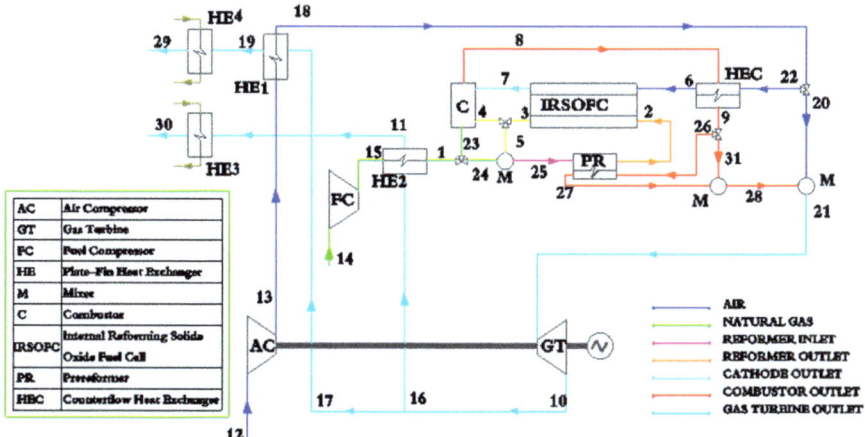

Figure 13.17 SOFC/GT hybrid system: internal reforming and anode recirculation, DIR arrangement.[50,68]

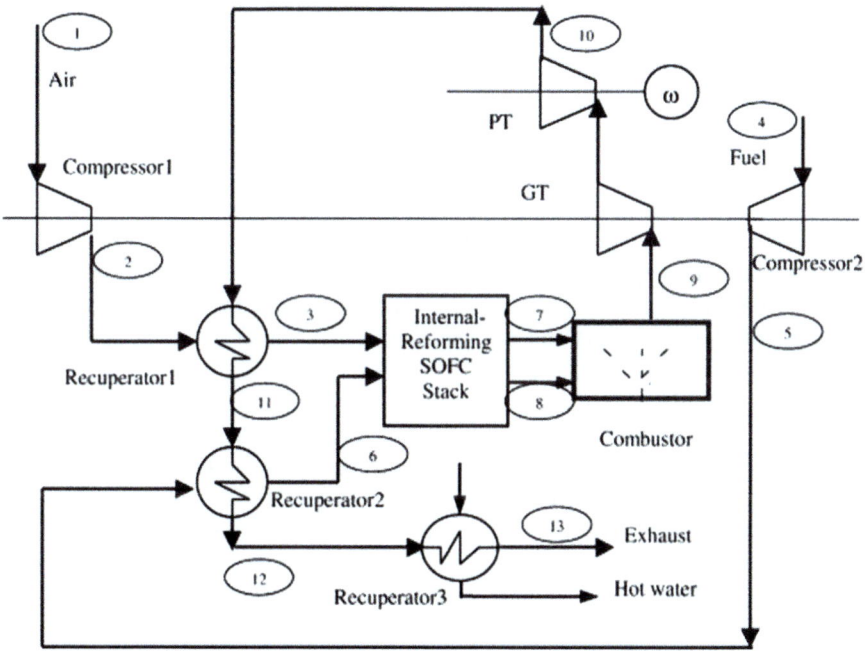

Figure 13.18 SOFC/GT hybrid system: internal reforming and anode recirculation, DIR arrangement, Free Power Turbine.[116]

with an intermediate intercooling process, using external air as cooling fluid; (ii) the pre-reformer is not used;[77] (iii) the expansion is accomplished by two separate turbines.[77] The simulation of the hybrid system is based on the operation map of the Northrop Grumman/Rolls Royce WR-21 gas turbine. The authors of this

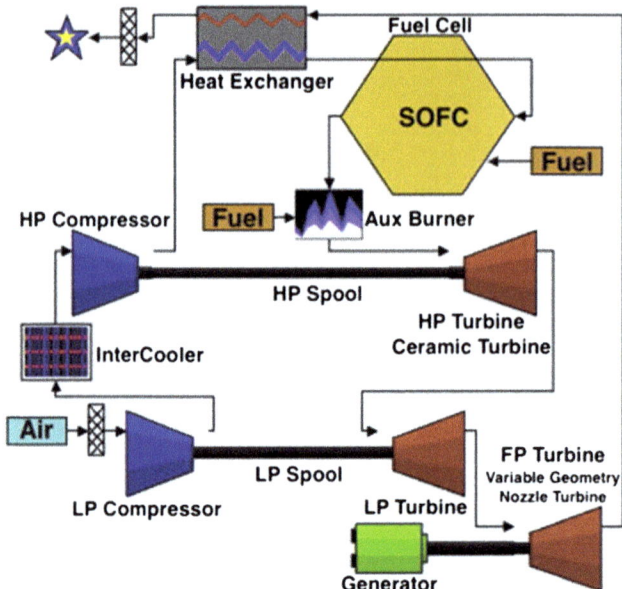

Figure 13.19   SOFC/GT hybrid system: internal reforming and anode recirculation, DIR arrangement, Free Power Turbine, Intercooling.[77]

study aimed to allow a large turndown at constant SOFC exit temperature and constant fuel utilization factor (85%), avoiding any cathode bypass or bleed. The high-pressure turbine is ceramic, due to its high operating temperature. The free turbine is a variable-geometry nozzle turbine. The system is manufactured using two separate single shaft turbochargers.[77] The nominal power of the system was close to 640 kW, whereas the electrical efficiency varies from 50% to 63%, depending on the part load ratio.[77]

A further study regarding pressurized SOFC/GT power plants, based on anode recirculation, internal reforming and free power turbine was presented by Haseli *et al.*[121,122] (Figure 13.20). In this study, the fuel utilization factor was set at 85%; the steam to carbon ratio was 2.5. The authors found the electrical efficiency of 60.55%. They also investigated the sources of irreversibilities in the system using both entropy and exergy. They found that increasing the TIT or compression ratio leads to a higher rate of entropy generation within the plant. It was found that the combustor and SOFC contribute predominantly to the total irreversibility of the system. About 60% of the irreversibility takes place in the following components at typical operating conditions: 31.4% in the combustor and 27.9% in the SOFC.[121,122]

13.4.3   Heat Recovery Steam Generator (HRSG)

As mentioned in the previous section, a possible alternative to the anode recirculation arrangement may consist in the use of a Heat Recovery Steam

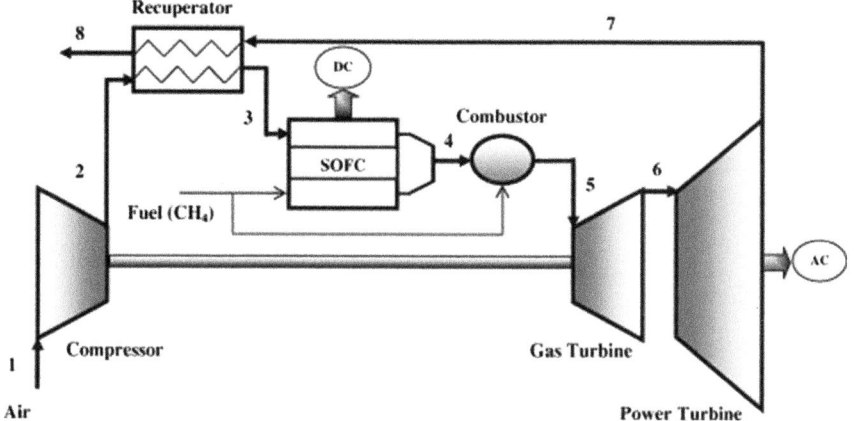

Figure 13.20 SOFC/GT hybrid system: internal reforming and anode recirculation, Free Power Turbine.[121,122]

Generator (HRSG), producing externally the steam required to support the SMR reaction, using the heat of the exhaust gases. This option was diffusely investigated in literature by a number of researchers presenting numerical models of internally reformed SOFC/GT power plants, including also a HRSG for steam production.[3,46,52–54,71,101–104]

A representative layout of a hybrid internally reformed SOFC/GT power plant, equipped with a HRSG, was presented by Chan *et al.*[103] (Figure 13.21). The authors of this study considered a DIR SOFC integrated in a recuperated Bryton cycle. In particular, two recuperators in series were considered. The high-temperature recuperator was used to pre-heat the air entering the SOFC stack, whereas the low-temperature recuperator was designed in order to preheat the fuel entering the anode compartment of the SOFC. The exhaust gases exiting from this second recuperator were finally employed in a HRSG to produce the steam required to support the SMR process occurring in the SOFC stack. Note also that the expansion is split in two turbines. The first one (high pressure) drives the air compressor and the low-pressure turbine drives the electrical generator. The authors of this study considered a 1.3 MW SOFC/GT hybrid plant. SOFC operating temperature and TIT were set, respectively, at 1000 °C and 1200 °C. The performance of the turbomachines was calculated on the basis of their operating map. Pressure ratio and fuel utilization factor were, respectively, set at 5.0 and 0.85. The authors calculated an electrical efficiency of 61.9%, whereas the global efficiency (electrical + thermal) was 86.4%.[103] Such values are in the same order of magnitude as those reported by other studies regarding the anode recirculation arrangement and discussed in the previous section.

Calise *et al.* also investigated a 1.5 MW system layout very similar to the one shown in Figure 13.21.[102] In this study, steam to carbon ratio and fuel utilization factor were set respectively at 2 and 0.85. Pressure ratio was 7.00. Results showed that electrical and global efficiencies were, respectively, 55%

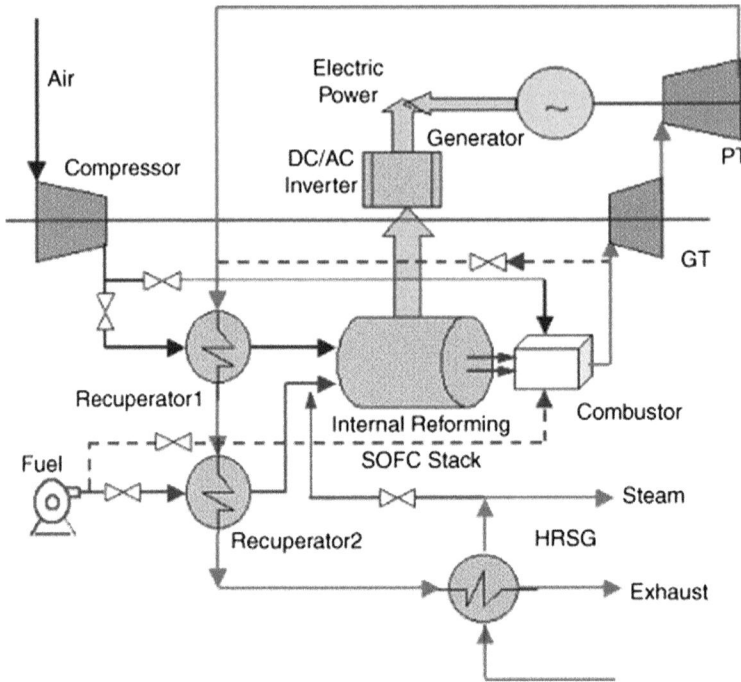

Figure 13.21 SOFC/GT hybrid system: internal reforming, HRSG, Free Power Turbine.[103]

and 65%. The authors performed also a comprehensive sensitivity analysis with the scope to analyze the effects of the variations of the main design parameters on the energetic and exergetic performance of the system.[102]

The authors concluded that both global and electrical efficiencies decrease for higher operating current densities (Figure 13.22). Obviously, the lower the current density, the higher the plant overall exergetic efficiency. This circumstance is due basically to the reduction of the exergy destruction rate in the SOFC, as a result of the reduction of the amount of its overvoltages. Note also that efficiency defects of fuel compressor, fuel-gas heat exchanger and of the pumps are always negligible. This is due to the low values of fuel and steam mass flow rates (Figure 13.22).[102]

The authors also showed (Figure 13.23) that it is possible to improve the electrical and global plant efficiencies by increasing the SOFC pressure, thereby lowering the irreversibilities in the heat exchangers, and mostly in the HRSG. In particular, the higher the operating pressure, the higher the cell voltage, determining a remarkable improvement of the electrical efficiency of the plant.[102]

Furthermore, authors also showed (Figure 13.24) that the Steam to Carbon (SC) ratio should be kept at lower levels because it determines lower plant electrical and global efficiencies. In particular, the slope of the global efficiency curve is higher with respect to the electrical efficiency curve: in fact, the higher

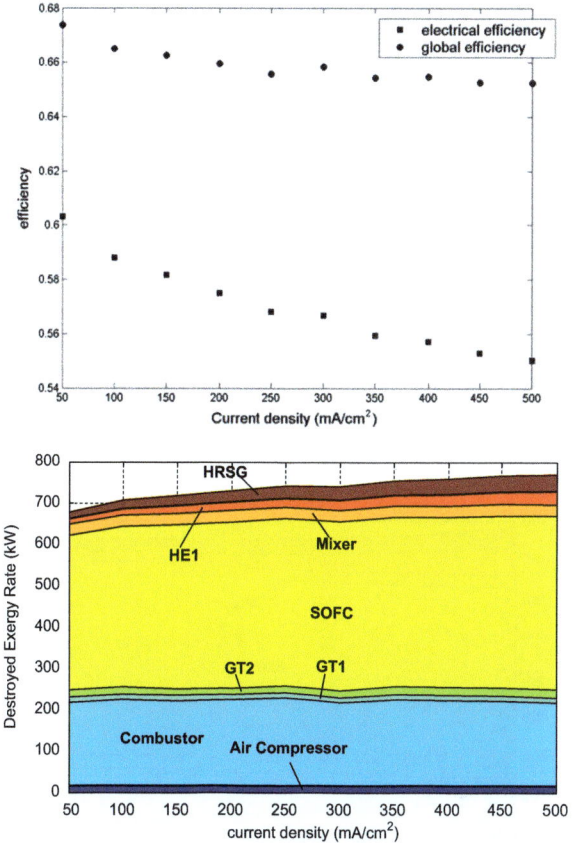

Figure 13.22 SOFC/GT hybrid plant: energetic and exergetic performance *vs.* current density.[102]

SC, the higher the amount of heat required by the HRSG, lowering plant thermal efficiency. Electrical efficiency slightly decreases since higher steam partial pressures cause higher Nernst overvoltages. The components' efficiency defects only slightly depend on the steam to carbon ratio.[102]

Finally, the authors also concluded that higher fuel utilization factors determine better overall global and electrical efficiencies (Figure 13.25). When the fuel utilization factor is very high, almost all the amount of hydrogen produced by the internal reforming reactions is consumed within the fuel cell by the anode electrochemical semi-reaction. Thus, the electricity produced by the SOFC increases, causing a rise in its electrochemical rate of reaction and consequently of its chemical exergy destruction rate. On the other hand, the fuel molar flow rates to the catalytic burner is reduced, thus determining lower local chemical exergy conversion rates. The higher the fuel utilization factor, the higher the plant exergetic efficiency since the amount of fuel converted via the electrochemical process, with respect to the amount reacted into the combustor, is higher. In fact, from the exergetic point of view, the former process is much

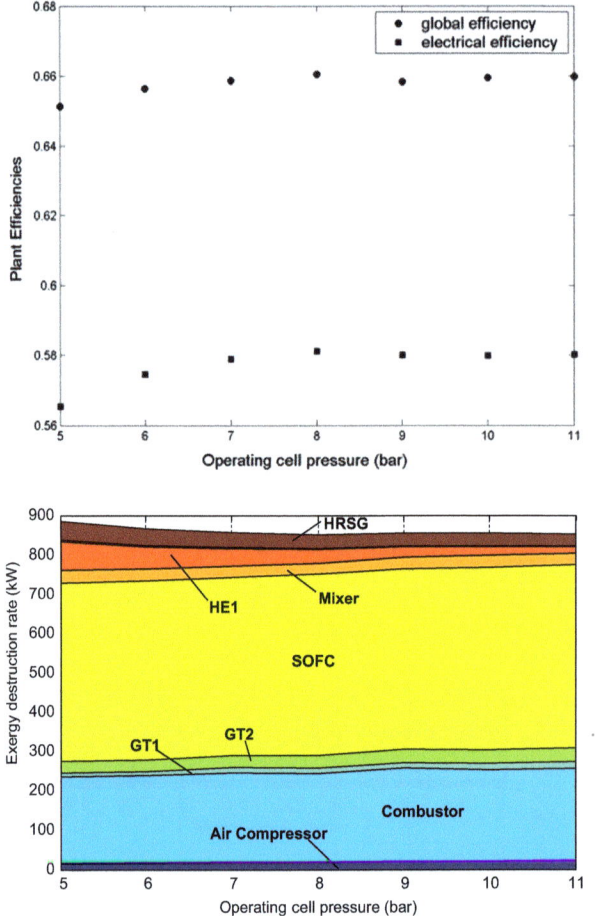

Figure 13.23 SOFC/GT hybrid plant: energetic and exergetic performance *vs.* SOFC pressure.[102]

more efficient than the latter. Furthermore, the higher the fuel utilization factor, the lower the chemical exergy flow available for the combustor, thereby determining a decrease of the corresponding efficiency defect. The same phenomena also determine the trend of the SOFC efficiency defect.[102]

The layout investigated by Calise *et al.*[102] and by Chan *et al.*[103] was slightly modified by Bavarsad *et al.*, by including a pre-reformer; furthermore, a single turbine was considered (Figure 13.26).[101] The authors assumed a 85% fuel utilization factor, a 30% pre-reforming degree, a 2.5 steam to carbon ratio and a pressure ratio equal to 3.0. In this configuration, the electrical efficiency calculated was about 65%. However, this value rapidly increased for higher pressure ratio (at pressure ratio of 7, the efficiency was around 72%). It was also concluded that efficiency decreases for lower values of fuel and air flow rates.[101]

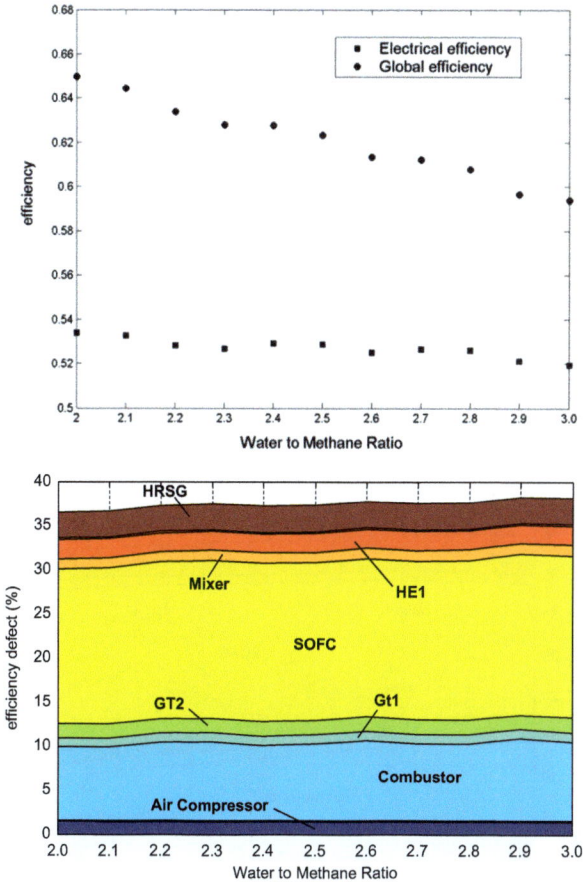

Figure 13.24 SOFC/GT hybrid plant: energetic and exergetic performance *vs.* steam to carbon ratio.[102]

The intercooled SOFC/GT hybrid system, already shown in Figure 13.19 in case of anode recirculation arrangement, can be also combined with the HRSG, as demonstrated by Yi *et al.*[85] The system layout investigated in this paper is shown in Figure 13.27. Here, the compression is divided into two stages, separated by a water cooled heat exchanger. The expansion also occurs in two separate turbines (HPT and LPT) although no reheating is considered. The stack is an internally reformed SOFC, based on Siemens tubular fuel cell. The steam required for the SMR is produced by a HRSG heated by the turbine exhaust gases. No air preheating is considered. The fuel utilization factor was set at 85%, whereas the pressure ratio in the optimal configuration was 50. The total power of the system was around 630 kW. In the basic configuration, an electrical efficiency of 69% was found. The authors calculated a slight increase of this efficiency for higher moisture content in the gas outlet of the humidifier. Conversely, results showed a slight reduction of system efficiency for higher

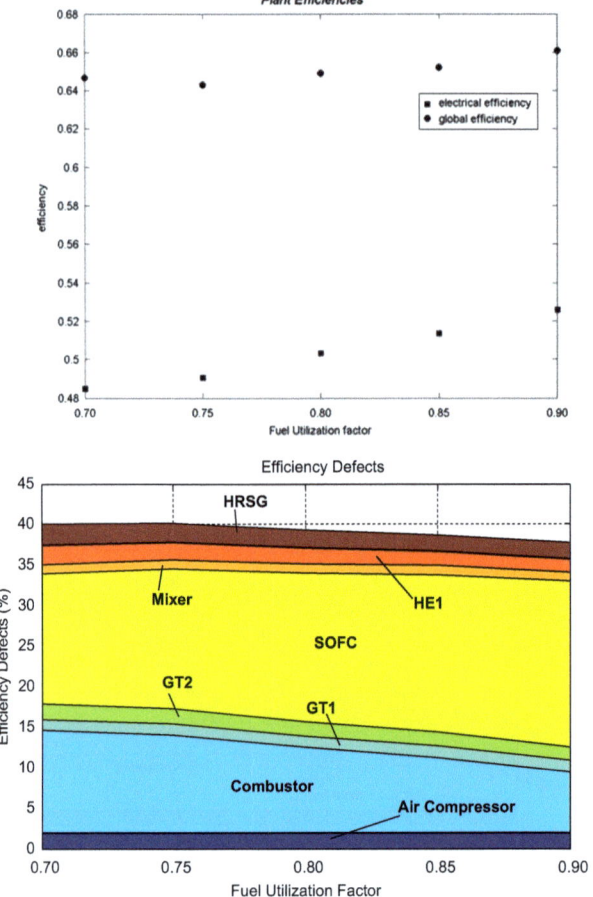

Figure 13.25 SOFC/GT hybrid plant: energetic and exergetic performance *vs.* fuel utilization factor.[102]

values of excess air. Finally, the authors also concluded that system efficiency dramatically increases for higher pressure ratio.

The HRSG configuration was also investigated by Granovskii *et al.*[104] in a paper in which the authors compared the performance of an internally reformed SOFC/GT power plant, equipped with a HRSG, with the one of an internally reformed SOFC/GT power plant, in an anode recirculation arrangement, coupled with a Rankine cycle. Figure 13.28 shows that the first layout (a) was similar to the previous one with some minor differences (recuperators are in parallel, water is recycled from exhaust gases). In the second layout (b) a conventional internally reformed SOFC/GT power plant, with an anode recirculation arrangement, is thermally coupled with a Rankine cycle, by a Heat Recovery Steam Generator which used GT exhaust to produce the superheated steam required for the Rankine cycle.[104] The authors of this study concluded that the second layout showed higher values of both electrical and

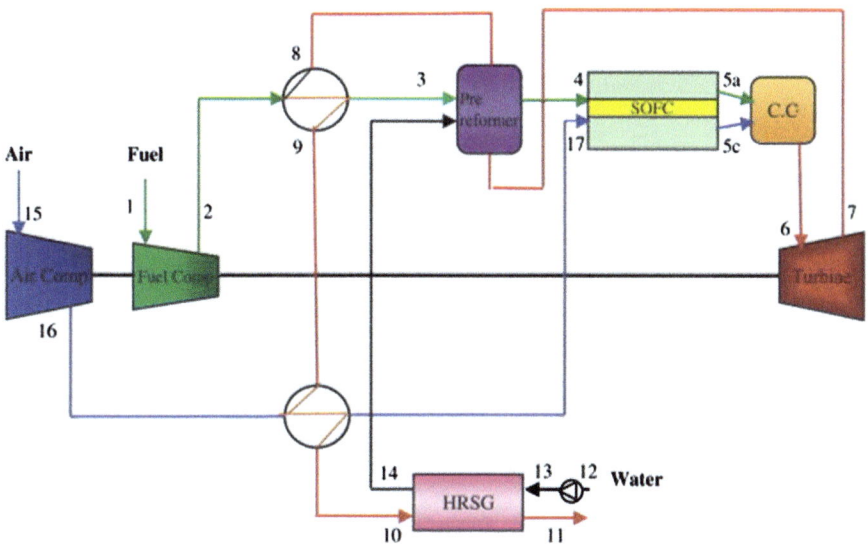

Figure 13.26 SOFC/GT hybrid system: internal reforming, HRSG, pre-reformer.[101]

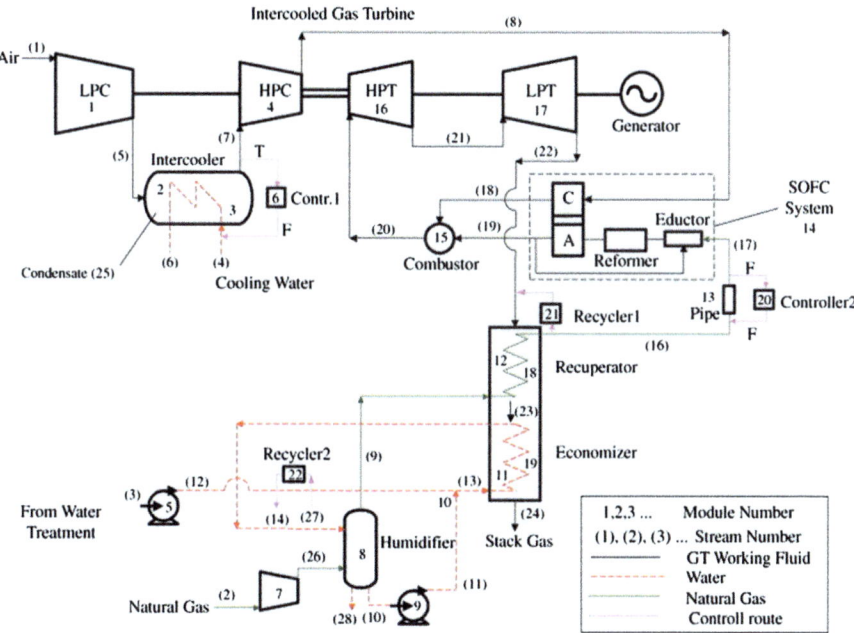

Figure 13.27 SOFC/GT hybrid system: internal reforming, HRSG, intercooling.[85]

exergy efficiencies, whereas the first layout showed a higher capability of power generation.[104]

Finally, the technology analyzed in this section has been also investigated as possible retrofitting strategy of an existing GT power plant by Cheddie.[125]

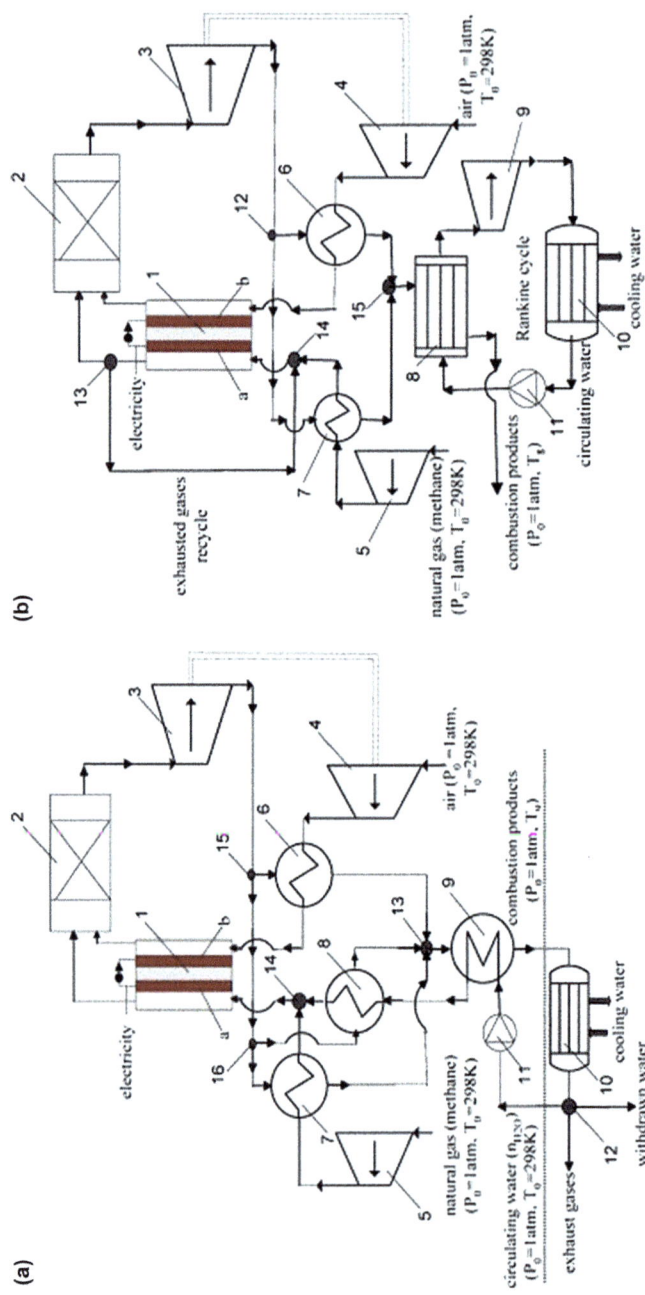

Figure 13.28 SOFC/GT hybrid system: internal reforming, HRSG (a); internal reforming, anode recirculation, Rankine cycle (b).[104]

The author of this study performed a thermo-economic optimization with the scope to design an optimized retrofitted internally reformed SOFC/GT hybrid plant, equipped with a HRSG. The electrical efficiency of the retrofitted pant was 66.2%, which was significantly higher than the one of the initial GT power plant. The cost of this retrofitting was estimated in about 24 M\$, which could generate a Net Present Value of about 35 M\$, assuming a future massive commercialization of SOFC technology.[125]

13.4.4 Externally Reformed SOFC/GT Cycles

The majority of the papers available in the open literature investigate internally reformed SOFC/GT hybrid cycles since this option is especially attractive due to the consequent possibility to reduce capital costs and improve system efficiency.[3,46,52] In fact, the externally reformed SOFC/GT hybrid power plants are very rare. The majority of systems equipped with external reformers are typically fed by more complex types of fuels (biogas, syngas, liquids, *etc.*), which cannot be safely supplied directly to the SOFC stack.[3,52,70,72,126]

In this framework, an interesting study was presented by Yang *et al.*[127] aiming at comparing internal and external SOFC/GT power plants. The authors investigated two power plants equipped with anode recirculation arrangements. In the first case (a) an IIR arrangement is considered, thermally coupling the SOFC and the reformer. In the second case (b) the reformer is separated by the SOFC stack and it is heated by a heat exchanger by the SOFC cathode exhaust gases (Figure 13.29). The authors of such work consider this method for sustaining the endothermic SMR process as one of the most efficient.[127]

The analysis was carried out considering Steam to Carbon Ratio of 3; the pressure ratio and the utilization factor were set respectively at 3.5 and 0.70. The allowed temperature of the SOFC ranged between 700 °C and 1000 °C. TIT varied from 750 °C to 1050 °C. The results of this study showed that the internally reformed SOFC/GT power plant shows a better performance than the externally reformed one. In fact, in the first case the efficiency ranges between 42% and 70% in the above-mentioned range of variation of SOFC temperature and TIT. In the same range, the efficiency of the externally reformed power plant varies between 32% and 60%. The authors concluded that the external reforming arrangement is penalized by a more complex thermal management, requiring additional amounts of fuels in order to achieve the desired SOFC and TIT temperatures and to maintain SOFC temperature difference within the range of acceptability.[127]

13.4.5 Hybrid SOFC/GT-Cheng Cycles

As mentioned before, a possible SOFC/GT hybrid configuration is based on a HRSG to produce the steam required for the SMR process. The steam produced by the HRSG can be further used for other purposes, such as thermal energy at medium temperature levels. A possible further utilization of this

(a)

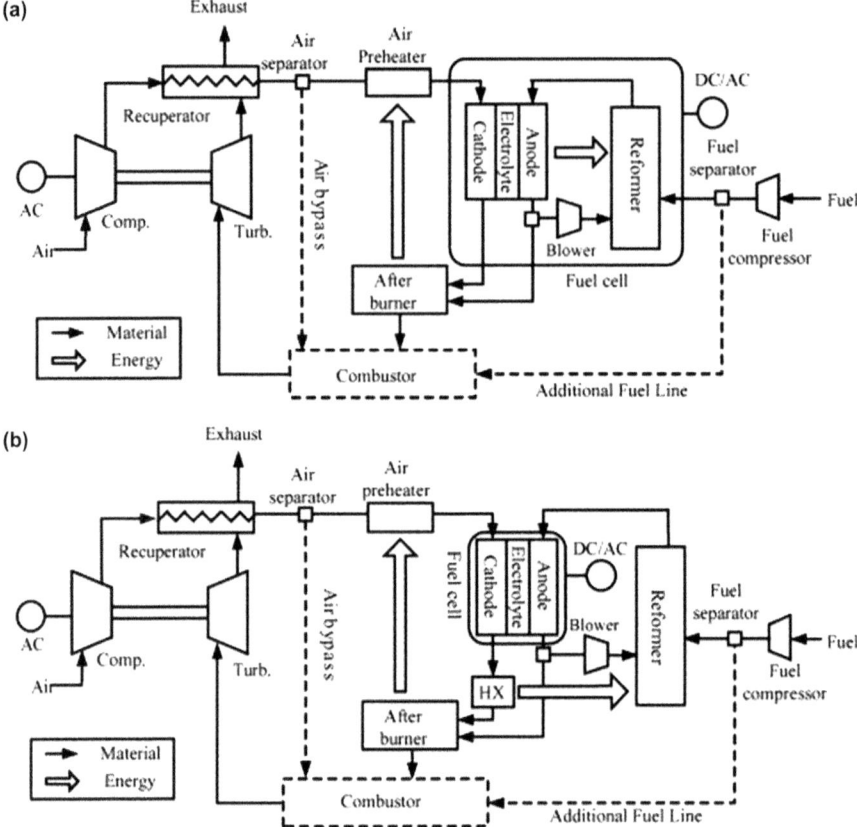

(b)

Figure 13.29 SOFC/GT power plants: internal *vs.* external reforming.[127]

steam lies in its injection in the gas turbine in order to improve its power production. In a conventional Bryton cycle, this arrangement is usually known as Cheng cycle. This arrangement allows one to increase the mass flow rate in the gas turbine, determining a consequent increase of the related power production. Cheng cycle is a good option to improve the flexibility of the hybrid system. In fact, the steam flow can be supplied to the gas turbine when the electrical demand is high; conversely, in case of increase of thermal demand, the steam can be used for thermal purposes. In case of SOFC/GT hybrid power systems, the Cheng cycle can be simply accomplished partially using the steam produced by the HRSG directly into the gas turbine. When the anode recirculation arrangement is not considered, the Cheng cycle does not require significant additional costs, since the HRSG is required in any case in order to supply steam to the SMR process. Cheng arrangement is expected to increase by 1–3% of the net electrical efficiency of the system.[3,46,52,128–130]

Nevertheless, the hybrid SOFC/GT-Cheng cycle was also studied when the SOFC is equipped with an anode recirculation arrangement. For example, the layout investigated by Mothar (Figure 13.30), considered a conventional

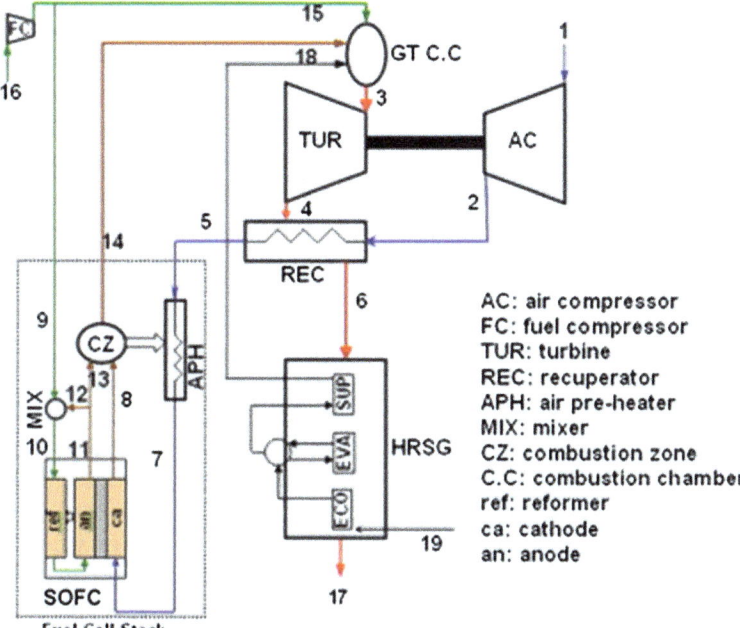

Figure 13.30   SOFC/GT-Cheng cycle: internal reforming and anode recirculation.[129]

recuperated Cheng cycle in which an internally reformed pressurized SOFC stack (with anode recirculation) plays the role of the combustor. Here, the turbine exhaust gases first preheat the air entering the SOFC stack, and subsequently supply heat to the HRSG, producing the superheated steam to be injected in the gas turbine. The authors of this study assumed steam to carbon ratio, pressure ratio and fuel utilization factor, respectively, of 2.5, 9.9 and 0.85. They compared the results of the SOFC/GT cycle with or without steam injection. The Cheng cycle determines an increase of the overall power production (13.7 MW *vs.* 1.6 MW) and an improvement of the system net electrical efficiency (66.12% *vs.* 58.87%). They also concluded that the use of the HRSG for steam injection determined a significant reduction of exergy losses of the exhausts. This circumstance determines a dramatic increase of the exergy efficiency of the Cheng cycle (65.34%) with respect to the case without steam injection (58.28%).[129]

A similar study was also performed by Kuchonthara *et al.*[128] comparing internally reformed pressurized SOFC/GT power plants with steam and heat recuperation. In particular, the base cycle was approximately the one shown in Figure 13.16, which was compared with a cycle in which exhaust gases supply heat both to the recuperator and to a HRSG, the latter producing the steam to be injected in the gas turbine. The authors of this study also concluded that the Cheng cycle determines lower values of the TIT, due to the steam dilution effect. However, the power production is higher due to the higher mass flow rate. This circumstance allows one to increase system efficiency specially in case of high steam to carbon ratio.[128]

13.4.6 Hybrid SOFC/Humidified Air Turbine (HAT)

In a Cheng cycle, the steam produced by the HRSG is directly injected in the combustion chamber and then expands in the GT with combustion gases. A similar approach is adopted in a SOFC-Humidified Air Turbine (HAT) cycle where the steam is included in the gas turbine flow by a saturation process, avoiding the use of HRSG as occurs in Cheng cycle. In other words, the air flow before entering the SOFC and/or the GT is adiabatically saturated by liquid water. The resulting stream subsequently expands in an appropriately designed turbine which is able to operate with such fluids, showing very high humidity rates. The SOFC–HAT system was investigated by Kuchonthara *et al.*[131] presenting the layout shown in Figure 13.31. Here, the air is compressed by a two-stage intercooled compressor. The intercooling is supported by liquid water, which is also used to additionally cool the stream exiting from the second compressor in an aftercooler. Therefore, the temperature of the water exiting from the intercooler and the aftercooler is increased. An additional amount of water is heated in an economizer using the heat of the system exhaust gases. All these hot liquid water streams are brought to the air saturator where the air coming from the compressors is supplied. In such a component the air is saturated by the liquid water. The excess water may be recirculated in the process, whereas the humidified air is first preheated in a recuperator and then enters the SOFC stack. Therefore, the stream expanding in gas turbine also includes the amount of water which was included in the air stream by the air saturator. This system was analyzed by Kuchonthara *et al.* and compared with a SOFC/GT Cheng cycle and a SOFC/GT Steam Turbine combined cycle.

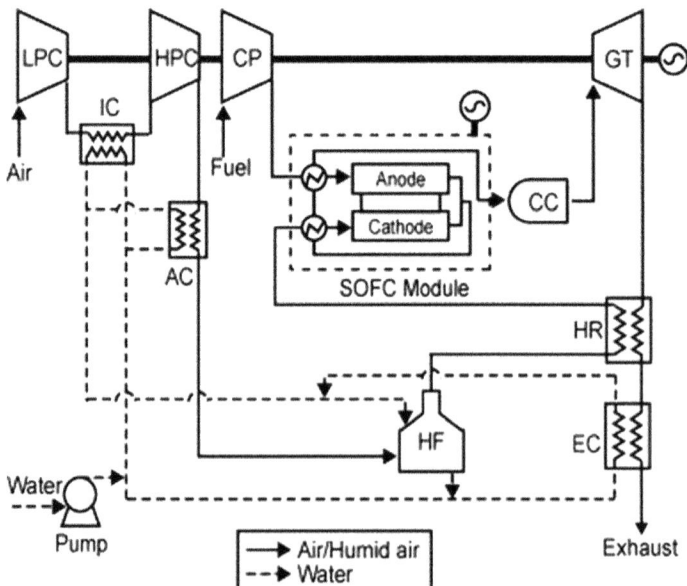

Figure 13.31 Hybrid SOFC/HAT cycle.[131]

The efficiency of the SOFC/HAT cycle varied approximately between 60% and 67% depending on several parameters. The authors also concluded that the SOFC–HAT system gives the lowest specific work at low-pressure ratios, corresponding to the lowest water consumption. It is worth noting that the SOFC–HAT system is a promising system with high efficiency and specific work output when the system operates at high TIT and pressure ratio conditions.[131]

13.4.7 Hybrid SOFC/GT-ITSOFC Cycles

A further possible SOFC/GT layout arrangement may consist of the integration of a high-temperature SOFC with an intermediate-temperature one (ITSOFC).[65,132] An example of this arrangement was investigated by Musa *et al.*[133] The authors of this study considered two different SOFC/GT system layouts. The first one is defined "single-staged" since only one type of SOFC stack is included in the system (high-temperature SOFC, HTSOFC, or intermediate-temperature SOFC, ITSOFC). In this case, the arrangement of the system layout is basically the one shown in Figure 13.18, differing from that only for the use of a single gas turbine. In this case, the SOFC stack may be HTSOFC or IT SOFC. In this layout, the fuel may also bypass the SOFC stack, entering directly the combustor. Furthermore, the anode recirculation arrangement is considered. The second configuration is defined as two-staged since the layout includes two SOFC stacks (ITSOFC+HTSOFC or ITSOFC+ITSOFC). Figure 13.32 shows the two-staged configuration including an ITSOFC and a HTSOFC. Here, the two stacks are supplied in parallel by the compressed and pre-heated fuel. Both stacks are equipped with anode recirculation arrangement. At the cathode side, the two stacks are connected in series: the compressed air is first preheated by the heat exchanger H/HE1 and then supplies the cathode compartment of the ITSOFC. Then, ITSOFC cathode exhaust gases supply the cathode side of the HTSOFC. Unreacted fuel and HTSOFC cathode exhaust stream are burnt in the combustor. Its exhaust

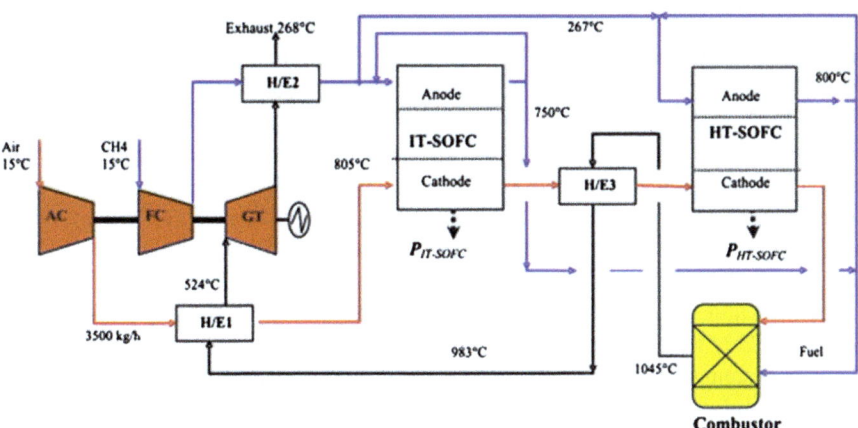

Figure 13.32 ITSOFC/HTSOFC/GT hybrid cycle.[133]

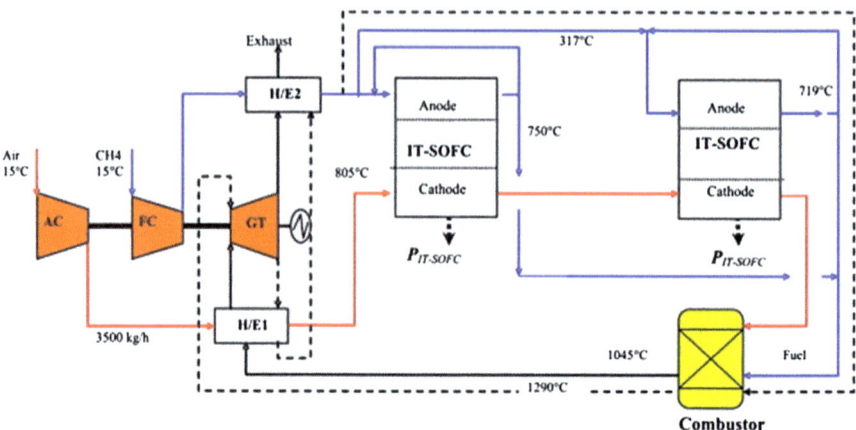

Figure 13.33   ITSOFC/ITSOFC/GT hybrid cycle.[133]

gases first preheat the cathode flow entering the HTSOFC and then preheat the air. Finally, they expand in the gas turbine. Note that in this layout the recuperative heat exchanger is placed upstream the GT, determining a significant decrease of the TIT. Figure 13.33 shows the two-staged layout including two ITSOFC. This layout is basically similar to the previous one, differing from it in a minor point: the heat exchanger H/E3 is not used since no further preheat is required due to lower operating temperature (ITSOFC) of the second-stage SOFC.[133]

The following values were assumed for the design parameters: steam to carbon ratio: 2.0; fuel utilization factor (total): 0.85; pressure ratio: 6.0; ITSOFC operating temperature: 750 °C; HTSOFC operating temperature: 750 °C. The results showed that the performance of the single-stage layout is dramatically affected by the cell operating temperature. In particular, in case of ITSOFC, the operating temperature was varied between 650 °C and 750 °C, resulting in an increase of system efficiency from 58% to 63%. In case of HTSOFC, the operating range was 750 °C–850 °C, resulting in an increase of efficiency from 53% to 58%. For the single-stage system, the authors performed also a parametric analysis, showing that for both ITSOFC and HTSOFC single-staged system the optimal cell operating pressure was around six bars and that the system's overall efficiency decreases for higher current density (especially in the case of single-staged ITSOFC cycles). The electrical efficiency of the two-staged system was higher by about 10% than the value reached by the corresponding single-staged system. In particular, the highest electrical efficiency is achieved by the two staged ITSOFC/ITSOFC cycle, showing values varying from 67% to 78%, depending on the operating current density. This result can be easily interpreted if one considers that in a two-staged system the ratio of the power produced by the fuel cell(s) versus the system total power is higher. Therefore, considering that the conversion efficiency of the SOFC is significantly higher than the one of GT, the overall system efficiency increases with the above-mentioned ratio.[133]

13.4.8 Hybrid SOFC/GT-Rankine Cycles

When the GT outlet temperature of a hybrid SOFC/GT system is sufficiently high, GT exhaust gases may be used to produce the superheated steam required to drive a Steam Turbine (ST) in a conventional Rankine cycle; so, a SOFC/GT/ST combined cycle can be obtained, with possible ultra-high electrical efficiencies.

Such an opportunity was thoroughly investigated by Arsalis *et al.*,[55,80,134] who analyzed four configurations, in which the following differences regarding the steam turbine (ST) bottoming cycle were assumed: single pressure level, dual pressure level, triple pressure level with and without reheat. The purpose of using multiple pressure levels is to achieve a higher power output from the steam turbine. The SOFC stack is based on the internally reformed Siemens tubular configuration, equipped with a pre-reformer and an anode recirculation arrangement. The layout of the hybrid SOFC/GT/ST power plant is shown in Figure 13.34. With regard to the topping cycle (SOFC/GT), the operating principle is basically the same of the one shown in Figure 13.17. The difference lies in the two cogenerative heat exchangers (HE3 and HE4) and the two

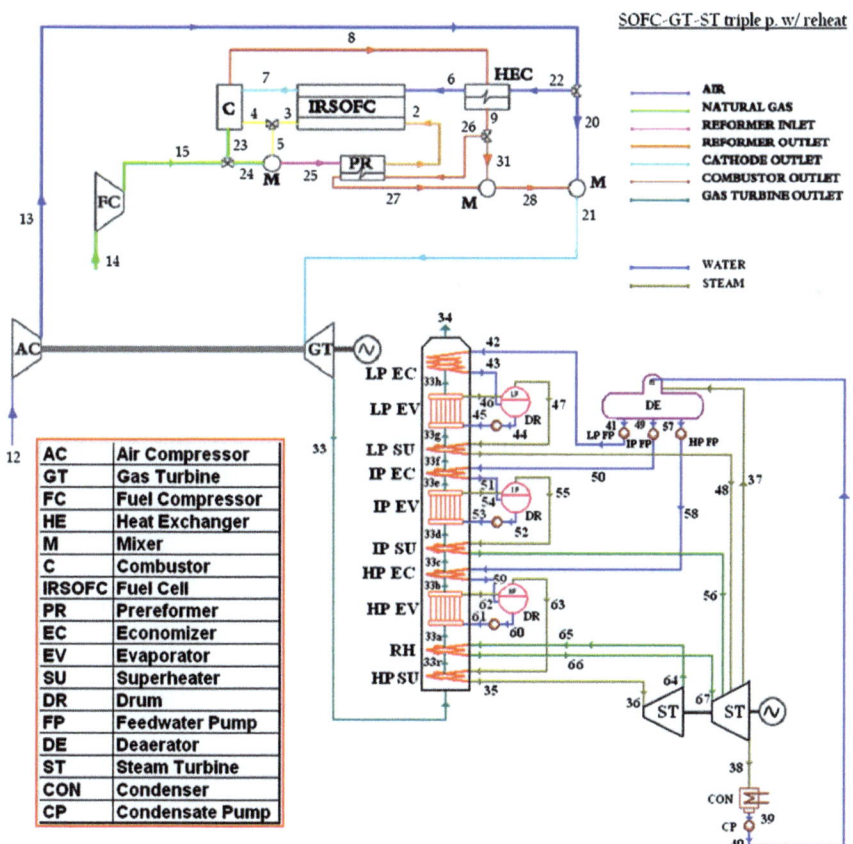

Figure 13.34 SOFC/GT/ST combined cycle.[55,80,134]

recuperative heat exchangers (HE1 and HE2), which are not employed in this system layout. In fact, in this combined cycle all the heat available from GT exhaust must be provided to the bottoming Rankine cycle. The operation of the SOFC/GT topping cycle is basically similar as the one discussed in previous sections. Air is compressed by the air compressor (AC) up to the fuel cell operating pressure. The air is then brought to the cathode inlet of the SOFC stack (state point 18). Similarly, fuel is compressed by the fuel compressor (FC) and then brought to the anode compartment of the stack (state point 1). Both fuel and air can bypass the fuel cell, *i.e.* a certain amount of fuel can flow directly to the combustor (C), bypassing the electrochemical reaction occurring within the stack (state point 23), while excess air can flow to the GT (state point 20). At the stack, fuel (state point 24) is mixed with the anode recirculation stream (state point 5) in order to support the steam reforming reaction in the pre-reformer and in the anode compartment of the fuel cell. The mixture at state point 25 consists of methane and steam. Thus, in the pre-reformer (PR), the first step in the fuel reforming process occurs. The energy required to support the pre-reforming reaction is derived from the hot stream at state point 26. The non-reacted fuel at state point 2 is involved in the internal reforming reaction within the anode compartment of the SOFC stack. Here, it is converted into the hydrogen that participates in the electrochemical reaction. On the cathode side, air is first preheated by a counter-flow heat exchanger air injection pipe (HEC) and then brought into the annulus (air pipe) of the SOFC where, at the three-phase boundaries, the cathode electrochemical reaction occur.[1,2,7,8] The electrochemical reactions, occurring in the fuel cell, produce DC electrical current and release thermal energy. The first of these is converted into AC current by the inverter; the latter is used by the internal reforming reaction and to heat up the fuel cell stack. The high energy flow rate at state point 8 is first used to preheat air in the counter-flow heat exchanger and then to supply energy to the pre-reforming reaction. This stream at state point 21 enters the gas turbine. The expansion in the GT supplies mechanical power which in turn is converted into electric power.

The operation of the steam turbine bottoming cycle, *e.g.*, the triple pressure with reheat variation (see Figure 13.1, bottom part), can be summarized as follows.[55,80,134]

- The GT exhaust stream (state point 33) flows to the heat recovery steam generator (HRSG). The gas mixture side of the HRSG passes through the 10 heat exchanger sections – high-pressure (HP) superheater (SU), reheater (RH), HP evaporator (EV), HP economizer (EC), intermediate-pressure (IP) SU, IP EV, IP EC, low-pressure (LP) SU, LP EV, and LP EC – and is exhausted at state point 34.

- The superheated steam produced by the HP SU (state point 35) is supplied to the HP stage of the steam turbine. After expansion the cold reheat (state point 64) at an intermediate pressure returns to the HRSG and there by means of a reheater is superheated (state point 66) and returned to the IP/LP steam turbine stage. Also the IP SU (state point 56) and the LP SU (state point 48) supply superheated steam to the double-admission IP/LP ST which during expansion produces mechanical power which in turn is converted into electric power in a generator. A small fraction of super-heated steam at low pressure is extracted (state point 37) to the deaerator (DE) to be used later on for feedwater preheating.

- The wet steam (state point 38) is then condensed in the condenser (CON). The condensate (state point 39) enters the condensate pump (CP) and is then pumped to the DE at state point 40.
- In the DE, any air oddments and impurities contained by the water are removed while the water is preheated at 60 °C. The preheated water (state points 57, 49, 41) enters the HP FP (feedwater pump), IP FP, and LP FP, and is then pumped to the HP EC, IP EC and LP EC at state points 58, 50, and 42, respectively.
- In the economizers, water is heated up to the saturated liquid point. Then it is evaporated at constant temperature/pressure in the evaporators.
- Water and saturated steam are separated in the drums, and the steam is supplied to the superheaters where it is superheated to the desired live-steam temperatures and fed to the ST to repeat the cycle.

The authors of this study considered three different hybrid plant sizes: $1.5\,MW_e$, $5\,MW_e$, and $10\,MW_e$. Also, two different SOFC sizes are considered based on current density. The operating current density for the selected fuel cell operates from $100\,mA/cm^2$ to $650\,mA/cm^2$. The systems were optimized, varying the main synthesis/design parameters. For example, SOFC temperature was varied in the range 950 °C–1100 °C, fuel utilization factor between 0.75 and 0.90, steam to carbon ratio between 2 and 3.5. The performance of each individual system is analyzed at full and part load conditions to determine the average and total efficiencies and total operating cost. Results of the thermo-economic optimizations for the various configurations under investigation are summarized in Table 13.1. The authors concluded that the hybrid SOFC/GT/ST configuration could be considered as an excellent candidate for ultra-efficient power production. For instance, the 10 MW_e SOFC/GT-ST hybrid triple pressure with reheat system exhibits efficiencies (maximum efficiency of 73.7%, an average efficiency of 65.3%, and a total efficiency of 68.4%). Furthermore, the SOFC/GT-ST hybrid power plant shows high efficiencies at off-design conditions as well. For a realistic system, a 1.5 MWe SOFC/GT-ST is not as attractive and efficient as a 5 MW_e or a 10 MW_e system because the gas turbine and especially the steam turbine are very inefficient at small sizes, resulting in lower overall system efficiencies.

13.4.9 Hybrid SOFC/GT with Air Recirculation or Exhaust Gas Recirculation (EGR)

The thermal design of a pressurized direct-coupled SOFC/GT hybrid system is very complex, due to various constraints regarding the SOFC stack in terms

Table 13.1 SOFC/GT/ST hybrid power plant: Cost and efficiency breakdown for all optimal models.[55,80,134]

Configuration	Description	Total cost	Operating cost	Capital cost	η_{ave}	η_{tot}	η_{max}
1	1.5 MWe single-pressure ST small SOFC	604 743	438 971	165 772	0.5696	0.6022	0.6638
2	1.5 MWe single-pressure ST large SOFC	613 655	405 328	208 327	0.6235	0.6525	0.6971
3	1.5 MWe dual-pressure ST small SOFC	602 033	435 146	166 887	0.5705	0.6031	0.6647
4	1.5 MWe dual-pressure ST large SOFC	611 026	401 575	209 451	0.6243	0.6533	0.6978
5	1.5 MWe triple-pressure ST small SOFC	599 321	431 428	167 893	0.5712	0.6035	0.6654
6	1.5 MWe triple-pressure ST large SOFC	608 046	397 831	210 215	0.6251	0.6536	0.6987
7	1.5 MWe triple RH-pressure ST small SOFC	598 691	430 598	168 093	0.5716	0.6038	0.6657
8	1.5 MWe triple RH-pressure ST large SOFC	607 446	397 031	210 415	0.6255	0.6539	0.6992
9	5 MWe single-pressure ST small SOFC	1 887 264	1 437 339	449 925	0.5791	0.6131	0.6795
10	5 MWe single-pressure ST large SOFC	1 935 053	1 340 421	594 632	0.6264	0.6573	0.7115
11	5 MWe dual-pressure ST small SOFC	1 872 891	1 411 966	460 925	0.5808	0.6149	0.6815
12	5 MWe dual-pressure ST large SOFC	1 920 649	1 316 017	604 632	0.6279	0.6589	0.7132
13	5 MWe triple-pressure ST small SOFC	1 871 891	1 405 966	465 925	0.5814	0.6155	0.6825
14	5 MWe triple-pressure ST large SOFC	1 924 649	1 310 017	614 632	0.6286	0.6596	0.7142
15	5 MWe triple RH-pressure ST small SOFC	1 869 891	1 402 966	466 925	0.5817	0.6159	0.6837
16	5 MWe triple RH-pressure ST large SOFC	1 923 649	1 307 017	616 632	0.6289	0.6601	0.7157
17	10 MWe single-pressure ST small SOFC	3 630 071	2 831 362	798 710	0.5876	0.6222	0.6922
18	10 MWe single-pressure ST large SOFC	3 717 621	2 603 916	1 113 705	0.6473	0.6766	0.7272
19	10 MWe dual-pressure ST small SOFC	3 630 067	2 819 358	810 710	0.5880	0.6228	0.6943
20	10 MWe dual-pressure ST large SOFC	3 717 624	2 591 919	1 125 705	0.6478	0.6772	0.7292
21	10 MWe triple-pressure ST small SOFC	3 640 067	2 805 358	834 710	0.5884	0.6234	0.6961
22	10 MWe triple-pressure ST large SOFC	3 715 624	2 577 919	1 137 705	0.6480	0.6774	0.7292
23	10 MWe triple RH-pressure ST small SOFC	3 606 144	2 822 572	783 573	0.5891	0.6242	0.6973
24	10 MWe triple RH-pressure ST large SOFC	3 697 652	2 577 948	1 119 703	0.6529	0.6836	0.7370

of operating temperature and temperature gradients. In fact, when the operating temperature of the fuel cell is high, the inlet cathode stream must be sufficiently hot in order to avoid excessive temperature gradients along SOFC stack, which may determine serious systems fails. To this scope, an air recuperator is often used for preheating the inlet air by the hot exhaust gases. However, in some circumstances, this recuperator might be insufficient, especially when the pressure ratio is high. In fact, in this case GT outlet temperature may be too low to preheat the inlet air, and special layouts must be considered.

In general, two possible arrangements can be used. The first one is based on the use of a Recuperative Heat Exchanger (RHE) which preheats the air entering the cathode compartment of the fuel cell, using the exhaust gases from the combustor. This configuration is quite different from the common ones shown in the previous sections, since in the majority of SOFC/GT layouts available in literature, this recuperative air heat exchanger is supplied by GT exhausts. Obviously, combustor exhaust gases' temperature is higher than the one of GT exhausts, allowing a better preheating of the air entering the fuel cell. On the other hand, it must be considered that such configuration determines a decrease of the TIT since combustor outlet stream is cooled in RHE before entering the GT. This circumstance may seriously affect the efficiency of the turbomachinery. A second alternative approach is based on the Exhaust Gas Recirculation (EGR) method, which recirculates part of combustor outlet stream to the cathode inlet, determining an increase of cathode inlet temperature.

A comprehensive comparison between RHE and EGR methods was presented by Zhang *et al.*[108] The authors of this work compared two SOFC/GT hybrid systems. The first one (RHE) is equipped with two recuperative air heat exchangers, HE2 and HE3, respectively, downstream and upstream with respect to the GT. HE3 may also be bypassed (Figure 13.35). The second system layout (EGR) is equipped with a high-temperature gas blower which recirculates combustor outlet gases to a mixer, placed upstream to the cathode inlet, which mixes this stream with the air exiting from the recuperator HE2 (Figure 13.36). The authors of this study assumed a steam to carbon ratio of 2.5, a fuel utilization factor of 0.85, a pressure ratio of 5.2. TIT was 1300 K and 1400 K, respectively, in the case of RHE and EGR. The authors of this paper analyzed the performance of both RHE and EGR power plants under different operating conditions in terms of: current density, oxygen utilization, fuel utilization and operating temperature. Results of their calculation, in both configurations (EGR and RHE), showed that their SOFC electrical efficiencies are very similar. Conversely, in the RHE configuration, GT performance decreases as a consequence of the reduction of the TIT. Therefore, the net electrical efficiency of the EGR configuration was remarkably higher than the one of the EHR system. For example, at 4.0 kA/m^2 the efficiencies of EGR and RHE configurations are 66.5% and 62.6%, respectively.

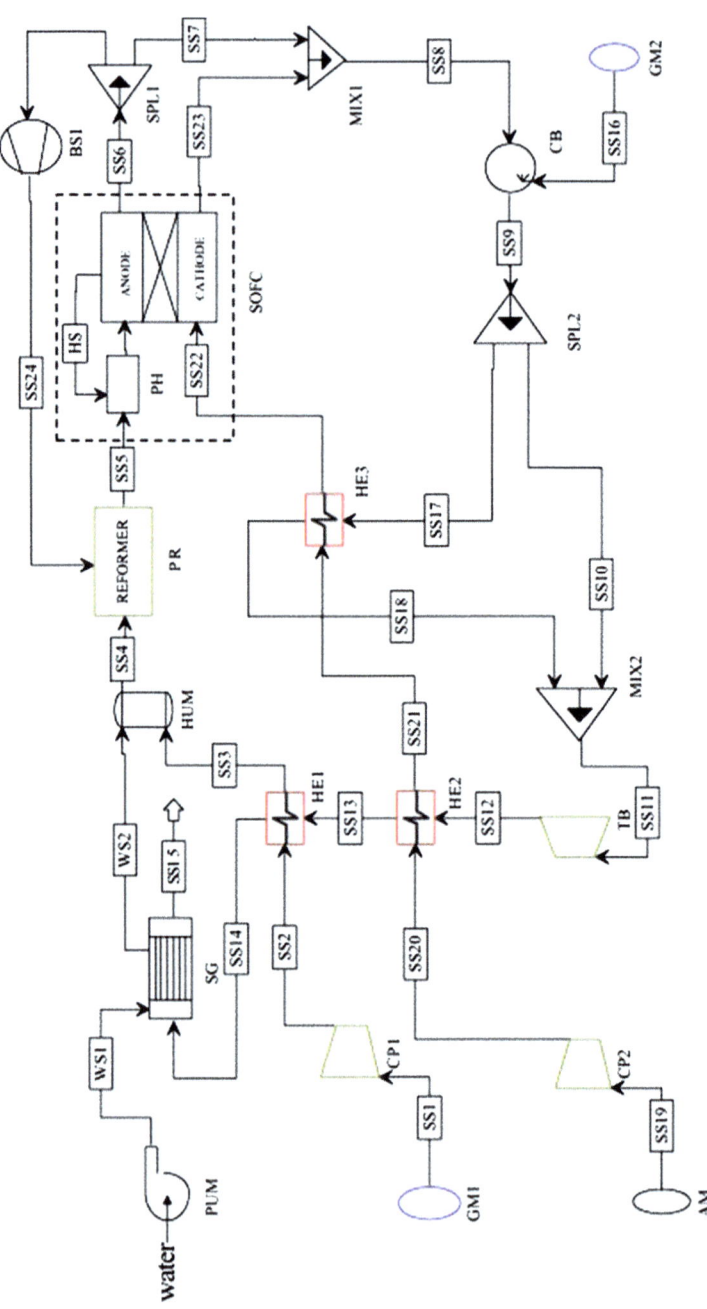

Figure 13.35 SOFC/GT with Recuperative Heat Exchanger (RHE).[135]

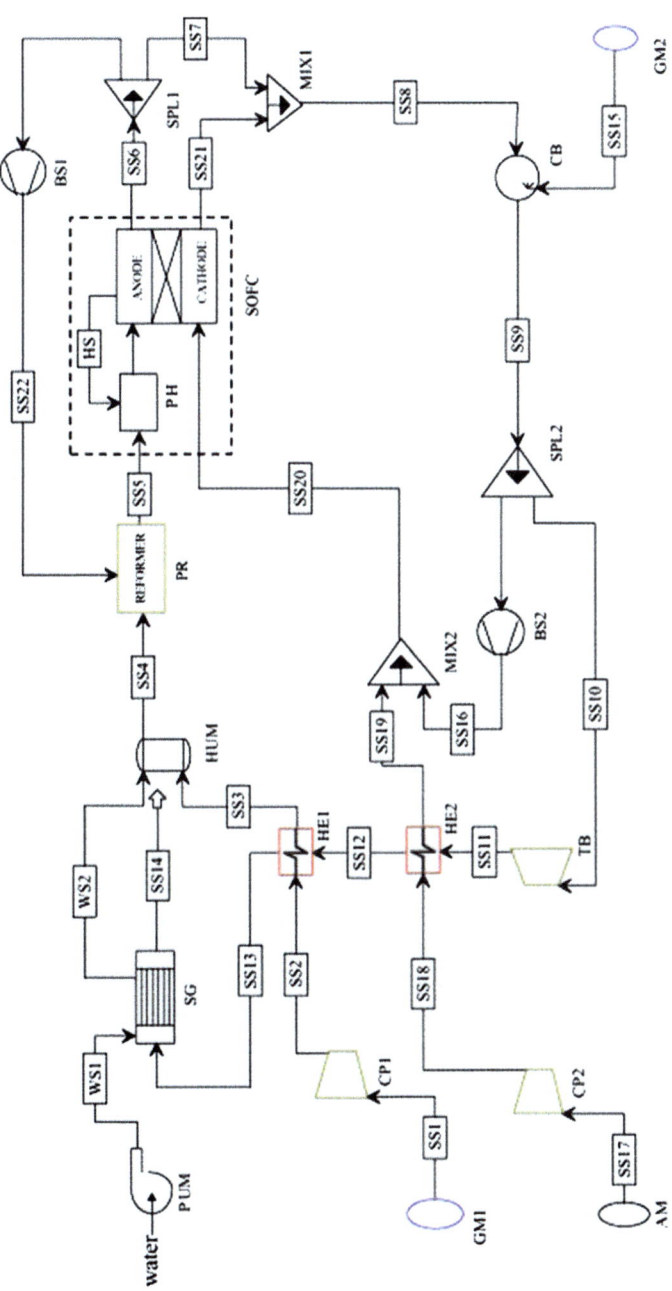

Figure 13.36 SOFC/GT with Exhaust Gas Recirculation (EGR).[135]

13.5   SOFC/GT Atmospheric Cycles

As mentioned above, the majority of the papers dealing with SOFC/GT hybrid power plants are focused on pressurized cycles, since this configuration is often considered more attractive than the atmospheric one from both energetic and economic points of view. In fact, when the SOFC operates at ambient pressure it cannot be directly coupled with the Bryton cycle, which is intrinsically pressurized. In this case, only an indirect coupling arrangement is allowed where the SOFC is operated at ambient pressure and its exhaust gases are used to heat the air exiting from the compressor and entering the gas turbine. This arrangement is typically less efficient than the pressurized one since: (i) the efficiency of the SOFC increases with its operating pressure; (ii) the TIT is lower with respect to the pressurized configuration due to the operation of the intermediate heat exchanger between SOFC exhausts and Bryton cycle. Furthermore, from the economic point of view, it must be considered that in the atmospheric configuration an additional cost must be considered due to the intermediate heat exchanger cost. This heat exchanger is typically very large, due to the low heat exchange coefficients of the primary and secondary fluids (both gaseous). On the other hand, the indirect coupling between atmospheric SOFC systems and Bryton cycles is much simpler to accomplish with respect to the direct coupling occurring with the pressurized SOFC stack. Furthermore, the atmospheric configuration may be also attractive for some special types of fuels which can be tolerated by the SOFC, being potential source of faults for the gas turbine.[52,89,109,110,125,136–140]

A very basic layout of an atmospheric SOFC/GT power plant is provided by Cheddie et al.,[89] presenting an internally reformed SOFC stack, equipped with anode recirculation arrangement, coupled with a basic Bryton cycle (Figure 13.37). Note that in an ambient pressure SOFC/GT system two different exhaust streams are available: the first from the SOFC subsystem and the second one by GT outlet. Therefore, special attention must be paid to the design of the heat exchanger network. For example, in this study the authors assumed to use SOFC exhaust first to preheat cathode inlet and then to supply the heat to the bottoming Bryton cycle. In addition, in this study, GT exhausts are used to preheat both air and fuel entering the SOFC stack, whereas no recuperative heat exchanger is considered inside the Bryton cycle. However, several alternative arrangements may be considered. For example, in reference[137] the authors considered an alternative layout in which the stream exiting from Bryton heat exchanger (state point 22) was mixed with the one exiting from the GT (state point 8) in order to enhance the thermal recovery. The net electrical efficiency calculated by the authors was 48.5%, which is significantly lower than the one reported for pressurized direct coupled SOFC/GT power plants. Furthermore, the additional layout investigated in reference[137] determines an increase of the electrical efficiency by 0.7%.

A comparison between pressurized and atmospheric SOFC/GT cycles was presented by Park et al.[138] The authors of this study considered two similar SOFC/GT hybrid systems. The first pressurized and the second at ambient

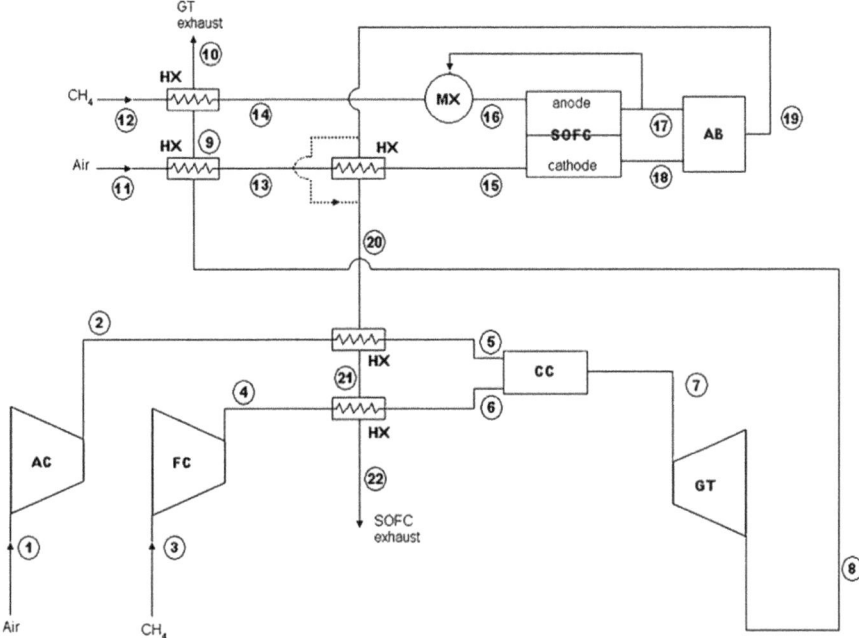

Figure 13.37 SOFC/GT atmospheric cycle: internal reforming and anode recirculation arrangement.[89]

pressure. The layouts of these two systems are basically similar to the ones discussed in this section. In particular, in the case of atmospheric systems, the SOFC exhaust gases are used to bottom the Bryton cycle, whereas GT exhausts heat the cathode air entering in the SOFC stack. Both pressurized and atmospheric configurations consider internal reforming and anode recirculation. In their calculations the authors considered steam to carbon ratio and fuel utilization of 3.0 and 0.7, respectively. The authors of this study calculated a decrease of system efficiency for increasing pressure ratios. Conversely, efficiency increases for higher cell temperatures. The authors also concluded that the efficiency of the ambient SOFC/GT system is 5%–10% lower than the one achieved by the respective pressurized system. They also detected a significant decrease of TIT for higher pressure ratios, especially in case of atmospheric systems. In fact, the atmospheric system is not feasible when high-pressure ratios are considered since the corresponding TIT is too low. In this case, special arrangements must be implemented such as: air bypass, additional fuel supply, *etc.* (Figure 13.38).[138]

Starting from the basic scheme of the atmospheric SOFC/GT power plant, several possible alternative layouts may be established. In fact, the layout can be modified handling: the type of the SOFC stack, the heat exchanger network, the type of reforming process, *etc.* One of these alternative layouts was presented by Roberts *et al.*,[139] investigating a possible modification with respect to the arrangements shown in previous figures. In fact, the authors of this study

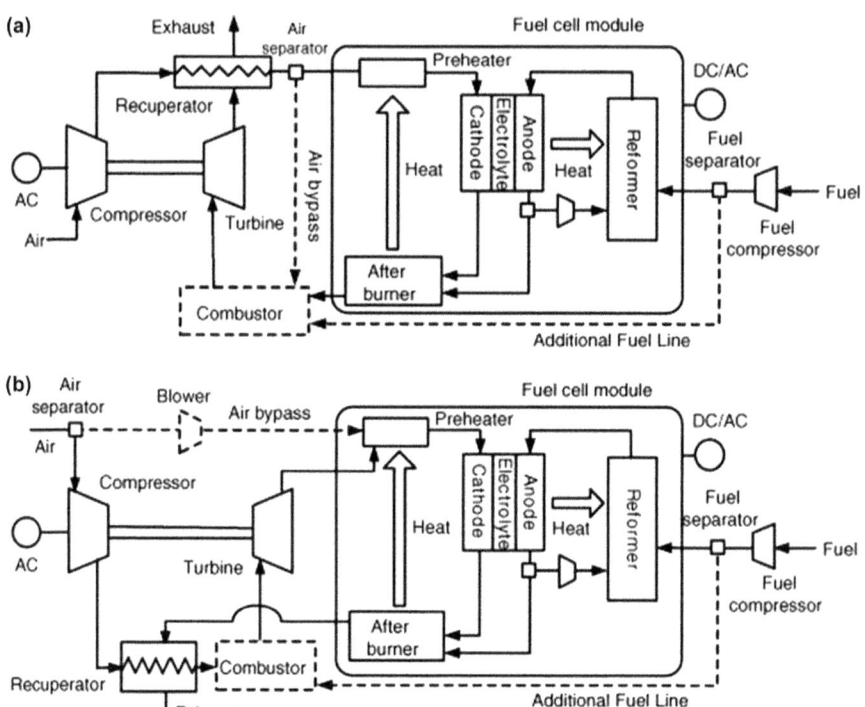

Figure 13.38   SOFC/GT cycle pressurized *vs.* ambient: internal reforming and anode recirculation.[138]

evaluated the possibility to use GT exhaust gases (atmospheric pressure) as cathode stream of the fuel cell. In fact, in an atmospheric SOFC/GT power plant, the turbine operates with air and not with the exhaust gases of the combustion process as occurs in conventional Bryton cycles. Also, considering a possible combustion, the content in oxygen of GT outlet stream is sufficiently high for the requirements of the cathode of the fuel cells. In fact, such a system always operates with very high excess air ratios. On the basis of this idea, the authors developed the system layout shown in Figure 13.39. Here, it is clearly shown that the Bryton cycle is powered by the heat provided by the atmospheric SOFC exhaust gases. The stream exiting from this heat exchanger is subsequently used to preheat the fuel entering the fuel cell. In this case the heat exchanger used for cathode air preheating is not required. In fact, the cathode flow comes directly from the GT outlet. Therefore, it is sufficiently hot and no further preheating is necessary. The dynamic thermal management of the system can be accomplished by the high pressure auxiliary combustor or by a cathode bypass (not shown in figure). The SOFC operates around 750 °C with a fuel utilization factor of 85%. The total power of the system is 350 kW and the maximum calculated efficiency is 66%.[139] This value is significantly higher than the ones rated for other similar atmospheric SOFC/GT systems. Such value is comparable with those achieved by pressurized SOFC/GT hybrid systems.

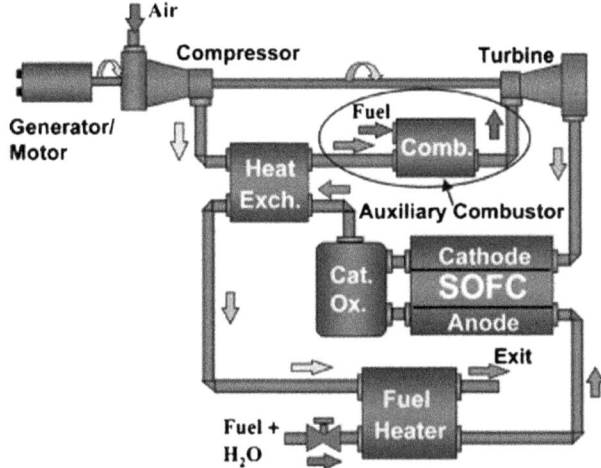

Figure 13.39 SOFC/GT atmospheric cycles: an alternative layout.[139]

13.6 SOFC/GT Power Plant: Control Strategies

One of the main advantages claimed for fuel cell systems lies in their modularity and in their capability to achieve high energy conversion efficiencies independently from their size. Theoretically, a fuel cell can show exactly the same efficiency at any size, ranging from micro-scale to multi-MW systems. In fact, the fuel cell is a battery, and larger powers can be achieved simply arranging a large number of fuel cells. This is true only from a theoretical point of view. In real systems, the efficiency increases for larger systems since the efficiency of their balance of plant increases is higher. In fact, the existing prototypes showed efficiency growing with the size of the system.[46] Furthermore, the modularity of the fuel cells is often also considered as a peculiarity which allows one to achieve electrical efficiencies independent from the load demand. In fact, in theory, when the electrical load demand decreases, the fuel cell may reduce its power production by simply deactivating a part of the modules, while the remaining ones operate at full load. In this way, the overall system efficiency might be always at its rated value. However, this approach is not always feasible since activation and deactivation of SOFC modules require long start-up and shut down times. Therefore, during the part-load operation it is common practice to reduce the power production of all the modules included in the SOFC stack simultaneously. Therefore, the existing SOFC prototypes show that their efficiency typically varies during the partial load operation.

This circumstance is much more stringent when the SOFC stack is coupled with a gas turbine in a hybrid cycle. In fact, in such hybrid power plants the efficiency is not independent from the system capacity. This is due to the fact that the efficiency of the SOFC stack typically slightly increases in case of higher capacities. In addition, the efficiency of the bottoming Bryton cycle dramatically increases for higher rated powers. In fact, when the system capacity is low, a micro-GT may be used, showing conversion efficiencies ranging from 20% to 25%. Conversely, when the size of the GT is higher, its electrical efficiency may increase up to 45%. Therefore, it must be concluded that the efficiency of hybrid SOFC/GT power plants is dramatically affected by the system capacity.

Furthermore, the operation at partial load of such SOFC/GT hybrid power plants is also very complex. As previously shown, the design of a pressurized direct-coupled SOFC/GT power plant is a very hard task, due to the difficulties in coupling the gas turbine and the SOFC stack. In fact, the domain of feasible operating points of a hybrid SOFC/GT power plant is very restricted, due to a large number of constraints regarding: SOFC operating temperature, SOFC operating pressure, maximum SOFC temperature difference, maximum SOFC pressure difference, TIT, matching between compressor and turbine, surge line, *etc.* This circumstance also dramatically affects the partial load operation of SOFC/GT power plants, determining a significant decrease of the electrical efficiency when the demanded electrical load is lower. Furthermore, it is also expected that the range of operation of a hybrid SOFC/GT power plant is significantly lower than the one of a conventional non-hybridized SOFC stack.

The problem of the part-load performance of SOFC/GT power plants was diffusely investigated, and a large number of papers can be found, presenting different control strategies for maximizing the performance of such hybrid systems also during the part-load operation.[59,75,100,106,113–115,141–146]

A simple method to operate at part-load the, SOFC/GT hybrid system may consist of the reduction of fuel and/or air mass flow rates, at constant turbomachinery speed. However, such a technique affects the system performance, since the operating point for turbomachinery moves far from the design condition, corresponding to the maximum efficiency. A more efficient partialization technique consists of reducing both fuel and air flow rates by a reduction of the rotational speed of turbomachinery. Such an approach leads to higher part-load efficiencies, due to a better operation of the turbomachinery. From typical operating maps of air compressors (Figure 13.40) and gas turbines (Figure 13.41) are considered, it is clear that their isentropic efficiency, at constant rotational speed, may dramatically decrease when the operating mass flow rate moves far from its design point, especially when close to the surge or the stall lines. Therefore, in the first approach, the efficiency of the hybrid system mainly decreases since the isentropic efficiency of turbomachinery decreases due to the reduction of their mass flow rate. Conversely, when a variable speed control is employed, a reduction of the rotational speed allows one to decrease the mass flow rate (and consequently the power) of the turbomachinery, with negligible variation in the efficiency.

The first partialization strategy was investigated by Calise *et al.*[115,147] In particular, in reference[147] the authors investigated the part-load operation of the SOFC/GT hybrid system shown in Figure 13.17. Two different strategies for part-load operation are investigated in this work. In the first one, part-load operation was achieved by reducing the fuel mass flow rate to the Internally Reformed SOFC (IRSOFC) and fixing all the other operating parameter values

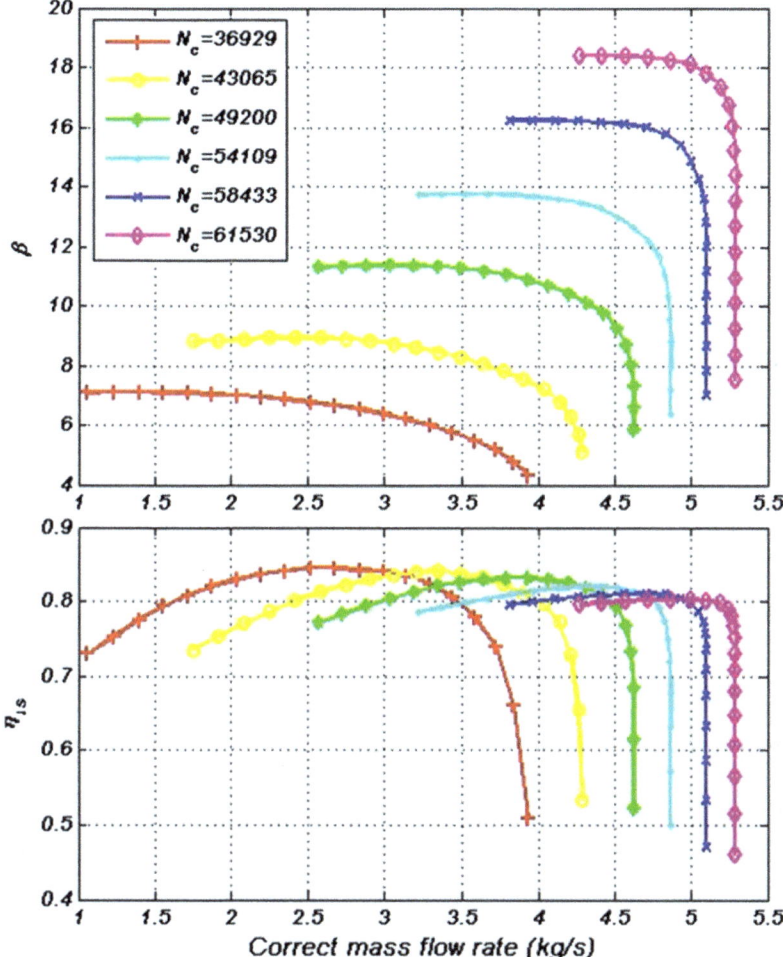

Figure 13.40 Centrifugal air compressor correct map.[115]

(Strategy A). Obviously, such an approach is very simple in terms of managing the operation of the plant, since it can be achieved by just operating the valve downstream of the fuel compressor. However, it also involves a lack of thermal balance in the plant, since the cooling effect of the amount of air supplying the IRSOFC becomes dominant over the energy released by the electrochemical reaction. Thus, a different part-load operating strategy (Strategy B) was investigated, in order to achieve a better thermal balance. According to this strategy, part-load operation is achieved by reducing both the fuel and air mass flow rates, while maintaining fixed their ratio.[115,147]

The partialization performed through strategy A (the fuel flow rate is reduced at constant air flow rate) determines a dramatic decrease of the system temperatures, due to the increased air cooling effect (Figure 13.42). Such a reduction causes a global decrease of all the plant temperatures state points. In

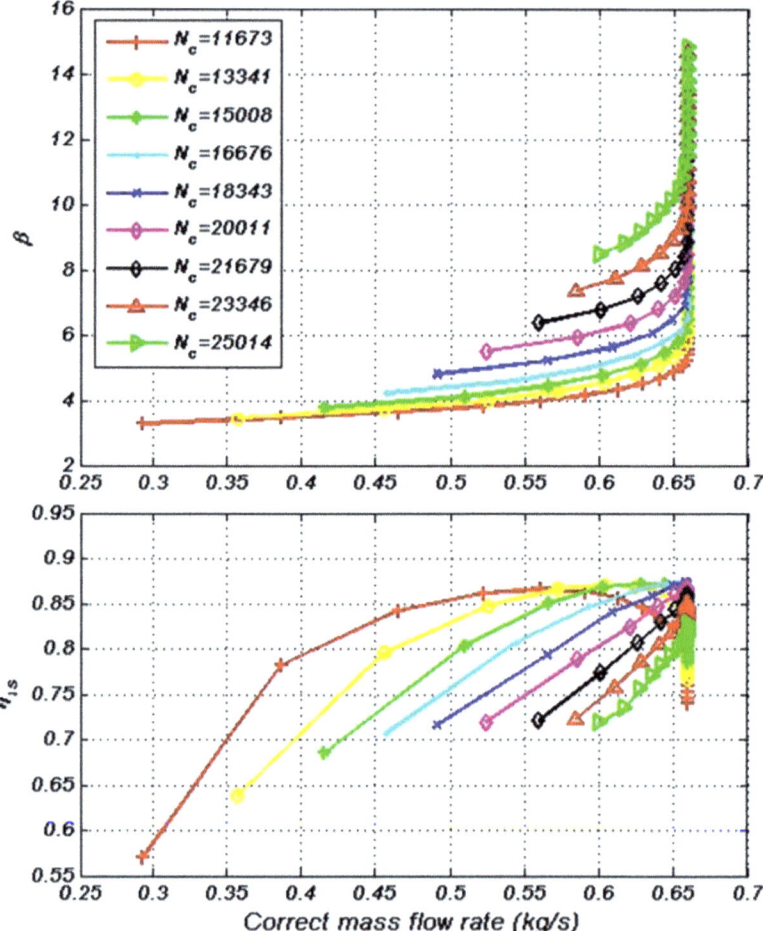

Figure 13.41   Radial gas turbine correct map.[115]

particular, the lower the electrical power, the lower the Turbine Inlet
Temperature (TIT) and Turbine Outlet Temperature (T10). This circumstance
also results in a dramatic reduction in the amount of electricity produced by the
SOFC. Part-load operation also results in a reduction of GT mechanical power,
mainly due to the decrease in the TIT, determining a strong reduction of its
isentropic efficiency (Figure 13.43).[147] The final result of Strategy A is shown in
Figure 13.44: the plant electrical efficiency significantly decreases during part-load
operation, due to the above-mentioned irreversibilities. The exergetic efficiency
shows the same trend. Finally, the global efficiency of the plant is only slightly
dependent on the load, due to an increase in the thermal efficiency of the plant.[147]

In the same work,[147] a further part-load operating strategy (Strategy B) is
introduced, aimed at improving the part-load performance of the hybrid
SOFC/GT power plant, by maintaining an optimal thermal balance of the
plant even during part-load operation. This goal is achieved by reducing the air

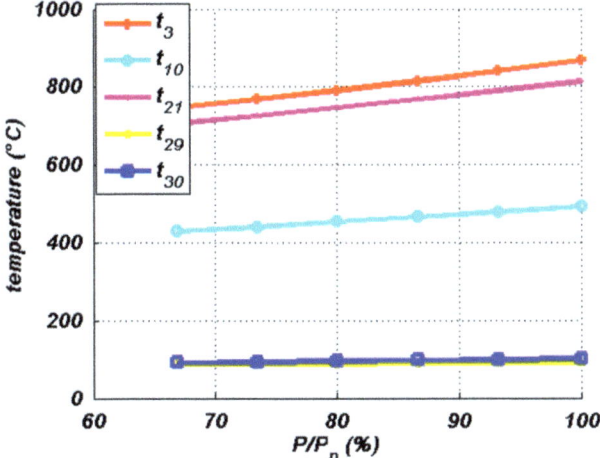

Figure 13.42 SOFC/GT part-load performance: temperatures, decrease fuel flow rate, constant air flow rate.[147]

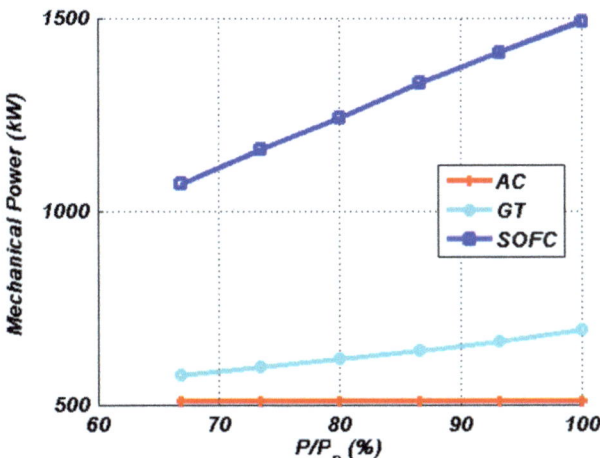

Figure 13.43 SOFC/GT part-load performance: mechanical power, decrease fuel flow rate, constant air flow rate.[147]

cooling effect, *i.e.* by simultaneously reducing the air and fuel mass flow rate, so that their ratio is kept constant.[147] In this case (strategy B), the stack and TIT increase at lower load factors. In particular, a significant increase of these temperatures is achieved as a consequence of the increased irreversibilities in the SOFC. The higher thermal level is also reflected in the turbine outlet temperature, further increasing the availability of energy for pre-heating and cogeneration. In contrast, the temperatures of the outlet streams of the plant (T29 and T30) are only slightly dependent on load (Figure 13.45). Furthermore, this strategy involves a higher power output from the GT (TIT and isentropic

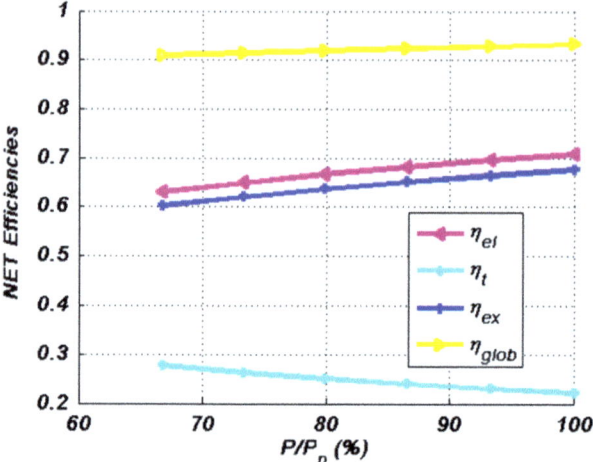

Figure 13.44   SOFC/GT part-load performance: efficiencies, decrease fuel flow rate, constant air flow rate.[147]

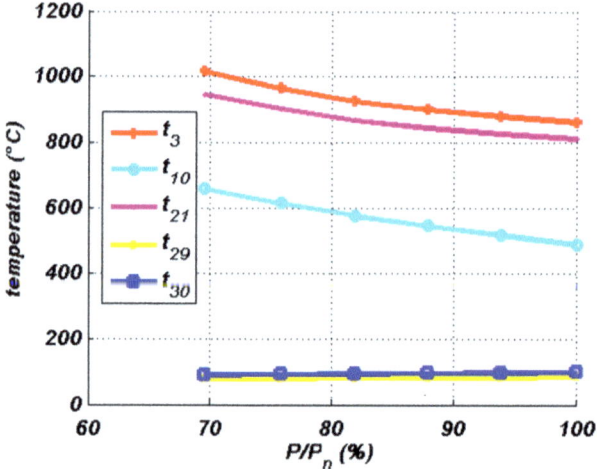

Figure 13.45   SOFC/GT part-load performance: temperatures, decrease of fuel and air flow rates.[147]

efficiency increase) and to reduce the mechanical power demanded by the air compressor, due to the reduction of the air flow rate (Figure 13.46). In summary, this second strategy results in very significant improvements in the electrical, thermal, exergetic, and global efficiencies of the plant (Figure 13.47), primarily due to the better thermal balance, that involves a reduction in the SOFC over-voltage and an increase in the isentropic efficiency of turbomachinery.[147]

A similar part-load strategy was also investigated by Song *et al.*[59] for a multi MW, internally reformed and recuperated SOFC/GT hybrid system, with anode recirculation arrangement, based on a commercial gas turbine (4.6 MW

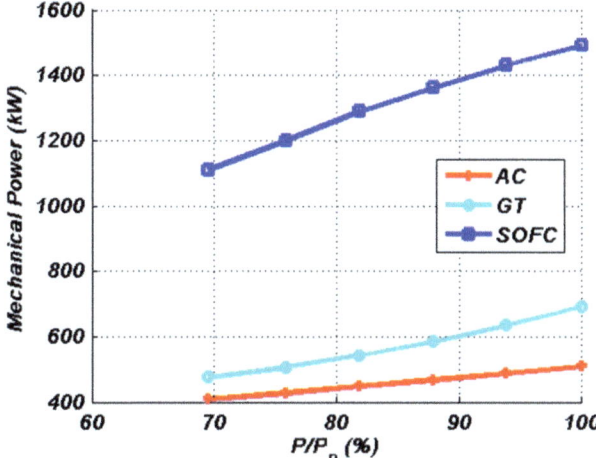

Figure 13.46 SOFC/GT part-load performance: mechanical powers, decrease of fuel and air flow rates.[147]

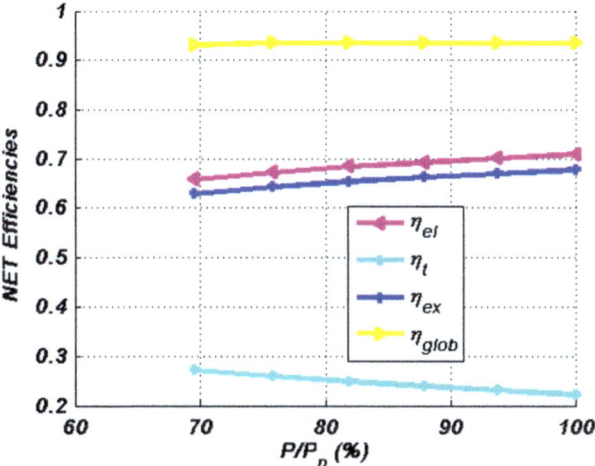

Figure 13.47 SOFC/GT part-load performance: efficiencies, decrease of fuel and air flow rates.[147]

Mercury 50). The authors of this study achieved a minimum part-load ratio of 65%. When the simultaneous reduction of both air and fuel flow rates are considered, in part-load conditions the efficiency increases from 59% to 61%. Conversely, when only the fuel to the GT or to the SOFC stack is reduced, the net efficiency decreases to 56% and 53%, respectively.[59]

The part-load operation of the system layout shown in Figure 13.21 was investigated by Chan *et al.*[113] The authors of this study proposed a control strategy based on the simultaneous reduction of the fuel and air flow rates supplying the SOFC stack. Additional fuel is also burnt in an external combustor in order to maintain the TIT at its design value. According to this strategy, the authors calculated that during the part-load operation stack temperature slightly decreases (by 100 °C), whereas TIT increase from 1300 K to 1350 K. The minimum part load ratio was about 57%. The system net efficiency varied from 62% (full load) to 38% (minimum load).[113]

The variable-speed control strategy was also analyzed by several authors.[75,92,114,142,144,146] For example, Costamagna et al.[141] investigated the design and part-load performance of a recuperated internally reformed SOFC/GT power plant with anode recirculation arrangement, similar to the one shown in Figure 13.11. Such analysis was carried out on the basis of a deterministic model of the SOFC stack and using the performance maps of turbomachinery. The authors of this study assumed constant fuel utilization factor of 0.85. The authors analyzed the part-load performance of the system in two cases: (i) fixed turbomachinery rotational speed, variable fuel flow rate; (ii) variable turbomachinery rotational speed. In the first case, they found that the hybrid system could be partialized up to 70% of the nominal capacity. Lower part load ratios could not be achieved due to the constraints in terms of SOFC temperature and TIT. In fact, a reduction of power production also determines a decrease of both SOFC and Turbine Inlet temperatures. The overall efficiency of the hybrid system decreased during the part-load operation from about 62% to 57%. In the second case, a better performance is achieved. In fact, varying the GT rotational speed, lower part load ratios might be achieved. For example, when the speed is decreased at 65 000 RPM (85 000 RPM at design point) the part load ratio is around 30% and the net system efficiency is slightly higher than 50%. Furthermore, decreasing the power load from 100% to 70%, no significant reduction of system efficiency is detected.[141]

The variable-speed control strategy was also investigated by Komatsu et al.[142] The authors of this study also focused their analysis on a recuperated internally reformed (IIR) SOFC/GT hybrid system, equipped with anode recirculation and pre-reformer, very similar to the 220 kW Siemens prototype. The analysis was carried out using turbomachinery maps and a deterministic model for the SOFC stack. The part-load operation was obtained reducing the rotational speed of the turbomachineries. The authors observed that pressure ratio, cell current density, cell voltage, air flow rate and fuel flow rate decrease during the part-load operation. They also calculated a reduction of the cell average temperature which varies between 930 °C (100% load) and 810 °C (60% load). This decrease determines a consequent reduction of both GT and SOFC efficiencies, resulting in an overall decrease of the electrical efficiency of the hybrid system. In fact, such efficiency was about 61% at design point, decreasing at 52% at the minimum calculated part-load ratio (60%).[142]

Another option for the part-load operation of SOFC/GT hybrid systems consists in the use of Variable Inlet Guide Vane (VIGV) control, as demonstrated by Yang et al.[146] In fact, the authors of this study pointed out an important issue: the variable-speed control is a strategy that can be applied only to small-scale systems, since they operate at high rotational speed which can be reduced during the part-load operation. Such small systems are typically equipped with high speed alternators and inverters. Conversely, in case of large gas turbine the rotational speed is lower and they are equipped with gear boxes coupled with a synchronous generator which requires constant rotational speed. In such large gas turbines, the part-load operation is typically achieved using the VIGV which is installed in front of the compressors and varies the amount of air supplied to the gas turbine.[146] The authors of this study focused their investigation on the 220 kW Siemens hybrid SOFC/GT prototype, presented in the previous section. They compared three different part-load strategies: (i) only reduction of fuel flow rate; (ii) variable speed control; (iii) VIGV. As regards the first control strategy, their results are in accordance with the previously presented works, assessing that the sole reduction of fuel flow rate causes a dramatic decrease of both SOFC temperature and TIT, determining a consequent significant decrease of the efficiency of the system. According to this strategy the efficiency of the system drops to 51% (at 65% part load ratio) when at the design point the efficiency was around 60%. Conversely, the variable speed control strategy determines an increase of the efficiency by 2–3% during the part load operation. This increase is due to the fact that, reducing the power demand, the relative amount of power produced by the gas turbine decreases and the one produced by the SOFC increases. Since the efficiency of the SOFC is higher than the one of the GT, this circumstance determines this increase in system efficiency. Finally, in case of VIGV only a slight decrease of the efficiency is detected (1–2%). This lower efficiency is related to the pattern of the compressor isentropic efficiency which decreases during the part-load operation. Therefore, the authors concluded that the VIGV may be a good alternative to the variable speed control mode, especially in large systems which need to be operated at constant speed.[146]

The control strategy of the 220 kW Siemens SOFC/GT system layout was investigated by Stiller *et al.*[75,92,114] aiming at evaluating the transient behaviour of the SOFC/GT hybrid systems. This system is subject to two degrees of freedom, namely the fuel flow rate and the air flow rate. The authors observed that the part-load operation of this system is affected by unstable regimes, especially in case of high fuel flow and low air flow. They also concluded that it is possible to control the SOFC mean temperature within a wide range of operation, but this temperature is very unstable at low load regimes. Therefore a feedback controller is required for controlling the SOFC temperature. The authors designed a multi-loop feedback controller manipulating some variables (namely: cell current, fuel flow and generator power) on the basis of some calculable/measurable input (power, air flow, fuel utilization factor, SOFC outlet temperature). This controller was implemented in the dynamic operation of the hybrid systems. The results showed that small changes in the load (few kW) are followed closely (<1 s), whereas large variations require longer response times (10–60 s). In these tests, only a modest variation of the SOFC operating temperature (approximately 40–65 K) was detected. The authors also evaluated possible malfunctions of the control system under investigation. In particular, they observed that eventual malfunctions of the fuel flow meters may be very severe, since they can determine a significant deviation of cell operating temperature. Furthermore, they also concluded that the cell degradation determines an increase of the operating temperature. This circumstance may be mitigated by periodically readjusting the control system in order to achieve the desired operating temperature.[114]

13.7 Hybrid SOFC/GT Systems Fed by Alternative Fuels

One of the most attractive peculiarities of SOFC systems is the flexibility in fuel selection. In fact, SOFC can be fed by a number of types of fuels (hydrogen, methane, carbon monoxide, biogas, syngas, *etc.*), whereas the other types of fuel cells are more restrictive. For example, PEM fuel cells can be fed only by highly pure hydrogen, and MCFC can be poisoned by a number of fuels that can be accepted by SOFC.[46] This makes SOFC systems suitable for integration in a number of applications, such as: landfill gas, biomass gasification, reformed liquid fuels, *etc.* Obviously, such peculiarity is emphasized when the SOFC is coupled with a Gas Turbine (GT) in a hybrid cycle, since in this case it is possible to achieve conversion efficiencies that are not feasible for the majority of conventional plants, when fed by the same kind of fuels. Therefore, the possibility to use alternative fuels for hybrid SOFC/GT power plants has been diffusely investigated in the literature by different research groups, analyzing several layout arrangements and several possible types of non-conventional fuels.[3,52,70,72,76,87,91,110,111,126,148–154] Among the possible alternative fuels for SOFC/GT systems (syngas from biomass gasification, methanol, dimethylether, kerosene, ammonia, coal syngas, *etc.*), the best candidates for this application are commonly considered the synthesis gases (syngas) obtainable from biomass[69,155–157] and coal,[107,158,159] due to their high availability and their low cost.

A comprehensive study regarding the use of gasified biomass in a hybrid SOFC/GT power plant was presented by Toonssen *et al.*[70] The authors of this paper investigated the influence of gasification technology, gas-cleaning technology and system scale on the overall performance of the system. Four different system configurations were investigated, namely: (S1) 30 MW system with indirect atmospheric steam gasification and low temperature gas cleaning; (S2) 30 MW system with direct pressurized air gasification and low-temperature gas cleaning; (S3) 30 MW system with direct pressurized air gasification and high-temperature gas cleaning; (S4) 100 kW system with direct pressurized air gasification and high-temperature gas cleaning. All these systems are based on a recuperated internally reformed SOFC/GT hybrid system equipped with both anode and cathode recirculation arrangements. The anode recirculation is used to provide the steam required for the internal SMR reaction, whereas the cathode recirculation is used for air preheating. The cathode recirculation arrangement is similar to that used in the EGR configuration discussed in the previous section. In fact, in the EGR configuration the combustor outlet gases are recirculated to preheat the cathode inlet stream. Here, instead, only the cathode outlet stream is recirculated to obtain the same heating effect. Obviously, in this case, the temperature of the hot stream is lower with respect to the EGR configuration. The operating pressure of the SOFC is around 8 bar (6 bar only in case of S4), whereas its average temperature is approximately 950 °C. In particular, in system S1 (Figure 13.48), the gasification is based on the fast internal circulating fluidized bed (FICB) gasifier. Such gasifier consists of two circulating fluidized beds. The gasification occurs in the first one (a bubbled fluidized bed), whereas the second one is used to combust char and a small amount of syngas. The heat provided by this second device is used to support the gasification in the first one by circulating the bed material. The operating temperature of the gasifier and combustor are 800–900 °C and 1000–1100 °C, respectively, whereas both operate at atmospheric pressure. The cleaning is obtained by cooling the gas to 120 °C, condensing the alkali metal compounds. Then, the gas is filtered for removing particulates. Subsequently, the gas passes through a water scrubber quenching the gas and removing the halogens, tars and residual alkalis. Finally, the gas is compressed, desulfurized, filtered and preheated for supplying the anode compartment of the fuel cell.[70]

In the scheme S2 (Figure 13.49), the gasification technology is based on direct pressurized air gasification (circulating fluidized bed) operating at 950–1000 °C and 8 bar. Biomass gas is produced in the top side of the gasifier. Then it is separated from bed material and char (recycled to the bottom of the gasifier). The heat required to support the endothermic gasification is supplied by the combustion of char. The scheme S3 (Figure 13.50) is equipped with a high-temperature cleaning system. Here, the gases exiting from the gasifier subsystem are filtered by a ceramic filter in order to remove particulates. Then, the tar is cracked by using a catalytic process at 950 °C. Then, again, the steam produced by a Heat Recovery Steam Generator is injected in the stream in order to avoid carbon deposition and to quench the gas. Subsequently, the stream is cooled to 700 °C. Thus it flows through a sorbent bed and a packed bed in order to remove alkali and hydrogen sulphide.

In the scheme S4 (Figure 13.51), the gasification is based on fixed bed downdraft gasifier, obtained by a modification of the commercially available atmospheric version.

A fuel utilization factor of 80% and a TIT ranging between 1060 °C and 1120 °C were assumed The following electrical efficiencies were calculated for the four systems under evaluation: 49.3%, 49.4%, 49.9% and 46.0%, respectively. The majority of this electricity is provided by the fuel cell, being the ratio between SOFC and GT powers around 6.0. Furthermore, it is also worth noting that a significant loss occurs in the gasifier (varying between 15% and 22%), whereas the loss in the gas cleaning subsystem is lower (ranging between 1.5% and 5%). Therefore, the authors of the study concluded that the gasification technology has hardly an influence on the overall system performance, whereas the high temperature cleaning system shows a better efficiency with respect to the low temperature one. Finally, it was also found that small-scale systems are less efficient than larger ones, due to a lower operating pressure and to higher losses in the components.[70]

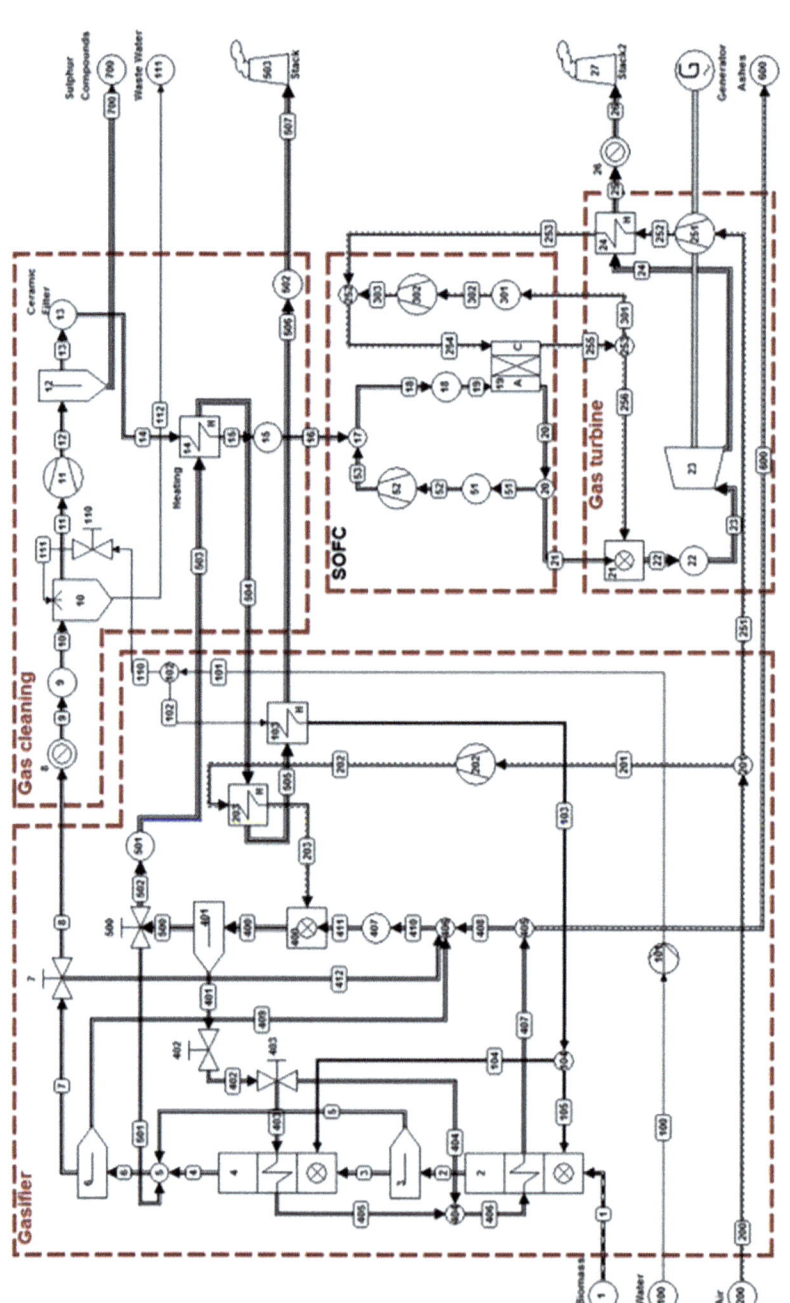

Figure 13.48 SOFC/GT hybrid system fed by gasified biomass: 30 MW system with indirect atmospheric steam gasification and low temperature gas cleaning.[70]

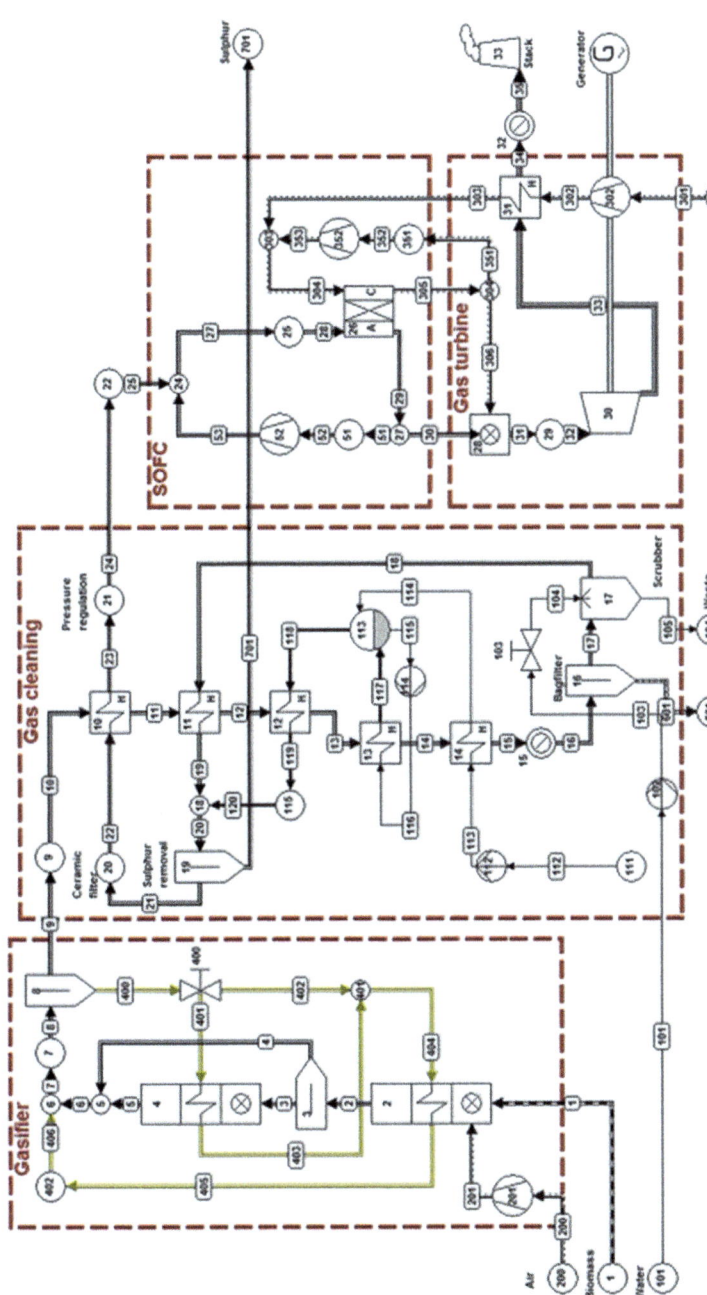

Figure 13.49 SOFC/GT hybrid system fed by gasified biomass: 30 MW system with direct pressurized air gasification and low temperature gas cleaning.[70]

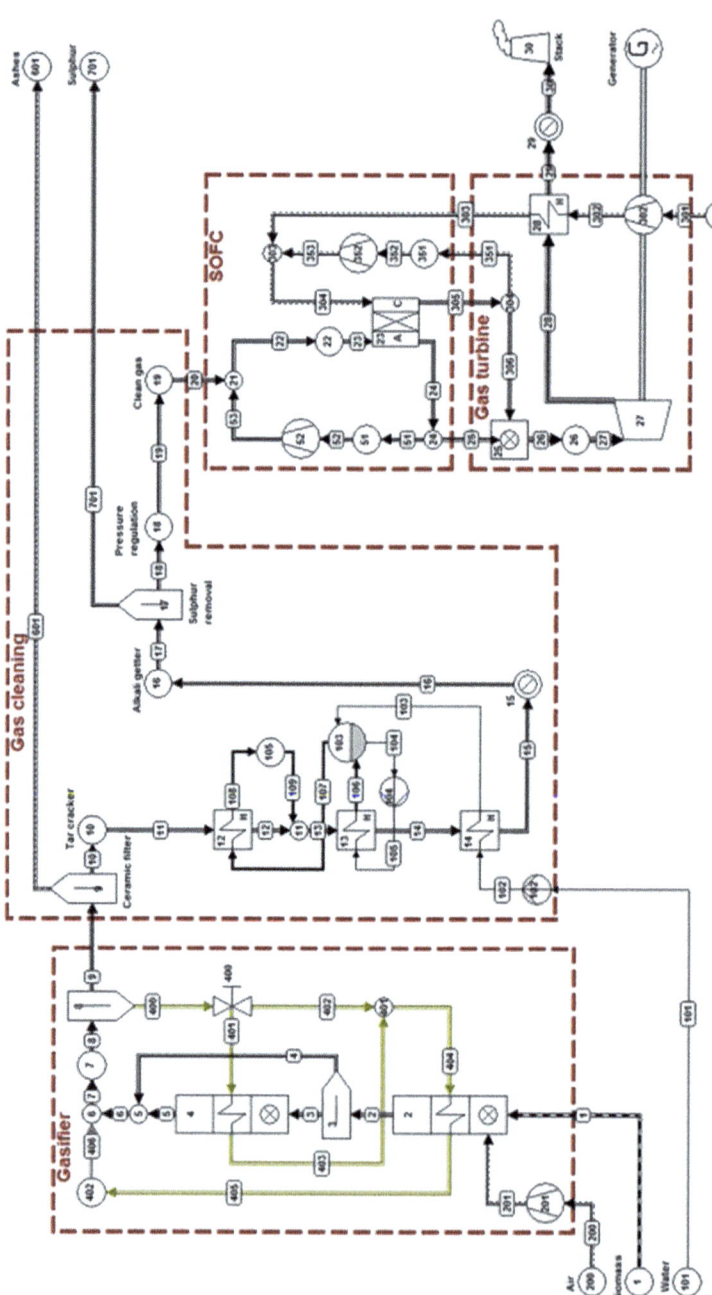

Figure 13.50 SOFC/GT hybrid system fed by gasified biomass: 30 MW system with direct pressurized air gasification and high temperature gas cleaning.[70]

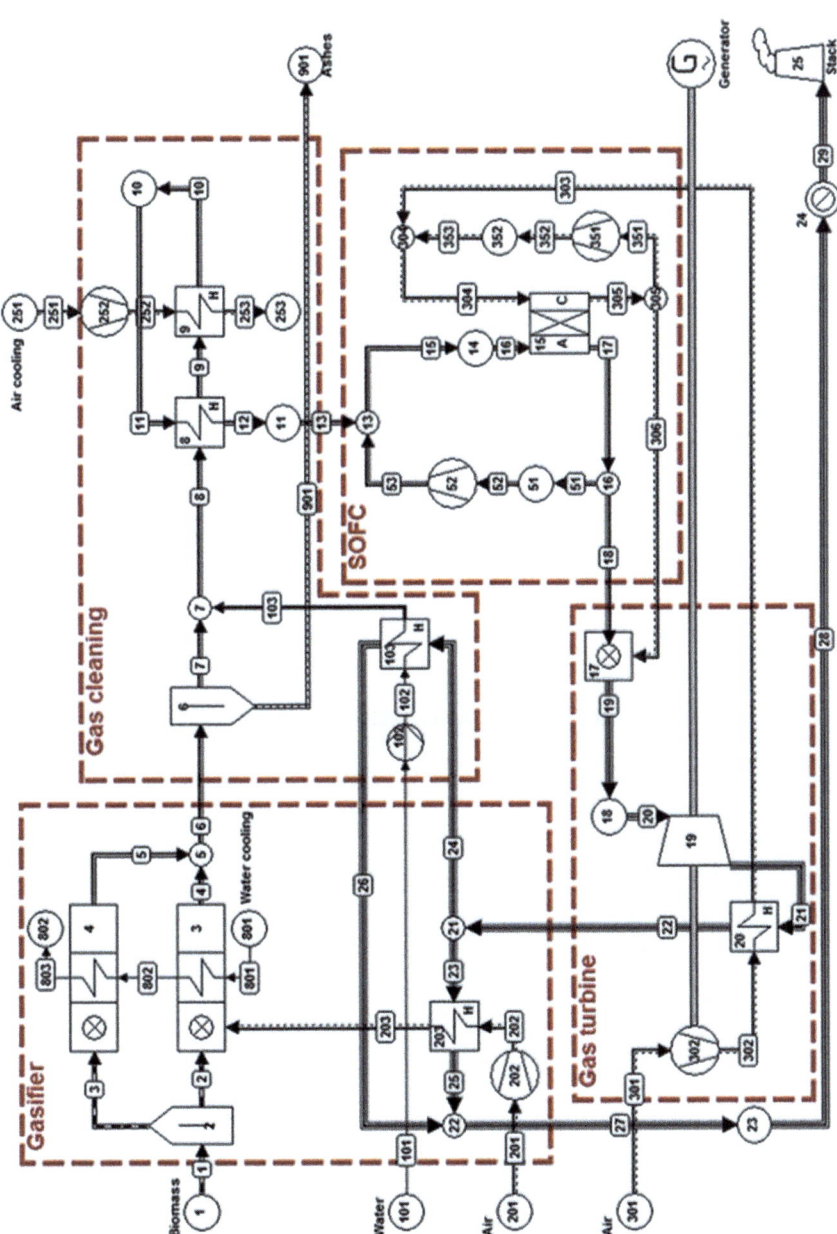

Figure 13.51 SOFC/GT hybrid system fed by gasified biomass: 100 kW system with direct pressurized air gasification and high temperature gas cleaning.[70]

Bang-Møller *et al.* investigated the integration of a SOFC-MGT hybrid system with an up-scaled version of the Viking low tar gasifier.[87,154] The system, shown in Figure 13.52, consists of an internally reformed recuperated SOFC/GT hybrid system fed by the syngas produced by a wt wood gasifier. Here, the wet wood is dried using the heat transfer from the hit syngas. The steam separated in the dryer is brought to the gasifier, together with the dried wood and the preheated air. Such air is preheated by the hot raw syngas exiting from the gasifier. This hot stream also supplies heat to the dryer and then it is

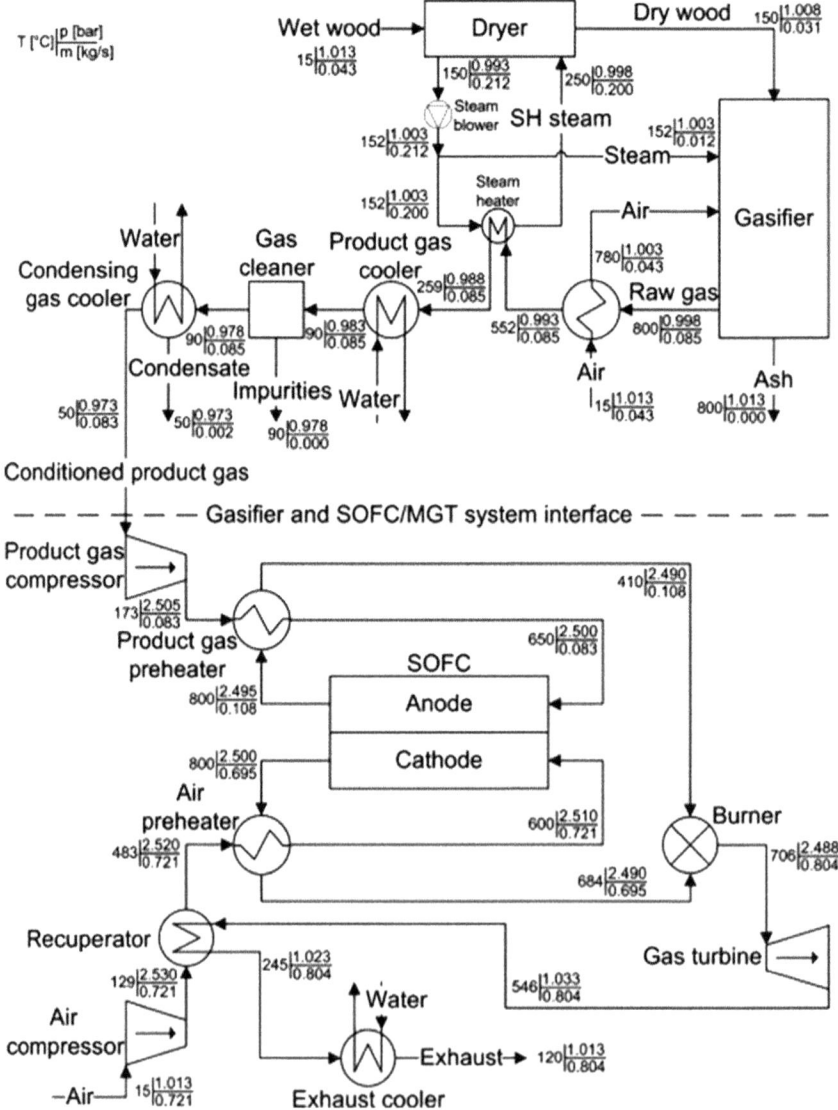

Figure 13.52 SOFC-MGT hybrid system integrated with the Viking gasifier.[87,154]

cooled by water to 90 °C. The hot water produced in this heat exchanger is used for district heating. The next step is the gas cleaning, removing syngas impurities. Then, the gas flow is cooled to 50 °C in order to separate the water included in the flow. Finally, the cleaned syngas is compressed and supplies the recuperated internally reformed hybrid SOFC/GT power plant. The gasifier operates at 800 °C and ambient pressure, the SOFC operating temperature is 800 °C and its fuel utilization factor is 0.85. TIT is 900 °C. On the basis of these data, the authors calculated a maximum electrical efficiency around 50% for a pressure ratio around 2.0. In fact, the plot of the electrical efficiency *vs.* the pressure ratio shows a bimodal trend, resulting in the efficiency decreasing for pressure ratio higher and lower than 2.0. The authors also calculated an increase of the electrical efficiency for higher SOFC temperature and TIT. Similarly, the efficiency increases for lower operating current density, due to the lower SOFC overvoltages. For example, when SOFC temperature is 1000 °C, the efficiency raises up to 56%. The authors also compared the gasifier/ SOFC/GT systems with two alternative systems, namely: Gasifier/GT and Gasifier/SOFC. They found that the gasifier/SOFC/GT allows one to achieve electrical efficiency by 14%–22% higher.[154] The same system layout was also analyzed from an exergetic point of view in order to calculate components' irreversibilities. The authors calculated the maximum exergy destruction in the gasifier, being about 78 kW. Significant exergy losses were also detected in the flue gases, burner and exhaust cooler (varying from 18 to 31 kW). SOFC and GT irreversibilities were lower (12 and 10 kW, respectively). On the basis of the results of the exergy analysis the authors proposed an alternative system layout (Figure 13.53), where the hot raw gases exiting from air preheated are now employed to preheat the syngas exiting from the fuel compressor (and not the dryer as in the first scheme). The dryer is instead heated by the GT recuperator exhaust gases. Finally, such exhaust gases, exiting from the dryer, pass through the exhaust cooler. This optimized system showed an efficiency increase from 55.0% to 58.2%, mainly due to the increased power production of the MGT.

A small scale SOFC/GT system fed by syngas from biomass was investigated by Kaneko *et al.*[111] An internally-reformed recuperated SOFC/GT power plant was analyzed, with a rated power of 35 kW. The gas turbine is an existing 5 kW model, with two shafts. The authors did not analyze the gasifier technology nor the clean-up system. Therefore, possible interactions among the gasifier, the clean-up system and the power system are not analyzed. A system efficiency of 56.3% was calculated in design conditions. The authors also observed a fluctuating power output when the system is fed by biomass syngas, due to small variations in the fuel compositions. Such fluctuations also affect, in a minor amount, the power produced by the gas turbine.[111]

A further possible alternative fuel for SOFC/GT systems is the coal syngas. This possibility was investigated by Zhao *et al.*[110]. In particular, the authors of this study assumed an internally reformed atmospheric SOFC stack, fed by coal syngas, indirectly coupled with a Bryton cycle. As in the previous case, the authors did not analyze the syngas production process nor the clean-up system, and therefore possible interactions among fuel processing power subsystems

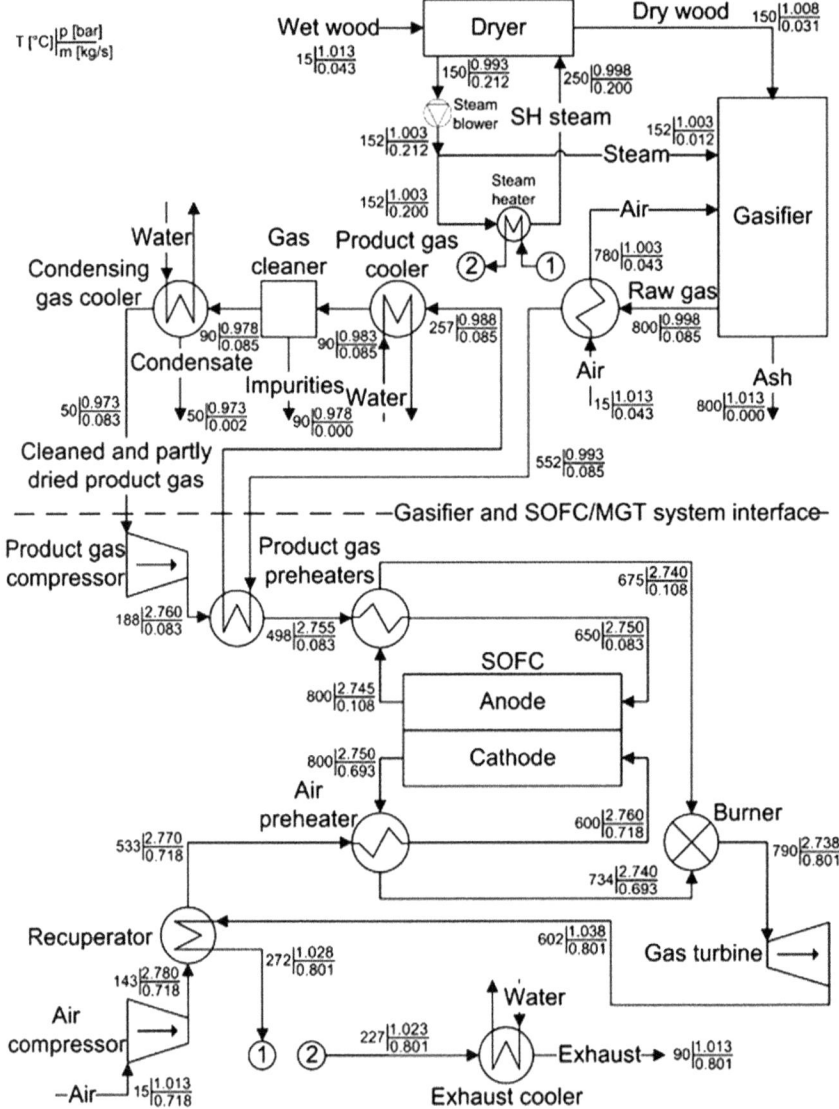

Figure 13.53 Optimized SOFC-MGT hybrid system integrated with the Viking gasifier.[87,154]

are not analyzed. The authors focused on the atmospheric arrangement rather than the pressurized one, since they expected lower system complexity and higher flexibility in terms of possible turbines to be integrated in the system. The system efficiency ranged between 48% and 56%, depending on the operating current density and cell temperature. In particular, the efficiency increase for lower current density. Conversely the trend of the efficiency *vs.* the temperature shows a bimodal trend (for example at 2000 A/m², the maximum occurs at about 970 K) (Figure 13.54).

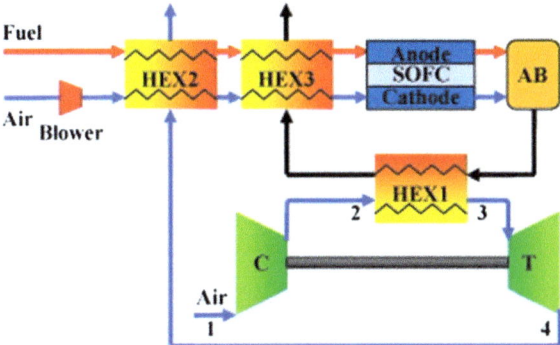

Figure 13.54 Atmospheric SOFC/GT system fed by coal syngas.[110]

As mentioned above, the majority of the studies regarding alternative fuels for SOFC/GT hybrid systems focus their analysis on gasified biomass gas and coal syngas. However, several additional fuels were also analyzed in further studies. For example, Cocco *et al.* investigated the possibility to use methanol and dimethilester (DME) as fuels in SOFC/GT hybrid plants. In particular, the authors focus on such fuels since they are expected to reduce the efficiency by only 0.5%–3.0%, whereas the power output decreases by 15%. On the other hand, their reforming temperature is very low (around 200–300 °C), allowing one to collocate the reformer outside the stack, improving the heat recovery from low-temperature exhaust. Moreover, the external reforming arrangement also involves an increase in the stack efficiency, since the average hydrogen partial pressure is higher with respect to the internal reforming arrangement. Therefore, the authors focused on an externally-reformed pressurized and recuperated SOFC/GT power plant. The simplified system layout is shown in Figure 13.55. A mixture of fuel (methanol or DME) is vapourized in the Fuel Vaporizer (FV) using the exhaust gases exiting from the recuperator (RC). Then, the fuel/water mixture is heated to the reforming required temperature in the superheater (SH). The next step is the reforming (RF). Finally, the fuel supplies the anode compartment of the SOFC stack. Here, the heat produced by the electrochemical and post-combustion reactions is further employed to preheat the pressurized air entering the stack. The following values were assumed for the operation parameters: current density: 300 mA/cm^2; pressure ratio: 4; fuel utilization factor: 0.85. The authors observed that the efficiency of the power plant is very sensitive to the reforming and stack temperatures. In particular, the efficiency shows a bimodal trend versus the reforming operating temperature. It was found that the optimal reforming temperature depended on the type of fuel and on the SOFC temperature. In particular, the results showed that in case of methanol, the optimal reforming temperature is around 240 °C, corresponding to an electrical efficiency around 73% when the SOFC operates at 1000 °C. When DME is used as fuel, the optimal reforming temperature is around 280 °C and the corresponding efficiency (with the SOFC operating at 1000 °C) is approximately 69%. The authors also suggested that Steam to

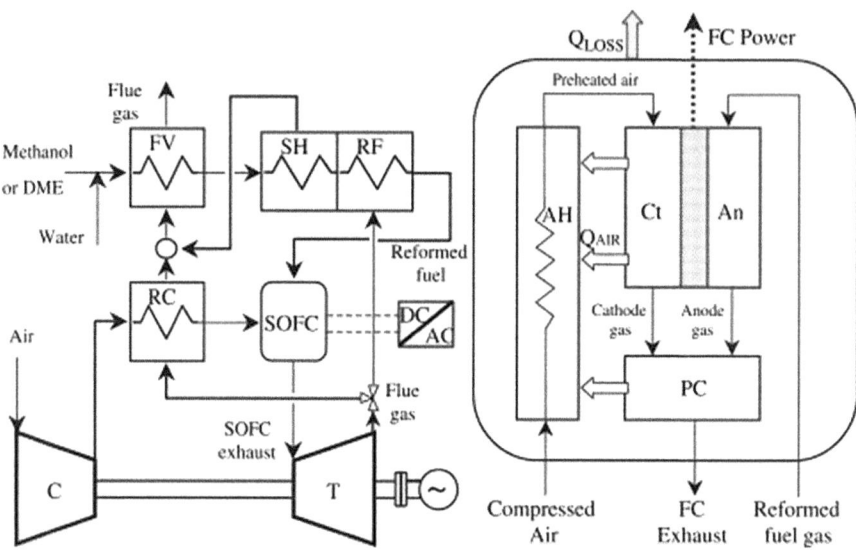

Figure 13.55   Externally reformed SOFC/GT fed by methanol or DME.[72]

Carbon ratios around 1.5–2 should be used, as a good compromise between prevention of carbon deposition and maximization of system efficiency.[72]

A further possible fuel for hybrid SOFC/GT power plant is ammonia. This opportunity was investigated by Ishak *et al.*[126] In particular, the authors based their analysis on a Direct Ammonia SOFC (DASOFC), which is a special type of SOFC capable to directly oxidize NH_3. In particular, NH_3 is supplied to the anode compartment of the SOFC where is subjected to catalytic dehydrogenation, producing N_2 and H_2. The rate of this reaction depends on the operating temperature and pressure. The nitrogen produced at the anode compartment acts as a diluent reducing the H_2 partial pressure, so determining a decrease in the Nernst reversible voltage. Two types of electrolytes can be used for DASOFC systems. The first one (SOFC-O) is an oxygen-ion conducting electrolyte which allows the transport of the ion O^{-2}. In this case, H_2O is produced at anode compartment. The second case (SOFC-H) is the proton conducting electrolyte in which the H^+ ion is allowed to pass through the electrolyte. In this case, water is produced at cathode side. This second configuration allows one to achieve higher Nernst voltage due to the higher H_2 partial pressure (steam vapour is produced at the cathode side). The hybrid SOFC/GT power plant fed by ammonia is shown in Figure 13.56. The ammonia is stored in a pressurized tank. When the ammonia is extracted by the tank, its temperature and pressure decrease. The temperature and pressure can be restored to the initial values by a coil installed inside the tank providing heat to the ammonia. This heating effect occurs at very low temperature. Therefore, the cooling fluid can be used as refrigerant for any cooling process demanded by the user. Then the ammonia is heated in heat exchangers HX1 and HX2 by the SOFC exhaust gases. Similarly, compressed air is preheated in heat exchanger HX3. Note that turbine outlet stream is split in

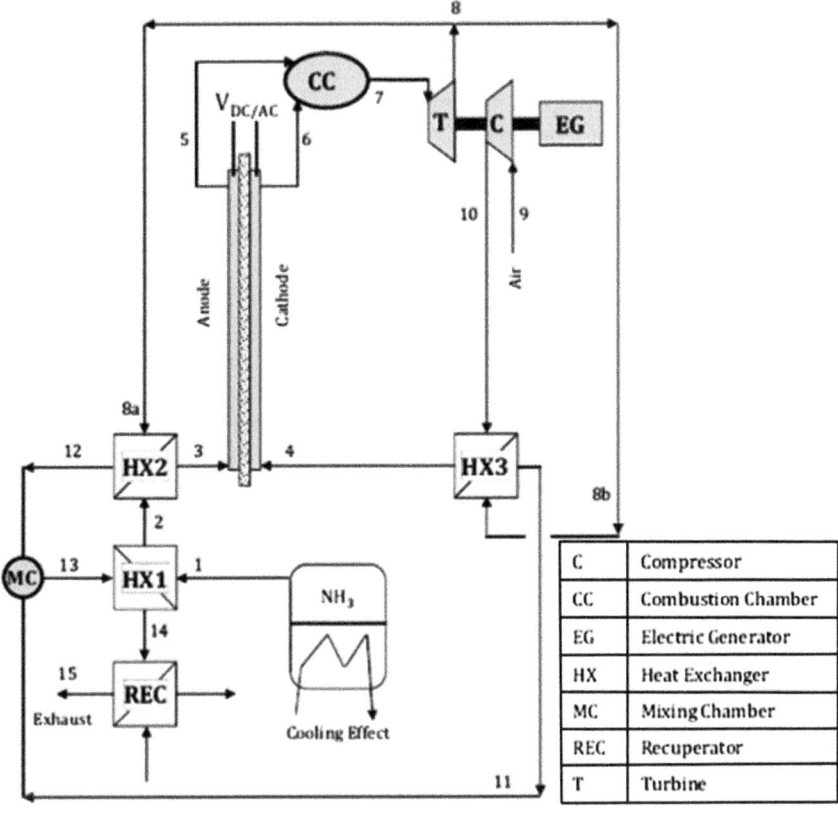

Figure 13.56 SOFC/GT hybrid system fed by ammonia.[126]

two flows, for air and fuel preheating, respectively. Finally, cogenerative heat is produced by the recuperator REC. The authors of this study investigated both types of DASOFC, their main assumptions were as follows: fuel utilization factor: 0.85; air utilization factor: 0.25; stack temperature: 1073 K; pressure ratio: 5 bar. First, authors calculated the polarization curves for both SOFC-O and SOFC-H, concluding that the SOFC-H performance was significantly better, showing a maximum power density 144% higher than SOFC-O. Also the Open Circuit Voltage of the SOFC-H was higher than the one of SOFC-O by more than 1.0 V. Such difference also affected the overall results of the hybrid SOFC/GT power plant. In fact, the electrical efficiency of the system based on SOFC-H is 69.3% whereas the one of SOFC-O is only 64.3%. The authors also performed some sensitivity analyses showing that efficiency decrease for higher ammonia utilization factors and stack temperatures and pressures.[126]

13.8 IGCC SOFC/GT Power Plants

It is well known that coal is one of the most abundant fossil fuels, especially for some of the emerging Countries (*e.g.* China). Unfortunately, the conventional energetic conversion of coal determines some major issues from the environmental point of view. However, in the last few years some innovative and environmental-friendly technologies for the energy conversion of coal have been developed. In this field, the most efficient technology is probably the Integrated Gasification Combined Cycle (IGCC), which consists of a conventional combined cycle (Brtyton + Rankine cycle) fed by the coal syngas produced by gasification. IGCC technology is also often coupled with the CO_2 capture technology (pre-combustion, post-combustion, oxy-fuel combustion) in order to further improve its environmental performance. A further possible improvement of IGCC technology may lie in its integration with a SOFC stack. In particular, this integration can be accomplished when the combustor of the Bryton topping cycle is replaced by a SOFC stack. This configuration is very attractive, since it promises simultaneously ultra-high electrical efficiencies and low environmental impact, using a cheap fuel. Nevertheless, it is worth noting that IGCC technology is available for very large systems. Therefore, the SOFC to be integrated should be sufficiently large as well (>100 MW). Unfortunately, for economic reasons, such large SOFC systems have never been experimented with. Therefore, IGCC SOFC/GT hybrid power plants are developed only from a theoretical point of view, whereas no experimental test has ever been performed.[1,3,52,76,160–162]

An example of IGCC SOFC/GT power plant was investigated by Park *et al.*[76] The system also includes pre-combustion CO_2 capture and is based on the state of the art of IGCC technology. The plant layout consists of three subsystems: gasifier, air separation unit and power block. The power block consists of: gas turbine, heat recovery steam generator, steam turbine and SOFC stack. The authors analyzed different plant configurations: (I) cathode recuperative heat exchanger (Figure 13.57); (II) cathode recirculation (Figure 13.58); (III) cathode recirculation and oxy-combustion (Figure 13.59). For the first two systems, two possible options are also evaluated: CO_2 capture and no CO_2 capture. The main design parameters assumed by the authors are listed in Table 13.2.

The configuration of the system is basically similar to the one discussed in case of SOFC/GT/ST combined cycle discussed earlier, with the difference that in this case the fuel comes from a coal gasifier. The difference from the first and the second configuration lies in the type of air preheating. In the first configuration, a conventional recuperative heat exchanger is used, whereas the second and the third configurations use the cathode recirculation arrangement. Both these arrangements were diffusely discussed in the previous sections.

Results of the calculations showed a system net power varying from 480 MW (case I with pre-combustion CO_2 capture) to 699 MW (case III). The minimum electrical efficiency (37.5%) was achieved in case of recuperative heat exchanger and pre-combustion CO_2 capture). The same system, without CO_2 capture, showed an increase of the electrical efficiency by 12.5%. The efficiency of the case II with or without pre-combustion CO_2 capture were 40.25% and 52.99%, respectively. Finally, the efficiency of case III (oxy-combustion CO_2 capture) was 46.6%. Therefore, it could be concluded that the best performance was

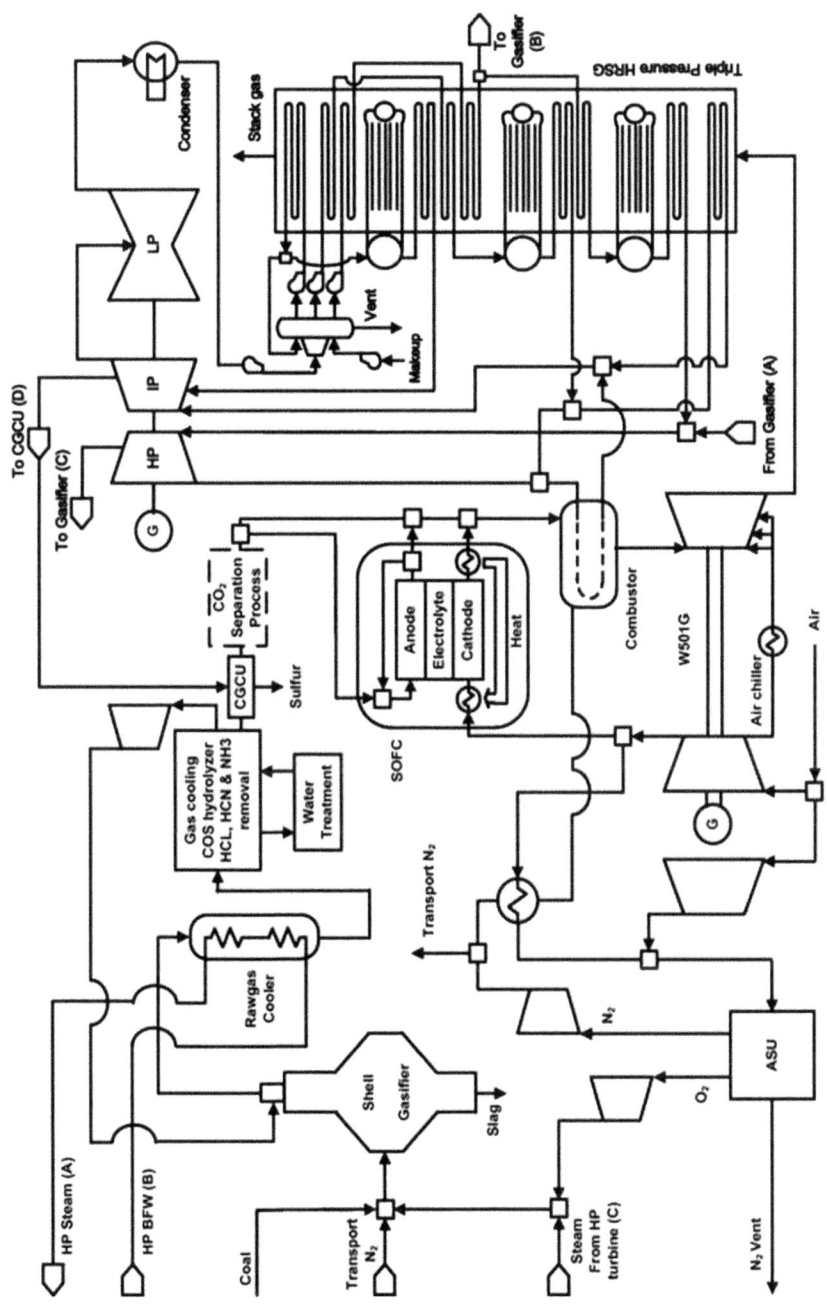

Figure 13.57 IGCC SOFC/GT power plant: cathode heat exchange.[76]

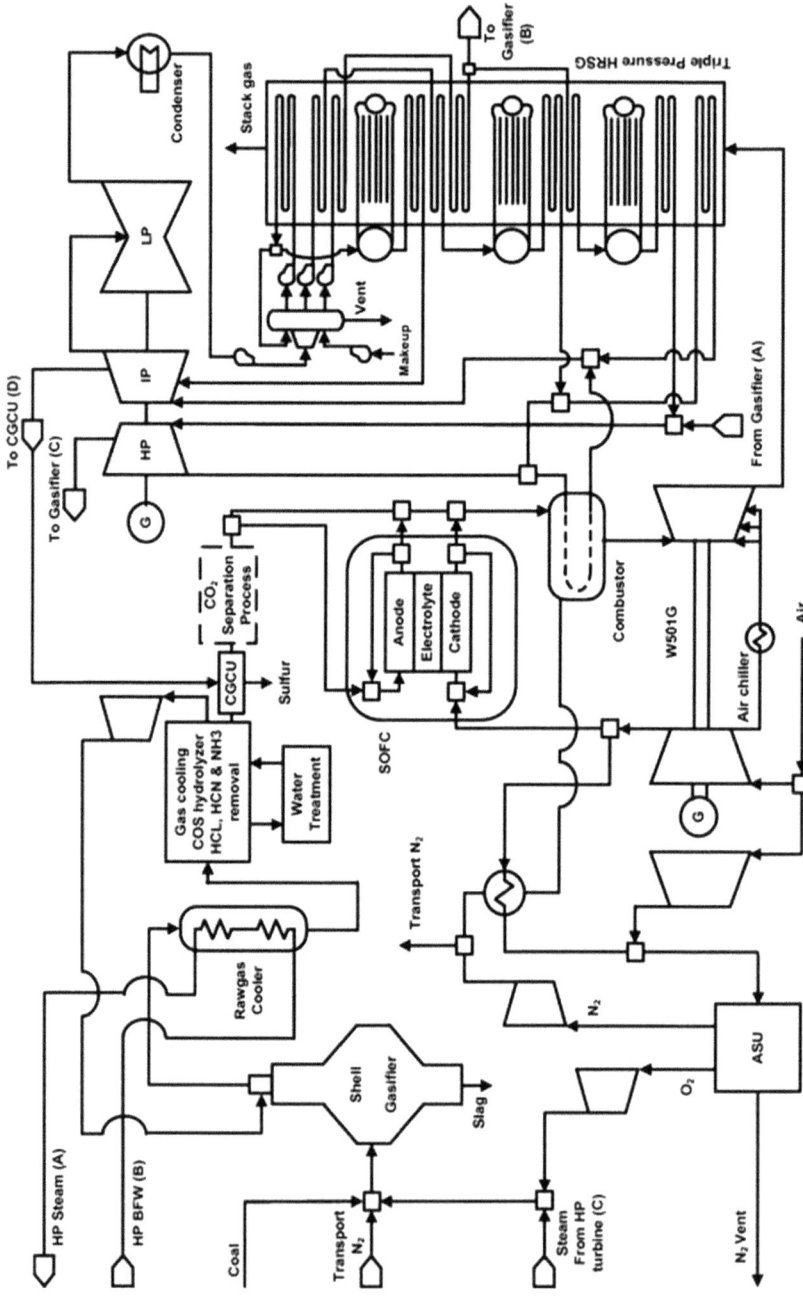

Figure 13.58   IGCC SOFC/GT power plant: cathode gas recirculation.[76]

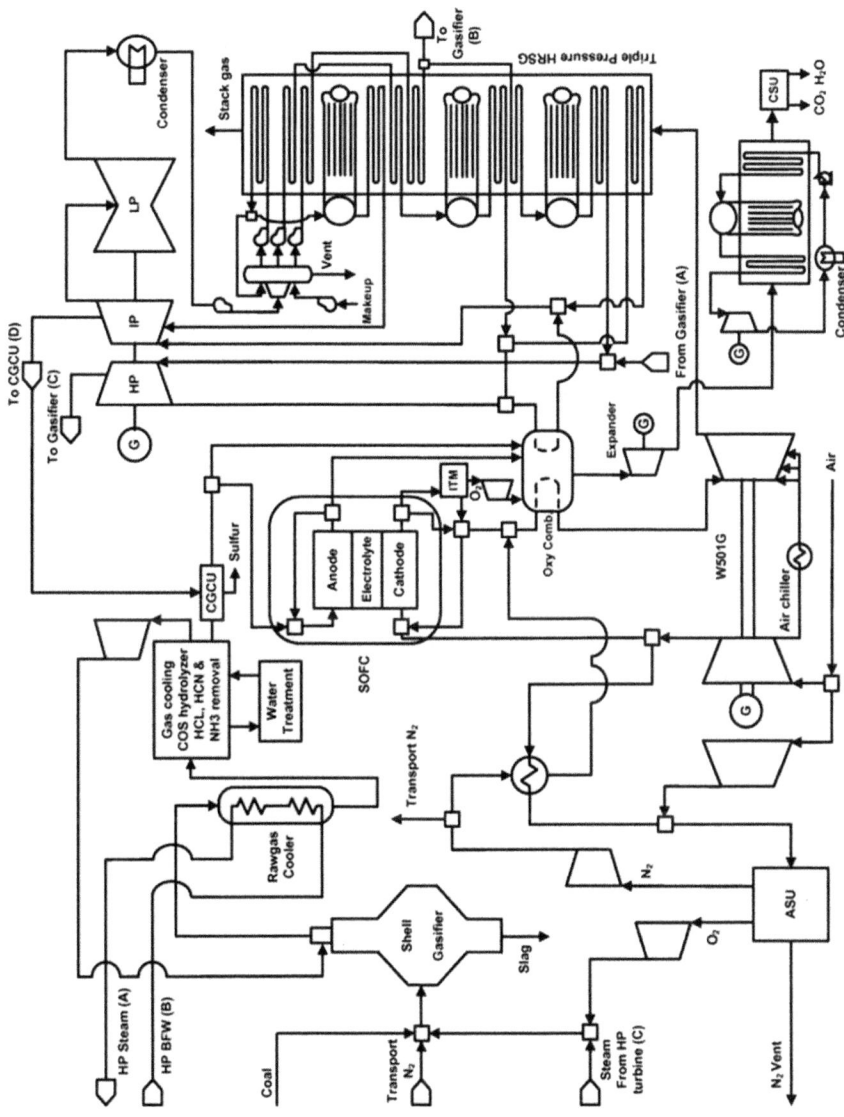

Figure 13.59 IGCC SOFC/GT power plant: cathode gas recirculation and oxy-combustion CO_2 capture.[76]

Table 13.2 IGCC SOFC/GT power plant: main design parameters.[76]

Parameter	Value
Gasifier exit temperature	1006.5 °C
Gasifier exit pressure	2430.4 kPa
TIT	1500 °C
GT pressure ratio	19.1
HRSG condensing pressure	5 kPa
SOFC temperature	900 °C
SOFC fuel utilization factor	0.70

achieved in case of cathode recirculation arrangement. In addition, the CO_2 capture option dramatically affects the system efficiency determining a drop around 13%. Finally, among the different CO_2 capture options, the oxy-combustion technology showed the best efficiency. It must be also noted that the integration of the SOFC into conventional IGCC systems determines a significant increase in both power capacity and electrical efficiency. In fact, the authors also analyzed the performance of the same IGCC system without the SOFC. In the case of pre-combustion CO_2 capture, an electrical efficiency of 33.4% and a net power of 337 MW were calculated. Without CO_2 capture, the efficiency was 46.3%, with a net power of 430 MW.[76]

References

1. IEA - International Energy Agency, *World Energy Outlook*, 2011. Available at: http://www.worldenergyoutlook.org.
2. N. Dui, Z. Guzovi, V. Kafarov, J. J. Klemea, B. vad Mathiessen, J. Yan, Sustainable development of energy, water and environment systems, *Appl. Energy*, In Press, Corrected Proof, 2012.
3. F. Zabihian and A. Fung, A Review on Modeling of Hybrid Solid Oxide Fuel Cell Systems, *Int. J. Eng.*, 2009, **3**, 85–119.
4. M. E. Biresselioglu and Y. Z. Karaibrahimoglu, The government orientation and use of renewable energy: Case of Europe, *Renewable Energy*, 2012, **47**, 29–37.
5. S. Keyuraphan, P. Thanarak, N. Ketjoy and W. Rakwichian, Subsidy schemes of renewable energy policy for electricity generation in Thailand, *Procedia Engineering*, 2012, **32**, 440–448.
6. N. I. Meyer, Renewable energy policy in Denmark, *Energy Sustainable Dev.*, 2004, **8**, 25–35.
7. Z. Peidong, Y. Yanli, S. Jin, Z. Yonghong, W. Lisheng, L. Xinrong, *Opportunities and Challenges for Renewable Energy Policy in China*, 2 ed, 2009, pp. 439–449.
8. Z. Xingang, L. Xiaomeng, L. Pingkuo, F. Tiantian, *The Mechanism and Policy on the Electricity Price of Renewable Energy in China*, 9 ed, 2011, pp. 4302–4309.

9. R. K. Dixon, E. McGowan, G. Onysko and R. M. Scheer, US energy conservation and efficiency policies: Challenges and opportunities, *Energy Policy*, 2010, **38**, 6398–6408.

10. T. A. Adams, II and P. I. Barton, Clean Coal: A new power generation process with high efficiency, carbon capture and zero emissions, *Comput.-Aided Chem. Eng.*, 2010, **28**, 991–996.

11. T. A. Adams, II and P. I. Barton, High-efficiency power production from natural gas with carbon capture, *J. Power Sources*, 2010, **195**, 1971–1983.

12. Stefano Campanari. Carbon dioxide separation from high temperature fuel cell power plants. *J. Power Sources*, 2002, **112**, 273–289.

13. M. W. Coney, C. Linnemann and H. S. Abdallah, A thermodynamic analysis of a novel high efficiency reciprocating internal combustion engine the isoengine, *Energy*, 2004, **29**, 2585–2600.

14. C.-C. Cormos, Hydrogen and power co-generation based on coal and biomass/solid wastes co-gasification with carbon capture and storage, *Int. J. Hydrogen Energy*, 2012, **37**, 5637–5648.

15. M. Liu, N. Lior, N. Zhang and W. Han, Thermoeconomic analysis of a novel zero-CO_2-emission high-efficiency power cycle using LNG coldness, *Energy Convers. Manage.*, 2009, **50**, 2768–2781.

16. P. Bajpai, V. Dash, *Hybrid Renewable Energy Systems for Power Generation in Stand-alone Applications: A review*, 5 ed, 2012, pp. 2926–2939.

17. C. Epp and M. Papapetrou, Co-ordination action for autonomous desalination units based on renewable energy systems ADU-RES, *Desalination*, 2004, **168**, 89–93.

18. O. Erdinc, M. Uzunoglu, *Optimum Design of Hybrid Renewable Energy Systems: Overview of Different Approaches,* 3 ed, 2012, pp. 1412–1425.

19. H. Polatidis and D. A. Haralambopoulos, Renewable energy systems: A societal and technological platform, *Renewable Energy*, 2007, **32**, 329–341.

20. M.-H. Chiang, A novel pitch control system for a wind turbine driven by a variable-speed pump-controlled hydraulic servo system, *Mechatronics*, 2011, **21**, 753–761.

21. S. McTavish, D. Feszty and T. Sankar, Steady and rotating computational fluid dynamics simulations of a novel vertical axis wind turbine for small-scale power generation, *Renewable Energy*, 2012, **41**, 171–179.

22. F. A. Al-Sulaiman, F. Hamdullahpur and I. Dincer, Performance assessment of a novel system using parabolic trough solar collectors for combined cooling, heating, and power production, *Renewable Energy*, 2012, **48**, 161–172.

23. F. Cao, L. Zhao, H. Li and L. Guo, Performance analysis of conventional and sloped solar chimney power plants in China, *Appl. Therm. Eng.*, 2013, **50**, 582–592.

24. S. A. Kalogirou, Solar thermoelectric power generation in Cyprus: Selection of the best system, *Renewable Energy*, 2013, **49**, 278–281.

25. J. Li, P. Guo and Y. Wang, Effects of collector radius and chimney height on power output of a solar chimney power plant with turbines, *Renewable Energy*, 2012, **47**, 21–28.

26. V. Boopathi Raja, V. Shanmugam, *A review and New Approach to Minimize the Cost of Solar Assisted Absorption Cooling System*, 9 ed, 2012, pp. 6725–6731.

27. F. Calise, Design of a hybrid polygeneration system with solar collectors and a Solid Oxide Fuel Cell: Dynamic simulation and economic assessment, *Int. J. Hydrogen Energy*, 2011, **36**, 6128–6150.

28. F. Calise, M. Dentice d Accadia, L. Vanoli, Design and dynamic simulation of a novel solar trigeneration system based on hybrid photovoltaic/thermal collectors (PVT), *Energy Convers. Manage.*, 2012, **60**, 214–225.

29. F. Calise, A. Palombo and L. Vanoli, A finite-volume model of a parabolic trough photovoltaic/thermal collector: Energetic and exergetic analyses, *Energy*, 2012, **46**, 283–294.

30. X. Q. Zhai, M. Qu, Yue Li, R. Z. Wang, *A Review for Research and New Design Options of Solar Absorption Cooling Systems*, 9 ed, 011. pp. 4416–4423.

31. A. M. Bassily, Numerical cost optimization and irreversibility analysis of the triple-pressure reheat steam-air cooled GT commercial combined cycle power plants, *Appl. Therm. Eng.*, 2012, **40**, 145–160.

32. A. G. Kaviri, M. N. Mohd Jaafar and T. M. Lazim, Modeling and multi-objective exergy based optimization of a combined cycle power plant using a genetic algorithm, *Energy Convers. Manage.*, 2012, **58**, 94–103.

33. A. Tic, H. Guéguen, D. Dumur, D. Faille and F. Davelaar, Design of a combined cycle power plant model for optimization, *Appl. Energy*, 2012, **98**, 256–265.

34. P. BrSder and H. F. Svendsen, Capacity and Kinetics of Solvents for Post-Combustion CO_2 Capture, *Energy Procedia*, 2012, **23**, 45–54.

35. D. Chinn, G. N. Choi, R. Chu and B. Degen, Cost efficient amine plant design for post combustion CO_2 capture from power plant flue gas, *Greenhouse Gas Control Technologies*, 2005, **7**, 1133–1138.

36. Eric S. Giovanna Fiandaca and S. Fraga, Brandani, Development of a Flowsheet Design Framework of Multi-Step PSA Cycles for CO_2 Capture, *Comput.-Aided Chem. Eng.*, 2009, **27**, 849–854.

37. J. Li, X. Liang and J. Gibbins, Early Opportunity for CO_2 Capture from Gasification Plants in China, *Energy Procedia*, 2012, **14**, 1451–1457.

38. Y.-J. Lin, C.-C. Chang, D. S.-H. Wong, S.-S. Jang and J.-J. Ou, Control strategies for flexible operation of power plant integrated with CO_2 capture plant, *Comput.-Aided Chem. Eng.*, 2012, **31**, 1366–1371.

39. Y. Peng, B. Zhao and L. Li, Advance in Post-Combustion CO_2 Capture with Alkaline Solution: A Brief Review, *Energy Procedia*, 2012, **14**, 1515–1522.

40. Z. Xu, S. Wang, J. Liu and C. Chen, Solvents with Low Critical Solution Temperature for CO_2 Capture, *Energy Procedia*, 2012, **23**, 64–71.

41. P. A. Marchioro Ystad, O. Bolland and M. Hillestad, NGCC and Hard-Coal Power Plant with CO_2 Capture Based on Absorption, *Energy Procedia*, 2012, **23**, 33–44.

42. D. A. Bell, B. F. Towler and M. Fan, Chapter 7 - Hydrogen Production and Integrated Gasification Combined Cycle (IGCC), *Coal Gasification and Its Applications*, 2011, 137–156.

43. F. Emun, M. Gadalla, T. Majozi and D. Boer, Integrated gasification combined cycle (IGCC) process simulation and optimization, *Comput. Chem. Eng.*, 2010, **34**, 331–338.

44. C. Kunze, K. Riedl and H. Spliethoff, Structured exergy analysis of an integrated gasification combined cycle (IGCC) plant with carbon capture, *Energy*, 2011, **36**, 1480–1487.

45. K. Park, K. Han and E. S. Yoon, Analysis of Integrated Gasification Combined Cycle (IGCC) Power Plant Based on Climate Change Scenarios with Respect to CO_2 Capture Ratio, *Comput.-Aided Chem. Eng.*, 2011, **29**, 1919–1923.

46. EG&G Technical Services Incorporated, *Fuel Cell Handbook*, 7th Edition, US Department of Energy, 2004. Available at http://www.osti.gov/bridge/servlets/purl/834188/834188.pdf.

47. A. J. Appleby, Fuel cell technology: Status and future prospects, *Energy*, 1996, **21**, 521–653.

48. S. J. McPhail, A. Aarva, H. Devianto, R. Bove and A. Moreno, SOFC and MCFC: Commonalities and opportunities for integrated research, *Int. J. Hydrogen Energy*, 2011, **36**, 10337–10345.

49. S. Mekhilef, R. Saidur, A. Safari, *Comparative Study of Different Fuel Cell Technologies*, 1 ed, 2012, pp. 981–989.

50. F. Calise, M. Dentice d Accadia, L. Vanoli and Michael R. von Spakovsky, Full load synthesis/design optimization of a hybrid SOFC GT power plant, *Energy*, 2007, **32**, 446–458.

51. S. Ahmad Hajimolana, M. Azlan Hussain, W. M. Ashri Wan Daud, M. Soroush and A. Shamiri, Mathematical modeling of solid oxide fuel cells: A review, *Renewable Sustainable Energy Rev.*, 2011, **15**, 1893–1917.

52. X. Zhang, S. H. Chan, Guojun Li, H. K. Ho, J. Li and Z. Feng, A review of integration strategies for solid oxide fuel cells, *J. Power Sources*, 2010, **195**, 685–702.

53. J. Larminie and A. Dicks, *Fuel Cell System Explained*, 2004.

54. S. C. Singhal, Kendall, K, *High Temperature Solid Oxide Fuel Cells*, 2003.

55. A. Arsalis, Thermoeconomic modeling and parametric study of hybrid SOFC gas turbine steam turbine power plants ranging from 1.5 to 10 MWe, *J. Power Sources*, 2008, **181**, 313–326.

56. M. Calì, M. G. L. Santarelli and P. Leone, Computer experimental analysis of the CHP performance of a 100 kW e SOFC Field Unit by a factorial design, *J. Power Sources*, 2006, **156**, 400–413.

57. T.-H. Lim, R.-H. Song, D.-R. Shin, J.-I. Yang, H. Jung, I. C. Vinke and S.-S. Yang, Operating characteristics of a 5 kW class anode-supported planar SOFC stack for a fuel cell/gas turbine hybrid system, *Int. J. Hydrogen Energy*, 2008, **33**, 1076–1083.

58. A. S. Martinez, J. Brouwer and G. Scott Samuelsen, Feasibility study for SOFC-GT hybrid locomotive power: Part I. Development of a dynamic 3.5 MW SOFC-GT FORTRAN model, *J. Power Sources*, 2012, **213**, 203–217.

59. T. W. Song, J. L. Sohn, T. S. Kim and S. T. Ro, Performance characteristics of a MW-class SOFC/GT hybrid system based on a commercially available gas turbine, *J. Power Sources*, 2006, **158**, 361–367.
60. H. Xu, Z. Dang and B.-F. Bai, Analysis of a 1 kW residential combined heating and power system based on solid oxide fuel cell, *Appl. Therm. Eng.*, 2013, **50**, 1101–1110.
61. Y. Zhe, L. Qizhao and B. Zhu, Thermodynamic analysis of ITSOFC hybrid system for polygenerations, *Int. J. Hydrogen Energy*, 2010, **35**, 2824–2828.
62. X. D. Zhou, S. C. Singhal, FUEL CELLS SOLID OXIDE FUEL CELLS | Overview, *Encyclopedia of Electrochemical Power Sources*, 2009, 1–16.
63. P. Aguiar, D. J. L. Brett and N. P. Brandon, Solid oxide fuel cell/gas turbine hybrid system analysis for high-altitude long-endurance unmanned aerial vehicles, *Int. J. Hydrogen Energy*, 2008, **33**, 7214–7223.
64. A. V. Akkaya, B. Sahin and H. H. Erdem, An analysis of SOFC/GT CHP system based on exergetic performance criteria, *Int. J. Hydrogen Energy*, 2008, **33**, 2566–2577.
65. T. Araki, T. Ohba, S. Takezawa, K. Onda and Y. Sakaki, Cycle analysis of planar SOFC power generation with serial connection of low and high temperature SOFCs, *J. Power Sources*, 2006, **158**, 52–59.
66. C. Bao, Y. Shi, C. Li, N. Cai and Q. Su, Multi-level simulation platform of SOFC GT hybrid generation system, *Int. J. Hydrogen Energy*, 2010, **35**, 2894–2899.
67. L. Blum, E. Riensche, FUEL CELLS SOLID OXIDE FUEL CELLS | Systems, *Encyclopedia of Electrochemical Power Sources*, 2009, 99–119.
68. F. Calise, M. Dentice d Accadia, L. Vanoli, M. R. von Spakovsky, Single-level optimization of a hybrid SOFC GT power plant, *J. Power Sources*, 2006, **159**, 1169–1185.
69. L. Fryda, K. D. Panopoulos and E. Kakaras, Integrated CHP with autothermal biomass gasification and SOFC MGT, *Energy Convers. Manage.*, 2008, **49**, 281–290.
70. R. Toonssen, S. Sollai, P. V. Aravind, N. Woudstra and A. H. M. Verkooijen, Alternative system designs of biomass gasification SOFC/GT hybrid systems, *Int. J. Hydrogen Energy*, 2011, **36**, 10414–10425.
71. P. Chinda and P. Brault, The hybrid solid oxide fuel cell (SOFC) and gas turbine (GT) systems steady state modeling, *Int. J. Hydrogen Energy*, 2012, **37**, 9237–9248.
72. D. Cocco and V. Tola, Externally reformed solid oxide fuel cell micro-gas turbine (SOFC MGT) hybrid systems fueled by methanol and di-methyl-ether (DME), *Energy*, 2009, **34**, 2124–2130.
73. A. Hahn, *Modeling and Control of Solid Oxide Fuel Cell Gas Turbine Power Plant Systems*, 2004.
74. Y. Inui, T. Matsumae, H. Koga and K. Nishiura, High performance SOFC/GT combined power generation system with CO_2 recovery by oxygen combustion method, *Energy Convers. Manage.*, 2005, **46**, 1837–1847.

75. R. Kandepu, L. Imsland, B. A. Foss, C. Stiller, B. Thorud and O. Bolland, Modeling and control of a SOFC-GT-based autonomous power system, *Energy*, 2007, **32**, 406–417.
76. S. K. Park, J.-H. Ahn and T. S. Kim, Performance evaluation of integrated gasification solid oxide fuel cell/gas turbine systems including carbon dioxide capture, *Appl. Energy*, 2011, **88**, 2976–2987.
77. W. Burbank, Jr., D. Witmer and F. Holcomb, Model of a novel pressurized solid oxide fuel cell gas turbine hybrid engine, *J. Power Sources*, 2009, **193**, 656–664.
78. D. McLarty, Y. Kuniba, J. Brouwer and S. Samuelsen, Experimental and theoretical evidence for control requirements in solid oxide fuel cell gas turbine hybrid systems, *J. Power Sources*, 2012, **209**, 195–203.
79. M. J. Moran, H. N. Shapiro, D. D. Boettner and M. B. Bailey, *Fundamentals of Engineering Thermodynamics*, 2011, John Wiley & sons, Inc.
80. A. Arsalis, *Thermoeconomic Modeling and Parametric Study of Hybrid Solid Oxide Fuel Cell – Gas Turbine – Steam Turbine Power Plants Ranging*, 2007.
81. F. A. Al-Sulaiman, I. Dincer and F. Hamdullahpur, Exergy analysis of an integrated solid oxide fuel cell and organic Rankine cycle for cooling, heating and power production, *J. Power Sources*, 2010, **195**, 2346–2354.
82. M. Rokni, Thermodynamic analysis of an integrated solid oxide fuel cell cycle with a rankine cycle, *Energy Convers. Manage.*, 2010, **51**, 2724–2732.
83. F. Ghirardo, M. Santin, A. Traverso and A. Massardo, Heat recovery options for onboard fuel cell systems, *Int. J. Hydrogen Energy*, 2011, **36**, 8134–8142.
84. D. Sánchez, J. M. Muñoz de Escalona, B. Monje, R. Chacartegui, T. Sánchez, Preliminary analysis of compound systems based on high temperature fuel cell, gas turbine and Organic Rankine Cycle, *J. Power Sources*, 2011, **196**, 4355–4363.
85. Y. Yi, A. D. Rao, J. Brouwer and G. Scott Samuelsen, Analysis and optimization of a solid oxide fuel cell and intercooled gas turbine (SOFC–ICGT) hybrid cycle, *J. Power Sources*, 2004, **132**, 77–85.
86. S. Samuelsen, J. Brouwer, APPLICATIONS STATIONARY | Fuel Cell/Gas Turbine Hybrid, *Encyclopedia of Electrochemical Power Sources*, 2009, 124–134.
87. C. Bang-Møller, M. Rokni and B. Elmegaard, Exergy analysis and optimization of a biomass gasification, solid oxide fuel cell and micro gas turbine hybrid system, *Energy*, 2011, **36**, 4740–4752.
88. M. Burer, K. Tanaka, D. Favrat and K. Yamada, Multi-criteria optimization of a district cogeneration plant integrating a solid oxide fuel cell gas turbine combined cycle, heat pumps and chillers, *Energy*, 2003, **28**, 497–518.
89. D. F. Cheddie, Thermo-economic optimization of an indirectly coupled solid oxide fuel cell/gas turbine hybrid power plant, *Int. J. Hydrogen Energy*, 2011, **36**, 1702–1709.
90. B. F. Möller, J. Arriagada, M. Assadi and I. Potts, Optimisation of an SOFC/GT system with CO_2 capture, *J. Power Sources*, 2004, **131**, 320–326.

91. M. Santin, A. Traverso, L. Magistri and A. Massardo, Thermoeconomic analysis of SOFC-GT hybrid systems fed by liquid fuels, *Energy*, 2010, **35**, 1077–1083.
92. C. Stiller, *Design, Operation and Control Modelling of SOFC/GT Hybrid Systems*, 2006.
93. T. P. Smith, *Hardware Simulation of Fuel Cell/Gas Turbine Hybrids*, 2007.
94. California Energy Commission, *220 kWe Solid Oxide Fuel Cell/ microturbine generator hybrid proof of concept demonstration report*, 2001. Available at http://www.energy.ca.gov/reports/2002-01-11_600-01-009.pdf.
95. S. E. Veyo, L. A. Shockling, J. T. Dederer, J. E. Gillett and W. L. Lundberg, Tubular solid oxide fuel cell/gas turbine hybrid cycle power systems: Status, *J. Eng., Gas Turbines Power*, 2002, **124**, 845–849.
96. R. A. George, Status of tubular SOFC field unit demonstrations, *J. Power Sources*, 2000, **86**, 134–139.
97. K. Huang and S. C. Singhal, Cathode-supported tubular solid oxide fuel cell technology: A critical review, *J. Power Sources*, 2013, **237**, 84–97.
98. T. Gengo, Y. Kobayashi, Y. Ando, N. Hisatome, T. Kabata, K. Kosaka, Development of 200kW Class SOFC Combined Cycle System and Future View, Technical Review, Mitsubishi Heavy Industries, Ltd. 45 (Mar. 2008).
99. S. Seidler, M. Henke, J. Kallo, W. G. Bessler, U. Maier and K. A. Friedrich, Pressurized solid oxide fuel cells: Experimental studies and modeling, *J. Power Sources*, 2011, **196**, 7195–7202.
100. X. Zhang, J. Li, G. Li and Z. Feng, Dynamic modeling of a hybrid system of the solid oxide fuel cell and recuperative gas turbine, *J. Power Sources*, 2006, **163**, 523–531.
101. P. G. Bavarsad, Energy and exergy analysis of internal reforming solid oxide fuel cell gas turbine hybrid system, *Int. J. Hydrogen Energy*, 2007, **32**, 4591–4599.
102. F. Calise, M. Dentice d Accadia, A. Palombo, L. Vanoli, Simulation and exergy analysis of a hybrid Solid Oxide Fuel Cell (SOFC) Gas Turbine System, *Energy*, 2006, **31**, 3278–3299.
103. S. H. Chan, H. K. Ho and Y. Tian, Multi-level modeling of SOFC gas turbine hybrid system, *Int. J. Hydrogen Energy*, 2003, **28**, 889–900.
104. M. Granovskii, I. Dincer and M. A. Rosen, Performance comparison of two combined SOFC gas turbine systems, *J. Power Sources*, 2007, **165**, 307–314.
105. O. Maurstad, R. Bredesen, O. Bolland, H. M. Kvamsdal and M. Schell, SOFC and gas turbine power systems Evaluation of configurations for CO_2 capture, *Greenhouse Gas Control Technologies*, 2005, **7**, 273–281.
106. S. K. Park, K. S. Oh and T. S. Kim, Analysis of the design of a pressurized SOFC hybrid system using a fixed gas turbine design, *J. Power Sources*, 2007, **170**, 130–139.
107. M. C. Romano, V. Spallina and S. Campanari, Integrating IT-SOFC and gasification combined cycle with methanation reactor and hydrogen firing for near zero-emission power generation from coal, *Energy Procedia*, 2011, **4**, 1168–1175.

108. X. Zhang, J. Li, G. Li and Z. Feng, Cycle analysis of an integrated solid oxide fuel cell and recuperative gas turbine with an air reheating system, *J. Power Sources*, 2007, **164**, 752–760.

109. X. Zhang, S. Su, J. Chen, Y. Zhao and N. Brandon, A new analytical approach to evaluate and optimize the performance of an irreversible solid oxide fuel cell-gas turbine hybrid system, *Int. J. Hydrogen Energy*, 2011, **36**, 15304–15312.

110. Y. Zhao, J. Sadhukhan, A. Lanzini, N. Brandon and N. Shah, Optimal integration strategies for a syngas fuelled SOFC and gas turbine hybrid, *J. Power Sources*, 2011, **196**, 9516–9527.

111. T. Kaneko, J. Brouwer and G. S. Samuelsen, Power and temperature control of fluctuating biomass gas fueled solid oxide fuel cell and micro gas turbine hybrid system, *J. Power Sources*, 2006, **160**, 316–325.

112. M. S. Koyama, K. Kraines, C. Tanaka, K. Wallace, K. Yamada and H. Komiyama, *International Journal of Energy Research*, 2004, **28**, 13–30.

113. S. H. Chan, H. K. Ho and Y. Tian, Modelling for part-load operation of solid oxide fuel cell gas turbine hybrid power plant, *J. Power Sources*, 2003, **114**, 213–227.

114. C. Stiller, B. Thorud, O. Bolland, R. Kandepu and L. Imsland, Control strategy for a solid oxide fuel cell and gas turbine hybrid system, *J. Power Sources*, 2006, **158**, 303–315.

115. F. Calise, A. Palombo and L. Vanoli, Design and partial load exergy analysis of hybrid SOFC GT power plant, *J. Power Sources*, 2006, **158**, 225–244.

116. S. H. Chan, H. K. Ho and Y. Tian, Modelling of simple hybrid solid oxide fuel cell and gas turbine power plant, *J. Power Sources*, 2002, **109**, 111–120.

117. T. W. Song, J. Lak Sohn, J. H. Kim, T. S. Kim, S. T. Ro and K. Suzuki, Performance analysis of a tubular solid oxide fuel cell/micro gas turbine hybrid power system based on a quasi-two dimensional model, *J. Power Sources*, 2005, **142**, 30–42.

118. F. Calise, G. Ferruzzi and L. Vanoli, Parametric exergy analysis of a tubular Solid Oxide Fuel Cell (SOFC) stack through finite-volume model, *Appl. Energy*, 2009, **86**, 2401–2410.

119. S. Campanari, Thermodynamic model and parametric analysis of a tubular SOFC module, *J. Power Sources*, 2001, **92**, 26–34.

120. Y. Yi, A. D. Rao, J. Brouwer and G. Scott Samuelsen, Fuel flexibility study of an integrated 25 kW SOFC reformer system, *J. Power Sources*, 2005, **144**, 67–76.

121. Y. Haseli, I. Dincer and G. F. Naterer, Thermodynamic analysis of a combined gas turbine power system with a solid oxide fuel cell through exergy, *Thermochimica Acta*, 2008, **480**, 1–9.

122. Y. Haseli, I. Dincer and G. F. Naterer, Thermodynamic modeling of a gas turbine cycle combined with a solid oxide fuel cell, *Int. J. Hydrogen Energy*, 2008, **33**, 5811–5822.

123. W.-H. Lai, C.-A. Hsiao, C.-H. Lee, Y.-P. Chyou and Y.-C. Tsai, Experimental simulation on the integration of solid oxide fuel cell

and micro-turbine generation system, *J. Power Sources*, 2007, **171**, 130–139.

124. C. Bao, Y. Shi, E. Croiset, C. Li and N. Cai, A multi-level simulation platform of natural gas internal reforming solid oxide fuel cell gas turbine hybrid generation system: Part I. Solid oxide fuel cell model library, *J. Power Sources*, 2010, **195**, 4871–4892.

125. D. F. Cheddie, Integration of A Solid Oxide Fuel Cell into A 10 MW Gas Turbine Power Plant, *Energies*, 2010, **3**, 754–769.

126. F. Ishak, I. Dincer and C. Zamfirescu, Energy and exergy analyses of direct ammonia solid oxide fuel cell integrated with gas turbine power cycle, *J. Power Sources*, 2012, **212**, 73–85.

127. W. J. Yang, S. K. Park, T. S. Kim, J. H. Kim, J. L. Sohn and S. T. Ro, Design performance analysis of pressurized solid oxide fuel cell/gas turbine hybrid systems considering temperature constraints, *J. Power Sources*, 2006, **160**, 462–473.

128. P. Kuchonthara, S. Bhattacharya and A. Tsutsumi, Energy recuperation in solid oxide fuel cell (SOFC) and gas turbine (GT) combined system, *J. Power Sources*, 2003, **117**, 7–13.

129. S. Motahar and A. A. Alemrajabi, Exergy based performance analysis of a solid oxide fuel cell and steam injected gas turbine hybrid power system, *Int. J. Hydrogen Energy*, 2009, **34**, 2396–2407.

130. K. Onda, T. Iwanari, N. Miyauchi, K. Ito, T. Ohba, Y. Sakaki and S. Nagata, Cycle analysis of combined power generation by planar SOFC and gas turbine considering cell temperature and current density distributions, *J. Electrochem. Soc.*, 2003, **150**, A1569–A1576.

131. P. Kuchonthara, S. Bhattacharya and A. Tsutsumi, Combinations of solid oxide fuel cell and several enhanced gas turbine cycles, *J. Power Sources*, 2003, **124**, 65–75.

132. T. Araki, T. Taniuchi, D. Sunakawa, M. Nagahama, K. Onda and T. Kato, Cycle analysis of low and high H_2 utilization SOFC/gas turbine combined cycle for CO_2 recovery, *J. Power Sources*, 2007, **171**, 464–470.

133. A. Musa and M. De Paepe, Performance of combined internally reformed intermediate/high temperature SOFC cycle compared to internally reformed two-staged intermediate temperature SOFC cycle, *Int. J. Hydrogen Energy*, 2008, **33**, 4665–4672.

134. A. Arsalis, M. R. Von Spakovsky and F. Calise, Thermoeconomic modeling and parametric study of hybrid solid oxide fuel cell-gas turbine-steam turbine power plants ranging from 1.5 MWe to 10 MWe, *J. Fuel Cell Sci. Technol.*, 2009, **6**, 0110151–01101512.

135. J. Wu and X. Liu, Recent Development of SOFC Metallic Interconnect, *J. Mater. Sci. Technol. (Shenyang, China)*, 2010, **26**, 293–305.

136. D. F. Cheddie and R. Murray, Thermo-economic modeling of an indirectly coupled solid oxide fuel cell/gas turbine hybrid power plant, *J. Power Sources*, 2010, **195**, 8134–8140.

137. D. F. Cheddie and R. Murray, Thermo-economic modeling of a solid oxide fuel cell/gas turbine power plant with semi-direct coupling and anode recycling, *Int. J. Hydrogen Energy*, 2010, **35**, 11208–11215.

138. S. K. Park and T. S. Kim, Comparison between pressurized design and ambient pressure design of hybrid solid oxide fuel cell gas turbine systems, *J. Power Sources*, 2006, **163**, 490–499.

139. R. Roberts, J. Brouwer, F. Jabbari, T. Junker and H. Ghezel-Ayagh, Control design of an atmospheric solid oxide fuel cell/gas turbine hybrid system: Variable versus fixed speed gas turbine operation, *J. Power Sources*, 2006, **161**, 484–491.

140. Y. Zhao, N. Shah and N. Brandon, Comparison between two optimization strategies for solid oxide fuel cell gas turbine hybrid cycles, *Int. J. Hydrogen Energy*, 2011, **36**, 10235–10246.

141. P. Costamagna, L. Magistri and A. F. Massardo, Design and part-load performance of a hybrid system based on a solid oxide fuel cell reactor and a micro gas turbine, *J. Power Sources*, 2001, **96**, 352–368.

142. Y. Komatsu, S. Kimijima and J. S. Szmyd, Performance analysis for the part-load operation of a solid oxide fuel cell micro gas turbine hybrid system, *Energy*, 2010, **35**, 982–988.

143. F. Leucht, W. G. Bessler and J. Kallo, K. Andreas Friedrich, H. Müller-Steinhagen, Fuel cell system modeling for solid oxide fuel cell/gas turbine hybrid power plants, Part I: Modeling and simulation framework, *J. Power Sources*, 2011, **196**, 1205–1215.

144. J. Milewski, A. Miller and J. SaBaciDski, Off-design analysis of SOFC hybrid system, *Int. J. Hydrogen Energy*, 2007, **32**, 687–698.

145. X.-J. Wu, Q. Huang and X.-J. Zhu, Power decoupling control of a solid oxide fuel cell and micro gas turbine hybrid power system, *J. Power Sources*, 2011, **196**, 1295–1302.

146. J. S. Yang, J. L. Sohn and S. T. Ro, Performance characteristics of a solid oxide fuel cell/gas turbine hybrid system with various part-load control modes, *J. Power Sources*, 2007, **166**, 155–164.

147. F. Calise, M. Dentice d'Accadia, L. Vanoli, M. R. von Spakovsky, Multipoint energy and exergy analysis of a 1.5 MWe hybrid SOFC-GT power plant, ESDA 2006, Turin, Italy, 2006.

148. V. Alderucci, P. L. Antonucci, G. Maggio, N. Giordano and V. Antonucci, Thermodynamic analysis of SOFC fuelled by biomass-derived gas, *Int. J. Hydrogen Energy*, 1994, **19**, 369–376.

149. C. Ozgur Colpan, F. Hamdullahpur, I. Dincer and Y. Yoo, Effect of gasification agent on the performance of solid oxide fuel cell and biomass gasification systems, *Int. J. Hydrogen Energy*, 2010, **35**, 5001–5009.

150. F. P. Nagel, T. J. Schildhauer and S. M. A. Biollaz, Biomass-integrated gasification fuel cell systems Part 1: Definition of systems and technical analysis, *Int. J. Hydrogen Energy*, 2009, **34**, 6809–6825.

151. F. P. Nagel, T. J. Schildhauer, N. McCaughey and S. M. A. Biollaz, Biomass-integrated gasification fuel cell systems Part 2: Economic analysis, *Int. J. Hydrogen Energy*, 2009, **34**, 6826–6844.

152. A. O. Omosun, A. Bauen, N. P. Brandon, C. S. Adjiman and D. Hart, Modelling system efficiencies and costs of two biomass-fuelled SOFC systems, *J. Power Sources*, 2004, **131**, 96–106.

153. M. Sucipta, S. Kimijima and K. Suzuki, Performance analysis of the SOFC MGT hybrid system with gasified biomass fuel, *J. Power Sources*, 2007, **174**, 124–135.

154. C. Bang-Møller and M. Rokni, Thermodynamic performance study of biomass gasification, solid oxide fuel cell and micro gas turbine hybrid systems, *Energy Convers. Manage.*, 2010, **51**, 2330–2339.

155. P. V. Aravind, Wiebren de Jong, *Evaluation of High Temperature Gas Cleaning Options for Biomass Gasification Product Gas for Solid Oxide Fuel Cells*, 6 ed, 2012. pp. 737–764.

156. E. Facchinetti, M. Gassner, M. D. Amelio, F. Marechal and D. Favrat, Process integration and optimization of a solid oxide fuel cell Gas turbine hybrid cycle fueled with hydrothermally gasified waste biomass, *Energy*, 2012, **41**, 408–419.

157. S. Kempegowda Rajesh, S. Øyvind, T. Khanh-Quang, *Cost modeling Approach and Economic Analysis of Biomass Gasification Integrated Solid Oxide Fuel Cell Systems*, 2012.

158. S. Chen, Z. Xue, D. Wang and W. Xiang, An integrated system combining chemical looping hydrogen generation process and solid oxide fuel cell/gas turbine cycle for power production with CO_2 capture, *J. Power Sources*, 2012, **215**, 89–98.

159. S. Wongchanapai, H. Iwai, M. Saito and H. Yoshida, Performance evaluation of a direct-biogas solid oxide fuel cell-micro gas turbine (SOFC-MGT) hybrid combined heat and power (CHP) system, *J. Power Sources*, 2013, **223**, 9–17.

160. R. S. El-Emam, I. Dincer and G. F. Naterer, Energy and exergy analyses of an integrated SOFC and coal gasification system, *Int. J. Hydrogen Energy*, 2012, **37**, 1689–1697.

161. M. Li, A. D. Rao, J. Brouwer and G. Scott Samuelsen, Design of highly efficient coal-based integrated gasification fuel cell power plants, *J. Power Sources*, 2010, **195**, 5707–5718.

162. E. Grol, Technical assessment of an integrated gasification fuel cell combined cycle with carbon capture, *Energy Procedia*, 2009, **1**, 4307–4313.

CHAPTER 14

Modelling and Control of Solid Oxide Fuel Cell

XIN-JIAN ZHU,*[a] HAI-BO HUO,[b] XIAO-JUAN WU[c] AND
BO HUANG*[a]

[a] Shanghai Jiaotong University, Institute of Fuel Cell, Shanghai 200240,
P.R. China; [b] Shanghai Ocean University, Ocean Engineering Research
Institute, Shanghai 201306, P.R. China; [c] University of Electronic Science and
Technology of China, School of Automation, Chengdu 610054, P.R. China
*Email: xjzhu803@163.com; huangbo2k@hotmail.com

An SOFC consists of an interconnect structure and a three-layer region composed of two ceramic electrodes, anode and cathode, separated by a dense ceramic electrolyte (often referred to as the PEN-Positive-electrode/Electrolyte/Negative-electrode). In this cell, the oxygen ions formed at the cathode migrate through the ion-conducting solid ceramic electrolyte to the anode/electrolyte interface where they react with the hydrogen and carbon monoxide contained in (and/or produced by) the fuel, producing water and carbon dioxide while liberating electrons that flow back to the cathode/electrolyte interface via an external circuit.[1] A single cell produces an open-circuit voltage of approximately 1V. Cells have to be connected together in a series arrangement to form a cell stack that delivers higher voltages suited to static converters.[2]

An important tool in fuel cell development is mathematical modelling, which is particularly appropriate for SOFCs, where localized experimental measurements are difficult due to the high operating temperature.[3] The results obtained from a reliable and effective model can be very useful to guide future research for fuel cell improvements and optimization.

RSC Energy and Environment Series No. 7
Solid Oxide Fuel Cells: From Materials to System Modeling
Edited by Meng Ni and Tim S. Zhao
© The Royal Society of Chemistry 2013
Published by the Royal Society of Chemistry, www.rsc.org

14.1   Static Identification Model

It is well known that the SOFC system is sealed, and works in a complicated high-temperature (600–1000 °C) environment. As a nonlinear multi-input and multi-output system, SOFC is hard to model using traditional methodologies. In the last several decades, fruitful results from SOFC stack modelling have been obtained.[4-6] However, these existing models focus on the design of the SOFC instead of its applications. What matters most to SOFC users, however, are not its relevant internal details but its performance under different operating conditions. What they really need are behavioural models, with which they can predict the SOFC behaviour under various operating conditions.

Motivated by this need, some researchers have attempted to establish novel SOFC models. System identification is a process which constructs a mathematical model by input and output data for a system undergoing testing. It is usually applied in the case when theoretical modelling is too complex. So, in this section, two kinds of static modelling methods for the SOFC are introduced.

14.1.1   Nonlinear Modelling Based on LS-SVM

A black-box identification technique such as the artificial neural network (ANN) has been used to derive a SOFC model from the experimental data quickly.[7] Although this ANN model shows a high accuracy and is much faster and easier to use, its practical design suffers from drawbacks such as the existence of local minima and over-fitting, choice of the number of hidden units, *etc.* So a new modelling approach is needed to provide a better solution. In this work, a least squares support vector machine (LS-SVM) is presented to establish a black-box model for the SOFC.

LS-SVM proposed by Suykens and Vandewalle[8] is a modification of the standard SVM. Unlike ANN, LS-SVM possesses prominent advantages: over-fitting is unlikely to occur by adopting the structural risk minimization (SRM) principle, and the global optimal solution can be uniquely obtained by solving a set of linear equations. A number of structures and algorithms for modelling using LS-SVM have been proposed.[9-11] However, the concrete study of modelling SOFC with LS-SVM cannot be found in prior papers.

As we know, cell voltage calculation is the core of any fuel cell modelling. For a given SOFC stack, the output voltage is influenced by many operating parameters such as temperature, pressure, fuel utilization, flow rate, *etc.* However, due to the high number of operating variables, a complete experimental database of SOFC under the different operating conditions is difficult to obtain and no data are available in the open literature yet.[12] Up to now, almost no model has ever been able to accommodate all these operating variables. Our LS-SVM model is no exception. Fuel utilization is one of the most important operating parameters for fuel cells and has significant effects on the cell voltage. In order to analyze the effects of different fuel utilization on output voltage, we choose current, which is determined by external load, and fuel utilization as variables, while holding other operating parameters as constant. Based on the

LS-SVM approach, we present a voltage-fuel utilization model under different current in this section. Furthermore, like the standard SVM, LS-SVM also has better generalization performance and this ability is independent on the dimensionality of the input data. So our LS-SVM model, obtained with the two variables, can predict stack voltage as precisely as a model considering more variables does. Besides, by adding more variables to our LS-SVM model and training it again, the new multi-dimensional model can be obtained easily.

14.1.1.1 LS-SVM for Nonlinear System Modelling

In the following, we briefly introduce LS-SVM algorithm for nonlinear system modelling.[8,13]

Assume a set of training data is given

$$(x_1, y_1), \cdots\cdots, (x_N, y_N) \in R^n \times R. \tag{1}$$

The nonlinear function $\psi(\cdot)$ is employed to map the original input space R^n to high dimensional feature space $\psi(x) = (\varphi(x_1), \varphi(x_2), \cdots, \varphi(x_N))$. Then the linear decision function $y(x_i) = w^T \cdot \varphi(x_i) + b$ is constructed in this high dimensional feature space. Thus, nonlinear function estimation in original space becomes linear function estimation in feature space.

The quadratic loss function is selected in LS-SVM. Then the optimization problem of LS-SVM is formulated as

$$\min_{w,b,e} J(w, e) = \frac{1}{2} w^T \cdot w + \gamma \frac{1}{2} \sum_{i=1}^{N} e_i^2, \quad \gamma > 0 \tag{2}$$

subject to the equality constraints

$$y_i = w^T \cdot \phi(x_i) + b + e_i, i = 1, \cdots, N \tag{3}$$

We construct the Lagrangian as

$$L(w, b, e, \alpha) = J(w, e) - \sum_{i=1}^{N} \alpha_i \{w^T \phi(x_i) + b + e_i - y_i\} \tag{4}$$

where α_i $(i = 1, \cdots, N)$ are the Lagrange multipliers. The conditions for optimality are given by

$$\frac{\partial L}{\partial w} = 0 \rightarrow w = \sum_{i=1}^{N} \alpha_i \phi(x_i)$$

$$\frac{\partial L}{\partial b} = 0 \rightarrow \sum_{i=1}^{N} \alpha_i = 0$$

$$\frac{\partial L}{\partial e_i} = 0 \rightarrow \alpha_i = \gamma e_i, i = 1 \cdots N$$

$$\frac{\partial L}{\partial \alpha_i} = 0 \rightarrow y_i = w^T \cdot \phi(x_i) + b + e_i, i = 1 \cdots N$$

With solution

$$
\begin{bmatrix}
0 & 1 & \cdots & 1 \\
1 & K(x_1, x_1) + 1/\gamma & \cdots & K(x_1, x_N) \\
\vdots & \vdots & \ddots & \vdots \\
1 & K(x_N, x_1) & \cdots & K(x_N, x_N) + 1/\gamma
\end{bmatrix}
\begin{bmatrix}
b \\
\alpha_1 \\
\vdots \\
\alpha_N
\end{bmatrix}
=
\begin{bmatrix}
0 \\
y_1 \\
\vdots \\
y_N
\end{bmatrix}
\tag{6}
$$

The resulting LS-SVM model for nonlinear system becomes

$$
y(x) = \sum_{i=1}^{N} \alpha_i K(x, x_i) + b \tag{7}
$$

Where, α_i, b are the solution to the linear system. Using the normal linear equations programme method, we can get the parameters α_i of (6.6). By the Karush-Kuhn-Tucker (KKT) conditions, the parameter b can be calculated, so the LS-SVM model for nonlinear system can be obtained.

The kernel function $K(x, x_i)$ is any symmetric function that satisfies Mercer's condition. The typical examples of kernel function include linear, polynomial, radial basis function (RBF) kernel.

$$
\text{Linear}: K(x_1, x_2) = x_1^T x_2 \tag{8}
$$

$$
\text{Polynomial}: K(x_1, x_2) = (x_1^T x_2 + 1)^p, \ p \in N \tag{9}
$$

$$
\text{RBF}: K(x_1, x_2) = \exp(-\|x_1 - x_2\|^2 / 2\sigma^2) \tag{10}
$$

The selection of kernel function needs some knowledge in advance, there is no common conclusion currently. In this section, the RBF function is used as the kernel function of LS-SVM because RBF kernels tend to give good performance under general smoothness assumptions.

14.1.1.2   Modelling SOFC Based on LS-SVM

A LS-SVM can be regarded as a black box which can produce certain output data as a response to the specific input data. In this modelling procedure, the relationship between input and output of SOFC can be emphasized while the sophisticated inner structure is ignored. In order to establish the expected nonlinear model of SOFC, we choose fuel utilization and cell current as the model inputs, and cell voltage as the output. In the following, identification structure of SOFC stack based on LS-SVM is given firstly, and then the processes of training and testing the LS-SVM model are presented.

14.1.1.2.1   Identification Structure of SOFC Stack Based on LS-SVM.   In general, a wide class of nonlinear systems can be described by nonlinear autoregressive model with exogenous inputs (NARX). So in this section the

SOFC nonlinear system with two inputs and one output can be described as follows:

$$V(k+1) = f[V(k), V(k-1), \cdots, V(k-n), u(k), u(k-1), \cdots u(k-m), I(k)]$$
(11)

Supposing there is a series of inputs $u(k-m), u(k-m+1), \cdots, u(k), I(k)$ and outputs
$V(k-n), V(k-n+1), \cdots, V(k)$, then the corresponding output $V(k+1)$ can be obtained from (11). And providing that

$$X(k) = (V(k), V(k-1), \cdots, V(k-n), u(k), u(k-1), \cdots, u(k-m), I(k))$$

$$k = 1, 2, \cdots, N$$
(12)

then

$$V(k+1) = f(X(k))$$
(13)

We firstly construct the training sample $set(X(k), V(k+1))$, and then the nonlinear sample data can be mapped as the linear outputs in high dimensional feature space by using LS-SVM.
Namely

$$\hat{V}(k+1) = \sum_{i=1}^{N} \alpha_i K(X(k), X(i)) + b$$
(14)

The identification structure of SOFC stack based on LS-SVM is shown in Figure 14.1, where TDL is the tapped delay line, and the predictive error $e(k+1) = V(k+1) - \hat{V}(k+1)$.

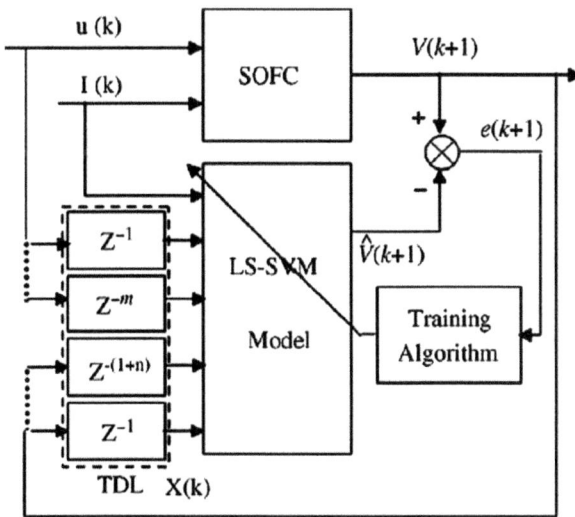

Figure 14.1 Identification structure of SOFC stack based on LS-SVM.[14]

14.1.1.2.2 Training Process of LS-SVM. In general, steps used in training LS-SVM include: training data choosing and preprocessing, selection of the optimal LS-SVM parameters, testing data choosing and preprocessing.

Training data choosing and preprocessing. In our study, a mathematical model[15] is used to generate the data required for the training of the LS-SVM model. The mathematical model has been developed to research the steady-state feasible operating regime of the SOFC. Here three groups of fuel utilization and cell voltage data at 100A, 200A, and 300A are chosen as training data, and each group has 101 pairs of data. Main operational parameters of SOFC are varied, such as fuel utilization (0.4–0.9), stack current (100–300A) and voltage in ranges that correspond to the fuel utilization and stack current as shown in Figure 14.2. Some parameters of the SOFC stack used in the LS-SVM modelling are shown in Table 14.1.

In most cases, all given training data are normalized to [0, 1] or [−1, 1] in order to increase the training speed, facilitate modelling and predicting. In this section, we normalize each group of training data, including fuel utilization, stack current and voltage, to [0, 1] by

$$x' = \frac{x_i - x_{\min}}{x_{\max} - x_{\min}} \tag{15}$$

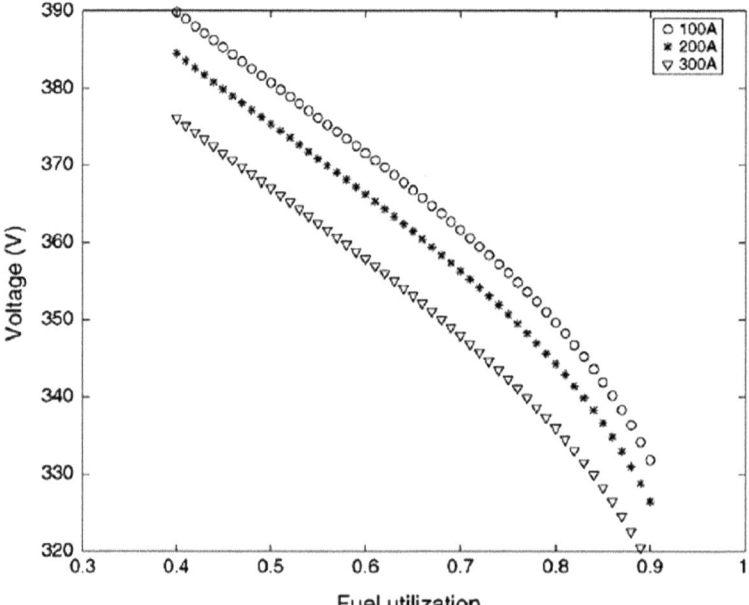

Figure 14.2 Training data: stack voltage (V) versus fuel utilization (u) at 100A, 200A and 300A.[14]

Table 14.1 Parameters of the SOFC stack used in the LS-SVM modelling.[14]

Item	Value
N_0	384
T	1273K
P	100kW
I	100–300A
V	Variable
U	0.4-0.9
qH_2	1.2e-3kmol/s
qO_2	2.4e-3kmol/s

Selection of the optimal LS-SVM parameter. The precision and convergence of LS-SVM are affected by regularization parameter γ and kernel width σ. So in order to obtain high level SOFC model, γ and σ in the LS-SVM have to be tuned.

1) γ, which determines the trade-off between minimizing training errors and minimizing model complexity, is important to increase the generalization performance of LS-SVM model.
2) σ influences directly the number of initial eigenvalues/eigenvectors. Small values of σ yield a large number of regressors, and eventually it can lead to over fitting. On the contrary, a large value of σ can lead to a reduced number of regressors, making the model more parsimonious, but eventually not so accurate.[16]

Several researchers have presented some methods for determining these two parameters, such as bootstrapping, Bayesian methods and so on. However, most of the available methods can be very expensive in terms of computation time and/or training data. For the industrial application of LS-SVM, there is a need for a fast and robust method to estimate these two parameters. Fortunately, we can rapidly tune these two parameters with a 10-fold cross-validation procedure and a grid search mechanism by LS-SVM toolbox.[17] In the final optimal LS-SVM parameters are: $\gamma = 419.0603$, $\sigma = 0.4080358$.

Testing data choosing and preprocessing. Testing data should be different from the data used for training. If testing data are identical to training data, then the LS-SVM is just interpolating points on a line, which is not what we expect the LS-SVM to do.[18] In our study, the testing data chosen for this work are also provided by the above mentioned mathematical model.[14] A group of fuel utilization and stack voltage data at 280A are chosen as testing data. Preprocessing of testing data is done in the same way as training data.

14.1.1.3   Predicting with the LS-SVM Model

The criterion of training LS-SVM is to minimize sum squared error (SSE). After training, a LS-SVM model is obtained, which can be used to predict new input data. Now the trained LS-SVM model is used to predict stack voltage at 280A with different fuel utilization. The comparison of predicted and experimental voltage-fuel utilization curve at 280A is then made to evaluate the LS-SVM model's prediction precision (as shown in Figure 14.3). At the same time, RBFNN model is also used to predict the stack voltage at 280A, and the predicted result is shown in Figure. 14.4. From Figures 14.3 and 14.4, we can see the LS-SVM is superior to the conventional RBFNN in predicting stack voltage with different fuel utilization. These indicate LS-SVM is a powerful tool for modelling SOFC and our LS-SVM model presented in this section is accurate and valid.

14.1.2   Nonlinear Modelling Based on GA-RBF

Neural networks are considered as an attractive structure to establish the mathematical relationship of the dynamic system based on the input–output data. A RBF neural network is a feed-forward neural network with one hidden layer and can uniformly approximate any continuous function to a prospected accuracy.[19] However, a key problem by using the RBF neural network approach is about how to choose the optimum initial values of the following three parameters: the output weights, the centres and widths of the hidden unit.

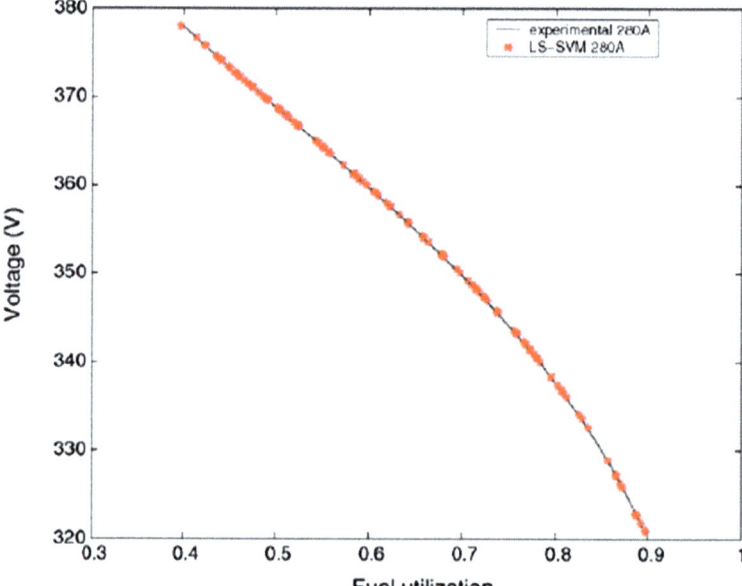

Figure 14.3   Voltage-fuel utilization characteristic: predicted by LS-SVM model and experimental at 280A.[14]

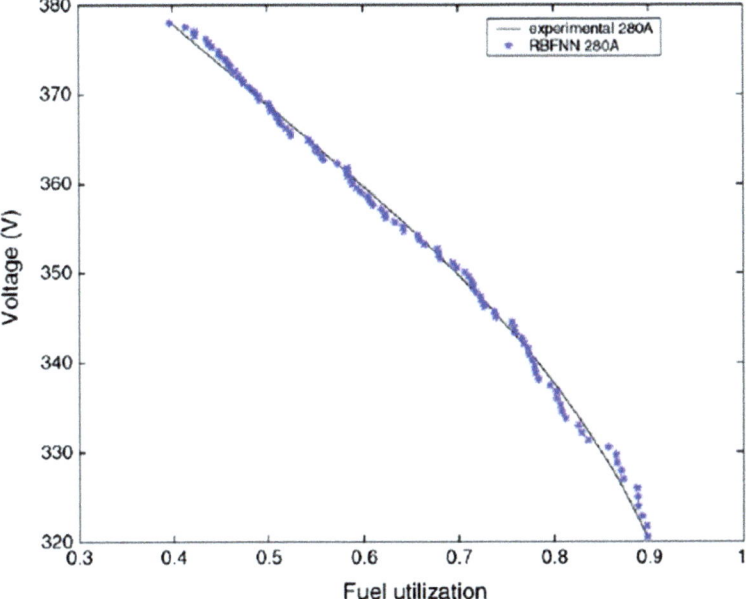

Figure 14.4 Voltage-fuel utilization characteristic: predicted by RBFNN model and experimental at 280A.[14]

If they are not appropriately chosen, the RBF neural network may degrade validity and accuracy of modelling. To assure the optimal performance of the RBF neural network approach for SOFC modelling, we consider applying a genetic algorithm to optimize the RBF neural network parameters in this study. Genetic algorithms are a kind of self-adaptive global searching optimization algorithm based on the mechanics of natural selection and natural genetics.[20] Different from conventional optimization algorithms, genetic algorithms are based on population, in which each individual is evolved parallel, and the ultimate result is included in the last population.

14.1.2.1 GA-RBF Neural Network for Nonlinear System Modelling

A RBF neural network has an input layer, a nonlinear hidden layer and a linear output layer. The nodes within each layer are fully connected to the previous layer nodes. The input variables are each assigned to nodes in the input layer and connected directly to the hidden layer without weights. The hidden layer nodes are RBF units. The nodes calculate the Euclidean distances between the centres and the network input vector, and pass the results through a nonlinear function.[21] The output layer nodes are weighted linear combinations of the RBF in hidden layer. The structure of a RBF neural network with n inputs, one output and q hidden nodes is given in Figure 14.5.

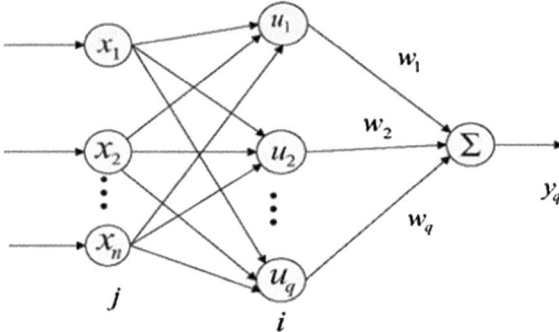

Figure 14.5   The structure of RBF neural networks.[22]

Where, input $x = [x_1, x_2, \cdots x_n]^T$ and $\omega = [\omega_1, \omega_2, \cdots, \omega_q]^T$ is the neural network weight. u_i is a nonlinear function and here, it is chosen as a Gaussian activation function

$$u_i = \exp\left[-\frac{(x - c_i)^T (x - c_i)}{2b_i^2}\right], (i = 1, 2 \cdots q) \tag{16}$$

Where $c_i = (c_{i1}, c_{i2}, \cdots c_{ij})^T$, $j = 1, 2, \cdots n$ is the centre of the ith RBF hidden unit, and b_i is the width of the ith RBF hidden unit. Then the ith RBF network output can be represented as a linearly weighted sum of q basis functions

$$y_q(k) = \sum_{i=1}^{q} \omega_i u_i = \sum_{i=1}^{q} \omega_i \exp\left[-\frac{(x - c_i)^T (x - c_i)}{2b_i^2}\right] \tag{17}$$

Let $y(k)$ represent the target value of the network at time k. The error of the network at time k is defined as:

$$e(k) = y(k) - y_q(k) \tag{18}$$

The cost function of the network is the squared error between the target and the predicted values, which is given by the following equation:

$$E(k) = \frac{1}{2}[e(k)]^2 \tag{19}$$

The learning algorithm aims to minimize the squared error using a gradient descent procedure. Hence, the change of the output weight ω_i the centres c_{ij} and the widths b_i is determined according to the following equation:

$$\omega_i(k) = \omega_i(k - 1) + \eta \Delta \omega_i + \alpha(\omega_i(k - 1) - \omega_i(k - 2)) \tag{20}$$

$$c_{ij}(k) = c_{ij}(k - 1) + \eta \Delta c_{ij} + \alpha(c_{ij}(k - 1) - c_{ij}(k - 2)) \tag{21}$$

$$b_i(k) = b_i(k - 1) + \eta \Delta b_i + \alpha(b_i(k - 1) - b_i(k - 2)) \tag{22}$$

where α is the momentum term and η is the learning rate, $\alpha \in [0,1], \eta \in [0,1]$. The term ω_i, c_{ij} and b_i are defined as:

$$\Delta\omega_i = \frac{\partial E}{\partial \omega_i} = \left(y(k) - y_q(k)\right)u_i = \left(y(k) - y_q(k)\right)\exp\left[-\frac{(x-c_i)^T(x-c_i)}{2b_i^2}\right] \quad (23)$$

$$\Delta c_{ij} = \frac{\partial E}{\partial c_{ij}} = \left(y(k) - y_q(k)\right)\omega_i\frac{x_j - c_{ij}}{b_i^2} \quad (24)$$

$$\Delta b_i = \frac{\partial E}{\partial b_i} = \left(y(k) - y_q(k)\right)\omega_i u_i \frac{\|x - c_i\|^2}{b_i^3} \quad (25)$$

When we programme to realize the RBF algorithm, how to choose the optimum initial values of the following three parameters in Eqs. (20)–(22): the output weight ω_i, the centres c_{ij} and the widths b_i, is very important. If they are not appropriately chosen, the RBF neural network may degrade validity and accuracy of modelling. So a genetic algorithm is used to optimize the RBF neural network parameters.

A genetic algorithm is an interactive procedure that maintains a population of strings which constitute the set of candidate solutions to the specific problem.[23] During each generation, the
strings in the current population are rated for their effectiveness as solutions. On the basis of these evaluations, a new population of candidate solutions is formed by using genetic operations, such as selection, crossover and mutation. There are four major steps required to use the genetic algorithm to solve a problem, including coding, evaluation of fitness, genetic operations and the terminate criterion.

14.1.2.2 *Modelling SOFC by GA-RBF*

For a given SOFC stack, the relation between terminal voltage U and current density I is influenced by many operating parameters, such as cell temperature, air flow rate, hydrogen flow rate, air pressure, hydrogen pressure, *etc*. However, due to the high number of operating variables, a complete experimental database of SOFC under the different operating conditions is difficult to obtain and no data are available in the open literature yet.[24] Up to now, almost no model has ever been able to accommodate all these operating variables. Our GA-RBF model is no exception. Temperature is one of the most operating parameters for the fuel cell and has a significant effect on the fuel cell. In order to analyze the effects of different temperatures on output voltage, we choose current density I, which is decided by the uncontrollable load, and cell temperature T as variables.

In general, a wide class of nonlinear systems can be described by nonlinear autoregressive model with exogenous inputs (NARX).[25] So in this section the

SOFC nonlinear system with two inputs and one output can be described as follows:

$$U(k+1) = f[U(k), U(k-1), \cdots U(k-n), I(k), I(k-1), \cdots I(k-m), T(k)]$$
$$(26)$$

Supposing there is a series of inputs $I(k), I(k-1), \cdots I(k-m), T(k)$ and outputs $U(k), U(k-1), \cdots U(k-n)$. The identification structure based on GA-RBF is shown in Figure 14.6, where TDL is the tapped delay line. The aim of our study is thus, to find a GA-RBF model that approximates Eq. (26) and it requires three steps to build an efficient GA-RBF model: preparing training data, training the data to obtain a GA-RBF model and predicting the new input data with the obtained model.

14.1.2.2.1 Preparing Training Data. In our study, a model[26] is used to generate the data required for the training of the GA-RBF model. Here, two groups of current density and cell voltage data at 800 °C and 1000 °C are chosen as training data, and each group has 701 pairs of data. Main operational parameters of SOFC are varied, such as temperatures (600–1000 °C), stack current density (0–700 m A cm^{-2}). In most cases, training data should be scaled, normally linearly, to [0, 1] or [−1, +1]. An example of scaled current density is shown in Table 14.2. Scaling can increase the training

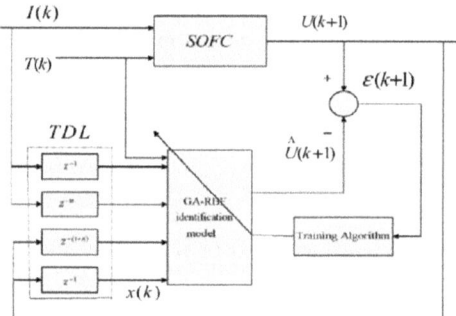

Figure 14.6   Identification structure of SOFC stack based on GA-RBF.[22]

Table 14.2   An example of cell current density scaled to [0,1].[22]

Unscaled(mAcm^{-2})	Scaled
0	0
100	0.143
150	0.214
300	0.429
450	0.643
550	0.786
700	1

speed and assist in selecting GA-RBF parameters. In this section, all the data, including, cell voltage, current density and temperature, are scaled to [0, 1] by Eq. (27).

$$x' = \frac{x_i - x_{\min}}{x_{\max} - x_{\min}} \tag{27}$$

14.1.2.2.2 Selection of the Optimal GA-RBF Parameters. In order to reduce the number of the parameters and improve the speed of program debug, the hidden layer of the RBF neural network is chosen three nodes. There are two inputs (current density I and cell temperature T) and one output (voltage U), so the structure of the RBF neural network is chosen 2-3-1. *i.e.* let the RBF neural network consists of input layer with 2 nodes, 1 hidden layer with 3 nodes and output layer with 1 node.

Coding structure. Coding aims to build the relationship between the problem and the individual in genetic algorithms. If the problems are expressed by coding strings, these strings are called an individual or a chromosome. A population contains a number of individuals. Generally, the population size n is chosen from 30 to 100. In order to save the running time, here, the size n is chosen 30. Each individual represents a variable or a part of the problem which is needed to be optimized. In this section, the parameters of RBF neural networks are needed to be optimized by the GA. So these individuals represent the widths and centres of the Gaussian function and the output weights, and the representation of an individual is

$$p = [b_1 b_2 b_3 c_{11} c_{12} c_{13} c_{21} c_{22} c_{23} \omega_1 \omega_2 \omega_3] \tag{28}$$

Here, each individual consists of 12 parameters, *i.e.* three widths and six centres of the hidden unit of the RBF neural network and three connection weights. Each parameter in the individual is expressed by a two-decimal coding of 10 bits.

All the width of the RBF hidden are chosen on the interval [0.1, 3] and the centres of the RBF hidden unit are chosen on the interval $[-3, 3]$. And all weights are chosen on the interval $[-1, 1]$.

Fitness function evaluation. All individuals of one generation are evaluated by a fitness function. When using a genetic algorithm to solve a problem, the problem is represented by a string and an evaluation function is defined. The evaluation function uses the value of the string as a parameter to evaluate the results of the problem. Each string is evaluated through the evaluation function and the new generation is formed by using the specific genetic operators. Here, an RBF neural network is used to model a SOFC stack. The value of the goal function is littler, and then the precision is higher. To get a higher regression precision, the goal function is defined as follows:

$$J = 50 \sum_{i=1}^{N} |e(i)| \tag{29}$$

Here, $e(i)$ is the error between the experimental output and the model output. Generally, the fitness function is defined as the reciprocal of the goal function, so we adopt the fitness function as below:

$$f = \frac{1}{j} = \frac{1}{50 \sum_{i=1}^{N} |e(i)|} \tag{30}$$

Genetic operations. There are mainly three genetic operations, including selection, crossover and mutation operations. These genetic operations have key effects on the performances of the genetic algorithm. In this study, we use the roulette-wheel selection method – a simulated roulette is spun – for this selection process. The response fitness value of every individual is p_i ($i = 1, 2, \ldots, N$). According to the p_i, a roulette wheel is divided into N parts. In the selection operation, spinning the roulette wheel, if a consulted point lies in the ith sector, we will choose the ith individual. Obviously the area of the sector is larger, and then the probability that the consulted point lies in the sector, is more. This indicates that the better an individual's fitness is, the more likely it is to be selected. An individual is probabilistically selected from the population on the basis of its fitness and the selected individual is then copied into the next generation of the population without any change.

Selection directs the search toward the best existing individuals but does not create any new individuals. In nature, an offspring has two parents and inherits genes from both. The main operator working on the parents is the crossover operator, the operation of which occurred for a selected pair with a crossover rate p_c that was set to 0.8 in this study. In each new population, there are $p_c \times n$ individuals which are needed crossover operations. Here, n is the population size. In the crossover step, we also keep the same number of chromosomes for each group. After this operation, the individuals with poor performances are replaced by the newly produced offspring.

Although selection and crossover will produce many new strings, they do not introduce any new information to the population at the site of an individual. Mutation is an operator that randomly alters the allele of a gene. With mutation, new genetic materials can be introduced into the population. In each new population, there are $p_m \times n \times L$ individuals which are needed mutation operations. Here, p_m is the mutation probability, n the population size and L is the string length. According to the above analyses, we know that every individual consists of 12 parameters and each parameter in the individual is expressed by a two-decimal coding of 10 bits. So L is 120. In the section, mutation probability p_m is chosen $0.003 - [1:1:size] \times 0.003/size$. This indicates that the smaller an individual's fitness is, the more likely it is to be mutated.

The terminate criterion. There are usually two criterions for terminating a run. The first criterion is deciding the maximum generation previously, and the second is that the process continues until the fitness function has no change. Here, we choose the first criterion and the maximum generation is chosen 100.

Table 14.3 The optimized widths and centres of the Gaussian function and the optimized output weights.[22]

b_1	b_2	b_3	c_{11}	c_{12}	c_{13}
1.9880	1.5146	2.8724	0.3724	−0.8416	−0.5073
c_{21}	c_{22}	c_{23}	ω_1	ω_2	ω_3
−1.7918	1.6276	−2.788	−0.4174	0.8397	0.8690

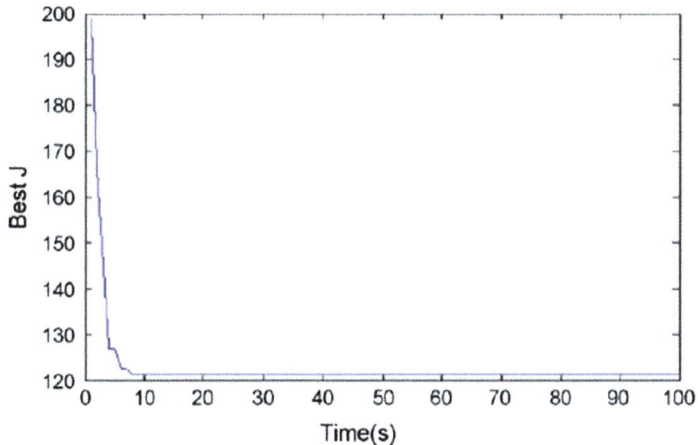

Figure 14.7 The goal function curve of the best individual with the population evolving.[22]

After 100 times genetic, the optimized initial values of the parameters are shown in Table 14.3. The optimization process of the goal function J in Eq. (29) is shown in Figure. 14.7 and the optimal value J is 121.3089.

After the optimized initial values of the three parameters are obtained, we utilize the gradient descent learning algorithms to adjust them, which can be seen in Eqs. (20)–(25). By tuning, the momentum term α is chosen 0.6 and the learning rate η is chosen 0.53.

14.1.2.2.3 Predicting with the GA-RBF Model. After training, a GA-RBF model is obtained, which can be used to predict new input dates. In our study, the testing data is also chosen from the above-mentioned model.[26] The cell voltage at 900 °C with the current density in the range from 0 to 700 m A cm^{-2} is predicted. And a comparison between the predicted data and the experimental data is made to evaluate the model's prediction precision, which is shown in Figure 14.8. At the same time, a BP neural network model is also used to predict the stack voltage at 900 °C. Via the toolbox of MATLAB 7.0, the predicted result is shown in Figure. 14.9. Comparing

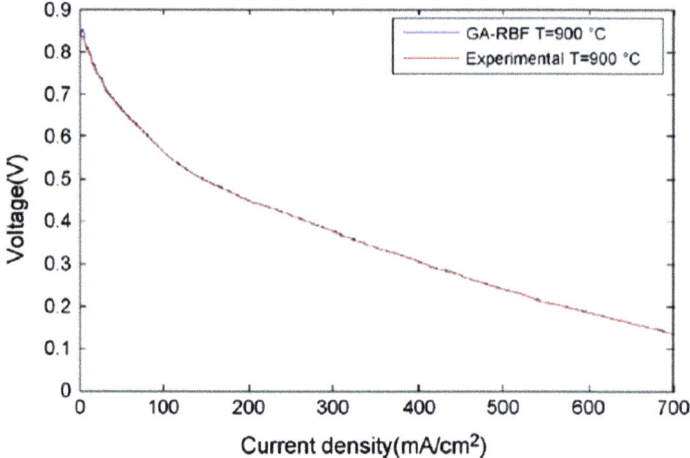

Figure 14.8   Voltage–current density characteristics: predicted by GA-RBF model and experimental at T = 900 °C.

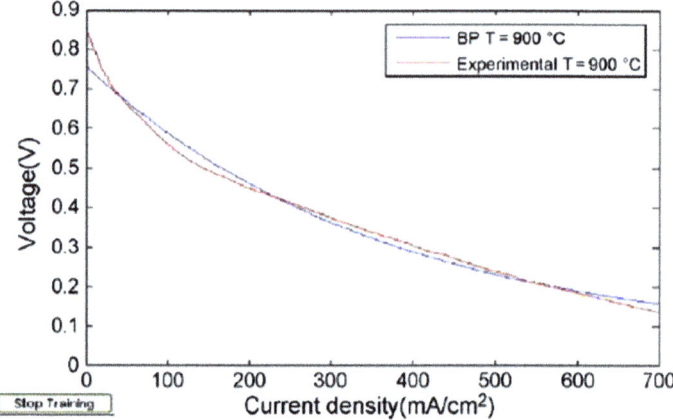

Figure 14.9   Voltage–current density characteristics: predicted by BP model and experimental at T = 900 °C.[22]

Figures 14.8 and 14.9, we can see clearly that the precision is greatly improved. It indicates that GA-RBF is a powerful tool for modelling SOFC and our GA-RBF model presented in this section is accurate and valid.

14.2   Dynamic Identification Modelling for SOFC

As a kind of nonlinear multi-input–multi-output (MIMO) system, SOFC is hard to model by the traditional methodologies. Although there have been many investigations into all aspects of mathematical modelling of the SOFC, most of them are concerned with static performance.[27–29] The SOFC, however, is a dynamic device which will affect the dynamic behaviour of the power system to which it is connected. Analysis of such a behaviour requires an accurate dynamic model. In the last several decades, fruitful results on modelling the nonlinear dynamics of the SOFC have been proposed.[30–33] However, most of these models emphasized the detailed description of cell internal processes, such as component material balance, energy balance and electrochemical kinetics, *etc.* These models are very useful for analyzing the transient characteristics of the SOFC, but they are too complicated to be used in a control system design. To develop effective control strategies, an autoregressive with exogenous input (ARX) identification model for a SOFC has been presented.[34] However, the performance of this model may be poor due to the highly nonlinear behaviour of the system. Therefore, a new nonlinear modelling approach is needed to provide a better solution. In this section, an adaptive neural-fuzzy inference system (ANFIS) model and a Hammerstein model are adopted to describe the nonlinear dynamic properties of the SOFC separately.

14.2.1 ANFIS Identification Modelling

An adaptive neural-fuzzy inference system (ANFIS)[35] was put forward by Jang in 1993. It integrates the advantages of both neural networks and fuzzy system, which not only has good learning capability, but also can be interpreted easily. The ANFIS has the ability to approximate a large class of dynamical nonlinear systems, which is widely applied in fields such as intelligent control and time series prediction[36,37] in recent years. However, according to our knowledge, practical application of ANFIS to build the dynamic model of SOFC stacks cannot be found in prior papers. In this section, a physical model of a 100 kW SOFC stack is used to generate the data required for the training and predicting of the ANFIS model.

14.2.1.1 SOFC Dynamic Physical Model

In this section, we assume that all the SOFCs in the SOFC stack operate at identical conditions. In addition, the following main assumptions have been made in developing the model.

1) All gases are assumed to be ideal gases.
2) The stack is fed with hydrogen and oxygen. If natural gas instead of hydrogen is used as fuel, the dynamics of the fuel processor must be considered in the model. The effect of the fuel processor in the model will be tested in the future.
3) The exhaust of each channel is via a single orifice. The ration of pressures between the interior and exterior of the channel is large enough to consider that the orifice is choked.
4) The channels that transport gases along the electrodes have a fixed volume, but their lengths are small. Thus it is only necessary to define one single pressure value in the interior.

5) The temperature is stable at all times.
6) The Nernst equation can be applied to calculate the voltage.
7) Ohmic, concentration, and activation losses are taken into account.

14.2.1.1.1 Electrochemical Model. The change in concentration of each specie that appears in the chemical reactions can be written in terms of input and output flow rates and exit molarity due to the following chemical reaction:[38,39]

$$\frac{dx_i}{dt} = \frac{RT}{V}\left(q_i - q_i^o - q_i^r\right) \tag{31}$$

According to the basic electrochemical relationships, the mole flow that reacts q_i^r can be calculated as:

$$q_i^r = \frac{NI}{2F} = 2K_r I \tag{32}$$

Replacing the mole flow that reacts by Eq. (32), we can get the following expression of the hydrogen partial pressure:

$$\frac{d}{dt}x_{H_2} = \frac{RT}{V}\left(q_{H_2} - q_{H_2}^o - 2K_r I\right) \tag{33}$$

For orifice that is choked, it could be considered that the molar flow of any gas through the valve is proportional to its partial pressure inside the channel, according to the expression

$$\frac{q_{H_2}}{x_{H_2}} = K_{H_2} \tag{34}$$

Replacing the output flow by Eq. (34), applying the Laplace transformation to the Eq. (33) and isolating the hydrogen partial pressure, yields the following expressions:

$$x_{H_2}(s) = \frac{\frac{1}{K_{H_2}}}{1 + \tau_{H_2}s}\left(q_{H_2} - 2K_r I\right) \tag{35}$$

Where, $\tau_{H_2} = \frac{V}{K_{H_2}RT}$.

Similarly component balances for O_2 and H_2O lead to the following set of equations:

$$x_{O_2}(s) = \frac{\frac{1}{K_{O_2}}}{1 + \tau_{O_2}s}\left(q_{O_2} - K_r I\right) \tag{36}$$

$$x_{H_2O}(s) = 2K_r I\frac{\frac{1}{K_{H_2O}}}{1 + \tau_{H_2O}s} \tag{37}$$

14.2.1.1.2 Operating Cell Voltage. The actual cell potential is decreased from its ideal potential because of several types of irreversible losses, such as activation, concentration and ohmic losses.

Activation losses. Activation losses are caused by sluggish electrode kinetics. In the case of electrochemical reactions with $\eta_{act} \geq 50 - 100mV$, it is possible to approximate the voltage drop due to activation polarization by a semi empirical equation, called the Tafel equation. The equation for activation polarization is shown as follows:

$$\eta_{act} = \frac{RT}{\alpha nF} \ln \frac{I}{I_0} \qquad (38)$$

The usual form of the Tafel equation that can be easily expressed by a Tafel plot is

$$\eta_{act} = \partial + \beta \log I \qquad (39)$$

Concentration losses. As a reactant is consumed at the electrode by electrochemical reaction, there is a loss of potential due to the inability of the surrounding material to maintain the initial concentration of the bulk fluid. Concentration loss equation is given as follows:

$$\eta_{conc} = \frac{RT}{nF} \ln \left(1 - \frac{I}{I_L} \right) \qquad (40)$$

Ohmic resistance. Ohmic losses occur because of resistance to the flow of ions in the electrolyte and resistance to flow of electrons through the electrode materials. Because both the electrolyte and fuel cell electrodes obey Ohm's law, the ohmic losses can be expressed by the Eq. (41)

$$\eta_{ohmic} = IR_{ohmic} \qquad (41)$$

The stack output voltage. Applying Nernst's equation and Ohm's law (taking into account ohmic, concentration, and activation losses), the stack output voltage is represented as follows:

$$U = E - \eta_{ohmic} - \eta_{conc} - \eta_{act} \qquad (42)$$

The dynamic physical model replaces the real SOFC stack to generate the simulation data required for the identification of the ANFIS model. The parameters of this fuel cell are given in Table 14.4. The data sources blocks is shown in Figure 14.10.

14.2.1.2 An Adaptive Neural-fuzzy Inference System

14.2.1.2.1 Problem Formulation. For a given SOFC stack, the relation between terminal voltage U and current I is influenced by many operating parameters, such as cell temperature, air flow, hydrogen flow, air pressure, hydrogen pressure, *etc.* However, due to the high number of operating variables, a complete experimental database of SOFC under the different operating conditions is difficult to obtain and no data are available in the

Table 14.4 Parameters in the SOFC dynamic model.[40]

Parameter	Unit	Value
T	K	1273
F	C/mol	96485
R	J/(mol K)	8.314
N	–	384
E^0	V	1
K_r	mol/(s A)	0.996×10^{-3}
K_{H_2}	mol/(s Pa)	8.32×10^{-6}
K_{H_2O}	mol/(s Pa)	2.77×10^{-6}
K_{O_2}	mol/(s Pa)	2.49×10^{-5}
τ_{O_2}	s	2.91
τ_{H_2}	s	26.1
τ_{H_2O}	s	78.3
R_{ohmic}	Ω	0.126
q_{H_2}	mol/s	Variable
q_{O_2}	mol/s	Variable
β	–	0.11
∂	–	0.05
I_L	A	800
U	V	Variable
I	A	Variable
P_{dc}	kW	100

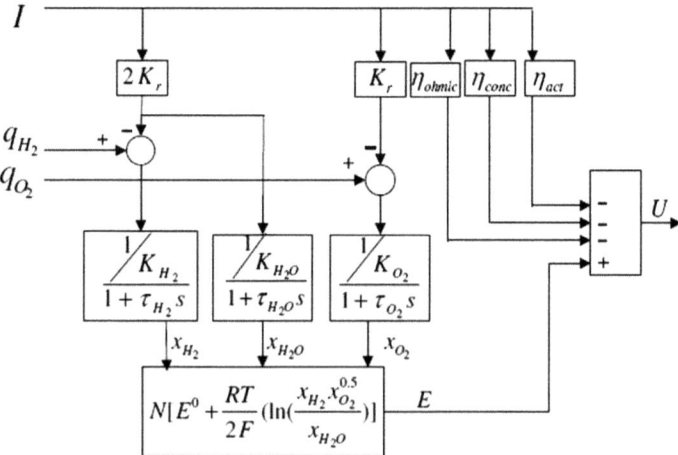

Figure 14.10 Data sources of ANFIS modelling.[40]

open literature yet.[41] Up to now, almost no model has ever been able to accommodate all these operating variables. Our ANFIS model is no exception. In our experiment, current I, which is decided by the uncontrollable load and air flow qO2 , hydrogen flow qH2 , are taken as variables.

In general, a wide class of nonlinear systems can be described by nonlinear autoregressive model with exogenous inputs (NARX).[42] So in this section the SOFC nonlinear system with three inputs and one output can be described as follows:

$$Y(k) = F\big(Y(k-1), \cdots, Y(k-n_y), V(k-1), \cdots, V(k-n_v)\big) \quad (43)$$

where, $Y(k) = U(k)$ denotes the output vector, $V(k) = \big[q_{H_2}(k), q_{O_2}(k), I(k)\big]$ the input vector, n_y and n_v the lags of the output and input, respectively, and $F(\cdot)$ a nonlinear function. The aim of our study is thus to find an ANFIS model that approximates Eq. (43).

14.2.1.2.2 ANFIS Theory. ANFIS is a neural network implementation of a T–S (Takagi–Suguno) fuzzy inference system.[43,44] In this section, with the five-layered network structure of the ANFIS, we shall define the function of each node.

Layer 1: Each node in this layer corresponds to one linguistic label (small, large, *etc.*) of one of the input variables. In other words, the membership value μ_i^j which specifies the degree to which an input value belongs to a fuzzy set, is calculated in this layer. For an external input $x_i = \big[q_{H_2}(k), q_{O_2}(k), I(k)\big]^T$ the following Gaussian membership function is assumed to be used:

$$\mu_i^j = \exp\left[-\left(\frac{x_i - c_{ij}}{b_i^j}\right)^2\right], (i = 1, 2, 3, \ j = 1, 2 \cdots c_i) \quad (44)$$

where, parameter $\{c_{ij}, b_{ij}\}$ is referred to as antecedent parameters.
Layer 2: Each node in layer 3 represents a possible IF-part for fuzzy rules and the layer aims to calculate the fitness value of every rule. Here the function of each rule is

$$\alpha_j = \mu_1^{i_1} \mu_2^{i_2} \mu_3^{i_3} \left(i_1 \in \{1, 2, \cdots, c_1\}, \cdots, i_3 \in \{1, 2, \cdots, c_3\}, \ j = 1, 2 \cdots, c, c = \prod_{i=1}^{3} c_i\right) \quad (45)$$

Layer 3: The jth node in this layer calculates the ratio of the jth rule's firing strength to the sum of all rules' firing strengths:

$$\bar{\alpha}_j = \frac{\alpha_j}{\sum_{i=1}^{c} a_i} (i = 1, 2, \cdots c) \quad (46)$$

Layer 4: The output of layer 3 is multiplied with the consequent function to obtain o_j:

$$O_j = \bar{\alpha}_j \big(p_j q_{H_2} + l_j q_{O_2} + m_j I + r_j\big)(i = 1, 2, \cdots c) \quad (47)$$

Layer 5: Each node in the layer represents an output variable which is the summation of incoming signals:

$$U = \sum_{j=1}^{c} O_j \qquad (48)$$

In this section, a hybrid learning algorithm combining back propagation (BP) and least squares estimate (LSE) is adopted to identify linear and nonlinear parameters in the ANFIS. The consequent parameters are identified by the LSE and the antecedent parameters are identified by the BP. Hybrid learning rule can be derived from Jang's research[35] on adaptive network. Jang observed that the output of a certain adaptive network is linear in some of the network parameters, which thereby can be identified by the least squares method. And then the combination of the gradient method and the LSE forms the so-called hybrid learning rule. The method of least squares assumes that the best-fit curve of a given type is the curve that has the minimal sum of the deviations squared (least square error) from a given set of data. Suppose that the data points are $(x_1, y_1), (x_2, y_2), \cdots, (x_n, y_n)$ where x the independent variable is and y is the dependent variable. The fitting curve $f(x)$ has the deviation d from each data point, i.e., $d_1 = y_1 - f(x_1), d_2 = y_2 - f(x_2), \cdots, d_n = y_n - f(x_n)$. According to the method of least squares, the best-fitting curve has the property that:

$$d_1^2 + d_2^2 + \cdots d_n^2 = \sum_{i=1}^{n} d_i^2 = [y_i - f(x_i)]^2 = \text{mininum} \qquad (49)$$

14.2.1.3 Model Training and Simulation Test

14.2.1.3.1 Preparing Simulation Data. For the purpose of identification, the dynamic physical model developed in Section 2 is excited with uniformly random input signals including the fuel flow (0–1.7 mol/s), the air flow (0–1.6 mol/s), and the current (0–800 A). To obtain values at integer time points, the fourth-order Runge–Kutta method was used to find the numerical solution to the dynamic physical model in the simulation. A set of 5000 data was collected from the simulation. The first 2000 data were used for the identification of an ANFIS model, while the remaining 3000 data were used for validation purposes (see Figure 14.11).

14.2.1.3.2 Predicting with the ANFIS Model. In this section, the presented ANFIS model has three inputs (hydrogen flow q_{H_2}, oxygen flow q_{O_2} and current I) and one output (voltage U). The toolbox of MATLAB 7.0 is used to build the ANFIS model. Root mean square error (RMSE) is employed here to evaluate modelling results, which is calculated by

$$RMSE = \sqrt{\frac{1}{N} \sum_{k=1}^{N} (y(k) - \hat{y}(k))^2} \qquad (50)$$

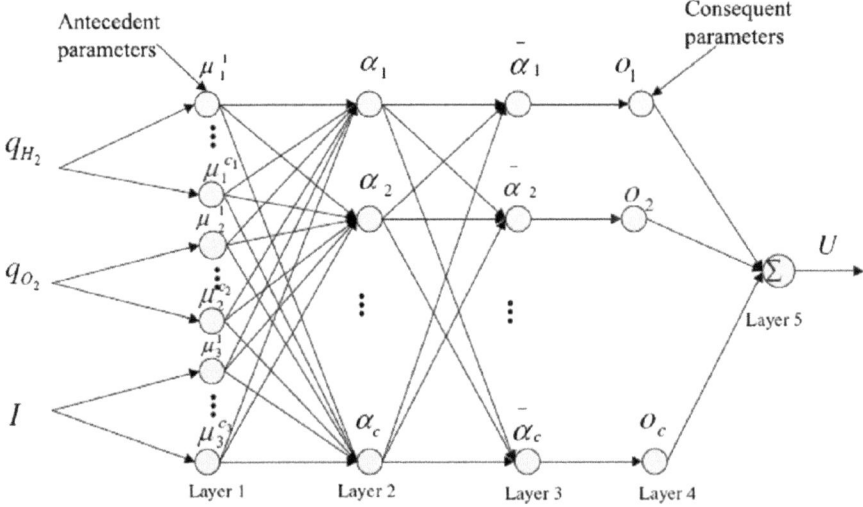

Figure 14.11 The structure of a self-adaptive fuzzy neural network.[40]

where N is the number of sample data from the dynamic physical model of SOFC, $y(k)$ is the predictive output of ANFIS model, $\hat{y}(k)$ is the output of the dynamic physical model. Based on satisfying the minimal RMSE conditions in Eq. (50) and shorter training time, 12 generalized bell membership functions are used in layer 1 by trial and error. An initial FIS (fuzzy inference system) structure is generated by the function "GENFIS1". Setting the parameters of training, and then the function "ANFIS" is used to train the ANFIS model. After 50 training epochs, an ANFIS model is obtained, which can be used to predict new input data. Under various fuel and air flow, comparing the discrete voltage values calculated by the ANFIS with actual sampled voltage data obtained from the dynamic model of SOFC stack, we obtained the initial and output membership function shown in Figure. 14.12 (For the space limitation, we only give the change of q_{H_2} membership function). Step changes in the stack current (from 200 A to 400 A in 100s and from 500 A to 300 A in 200s) are applied. The comparison between predicted (ANFIS model) and experimental (physical model) output voltage dynamic curves is represented in Figures 14.13 and 14.14. The RMSE of output voltage obtained is 0.4637 and 0.5126, respectively. From Figures 14.13–14.14, we can see that the obtained ANFIS model can approximate the dynamic behaviour of the physical SOFC model with good accuracy.

The simulation results reveal that it is feasible to set up the model of the SOFC stack based on ANFIS identification. The most important is that the modelling process avoids using complicated differential equations groups to describe the stack, and the inputs–outputs dynamic characteristics of the SOFC stack can be predicted. In the future, based on the ANFIS model, some control scheme studies such as predictive control and robust control can be developed.

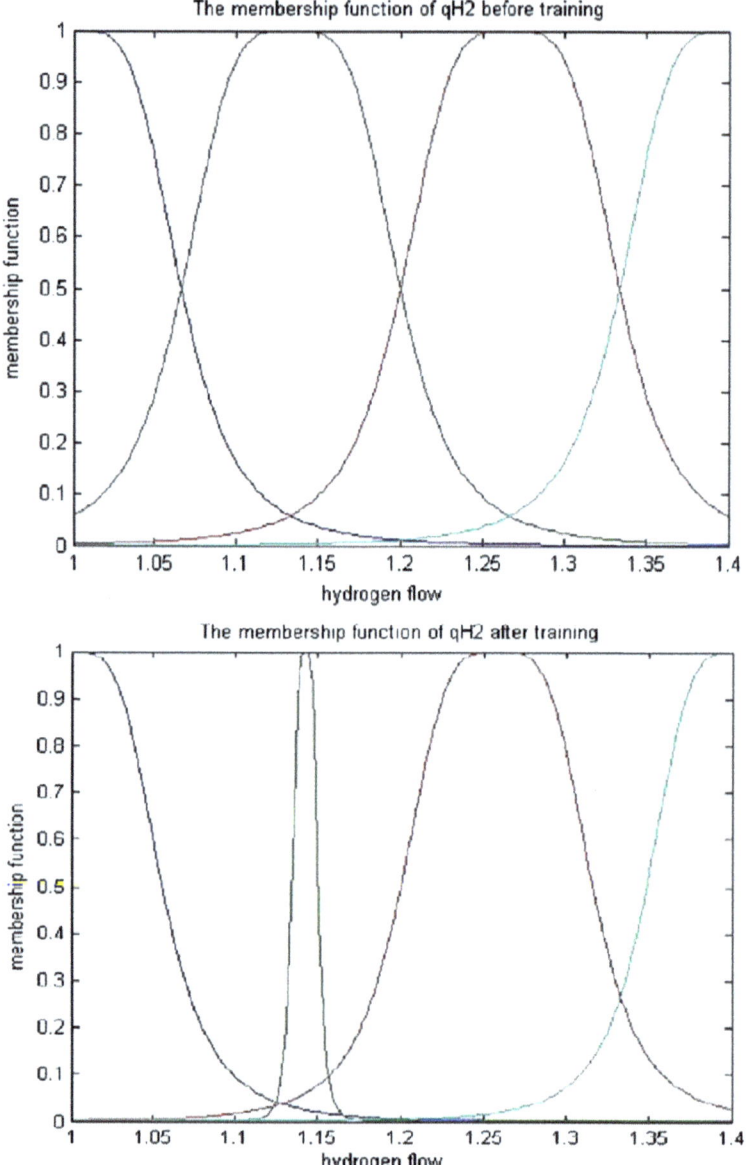

Figure 14.12   Change of the membership function.[40]

In addition, the ANFIS approach is finding relationships between the input and output variables of the system without explicit knowledge of its physical behaviour. Therefore, in the future it will also be applied to the other type of fuel cells, as long as we get the abundant input–output data of the system. If natural gas instead of hydrogen is used as fuel in the SOFC, the effect of the fuel processor in the model will be tested in the future.

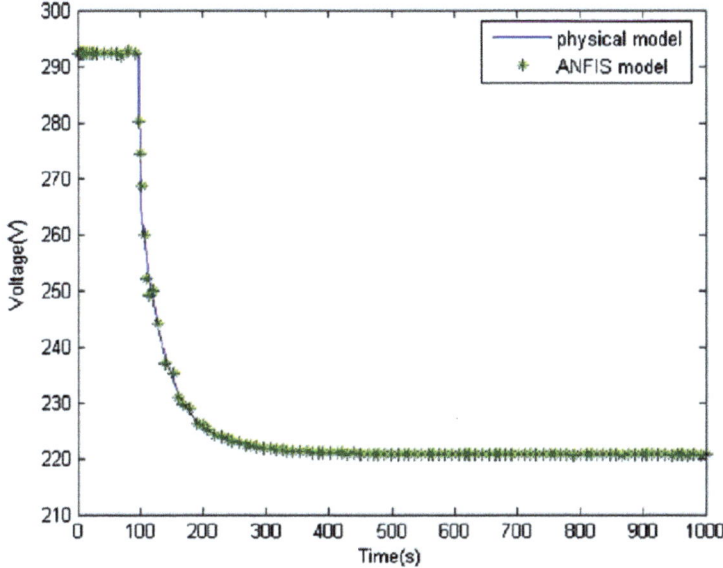

Figure 14.13 Output voltage dynamic response (from 200 A to 400 A).[40]

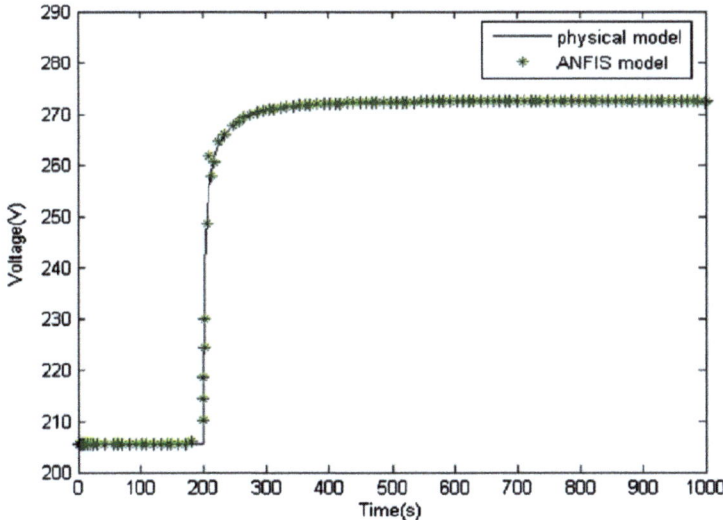

Figure 14.14 Output voltage dynamic response (from 500 A to 300 A).[40]

14.2.2 Hammerstein Identification Modelling

In this section, a Hammerstein model, consisting of a radial basis function neural network (RBFNN) in cascade with an ARX model, is adopted to describe the nonlinear dynamic properties of the SOFC.

The Hammerstein model consists of a static nonlinear part followed in series by a dynamic linear part. It is a type of commonly used nonlinear model, and has been successfully used to model a class of nonlinear systems.[45–48] The identification of the Hammerstein system involves estimating both the nonlinear and the linear parts from the input-output observations. Jurado[49] has presented a Hammerstein model for SOFC, in which the base functions were used for the representation of the linear and nonlinear blocks in the Hammerstein model. But this kind of identification method needs prior information of the system.[50] To overcome the aforementioned deficiencies, Narendra and Parthasarathy.[51] have pointed out that a neural network could be used as the nonlinear operator in a Hammerstein model. Due to a number of advantages of RBFNNs compared with other types of artificial neural networks (ANNs), such as better approximation properties, simpler network structures and faster learning algorithms,[52] a Hammerstein model of the SOFC is presented in this section, in which the nonlinear static part is approximated by a RBFNN and the linear dynamic part is modelled by an ARX model.

In addition, one of the most important cell performance variables, fuel utilization, has not been examined[49] when they explored the SOFC dynamic response after the disturbances. Furthermore, they have not included the SOFC fuel processor in their investigation. So in this section, the fuel processor is included and the operating issue about the fuel utilization is considered specifically.

14.2.2.1 Theory for the SOFC Dynamic Model

Based on the work reported,[53–55] the SOFC dynamic model is briefly reviewed in this section. The SOFC dynamic model including the fuel processor adopted in this section is shown in Figure 14.15.

14.2.2.1.1 The Balance of Plant (BOP). The BOP consists of the natural gas fuel storage, fuel valve controlled by its controller, and the fuel processor that reforms the natural gas input q_f to the hydrogen-rich fuel $q_{H_2}^{in}$. The authors[53] introduced a simple model of a fuel processor that converts fuels such as natural gas to hydrogen and by-product gases. The model is a first-order transfer function with time constant τ_f. Hence, the fuel processor is represented simply by this first-order model in this study.

Although CO can be a fuel in a SOFC, we suppose all CO will take part in the CO-shift reaction if the gas contains water.[53] Thus, the overall cell reaction of the SOFC is:

$$H_2 + \frac{1}{2}O_2 \rightarrow H_2O \tag{51}$$

From Eq. (51), it is seen that the stoichiometric ratio between hydrogen and oxygen is 2 to 1. Oxygen excess is always taken in to let hydrogen react with

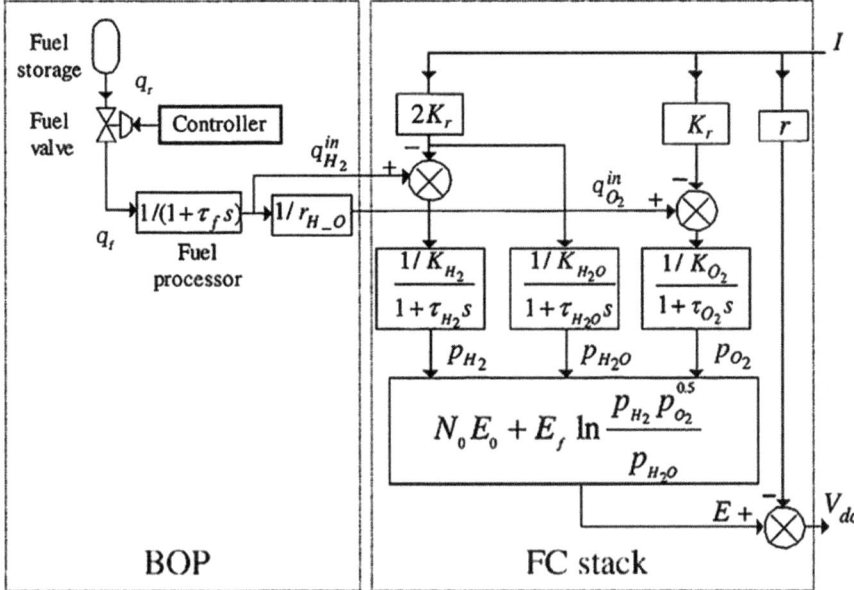

Figure 14.15 SOFC dynamic model.[54]

oxygen more completely. So, the flow ratio of hydrogen to oxygen is kept at 1.145 in this section.[54]

14.2.2.1.2 Solid Oxide Fuel Cell. The SOFC consists of hundreds of cells connected in series or in parallel. By regulating the fuel valve, the amount of fuel into the SOFC stack can be adjusted, and the output voltage of the SOFC can be controlled.

The Nernst's equation and Ohm's law determine the average voltage magnitude of the fuel cell stack. So, applying Nernst's equation and Ohm's law (taking into account ohmic losses), the output voltage of the SOFC can be modelled as follows:[53–56]

$$V_{dc} = E - rI \tag{52}$$

$$E = N_0 E_0 + \frac{N_0 RT}{2F} \ln \frac{p_{H_2} p_{O_2}^{0.5}}{p_{H_2O}} \tag{53}$$

where

$$p_{H_2} = \frac{1/K_{H_2}}{1 + \tau_{H_2}s}(q_{H_2}^{in} - 2K_r I) \tag{54}$$

$$p_{O_2} = \frac{1/K_{O_2}}{1 + \tau_{O_2}s}(q_{O_2}^{in} - K_r I) \tag{55}$$

$$p_{H_2O} = \frac{1/K_{H_2O}}{1 + \tau_{H_2O}s} 2K_r I \tag{56}$$

14.2.2.1.3 Fuel Utilization. Fuel utilization is one of the most important operating variables that may affect the performance of FC. It is defined as

$$u_f = \frac{q_{H_2}^{in} - q_{H_2}^{out}}{q_{H_2}^{in}} = \frac{q_{H_2}^{r}}{q_{H_2}^{in}} = \frac{N_0 I}{2 F q_{H_2}^{in}} \tag{57}$$

where $q_{H_2}^{r}$ is the hydrogen reacted flow rate. For protecting the SOFC stack, the desired range of fuel utilization is from 0.7 to 0.9. An overused-fuel condition ($u_f > 0.9$) could lead to permanent damage to the cells due to fuel starvation whereas an underused-fuel condition ($u_f < 0.7$) results in a rapid rise of the cell voltage.[57]

To protect the SOFC stack and expect small deviations in the terminal voltage due to changes in stack current, we will hold the utilization as constant. According to Eq. (57), the operation of the SOFC stack with a fuel input proportional to the stack current results in a constant utilization factor in the steady state. Thus, the SOFC stack is operated with constant steady-state utilization by controlling the natural gas input flow to the stack as:[55]

$$q_f = \frac{N_0 I}{2 F u_s} \tag{58}$$

where u_s is the desired utilization in steady state. Furthermore, because the fuel processor is specially considered, the relationship between a small change of stack current ΔI and a small change of hydrogen input $\Delta q_{H_2}^{in}$ fed to the SOFC stack can be derived as:[55]

$$\Delta q_{H_2}^{in} = \frac{N_0}{2 F u_s (1 + \tau_f s)} \Delta I \tag{59}$$

14.2.2.2 Hammerstein Model

14.2.2.2.1 Problem Statement. In this section, we will consider the problem of estimating a model for a single-input-single-output (SISO) Hammerstein system based on the input-output data, *i.e.*, $\{u_i\}_{i=1,\dots,n}$ and $\{y_i\}_{i=1,\dots,n}$. The Hammerstein model consists of a RBFNN for identification of the static nonlinear part, in series with an ARX model for identification of the linear part. The structure of the Hammerstein model adopted in this section is illustrated in Figure 14.16, where $u(k)$ and $y(k)$ are the input and output of the Hammerstein model at the kth sampling instant, respectively,

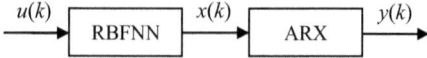

Figure 14.16 Hammerstein model.[58]

and $x(k)$ is the output of RBFNN which is unmeasurable. The output of the Hammerstein model can be expressed as

$$y(k) = -\sum_{i=1}^{n_a} a_i y(k-i) + \sum_{j=0}^{n_b} b_j x(k-j) \tag{60}$$

where $a_i(i=1,\cdots,n_a)$ and $b_j(j=0,\cdots,n_b)$ are the parameters of the ARX model, n_a and n_b are integers related to the model order and the function.

The static nonlinear part in the Hammerstein model is represented by using the RBFNN depicted in Figure 14.17 as

$$x(k) = \sum_{i=1}^{M} w_i \varphi_i(u(k)) \tag{61}$$

where

$$\varphi_i(u(k)) = \exp(-\frac{\|u(k) - c_i\|^2}{2d_i^2}) \tag{62}$$

is the Gaussian function. M is the number of hidden node, c_i and d_i are the centres and widths of the ith RBF hidden unit, respectively. w_i are the connection weights from the ith hidden node to the output, $\|\cdot\|$ denotes the Euclidean norm.

14.2.2.2.2 Identification of the Hammerstein Model. The identification of the Hammerstein model involves estimating the hidden centres, the radial basis function widths and the connection weights of the RBFNN and the orders and parameters of the ARX model. The objective is to develop training algorithms by which we can adjust the above parameters so that the application of a set of inputs produces a desired set of outputs.

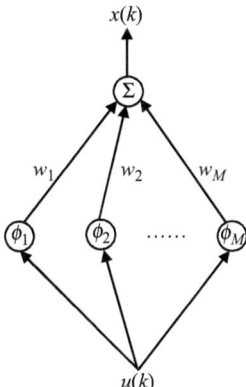

Figure 14.17 RBF neural network.[58]

Identification of the RBFNN. For a neural network, gradient-based learning algorithm is commonly used in network training. In this research, a new gradient descent algorithm is derived to update the hidden centres, the radial basis function widths and the connection weights of the RBFNN in the Hammerstein model.

In order to adjust the hidden centres, the radial basis function widths and the connection weights of the RBFNN by using gradient-based learning algorithm, a problem arises in determining a measure of the error at the output of the RBFNN. One knows the desired output ($y(k)$) of the Hammerstein model, but the desired output ($x(k)$) for the RBFNN is unknown in advance. Intuitively, the error at the output of the RBFNN must be related to the error at the output of the ARX model.[59] This idea is used to derive the updating laws for the connection weights, the hidden centres and the radial basis function widths of the RBFNN by defining the following error at the output of the ARX model:

$$E(k) = \frac{1}{2}(y_d(k) - y(k))^2 \tag{63}$$

where $y_d(k)$ is the desired output of the system. Combining Eqs. (60), (61) and (63), one gets

$$E(k) = \frac{1}{2}(y_d(k) + a_1 y(k-1) + \cdots + a_{n_a} y(k - n_a) - b_0 x(k) - \cdots - b_{n_b} x(k - n_b))^2$$

$$= \frac{1}{2}(y_d(k) + a_1 y(k-1) + \cdots + a_{n_a} y(k - n_a) - b_0 \left(\sum_{i=1}^{M} w_i \phi_i(u(k)) \right)$$

$$- \cdots - b_{n_b} x(k - n_b))^2 \tag{64}$$

To minimize the error $E(k)$, the connection weights w_i, the hidden centres c_i and the radial basis function widths d_i should be updated in the negative direction of the gradient of $E(k)$, $\nabla E(k)$. Thus, the updating laws of w_i, c_i and d_i are derived as follows:

$$\nabla w_i(k) = -\eta \frac{\partial E(k)}{\partial w_i(k)} = b_0 \eta (y_d(k) - y(k)) \varphi_i(u(k)) \tag{65}$$

$$\nabla w_i(k) = -\eta \frac{\partial E(k)}{\partial w_i(k)} = b_0 \eta (y_d(k) - y(k)) \varphi_i(u(k)) \tag{66}$$

$$\nabla c_i(k) = -\eta \frac{\partial E(k)}{\partial c_i(k)} = b_0 \eta w_i(k)(y_d(k) - y(k)) \varphi_i(u(k)) \frac{u(k) - c_i(k)}{(d_i(k))^2} \tag{67}$$

$$c_i(k+1) = c_i(k) + b_0 \eta w_i(k)(y_d(k) - y(k)) \varphi_i(u(k)) \frac{u(k) - c_i(k)}{(d_i(k))^2} \tag{68}$$

$$\nabla d_i(k) = -\eta \frac{\partial E(k)}{\partial d_i(k)} = b_0 \eta w_i(k)(y_d(k) - y(k)) \varphi_i(u(k)) \frac{\|u(k) - c_i(k)\|^2}{(d_i(k))^3} \tag{69}$$

$$d_i(k+1) = d_i(k) + b_0 \eta w_i(k)(y_d(k) - y(k))\varphi_i(u(k))\frac{\|u(k) - c_i(k)\|^2}{(d_i(k))^3} \quad (70)$$

where η is the learning rate.

Identification of the ARX model. An important step, which precedes the parameter estimation of the ARX model, is to determine the model structure which is entirely defined by the integers n_a and n_b. Many statistical model selection criteria have been developed. One popular criterion is the Akaike Information Criterion (AIC).[60] So in this study, the best orders of the ARX model are determined by minimizing AIC, which is defined as

$$AIC = -N \log(E) + 2(n_a + n_b) \quad (71)$$

$$E = \frac{1}{N}\sum_{k=1}^{N}(y(k) - \hat{y}(k))^2 \quad (72)$$

where, $\hat{y}(k)$ is the estimated output at time step "k" of the ARX model, N is the number of sample points, E is the mean square error between actual output and estimated output of the ARX model.

In the sequel, the least squares (LS) algorithm is adopted to estimate the parameters of the ARX model, *i.e.*

$$\hat{\theta} = [H^T H]^{-1}H^T Y \quad (73)$$

where $\hat{\theta} = [\hat{a}_1, \hat{a}_2, \cdots, \hat{a}_{n_a}, \hat{b}_0, \hat{b}_1, \cdots, \hat{b}_{n_b}]^T$ is the estimated parameters of the ARX model and

$$H = [h(1), h(2), \cdots, h(N)]^T_{N \times (n_a + n_b + 1)} \quad (74)$$

$$Y = [y(1), y(2), \cdots, y(N)]^T_{N \times 1} \quad (75)$$

$$h(k) = [-y(k-1), -y(k-2), \cdots, -y(k-n_a), x(k), x(k-1), \cdots, \\ x(k-n_b)]^T_{(n_a + n_b + 1) \times 1} \quad (76)$$

14.2.2.3 Results and Discussion

The main aim of this section is to construct a dynamic model for the output voltage V_{dc} of the SOFC, as a function of the natural gas input flow to the stack N_f. For the purpose of identification, the white-box model described in Section 14.2 is used to generate the modelling data. In the following condition, it is assumed that the load resistor has the following variation. The SOFC is operating at its rated operation point initially. The nominal operating conditions of the SOFC are given in Table 14.5.[53,54] At $t=150s$, a load disturbance causes the stack current to have a step change (from 300 A to 270 A). In this situation, the variation of the natural gas input flow to the

Table 14.5   SOFC operating point data.[58]

Item	Value
N_0	384
T	1273 K
I_{rate}	300 A
u_s	0.8
E_0	1.18 V
K_{H_2}	0.843 mol/(s.atm)
K_{O_2}	2.52 mol/(s.atm)
K_{H_2O}	0.281 mol/(s.atm)
τ_{H_2}	26.1 s
τ_{O_2}	2.91 s
τ_{H_2O}	78.3 s
τ_f	5 s
r	0.126 Ω
r_{H_O}	1.145

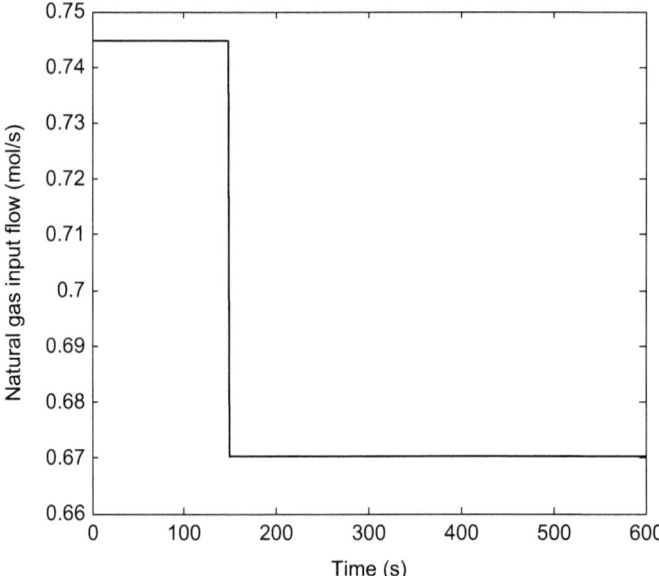

Figure 14.18   Variation of natural gas input flow due to step change of current from 300A to 270A.[58]

stack is depicted in Figure 14.18. In order to establish the Hammerstein model, a record of 600 experimental samples of the output voltage is collected due to this change of the natural gas input flow.

Based on these data, we firstly determine the structure of the ARX model in the Hammerstein model. To determine the orders of the ARX model, the sampled input-output data are used to identify a linear ARX model for the

SOFC.[49] By minimizing AIC, the structure of the ARX model can be determined as follows:

$$V_{dc}(k) = -a_1 V_{dc}(k-1) - a_2 V_{dc}(k-2) - a_3 V_{dc}(k-3) - a_4 V_{dc}(k-4)$$

$$- a_5 V_{dc}(k-5) - a_6 V_{dc}(k-6) - a_7 V_{dc}(k-7) - a_8 V_{dc}(k-8)$$

$$- a_9 V_{dc}(k-9) + b_0 x(k) + b_1 x(k-1) + b_2 x(k-2) + b_3 x(k-3)$$

$$+ b_4 x(k-4) + b_5 x(k-5) + b_6 x(k-6) \tag{77}$$

Thus, the structure of the Hammerstein model selected in this section will consist of a RBFNN with 6 RBF neurons in the hidden layer as shown in Figure 14.17, and an ARX model with the structure as shown in Eq. (77).

Secondly, the connection weights w_i, the hidden centres c_i and the radial basis function widths d_i of the RBFNN are initialized with small random values. Thus, one can convert the natural gas input flow, $N_f(k)$, into the intermediate variable, $x(k)$, using Eq. (11). After selecting the learning rate $\eta = 0.04$, the LS algorithm (*i.e.*, Eq. (73)) and the derived gradient descent learning algorithm (*i.e.*, Eqs. (75)–(80)) are used to adjust the parameters of the ARX model and the RBFNN, respectively. As a result, the actual output voltage of the SOFC and the identified output voltage of the Hammerstein model are shown in Figure 14.19. Figure 14.19 indicates that the proposed Hammerstein modelling method is applicable to describe the nonlinear dynamic behaviours of the SOFC.

For the purpose of comparison, the same input-output data are used to identify a RBFNN model, which performs at the same structure and initial

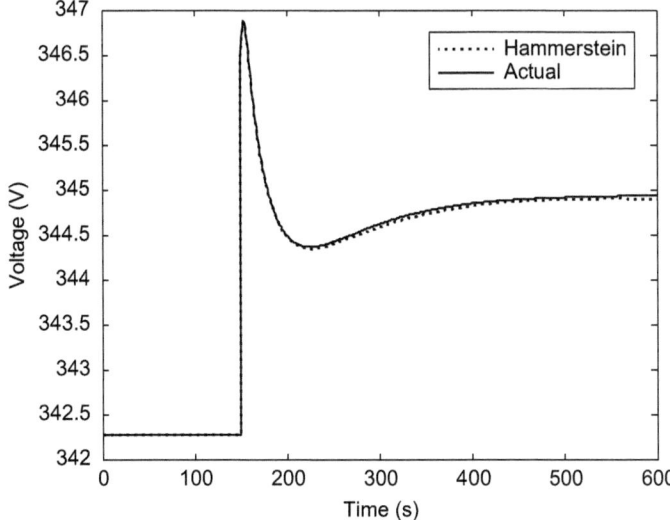

Figure 14.19 Output voltage of the actual and identified Hammerstein models.

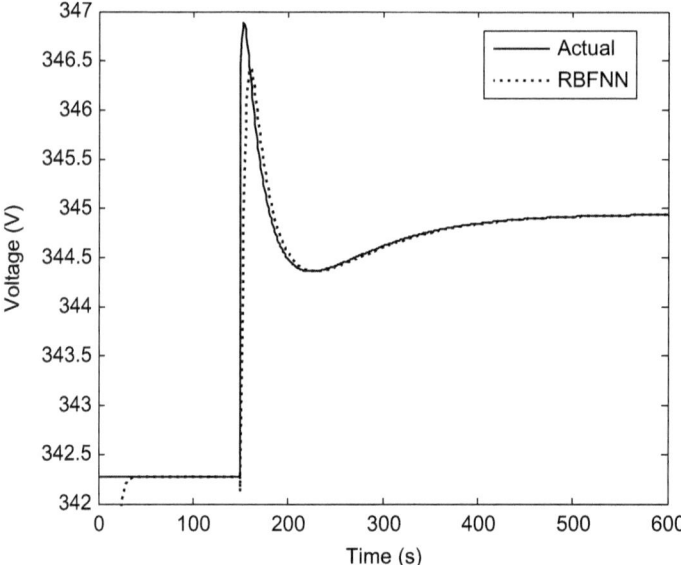

Figure 14.20   Output voltage of the actual and identified RBFNN models.

values of the RBFNN parameters as in the Hammerstein model. In this study, the traditional gradient descent algorithm is used to adjust the RBFNN parameters. Finally, the actual output voltage of the SOFC and the estimated output voltage of the RBFNN model are depicted in Figure. 14.20. Comparing the identification results in Figures 14.19 and 14.20, one will notice that the Hammerstein model yields higher modelling accuracy. These indicate the Hammerstein model is a powerful tool for describing the nonlinear dynamic properties of the SOFC and the Hammerstein model presented is accurate and valid.

14.3   Control Strategies of the SOFC

Transients in a load have a significant impact on the life of the SOFC. One of the reasons is that the fuel utilization changes drastically due to the load transient. Fuel utilization u_f is one of the most important operating variables of the SOFC system. For protecting the SOFC stack, the desired range of fuel utilization is from 0.7 to 0.9. The terminal voltage and the fuel utilization cannot be kept constant simultaneously when load changes.[61] So there are two control strategies, which can guarantee the fuel utilization to operate within a safe range. One is directly controlling the input hydrogen fuel in proportion to the stack current, and the constant utilization control can be accomplished, which is analyzed in reference.[62] The other one is to maintain a constant cell voltage at the SOFC terminal. One proper value for cell voltage can guarantee the fuel utilization within the desired safe rang when the load changes.

In this section, a nonlinear model predictive control (MPC) scheme is proposed to control the voltage and guarantee the fuel utilization within a safe range. The SOFC system is an uncertain nonlinear system and its structure and parameters vary with the change of operating point. So the controller should be robust to uncertainty and meet closed-loop objectives such as tracking, regulation and disturbance attenuation.

MPC is a feedback control strategy based on a predictive model and receding horizon optimization. The important advantage of MPC comes from the predictive model, which allows the design of controller in bigger horizon than the other control algorithms without the form of predictive model.[63] Therefore, the model is the key to determine the control quality. The traditional approach usually is based on approximate linearization theory, which is often difficult to identify an accurate mathematical model of the system and imposes serious restrictions on the structure of nonlinear systems.[64] Moreover, the robustness of the closed-loop system cannot be guaranteed, especially when the parameters of plant are uncertain or there is noise or disturbance in the process.[65] Therefore, specific nonlinear modelling approaches might be required. Neural networks are considered as an attractive structure to establish the mathematical relationship of the dynamic system based on the input–output data. A RBF neural network is a feed forward neural network with one hidden layer and can uniformly approximate any continuous function to a prospected accuracy.[66] However, a key problem by using the RBF neural network approach is about how to choose the optimum initial values of the following three parameters: the output weights, the centres and widths of the hidden unit. If they are not appropriately chosen, the RBF neural network may degrade validity and accuracy of modelling.[67] To assure the optimal performance of the RBF neural network approach for SOFC modelling, we consider applying a genetic algorithm (GA) to optimize the RBF neural network parameters in this study. Genetic algorithms are a kind of self-adaptive global searching optimization algorithm based on the mechanics of natural selection and natural genetic.[68] Different from conventional optimization algorithms, genetic algorithms are based on population, in which each individual is evolved parallel, and the ultimate result is included in the last population. In this section, a physical model of a 100kW SOFC system is used to generate the data required for the training and predicting of the GA-RBF model.

The dynamic physical model replaces the real SOFC stack to generate the simulation data required for the identification of the GA-RBF model. The data sources blocks developed in MATLAB is shown in Figure 14.21.

14.3.1 Constant Voltage Control

The basic frame of the proposed model predictive controller for the SOFC system is shown in Figure 14.22, where y_r is the reference track of the control system, y and y_R is the SOFC dynamic model output and the GA-RBF predictive model output, respectively.

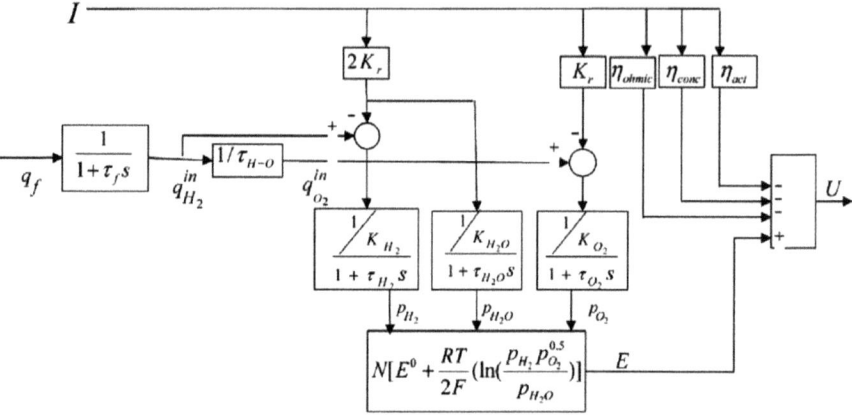

Figure 14.21   SOFC dynamic model.[69]

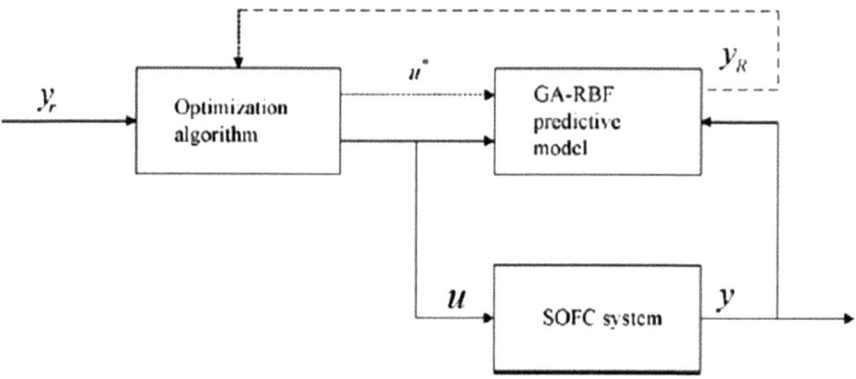

Figure 14.22   Block diagram of the proposed control system.[69]

A nonlinear model that the controller will use for the optimization is the first step in designing a nonlinear MPC system. This model should be as accurate as possible, while being simple enough to allow for repeated calculations during the optimization. If the dynamic model developed in Section 14.3 is applied in the predictive control scheme, it will consume much time to obtain the solutions. For this reason, a nonlinear offline voltage model of SOFC is built by a GA-RBF neural network.

14.3.1.1   Predictive Model Based on GA-RBF Neural Network

Suppose the operating temperature and pressure of the SOFC are kept constant in this section. The oxygen flow is expressed by using the hydrogen–oxygen flow ratio, so the terminal voltage

U is mainly influenced by the inlet hydrogen flow $q_{H_2}^{in}$ and current I. The following NARMAX model is used to describe the controlled voltage system

$$f(k) = f\left[y(k-1), y(k-2), \cdots y(k-n_y), u(k-1), u(k-2), \cdots u(k-n_u)\right] \quad (78)$$

where y is the SOFC terminal voltage, u is the hydrogen flow rate and current, n_y and n_u are the lags of the output and input, respectively, and $f(\cdot)$ is a nonlinear function. In this section we adopt a GA-RBF neural network to identify the nonlinear function $f(\cdot)$. The structure of the RBF neural network is shown in Figure 14.23.

The output of hidden layer is

$$u_i = \exp\left[-\frac{(x-c_i)^T(x-c_i)}{2b_i^2}\right], (i=1,2\cdots q) \quad (79)$$

Where $x = \left(I, q_{H_2}^{in}\right)^T$, $c_i = (c_{i1}, c_{i2})^T$ is the centre of the ith RBF hidden unit, and is the width of the ith RBF hidden unit.

The output is

$$y_R(k) = \sum_{i=1}^{q} \omega_i u_i \quad (80)$$

where $y_R = U$; $\omega = [\omega_1, \omega_2, \cdots, \omega_q]^T$ is the neural network weight that connects output y_R and neuron i in the hidden layer.

In our case, it is done by minimizing the following quadratic cost function of the output error. This error is calculated by comparing the output value of the network and the desired output value.

$$J = \frac{1}{2} \sum_{k=1}^{M} [y_R(k) - y(k)]^2 \quad (81)$$

where y_R is the output voltage calculated from neural network; y the output voltage of the dynamic physical model; M the number of training data. A gradient descent algorithm is adopted to minimize J. And the genetic

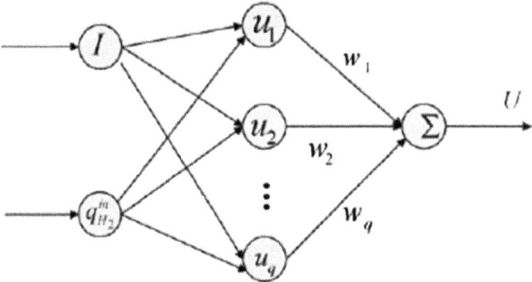

Figure 14.23   The structure of the RBF neural network.[69]

algorithm is used to obtain the optimum initial values of the following three parameters: the output weight w_i, the centres c_i and widths b_i.

The predictive output of p step can be obtained by the GA-RBF model as:[70]

$$y_R(k+p) = f_R[y(k+p-1), y(k+p-2), \cdots y(k+p-n_y),$$
$$u(k+p-1), u(k+p-2), \cdots u(k+p-n_u)] \tag{82}$$

And the error at instance k is defined as

$$e(k) = y(k) - y_R(k) \tag{83}$$

Therefore the predictive output of feedback system is

$$y(k+j) = y_R(k+j) + e(k) \tag{84}$$

where p is the predictive horizon. The method of the offline model is simple and convenient only if the experimental data or experience of the input and output variable is obtainable. It is not necessary to decide the coefficient of the SOFC material and structure of the mechanism model. Therefore it can be used in the control system.

14.3.1.2 Optimization Algorithm

At each sample time k, a series of future output values $y(k+j)$ is calculated through GA-RBF predictive model and it is compared to the reference output value $y_r(k)$. Referenced trajectories of output voltage are introduced to avoid excessive movement of the control input, which are defined as

$$y_r(k+j) = c^j y(k) + (1 - c^j)r, (1 \leq j \leq p, 0 < c < 1) \tag{85}$$

where y_r and y are the reference trajectories and predictive output, respectively. r is the system set value. The optimization problem for the model predictive controller is the minimization of the sum of squared errors between the referenced trajectory y_r and the predictive output y with an additional penalty imposed on rapid changes in the manipulated variables. Define objective function as

$$J(k) = \sum_{j=1}^{n} [y_R(k+j) - y_r(k+j)] - \sum_{i=1}^{m} \lambda u^2(k+i-1), u(k+i-1) \in U^* \tag{86}$$

where n and m are the predictive horizon and control horizon, respectively. $U^* = [u_{min}, u_{max}]$ is the control range. Golden mean method is used to obtain the optimal output of controller $u^*(k)$, the optimization process is presented as follows:[70]

- Step 1: Define the initial searching area $[\alpha_1, \beta_1] = [u_{min}, u_{max}]$; the initial number of steps $k = 1$; and define ζ ,it is positive and adequately small; the initial searching points are given as: $\lambda_1 = \alpha_1 + (1 - 0.618)(\beta_1 - \alpha_1), \lambda_1' = \alpha_1 + 0.618(\beta_1 - \alpha_1)$. And then, compute the values of objective function $J(\lambda_1)$ and $J(\lambda_1')$.

- Step 2: If $\beta_k - \alpha_k < \xi$, then $u(k) = (\beta_k + \alpha_k)/2$; Otherwise, if $J(\lambda_1) < J(\lambda_1')$, go to step 3; if $J(\lambda_1) > J(\lambda_1')$, go to step 4.
- Step 3: $\alpha_{k+1} = \lambda_k, \beta_{k+1} = \beta_k, \lambda_{k+1} = \lambda_k'$, then $\lambda_{k+1}' = \alpha_{k+1} + 0.618(\beta_{k+1} - \alpha_{k+1})$ and then, compute the value of objective function $J(\lambda_{k+1}')$, go to Step 5.
- Step 4: $\alpha_{k+1} = \alpha_k, \beta_{k+1} = \lambda_k', \lambda_{k+1} = \lambda_k'$, then $\lambda_{k+1}' = \alpha_{k+1} + (1 - 0.618)$ $(\beta_{k+1} - \alpha_{k+1})$ and then, compute the value of objective function $J(\lambda_{k+1}')$, go to Step 5.
- Step 5: $k = K + 1$, return to Step 2.

After iterations for several times, the optimal control moves $u^*(k)$ can be obtained.

14.3.2 Constant Fuel Utilization Control

According to Eq. (57), the operation of the SOFC stack with a fuel input proportional to the stack current results in a constant utilization factor in the steady-state. Thus, the SOFC is operated with a constant steady-state utilization factor by controlling the natural gas flow to the stack as[62]

$$q_f = \frac{NI}{2Fu_{fs}} \tag{87}$$

where u_{fs} is the desired utilization in steady-state. Furthermore, because the fuel processor is specially considered, the relationship between a small change of stack current I and a small change of hydrogen input in $\Delta q_{H_2}^{in}$ fed to the SOFC stack can be derived as:[62]

$$\Delta q_{H_2}^{in} = \frac{N}{2Fu_{fs}(1 + \tau_f s)} \Delta I = \frac{2K_r}{u_{fs}(1 + \tau_f s)} \Delta I \tag{88}$$

From Eq. (57), the small-signal relationship between Δu_f, ΔI, and $\Delta q_{H_2}^{in}$ about their initial steady-state values u_{fs}, I_0, and $q_{H_2,0}^{in}$ can be written as

$$\Delta I = \frac{q_{H_2,0}^{in}}{2K_r} \Delta u_f + \frac{u_{fs}}{2K_r} \Delta q_{H_2}^{in} \tag{89}$$

Under constant u_f control, substituting Eq. (88) into Eq. (89), and Eq. (89) can be rewritten as

$$\Delta u_f = \frac{u_{fs}}{I_0} \frac{\tau_f s}{1 + \tau_f s} \Delta I \tag{90}$$

where I_0 the initial stack current.

14.3.3 Simulation

In this section, we present numerical experiments to show the validation of the proposed nonlinear.

predictive control scheme based on the GA-RBF neural network model of the SOFC.

For the purpose of identification, the input signals of the dynamic physical model are uniformly random, including the hydrogen flow rate (0–1.2 mol s^{-1}) and the current (0–800 A). To obtain values at integer time points, the fourth-order Runge–Kutta method was used to find the numerical solution to the dynamic physical model in the simulation. A set of 3000 data was collected from the simulation. The first 1000 data were used for the identification of GA-RBF model, while the remaining 2000 data were used for validation purposes.

In order to reduce the numbers of the parameters and improve the speed of program debug, the hidden layer of the RBF neural network is chosen 2 nodes. After many trials, a population size of 50, a crossover probability of 0.4, and a mutation probability of 0.001 are used. The optimized initial values of the parameters are shown in Table 14.6. After the optimized initial values of the three parameters are obtained, we utilize the gradient descent learning algorithms to adjust them. Root mean square error (RMSE) is employed here to evaluate modelling results, which is calculated by

$$RMSE = \sqrt{\frac{1}{N}\sum_{k=1}^{N}(y(k) - \hat{y}(k))^2} \tag{91}$$

where N is the number of sample data from the dynamic physical model of SOFC, $y(k)$ is the predictive output of GA-RBF model, $\hat{y}(k)$ is the output of the dynamic physical model. Under various hydrogen flows and current, the voltage identification model is shown in Figure 14.24. The RMSE of output voltage obtained is 1.2584 and 1.1836, respectively. The result shows that the GA-RBF neural network can approximate the behaviour of the physical SOFC model with good accuracy.

In normal working conditions, the current of the SOFC system is 300 A. The steady output of the voltage is 330V. Assuming at $t = 200$ s, a load disturbance

Table 14.6   The optimized initial values of widths, centres and output weights.[69]

b_1	0.4856
b_2	1.1062
c_{11}	0.5678
c_{12}	-0.3561
c_{21}	-0.8409
c_{22}	1.032
w_1	0.4567
w_2	-0.2365

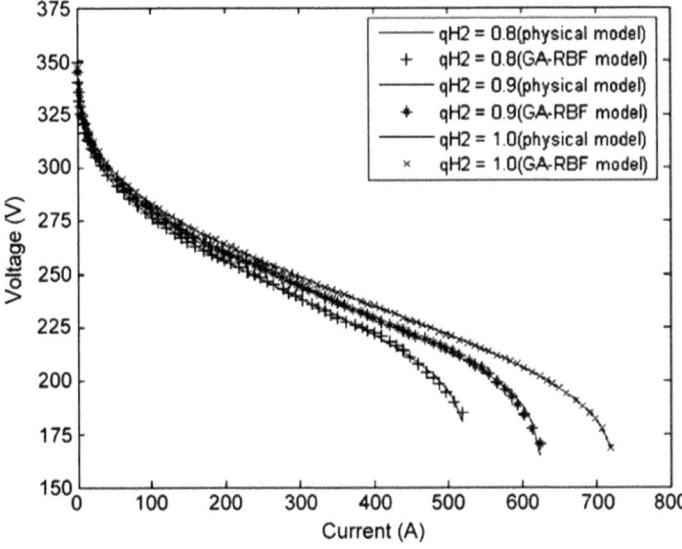

Figure 14.24 Cell terminal voltage under various current and hydrogen flow.[69]

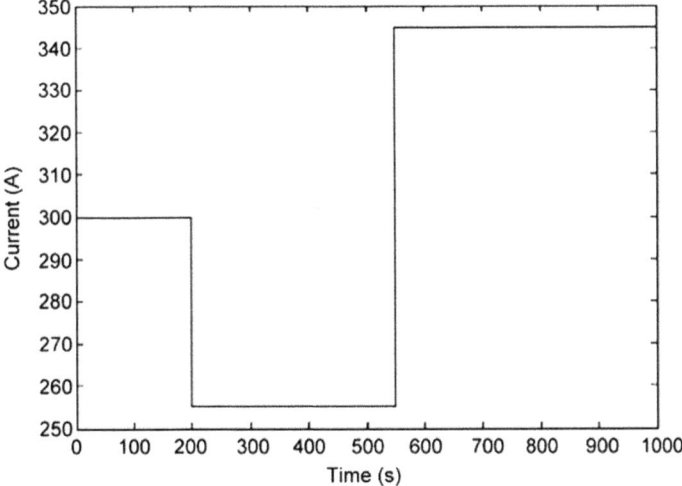

Figure 14.25 Step increase in cell current.[69]

causes the stack current to have a step change (from 300 to 255 A), and at $t = 550$ s, a load disturbance causes the stack current to have a step change (from 255 to 345 A). The series of step in stack current input are shown in Figure 14.25. The predictive controller is used to adjust the voltage to its steady value. In this section, let the predictive horizon be $m = 10$, and the control horizon $n = 4$. We can get the tracking curve of the controlled voltage of the SOFC system, which is shown in Figure 14.26. From Figure 14.26, we can get

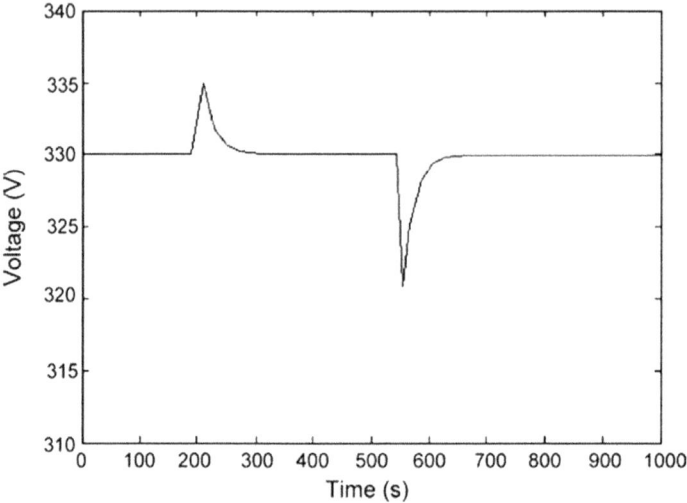

Figure 14.26 Trajectories of SOFC voltage by MPC.[69]

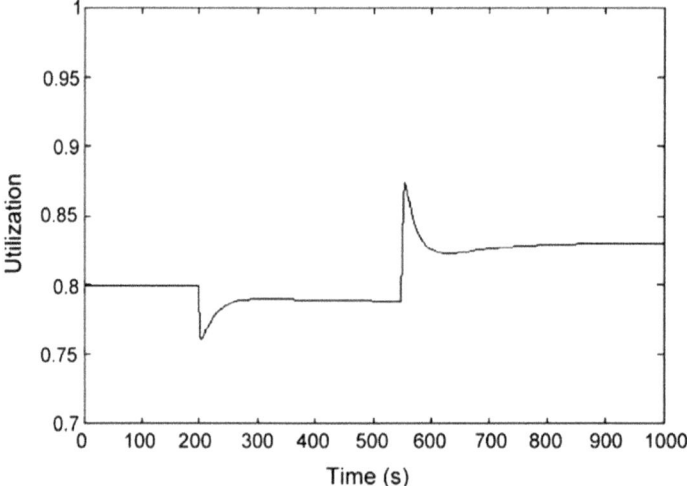

Figure 14.27 Fuel utilization response by MPC.[69]

when the stack current fluctuates the voltage is changed rapidly and then returns to the reference value. Based on the results, we can get the fuel utilization response curve of the SOFC system, which is shown in Figure 14.27. The result shows that the predictive control scheme can guarantee the fuel utilization of the SOFC operating within a safe range when load changes.

Now we use the same load disturbances as before and use the constant fuel utilization control method. The tracking curve of the controlled fuel utilization of the SOFC system is shown in Figure 14.28. And the manipulated variable q_f is shown in Figure 14.29. Comparing Figure 14.27 with Figure 14.28, we can

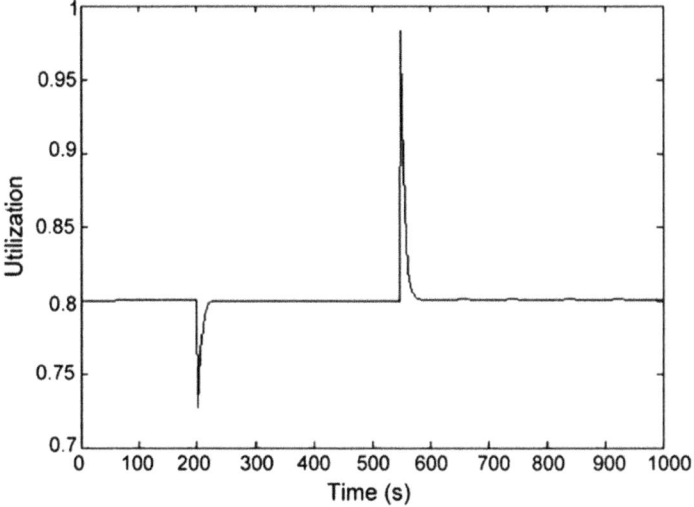

Figure 14.28 Fuel utilization response by constant fuel utilization control.[69]

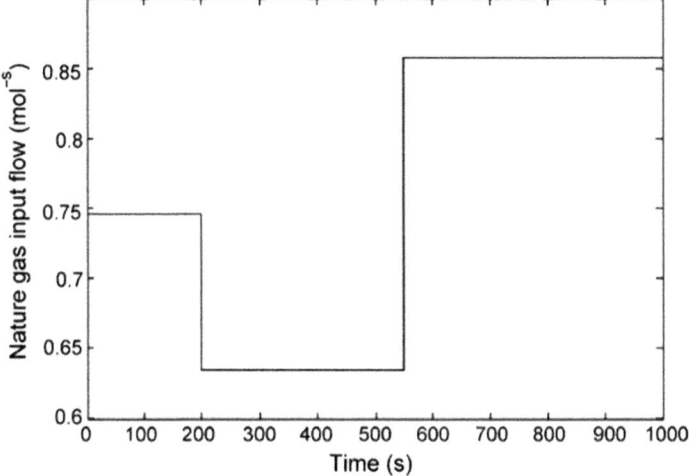

Figure 14.29 Curves of manipulated variable q_f.[69]

get the conclusions: the constant fuel utilization control method is simpler; however the fuel utilization u_f has a larger excursion from the safe range when the stack current changes from 255 to 345A at $t = 550$ s.

14.4 Conclusions

In this chapter, static and dynamic modeling studies of the SOFC stack are described. These models are much more time-efficient and economical for the researchers to understand the physical/chemical phenomena involved in the

SOFC operation. They also provide a tool to facilitate design, analysis and optimization of the SOFC. Furthermore, based on the established dynamic identification models in this chapter, several control strategies for the SOFC can be developed. In addition, the excellent voltage control effects of the SOFC by designing MPC controller is also revealed in this chapter

References

1. J. R. Ferguson, J. M. Fiard and R. Herbin, Three-dimensional numerical simulation for various geometries of solid oxide fuel cells, *J. Power Sources*, 1996, **58**, 109–122.
2. R. Saisset, C. Turpin, S. Astier, B. Lafage, Study of thermal imbalances in arrangements of solid oxide fuel cells by means of bond graph modelling, in: IEEE 33rd Annual Power Electronics Specialists Conference, 2002, **1**, pp. 327–332.
3. P. Iora, P. Aguiar, C. S. Adjiman and N. P. Brandon, Comparison of two IT DIR-SOFC models: Impact of variable thermodynamic, physical, and flow properties, Steady-state and dynamic analysis, *Chemical Engineering Science*, 2005, **60**, 2963–2975.
4. E. Achenbach, Three-dimensional and time dependent simulation of a planar solid oxide fuel cell stack, *J. Power Sources*, 1994, **49**, 333–348.
5. P. Aguiar, C. S. Adjiman and N. P. Brandon, Anode-supported intermediate temperature direct internal reforming solid oxide fuel cell. I: model-based steady-state performance, *J. Power Sources*, 2004, **138**, 120–136.
6. K. P. Recknagle, R. E. Williford, L. A. Chick, D. R. Rector and M. A. Khaleel, Three-dimensional thermo-fluid electrochemical modelling of planar SOFC stacks, *J. Power Sources*, 2003, **113**, 109–114.
7. J. Arriagada, P. Olausson and A. Selimovic, Artificial neural network simulator for SOFC performance prediction, J. Power Sources, 2002, **112**, 54–60.
8. J. A. K. Suykens and J. Vandewalle, Least squares support vector machine classifiers, *Neural Processing Letters*, 1999, **9**(3), 293–300.
9. T. Van Gestel, J. A. K. Suykens, D. E. Baestaens, A. Lambrechts, G. Lanckriet, B. Vandaele, B. D. Moor and J. Vandewalle, Financial time series prediction using least squares support vector machines within the evidence framework, *IEEE Transactions on Neural Networks*, 2001, **12**(4), 809–821.
10. C. Lu, T. Van Gestel, J. A. K. Suykens, S. Van Huffel, I. Vergote and D. Timmerman, Preoperative prediction of malignancy of ovarian tumors using least squares support vector machines, *Artificial Intelligence in Medicine*, 2003, **28**, 281–306.
11. B. J. Kim and I. K. Kim, An application of hybrid least squares support vector machine to environmental process modelling, *Lecture Notes in Computer Science*, 2004, **3320**, 184–187.

12. P. Costamagna, L. Magistri and A. F. Massardo, Design and part-load performance of a hybrid system based on a solid oxide fuel cell reactor and a micro gas turbine, *J. Power Sources*, 2001, **96**, 352–368.

13. M. G. Zhang, X. G. Wang, W. H. Li, Nonlinear system identification using least squares support vector machines, In: International Conference on Neural Networks and Brain, ICNN&B 05, vol.1, (13–15) 2005, pp. 414–418.

14. H. B. Huo, X. J. Zhu and G. Y. Cao, Nonlinear modelling of a SOFC stack based on a least squares support vector machine, *J. Power Sources*, 2006, **162**, 1220–1225.

15. Y. H. Li, S. S. Choi and S. Rajakaruna, An Analysis of the Control and Operation of a Solid Oxide Fuel-Cell Power Plant in an Isolated System, *IEEE Trans. Energy Convers*, 2005, **20**(2), 381–387.

16. M. Espinoza, J. A. K. Suykens, B. De Moor, Least Squares Support Vector Machines and Primal Space Estimation, in: Proceedings of the 42nd IEEE Conference on Decision and Control Maui, Hawaii USA, 2003, pp. 3451–3456.

17. C.-C. Chang, C.-J. Lin, LIBSVM: a library for support vector machines, 2001, Software available at http://www.csie.ntu.edu.tw/~cjlin/libsvm.

18. T. Hansen and C. J. Wang, Support vector based battery state of charge estimator, *J. Power Sources*, 2005, **141**, 351–358.

19. K.Warwick, Proceedings of the 35th Conference on Decision and Control, Kobe, Japan, December, 1996, pp. 464–469.

20. D. E. Goldberg, Genetic Algorithms in Search, Optimization and Machine Learning, Addison-Wesley, 1989.

21. A. Ai-Amoudi, L. Zhang and IEE Pro.-Gener, *Transm. Distrib.*, 2000, **147**(5), 310–316.

22. X. J. Wu, X. J. Zhu, G. Y. Cao and H. Y. Tu, Modelling a SOFC stack based on GA-RBF neural networks identification, *J. Power Sources*, 2007, **167**, 145–150.

23. Y. Gao, L. Shi, P. J. Yao, Proceedings of the 3rd World Congress on Intelligent Control and Automation June 28–July 2, Hefei, PR China, 2000.

24. P. Costamagna, L. Magistri and A. F. Massardo, *J. Power Sources*, 2001, **96**, 352–368.

25. J. Sjoberg, Q. H. Zhang and L. Ljung, *et al.*, *Automatica*, 1995, **31**(12), 1691–1724.

26. F. Calise, M. Dentice d'Accadia, A. Palombo and L. Vanoli, *Energy*, 2006, **31**, 3278–3299.

27. J. Arriagada, P. Olausson and A. Selimovic, Artificial neural network simulator for SOFC performance prediction, *J. Power Sources*, 2002, **112**(1), 54–60.

28. P. Aguiar, C. S. Adjiman and N. P. Brandon, Anode-supported intermediate temperature direct internal reforming solid oxide fuel cell. I: Model-based steady-state performance, *J. Power Sources*, 2004, **138**, 120–136.

29. H. B. Huo, X. J. Zhu and G. Y. Cao, Nonlinear modelling of a SOFC stack based on a least squares support vector machine, *J. Power Sources*, 2006, **162**(2), 1220–1225.
30. J. Padulles, G. W. Ault and J. R. McDonald, An integrated SOFC plant dynamic model for power systems simulation, J. Power Sources **86** *(1–2)*, 2000, 495–500.
31. D. J. Hall and R. G. Colclaser, Transient modelling and simulation of a tubular solid oxide fuel cell, *IEEE Trans. Energy Convers*, 1999, **14**(3), 749–753.
32. K. Sedghisigarchi and A. Feliachi, Dynamic and transient analysis of power distribution systems with fuel cells-part I: fuel-cell dynamic model, *IEEE Trans. Energy Convers*, 2004, **19**(2), 423–428.
33. Y. Zhu and K. Tomsovic, Development of models for analyzing the load-following performance of microturbines and fuel cells, *Elect. Power Syst. Res*, 2002, **62**(1), 1–11.
34. F. Jurado, Modelling SOFC plants on the distribution system using identification algorithms, *J. Power Sources*, 2004, **129**(2), 205–215.
35. J.-S. Roger Jang, ANFIS: Adaptive-network-based fuzzy inference system, *IEEE Trans. Syst. Man Cybernet*, 1993, **23**(03), 665–685.
36. Feng-Jie Lin Jeich Mar, An ANFIS controller for the car-following collision prevention system, *IEEE Trans. Vehicular Technol.*, 2001, **50**(4), 1106–1113.
37. Kok Chew Lee and Peter Gardner, Adaptive neuro-fuzzy inference system (ANFIS)digital predistorter for RF power amplifier linearization, *IEEE Trans. Vehicular Technol.*, 2006, **50**(1), 43–51.
38. J. Padulle's, G. W. Ault and J. R. McDonald, An integrated SOFC plant dynamic model for power system simulation, *J. Power Sources*, 2000, **86**, 495–500.
39. Ali Kourosh Sedghisigarchi, Feliachi, Dynamic and transient analysis of power distribution systems with fuel cells – Part I: Fuel-cell dynamic model, *IEEE Trans.*, *Energy Conversion*, 2004, **19**(2), 423–428.
40. X. J Wu, X. J Zhu, G. Y Cao and H. Y Tu, Nonlinear modelling of a SOFC stack based on ANFIS identification, *Simulation Modelling Practice and Theory*, 2008, **16**, 399–409.
41. P. Costamagna, L. Magistri and A. F. Massardo, Design and part-load performance of a hybrid system based on a solid oxide fuel cell reactor and a microgas turbine, *J. Power Sources*, 2001, **96**, 352–368.
42. J. Sjoberg, H. Q. Zhang and L. Ljung, *et al.*, Nonlinear black-box modelling in system identification: a unified overview [J], *Automatica*, 1995, **31**(12), 1691–1724.
43. T. Takagi and M. Sugeno, Fuzzy identification of systems and its applications to modelling and control, *IEEE Trans. Syst. Man Cybernet.*, 1985, **15**, 116–132.
44. M. Sugeno and G. T. Kang, Structure identification of fuzzy model, *Fuzzy Sets Syst.*, 1998, **28**, 15–33.
45. R. K. Pearson and M. Pottmann, Gray-box identification of block-oriented nonlinear models, *J. Process Control*, 2000, **10**(4), 301–315.

46. H. T. Su and T. J. McAvoy, Integration of multilayer perceptron networks and linear dynamic models: A Hammerstein modelling approach, *Ind. Eng. Chem. Res.*, 1993, **26**, 1927–1936.

47. D. K. Rollins, N. Bhandari, A. M. Bassily, G. M. Colver and S. T. Chin, A continuous-time nonlinear dynamic predictive modelling method for Hammerstein processes, *Ind. Eng. Chem. Res.*, 2003, **42**(4), 860–872.

48. A. Balestrino, A. Landi, M. Ould-Zmirli and L. Sani, Automatic nonlinear auto-tuning method for Hammerstein modelling of electrical drives, *IEEE Trans. Ind. Electron*, 2001, **48**(3), 645–655.

49. F. Jurado, A method for the identification of solid oxide fuel cells using a Hammerstein model, *J. Power Sources*, 2006, **154**(1), 145–152.

50. L. M. Sun, W. J. Liu, A. Sano, Least squares identification of Hammerstein model based on over-sampling scheme, in: Proceedings of the 1996 UKACC International Conference on Control, Exeter, UK,1996, pp. 240–245.

51. K. S. Narendra and K. Parthasarathy, Identification and control of dynamical systems using neural networks, *IEEE Trans. Neural Netw.*, 1990, **1**(1), 4–27.

52. H. Sarimveis, A. Alexandridis, S. Mazarakis and G. Bafas, A new algorithm for developing dynamic radial basis function neural network models based on genetic algorithms, *Comput. Chem. Eng.*, 2004, **28**(1–2), 209–217.

53. Y. Zhu and K. Tomsovic, Development of models for analyzing the load-following performance of microturbines and fuel cells, *Elect. Power Syst. Res.*, 2002, **62**(1), 1–11.

54. Y. H. Li, S. S. Choi and S. Rajakaruna, An analysis of the control and operation of a solid oxide fuel-cell power plant in an isolated system, *IEEE Trans. Energy Convers*, 2005, **20**(2), 381–387.

55. Y. H. Li, S. Rajakaruna and S. S. Choi, Control of a solid oxide fuel cell power plant in a grid-connected system, *IEEE Trans. Energy Convers*, 2007, **22**(2), 405–413.

56. J. Padulles, G. W. Ault and J. R. McDonald, An integrated SOFC plant dynamic model for power systems simulation, *J. Power Sources*, 2000, **86**(1–2), 495–500.

57. K. Sedghisigarchi and A. Feliachi, Impact of fuel cells on load-frequency control in power distribution systems, *IEEE Trans. Energy Convers*, 2006, **21**(1), 250–256.

58. H. B. Huo, Z. D. Zhong, X. J. Zhu and H. Y. Tu, Nonlinear dynamic modelling for a SOFC stack by using a Hammerstein model, *J. Power Sources*, 2008, **175**, 441–446.

59. H. Al-Duwaish and M. N. Karim, A new method for the identification of Hammerstein model, *Automatica*, 1997, **33**(10), 1871–1875.

60. L. Ljung, *System Identification: Theory for the User*, second ed., Prentice Hall, Englewood Cliffs, NJ, 1999.

61. Y. H. Li, S. S. Choi and S. Rajkaruna, *IEEE Trans. Energy Convers.*, 2005, **20**(2), 381–387.

62. Y. H. Li, S. Rajakaruna and S. S. Choi, *IEEE Trans. Energy Convers.*, 2007, **22**(2), 405–413.
63. G. Wei, H. Min, Proceedings of the 2006 American Control Conference Minneapolis, MN, USA, June 14–16, 2006, pp. 1569–1574.
64. T. Van Den Boom, M. A. Botto and J. S̈a Da Costa, *Int. J. Control*, 2003, **76**(18), 1783–1789.
65. K. Hwang, S. Tan and M. Tsai, *IEEE Trans. Syst. Man Cybern. B: Cybern.*, 2003, **33**(3), 514–521.
66. K.Warwick, Proceedings of the 35th Conference on Decision and Control, Kobe, Japan, December, 1996, pp. 464–469.
67. X.-J. Wu, X.-J. Zhu and G.-Y. Cao, *et al.*, *J. Power Sources*, 2007, **167**(1), 145–150.
68. D. E. Goldberg, *Genetic Algorithms in Search, Optimization and Machine Learning,* Addison-Wesley, 1989.
69. X. J Wu, X. J Zhu, G. Y Cao and H. Y Tu, Predictive control of SOFC based on a GA-RBF neural network model, *J. Power Sources*, 2008, **179**, 232–239.
70. J. Zhu, *Intelligent Predictive Control Technology and Application [M],* Zhejiang University Press, Hangzhou, 2006.

Subject Index

Ab inito thermodynamics, 228–230
"active two-phase boundary", 181
adaptive neural-fuzzy inference
    system (ANFIS), 479
AES, *see* Auger electron spectroscopy
    (AES)
alkaline fuel cell (AFC), 3
ammonia fuel, 100–101
anode materials, of PC-SOFCs,
    262–269
APU, *see* auxiliary power units (APU)
area specific resistance (ASR), 26
Arrhenius equation, 13
artificial neural network (ANN), 464
atomistic modelling, 220–222
Auger electron spectroscopy (AES),
    264
auxiliary power units (APU), 237

back-scattered electrons, 184
BCY15 electrolyte, 262
Beer–Lambert Law, 189
Bieberle's model, 214
biofuel cells (BFCs), 3
biomass-fueled power production
    system
        continuity equations, 15
        current density and carbon
            activity distributions, 20
        electrical and exergetic
            efficiencies, 20–21
        GHG emissions, 21–22
        heat transfer equations, 16–17
        input data, 19
        natural convection, 14

Reynolds number, 14
specific greenhouse gas
    emissions, 18–19
temperature distribution, 20–21
total exergy destructions, 17
biomass gas, 437
Bruggemann factor, 234
Bruggemann relationship, 234
Butler–Volmer electrochemistry, 237
Butler–Volmer equation, 13
Butler–Volmer type equations, 232
button cell experiment, 201–202

cathode gas recirculation, 450
cathode heat exchange, 449
cathode material development
    area-specific resistance, 57
    ohmic resistance, 56
    oxygen ion-conducting
        electrolyte, 57 (*see also*
        oxygen ion-conducting
        electrolyte)
    proton-conducting electrolyte
        cathode reaction
            mechanisms, 80–82
        electron-conducting
            cathodes, 73–74
        microstructure optimized
            cathodes, 78–79
        mixed electron–proton
            conducting cathodes,
            77–78
        mixed oxygen ion–electron
            conducting cathodes,
            74–77

cathode poisoning
    chromium poisoning, 310–313
    SO₂ poisoning, 313–315
cathode-support tubular cell, 298
CC, *see* current collector (CC)
CC theory, *see* concept of contiguity
    (CC) theory
cell-level models, 225
Center for X-ray Optics (CXRO), 192
centrifugal air compressor, 429
CFD, *see* computational fluid
    dynamics (CFD)
chromium poisoning, 310–313
coal syngas, 445
co-generation, using fuel cells,
    328–329
combined cooling, heat and power
    (CCHP)
    application characteristics and
        building integration, 331–332
        building integration,
            335–336
        commercial buildings,
            332–334
        operating strategies,
            336–338
        residential applications,
            334–335
    co- and tri-generation, using
        fuel cells, 328–329
    market barriers and challenges
        energy pricing, 371–372
        market barriers and
            environmental impact,
            373–376
        SOFC costs, 372
        technical barriers, 373
    modelling approaches, 342–344
        building energy demands,
            and heating/cooling
            systems, 345
        overview, 344
        simplified cell-level
            modelling, 350–353
        simplified stack modelling,
            353–355

SOFC system-level,
            345–347
        system performance
            metrics, 347–349
        techno-economic model
            formulations, system
            optimization using,
            355–356
    overview of, 329–331
    SOFC system, 338–339
        for CHP, 339–340
        large-scale CHP and,
            363–365
        micro-CHP, 356–363
        tri-generation, 340–342
combined heat and power (CHP)
    application characteristics and
        building integration, 331–332
        building integration,
            335–336
        commercial buildings,
            332–334
        operating strategies,
            336–338
        residential applications,
            334–335
    co- and tri-generation, using
        fuel cells, 328–329
    market barriers and challenges
        energy pricing, 371–372
        market barriers and
            environmental impact,
            373–376
        SOFC costs, 372
        technical barriers, 373
    modelling approaches, 342–344
        building energy demands,
            and heating/cooling
            systems, 345
        overview, 344
        simplified cell-level
            modelling, 350–353
        simplified stack modelling,
            353–355
        SOFC system-level,
            345–347

system performance
metrics, 347–349
techno-economic model
formulations, system
optimization using,
355–356
overview of, 329–331
SOFC system, 338–339
for CHP, 339–340
large-scale CHP and,
363–365
micro-CHP, 356–363
tri-generation, 340–342
commercial YSZ, 269, 271
composite cathode materials, 188
composite electrode microstructure,
180
computational fluid dynamics based
design, 226
computational fluid dynamics (CFD),
226
computational scheme, 211–212
computed tomography (CT), 189
concept of contiguity (CC) theory,
200
conductivity degradation behaviour,
for Ni-YSZ electrolyte, 306
"continuous particle size diameter",
195
continuum approach, 224
conventional tomography techniques,
182
cross-sectional area, tortuosity factors
against, 206
CT, *see* computed tomography (CT)
current collector (CC), 210
CXRO, *see* Center for X-ray Optics
(CXRO)

data analysis, 194–196
de Bore current density model, 212
DGM, *see* dusty gas model (DGM)
differential scanning calorimetry
(DSC), 277
dimension reduction, 235
direct coupling, 227

direct ethanol fuel cells (DEFCs), 3
direct formic acid fuel cells
(DFAFCs), 3
direct internal reforming (DIR)
configuration, 396, 401, 402
direct internal reforming (DIR-
SOFC), 6–7
direct methanol fuel cells (DMFCs),
2–3, 266
direct pressurized air gasification, 439,
440, 441
DIR-SOFC, *see* direct internal
reforming (DIR-SOFC)
DMFCs, *see* direct methanol fuel cells
(DMFCs)
DSC, *see* differential scanning
calorimetry (DSC)
dual beam systems, 184
dual-phase composite electrolyte,
45–46
dusty gas model (DGM), 207
dynamic identification modelling,
478–479
ANFIS
model training and
simulation test,
484–487
problem formulation,
481–483
SOFC dynamic physical
model, 480–481
theory, 483–484
Hammerstein identification
modelling, 487–488
identification of, 491–493
problem statement,
490–491
results and discussion,
493–496
SOFC dynamic model,
487–490

EGR, *see* exhaust gas recirculation
(EGR)
electrochemical reactions in H_2S–air
fuel cells, 280

electrochemistry, 228–232
    coupling of, 225
electrode
    cross-section, micrograph of,
        187
    microstructure of, 181
    reactivity, impurities and
        poisoning effects on, 292–294
electrolyte materials
    ASR, 26
    classification
        dual-phase composite
            electrolyte, 45–46
        oxygen-ion conducting
            electrolyte (*see* oxygen-
            ion conducting
            electrolyte)
        proton-conducting
            electrolyte (*see* proton-
            conducting electrolyte)
    ohmic resistance, 26
    requirements of, 27
    Westinghouse's cathode-
        supported tubular cell, 26
electron beam, interaction of, 185
electron imaging, 184
elementary kinetics, 230–232
exhaust gas recirculation (EGR),
    419–423
external reforming (IR) arrangement,
    411

fabrication
    operation procedures, 291
    technology, 289
FIB tomography, *see* focused ion
    beam (FIB) tomography
Fick's law of diffusion, 196
field emission electron, 185
flatten tubular cell, 291
focused ion beam (FIB) tomography,
    183
focused ion beam scanning electron
    microscopy (FIB-SEM), 188
    microstructure reconstruction
        using, 202

techniques, application of,
    186–188
focused ion beam scanning electron
    microscopy (FIB-SEM), 201
force field, 222
4-dimensional tomography, 193–194
free power turbine, 403, 404
fuel cell reactor set-up, 248
fuel vaporizer (FV), 445

GA-RBF system, nonlinear modelling
    based on, 470–471
    modelling SOFC by, 473–474
    neural network for, 471–473
    optimal GA-RBF parameters,
        selection of, 475–477
    predicting with, 477–478
    predictive model based on,
        498–500
    preparing training data,
        474–475
gas injection systems, 184
gas turbine systems
    SOFC
        by alternative fuels,
            436–447
        anode recirculation,
            396–402
        atmospheric cycles,
            424–427
        control strategies, 427–436
        EGR, 419–423
        externally reformed, 411
        HAT, 414–415
        HRSG, 402–411
        hybrid SOFC/GT-Cheng
            cycle, 411–413
        hybrid SOFC/GT-
            ITSOFC cycles,
            415–416
        hybrid SOFC/GT-rankine
            cycles, 417–419
        IGCC power plants,
            447–452
        internal reformed,
            395–396

layouts classification,
392–394
pressurized cycles,
394–395
prototypes, 385–392
global kinetics, 232
Goldschmidt tolerance factor,
39–40

HAT, *see* humidified air turbine
(HAT)
heat recovery steam generator
(HRSG), 402–411
high temperature gas cleaning, 440
high temperature proton-conducting
materials (HTPCs), 37
homogenization, 234
horizontal coupling, 227
HRSG, *see* heat recovery steam
generator (HRSG)
humidified air turbine (HAT),
414–415
hybrid power plant, SOFC system for,
237
hydrocarbon fuels, 94–95
hydrogen fuel, 89–90
hydrogen oxidation reaction, 178

IGCC, *see* integrated gasification
combined cycle (IGCC)
IIR-SOFC, *see* indirect internal
reforming (IIR-SOFC)
image based modelling, 196
indirect atmospheric steam
gasification, 438
indirect coupling, 227
indirect internal reforming (IIR)
arrangement, 396
indirect internal reforming solid
oxide fuel cell (IIR-SOFC), 6–7
integrated coal gasification fuel cell
combined cycle (IGFC), 319
integrated gasification combined cycle
(IGCC), 447–452
interconnect materials
demands to, 106–107

LaCrO$_3$ (*see* lanthanum
chromites (LaCrO$_3$))
mechanical support, 106
metallic alloys
advantages, 116
alloy groups, properties,
120–121
chromium based alloys,
116–117
chromium poisoning,
122–123
coating material,
attributes, 123
CrFe PM alloy, 129–130
electrophoretic deposition,
127–128
Fe–Cr-based alloys,
117–119
ferritic stainless steel,
120–121
metal organic chemical
vapour deposition, 128
(Mn,Co)$_3$O$_4$, 126–129
Ni–Cr-based alloys,
119–120
nitride coatings,
123–124
oxidation resistance, Cr/Al,
116
oxide scale, growth and
spallation, 121–122
perovskite materials,
124–125
spinels, 125–126
TEC, 120
Wagner's theory,
oxidation, 120
internal reforming (IR)
and anode recirculation, 400
arrangement, 395–396

Karush–Kuhn–Tucker (KKT), 466
KIER hybrid layout, 391
kinetic Monte Carlo, 222
kinetic reaction mechanism, 223
kyocera cells, 293

lab-based sources, 190
lanthanum chromites (LaCrO₃)
    chemical compatibility, 113–114
    conductivity
        activation energy, 110
        calcium substituted
            lanthanum chromites,
            109
        electronic blocking
            electrochemical method,
            109–110
        oxygen partial pressure,
            108–109
        p-type nonstoichiometric
            reaction, 108
        substitution reaction, 108
        temperature dependence,
            110–111
    Cr₂O₃ formation, 113
    flattened tube-type cells, 115
    high-surface-area powders, 113
    ionic radii, 108
    Rolls-Royce fuel cells, 116
    Siemens-Westinghouse tubular
        cells design, 114–115
    structure of, 107
    TEC, 111–113
    yttrium chromite, 114
lanthanum strontium cobalt ferrite
    (LSCF), 309
    sulfur distribution in, 315
lanthanum strontium manganite
    (LSM), 293, 296, 309
Lattice Bhatnagar-Gross-Krook
    (LBGK) model, 211
Lattice Boltzmann method (LBM),
    196, 201
    simulations, 191
LBGK model, *see* Lattice Bhatnagar-
    Gross-Krook (LBGK) model
LBM, *see* Lattice Boltzmann method
    (LBM)
least squares support vector machine
    (LS-SVM), 464–470
    modelling SOFC based on,
        466–469

predicting with, 470
    training process of, 468–469
linear parameter varying (LPV)
    method, 240
liquefied natural gas (LNG), 390
liquid fuels, 98–100
liquid metal ion source (LMIS),
    184
LMIS, *see* liquid metal ion source
    (LMIS)
long term operating stability
    cathode and anodes,
        performance degradations of
        cathode poisoning,
            309–315
        Ni cermet anodes,
            sintering of, 316–320
    electrolytes, deteriorations of
        Mn dissolved YSZ,
            destabilization of,
            296–302
        Ni dissolved YSZ,
            conductivity decrease
            in, 302–309
    stacks/systems, durability of
        electrode reactivity,
            impurities and
            poisoning effects on,
            292–294
        performance degradation
            and materials
            deteriorations, 289–292
        stack performance,
            determination of, 289
low temperature gas cleaning, 438,
    439
LPV method, *see* linear parameter
    varying (LPV) method

Maxwellian local equilibrium
    distribution, 211
MEA, *see* membrane electrode
    assembly (MEA)
mean-field approach, 222
mean-field elementary kinetics,
    electrochemistry with, 222–223

membrane electrode assembly
(MEA), 248
metal sulfide catalysts, 276–277
methane fuel
  alkaline earth oxides, 93
  anode-electrolyte assemblies, 94
  bi-layer anode, 93
  carbon deposition, 91
  catalytic carbon, 90–91
  chemical vapour deposition
    method, 93
  gas-phase reactions, 91
  Ni/YSZ anodes, 91–92
  perovskite anodes, 92
micro-Raman spectroscopy, 307
microstructural evolution processes,
  182
microstructural parameters,
  quantification of, 202–207
microstructural requirements, 181
microstructure-resolved approaches,
  223
MIEC material, *see* mixed
  ionic-electronic conducting
  (MIEC) material
Mitsubishi module, 389, 390
mixed ionic and electronic conductor
  (MIEC), 310
mixed ionic-electronic conducting
  (MIEC) material, 178
Mn dissolved
  YSZ, destabilization of,
    296–302
mobile APU applications
  SOFC system for, 237–240
modelling approaches, 342–344
  building energy demands, and
    heating/cooling systems, 345
  overview, 344
  simplified cell-level modelling,
    350–353
  simplified stack modelling,
    353–355
  SOFC system-level, 345–347
  system performance metrics,
    347–349

techno-economic model
  formulations, system
  optimization using,
  355–356
model-predictive control (MPC), 240
model predictive control (MPC)
  scheme, 497
molecular dynamics, 222
momentum conservation, 225
motivation, 219–220
multi-scale methods, 227
multi-scale modelling approach, 237

"nano-CT" system, 190
nano-structured electrodes
  anodes
    ceramic oxide scaffolds,
      152–155
    Cu-infiltrated anodes,
      150–151
    H_2 oxidation reaction, 150
    Ni and Ni/YSZ scaffolds,
      151–153
  cathodes
    catalytic active
      components, 145
    electronic and oxygen
      ionic conductivity, 143
    electronic conducting
      scaffolds, 145–146
    ionic conducting scaffolds,
      146–147
    LSM, 143
    mixed ionic and electronic
      conducting scaffolds,
      147–150
  electrocatalytic effects,
    infiltrated nanoparticles
    activation energy, 162–163
    H_2 reduction reaction,
      164–165
    impedance responses,
      165–166
    inert materials, 167
    iR-free polarization
      curves, 167–168

nano-structured electrodes
  (*continued*)
    O₂ reduction reaction,
      163–164
    oxygen surface exchange
      coefficient, 166–167
    Pd/PdO redox couple, 164
    promotion effect, 162
  electrochemical performance,
    142
  electrode/electrolyte interfacial
    bonding, 136
  infiltration process, 169
    electronic conducting
      electrode materials, 139
    factors, 140–142
    low melting point phase,
      137
    metallic nanoparticles,
      138
    metal salt nitrate solution/
      nanoparticle
      suspension, 137
    porous BSCF cathode,
      138–139
    pre-sintered porous
      electrode/electrolyte
      scaffold/skeleton, 139
    Pt-based nano-structured
      electrocatalysts, 136
    total conductivity,
      139–140
  microstructural stability
    ASR, 160–161
    glass and glass-ceramics,
      161
    Mn and Pd co-infiltration,
      159–160
    operation temperature,
      160
    poisoning effect, boron,
      161–162
    sintering and grain
      growth, 158
    thermal and performance
      stability, 159

  microstructure effect
    highly porous YSZ
      scaffold, 156–157
    LSM-YSZ composite
      cathode, 157
    finite element calculation,
      157
    pre-sintered LSM
      electrode scaffold,
      155–156
    sphere particles and
      electronic/ionic
      conducting scaffold,
      157, 159
  promotion factor, 143–144
  three phase boundaries, 136
Navier–Stokes equations, 226
Ni cermet anodes, sintering of,
    316–320
nickel oxide, 302
Ni-dissolved YSZ, conductivity
    decrease in, 302–309
NiO dissolution, effect of, 309
NiO-8YSZ anode, 201
Ni-S-O-H-C system, 318
Ni/YSZ, *see* Ni–yttria-stabilized-
    zirconia (Ni/YSZ)
Ni-yttria-stabilized-zirconia (Ni/
    YSZ), 248–249
nonlinear autoregressive model with
    exogenous inputs (NARX), 466

ohmic effects, 182
optimization algorithm, 500–501
oxygen ion-conducting electrolyte
    anion vacancy concentration, 29
    bismuth oxide compositions,
      28–29
oxygen ion-conducting electrolyte
    cathode reaction mechanisms
      cathode/electrolyte
        interface, 72
      current density, 72
      electrochemical reactions,
        70
      MIEC-based cathodes, 73

oxygen reduction paths, 71
oxygen surface exchange
    coefficient, 73
oxygen-ion conducting electrolyte
    ceria-based ceramics, 34
    ceria oxides, 32–33
    chemical compatibility, 37
    defective clustering, 36
    electron conducting cathodes
        cathode/electrolyte
            interface, 60
        electronic conductivity,
            58–59
        LSM-YSZ composites,
            electrochemical
            performances, 60–61
        perovskite-type
            manganites, 57
    fifluorite-type oxide-ion
        conductor, 32
    GDC and SDC, 32
    grain boundary conductivity, 34
    ionic conductivity, 28
    Kroger–Vink notation, 30, 36
    lanthanum-doped ceria, 33
    lanthaum gallate, 35–36
    microstructure optimized
        cathodes
        ASR, 67–68, 70
        electrode microstructure, 67
        infiltration/impregnation
            method, 68–69
        ionic conducting phase, 65
        thermal expansion
            coefficient, 67
        Weibull theory, 69
    mixed oxygen ion-electron
        conducting cathodes
        Arrhenius conductivity
            plots, 64
        bivalent elements, 61–62
        cobalt ions, 61
        electrical conductivity, 61
        electrochemical
            performance, LSCF,
            65–66

gadolinia-doped ceria, 65
hole-conduction
    mechanism, 64
ionic conductivity, 61–62
metal-insulator transition,
    61
oxygen partial pressure
    dependence, 63
strontium-doped
    lanthanum ferrites, 63
open circuit voltage, 35
oxygen vacancy concentration,
    30
perovskite oxides, 35
samarium and strontium, 33
scavenging effect, 34
ScSZ, 30–32
sol–gel technique, 36
ultra-thin interfacial electrolyte
    layer, 35
yttrium-stabilized zirconia, 30
ZrO_2, 30

PC-SOFCs, *see* proton conducting
    solid oxide fuel cells (PC-SOFCs)
Pechini method, 269
PEMFCs, *see* proton exchange
    membrane fuel cells (PEMFCs)
performance degradation, and
    materials deteriorations, 289–292
platinum meshes, 201
platinum paste, 253
PMMA, *see* polymethylmethacrylate
    (PMMA)
polarization simulation, equations
    for, 207–210
polymer electrolyte fuel cells
    (PEFCs), 288
polymethylmethacrylate (PMMA),
    250
pore tortuosity factor, 207
porous transport coefficients, 233–234
proton-conducting electrolyte
    barium carbonate and
        hydroxide, 42
    barium cerate, 41

proton-conducting electrolyte
    (*continued*)
    bulk conductivity, 44
    cathode reaction mechanisms,
        80–82
    chemical stability, 41
    doping barium zirconate, 44
    electron-conducting cathodes,
        73–74
    Goldschmidt tolerance factor,
        39–40
    Grotthuss mechanism, 39
    HTPCs, 37
    ionic conductivity, 43
    Kroger–Vink notation, 37–38
    microstructure optimized
        cathodes, 78–79
    mixed electron-proton
        conducting cathodes, 77–78
    mixed oxygen ion-electron
        conducting cathodes,
        74–77
    perovskite-type materials, 37
    proton defects, 37
    proton transport mechanism,
        38–39
    synergetic effect, 42–43
    total conductivity, 40–41, 45
    ultrafine barium zirconate
        powders, 43–44
proton conducting solid oxide fuel
    cells (PC-SOFCs), 256
    anode materials of, 262–269
    electrolyte of, 258–262
proton exchange membrane fuel cells
    (PEMFCs), 2–3, 266

quantum chemistry, 221–222

radial gas turbine, 430
Raman spectra, of cell components,
    304
random resistor network models, 225
reactive force fields, 222
recuperative heat exchanger (RHE),
    422

renewable energy sources, 384
representative repeat elements,
    235–237

scandia stabilized zirconia (ScSZ),
    299
sealless tubular cell, 291
secondary ion mass spectrometry
    (SIMS), 292, 293
selected area electron diffraction
    (SAED), 253
Siemens–Westinghouse
        220 kW SOFC/GT unit, 388
        tubular SOFC bundle, 386
        tubular SOFC cross section, 387
SOFC, *see* solid oxide fuel cells
    (SOFC)
SOFC-CHP system, commercial
    developments of
        commercialization efforts,
            366–367
        demonstrations, 367–371
SOFCs
solid oxide fuel cells (SOFC), 200,
    219, 247, 288–289
    advantages, 4
    anode materials
        ammonia fuel, 100–101
        biomass-actual gas, 97
        biomass gasification
            syngas, 95
        biomass-simulated gas,
            96–97
        butane, 95
        hydrogen fuel, 89–90
        liquid fuels, 98–100
        methane fuel (*see* methane
            fuel)
        propane, 94–95
        properties of, 88–89
        tar sample, elemental
            analysis, 96
    biomass feed stocks, 7–8
    bio-oil, 8–9
    catalytic cracking, ammonia
        reaction, 7–8

cathode material development
(*see* cathode material
development)
on C_2H_6 and C_3H_8, 256–257
CHP/CCHP, 338–339
for CHP, 339–340
large-scale CHP and,
363–365
micro-CHP, 356–363
tri-generation, 340–342
classification
anode-supported,
cathode-supported and
electrolyte-supported
type, 6
cell and stack designs, 5–6
external reforming, 6–7
flow configurations, 6
internal reforming, 6–7
self-supporting
configuration, 6
temperature levels, 5
cleanup process, types, 8
control strategies of, 496–497
constant fuel utilization
control, 501
constant voltage control,
497–501
simulation, 502–505
electrodes, 178
electrolyte materials (*see*
electrolyte materials)
on H_2S, 276–281
hydrogen sulfide, 7
integrated SOFC
vs. biomass-fueled power
production system (*see*
biomass-fueled power
production system)
biomass gasification
system, 11
gasification systems, 10
gas turbine systems, 9
interconnect materials (*see*
interconnect materials)
materials, 186–188

materials usage, 4
materials used, 135
military applications, 4
modelling
Arrhenius equation, 13
Butler–Volmer equation,
13
electrical efficiency, 14
fuel and air utilization
ratios, 12
Nernst voltage, 12
ohmic polarization,
12–13
power output, 13
zero-dimensional SOFC
model, 12
nano-structured electrodes
(*see* nano-structured
electrodes)
operation principle, 4
oxygen ion conducting
electrolyte, 135
polarisation curve, 181
on sourgas, 248–256
on syngas containing H_2S,
269–276
system simulation, multi-scale
models for, 237
water-gas shift reaction, 7
solid-state reactions, 259
stack performance
determination of, 289
static identification model
GA-RBF, nonlinear modelling
based on, 470–471
modelling SOFC by,
473–474
neural network for,
471–473
optimal GA-RBF
parameters, selection of,
475–477
predicting with,
477–478
preparing training data,
474–475

static identification model (*continued*)
    LS-SVM, nonlinear modelling
        based on, 464–466
            modelling SOFC based
                on, 466–469
            predicting with, 470
stationary fuel cells
    components, 1–2
    low-temperature fuel cells
        AFC, 3
        BFCs, 3
        DEFCs, 3
        DFAFCs, 3
        types, 2–3
    SOFCs (*see* solid oxide fuel cells
        (SOFCs))
    stacking process, 2
steam methane reforming (SMR)
    reaction, 396, 397
steam to carbon (SC) ratio, 404
stereology
    limitations of, 179–184
structure-resolved approach, 224–225
sulfur, 316
superheater (SH), 445
surface transport coefficients, 232–233
synchrotron radiation, 190
syngas, 269

TEC, *see* thermal expansion
    coefficient (TEC)
temperature programmed desorption
    (TPD) experiments, 231
temperature programmed oxidation
    (TPO) technique, 262–263
temperature programmed reaction
    (TPR), 249
tetragonal precipitations, 305
TGA, *see* thermogravimetric analysis
    (TGA)
"theatre curtain effect", 185, 186
thermal expansion coefficient (TEC),
    111–113
thermodynamic calculation,
    transition metals/YSZ and
    solubility of, 299

thermogravimetric analysis (TGA),
    271
3D characterisation, importance of,
    179–184
three-dimensional parameters, 182
three phase boundary (TPB) theory,
    201, 203, 204, 277
TMCs, *see* transition metal carbides
    (TMCs)
Tortuosity factor, 206
TPB theory, *see* three phase boundary
    (TPB) theory
TPD experiments, *see* temperature
    programmed desorption (TPD)
    experiments
TPO technique, *see* temperature
    programmed oxidation (TPO)
    technique
TPR, *see* temperature programmed
    reaction (TPR)
transition metal carbides (TMCs),
    266
transmission X-ray microscopy,
    189
transport coefficients, 233
tri-generation, using fuel cells,
    328–329
triple phase boundary (TPB), 178,
    180, 181, 266
turbine inlet temperature (TIT), 396,
    397, 430
turbine outlet temperature (TOT), 430

variable inlet guide vane (VIGV),
    434–435
variable-speed control strategy, 434
vertical coupling, 227
Viking gasifier, 442
VoF method, *see* volume of fluid
    (VoF) method
volume of fluid (VoF) method, 225

working electrode (WE), 209

XCT, *see* X-ray computed
    tomography (XCT)

Xradia instrument, 192
X-ray computed tomography (XCT), 201
X-rays
    instruments
        lab, 191–192
        synchrotron, 192–193

microscopy and tomography, 189–191
microstructural characterisation using, 189

yttria stabilized zirconia (YSZ), 294–309